Physical Properties of High-Temperature Superconductors

Wiley Series in Materials for Electronic and Optoelectronic Applications

www.wiley.com/go/meoa

Series Editors

Professor Arthur Willoughby, *University of Southampton, Southampton, UK*

Dr Peter Capper, *SELEX Galileo Infrared Ltd, Southampton, UK*

Professor Safa Kasap, *University of Saskatchewan, Saskatoon, Canada*

Published Titles

Bulk Crystal Growth of Electronic, Optical and Optoelectronic Materials, Edited by P. Capper

Properties of Group-IV, III–V and II–VI Semiconductors, S. Adachi

Charge Transport in Disordered Solids with Applications in Electronics, Edited by S. Baranovski

Optical Properties of Condensed Matter and Applications, Edited by J. Singh

Thin Film Solar Cells: Fabrication, Characterization, and Applications, Edited by J. Poortmans and V. Arkhipov

Dielectric Films for Advanced Microelectronics, Edited by M. R. Baklanov, M. Green, and K. Maex

Liquid Phase Epitaxy of Electronic, Optical and Optoelectronic Materials, Edited by P. Capper and M. Mauk

Molecular Electronics: From Principles to Practice, M. Petty

CVD Diamond for Electronic Devices and Sensors, Edited by R. S. Sussmann

Properties of Semiconductor Alloys: Group-IV, III–V, and II–VI Semiconductors, S. Adachi

Mercury Cadmium Telluride, Edited by P. Capper and J. Garland

Zinc Oxide Materials for Electronic and Optoelectronic Device Applications, Edited by C. Litton, D. C. Reynolds, and T. C. Collins

Lead-Free Solders: Materials Reliability for Electronics, Edited by K. N. Subramanian

Silicon Photonics: Fundamentals and Devices, M. Jamal Deen and P. K. Basu

Nanostructured and Subwavelength Waveguides: Fundamentals and Applications, M. Skorobogatiy

Photovoltaic Materials: From Crystalline Silicon to Third-Generation Approaches, G. Conibeer and A. Willoughby

Glancing Angle Deposition of Thin Films: Engineering the Nanoscale, Matthew M. Hawkeye, Michael T. Taschuk, and Michael J. Brett

Physical Properties of High-Temperature Superconductors

RAINER WESCHE

WILEY

This edition first published 2015

Registered office
John Wiley & Sons Ltd, The Atrium, Southern Gate, Chichester, West Sussex, PO19 8SQ, United Kingdom

For details of our global editorial offices, for customer services and for information about how to apply for permission to reuse the copyright material in this book please see our website at www.wiley.com.

Library of Congress Cataloging-in-Publication Data applied for.

Wesche, Rainer, 1956–
Physical properties of high-temperature superconductors / Rainer Wesche.
pages cm. – (Wiley series in materials for electronic and optoelectronic applications)
Includes bibliographical references and index.
ISBN 978-1-119-97881-7 (cloth)
1. High temperature superconductors. I. Title.
QC611.98.H54W47 2015
621.3′5 – dc23

2015009531

A catalogue record for this book is available from the British Library.

ISBN: 9781119978817

Typeset in 10/12pt TimesLTStd by Laserwords Private Limited, Chennai, India

1 2015

Contents

About the Author

Rainer Wesche studied physics at the University of Constance, Germany (M.Sc. in 1984) and received his Ph.D. in physics from the University of Constance, Germany, in 1988. From 1989 to 1993, he was a research scientist at the Paul Scherrer Institute, Switzerland, where he led an experimental study of high-current applications of high-T_c superconductivity funded by the Swiss National Science Foundation. Since 1994, he has been a research scientist at the Swiss Federal Institute of Technology Lausanne (Centre de Recherches en Physique des Plasmas (CRPP)). His present research is in the field of applied superconductivity.

Series Preface

Wiley Series in Materials for Electronic and Optoelectronic Applications

This book series is devoted to the rapidly developing class of materials used for electronic and optoelectronic applications. It is designed to provide much-needed information on the fundamental scientific principles of these materials, together with how these are employed in technological applications. The books are aimed at (postgraduate) students, researchers, and technologists, engaged in research, development, and the study of materials in electronics and photonics, and industrial scientists developing new materials, devices, and circuits for the electronic, optoelectronic, and communications industries.

The development of new electronic and optoelectronic materials depends not only on materials engineering at a practical level, but also on a clear understanding of the properties of materials, and the fundamental science behind these properties. It is the properties of a material that eventually determine its usefulness in an application. The series therefore also includes such titles as electrical conduction in solids, optical properties, thermal properties, and so on, all with applications and examples of materials in electronics and optoelectronics. The characterization of materials is also covered within the series in as much as it is impossible to develop new materials without the proper characterization of their structure and properties. Structure-property relationships have always been fundamentally and intrinsically important to materials science and engineering.

Materials science is well known for being one of the most interdisciplinary sciences. It is the interdisciplinary aspect of materials science that has led to many exciting discoveries, new materials, and new applications. It is not unusual to find scientists with a chemical engineering background working on materials projects with applications in electronics. In selecting titles for the series, we have tried to maintain the interdisciplinary aspect of the field, and hence its excitement to researchers in this field.

Arthur Willoughby
Peter Capper
Safa Kasap

Preface

The now more than 100-year-long history of superconductivity started with the discovery of the phenomenon by H. Kammerlingh Onnes in 1911 at the University of Leiden, The Netherlands. The discovery of J.G. Bednorz and K.A. Müller in 1986 that the superconducting state can exist in complex oxides above 30 K revitalized the field of superconductivity. This breakthrough started a race to find cuprate high-T_c superconductors with higher and higher critical temperatures. As early as the following year, $YBa_2Cu_3O_{7-x}$, the first superconductor with a critical temperature above 77 K, the boiling point of liquid nitrogen, was discovered. Today a large number of cuprate high-T_c superconductors with critical temperatures well above this are known.

The discovery of J.G. Bednorz and K.A. Müller opened up the search for the pairing mechanism in the cuprate high-T_c superconductors. Furthermore, the investigation of the phase diagrams and the processing conditions of complex cuprate superconductors led to progress in the materials science of complex multicomponent compounds. Last but not least, the potential for reduced operation costs resulting from the use of liquid nitrogen as coolant renewed interest in power applications of superconductivity.

This book provides an overview of the known cuprate- and iron-based high-T_c superconductors and their physical properties. In addition, the special case of the intermediate-temperature superconductor MgB_2 is considered. Further aspects presented are the synthesis of these materials, the manufacture of superconducting wires and tapes, and the deposition of superconducting films. The book should be suitable for use in graduate-level courses on superconductivity. A large number of figures, tables, and references illustrate the status of research and development in the field in mid-2014.

In Chapter 1, the milestones in the history of superconductivity are briefly described. A special aspect of importance is the development of the maximum known critical temperatures in metals, oxides, molecular, and iron-based superconductors. The fundamental physical principles of normal-state electrical conductivity and the well-known characteristics of metallic superconducting elements (Type I superconductors) are presented in Chapter 2. Because the superconducting state in these materials can be destroyed by magnetic fields as small as 100 mT, they are not suitable for magnet applications. The main results of the Bardeen–Cooper–Schrieffer (BCS) theory, the microscopic quantum theory of superconductivity in conventional metallic superconductors, are briefly described in Chapter 3. In addition, it is shown that superconductivity is a macroscopic quantum phenomenon, which is reflected in flux quantization and tunneling effects. In Chapter 4, the properties of Type II superconductors are presented. Because normal and superconducting regions can coexist in

Type II metallic superconductors, the magnetic fields required to destroy superconductivity can exceed 20 T. In contrast, the high-T_c materials are extreme Type II superconductors with upper critical fields of the order of 100 T.

An overview of the most important families of cuprate high-T_c superconductors is given in Chapter 5. The crystal structures of cuprate superconductors are described in Chapter 6. Empirical rules for the critical temperature of cuprate high-T_c superconductors are discussed in Chapter 7. The generic phase diagram of hole-doped cuprate superconductors is presented in Chapter 8. Aspects to be discussed are how high-T_c superconductivity can come into existence in an insulating antiferromagnetic parent compound, in which mobile holes are doped into the CuO_2 planes, the superconducting order parameter has *d*-wave symmetry, and a pseudogap exists. In Chapters 9–13, the physical properties of the cuprate high-T_c superconductors are described. Chapter 14 focuses on the synthesis of cuprate superconductor powders and the manufacture of bulk material. In Chapter 15, the manufacture and the performance of first- (Ag/Bi-2212 round wires and tapes, Ag/Bi-2223 tapes) and second-generation (biaxially textured RE-123 (where RE is yttrium or another rare earth element) coated conductors) cuprate high-T_c superconductor wires and tapes are discussed. Chapter 16 is devoted to cuprate high-T_c superconductor films deposited on single-crystal substrates, and to the achieved critical temperatures and transport critical current densities.

Chapter 17 provides an overview of the physical properties of MgB_2, the status of the development of MgB_2 wires and tapes, and the preparation of MgB_2 films. An overview of the recently discovered iron-based superconductors and their properties is provided in Chapter 18.

Finally, an outlook on future research and development is given in Chapter 19. Future research is expected to focus in two directions, namely the pairing mechanisms in cuprate and iron-based superconductors, and the development of high-T_c superconductors and MgB_2 for magnet and power applications.

Rainer Wesche

Acknowledgment

I wish to thank P. Bruzzone, the head of the superconductivity section of CRPP, for his support and encouragement. The careful reviewing and many constructive suggestions on the manuscript by J.F. Crawford were especially appreciated. The kind permission from Springer Science+Business Media B.V. to reprint parts of my previous monograph "High-Temperature Superconductors: Materials, Properties, and Applications" (ISBN 0-7923-8386-9) published by Kluwer Academic Publishers in 1998 is gratefully acknowledged.

List of Tables

Nomenclature

Symbol	Meaning	Section first used
$2a$	width of a slab	2.4
a	lattice spacing	2.2
a	wire radius	14.3.3
a, b	contact dimensions	3.3
a, b, c	lattice parameters	6.2.3
A	cross-sectional area	11.3
$\mathbf{A}$, A	vector potential	2.4
A_J, A_L	single contact area, loop area	3.3
A_{ph}	constant defined by $A_{\mathrm{ph}} = 234R/\theta_D{}^3$	2.5
A_{sc}	cross-section of superconductor	15.3.3
A_t	area of single tunneling contact	3.3
A_{tot}	total tape cross-section	15.3.3
A_v	area of a single vortex	4.3
$\mathbf{B}$, B	magnetic induction	2.3
B_0	magnetic induction at superconductor surface	2.3
B_a	applied magnetic field	7.3
B_c	critical magnetic field	2.3
B_{c1}	lower critical field	4.3
$B_{c1,ab}$, $\mathrm{B}_{c1,c}$	lower critical fields for $B \parallel ab$, $B \parallel c$	4.5
B_{c2}	upper critical field	4.3
$B_{c2,ab}$, $B_{c2,c}$	upper critical fields for $B \parallel ab$, $B \parallel c$	4.5
$B_{c2}{}^{\mathrm{clean}}$	upper critical field in clean limit	17.2
B_{fj}	flux jump field	17.4
B_i	magnetic induction inside the superconductor	2.3
B_{irr}	irreversibility field	9.4
B_n	magnetic field at the nucleus	7.3
B_p	penetration field	14.3.3
B_{Pauli}	Pauli limiting field	4.3
B_{sc}	scaling field	9.5.4
c_s	velocity of sound	3.2
C	specific heat	2.5
C_{el}	electron specific heat	2.5

Symbol	Meaning	Section first used
C_{ji}	components of the elastic tensor	7.4
C_n	specific heat in normal state	2.5
C_{ne}	electron specific heat in normal state	2.5
C_{ph}	phonon specific heat	2.5
C_s	specific heat in the superconducting state	2.5
C_{se}	electron specific heat in the superconducting state	2.5
C_{ST}	concentration of $Na_2S_2O_3$	5.3
d	average separation of vortices	4.3
d	distance of lattice planes	6.2.1
d	wire diameter	15.2.2
d_{bulk}, d_{strained}	bulk lattice parameter, strained film lattice parameter	16.4
dQ/dt	heat load	1.2.4
d_r	deposition rate	15.3.2
$D(E)$	density of energy states	2.2
$D_A(E')$, $D_B(E')$	electron density of states in metals A and B	3.3
D_{An}, D_{Bs}	electron density of states in normal conductor A, superconductor B	3.3
$D_n(E)$	electron density of states in normal state	3.2
$D_n(E_F)$	electron density of states in normal state at E_F	2.5
$D_{ph}(\omega)$	phonon density of states	3.2
D_R	roller diameter	15.2.3
$D_s(E)$	single electron density of states in superconductor	3.2
e	electron charge	1.2.2
E	energy	2.2
E'	$E' = E - E_F$	3.3
E_b	height of energy barrier	3.3
E_c	condensation energy	4.2
$\mathbf{E_e}$, E_e	electric field	2.2
E_{ec}	electric field criterion used to define I_c	9.5.1
E_F	Fermi energy	2.2
E_{fl}	energy for formation of a flux line	10.2
E_g	energy gap	2.2
E_k	eigenvalue of energy	2.2
$f(E)$	Fermi–Dirac distribution function	2.2
f_{ip}, f_{op}	fraction of holes in inner, outer CuO_2 planes	7.3
f_j	atomic form factor	6.2.4
$F(hkl)$	structure factor	6.2.4
$\mathbf{F_L}$	Lorentz force	4.4
G	Gibbs free energy	2.5
G_{hkl}	reciprocal lattice vector	6.2.5
G_n	Gibbs free energy in the normal state	2.5

Symbol	Meaning	Section first used
G_s	Gibbs free energy in the superconducting state	2.5
h	Planck's constant	1.2.2
$\hbar$	$h/2\pi$	2.2
$h\,k\,l$	Miller indices	6.2.2
$h\,k\,i\,l$	Miller–Bravais indices	6.2.2
$\mathbf{H}$, H	magnetizing field	2.3
I	current	2.3
I	nuclear spin	7.3
I_0	initial current	2.3
I_{AC}	Josephson AC current	3.3
I_c	critical current	3.3
$I_{c,field}$, $I_{c,offset}$	critical current definitions (see text)	9.5.1
I_{NIN}, I_{NIS}	tunneling currents	3.3
I_s^{ij}	supercurrent across junction ij	8.2.2
$\mathbf{j}$, j	current density	2.4
j_{AB}, j_{BA}	number density of electrons flowing from $A \to B$, $B \to A$	3.3
j_c	critical current density	4.4
j_{cg}	critical current density within grains	9.5.2
j_{cgb}	critical current density across grain boundary	9.5.2
j_c^{perp}	j_c for $B \perp ab$	9.5.4
j_{ct}	transport critical current density	9.5.2
j_e	engineering critical current density	15.1
j_Q	thermal current density	11.3
j_s	screening current density	2.4
j_{s0}	screening current density at superconductor surface	2.4
J_c	critical current density	3.3
$\mathbf{k}$	wave vector	2.2
k	wave number	2.2
k	thermal conductivity	11.3
k_{ab}	thermal conductivity in the crystallographic ab planes	11.3
k_B	Boltzmann constant	2.2
k_c	thermal conductivity along the crystallographic c-axis	11.3
k_e	elastic constant	3.2
k_{el}	electron contribution to thermal conductivity	11.3
k_F	Fermi wave number	2.2
k_{ph}	phonon contribution to thermal conductivity	11.3
k_x, k_y, k_z	components of wave vector	2.2
K	Knight shift	7.3

Symbol	Meaning	Section first used
l_{fl}	length of flux line	10.2
L	inductance	2.3
L	length	2.2
L	conductor length	15.3.4
L	Lorenz number	11.3
L	mean free path	17.2
L_{el}	electron mean free path	11.3
L_{ph}	phonon mean free path	11.3
$\mathbf{m}$	magnetic moment	2.5
m	electron mass	2.2
m^*	effective electron mass	2.2
m_{ab}, m_c	effective electron mass for currents along ab, c	4.5
m_C	mass of a Cooper pair	2.4
m_n	neutron mass	6.2.5
$\mathbf{M}$, M	magnetization	2.3
M	isotopic mass	3.2
M	magnetic quantum number	7.3
M_0	initial magnetization	10.4
M_{rem}	remanent magnetization	10.4
M_{sc}	molar weight	5.3
$\mathbf{n}$	unit vector perpendicular to interface	4.2
n	number of CuO_2 planes in a copper oxide block	6.3.6
n	n factor	9.5.1
n_C	number density of Cooper pairs	2.4
n_{CA}, n_{CB}	number densities of Cooper pairs in superconductors A, B	3.3
n_{Ci}	number density of Cooper pairs deep inside the superconductor	4.2
n_e	number density of electrons $n_e = N_e/V$	2.2
n_h	number density of holes	11.2
N	number of atoms	2.2
N	average electron number density	7.3
N_A	Avogadro's number	2.5
N_e	number of electrons	2.2
N_E	Einstein constant	13.2
N_h	number of holes per CuO_2 unit	7.2
N_{opt}	number of holes leading to the highest T_c	7.2
N_v	number of vortices	4.3
N_ϕ	integer number of flux quanta	3.2
p	pressure	2.5
p	chamber pressure	16.2.2
$\mathbf{p}$	momentum	2.2

Symbol	Meaning	Section first used
p	formal valence of copper $2 + p$	5.3
$\mathbf{p}_1$, $\mathbf{p}_2$, $\mathbf{p}_1'$, $\mathbf{p}_2'$	momenta	3.2
p_b	chamber base pressure	17.5
p_O	oxygen partial pressure	15.3.2
pO_2	oxygen partial pressure	11.2
P_{in}	refrigerator input power	1.2.4
P_t	probability of electron tunneling	3.3
q	electric charge	5.3
$\mathbf{q}$	phonon wave vector	6.2.5
q_C	charge of a Cooper pair	2.4
$\mathbf{q}$	phonon momentum	3.2
Q	heat	1.2.4
Q_P	Peltier heat	12.1
r	radius	4.4
$\mathbf{r}$	position vector	2.2
R	gas constant	2.5
R	resistance	2.3
R_H	Hall coefficient	5.3
R_r	reduction ratio	15.2.3
RRR	residual resistivity ratio, $RRR = \rho(273\text{ K})/\rho(4\text{ K})$	2.2
s_1, s_2	electron spin	3.2
S	entropy	2.5
S	relaxation rate	10.4
S	Seebeck coefficient	12.1
S_e, S_h	Seebeck coefficients for electrons, holes	12.2
S_n	entropy in normal state	2.5
S_s	entropy in superconducting state	2.5
t	time	1.2.4
t	barrier thickness	8.2.2
t_i, t_f	initial, final tape thickness	15.2.3
T	temperature	2.2
T^*	pseudogap temperature	8.2.1
T_a	annealing temperature	16.2.2
T_c	critical temperature, transition temperature	1.1
T_c (end)	T_c defined by the end of the superconducting transition	2.3
T_c (mid)	T_c defined by the midpoint of the superconducting transition	2.3
T_c (onset)	T_c defined by the first deviation from normal-state resistivity	2.3
T_c^{clean}	critical temperature in clean limit	17.2
$T_{c,\text{opt}}$	optimum critical temperature	7.2

Symbol	Meaning	Section first used
T_E	Einstein temperature	13.2
T_F	Fermi temperature	2.2
T_g	glass temperature	10.5
T_M	melting temperature (flux line lattice)	9.4
T_N	Néel temperature	6.1
T_{op}	operation temperature	1.2.4
T_s	substrate temperature	15.3.2
T_s	temperature at structural transition	18.2
U	internal energy	2.5
U	thermovoltage	12.1
U, U_1	voltage	3.3
$\mathbf{v}$, v	velocity	2.2
v_F	Fermi velocity	11.3
v_{ph}	phonon velocity	11.3
V	volume	2.2
V	voltage	10.5
V_1	volume	5.3
V_m	molar volume	7.4
V_p	Electron–phonon interaction constant	3.2
V_p, V_{p0}	pinning potential with and without current	10.4
w	width	8.2.2
W_1	sample weight	5.3
x, y, z	coordinates	2.2
x_L	lattice vector	2.2
α	inclination of evaporation direction to surface normal	15.3.2
$\alpha(\omega)$	electron–phonon coupling strength	3.2
α, β, γ	angles between crystallographic axes	6.2.3
α_{xy}	off-diagonal element of the Peltier conductivity tensor	12.1
β	isotope effect exponent	3.2
β_B, β_{Mg}	partial boron, magnesium isotope effect exponents	17.2
γ	constant defined by $C_{el} = \gamma T$	2.5
γ	gyromagnetic ratio	7.3
γ_a	electronic anisotropy	4.5
δ	phase difference	3.3
Δ	half-width of the superconducting energy gap	3.2
Δ_{max}	maximum half-width of superconducting energy gap	8.2.2
$\Delta\alpha_i$	mean-field jump in thermal expansion coefficient (i: lattice vectors a, b, c)	7.4

Symbol	Meaning	Section first used
ΔT_c	transition width	2.3
Δ_σ	half-width of σ-band superconducting energy gap	17.2
Δ_π	half-width of π-band superconducting energy gap	17.2
ε	dielectric constant	9.6
ε_{ab}, ε_c	strain in *ab* plane, strain along the *c*-axis	16.4
ε_i	strain along direction *i*, *i*: lattice vectors *a*, *b*, *c*	7.4
η_C	Carnot efficiency	1.2.4
θ	angles defined in text	6.2.1
θ	Hall angle	12.1
θ_0	grain boundary misalignment angle leading to 63% reduction of j_c	9.6
θ_D	Debye temperature	2.5
κ	Ginzburg–Landau parameter	4.2
κ_{ab}, κ_c	κ for $B \parallel ab, c$	4.5
λ	wavelength	6.1
λ_0	penetration depth at $T = 0$	9.2
λ_{ab},λ_c	penetration depths for screening currents along *ab*, *c*	4.5
λ_e	electron wavelength	3.3
λ_{ep}	electron–phonon coupling parameter	3.2
λ_L	penetration depth	2.3
λ_{Lo}	London penetration depth	2.4
λ_{TF}	Thomas–Fermi screening length	9.6
μ	chemical potential	2.2
μ	magnetic moment (electron)	4.3
μ_0	permeability of free space	2.3
μ_B	Bohr magneton	4.3
$\mu_c{}^*$	screened Coulomb interaction parameter	3.2
μ_r	relative permeability	2.5
μ_T	Thomson coefficient	12.1
ν	Nernst coefficient	8.2.3
ν, ν_0	frequency, attempt frequency	10.4
ν_D	Debye frequency	13.1
ν_J	Josephson frequency	3.3
ν_p	phonon frequency	3.2
ξ	coherence length	4.2
ξ_0	BCS value of the coherence length	17.2
ξ_{ab}, ξ_c	coherence length along *ab*, *c*	4.5
ξ_{Co}	diameter of a Cooper pair	3.2
Π	Peltier coefficient	12.1
ρ	electrical resistivity	2.2

Symbol	Meaning	Section first used
ρ_0, ρ_i	residual resistivity, intrinsic resistivity	2.2
ρ_a, ρ_b, ρ_c	resistivity for currents along crystallographic *a*-, *b*-, *c*-axis	11.2
ρ_{ab}	resistivity for currents in the *ab* plane	11.2
ρ_c	resistivity criterion (definition of I_c)	9.5.1
ρ_{ff}	flux flow resistivity	4.4
σ	electrical conductivity	2.2
σ_c	chemical shift	7.3
σ_e, σ_h	electrical conductivity for electron band, hole band	12.2
σ_{xy}	off-diagonal element of the conductivity tensor	12.1
τ	relaxation time	2.2
τ_0	relaxation time	10.4
φ	electrical potential	12.1
$\varphi(\mathbf{r})$	phase of the wave function of the Cooper pairs	2.4
φ_A, φ_B	phase of Cooper pair wave functions A, B	3.3
φ_i, φ_j	angles between crystallographic axes and junction interface	8.2.2
ϕ	magnetic flux	3.3
ϕ_0	flux quantum	3.2
ϕ_J, ϕ_L	magnetic flux in a single contact, in a loop	3.3
Φ	superconducting order parameter	4.2
χ	magnetic susceptibility	2.5
χ_P	paramagnetic susceptibility	4.3
ψ	collective wave function of the Cooper pairs	2.4
ψ, ψ_k	electron wave functions	2.2
ψ_1, ψ_2, ψ_3	wave functions	3.3
ψ_A, ψ_B	Cooper pair wave functions	3.3
ω	angular frequency	2.5
ω_D	Debye angular frequency	2.5

1
Brief History of Superconductivity

1.1 Introduction

In 1911, the phenomenon of superconductivity was discovered by Heike Kammerlingh Onnes at the University of Leiden (The Netherlands). The centenary of this discovery was celebrated with a joint conference of the European Applied Superconductivity Conference (EUCAS 2011), the International Superconductive Electronics Conference (ISEC 2011), and the International Cryogenic Materials Conference (ICMC 2011) on September 2011 in The Hague (The Netherlands) [1]. Because of this anniversary, it seems to be worthwhile to briefly describe the history of superconductivity. This overview of the most important events in the history of superconductivity allows us to consider the high-temperature superconductors (HTSs) and their properties in the broader context of superconductivity in general. The most important milestones in the field of superconductivity are listed in Table 1.1.

1.2 Milestones in the Field of Superconductivity

1.2.1 Early Discoveries

In 1908, the successful development of helium liquefaction techniques in the laboratory of H. Kammerlingh Onnes at the University of Leiden made temperatures accessible down to about 1 K [2]. One of the first aspects to be investigated was the electrical resistance of pure metals at very low temperatures. At that time, no proper theory of the electrical resistivity of metals existed. It was already known that the electrical resistance of metals decreases with decreasing temperatures. At the lowest temperatures, a nearly temperature-independent residual resistivity was observed for platinum and gold. This residual resistivity was found to decrease with increasing purity of the investigated sample. Mercury was selected for the further investigations because this low melting-point metal

Physical Properties of High-Temperature Superconductors, First Edition. Rainer Wesche.

Table 1.1 *Important milestones in the history of superconductivity*

Year	Milestone	Reference
1908	Successful liquefaction of He in the laboratory of Kammerlingh Onnes at the University of Leiden	[2]
1911	Onnes discovered superconductivity in Hg	[2]
1913	Onnes received the Nobel prize for the liquefaction of He	
1933	Meissner and Ochsenfeld found perfect diamagnetism for the superconducting state	[3]
1935	London theory of the superconductor electrodynamics	[4]
1941	Discovery of superconductivity in NbN ($T_c = 15$ K)	[5]
1950	Ginzburg and Landau developed a phenomenological theory of superconctivity	[6]
1953	Discovery of the first A15 superconductor V_3Si ($T_c = 17.1$ K)	[7]
1957	Bardeen, Cooper, and Schrieffer (nobel prize 1972) developed a quantum theory of superconductivity	[8, 9]
1960	Giaever measured the energy gap by electron tunneling	[10, 11]
1961	Experimental confirmation of flux quantization	[12, 13]
1962	Josephson predicted the possibility of Cooper pair tunneling between two superconductors separated by a thin insulating oxide layer	[14]
1973	Josephson received the Nobel prize for his theoretical prediction of Cooper pair tunneling through a thin insulating layer separating two superconductors	
1974	Record T_c-value of 23.2 K in Nb_3Ge	[15–17]
1980	Bechgaard and colleagues found the first organic superconductor	[18]
1986	Bednorz and Müller discovered superconductivity at 30 K in La–Ba–Cu–O	[19]
1987	Discovery of the first superconductor with a T_c well above the boiling point of liquid nitrogen ($YBa_2Cu_3O_{7-x}$, $T_c = 93$ K)	[20]
1987	Bednorz and Müller received the Nobel prize for the discovery of high-temperature superconductivity	
1991	Superconductivity in K_3C_{60} at 18 K	[21]
1993	Record T_c-value of 135 K in $HgBa_2Ca_2Cu_3O_{8+\delta}$	[22, 23]
2001	Akimitsu *et al.* found superconductivity at 39 K in MgB_2	[24]
2003	Abrikosov and Ginzburg received the Nobel prize for their contributions to the GLAG (Ginzburg–Landau–Abrikosov–Gor'kov) theory	
2008	Hosono and colleagues discovered superconductivity at 26 K in $La(O_{1-x}F_x)FeAs$	[25]

could be purified by repeated distillation. In 1911, Kammerlingh Onnes found that, at a temperature of approximately 4.2 K, the electrical resistivity of mercury suddenly drops to a value too small to be measured. The temperature at which the superconducting state is reached is called the transition or critical temperature (T_c). The resistance versus temperature data measured by Kammerlingh Onnes are shown in Figure 1.1. The loss of

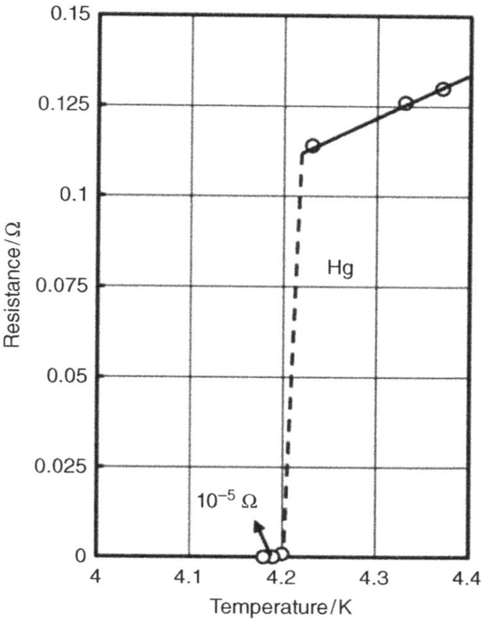

Figure 1.1 *Resistance vs. temperature plot obtained for mercury by Kammerlingh Onnes (adapted from [2])*

resistivity occurred within a temperature interval of 0.04 K. The unexpected phenomenon of superconductivity had been discovered [2]. Soon after this discovery, it was found in Leiden that tin (T_c < 3.8 K) and lead (T_c < 7.2 K) are also superconducting metals [2].

The most remarkable property of the superconducting state is field exclusion. In 1933, Meissner and Ochsenfeld observed that the magnetic field is expelled from the interior of a superconductor, which is cooled below the transition temperature in the presence of a small magnetic field [3]. This effect, nowadays known as the Meissner effect, cannot be explained by zero resistance and is in fact a second characteristic property of the superconducting state.

In 1935, the brothers Fritz London and Heinz London developed the first phenomenological theory of superconductivity [4]. The London equations (see Section 2.4) provide a theoretical description of the electrodynamics of superconductors, including the Meissner effect. In a thin surface layer, just inside the superconductor, screening currents flow without resistance, which cancel the applied magnetic field in the interior of the superconductor. The thickness of this layer, known as the London penetration depth, is a characteristic of the superconductor in question. In addition, London recognized that superconductivity is an example of a macroscopic quantum phenomenon. The behavior of a superconductor is governed by the laws of quantum mechanics like that of a single atom, but on a macroscopic scale [26, 27].

In 1941, Aschermann *et al.* found superconductivity in NbN with a transition temperature of 15 K [5]. Most superconductors discovered before 1941 were elemental metals or alloys. An interesting aspect is the relatively high critical temperature in a compound containing the nonmetallic element nitrogen.

1.2.2 Progress in the Understanding of Superconductivity

In 1950, Ginzburg and Landau developed a phenomenological theory of superconductivity [6], based on the principles of thermodynamics and empirical in nature. Abrikosov [28] solved the Ginzburg–Landau (GL) equations and found solutions explaining the penetration of magnetic flux into Type II superconductors (see Chapter 4), leading to the formation of a flux line lattice. The London theory of superconductivity is a special case of the GL theory [29].

In 1957, Bardeen, Cooper, and Schrieffer [8, 9] developed a microscopic quantum theory of superconductivity (BCS theory, see Section 3.2). The electron–phonon interaction leads to a weak attraction between two electrons. This leads to the formation of Cooper pairs consisting of two electrons of opposite spin and momentum. The total spin of a Cooper pair is therefore zero and as the consequence the paired electrons do no longer obey Pauli's exclusion principle. All the Cooper pairs condense into a single ground state.

Gor'kov demonstrated in 1959 [30] that the GL theory can be derived from the BCS theory, the microscopic theory of superconductivity. The GL theory is still a very useful tool in the field of applied superconductivity.

In 1960, Giaever used aluminum/aluminum oxide/lead sandwiches to perform electron tunneling experiments. The two metals were separated by an Al_2O_3 layer as thin as 1.5–2.0 nm. The tunneling current across the contact was measured in the temperature range of 1.6–4.2 K. Depending on the strength of the applied magnetic field, lead was in its normal or superconducting state, while aluminum ($T_c = 1.18$ K) was always in its normal state. The energy gap for the quasi-particles (unpaired electrons) in lead was determined from the current–voltage curves of the tunneling contact [10]. The existence of an energy gap for the quasi-particles in the superconducting state is one of the central predictions of the BCS theory.

Independent of each other, Deaver and Fairbank [12] and Doll and Näbauer [13] demonstrated experimentally in 1961 that the trapped flux in a superconducting hollow cylinder is quantized in units of $h/2e$, where h is Planck's constant and e is the charge of an electron. The fact that the flux is quantized in units of $h/2e$ instead of h/e indicates the existence of Cooper pairs in the superconductive state.

The possibility that Cooper pairs can tunnel through an insulating barrier separating two superconductors was predicted in 1962 by Josephson [14] (see Section 3.3).

1.2.3 Discovery of High-Temperature Superconductivity

The discovery of high-temperature superconductivity has to be considered in the context of the evolution of the record values of the critical temperature in different classes of superconductive materials (see Figure 1.2). In 1974, a record T_c-value of 23.2 K was reached in Nb_3Ge films [15–17]. This remained the highest known critical temperature until the discovery of superconductivity in the Ba–La–Cu–O system at a temperature as high as 30 K [19]. This discovery by Bednorz and Müller in 1986 started a race to higher and higher critical temperatures. In 1987, $YBa_2Cu_3O_{7-x}$, the first superconductor with a transition

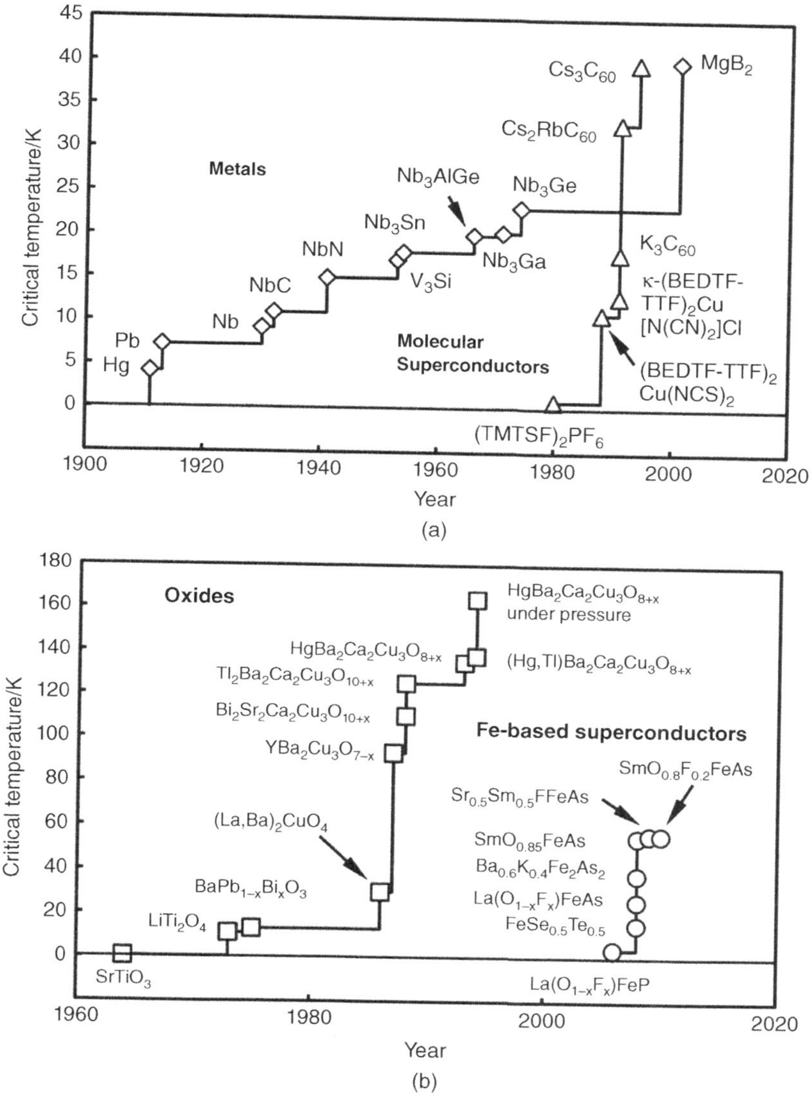

Figure 1.2 *Map of the highest known critical temperatures in different classes of superconducting materials. Panel (a): metals and molecular superconductors [2, 5, 7, 15–18, 21, 24, 29, 31–37]. Panel (b): HTSs (cuprates and iron-based superconductors) [19, 20, 22, 23, 25, 38–51]*

temperature well above the boiling point of liquid nitrogen (77 K), was discovered [20]. By 1993, the maximum critical temperature was pushed up to 135 K in $HgBa_2Ca_2Cu_3O_{8+x}$ (Hg-1223) [22, 23]. A slightly higher transition temperature of 138 K was reported for the partly thallium substituted $Hg_{0.8}Tl_{0.2}Ba_2Ca_2Cu_3O_{8+x}$ [44]. Under a large quasi-hydrostatic pressure of 31 GPa, a critical temperature as high as 164 K was measured in Hg-1223 [45].

A common feature of all HTSs is the presence of CuO_2 planes in the crystal structure; these materials are cuprates.

One of the merits of Bednorz and Müller was the search for higher transition temperatures in the oxides. It was already known that superconductivity can occur in oxides. However, the critical temperature of $SrTiO_3$, the first known oxide superconductor, was as low as 0.25 K [38]. In the seventies of the 20th century moderately high critical temperatures of 11 K [39] and 13 K [40] were found in the oxides $LiTi_2O_4$ and $BaPb_{1-x}Bi_xO_3$, respectively.

In 1980, Bechgaard and colleagues discovered the first organic superconductor $(TMTSF)_2PF_6$ (tetramethyl-tetraselenafulvalene) [18]. Other molecular superconductors are alkali-doped C_{60} molecules. In K_3C_{60}, superconductivity at 18 K [21] was found in 1991. The critical temperature of Cs_3C_{60} under a high pressure of 40 K [37] is considerably larger than that of Nb_3Ge and comparable to that of $La_{2-x}Ba_xCuO_4$, the first HTS discovered by Bednorz and Müller.

Akimitsu *et al.* found in 2001 that the simple binary compound MgB_2, a material known since the 1950s, is a superconductor with the extraordinarily high transition temperature of 39 K [24].

Hosono *et al.* discovered superconductivity in the iron pnictide $La(O_{1-x}F_x)FeAs$ at a temperature of 26 K in 2008 [25]. Very rapidly the critical temperature could be enhanced to a value as high as 55 K in the related iron pnictide $SmO_{0.85}FeAs$ [49]. Because they have transition temperatures well above 50 K, the iron pnictides are considered as the second family of HTSs. The discovery of the iron-based superconductors suggests that in addition to the cuprates and the iron pnictides their may exist other families of HTSs.

1.2.4 Importance of Higher Transition Temperatures for Applications

Most applications of superconductivity require critical current densities of at least $10^4\ \mathrm{A\,cm^{-2}}$ at the envisaged operation field and temperature [52]. Another aspect of importance is the cost of the cooling of the superconductive device. The efficiency of a refrigeration cycle is closely related to the operation temperature. Considering an ideal, reversible cooling cycle the maximum possible refrigeration efficiency is given by the Carnot efficiency

$$\eta_C = \frac{T_{\mathrm{op}}}{300\ \mathrm{K} - T_{\mathrm{op}}} \tag{1.1}$$

where T_{op} is the operating temperature. The input power (P_{in}), required by an ideal, reversible refrigerator, to remove the heat load dQ/dt is proportional to the reciprocal value of the Carnot efficiency:

$$P_{\mathrm{in}} = \frac{1}{\eta_C}\frac{dQ}{dt} \tag{1.2}$$

The reciprocal values of the Carnot efficiency as a function of operating temperature are presented in Figure 1.3. For example, the input power needed to remove a heat load of 1 W at operation temperatures of 4.2, 20, and 77 K is 70.4, 14, and 2.9 W, respectively. However, the efficiency of a real cooling cycle is considerably lower than the Carnot efficiency. In general, large refrigerators are more efficient than smaller ones [53, 54]. For large refrigerators with a heat removal capacity well above 10 kW, the efficiency can reach

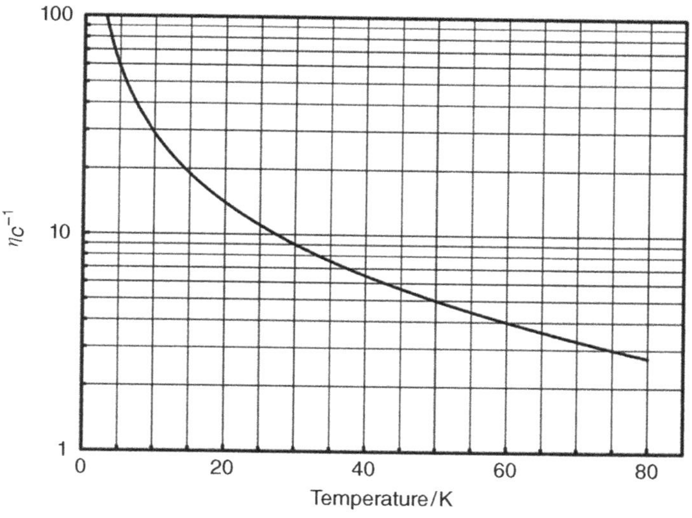

Figure 1.3 The reciprocal value of the Carnot efficiency as a function of the operation temperature indicates that for the removal of a heat load of 1 W at an operation temperature of 4.2 K a power of ≈70 W is required, while this power is only 2.9 W at an operation temperature of 77 K

≈30% of the Carnot efficiency. On the other hand, the efficiency of refrigerators with a heat removal capacity of 1 W is only a few per cent of the Carnot efficiency [53, 54].

The considerations of refrigeration efficiency clearly indicate that the ability to operate superconductive devices at 77 K instead of 4.2 K provides a drastic reduction in the cost of the cooling. The discovery of the cuprate HTSs therefore renewed interest in power applications of superconductivity, e.g. in power transmission cables, transformers, fault current limiters, and generators. For example, the technical feasibility of superconducting transmission cables, made of low-temperature superconductors requiring operation temperatures close to 4 K, was already demonstrated in the 1970s and the 1980s [55–59]. However, these cables were not economically competitive with conventional technology because of the high cost of cooling. As soon as industrially manufactured HTS tapes became available, several demonstrators of power transmission cables have been made of them [60–65]. Operation of HTS cables has been demonstrated with liquid nitrogen cooling; the cooling cost is considerably reduced as compared to operation at ≈4 K. In addition, high critical current density and sufficiently low superconductor cost are required if future HTS power transmission cables are to become economically competitive with conventional technology. The critical current density of HTS tapes is still increasing, and the cost is expected to be further reduced.

References

1. EUCAS–ISEC–ICMC, *Superconductivity Centennial Conference*, The Hague, The Netherlands, September 18–23, 2011, http://www.eucas2011.org/.

2. R. de Bruyn Ouboter, Superconductivity: Discoveries during the early years of low temperature research at Leiden 1908–1914, *IEEE Trans. Magn.*, **23**, 355–370 (1987).
3. W. Meissner and R. Ochsenfeld, Ein neuer Effekt bei Eintritt der Supraleitfähigkeit, *Die Naturwissenschaften*, **21**, 787–788 (1933).
4. F. London and H. London, Supraleitung und Diamagnetismus, *Physica*, **2**, 341—354 (1935).
5. G. Aschermann, E. Friedrich, E. Justi, and J. Kramer, Supraleitfähige Verbindungen mit extrem hohen Sprungtemperaturen (NbH und NbN), *Phys. Z.*, **42**, 349–360 (1941).
6. V.L. Ginzburg, Nobel lecture: On superconductivity and superfluidity (what I have and have not managed to do) as well as on the "physical minimum" at the beginning of the XXI century, *Rev. Mod. Phys.*, **76**, 981–998 (2004).
7. G.F. Hardy and J.D. Hulm, Superconducting silicides and germanides, *Phys. Rev.*, **89**, 884 (1953).
8. J. Bardeen, L.N. Cooper, and J.R. Schrieffer, Microscopic theory of superconductivity, *Phys. Rev.*, **106**, 162–164 (1957).
9. J. Bardeen, L.N. Cooper and J.R. Schrieffer, Theory of superconductivity, *Phys. Rev.*, **108**, 1175–1204 (1957).
10. I. Giaever, Energy gaps in superconductors measured by electron tunneling, *Phys. Rev. Lett.*, **5**, 147–148 (1960).
11. I. Giaever, Electron tunneling between two superconductors, *Phys. Rev. Lett.*, **5**, 464–466 (1960).
12. B.S. Deaver Jr., and W.M. Fairbank, Experimental evidence for quantized flux in superconducting cylinders, *Phys. Rev. Lett.*, **7**, 43–46 (1961).
13. R. Doll and M. Näbauer, Experimental proof of magnetic flux quantization in a superconducting ring, *Phys. Rev. Lett.*, **7**, 51–52 (1961).
14. B.D. Josephson, Possible new effects in superconductive tunnelling, *Phys. Lett.*, **1**, 251–253 (1962).
15. L.R. Testardi, J.H. Wernick, and W.A. Royer, Superconductivity in Nb_3Ge films above 22 K, *Appl. Phys. Lett.*, **23**, 480–482 (1973).
16. L.R. Testardi, J.H. Wernick, and W.A. Royer, Superconductivity with onset above 23 K in Nb-Ge sputtered films, *Solid State Commun.*, **15**, 1–4 (1974).
17. J.R. Gavaler, M.A. Janocko, and C.K. Jones, Preparation and properties of high-T_c Nb-Ge films, *J. Appl. Phys.*, **45**, 3009–3013 (1974).
18. D. Jérome, A. Mazaud, M. Ribault, and K. Bechgaard, Superconductivity in a synthetic organic conductor $(TMTSF)_2PF_6$, *J. Phys. Lett.*, **41**, 201–204 (1980).
19. J.G. Bednorz and K.A. Müller, Possible high T_c superconductivity in the Ba–La–Cu–O system, *Z. Phys. B*, **64**, 189–193 (1986).
20. M.K. Wu, J.R. Ashburn, C.J. Torng, P.H. Hor, R.L. Meng, L. Gao, Z.J. Huang, Y.Q. Wang, and C.W. Chu, Superconductivity at 93 K in a new mixed-phase, Y–Ba–Cu–O compound system at ambient pressure, *Phys. Rev. Lett.*, **58**, 908–910 (1987).
21. A.F. Hebard, M.J. Rosseinsky, R.C. Haddon, D.W. Murphy, S.H. Glarum, T.T.M. Palstra, and A.P. Ramirez, Superconductivity at 18 K in potassium-doped C_{60}, *Nature*, **350**, 600–601 (1991).
22. A. Schilling, M. Cantoni, J.D. Guo, and H.R. Ott, Superconductivity above 130 K in the Hg–Ba–Ca–Cu–O system, *Nature*, **363**, 56–58 (1993).

23. J.J. Capponi, J.L. Tholence, C. Chaillout, M. Marezio, P. Bordet, J. Chenavas, S.M. Loureiro, E.V. Antipov, E. Kopnine, M.F. Gorius, M. Nunez-Regueiro, B. Souletie, P. Radaelli, and F. Gerhards, High pressure synthesis and properties of the $HgBa_2Ca_{n-1}Cu_nO_{2n+2+x}$ ($n = 1$–6), *Physica C*, **235–240**, 146–149 (1994).
24. J. Nagamatsu, N. Nakagawa, T. Muranaka, Y. Zenitani, and J. Akimitsu, Superconductivity at 39 K in magnesium diboride, *Nature*, **410**, 63–64 (2001).
25. Y. Kamihara, T. Watanabe, M. Hirano, and H. Hosono, Iron-based layered superconductor $La[O_{1-x}F_x]FeAs$ ($x = 0.05$–0.12) with $T_c = 26$ K, *J. Am. Chem. Soc.*, **130**, 3296–3297 (2008).
26. P.W. Anderson, Thinking big – Fritz London's single-minded thinking led him to surpass even Einstein, as he believed correctly that quantum mechanics was right at all scales, including the macroscopic, *Nature*, **437**, 625 (2005).
27. S. Blundell, The forgotten brothers, *Physics World*, April 2011, pp. 26–29.
28. A.A. Abrikosov, On the magnetic properties of superconductors of the second group, *Sov. Phys.– JETP*, **5**, 1174–1182 (1957).
29. C.P. Poole, Jr.,, H.A, Farach, and R.J. Creswick, *Superconductivity*, Academic Press, San Diego, CA, 1995.
30. L.P. Gor'kov, Microscopic derivation of the Ginzburg-Landau equations in the theory of superconductivity, *Sov. Phys.– JETP*, **9**, 1364–1367 (1959).
31. D.K. Finnemore, Superconducting materials, In *Encyclopedia of Physics*, R.G. Lerner and G.L. Trigg (eds.), VCH Publishers, New York, 1991.
32. B.T. Matthias, T.H. Geballe. S. Geller, and E. Corenzwit, Superconductivity of Nb_3Sn, *Phys. Rev.*, **95**, 1435 (1954).
33. G.W. Webb, J.L. Vieland, R.E. Miller, and A. Wicklund, Superconductivity above 20 K in stoichiometric Nb_3Ga, *Solid State Commun.*, **9**, 1769–1773 (1971).
34. K. Oshima, H. Urayama, H. Yamochi, and G. Saito, A new ambient pressure organic superconductor $(BEDT\text{-}TTF)_2Cu(NCS)_2$ with T_c above 10 K, *Physica C*, **153–155**, 1148–1152 (1988).
35. J.M. Williams, A.J. Schultz, U. Geiser, K.D. Carlson, A.M. Kini, H.H. Wang, W.-K. Kwok, M.-H. Whangbo, and J.E. Schirber, Organic superconductors – New benchmarks, *Science*, **252**, 1501–1508 (1991).
36. K. Lüders, Supraleitende fullerene, *Phys. Bl.*, **50**, 166–167 (1994).
37. T.T.M. Palstra, O. Zhou, Y. Iwasa, P.E. Sulewski, R.M. Fleming, and B.R. Zegarski, Superconductivity at 40 K in cesium doped C_{60}, *Solid State Commun.*, **93**, 327–330 (1995).
38. J.F. Schooley, W.R. Hosler, and M.L. Cohen, Superconductivity in semiconducting $SrTiO_3$, *Phys. Rev. Lett.*, **12**, 474–475 (1964).
39. D.C. Johnston, H. Prakash, W.H. Zachariasen, and R. Viswanathan, High temperature superconductivity in the Li-Ti-O ternary system, *Mater. Res. Bull.*, **8**, 777–784 (1973).
40. A.W. Sleight, J.L. Gillson, and P.E. Bierstedt, High-temperature superconductivity in the $BaPb_{1-x}Bi_xO_3$ system, *Solid State Commun.*, **17**, 27–28 (1975).
41. H. Maeda, Y. Tanaki, M. Fukutomi, and T. Asano, A new high-T_c oxide superconductor without a rare earth element, *Japan. J. Appl. Phys.*, **27**, 303–304 (1988).
42. Z.Z. Sheng and A.M. Herrmann, Superconductivity in the rare-earth-free Tl–Ba–Cu–O system above liquid nitrogen temperature, *Nature*, **332**, 55–58 (1988).

43. S.S.P. Parkin, V.Y. Lee, E.M. Engler, A.I. Nazzal, T.C. Huang, G. Gorman, R. Savoy, and R. Beyers, Bulk superconductivity at 125 K in $Tl_2Ca_2Ba_2Cu_3O_x$, *Phys. Rev. Lett.*, **60**, 2539–2542 (1988).
44. G.F. Sun, K.W. Wong, B.R. Xu, Y. Xin, and D.F. Lu, T_c enhancement of $HgBa_2Ca_2Cu_3O_{8+\delta}$ by Tl substitution, *Phys. Lett. A*, **192**, 122–124 (1994).
45. L. Gao, Y.Y. Xue, F. Chen, Q. Xiong, R.L. Meng, D. Ramirez, C.W. Chu, J.H. Eggert, and H.K. Mao, Superconductivity up to 164 K in $HgBa_2Ca_{m-1}Cu_mO_{2m+2+\delta}$ under quasihydrostatic pressures, *Phys. Rev. B*, **50**, 4260–4263 (1994).
46. Y. Kamihara, H. Hiramatsu, M. Hirano, R. Kawamura, H. Yanagi, T. Kamiya, and H. Hosono, Iron-based layered superconductor: LaOFeP, *J. Am. Chem. Soc.*, **128**, 10012–10013 (2006).
47. K.-W. Yeh *et al.*, Tellurium substitution effect on superconductivity of the α-phase iron selenide, *Europhys. Lett.*, **84**, 37002 (2008).
48. M. Rotter, M. Tegel, and D. Johrendt, Superconductivity at 38 K in the iron arsenide $(Ba_{1-x}K_x)Fe_2As_2$, *Phys. Rev. Lett.*, **101**, 107006 (2008).
49. Z.-A. Ren, G.-C. Che, X.-L. Dong, J. Yang, W. Lu, W. Yi, X.-L. Shen, Z.-C. Li, L.-L. Sun, F. Zhou, and Z.-X. Zhao, Superconductivity and phase diagram in iron-based arsenic-oxides $ReFeAsO_{1-\delta}$ (Re = rare earth metal) without fluorine doping, *Europhys. Lett.*, **83**, 17002 (2008).
50. G. Wu, Y.L. Xie, H. Chen, M. Zhong, R.H. Liu, B.C. Shi, Q.J. Li, X.F. Wang, T. Wu, Y.J. Yan, J.J. Ying, and X.H. Chen, Superconductivity at 56 K in samarium-doped SrFeAsF, *J. Phys.: Condens. Matter*, **21**, 142203 (2009).
51. C. Wang, Z. Gao, L. Wang, Y. Qi, D. Wang, C. Yao, Z. Zhang, and Y. Ma, Low-temperature synthesis of $Sm_{0.8}F_{0.2}FeAs$ superconductor with $T_c = 56.1$ K, *Supercond. Sci. Technol.*, **23**, 55002 (2010).
52. D. Larbalestier, Critical currents and magnet applications of high-T_c superconductors, *Physics Today*, June 1991, pp. 74–82.
53. P. Komarek, *Hochstromanwendung der Supraleitung*, B.G. Teubner, Stuttgart, 1995.
54. T.R. Strobridge, Refrigeration for superconducting and cryogenic systems, *IEEE Trans. Nucl. Sci.*, **16**, 1104–1108 (1969).
55. G. Bogner, Power transmission studies in Europe, *Cryogenics*, **15**, 79–87 (1975).
56. B.C. Belanger, Superconducting power transmission programme in the USA, *Cryogenics*, **15**, 88–90 (1975).
57. T. Horigome, The present state of R&D for superconducting transmission in Japan, *Cryogenics*, **15**, 91–94 (1975).
58. P.A. Klaudy and J. Gerhold, Practical conclusions from field trials of a superconducting cable, *IEEE Trans. Magn.*, **19**, 656–661 (1983).
59. E.B. Forsyth and R.A. Thomas, Performance summary of the Brookhaven superconducting power transmission system, *Cryogenics*, **26**, 599–614 (1986).
60. J. Rieger, M. Leghissa, J. Wiezoreck, H.-P. Krämer, G. Ries, and H.W. Neumüller, Development of a 10 m long superconducting multistrand conductor for power transmission cables, *Supercond. Sci. Technol.*, **11**, 902–908 (1998).
61. G. Pasztor, A. Anghel, A.M. Fuchs, B. Jakob, D. Trajkovic, G. Vécsey, R. Wesche, and R. Schindler, Manufacture and testing of a neon gas-cooled HTS prototype cable for power transmission, *Inst. Phys. Conf. Ser.*, **167**, 1119–1122 (2000).

62. P.W. Fisher, M.J. Cole, J.A. Demko, C.A. Foster, M.J. Gouge, R.W. Grabovickic, J.W. Lue, J.P. Stovall, D.T. Lindsay, M.L. Roden, and J.C. Tolbert, Design, analysis, and fabrication of a tri-axial cable system, *IEEE Trans. Appl. Supercond.*, **13**, 1938–1941 (2003).
63. S.H. Sohn, H.S. Choi, H.R. Kim, O.B. Hyun, S.W. Yim, T. Masuda, K. Yatsuka, M. Watanabe, H.S. Ryoo, H.S. Yang, D.L. Kim, and S.D. Wang, Field test of 3 phase, 22.9 kV, 100 m HTS cable system in KEPCO, *Inst. Phys. Publishing J. Phys.: Conf. Ser.*, **43**, 885–888 (2006).
64. M. Yagi, S. Mukoyama, M. Ichikawa, T. Takahashi, H. Suzuki, and A. Kimura, Demonstration of a 500 m HTS power cable in the Super-ACE project, *Inst. Phys. Publishing J. Phys.: Conf. Ser.*, **43**, 849–852 (2006).
65. V.S. Vysotsky, K.A. Shutov, A.A. Nosov, N.V. Polyakova, S.S. Fetisov, V.V. Zubko, V.E. Sytnikov, W.L. Carter, S. Fleshler, A.P. Malozemoff, and G. Snitchler, AC loss of a 5 m 2G HTS power cable using wires with NiW substrate, *J. Phys.: Conf. Ser.*, **234**, 032061 (2010).

2

The Superconducting State

2.1 Introduction

In this chapter, we will first describe the superconducting state focusing on the physical properties of simple metallic superconductors. In spite of much higher transition temperatures and much more complex physics and chemistry of high-temperature superconductors (cuprates and iron-based superconductors), some aspects of the phenomenological description are very similar to those of simple metallic superconductors. Moreover, the knowledge of the properties of metallic low-temperature superconductors is required to recognize new phenomena characteristic of new classes of superconducting materials (cuprates, iron-based superconductors). In addition to a relatively high-transition temperature of 39 K, the chemically and structurally simple superconductor MgB_2 was found to show the special feature of two-gap superconductivity. As a basis for the description of the superconducting state, we will first briefly consider the electrical resistance of normal metals.

2.2 Electrical Resistance[1]

In contrast to the electrons of the completely filled inner atomic shells, the valence electrons from the uppermost atomic levels are not bound to individual atoms in a metal. These conduction electrons can move nearly freely in the background of the positive ions. As a first approximation, the conduction electrons may be regarded as free electrons. In this approximation, the electron–electron and the potential energy caused by the lattice of the

[1] Reprinted with minor changes from *"High-Temperature Superconductors: Materials, Properties, and Applications"* by Rainer Wesche, Kluwer Academic Publishers, 1998, ISBN 0-7923-8386-9, pp. 7–15 with kind permission from Springer Science+Business Media B.V.

Physical Properties of High-Temperature Superconductors, First Edition. Rainer Wesche.

positive ions are neglected. The Schrödinger equation for a free electron is

$$-\frac{\hbar^2}{2m}\left(\frac{\partial^2}{\partial x^2}+\frac{\partial^2}{\partial y^2}+\frac{\partial^2}{\partial z^2}\right)\psi_{\mathbf{k}}(\mathbf{r}) = E_{\mathbf{k}}\psi_{\mathbf{k}}(\mathbf{r}) \tag{2.1}$$

where $\hbar = 1.0546 \times 10^{-34}$ J s and m is the mass of the free electron. $\psi_{\mathbf{k}}(\mathbf{r})$ is the wave function of the electron and $E_{\mathbf{k}}$ is the corresponding eigenvalue of the energy. The use of periodic boundary conditions

$$\psi(x+L,y,z) = \psi(x,y+L,z) = \psi(x,y,z+L) = \psi(x,y,z) \tag{2.2}$$

leads, for an electron of momentum $\mathbf{p} = \hbar\mathbf{k}$, to the following wave function:

$$\psi_{\mathbf{k}}(\mathbf{r}) = \exp(i\mathbf{k}\cdot\mathbf{r}) \tag{2.3}$$

where $i = \sqrt{-1}$ and

$$k_x = 0, \pm\frac{2\pi}{L}, \pm\frac{4\pi}{L}, \ldots \quad k_y = 0, \pm\frac{2\pi}{L}, \pm\frac{4\pi}{L}, \ldots \quad k_z = 0, \pm\frac{2\pi}{L}, \pm\frac{4\pi}{L}, \ldots \tag{2.4}$$

Equations 2.2 and 2.4 are valid only for isotropic solids. The wave functions (2.3) correspond to plane waves propagating in the direction of the wave vector **k**. Inserting Equation 2.3 into Equation 2.1 leads to the following energy eigen values $E_{\mathbf{k}}$:

$$E_{\mathbf{k}} = \frac{\hbar^2}{2m}(k_x^2+k_y^2+k_z^2) = \frac{\hbar^2}{2m}k^2 \tag{2.5}$$

In a solid composed of N atoms, each atomic energy level splits into N levels. As a consequence of the large number of atoms in a solid, it is justifiable to replace the discrete values of the wave number k by a continuous variable. Figure 2.1 shows the energy of free electrons in a one-dimensional lattice as a function of the wave number k.

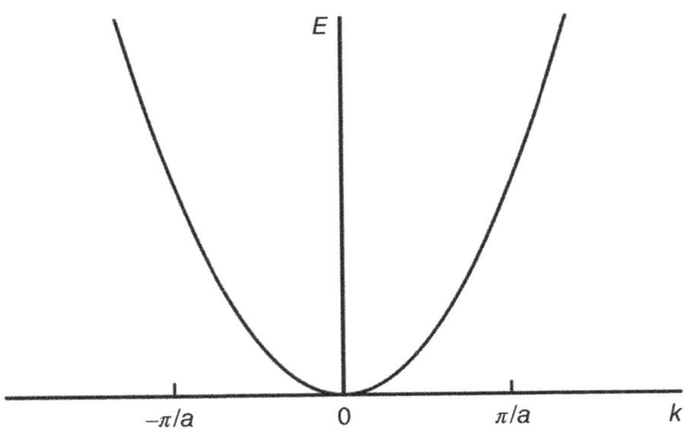

Figure 2.1 *Energy E of free electrons in a one-dimensional lattice versus wave number k; a is the lattice constant (Reprinted from "High-Temperature Superconductors: Materials, Properties, and Applications" by Rainer Wesche, Kluwer Academic Publishers, 1998, ISBN 0-7923-8386-9, Figure 2-1, p. 8 with kind permission from Springer Science+Business Media B.V.)*

In the ground state, the electrons in atoms and solids occupy the lowest energy levels allowed by Pauli's exclusion principle which states that no two electrons in an atom or solid may have the same set of quantum numbers. Because of the two possible directions of the electron spin, each energy level can be occupied by two electrons.

The Fermi–Dirac distribution law gives the probability that an energy state is occupied by the electrons in thermal equilibrium. The Fermi–Dirac distribution function $f(E)$ is

$$f(E) = \frac{1}{\exp\{(E-\mu)/k_B T\}+1} \tag{2.6}$$

where μ is the chemical potential, k_B is the Boltzmann constant, and T is the temperature. It is a fairly good approximation to replace the chemical potential μ in Equation 2.6 by the Fermi energy E_F.

$$\begin{aligned} \mu &= E_F \quad \text{for} \quad T = 0 \\ \mu &\approx E_F \quad \text{for} \quad T \neq 0. \end{aligned} \tag{2.7}$$

The Fermi energy E_F is by definition the highest occupied energy state at zero temperature. At $T = 0$, the Fermi–Dirac distribution function $f(E)$ is a step function with $f(E) = 1$ for $E < E_F$ and $f(E) = 0$ for $E > E_F$. Figure 2.2 shows the Fermi–Dirac distribution function $f(E)$ for low and high temperatures. The Fermi energy $E_F = k_B T_F$ typically corresponds to temperatures T_F of more than 50 000 K. Room temperature may, therefore, be considered as low.

Next, the density of energy states $D(E) = dN_e(E)/dE$ will be considered. The energy levels in the ground state can be represented by points in **k**-space. The occupied states are within a sphere of radius k_F, where the Fermi wave number k_F is defined by

$$E_F = \frac{\hbar^2}{2m} k_F^2. \tag{2.8}$$

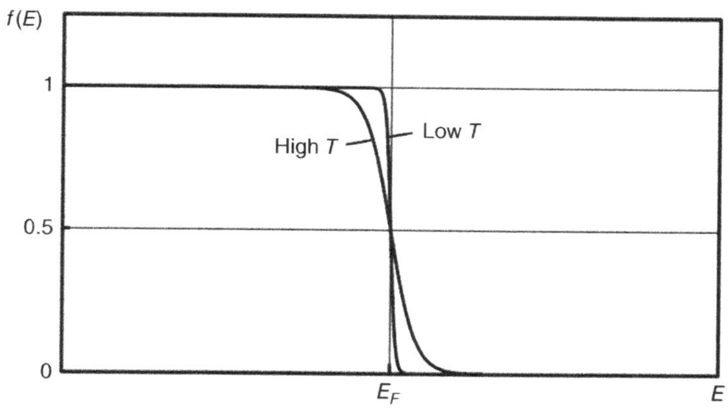

Figure 2.2 *Fermi–Dirac distribution for low and high temperatures (Reprinted from "High-Temperature Superconductors: Materials, Properties, and Applications" by Rainer Wesche, Kluwer Academic Publishers, 1998, ISBN 0-7923-8386-9, Figure 2-2, p. 9 with kind permission from Springer Science+Business Media B.V.)*

According to Equation 2.4, each energy state in an isotropic solid requires a volume of $(2\pi/L)^3$ in **k**-space. Taking into account that each of these states can be occupied by two electrons having opposite spin directions, the number of electrons N_e within the Fermi sphere of volume $(4/3)\pi k_F^3$ is

$$N_e = \frac{V}{3\pi^2} k_F^3 \tag{2.9}$$

where $V = L^3$ is the volume of the solid. Using Equation 2.9, the Fermi wave number can be related to the electron concentration $n_e = N_e/V$, and for the Fermi energy E_F, we obtain

$$E_F = \frac{\hbar^2}{2m}\left(\frac{3\pi^2 N_e}{V}\right)^{2/3} = \frac{\hbar^2}{2m}(3\pi^2 n_e)^{2/3} \tag{2.10}$$

Rearranging Equation 2.10 leads to

$$N_e(E) = \frac{V}{3\pi^2}\left(\frac{2mE}{\hbar^2}\right)^{3/2} \tag{2.11}$$

where $N_e(E)$ is the number of states below the energy E. The resulting density of states $D(E)$ is

$$D(E) = \frac{dN_e(E)}{dE} = \frac{V}{2\pi^2}\left(\frac{2m}{\hbar^2}\right)^{3/2} E^{1/2} \tag{2.12}$$

Figure 2.3 shows the density of states for a free electron gas. The number of occupied energy states between E and $E + dE$ is $D(E)f(E)$, as indicated in Figure 2.3.

As a consequence of the very high Fermi temperatures $T_F \approx 50\,000$ K, only about 1% of the conduction electrons in a metal can reach energy states above the Fermi energy at room temperature. This explains the very small contribution of the electron gas ($\approx 0.01 \times (3/2)k_B T$) to the specific heat of metals.

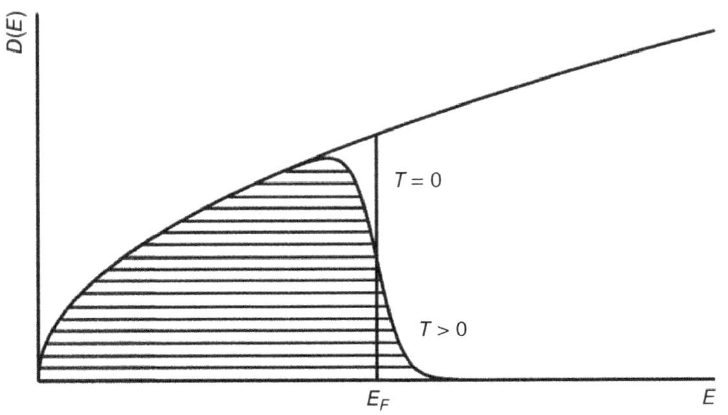

Figure 2.3 Density of states D(E) and occupation of the energy states for a free electron gas (Reprinted from "High-Temperature Superconductors: Materials, Properties, and Applications" by Rainer Wesche, Kluwer Academic Publishers, 1998, ISBN 0-7923-8386-9, Figure 2-3, p. 10 with kind permission from Springer Science+Business Media B.V.)

So far, the periodically varying potential of the positive ions has been neglected. Taking this potential into consideration, the solutions of Schrödinger's equation are Bloch functions

$$\psi(\mathbf{r}) = \exp(i\mathbf{k} \cdot \mathbf{r})\, u(\mathbf{r}) \tag{2.13}$$

where the modulating amplitude $u(\mathbf{r}) = u(\mathbf{r} + \mathbf{x}_\mathrm{L})$ reflects the periodicity of the lattice. Thus, the vector $\mathbf{x}_\mathrm{L}$ is a lattice vector. The motion of the electron waves in the crystal lattice is similar to that of electromagnetic waves. Scattering of X-rays is allowed only in the directions given by the Bragg condition. In a one-dimensional lattice of spacing a, the Bragg condition is satisfied by the electrons with wave vectors of $\pm\pi/a \pm 2\pi/a \pm 3\pi/a \pm \cdots$. These k-values are the boundaries of the Brillouin zones. Bragg scattering leads to the interference of the electron waves propagating in opposite directions. From this interference, for example for $k = \pi/a$, the following two different standing waves result:

$$\begin{aligned} \psi_+ &\propto \left\{\exp\left(ix\frac{\pi}{a}\right) + \exp\left(-ix\frac{\pi}{a}\right)\right\} = 2\cos\left(x\frac{\pi}{a}\right) \\ \psi_- &\propto \left\{\exp\left(ix\frac{\pi}{a}\right) - \exp\left(-ix\frac{\pi}{a}\right)\right\} = 2i\sin\left(x\frac{\pi}{a}\right) \end{aligned} \tag{2.14}$$

The probability of finding the electron in an interval dx around the position x is given by the probability density $\psi(x)\psi^*(x)$, where $\psi^*(x)$ is the complex conjugate of $\psi(x)$. The maximum probability density is at the positions of the positive ions for $\psi_+\psi_+*$, whereas it is minimum at the positions for $\psi_-\psi_-*$. Therefore, the potential energies for the wave functions ψ_+ and ψ_- are different, leading to an energy gap at the boundaries of the Brillouin zones. Figure 2.4 shows the dependence of the electron energy E on the wave number k for a one-dimensional periodic potential, together with the band structure resulting from the energy gaps E_g at the k-values of $\pm\pi/a$, $\pm 2\pi/a$, and $\pm 3\pi/a$. The energy of electrons in a periodic potential can be expressed by the free electron Equation (2.5) if the free electron mass is replaced by an effective mass m^*:

$$E_\mathbf{k} = \frac{\hbar^2}{2m^*}k^2 \tag{2.15}$$

where

$$\frac{1}{m^*(k)} = \frac{1}{\hbar^2}\frac{d^2E}{dk^2} \tag{2.16}$$

The difference between m and m^* is a consequence of the electron–lattice interaction. In the vicinity of the boundaries of the Brillouin zones, m^* may deviate considerably from the free electron mass m. A more detailed description of the band structure of solids can be found in [1].

The band structures of insulators, semiconductors, and metals are illustrated schematically in Figure 2.5. In insulators, the valence band is fully occupied by electrons, whereas the conduction band is completely empty. Because the energy gap E_g between the valence and conduction bands is typically larger than 2 eV for insulators, thermal energy is not sufficient to excite electrons into the conduction band. As a consequence of the exclusion principle and the lack of empty states in the uppermost occupied band, the electrons cannot be moved by an applied electric field.

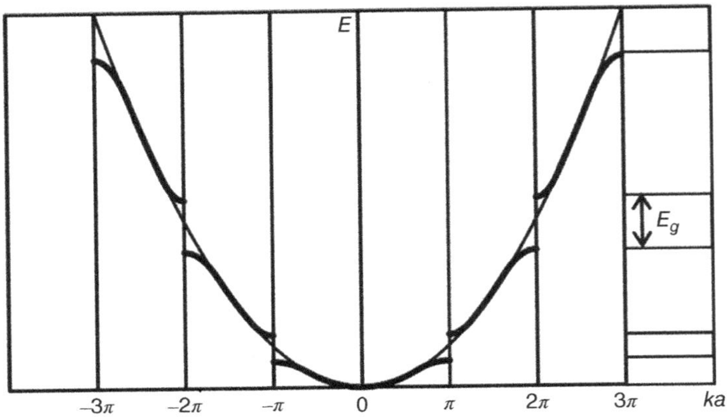

Figure 2.4 Energy versus wave number k for electrons in a one-dimensional periodic potential. The energy gaps at the boundaries of the Brillouin zones lead to the band structure shown on the right (Reprinted from "High-Temperature Superconductors: Materials, Properties, and Applications" by Rainer Wesche, Kluwer Academic Publishers, 1998, ISBN 0-7923-8386-9, Figure 2-4, p. 12 with kind permission from Springer Science+Business Media B.V.)

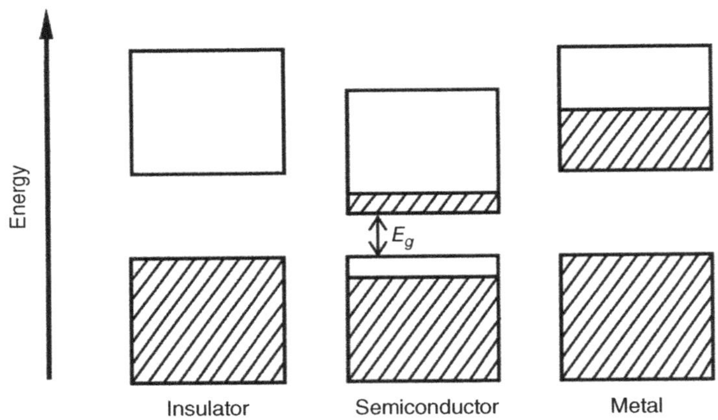

Figure 2.5 Schematic illustration of the band structures of insulators, pure semiconductors, and metals (Reprinted from "High-Temperature Superconductors: Materials, Properties, and Applications" by Rainer Wesche, Kluwer Academic Publishers, 1998, ISBN 0-7923-8386-9, Figure 2-5, p. 13 with kind permission from Springer Science+Business Media B.V.)

Pure semiconductors are also insulators at $T = 0$. However, the energy gap E_g of semiconductors is only around 1 eV or less. The relatively small energy gap allows the thermal excitation of a few electrons from the valence band into the conduction band. In this process, holes are created in the valence band, which behave like positive electrons. Both the electrons in the conduction band and the holes in the valence band contribute to the electrical conductivity of semiconductors. The number of electron–hole pairs is proportional to $\exp(-E_g/2k_BT)$. Thus, the number of electron–hole pairs increases with increasing temperature, leading to an enhanced intrinsic conductivity of semiconductors at higher temperatures.

In metals, the Fermi energy E_F is within the conduction band. Because of the empty energy states available in the conduction band, the electrons in a metal can be accelerated by an applied electric field $\mathbf{E_e}$. Figure 2.6 (left) shows the occupation of states for $\mathbf{E_e} = 0$. Equal numbers of electrons move in the x and $-x$ directions, and the resulting current is zero. On the other hand, for an electric field $\mathbf{E_e}$ applied in the $-x$ direction (right), more electrons travel in the $+x$ direction than in the $-x$ direction, and a net current flows.

Propagation without resistance would be expected for electrons in a perfectly periodic lattice. However, impurity atoms, lattice defects, and the thermally activated lattice vibrations (phonons) disturb the periodicity of the lattice. As a consequence, the electrons are scattered at these lattice imperfections, transferring energy from the electrons to the lattice. Between successive scattering processes, the electrons are accelerated by the applied electric field $\mathbf{E_e}$. The resulting drift velocity of the electrons is

$$\mathbf{v} = \frac{e\mathbf{E}_e}{m^*}\tau \qquad (2.17)$$

where the electron's charge e is understood to be negative and m^* is its effective mass. The relaxation time τ is of the order of the time between two successive collisions. The electrical

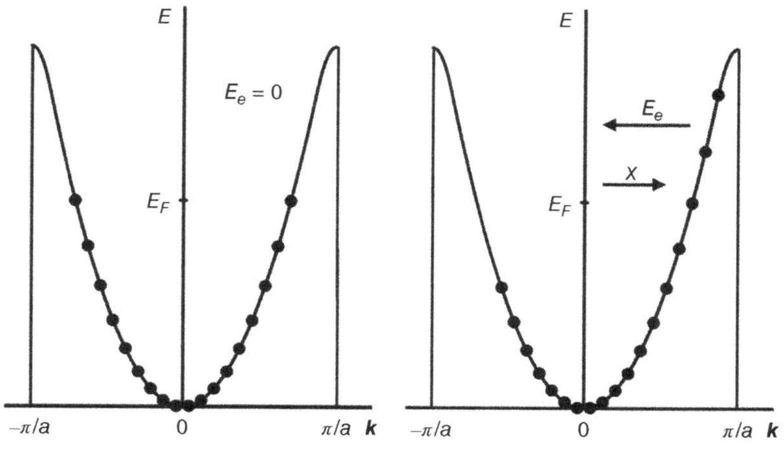

Figure 2.6 Occupation of energy states in a one-dimensional metal for $\mathbf{E_e} = 0$ (left) and an electric field $\mathbf{E_e}$ applied in the $-x$ direction (right) (Reprinted from "High-Temperature Superconductors: Materials, Properties, and Applications" by Rainer Wesche, Kluwer Academic Publishers, 1998, ISBN 0-7923-8386-9, Figure 2-6, p. 13 with kind permission from Springer Science+Business Media B.V.)

conductivity σ can be written as

$$\sigma = \frac{n_e e^2 \tau}{m^*} \tag{2.18}$$

where n_e is the number density of electrons. The resistivity of many metals can be well described by Matthieson's rule

$$\rho(T) = \rho_0 + \rho_i(T) \tag{2.19}$$

The temperature-independent residual resistivity ρ_0 is caused by impurities and lattice defects. Typically, 1 at% of impurities leads to a residual resistivity of 1 μΩ cm. The intrinsic resistivity $\rho_i(T)$ due to the electron–phonon interaction is a characteristic property of the host metal. At sufficiently large temperatures, the resistivity of metals is dominated by the phonon contribution $\rho_i(T)$, and a linear temperature dependence results in $\rho(T)$ above room temperature:

$$\rho(T) = \rho(300\ \text{K}) + k_1 T \tag{2.20}$$

At low temperatures, the intrinsic resistivity is typically proportional to T^5. Thus, the resistivity at very low temperatures is given by

$$\rho(T) = \rho_0 + k_2 T^5 \tag{2.21}$$

The constants k_1 and k_2 are characteristic of the metal in question. Figure 2.7 shows the temperature dependence of the electrical resistivity of copper [2] for the selected values of the residual resistivity ratio (RRR), which is defined as

$$\text{RRR} = \frac{\rho(273\ \text{K})}{\rho(4\ \text{K})} \tag{2.22}$$

RRR increases with increasing purity, reaching values above 10 000 in very pure metals.

[2]So far, we have considered the conduction electrons in a metal as a Fermi gas. In this model, the interactions among the electrons are neglected. In the expressions for the free electron gas, the interaction of the electrons with the atomic lattice has been taken into account by the use of an effective mass m^*, which may be considerably different from the mass of a free electron. The Fermi liquid theory, developed by Landau in 1957 [3, 4], takes into account the interactions of the electrons in a nonsuperconducting metal. The Landau–Fermi liquid theory indicates that there exist close correspondences between the physical properties of a Fermi gas and a Fermi liquid. Other examples of Fermi liquids are liquid ^{3}He, and the protons and neutrons in the atomic nuclei. The electrons (quasi-particles) in a Fermi liquid obey both Pauli's exclusion principle and Fermi statistics. The interacting fermions are replaced by noninteracting quasi-particles, characterized by the same spin, charge, and momentum of the original particles. The specific heat (electronic contribution linear in temperature), the compressibility, and the spin susceptibility, all show a similar dependence on temperature as a Fermi gas. However, the absolute values may be significantly different. The resistance at low temperatures is

[2] Next three paragraphs: Material added for the present monograph not reprinted from "*High-Temperature Superconductors: Materials, Properties, and Applications*" by Rainer Wesche, Kluwer Academic Publishers, 1998, ISBN 0-7923-8386-9.

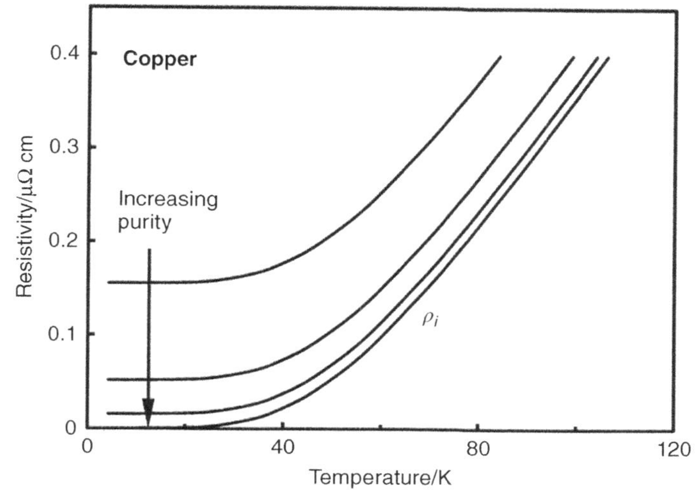

Figure 2.7 *Resistivity of copper for RRR-values of 10, 30, and 100. At very low temperatures, the intrinsic resistivity $\rho_i(T)$ due to the electron–phonon interaction approaches zero ($\rho_i(T)$ data from [2]) (Reprinted from "High-Temperature Superconductors: Materials, Properties, and Applications" by Rainer Wesche, Kluwer Academic Publishers, 1998, ISBN 0-7923-8386-9, Figure 2-7, p. 15 with kind permission from Springer Science+Business Media B.V.)*

dominated by electron–electron scattering in combination with Umklapp scattering, and a quadratic dependence on temperature can be expected for a Fermi liquid. In systems with strongly interacting, correlated electrons, the Fermi liquid behavior may break down. In the high-temperature cuprate superconductors, some evidence for a non-Fermi liquid behavior has been observed.

In addition to the band insulators, in which the unavailability of empty states and Pauli's exclusion principle are responsible for the insulating character of the material as has been already discussed, there exists a second class of insulators known as Mott insulators. The fundamental work in this field was performed by the British physicist Mott [5]. For example, the parent compounds of cuprate superconductors are Mott insulators with a half-filled conduction band. Band theory would predict metallic behavior for such a material. However, in a Mott insulator, the strong Coulomb repulsion of the electrons prevents the occupation of the same state by a second electron with an opposite spin direction. The energy barrier for this process is of the order of several electron volts. In spite of the availability of empty states, the material is insulating because of the strong Coulomb interaction among the electrons hindering their motion.

In general, changes of the distances of neighboring atoms can lead to a Mott transition. At sufficiently large inter-atomic distances, the broad conduction band of a metal can be separated into distinct atomic levels, and, hence, there occurs a transition from delocalized conduction electrons to electrons bound to single atoms [1, 6]. Doping or application of pressure can transform a Mott insulator to a metallic state. Examples of Mott insulators are NiO, FeO, and CoO.

2.3 Characteristic Properties of Superconductors[3]

In the previous section, the temperature dependence of the resistivity of metals in the normal state was described. The resistivity in the normal state decreases continuously and reaches a constant value ρ_0 at low temperatures. Figure 2.8 shows resistance versus temperature for a low-temperature superconductor. At the transition temperature T_c, the resistance drops abruptly to an immeasurably small value. This behavior is remarkably different from the steadily decreasing resistance of nonsuperconducting metals (see Figure 2.7) and suggests the existence of a physically different superconducting state. In pure metals, the zero resistance state can be reached within a temperature range of 1 mK. In the case of impure metals, the transition to the superconducting state may be considerably broadened. A transition width of $\approx$0.05 K has been observed for impure tin [7]. Above the critical temperature T_c, the metal is in the normal state and the resistance is proportional to T^5. In many metals, the exponent is between 2 and 6, considerably different from the value of 5 predicted by the Bloch theory [2].

Figure 2.9 shows resistance versus temperature for a single- and a multiphase high-temperature superconductor. Normal and superconducting metals are simple. Bi-, Tl-, and Hg-based cuprate superconductors are chemically complex materials, in which there may exist several superconducting phases in one specimen. A two-step transition reflects the presence of at least two superconducting phases. Generally, the transition to the superconducting state, even in single-phase materials, is less sharp than in metallic low-temperature superconductors. The transition width ΔT_c for single-phase high-temperature superconductors is typically $\approx$1 K. In epitaxial $YBa_2Cu_3O_7$ films, ΔT_c-values as small as 0.3 K have been achieved [8]. Typically, above T_c a linear dependence of resistivity on temperature is observed.

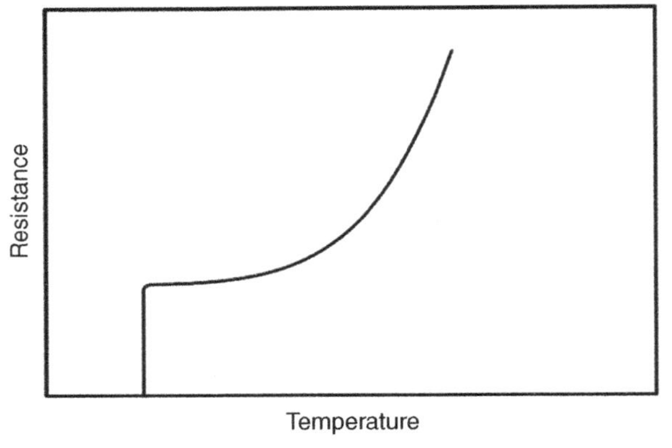

Figure 2.8 Resistance versus temperature for a low-temperature superconductor (Reprinted from "High-Temperature Superconductors: Materials, Properties, and Applications" by Rainer Wesche, Kluwer Academic Publishers, 1998, ISBN 0-7923-8386-9, Figure 2-8, p. 16 with kind permission from Springer Science+Business Media B.V.)

[3] Reprinted with minor changes from "*High-Temperature Superconductors: Materials, Properties, and Applications*" by Rainer Wesche, Kluwer Academic Publishers, 1998, ISBN 0-7923-8386-9, pp. 16–24 with kind permission from Springer Science+Business Media B.V.

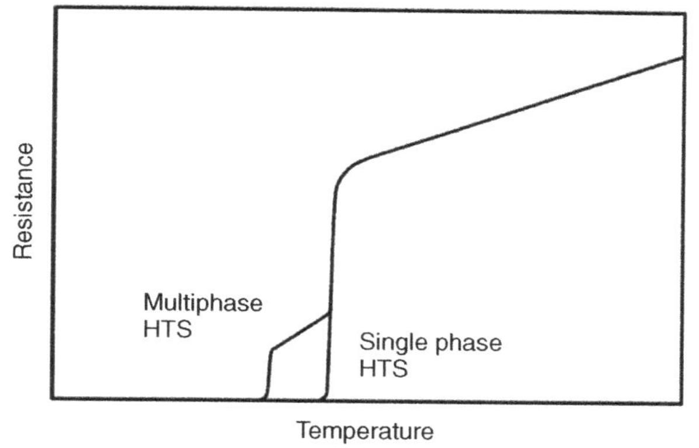

Figure 2.9 *Resistance versus temperature curves of a single- and a multiphase high-temperature superconductor (Reprinted from "High-Temperature Superconductors: Materials, Properties, and Applications" by Rainer Wesche, Kluwer Academic Publishers, 1998, ISBN 0-7923-8386-9, Figure 2-9, p. 17 with kind permission from Springer Science+Business Media B.V.)*

Figure 2.10 shows the resistance as a function of temperature for a multifilamentary Ag/Bi-2212 wire of 1 mm diameter. Some definitions of the critical temperature T_c used in the literature are illustrated in Figure 2.10. Because of the broadening of the superconducting transition in high-temperature superconductors, the T_c-value can be defined in several ways. The first deviation from the linear behavior in the normal state defines T_c(onset). Another way to define T_c is as the endpoint of the superconducting transition, where the resistance has fallen to an immeasurably small value. Generally, the determination of the onset or the endpoint of the transition is difficult. It is, therefore, more convenient to define T_c by the midpoint of the superconducting transition T_c(mid). The T_c-values corresponding to 10% and 90% of the normal state resistance can be used to define a transition width $\Delta T_c = T_c(90\%) - T_c(10\%)$. The width ΔT_c of the superconducting transition shown in Figure 2.10 is about 1.2 K. Above T_c, the electrical conductivity of the Ag/Bi-2212 multicore wire is determined mainly by the silver matrix.

Typically, a resistance drop of several orders of magnitude is observed at the transition temperature T_c. However, it is in principle impossible to prove experimentally that the resistance in the superconducting state is in fact zero. The most efficient way to determine an upper limit of the resistance is to induce a current in a superconducting loop and to detect the decay of the magnetic fields produced by the supercurrents. The time dependence of the current I in the loop is given by $I(t) = I_0 e^{-t/\tau}$, where I_0 is the initial value of the current and t is the time which has elapsed since the supercurrent has been induced. The resistance R and self-inductance L of the superconducting loop determine the time constant $\tau = L/R$ for the decay of the current. Several investigators have performed such experiments and established upper limits for the resistivity in the superconducting state [9–13]. An upper limit of 3.6×10^{-23} Ω cm has been determined for the resistivity in low-temperature Type I superconductors [9]. As a consequence of flux creep effects, slightly decaying supercurrents have been observed in Type II superconductors (see Chapter 4). However, the creep rates are so small that the supercurrents in Nb_3Sn, 3Nb–Zr and Nb powder would die out

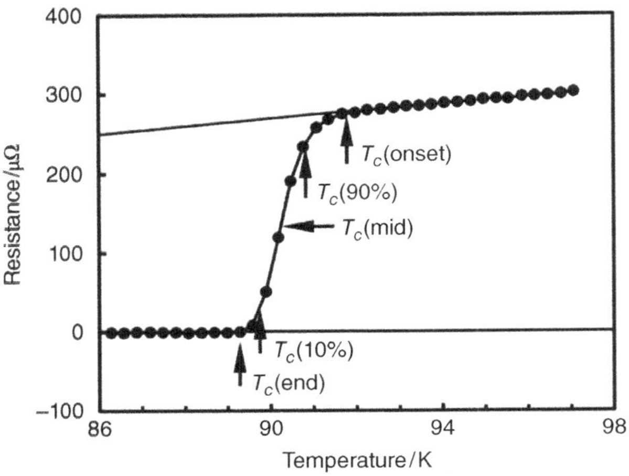

Figure 2.10 *Resistance versus temperature for a Ag/Bi-2212 multicore wire. The T_c-values resulting from different definitions of the critical temperature are indicated (Reprinted from "High-Temperature Superconductors: Materials, Properties, and Applications" by Rainer Wesche, Kluwer Academic Publishers, 1998, ISBN 0-7923-8386-9, Figure 2.10, p. 18 with kind permission from Springer Science+Business Media B.V.)*

after times of more than 3×10^{92} years [10]. Therefore, in any practical sense, the supercurrents can be considered as persistent currents. Upper resistivity limits of 2×10^{-18} [11] and 7×10^{-23} Ω cm [12] have been reported for YBaCuO at 77 K. As in conventional Type II superconductors, flux creep effects have been observed [11]. These values of the upper limit of the resistivity in the superconducting state have to be compared with the resistivity of annealed, very pure metals. The resistivity of aluminum with an RRR-value of 40 000 is $\approx 10^{-10}$ Ω cm at 4.2 K [2]. This value is several orders of magnitude larger than the upper limits found for a possible resistivity in the superconducting state. For all practical purposes, it is, therefore, justified to assume zero resistance below T_c.

In addition to resistanceless current transport, the superconducting state is characterized by perfect diamagnetism. The magnetic behavior of a superconductor is illustrated in Figure 2.11. Two different situations have to be distinguished. In the first case, the superconductor is cooled below T_c without an applied magnetic field (zero field-cooled ZFC, Figure 2.11, left). Below the transition temperature T_c, a magnetic field is applied. Due to the time variation of the magnetic field $dB/dt \neq 0$, persistent screening currents are induced in the surface layer of the superconductor. These currents generate a flux density opposite to that of the applied magnetic field. The magnetic flux density is, therefore, zero everywhere inside the superconductor. On the other hand, outside the superconducting sphere, the magnetic field is enhanced as a consequence of the superposition of the flux of the applied magnetic field and that of the screening currents flowing in the surface layer of the superconductor. A similar behavior would be expected for a perfect conductor with zero resistance. The superconductor is again unmagnetized when the applied magnetic field is removed.

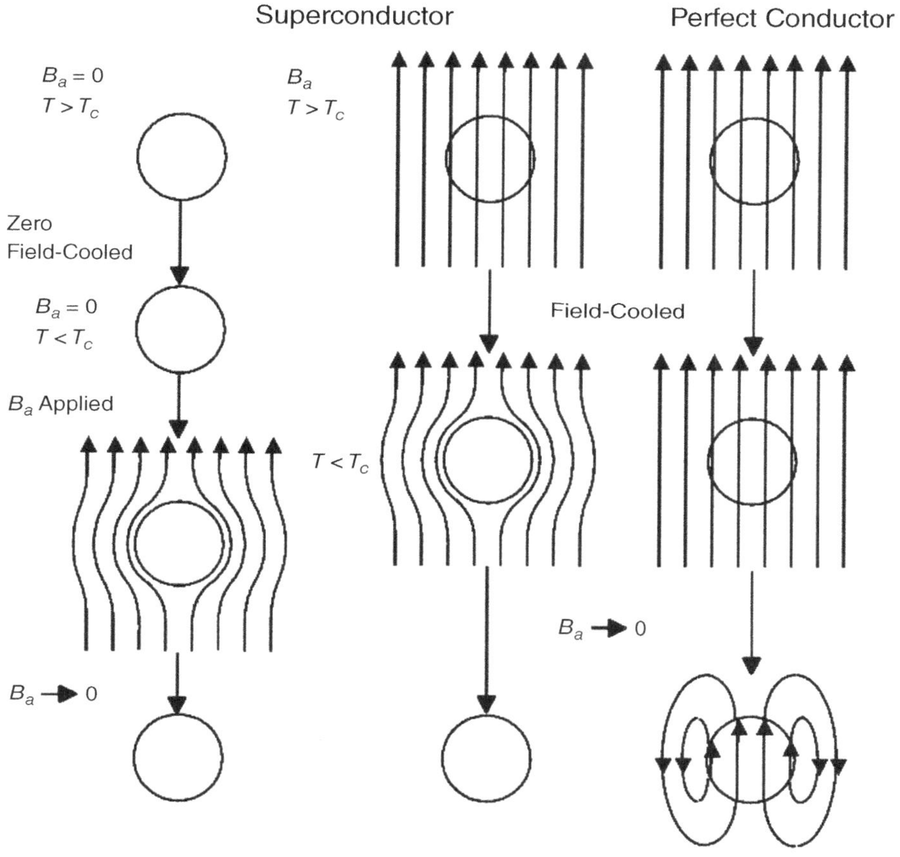

Figure 2.11 *The magnetic flux is excluded from the interior of a superconductor without field-cooling (left) as well as with it (center). In contrast to this behavior, a magnetic flux would exist in the interior of a field-cooled perfect conductor (right) (Reprinted from "High-Temperature Superconductors: Materials, Properties, and Applications" by Rainer Wesche, Kluwer Academic Publishers, 1998, ISBN 0-7923-8386-9, Figure 2-11, p. 20 with kind permission from Springer Science+Business Media B.V.)*

In the second case, the superconductor is cooled below T_c, while a magnetic field is applied (field-cooled FC, Figure 2.11, center). As soon as the temperature has fallen below T_c, the magnetic flux is excluded from the interior of the superconductor. This remarkable behavior of a superconductor is called the Meissner effect. Reduction of the magnetic field to zero leads to an unmagnetized superconductor. Independent of the way the superconducting state was reached there is no magnetic flux inside a superconductor. It is of importance to notice that perfect conductivity does not explain the Meissner effect. The different behavior of an FC perfect conductor is also shown in Figure 2.11 (right). The values of the relative magnetic permeability for nonferromagnetic metals are close to unity. Thus, the magnetic flux density inside the metal is nearly the same as that of the applied

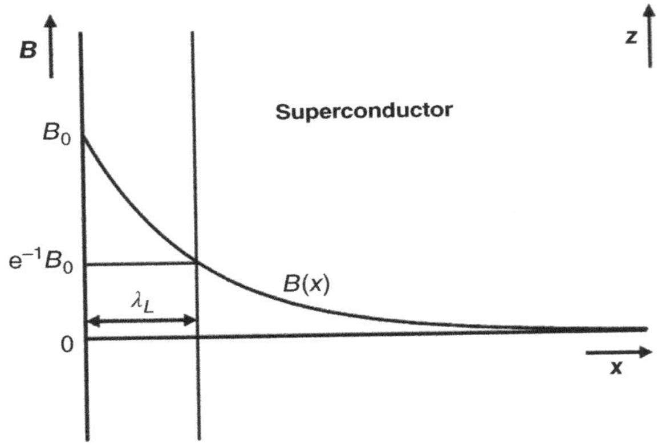

Figure 2.12 Penetration of an applied magnetic field into the surface layer of a semi-infinite superconducting plate (Reprinted from "High-Temperature Superconductors: Materials, Properties, and Applications" by Rainer Wesche, Kluwer Academic Publishers, 1998, ISBN 0-7923-8386-9, Figure 2-12, p. 21 with kind permission from Springer Science+Business Media B.V.)

magnetic field. Because $dB/dt = 0$, no screening currents are induced and the magnetic flux is, therefore, not removed from the interior of a perfect conductor at low temperatures. Finally, the magnetic field is removed ($dB/dt \neq 0$), and the perfectly conducting sphere is magnetized due to the induced currents.

Because of the limited critical current densities, the screening currents flow in a thin but finite layer, within which magnetic flux penetrates into the superconductor. Figure 2.12 shows the exponential decay of the magnetic field in the surface layer of a semi-infinite superconducting plate. The magnetic field at a depth x from the surface of the semi-infinite superconducting plate is

$$B(x) = B_0 \exp\left(-\frac{x}{\lambda_L}\right) \tag{2.23}$$

where B_0 is the magnetic field at the surface and λ_L is the penetration depth. Typical values of the penetration depth are 10^{-5}–10^{-6} cm at $T = 0$. The penetration depth λ_L is an important parameter in the London theory of superconductivity (see Section 2.4). The penetration depth λ_L depends on temperature and becomes infinite at the transition temperature T_c. The temperature dependence is given by

$$\lambda_L(T) = \lambda_L(0)\left(1 - \left(\frac{T}{T_c}\right)^4\right)^{-1/2} \tag{2.24}$$

where T is the temperature and $\lambda_L(0)$ is the penetration depth at zero temperature. Figure 2.13 shows the normalized penetration depth $\lambda_L/\lambda_L(0)$ as a function of reduced temperature T/T_c.

The magnetizing field **H**, the magnetic flux density **B**, and the magnetization **M** are related by the expression $\mathbf{B} = \mu_0(\mathbf{H} + \mathbf{M})$. For example, for a superconducting cylinder

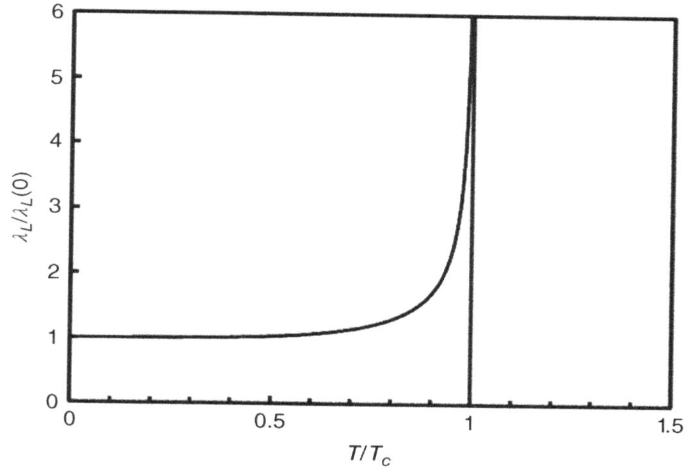

Figure 2.13 *Normalized penetration depth $\lambda_L/\lambda_L(0)$ versus reduced temperature (Reprinted from "High-Temperature Superconductors: Materials, Properties, and Applications" by Rainer Wesche, Kluwer Academic Publishers, 1998, ISBN 0-7923-8386-9, Figure 2-13, p. 21 with kind permission from Springer Science+Business Media B.V.)*

with H applied parallel to the axis, the B field, and the magnetization M are also parallel to the axis of the cylinder. The H fields inside and outside of the superconducting cylinder are equal ($H_a = H_i$). In the surface layer of the superconductor, the flux density decreases rapidly due to the magnetization generated by the screening currents. In the center of the cylinder, the magnetization cancels the applied field, i.e. $M_i = -H_i$, and, hence, $B_i = 0$. For simplicity, in the present book we will frequently denote the magnetic flux density B as the magnetic field.

The superconducting state can be destroyed by sufficiently large magnetic fields. The transition from the superconducting state occurs at a certain critical magnetic field B_c, which is the characteristic of the superconductor in question. The temperature dependence of the critical field B_c is typically well described by

$$B_c(T) = B_c(0)\left(1 - \left(\frac{T}{T_c}\right)^2\right) \tag{2.25}$$

where $B_c(0)$ is the critical field at $T = 0$. The resulting phase diagram is shown in Figure 2.14. The $B_c(T)$ curve separates the superconducting from the normal state. The Meissner effect ensures that the superconducting state is a state in the sense of thermodynamics. The properties at the point P_f in the superconducting state (see Figure 2.14) are independent of the order in which the final conditions of temperature and the applied magnetic field are reached. Figure 2.15 shows the critical field B_c as a function of temperature for selected superconducting elements. Table 2.1 gives the distribution of the superconducting elements in the periodic table, with the T_c- and B_c-values of elements that are superconducting at ambient pressure. These values and Equation 2.25 were used to

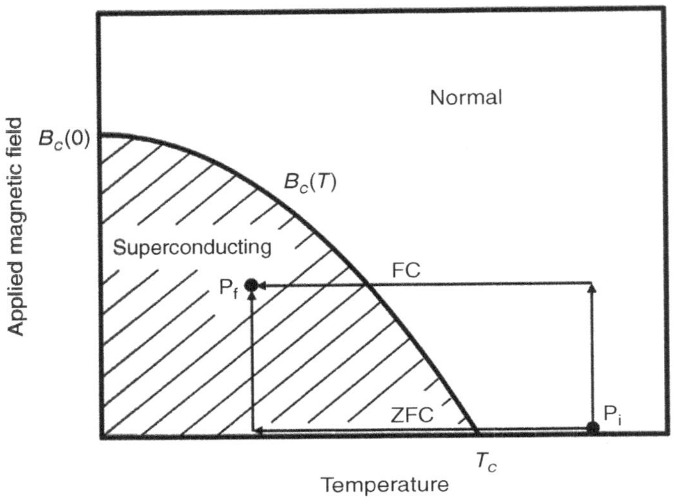

Figure 2.14 *Phase diagram for Type I superconductors (Reprinted from "High-Temperature Superconductors: Materials, Properties, and Applications" by Rainer Wesche, Kluwer Academic Publishers, 1998, ISBN 0-7923-8386-9, Figure 2-14, p. 22 with kind permission from Springer Science+Business Media B.V.)*

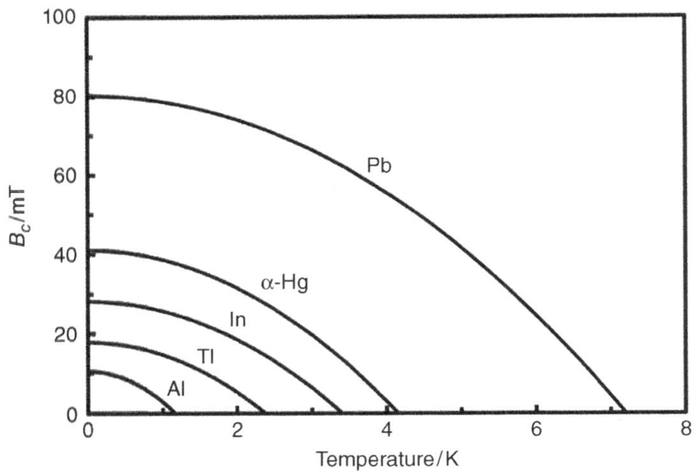

Figure 2.15 *Critical field B_c versus temperature for selected superconducting elements (Reprinted from "High-Temperature Superconductors: Materials, Properties, and Applications" by Rainer Wesche, Kluwer Academic Publishers, 1998, ISBN 0-7923-8386-9, Figure 2-15, p. 24 with kind permission from Springer Science+Business Media B.V.)*

Table 2.1 Distribution of superconductors in the periodic table (data taken from [14–16])

Superconducting Elements

T_c depending on crystal structure (shaded)

Superconductive under high pressure (hatched)

T_c (K)
B_c (mT)

Li	Be 0.026											B	C	N	O	F	Ne
Na	Mg											Al 1.175 10.5	Si	P	S	Cl	Ar
K	Ca	Sc	Ti 0.4 5.6	V 5.4 140.8	Cr	Mn	Fe	Co	Ni	Cu	Zn 0.85 5.4	Ga 1.083 5.83	Ge	As	Se	Br	Kr
Rb	Sr	Y	Zr 0.61 4.7	Nb 9.25 206	Mo 0.92 9.6	Tc 7.8 141	Ru 0.49 6.9	Rh $35\ 10^{-6}$	Pd	Ag	Cd 0.517 2.8	In 3.408 28.2	Sn w 3.722 30.5	Sb	Te	I	Xe
Cs	Ba	La β 6.00 110	Hf 0.128 1.27	Ta 4.47 82.9	W 0.015 0.12	Re 1.70 20.0	Os 0.66 7.0	Ir 0.113 1.6	Pt	Au	Hg α 4.154 41.1	Tl 2.38 17.8	Pb 7.196 80.3	Bi	Po	At	Rn
Fr	Ra	Ac															
			Ce	Pr	Nd	Pm	Sm	Eu	Gd	Tb	Dy	Ho	Er	Tm	Yb	Lu	
			Th 1.38 0.16	Pa 1.4	U α 0.2	Np 0.075	Pu	Am β 1.0	Cm	Bk	Cf	Es	Fm	Md	No	Lr	

Table 2.2 Critical temperatures T_c of selected elements superconductive only under high pressure[4]

Element	T_c(K)	Pressure (GPa)	Reference
Li	20	48	[17]
Ca	15	150	[18]
Ca	25	161	[19]
Fe	2	21	[21]
Si	8.5	12	[22, 23]
Ge	5.4	11.5	[15]
P	18	30	[24]
Bi	8.7	9	[25]
S	17	160	[26]
Se	7	13	[27]

calculate the $B_c(T)$ curves shown in Figure 2.15. In addition to the ferromagnetic metals, the noble metals and most of the alkali metals have been found to be nonsuperconducting in the temperature ranges investigated so far. Generally, both the transition temperature T_c and the critical field B_c depend on the crystal structure. For example, the T_c-values of hexagonal close-packed and face-centered cubic lanthanum are 4.88 K and 6.00 K, respectively.

Some of the elements are superconducting only at high pressure or in high pressure phases. In Table 2.2, the critical temperatures of selected elements superconductive only under high pressure are listed. The earth alkali metal calcium shows a critical temperature as high as 25 K at a pressure of 161 GPa, the highest T_c among the superconducting elements [19]. Other elements characterized by high-transition temperatures at high pressure are lithium (T_c = 20 K at 48 GPa) [15–17], sulfur (T_c = 17 K at 160 GPa) [26] and boron (T_c = 11.2 K at 250 GPa) [20]. At a pressure of 21 GPa, superconductivity has been observed even in iron [21]. At this pressure, iron crystallizes in a hexagonal close-packed structure and is supposed to be nonferromagnetic. Other elements show superconductivity only in certain allotropes. In the case of carbon, the allotropes diamond (boron doped) [28] and carbon nanotubes [29] have been found to be superconductive, whereas graphite is not superconducting. In some elements, superconductivity has been observed only in films. A chromium film embedded between two gold layers leading to a face-centered cubic crystal structure has been found to be superconductive below 3 K [30].

2.4 Superconductor Electrodynamics[5]

Two years after the discovery of the Meissner effect, a phenomenological theory of superconductivity was developed by London and London [31]. In the normal conducting

[4] Table 2.2 and following paragraph: Material added for the present monograph not reprinted from "*High-Temperature Superconductors: Materials, Properties, and Applications*" by Rainer Wesche, Kluwer Academic Publishers, 1998, ISBN 0-7923-8386-9.

[5] Reprinted with minor changes from "*High-Temperature Superconductors: Materials, Properties, and Applications*" by Rainer Wesche, Kluwer Academic Publishers, 1998, ISBN 0-7923-8386-9, pp. 51–56 with kind permission from Springer Science+Business Media B.V.

state, the current density $\mathbf{j}$ and the applied electric field $\mathbf{E_e}$ are connected by Ohm's law $\mathbf{j} = \sigma \mathbf{E_e}$, where σ is the conductivity. As a consequence of infinite conductivity in the superconducting state, Ohm's law has to be replaced by the first London equation

$$\frac{d\mathbf{j}}{dt} = \frac{1}{\mu_0 \lambda_{\mathrm{Lo}}^2} \mathbf{E_e} \tag{2.26}$$

where μ_0 is the permeability of free space and λ_{Lo} is a constant. The electric field in the superconductor is zero for constant current density. It may be noticed that time varying currents produce an electric field within the superconductor. The second London equation gives a relation between the current density $\mathbf{j}$ and the magnetic field $\mathbf{B}$:

$$\mathrm{curl}\, \mathbf{j} = -\frac{1}{\mu_0 \lambda_{\mathrm{Lo}}^2} \mathbf{B} \tag{2.27}$$

In quantum physics, the momentum $\mathbf{p}$ has to be replaced by the operator $-i\hbar\nabla$; the following expression results for the current density $\mathbf{j}$ (see also Chapter 3):

$$\mathbf{j} = q_C \psi^* \mathbf{v} \psi = \frac{n_C q_C}{m_C} (\hbar \nabla \varphi(\mathbf{r}) - q_C \mathbf{A}) \tag{2.28}$$

where $\mathbf{A}$ is the vector potential, $\mathbf{v}$ is the velocity, n_c is the concentration, m_c is the mass, q_C is the charge, and ψ is the collective wave function of the Cooper pairs, given by

$$\psi = n_C^{1/2} \exp(i\varphi(\mathbf{r})), \qquad \text{with} \quad n_C = \psi^* \psi \tag{2.29}$$

Here, $\varphi(\mathbf{r})$ is the phase of the wave function (see also Chapter 3). Rewriting Equation 2.28 and taking into account that $\nabla \times \nabla \varphi(\mathbf{r}) = 0$, the following expression is obtained:

$$\mathrm{curl}\, \mathbf{j} = -\frac{n_C q_C^2}{m_C} \mathbf{B} \tag{2.30}$$

A comparison of Equations 2.27 and 2.30 allows us to write

$$\lambda_{\mathrm{Lo}} = \sqrt{\frac{m_C}{\mu_0 n_C q_C^2}} \tag{2.31}$$

Next, it will be shown that the second London equation includes the Meissner effect. The London equations do not replace the Maxwell equations but are additional conditions which are met by the supercurrents. The electric and magnetic fields are connected by the Maxwell equation

$$\mathrm{curl}\, \mathbf{E_e} = -\frac{\partial \mathbf{B}}{\partial t} \tag{2.32}$$

Using Equation 2.26, the electric field $\mathbf{E_e}$ can be replaced by the current density $\mathbf{j}$ and the following expression results:

$$\mathrm{curl} \left(\mu_0 \lambda_{\mathrm{Lo}}^2 \frac{\partial \mathbf{j}}{\partial t} \right) = -\frac{\partial \mathbf{B}}{\partial t} \tag{2.33}$$

Equation 2.33 describes the behavior of a perfect conductor with infinite conductivity. The magnetic field in a perfect conductor depends on the initial conditions (see Figure 2.11). However, as a consequence of the Meissner effect, the magnetic field in the interior of a sufficiently large superconductor is zero regardless of the initial conditions. The second London equation 2.27 can be obtained by the time integration of Equation 2.33 under the condition that the integration constant is zero. Thus, the second London equation ensures that the magnetic field in the interior of a superconductor is zero whatever the initial conditions. In addition to the second London equation 2.27, the current density **j** and the magnetic field **B** are connected by the following Maxwell equation:

$$\text{curl}\,\mathbf{B} = \mu_0 \mathbf{j} \tag{2.34}$$

Combining Equations 2.27 and 2.34 leads to

$$\text{curl curl}\,\mathbf{B} = \mu_0\,\text{curl}\,\mathbf{j} = -\frac{1}{\lambda_{\text{Lo}}^2}\mathbf{B}. \tag{2.35}$$

Taking into consideration that $\nabla \cdot \mathbf{B} = 0$, the following differential equation results for the magnetic field:

$$\nabla^2\mathbf{B} = \frac{1}{\lambda_{\text{Lo}}^2}\mathbf{B} \tag{2.36}$$

Using Equation 2.36, the field distribution in a semi-infinite superconductor will be calculated. Assuming that the magnetic field is applied along the z direction, Equation 2.36 reduces to

$$\frac{d^2B(x)}{dx^2} - \frac{1}{\lambda_{\text{Lo}}^2}B(x) = 0 \tag{2.37}$$

A solution of (2.37) is

$$B(x) = B(0)\exp\left(-\frac{x}{\lambda_{\text{Lo}}}\right) \tag{2.38}$$

Thus, the magnetic field decays exponentially within the surface layer of the superconductor with a characteristic length λ_{Lo}, called the London penetration depth. Equation 2.31 indicates that λ_{Lo} is proportional to $n_C^{-1/2}$. The penetration depth increases rapidly in the vicinity of T_c (see Equation 2.24). This is a consequence of the decreasing number density of Cooper pairs at temperatures near T_c. Figure 2.16 shows the exponential decay of the magnetic field in the surface layer of a semi-infinite superconductor. It may be remarked that the decay of the magnetic field in real superconductors deviates considerably from the exponential law given by Equation 2.38.

A more general definition of the penetration depth is

$$\lambda_L = \frac{1}{B(0)}\int_0^\infty B(x)\,dx \tag{2.39}$$

Measured values of the penetration depth λ_L are typically a factor of two larger than the values predicted by the London theory [7].

Finally, field penetration in a thin superconducting slab will be considered. The magnetic field B is again applied along the z direction; the width of the slab is $2a$. This situation is described by Equation 2.37. In contrast to the case of a semi-infinite superconductor,

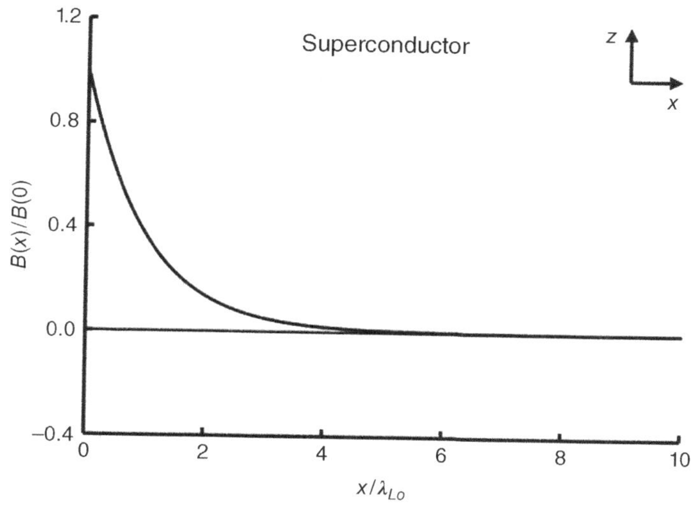

Figure 2.16 Exponential decay of an applied magnetic field in the surface layer of a semi-infinite superconductor (Reprinted from "High-Temperature Superconductors: Materials, Properties, and Applications" by Rainer Wesche, Kluwer Academic Publishers, 1998, ISBN 0-7923-8386-9, Figure 2-36, p. 54 with kind permission from Springer Science+Business Media B.V.)

both linearly independent solutions of this equation are required for the slab geometry. The general solution is

$$B(x) = B_1 \exp\left(-\frac{x}{\lambda_{\text{Lo}}}\right) + B_2 \exp\left(\frac{x}{\lambda_{\text{Lo}}}\right) \tag{2.40}$$

where the constants B_1 and B_2 depend on the boundary conditions. Taking into account that for the slab $B(-a) = B(a) = B_0$, the following solution results:

$$B(x) = B_0 \frac{\cosh \frac{x}{\lambda_{\text{Lo}}}}{\cosh \frac{a}{\lambda_{\text{Lo}}}} \tag{2.41}$$

Figure 2.17 shows calculated field profiles for selected values of the slab thickness $2a$. Even in the center of a thin superconducting slab, a considerable magnetic field is present. The field in the center reaches the values of $0.887B_0$, $0.648B_0$, and $0.266B_0$ for the slabs of thicknesses λ_{Lo}, $2\lambda_{\text{Lo}}$, and $4\lambda_{\text{Lo}}$, respectively.

Using Equation 2.34, the resulting screening current densities $j_s(x)$ can be calculated

$$j_s(x) = -\frac{1}{\mu_0}\frac{dB_z(x)}{dx} = -\frac{B_0 \sinh \frac{x}{\lambda_{\text{Lo}}}}{\mu_0 \lambda_{\text{Lo}} \cosh \frac{a}{\lambda_{\text{Lo}}}} \tag{2.42}$$

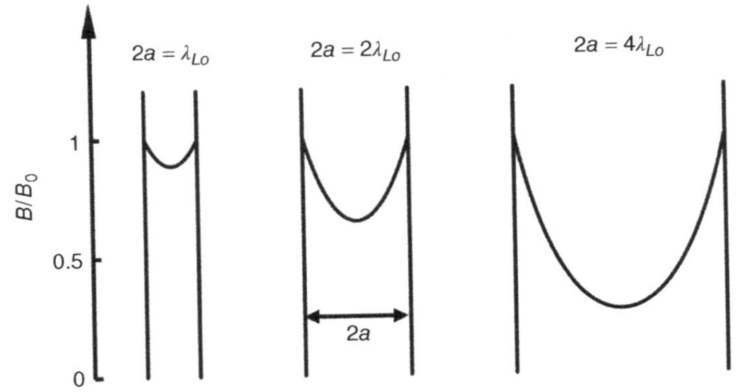

Figure 2.17 Field profiles in thin superconducting slabs with widths 2a of λ_{Lo}, $2\lambda_{Lo}$, and $4\lambda_{Lo}$. Even at the center of thin slabs a considerable magnetic field is present (Reprinted from "High-Temperature Superconductors: Materials, Properties, and Applications" by Rainer Wesche, Kluwer Academic Publishers, 1998, ISBN 0-7923-8386-9, Figure 2-37, p. 55 with kind permission from Springer Science+Business Media B.V.)

The screening currents flow in the y direction. The maximum screening current density j_{s0} at the surface of the slab is

$$j_{s0} = \frac{B_0}{\mu_0 \lambda_{\mathrm{Lo}}} \tanh \frac{a}{\lambda_{\mathrm{Lo}}}. \tag{2.43}$$

Inserting j_{s0} into Equation 2.42 leads to

$$j_s(x) = -j_{s0} \frac{\sinh \frac{x}{\lambda_{\mathrm{Lo}}}}{\sinh \frac{a}{\lambda_{\mathrm{Lo}}}} \tag{2.44}$$

The sign in Equation 2.44 gives the direction of the screening currents: a positive sign corresponds to a current in the $+y$ direction. Figure 2.18 shows the resulting screening current densities in superconducting slabs with widths $2a$ of λ_{Lo}, $2\lambda_{\mathrm{Lo}}$, and $4\lambda_{\mathrm{Lo}}$.

2.5 Thermodynamics of Superconductors[6]

In this section, the laws of thermodynamics will be used to gain deeper insights into the properties of the superconducting state. First, the temperature behavior of the specific heat $C_n(T)$ in the normal state will be briefly described. The specific heat $C_n(T)$ will then be used to derive expressions for the Gibbs free energy and the entropy of normal conductors. The phonon contribution C_{Ph} to the specific heat can be well described by the Debye approximation [1]

$$C_{\mathrm{Ph}} = 9Nk_B \left(\frac{T}{\theta_D}\right)^3 \int_0^{x_D} \frac{x^4 \exp x}{(\exp x - 1)^2} dx \tag{2.45}$$

[6] Reprinted with minor changes from "*High-Temperature Superconductors: Materials, Properties, and Applications*" by Rainer Wesche, Kluwer Academic Publishers, 1998, ISBN 0-7923-8386-9, pp. 56–65 with kind permission from Springer Science+Business Media B.V.

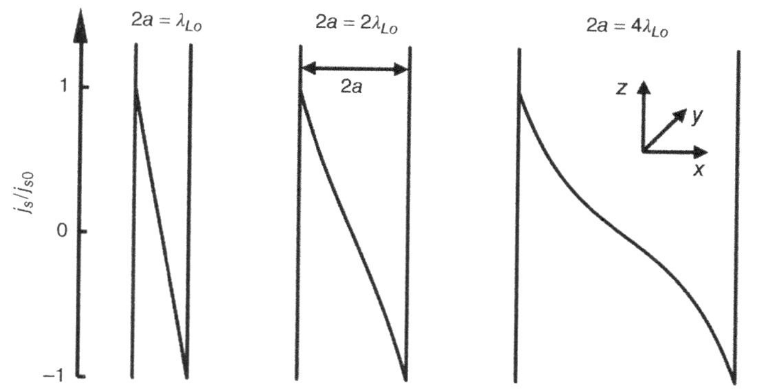

Figure 2.18 *Normalized screening current densities j_s/j_{s0} in thin superconducting slabs with widths 2a of λ_{Lo}, $2\lambda_{Lo}$, and $4\lambda_{Lo}$. Positive values of j_s correspond to screening currents flowing in the +y direction. A nearly linear dependence of j_s on the position in the x direction results for very thin slabs (Reprinted from "High-Temperature Superconductors: Materials, Properties, and Applications" by Rainer Wesche, Kluwer Academic Publishers, 1998, ISBN 0-7923-8386-9, Figure 2-38, p. 55 with kind permission from Springer Science+Business Media B.V.)*

where N is the number of atoms in the solid, k_B is Boltzmann's constant, θ_D is the Debye temperature, and $x_D = \theta_D/T$. The Debye temperature θ_D is connected to the Debye frequency ω_D by

$$\theta_D = \frac{\hbar\omega_D}{k_B} \tag{2.46}$$

Figure 2.19 shows the phonon-specific heat C_{Ph} as a function of temperature. Well above the Debye temperature θ_D, the specific heat C_{Ph} reaches the classical value of $3R$, where $R = N_A\,k_B$ ($N_A = 6.023 \times 10^{23}$) is the gas constant. On the other hand, the phonon contribution is proportional to T^3 for temperatures below $0.1\theta_D$. At low temperatures, the phonon-specific heat is found from Equation 2.45 to be [1]

$$C_{\text{Ph}} = 234Nk_B\left(\frac{T}{\theta_D}\right)^3 \tag{2.47}$$

In addition to the phonons, the free electron gas contributes to the specific heat of a conductor. However, this contribution is typically less than 1% of the phonon-specific heat at room temperature. As a consequence of Pauli's exclusion principle, thermal excitation is limited to those electrons that can reach empty energy states above the Fermi energy E_F (see Figures 2.2 and 2.3). Therefore, only a small fraction k_BT/E_F of the free electrons contribute to the specific heat of a conductor in the normal state. The electron-specific heat C_{el} can be written as [1]

$$C_{\text{el}} = \frac{1}{3}\pi^2 D(E_F)k_B^2 T \tag{2.48}$$

where $D(E_F)$ is the density of states at the Fermi energy. At low temperatures $T \ll \theta_D$, the specific heat of a conductor in the normal state can be written as

$$C_n = \gamma T + A_{\text{Ph}}T^3 \tag{2.49}$$

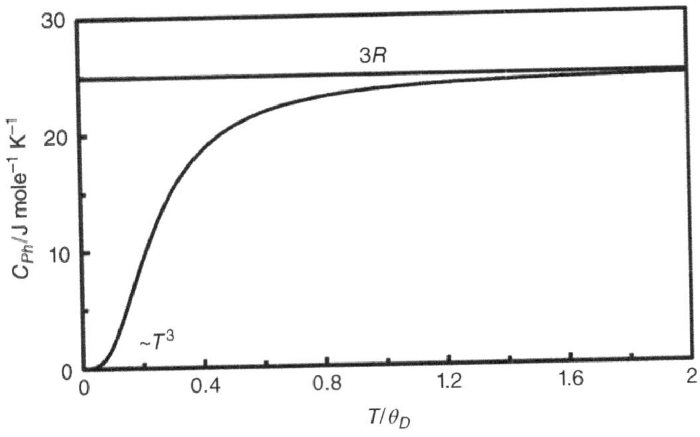

Figure 2.19 *Phonon-specific heat C_{Ph} according to the Debye approximation (Reprinted from "High-Temperature Superconductors: Materials, Properties, and Applications" by Rainer Wesche, Kluwer Academic Publishers, 1998, ISBN 0-7923-8386-9, Figure 2-39, p. 57 with kind permission from Springer Science+Business Media B.V.)*

where $A_{\text{Ph}} = 234R/\theta_D^3$. Because of the rapidly declining phonon-specific heat $C_{\text{Ph}} = A_{\text{Ph}}T^3$, the electron-specific heat $C_{\text{el}} = \gamma T$ is the dominating contribution to the specific heat at sufficiently low temperatures. Equation 2.49 can be rewritten as

$$\frac{C_n}{T} = \gamma + A_{\text{Ph}}T^2 \tag{2.50}$$

Figure 2.20 shows C_n/T versus T^2 for selected conductors in the normal state. The phonon contributions A_{Ph} correspond to the slopes of the straight lines. The γ-values for the electron-specific heat are given by the intercepts at $T = 0$.

Next, the Gibbs free energy will be used to describe the thermodynamics of the normal and the superconducting state. The Gibbs free energy is given by [32]

$$G = U - TS + pV - \mathbf{m} \cdot \mathbf{B} \tag{2.51}$$

where U is the internal energy, S is the entropy, p is the pressure, V is the volume, $\mathbf{m}$ is the magnetic moment, and $\mathbf{B}$ is the applied magnetic field. The differential of the internal energy is [32]

$$dU = TdS - pdV + \mathbf{B} \cdot d\mathbf{m} \tag{2.52}$$

The resulting differential for the Gibbs free energy is [32]

$$dG = -SdT + Vdp - \mathbf{m} \cdot d\mathbf{B} \tag{2.53}$$

The independent variables are the temperature T, the pressure p, and the magnetic field $\mathbf{B}$, which can be easily controlled in an experiment. Because the changes in volume are typically very small for solids, their specific heats at constant pressure C_p and at constant volume C_V are not very different. For the thermodynamically stable phase, the Gibbs free energy reaches a minimum.

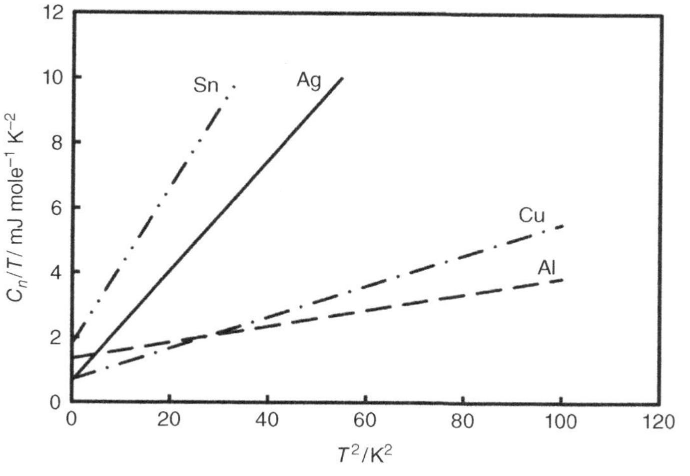

Figure 2.20 *Low-temperature-specific heat of various metals in the normal state (Reprinted from "High-Temperature Superconductors: Materials, Properties, and Applications" by Rainer Wesche, Kluwer Academic Publishers, 1998, ISBN 0-7923-8386-9, Figure 2-40, p. 59 with kind permission from Springer Science+Business Media B.V.)*

At the transition temperature T_c, the superconducting and normal phases coexist, and the Gibbs free energies of both are equal. The entropy S and the specific heat C are related by the second law of thermodynamics

$$dS = \frac{CdT}{T} \tag{2.54}$$

Using Equations 2.49 and 2.54, the normal state entropy S_n can be derived:

$$S_n = \int_0^T \frac{\gamma T + A_{\mathrm{Ph}} T^3}{T} dT = \gamma T + \frac{1}{3} A_{\mathrm{Ph}} T^3 \tag{2.55}$$

The Gibbs free energy at constant pressure can be found by the integration of Equation 2.53. As a consequence of the third law of thermodynamics, the entropy S_0 at $T = 0$ is zero independent of the volume and the pressure. The resulting Gibbs free energy for the normal state is

$$G_n = -\frac{\gamma}{2} T^2 - \frac{1}{12} A_{\mathrm{Ph}} T^4 \tag{2.56}$$

In the presence of a homogeneous magnetic field, the term $\mathbf{m} \cdot d\mathbf{B}$ in Equation 2.53 contributes to the Gibbs free energy. In a long cylindrical rod, demagnetization effects can be neglected and the magnetic moment $\mathbf{m}$ is given by

$$\mathbf{m} = \mathbf{M}V \tag{2.57}$$

where $\mathbf{M}$ is the magnetization and V is the volume of the rod. The magnetization can be written as

$$\mathbf{M} = \chi \frac{\mathbf{B}}{\mu_0} = (\mu_r - 1)\frac{\mathbf{B}}{\mu_0} \tag{2.58}$$

where χ is the magnetic susceptibility and μ_r is the relative permeability. Because μ_r is close to unity for nonferromagnetic metals, the magnetic energy $\mathbf{m} \cdot d\mathbf{B}$ can be neglected for the normal state; it is important in the superconducting state. Because of perfect diamagnetism $\chi = -1$ in the superconducting state.

For constant pressure and temperature, Equation 2.53 reduces to

$$dG = -\mathbf{m} \cdot d\mathbf{B} \tag{2.59}$$

Using Equation 2.59, the field dependence of the Gibbs free energy in the superconducting state can be derived. The following expression for the difference of the Gibbs free energies with and without the applied magnetic field is obtained

$$G_s(T,B) - G_s(T,0) = -\int_0^B \mathbf{m} \cdot d\mathbf{B} \tag{2.60}$$

where $\mathbf{m} = \chi\mathbf{B}V/\mu_0 = -\mathbf{B}V/\mu_0$. The resulting difference of the Gibbs free energies is

$$G_s(T,B) - G_s(T,0) = \frac{B^2 V}{2\mu_0} \tag{2.61}$$

where $B \le B_c(T)$ is the applied magnetic field. Obviously, the magnetic energy is independent of the temperature. At a certain temperature $T < T_c$, the maximum possible magnetic energy is reached for $B = B_c(T)$. Furthermore, the Gibbs free energies of both the normal and superconducting states are equal at the phase transition. The Gibbs free energy of the normal state is considered to be independent of the applied magnetic field B. Neglecting the very small change in volume at the phase transition, the difference of the Gibbs free energies of the normal and the superconducting state at the zero applied field can be expressed as

$$G_n(T) - G_s(T) = \frac{B_c^2(T)\, V}{2\mu_0} \tag{2.62}$$

Figure 2.21 shows the Gibbs free energies of both the superconducting and the normal state as functions of temperature. Using Equations 2.60 and 2.62, and the expression $B_c(T) = B_c(0)(1-(T/T_c)^2)$ (Equation 2.25), the Gibbs free energy of the superconducting state can be calculated. In the presence of an applied magnetic field B, the difference of the Gibbs free energies is

$$G_n(T,B) - G_s(T,B) = \frac{V}{2\mu_0}(B_c^2(T) - B^2) \tag{2.63}$$

The entropy difference between the normal and superconducting states can be found by differentiating the Gibbs free energy at the constant applied field.

$$S = -dG/dT \tag{2.64}$$

Combining Equations 2.63 and 2.64, the following expression for the entropy difference is obtained:

$$S_s(T) - S_n(T) = \frac{VB_c(T)}{\mu_0}\frac{dB_c}{dT} \tag{2.65}$$

Figure 2.22 shows the entropies of the normal and superconducting states as functions of temperature for the zero applied field. The curves shown in Figure 2.22 were calculated

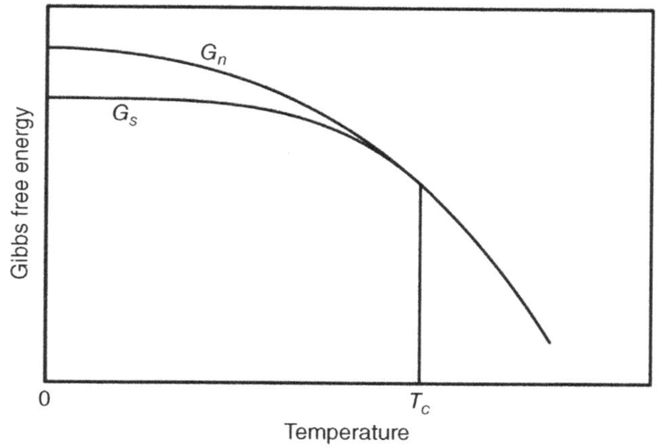

Figure 2.21 *Gibbs free energies of the superconducting and the normal states versus temperature for the zero applied field (Reprinted from "High-Temperature Superconductors: Materials, Properties, and Applications" by Rainer Wesche, Kluwer Academic Publishers, 1998, ISBN 0-7923-8386-9, Figure 2-41, p. 62 with kind permission from Springer Science+Business Media B.V.)*

using Equations 2.25, 2.55, and 2.65. At the critical temperature T_c, the critical field B_c is zero. Thus, the entropies of the superconducting and normal states are equal at the transition, and no latent heat is released at the phase transition for $B = 0$. This behavior is the characteristic of a second-order phase transition. On the other hand, the critical temperature $T_c(B)$ in the presence of an applied magnetic field is lower than $T_c(0)$. The resulting jump in the entropy for $B \neq 0$ releases latent heat at the transition. Thus, the transition from the normal to the superconducting state is of first order in the presence of an applied magnetic field. The entropy of the superconducting state is lower than that of the normal state, in agreement with the microscopic theory of superconductivity. Due to the formation of Cooper pairs, there exists a higher degree of order in the superconducting state.

Next, an expression for the specific heat C_s in the superconducting state will be derived. The specific heat C and the entropy S are connected by the expression

$$C = T\frac{\partial S}{\partial T} \tag{2.66}$$

where B is held constant. Using Equation 2.65, the difference of the specific heats in the superconducting and normal state can be written as

$$C_s - C_n = \frac{VT}{\mu_0}\left(\left(\frac{\partial B_c}{\partial T}\right)^2 + B_c\frac{\partial^2 B_c}{\partial T^2}\right) \tag{2.67}$$

Using Equation 2.25, the following expression is obtained:

$$C_s - C_n = \frac{2V}{\mu_0}B_c(0)^2\left(\frac{3T^2}{T_c^2} - 1\right)\frac{T}{T_c^2} \tag{2.68}$$

Figure 2.23 shows the specific heats of the superconducting and normal states as functions of temperature, calculated from Equations 2.49 and 2.68. Using Equation 2.67 and taking

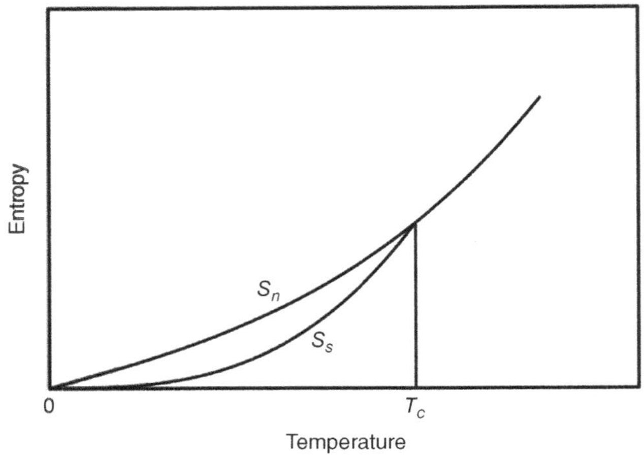

Figure 2.22 *Entropy versus temperature for zero applied field (Reprinted from "High-Temperature Superconductors: Materials, Properties, and Applications" by Rainer Wesche, Kluwer Academic Publishers, 1998, ISBN 0-7923-8386-9, Figure 2-42, p. 62 with kind permission from Springer Science+Business Media B.V.)*

into consideration that $B_c(T_c) = 0$, the following expression for the jump in the specific heat at the transition temperature T_c results:

$$C_s - C_n = \frac{VT_c}{\mu_0}\left(\frac{\partial B_c}{\partial T}\right)^2_{T=T_c} \tag{2.69}$$

Equation 2.69 is known as Rutger's formula. The Bardeen, Cooper, and Schrieffer (BCS) theory predicts for the jump in the electronic-specific heat [33]

$$\frac{C_{\rm se} - C_{\rm ne}}{C_{\rm ne}} = \frac{C_{\rm se} - \gamma T}{\gamma T} = 1.52 \tag{2.70}$$

The normal-state-specific heat $C_{\rm ne}$ can be determined experimentally by applying a magnetic field $B > B_c$ to the superconductor. Equation 2.68 indicates that the specific heats of the two states are equal for $T = T_c / \sqrt{3}$ (see Figure 2.23). Below this temperature, the specific heat in the normal state is larger than in the superconducting state.

So far, only the effects of temperature and the magnetic field on the free energy of the superconducting state have been taken into account. However, the transition temperature T_c and the critical field depend also on the applied pressure. Negative pressure derivatives of the critical temperature dT_c /dp are related to a decreasing critical field B_c under the applied pressure. At the transition from the normal to the superconducting state, very small changes in the sample volume occur [32, 34], which contribute to the Gibbs free energy. The application of hydrostatic pressure allows the study of the effects of a continuous variation of the lattice constants on the superconducting properties. According to Equation 3.8 (Chapter 3), the critical temperature of a superconductor obeying the BCS theory depends on the Debye temperature θ_D, the density of states $D_n(E_F)$ at the

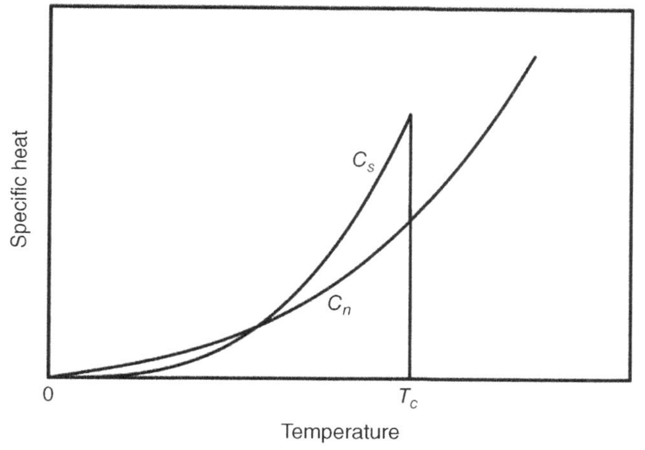

Figure 2.23 Specific heats in the superconducting and normal states versus temperature for $B = 0$ (Reprinted from "High-Temperature Superconductors: Materials, Properties, and Applications" by Rainer Wesche, Kluwer Academic Publishers, 1998, ISBN 0-7923-8386-9, Figure 2-43, p. 64 with kind permission from Springer Science+Business Media B.V.)

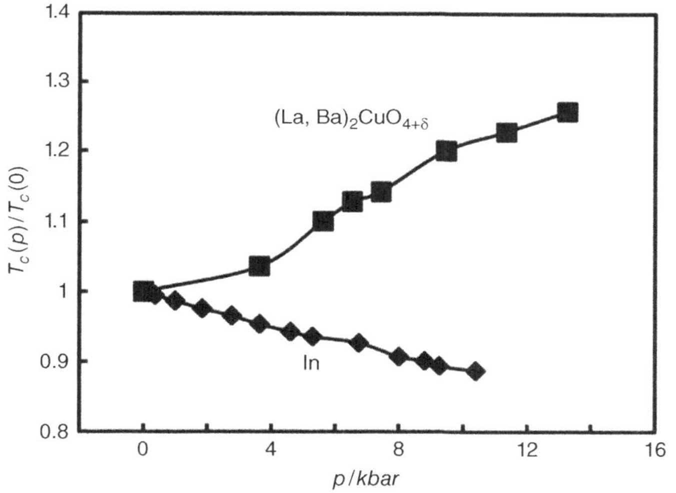

Figure 2.24 Normalized T_c-values versus applied pressure. The critical temperatures for indium and $(La,Ba)_2CuO_{4+\delta}$ at ambient pressure are 3.403 and 32 K, respectively (results from [35, 36]) (Reprinted from "High-Temperature Superconductors: Materials, Properties, and Applications" by Rainer Wesche, Kluwer Academic Publishers, 1998, ISBN 0-7923-8386-9, Figure 2-44, p. 64 with kind permission from Springer Science+Business Media B.V.)

Fermi energy E_F and the electron–phonon interaction parameter V_p. Pressure experiments, therefore, shed some light on the interplay of the lattice constants and these metal properties. Figure 2.24 shows normalized critical temperatures versus applied pressure for indium and the high-temperature superconductor $(La,Ba)_2CuO_{4+\delta}$. An applied pressure of 10 kbar causes a 10% reduction of the critical temperature of indium. On the other hand, the critical temperature of $(La,Ba)_2CuO_{4+\delta}$ is remarkably enhanced under the applied pressure. A similar effect can be achieved by a chemical pressure. Large ions can be replaced by smaller ions, and as a consequence, the lattice constants are reduced. The observed pressure dependence of the critical temperature of $(La,Ba)_2CuO_{4+\delta}$ led Chu and collaborators to the idea of replacing the large lanthanum ions by smaller yttrium ions. In this way the Y–Ba–Cu–O system with a transition temperature T_c of 93 K was discovered [37]. The investigation of pressure effects may, therefore, be a useful tool in the search for new superconducting materials with higher transition temperatures.

References

1. Ch. Kittel, *Introduction to Solid State Physics*, John Wiley & Sons, Inc., New York, 1976.
2. F.R. Fickett, Electrical properties, in *Materials at Low Temperatures*, R.P. Reed and A.F. Clark (eds.), American Society for Metals, Metals Park, OH, 1983.
3. L.D. Landau, Theory of Fermi-liquids, *Sov. Phys.– JETP*, **3**, 920 (1957).
4. L.D. Landau, Oscillations in a Fermi-liquid, *Sov. Phys.– JETP*, **5**, 101 (1957).
5. N.F. Mott, *Metal-Insulator Transitions*, Taylor & Francis, London, 1974.
6. C.P. Poole, Jr.,, H.A, Farach, and R.J. Creswick, *Superconductivity*, Academic Press, San Diego, 1995.
7. A.C. Rose-Innes and E.H. Rhoderick, *Introduction to Superconductivity*, Pergamon Press, Oxford, 1969.
8. B. Hopfengärtner, B. Hensel, and G. Saemann-Ischenko, Analysis of the fluctuation-induced excess dc conductivity of epitaxial $YBa_2Cu_3O_7$ films: Influence of a short-wavelength cutoff in the fluctuation spectrum, *Phys. Rev. B*, **44**, 741–749 (1991).
9. D.J. Quinn, W.B. Ittner, Resistance in a superconductor, *J. Appl. Phys.*, **33**, 748–749 (1962).
10. Y.B. Kim, C.F. Hempstead, and A.R. Strnad, Critical persistent currents in hard superconductors, *Phys. Rev. Lett.*, **9**, 306–309 (1962).
11. W.-J. Yeh, L. Chen, F. Xu, B. Bi, and P. Yang, Persistent current in Ba–Y–Cu–O in liquid nitrogen, *Phys. Rev. B*, **36**, 2414–2416 (1987).
12. F.J. Kedves, S. Mészáros, K. Vad, G. Halász, B. Keszei, and L. Mihály, Estimation of maximum electrical resistivity of high T_c superconducting ceramics by the Meissner effect, *Solid State Commun.*, **63**, 991–992 (1987).
13. J. File and R.G. Mills, Observation of persistent current in a superconducting solenoid, *Phys. Rev. Lett.*, **10**, 93–96 (1963).
14. L.I. Berger and B.W. Roberts, Properties of superconductors, *in Handbook of Chemistry and Physics*, D.R. Lide and H.P.R. Frederikse (eds.), CRC Press, Boca Raton, FL, 74th edition, 1993–1994.

15. C. Buzea and K. Robbie, Assembling the puzzle of the superconducting elements: a review, *Supercond. Sci. Technol.*, **18**, R1–R8 (2005).
16. N.W. Ashcroft, Putting the squeeze on lithium, *Nature*, **419**, 569–572 (2002).
17. K. Shimizu, H. Ishikawa, D. Takao, T. Yagi, and K. Amaya, Superconductivity in compressed lithium at 20 K, *Nature*, **419**, 597–599 (2002).
18. K.J. Dunn and F.P. Bundy, Electrical resistance of behaviour of Ca at high pressures and low temperatures, *Phys. Rev. B*, **24**, 1643–1650 (1981).
19. J. Schmalian, Unkonventionell und komplex, *Phys. J.*, **10** (6), 37–43 (2011).
20. M.I. Eremets, V.V. Struzhkin, H.K. Mao, and R.J. Hemley, Superconductivity in boron, *Science*, **293**, 272–274 (2001).
21. K. Shimizu, T. Kimura, S. Furomoto, K. Takeda, K. Kontani, Y. Onuki, and K. Amaya, Superconductivity in the non-magnetic state of iron under pressure, *Nature*, **412**, 316–318 (2001).
22. K.J. Chang, M. Dacorogna, M. Cohen, J. Mignot, G. Chouteau, and G. Martinez, Superconductivity in high-pressure metallic phases of Si, *Phys. Rev. Lett.*, **54**, 2375–2378 (1985).
23. M.A. Il'ina and E.S. Itskevich, Superconductivity of high-pressure phases of silicon in the pressure range up to 14 GPa, *Sov. Phys. – Solid State*, **22**, 1833–1835 (1980).
24. I. Shirotani, H. Kawamura, K. Tsuburaya, and K. Tachikawa, Superconductivity of phosphorus and phosphorus–arsenic alloy under high pressures, *Japan. J. Appl. Phys. Suppl.*, **26**, 921–922 (1987).
25. M.A. Il'ina, E.S. Itskevich, and E.M. Dizhur, Superconductivity of bismuth, barium, and lead at pressures exceeding 100 kbar, *Sov. Phys. – JETP*, **34**, 1263–1265 (1972).
26. V.V. Struzhkin, R.J. Hemley, H.K. Mao, and Y.A. Timofeev, Superconductivity at 10–17 K in compressed sulphur, *Nature*, **390**, 382–384 (1997).
27. Y. Akahama, M. Kobayashi, and H. Kawamura, Pressure induced superconductivity and phase transition in selenium and tellurium, *Solid State Commun.*, **84**, 803–806 (1992).
28. E.A. Ekimov, V.A. Sidorov, E.D. Bauer, N.N. Mel'nik, N.J. Curro, J.D. Thompson, and S.M. Stishov, Superconductivity in diamond, *Nature*, **428**, 542–545 (2004).
29. Z.K. Tang, L. Zhang, N. Wang, X.X. Zhang, G.H. Wen, G.D. Li, J.N. Wang, C.T. Chan, and P. Sheng, Superconductivity in 4 angstrom single-walled carbon nanotubes, *Science*, **292**, 2462–2465 (2001).
30. M.B. Brodsky, P. Marikar, R.J. Friddle, L. Singer, and C.H. Sowers, Superconductivity in Au/Cr/Au epitaxial metal film sandwiches (EMFS), *Solid State Commun.*, **42**, 675–678 (1982).
31. F. London and H. London, Supraleitung und Diamagnetismus, *Physica*, **2**, 341–354 (1935).
32. W. Buckel, *Supraleitung, Grundlagen und Anwendungen*, Wiley-VCH, Weinheim, Germany, 1994.
33. J. Bardeen, L.N. Cooper, and J.R. Schrieffer, Theory of superconductivity, *Phys. Rev.*, **108**, 1175–1204 (1957).
34. J.L. Olsen and H. Rohrer, The volume change at the superconducting transition, *Helv. Phys. Acta*, **30**, 49–66 (1957).
35. L.D. Jennings and C.A. Swenson, Effects of pressure on the superconducting transition temperatures of Sn, In, Ta, Tl, and Hg, *Phys. Rev.*, **112**, 31–43 (1958).

36. C.W. Chu, P.H. Hor, R.L. Meng, L. Gao, Z.J. Huang, and Y.Q. Wang, Evidence for superconductivity above 40 K in the La–Ba–Cu–O compound system, *Phys. Rev. Lett.*, **58**, 405–407 (1987).
37. M.K. Wu, J.R. Ashburn, C.J. Torng, P.H. Hor, R.L. Meng, L. Gao, Z.J. Huang, Y.Q. Wang, and C.W. Chu, Superconductivity at 93 K in a new mixed-pase Y–Ba–Cu–O compound system at ambient pressure, *Phys. Rev. Lett.*, **58**, 908–910 (1987).

3

Superconductivity: A Macroscopic Quantum Phenomenon

3.1 Introduction

In Chapter 2, the phenomenology of the superconducting state was considered. The microscopic causes of superconductivity in simple metals will be briefly described in this chapter. In 1957, a microscopic quantum theory of superconductivity was proposed by Bardeen, Cooper, and Schrieffer [1, 2], which is now well known as the BCS theory. In this theory, the electron–phonon interaction is responsible for the formation of the Cooper pairs. It seems to be unlikely that the very high transition temperatures in the cuprate superconductors can be explained in this way. While a derivation of the BCS theory is beyond the scope of this book, some fundamental ideas and results will be presented. A more detailed description can be found in Ref. [3].

An important aspect of the development of the BCS theory [1, 2] was the idea of F. London that superconductivity is a macroscopic quantum phenomenon. Cooper-pair tunneling and quantum interference effects in Josephson junctions (see Section 3.3) are consequences of the macroscopic quantum nature of superconductivity.

3.2 BCS Theory of Superconductivity[1]

A breakthrough toward a microscopic theory of superconductivity was the discovery that the interaction of two electrons via lattice vibrations (phonons) can lead to attraction between the two electrons. This theoretical result was found independently of each other by Fröhlich [4] and Bardeen [5]. The electron–phonon interaction is schematically

[1] Reprinted with minor changes from "*High-Temperature Superconductors: Materials, Properties, and Applications*" by Rainer Wesche, Kluwer Academic Publishers, 1998, pp. 24–33, ISBN 0-7923-8386-9, with kind permission from Springer Science+Business Media B.V.

Physical Properties of High-Temperature Superconductors, First Edition. Rainer Wesche.

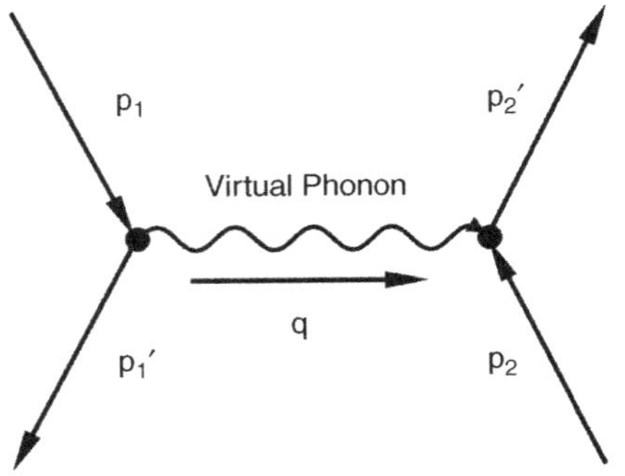

Figure 3.1 Interaction of two electrons via the exchange of a virtual phonon (Reprinted from "High-Temperature Superconductors: Materials, Properties, and Applications" by Rainer Wesche, Kluwer Academic Publishers, 1998, ISBN 0-7923-8386-9, Figure 2-16, p. 25 with kind permission from Springer Science+Business Media B.V.)

illustrated in Figure 3.1. In this process two electrons exchange a virtual phonon, which exists only for a very short time, and which can be emitted only in the presence of a second electron ready to absorb it. In Figure 3.1, $\mathbf{p_1}$, $\mathbf{p_2}$ and $\mathbf{p_1}'$, $\mathbf{p_2}'$ are the momenta of the two electrons before and after the exchange of the virtual phonon, respectively. The momentum $\mathbf{q}$ of the phonon is

$$|\mathbf{q}| = \frac{h\nu_p}{c_s} \tag{3.1}$$

where h is Planck's quantum, ν_p the frequency of the phonon, and c_s the velocity of sound. Equation 3.1 is analogous to the expression for the momentum of a photon. In the exchange momentum is conserved:

$$\mathbf{p_1} + \mathbf{p_2} = \mathbf{p}_1' + \mathbf{p}_2' \tag{3.2}$$

where $\mathbf{p_1}' = \mathbf{p_1} - \mathbf{q}$ and $\mathbf{p_2}' = \mathbf{p_2} + \mathbf{q}$. Furthermore, energy is conserved for the initial and final states. However, during the short period of time in which the virtual phonon exists, the energy of the system may be different from that of the initial state. This is a consequence of the uncertainty principle $\Delta E \Delta t \cong \hbar$ between energy and time. Short interaction times are connected with large energy uncertainties. If the energy difference between the initial and the intermediate states is smaller than $h\nu_p$, the electron–phonon interaction is attractive. This can be understood as a consequence of the polarization of the lattice by the other electron. The concentration of positive charges produced by the first electron allows a reduction of the potential energy of the second electron. Due to this attractive interaction, electrons with opposite momenta and spins form so-called Cooper pairs in a conventional superconductor. The electron–phonon interaction is typically effective over a distance of ~ 1 μm, which may be considered as the spatial extent ξ_{Co} of a Cooper pair. For tin, $\xi_{\mathrm{Co}} = 180$ nm, $n_C = 7.5 \times 10^{22}$ cm^{-3}, which leads to 4.35×10^8 Cooper pairs present within a volume $\xi_{\mathrm{Co}}{}^3$. The following pair correlation is valid for the two electrons in a Cooper pair:

$$\mathbf{p_1} = -\mathbf{p_2} \tag{3.3}$$

$$\mathbf{s_1} = -\mathbf{s_2} \tag{3.4}$$

Thus, the momentum and spin of a Cooper pair are both zero. The Cooper pairs have to be considered as new particles. Because their spin is zero, they do not obey Pauli's exclusion principle. Further, the occupation of energy states is described by the Bose–Einstein distribution law. The probability that a state is occupied increases with the number of particles populating the same quantum state. This behavior may be considered as a kind of anti-Pauli principle. Because they occupy the same quantum state, all Cooper pairs have the same momentum in the presence of an applied electric field. Thus, only scattering processes that break up a Cooper pair result in a change of momentum. Resistance occurs as soon as the energy taken from the electric field exceeds the binding energy of the Cooper pairs. This corresponds to the existence of a critical current density. The electron–phonon interaction leads to an energy gap of a few meV in the density of states for the single electrons. Figure 3.2 shows the density of states for the quasiparticles (single electrons) in the superconducting state around the Fermi energy. The energy states present in the normal state within the energy gap are shifted to the edges. Therefore, the density of states is strongly enhanced at the edges of the energy gap. The BCS expression for the single-electron density of states $D_s(E - E_F)$ in a superconductor in the vicinity of the Fermi energy is

$$D_s(E - E_F) = D_n(E_F)\frac{|E - E_F|}{\sqrt{(E - E_F)^2 - \Delta^2}} \tag{3.5}$$

where $|E - E_F| \geq \Delta$ and Δ is the half width of the energy gap. The variation of the density of states $D_n(E)$ in the normal state can be neglected because $E_F >> 2\Delta$.

Figure 3.3 shows the temperature dependence of the energy gap given by the BCS theory. At temperatures below $0.6T_c$, the width of the energy gap depends only weakly on

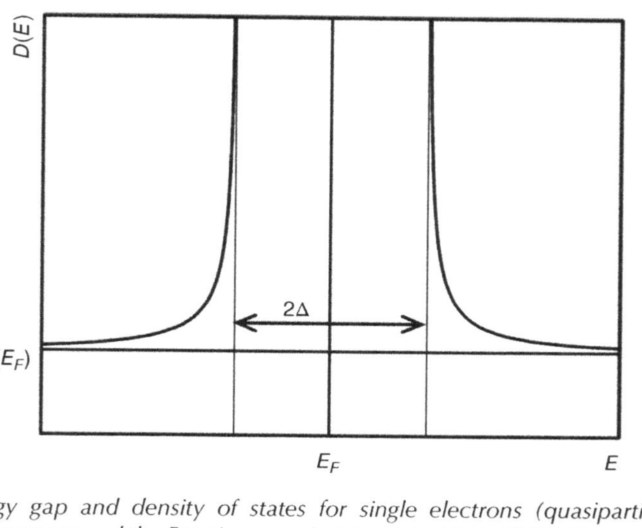

Figure 3.2 *Energy gap and density of states for single electrons (quasiparticles) in the superconducting state around the Fermi energy E_F (Reprinted from "High-Temperature Superconductors: Materials, Properties, and Applications" by Rainer Wesche, Kluwer Academic Publishers, 1998, ISBN 0-7923-8386-9, Figure 2-17, p. 27 with kind permission from Springer Science+Business Media B.V.)*

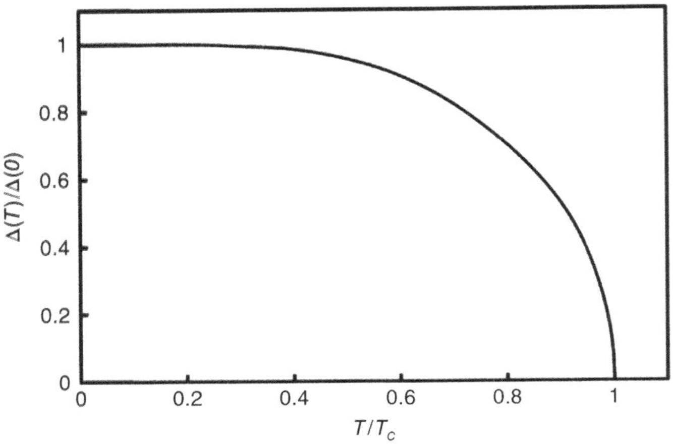

Figure 3.3 *Temperature dependence of the energy gap as given by the BCS theory (Reprinted from "High-Temperature Superconductors: Materials, Properties, and Applications" by Rainer Wesche, Kluwer Academic Publishers, 1998, ISBN 0-7923-8386-9, Figure 2-18, p. 27 with kind permission from Springer Science+Business Media B.V.)*

temperature. Near the critical temperature, the half width of the energy gap is approximately [2]

$$\Delta(T) = 3.2k_B T_c (1 - T/T_c)^{1/2} \tag{3.6}$$

where k_B is Boltzmann's constant.

An important prediction of the BCS theory is that the width of the energy gap $2\Delta(0)$ is closely connected to the transition temperature T_c:

$$2\Delta(0) = 3.5k_B T_c \tag{3.7}$$

Figure 3.4 shows the width of the energy gap $2\Delta(0)$ versus $k_B T_c$ for various superconducting elements. The solid line corresponds to the ratio $2\Delta(0)/k_B T_c = 3.5$ predicted from the BCS theory. Experimentally determined values of this ratio are typically between 3.2 and 4.6 for superconducting elements [3].

In addition, the BCS theory allows us to calculate the transition temperature T_c. Assuming that the attractive electron–phonon interaction can be described by the constant V_p, we obtain the following expression for the critical temperature T_c [2]:

$$T_c = 1.14 \frac{\hbar \omega_D}{k_B} \exp\left(\frac{-1}{D_n(E_F)V_p}\right) \tag{3.8}$$

where ω_D is the Debye frequency and $D_n(E_F)$ is the density of states in the normal state at the Fermi energy E_F. Equation 3.8 is valid for $D_n(E_F)V_p << 1$ (weak coupling limit). In more sophisticated expressions for the critical temperature T_c, the expression $D_n(E_F)V_p$ is replaced by the difference between the dimensionless parameters λ_{ep} and μ^*, which

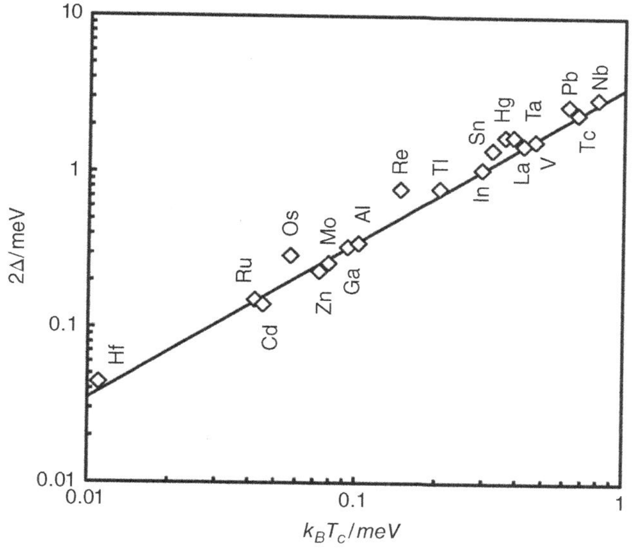

Figure 3.4 *Width of the energy gap $2\Delta(0)$ versus k_BT_c for various superconducting elements. The solid line corresponds to the ratio of $2\Delta(0)/k_BT_c = 3.5$ predicted by the BCS theory (data from [3]) (Reprinted from "High-Temperature Superconductors: Materials, Properties, and Applications" by Rainer Wesche, Kluwer Academic Publishers, 1998, ISBN 0-7923-8386-9, Figure 2-20, p. 29 with kind permission from Springer Science+Business Media B.V.)*

characterize the electron–phonon coupling and the screened Coulomb interaction, respectively [3]

$$T_c = 1.14\frac{\hbar\omega_D}{k_B}\exp\left(\frac{-1}{\lambda_{ep} - \mu_c^*}\right) \tag{3.9}$$

The dimensionless electron–phonon coupling constant λ_{ep} is defined as [3]

$$\lambda_{ep} = 2\int_0^{\infty}\frac{\alpha^2(\omega)D_{ph}(\omega)}{\omega}d\omega \tag{3.10}$$

where $\alpha^2(\omega)$ is the electron–phonon coupling strength and $D_{ph}(\omega)$ the phonon density of states. Depending on the magnitude of λ_{ep}, the superconductors are characterized as

$$\begin{array}{ll} \lambda_{ep} << 1 & \text{weak coupling} \\ \lambda_{ep} \approx 1 & \text{intermediate coupling} \\ \lambda_{ep} >> 1 & \text{strong coupling} \end{array} \tag{3.11}$$

Equations 3.8 and 3.9 indicate that the critical temperature T_c is proportional to the Debye frequency ω_D, the cutoff frequency of the phonon spectrum in a solid. The atoms in a solid

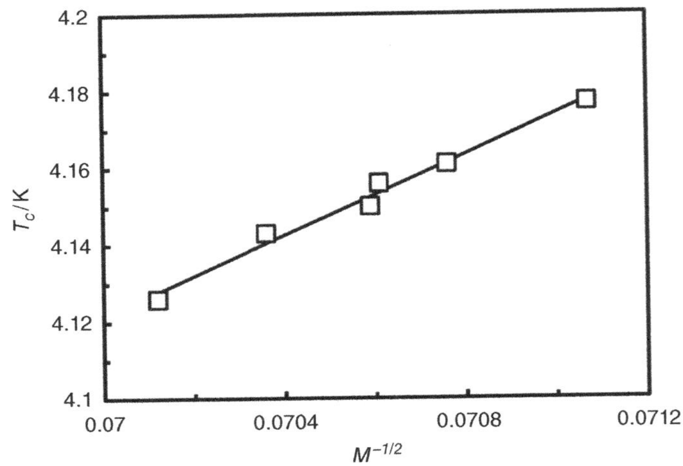

Figure 3.5 *Isotope effect for mercury (data from [6, 7]) (Reprinted from "High-Temperature Superconductors: Materials, Properties, and Applications" by Rainer Wesche, Kluwer Academic Publishers, 1998, ISBN 0-7923-8386-9, Figure 2-21, p. 30 with kind permission from Springer Science+Business Media B.V.)*

may be considered as harmonic oscillators of frequency $\omega = (k_e/M)^{1/2}$, where the elastic constant k_e results from the binding forces in the solid and is expected to be independent of the isotopic mass M. Therefore, both the Debye frequency ω_D and the critical temperature T_c should be proportional to $M^{-1/2}$ for different isotopes of the same element. In many superconductors, this isotope effect has been observed [6–9]. Figure 3.5 shows the T_c values for different mercury isotopes as a function of $M^{-1/2}$, showing good agreement with an $M^{-\beta}$ dependence where $\beta = 0.5$.

Figure 3.6 shows β-values of selected superconductors. As expected from Equation 3.8, the exponent β is close to 0.5 for many simple metals. Only for the transition metals have considerable deviations from $\beta = 0.5$ been found. In spite of a missing or reduced isotope effect for the transition metals, the electron–phonon interaction is responsible for superconductivity in these materials [8].

Next, it will be shown that superconductivity is a macroscopic quantum phenomenon. In the superconducting state, electrons with opposite spins and momenta form Cooper pairs. Because their total spin is zero, the Cooper pairs are not restricted by Pauli's exclusion principle; all can populate the same quantum state. The collective wave function for the Cooper pairs is

$$\psi = n_C^{1/2} \exp(i\varphi(\mathbf{r})) \tag{3.12}$$

where n_C is their number density. It then follows that $n_C = \psi^*\psi$. Because the phase coherence of the Cooper pairs is effective for large distances, the phase $\varphi(\mathbf{r})$ of the wave function given by Equation 3.12 is a macroscopically observable quantity. The momentum of a Cooper pair in the presence of a magnetic field is

$$\mathbf{p} = (m_C\mathbf{v} + q_C\mathbf{A}) \tag{3.13}$$

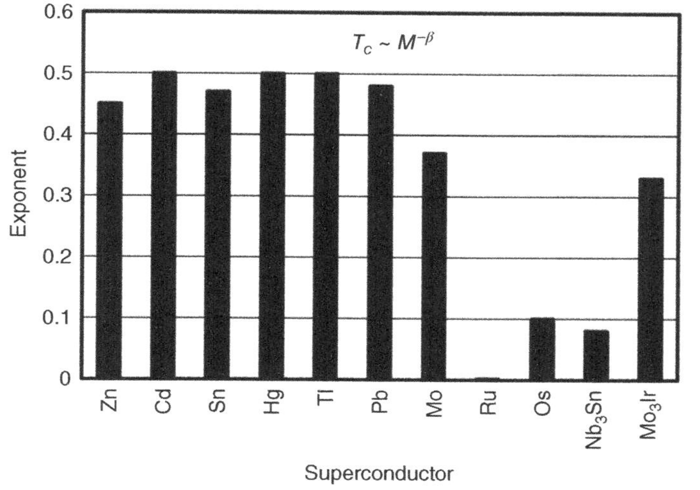

Figure 3.6 *Exponent β for selected superconductors (data from [8]) (Reprinted from "High-Temperature Superconductors: Materials, Properties, and Applications" by Rainer Wesche, Kluwer Academic Publishers, 1998, ISBN 0-7923-8386-9, Figure 2-22, p. 31 with kind permission from Springer Science+Business Media B.V.)*

where **A** is the vector potential, **v** the velocity, q_C the charge, and m_C the mass of a Cooper pair. The magnetic field **B** and the vector potential **A** are connected by

$$\mathbf{B} = \text{curl } \mathbf{A} \tag{3.14}$$

In quantum physics, the momentum $\mathbf{p} = m\mathbf{v}$ has to be replaced by the operator $-i\hbar\nabla$; the following expression results for the current density **j**:

$$\mathbf{j} = q_C \psi^* \mathbf{v} \psi = \frac{n_C q_C}{m_C}(\hbar \nabla \varphi(\mathbf{r}) - q_C \mathbf{A}) \tag{3.15}$$

From Equation 3.15, we obtain

$$\hbar \nabla \varphi(\mathbf{r}) = q_C \left(\mathbf{A} + \frac{m_C}{n_C {q_C}^2} \mathbf{j} \right) \tag{3.16}$$

The phase shift for a closed path C is

$$\oint_C \nabla \varphi(\mathbf{r}) d\mathbf{r} = 2\pi N_\phi \tag{3.17}$$

where N_ϕ must be an integer as a consequence of the definiteness of the macroscopically observable phase $\varphi(\mathbf{r})$. Combining Equations 3.16 and 3.17 leads to the following expression:

$$\oint_C \left(\mathbf{A} + \frac{m_C}{n_C {q_C}^2} \mathbf{j} \right) d\mathbf{r} = \frac{2\pi \hbar}{q_C} N_\phi \tag{3.18}$$

The quantity on the left-hand side of Equation 3.18 is called a fluxoid. Using Stokes' theorem, the following condition is obtained for the magnetic flux ϕ:

$$\oint_C \mathbf{A}\, d\mathbf{r} = \int_F \text{curl}\ \mathbf{A}\, d\mathbf{F} = \int_F \mathbf{B}\, d\mathbf{F} = \phi \tag{3.19}$$

where F is a surface defined by the contour C. For a closed path in the interior of the superconductor, $\mathbf{j}$ and $\mathbf{B}$ are zero because of the Meissner effect, and for the magnetic flux,

$$\phi = \int_F \mathbf{B}\, d\mathbf{F} = \frac{h}{q_C} N_\phi \tag{3.20}$$

Equation 3.20 indicates that the magnetic flux is quantized in units of $\phi_0 = h/q_C$, where ϕ_0 is called a flux quantum. In principle, the supercurrents flowing near the surface of the superconductor also contribute to the magnetic flux ϕ; however, this is typically only a small correction.

In 1961, two groups of researchers [10, 11] showed experimentally that magnetic flux is quantized. For these experiments, very small superconducting hollow cylinders were prepared by coating a nonsuperconducting material. The total flux trapped in such cylinders is always a discrete number of flux quanta irrespective of the value of the applied field used to establish the persistent currents. The trapped flux was measured in different ways. Doll and Näbauer [10] connected a torsion wire orthogonal to the cylinder axis of their sample. A second external field was used to establish a torsional vibration. In the experiment of Deaver and Fairbank [11], the cylinder was moving up and down with a frequency of 100 Hz. Two small pick-up coils were used to determine the magnetic flux trapped in the superconducting cylinder. The experimental results indicate that $q_C = 2e$ as expected for Cooper pairs. Thus, the flux quantum ϕ_0 has a value of 2.068×10^{-15} V s.

3.3 Tunneling Effects[2]

Tunneling, or barrier penetration, is a quantum mechanical process. Based on classical mechanics, a particle could pass through a potential barrier only if the kinetic energy E is larger than the height E_b of the barrier; otherwise it would be reflected. However, this classical picture is not correct because of the wave-like nature of the electron. The tunneling of electrons through a thin potential barrier is schematically illustrated in Figure 3.7. The wavelength of an electron with momentum $\mathbf{p} = m\mathbf{v}$ is

$$\lambda_e = \frac{h}{p} \tag{3.21}$$

where h is Planck's constant. This is the fundamental de Broglie relationship. The resulting wave functions for electrons traveling from left to right are also shown in Figure 3.7 (bottom). Within the barrier the wave function is

$$\psi_2(x) = C \exp(k'x) + D \exp(-k'x) \tag{3.22}$$

[2] Reprinted with minor changes from "*High-Temperature Superconductors: Materials, Properties, and Applications*" by Rainer Wesche, Kluwer Academic Publishers, 1998, pp. 34–51, ISBN 0-7923-8386-9, with kind permission from Springer Science+Business Media B.V.

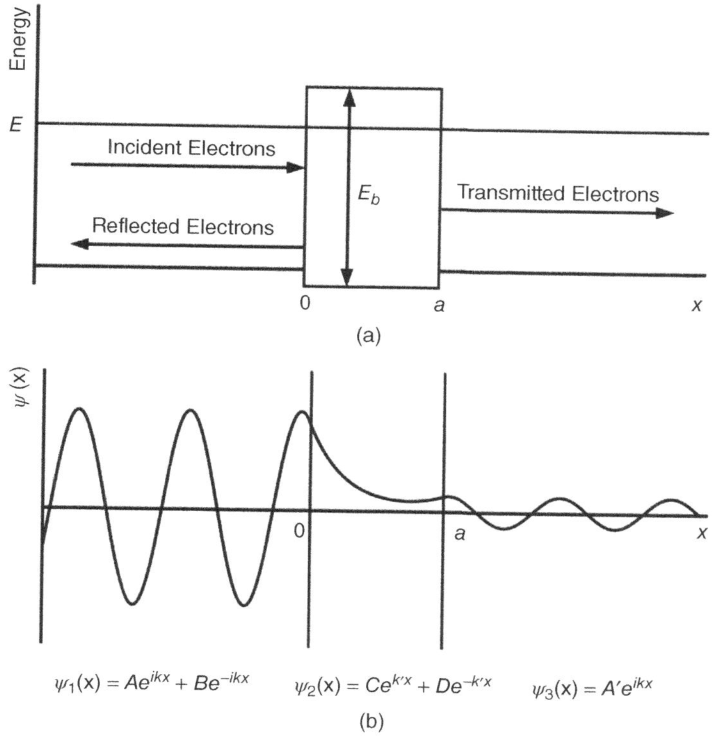

Figure 3.7 *Illustration of electron tunneling through a sufficiently thin potential barrier. The electron wave functions (bottom) indicate that a few electrons are able to pass through the barrier (Reprinted from "High-Temperature Superconductors: Materials, Properties, and Applications" by Rainer Wesche, Kluwer Academic Publishers, 1998, ISBN 0-7923-8386-9, Figure 2-24, p. 35 with kind permission from Springer Science+Business Media B.V.)*

where

$$k' = [2m(E_b - E)]^{1/2}/\hbar \tag{3.23}$$

The constants C and D depend on the boundary conditions. The wave function $\psi_2(x)$ decays nearly exponentially within the potential barrier. There exists a small but finite probability that electrons can pass through a thin potential barrier.

We now discuss single-electron tunneling between two normal conductors separated by a thin insulating oxide layer. The thickness of the oxide layer is typically less than 2 nm. Figure 3.8 shows the experimental setup for the measurement of the tunneling currents. In tunneling processes energy must be conserved. Moreover, tunneling can take place only if empty target energy states are available; otherwise the process is forbidden by the Pauli principle. The tunneling current is therefore proportional to both the number of incident electrons and the number of empty target states. Figure 3.9 shows the density of states $D(E)$ and their occupation at zero temperature, for two normal conductors. Applying a voltage U_1 leads to a difference eU_1 between the Fermi energies of the two metals; electrons can flow from left to right (see Figure 3.9). The resulting current voltage characteristic of the

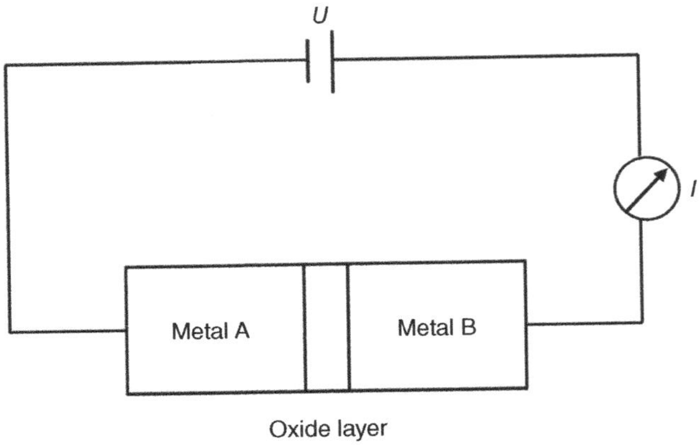

Figure 3.8 *Tunneling contact of two metals separated by a thin oxide layer (Reprinted from "High-Temperature Superconductors: Materials, Properties, and Applications" by Rainer Wesche, Kluwer Academic Publishers, 1998, ISBN 0-7923-8386-9, Figure 2-25, p. 35 with kind permission from Springer Science+Business Media B.V.)*

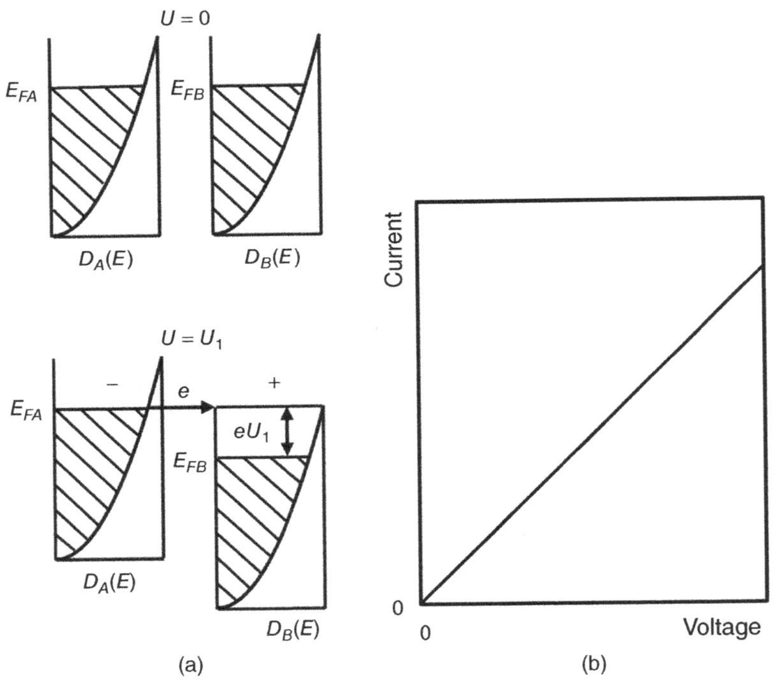

Figure 3.9 *Density of states for normal metals. The states in the hatched areas are occupied at* $T = 0$ *(left). The applied voltage leads to a difference* eU_1 *in the Fermi energies. The resulting N-I-N current–voltage characteristic is shown on the right (Reprinted from "High-Temperature Superconductors: Materials, Properties, and Applications" by Rainer Wesche, Kluwer Academic Publishers, 1998, ISBN 0-7923-8386-9, Figure 2-26, p. 37 with kind permission from Springer Science+Business Media B.V.)*

normal–insulator–normal (N-I-N) tunneling junction is also shown in Figure 3.9 (right) and will be discussed below.

The net tunneling current through the contact shown in Figure 3.8 is

$$I = e(j_{\mathrm{AB}} - j_{\mathrm{BA}})A_t \tag{3.24}$$

where A_t is the cross-sectional area of the contact and e the electron charge. The number densities of electrons flowing from A→B and B→A are j_{AB} and j_{BA}, respectively. Because their charge is negative, the current flows in the opposite direction to the electrons. The number densities j_{AB} and j_{BA} are proportional to the probability P_t of electron tunneling through the barrier. The resulting number density j_{AB} of electrons flowing from A→B is

$$j_{\mathrm{AB}} = \int P_t D_{\mathrm{A}}(E')f(E')D_{\mathrm{B}}(E' - eU_1)(1 - f(E' - eU_1))dE' \tag{3.25}$$

where $D_{\mathrm{A}}(E')$ and $D_{\mathrm{B}}(E')$ are the densities of states and $E' = E - E_F$. The number of occupied energy states in metal A is proportional to the Fermi function $f(E')$, whereas the number of empty energy states in metal B is proportional to $(1 - f(E' - eU_1))$. The corresponding number density of electrons passing through the barrier in the reverse direction is

$$j_{\mathrm{BA}} = \int P_t D_{\mathrm{B}}(E' - eU_1)f(E' - eU_1)D_{\mathrm{A}}(E')(1 - f(E'))dE' \tag{3.26}$$

The resulting net number density $j = j_{\mathrm{AB}} - j_{\mathrm{BA}}$ of electrons passing through the barrier is

$$j = \int P_t D_{\mathrm{A}}(E')D_{\mathrm{B}}(E' - eU_1)(f(E') - f(E' - eU_1))dE' \tag{3.27}$$

By inserting $E' = E - E_F$ in Equation 2.12 (Chapter 2), the following expression for the density of states is obtained:

$$D(E') = D(0)(E_F + E')^{1/2}/E_F^{1/2} \tag{3.28}$$

Generally, the energy shifts connected with the relatively small tunneling voltages are much smaller than the Fermi energy E_F and as a consequence

$$D(E') \cong D(0) \tag{3.29}$$

The Fermi function given in Equation 2.6 (Chapter 2) can be written as

$$f(E') = \frac{1}{\exp(E'/k_B T) + 1} \tag{3.30}$$

In Equation 3.27, the difference of the Fermi functions at $T = 0$ is

$$\begin{aligned} f(E') - f(E' - eU_1) &= 1 \quad \text{for} \quad 0 \le |E'| \le |eU_1| \\ f(E') - f(E' - eU_1) &= 0 \quad \text{for} \quad |E'| \ge |eU_1| \end{aligned} \tag{3.31}$$

For the polarity considered in Figure 3.9, E' and eU_1 are negative. Using Equations 3.27, 3.29, and 3.31 and taking into consideration that the probability P_t for electron tunneling through the barrier is typically independent of the applied voltage, we obtain the following expression for the net number density j of electrons passing through the junction:

$$j = P_t D_A(0) D_B(0) \int_{eU_1}^{0} dE' = -eU_1 P_t D_A(0) D_B(0) \tag{3.32}$$

where $-eU_1 > 0$. Equation 3.32 indicates that the N-I-N tunneling current is proportional to the applied voltage, as shown in Figure 3.9 (right).

We now consider the tunneling of single electrons in the case of normal–insulator–superconductor (N-I-S) and superconductor–insulator–superconductor (S-I-S) junctions. Tunneling experiments allow the measurement of the energy gap of superconductors [12–14]. In 1960 this method was first used by I. Giaever to determine the energy gap in lead [12]. Figure 3.10 shows the N-I-S tunneling process in a Bose condensation representation. Because their total spin is zero, the Cooper pairs do not obey Pauli's exclusion principle. As a consequence, all Cooper pairs populate the same quantum state at the Fermi energy E_F. The Fermi energy of a superconductor connected to the positive pole of a battery is shifted to lower values. For a voltage of $U = \Delta/e$, the empty energy states of the superconductor above the energy gap reach the same level as the Fermi energy of the normal conductor, and strong tunneling sets in. On the other hand, for a voltage $U = -\Delta/e$, pair breaking becomes possible. One of the electrons tunnels into the normal conductor and gains an energy Δ. This energy Δ is available for the excitation of the second electron to an empty state above the energy gap. Thus, energy is conserved in this process.

In the following section, a quantitative description of the N-I-S tunneling process will be given. Assuming that $U = -\Delta/e$ (see Figure 3.10, right), the net number density of

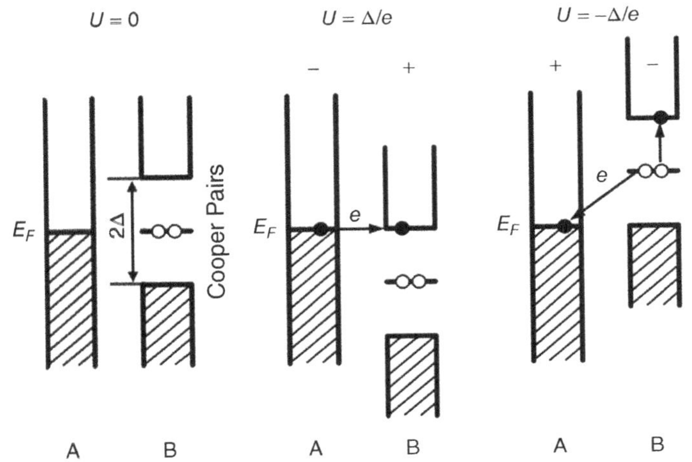

Figure 3.10 *Bose condensation representation of the N-I-S tunneling process. Strong tunneling sets in for $U \geq |\Delta/e|$ (Reprinted from "High-Temperature Superconductors: Materials, Properties, and Applications" by Rainer Wesche, Kluwer Academic Publishers, 1998, ISBN 0-7923-8386-9, Figure 2-27, p. 38 with kind permission from Springer Science+Business Media B.V.)*

electrons flowing through the barrier is

$$j = j_{BA} - j_{AB} = \int P_t D_{An}(E' - eU) D_{Bs}(E')(f(E') - f(E' - eU))dE' \tag{3.33}$$

where $D_{An}(E' - eU) \cong D_{An}(0)$. From Equation 3.5, the density of states in the superconductor B is

$$D_{Bs}(E') = D_{Bn}(0)\,|E'|/\sqrt{E'^2 - \Delta^2} \tag{3.34}$$

where $|E'| > \Delta$. Inserting Equations 3.31 and 3.34 into Equation 3.33 and taking into consideration that there are no energy states within the gap, the following expression for j is obtained:

$$j = \int_{eU}^{-\Delta} P_t D_{An}(0) D_{Bn}(0) \frac{|E'|}{\sqrt{E'^2 - \Delta^2}} dE' \tag{3.35}$$

Integration of Equation 3.35 leads to

$$j = -eP_t\, D_{An}(0) D_{Bn}(0) \sqrt{U^2 - \left(\frac{\Delta}{e}\right)^2} \tag{3.36}$$

where $-e > 0$, i.e., the electrons tunnel from B $\rightarrow$ A. A comparison of Equations 3.32 and 3.36 indicates that

$$I_{\mathrm{NIS}} = \frac{I_{\mathrm{NIN}}}{U} \sqrt{U^2 - \left(\frac{\Delta}{e}\right)^2} \tag{3.37}$$

The resulting N-I-S current–voltage characteristic at $T = 0$ and the semiconductor representation of the tunneling process are shown in Figure 3.11.

Finally, single-electron tunneling between two identical superconductors is briefly described. Figure 3.12 shows the Bose condensation representation of this S-I-S tunneling process. Strong tunneling sets in at an applied voltage $U \geq |2\Delta/e|$. This voltage is high enough to allow the empty energy states above the gap of the first superconductor, and the filled energy states below the gap of the second superconductor, to overlap; pair breaking processes become possible. Figure 3.13 (bottom) shows the resulting current–voltage characteristic of this S-I-S tunnel junction. At $T = 0$ the tunneling current increases rapidly as soon as the voltage exceeds $|2\Delta/e|$. However, even at voltages well below $|2\Delta/e|$, a weak tunneling current is measured for temperatures $T > 0$. The reason for this behavior is that a few electrons are thermally excited into states above the energy gap, as shown in the semiconductor representation of the S-I-S tunneling process in Figure 3.13 (top). These few electrons populating states above the energy gap are responsible for a weak tunneling current flowing at voltages below $|2\Delta/e|$.

So far, only single-electron tunneling has been considered. In 1962, B.D. Josephson predicted theoretically that Cooper-pair tunneling through a very thin insulating layer is possible [15, 16]. Figure 3.12 suggests that for $U = 0$ across the insulating layer Cooper pairs can tunnel through the barrier. It is difficult to imagine Cooper-pair tunneling if we consider the Cooper pairs as particles with a spatial extension ξ_{Co}. However, this process can be well understood as the tunneling of the wave function describing the collective of the Cooper pairs. Figure 3.14 shows the Josephson DC current resulting from Cooper-pair tunneling at zero voltage across the insulating layer. Below the critical current I_c of the

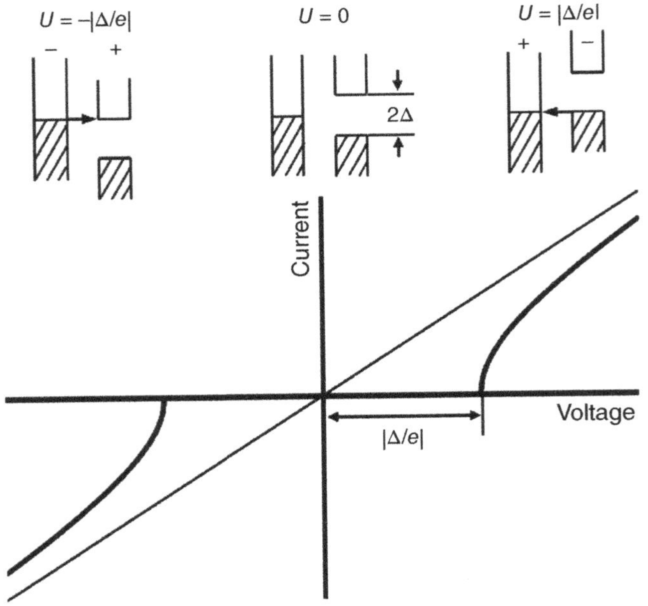

Figure 3.11 Semiconductor representation of N-I-S tunneling (top) and the corresponding current–voltage characteristic at $T = 0$ (bottom) (Reprinted from "High-Temperature Superconductors: Materials, Properties, and Applications" by Rainer Wesche, Kluwer Academic Publishers, 1998, ISBN 0-7923-8386-9, Figure 2-28, p. 40 with kind permission from Springer Science+Business Media B.V.)

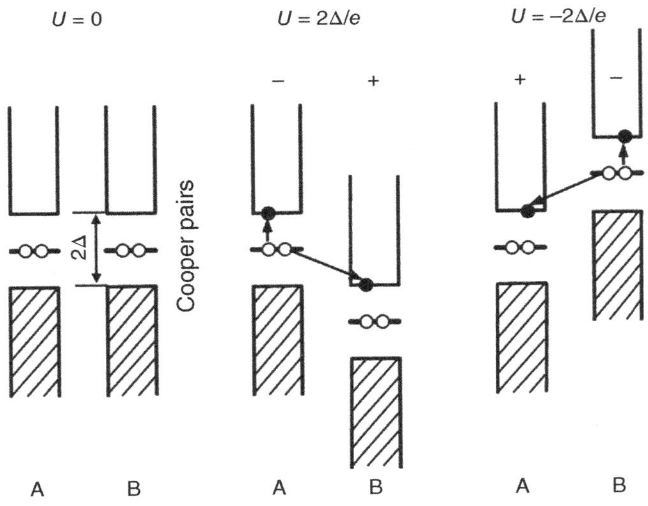

Figure 3.12 Bose condensation representation of single-electron tunneling between two identical superconductors. Strong tunneling sets in for $U \geq |2\Delta/e|$ (Reprinted from "High-Temperature Superconductors: Materials, Properties, and Applications" by Rainer Wesche, Kluwer Academic Publishers, 1998, ISBN 0-7923-8386-9, Figure 2-29, p. 41 with kind permission from Springer Science+Business Media B.V.)

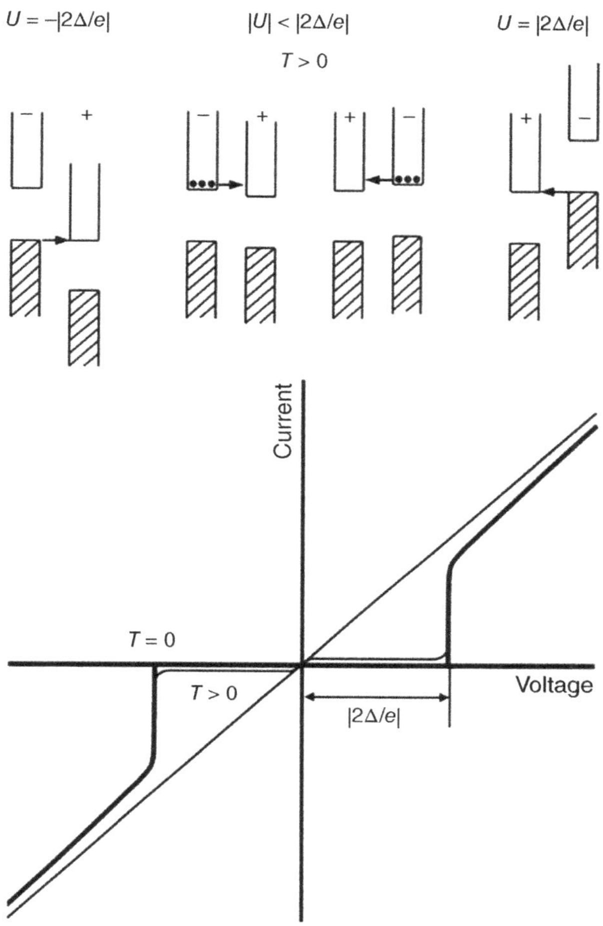

Figure 3.13 *Current–voltage characteristic of an S-I-S tunnel junction (bottom). Strong tunneling sets in for $U \geq |2\Delta/e|$. The semiconductor representation of the S-I-S tunneling process (top) shows that a few electrons are thermally excited into states above the gap at $T > 0$. These electrons are responsible for the weak tunneling current observed at low voltages (Reprinted from "High-Temperature Superconductors: Materials, Properties, and Applications" by Rainer Wesche, Kluwer Academic Publishers, 1998, ISBN 0-7923-8386-9, Figure 2-30, p. 42 with kind permission from Springer Science+Business Media B.V.)*

S-I-S junction, the Josephson current flows without resistance. As soon as the current exceeds I_c, voltage and current jump to values on the current–voltage characteristic for single-electron tunneling which is also shown in Figure 3.14. In the following section, a quantitative description of the DC Josephson current will be given.

To calculate the DC Josephson current flowing between two superconductors A and B separated by a thin oxide layer, the time-dependent Schrödinger equation has to be

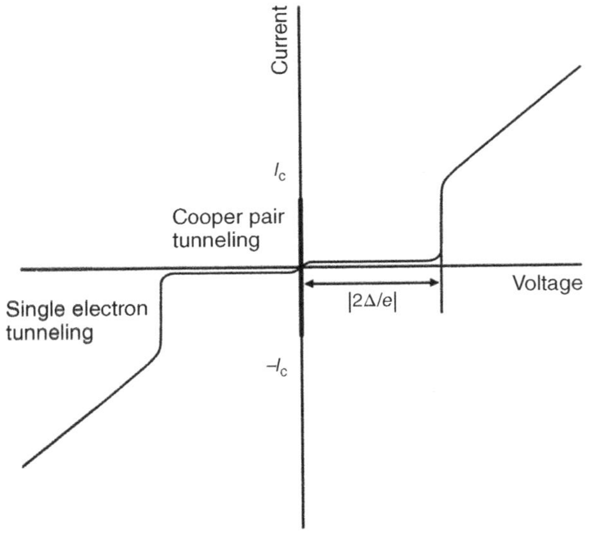

Figure 3.14 *Josephson DC current resulting from Cooper-pair tunneling at zero voltage across the oxide layer of the S-I-S junction. The corresponding current–voltage characteristic for single-electron tunneling is also shown (Reprinted from "High-Temperature Superconductors: Materials, Properties, and Applications" by Rainer Wesche, Kluwer Academic Publishers, 1998, ISBN 0-7923-8386-9, Figure 2-31, p. 43 with kind permission from Springer Science+Business Media B.V.)*

applied to the wave functions of both Cooper-pair systems. The resulting Schrödinger equations are

$$i\hbar\frac{\partial\psi_A}{\partial t} = E_A\psi_A + K_{\text{Co}}\psi_B$$
$$i\hbar\frac{\partial\psi_B}{\partial t} = E_B\psi_B + K_{\text{Co}}\psi_A \tag{3.38}$$

where K_{Co} is the coupling constant describing the Cooper-pair tunneling through the insulating barrier. The wave functions for the two systems of Cooper pairs can be written as

$$\psi_A = n_{CA}^{1/2}e^{i\varphi_A} \quad \text{and} \quad \psi_B = n_{CB}^{1/2}e^{i\varphi_B} \tag{3.39}$$

where n_{CA} and n_{CB} are the number densities of the Cooper pairs in the two superconductors. The phase difference between the two wave functions is

$$\delta = \varphi_B - \varphi_A \tag{3.40}$$

Inserting the wave functions 3.39 into Equation 3.38 leads to the following expressions:

$$\frac{1}{2}n_{CA}^{-1/2}e^{i\varphi_A}\frac{\partial n_{CA}}{\partial t} + in_{CA}^{1/2}e^{i\varphi_A}\frac{\partial\varphi_A}{\partial t} = -\frac{i}{\hbar}(E_A n_{CA}^{1/2}e^{i\varphi_A} + K_{\text{Co}}\psi_B) \tag{3.41}$$

$$\frac{1}{2}n_{CB}^{-1/2}e^{i\varphi_B}\frac{\partial n_{CB}}{\partial t} + in_{CB}^{1/2}e^{i\varphi_B}\frac{\partial\varphi_B}{\partial t} = -\frac{i}{\hbar}(E_B n_{CB}^{1/2}e^{i\varphi_B} + K_{\text{Co}}\psi_A) \tag{3.42}$$

Multiplying Equation 3.41 by ${n_{CA}}^{1/2}\exp(-i\varphi_A)$, Equation 3.42 by ${n_{CB}}^{1/2}\exp(-i\varphi_B)$, and using Equation 3.40, leads to

$$\frac{1}{2}\frac{\partial n_{CA}}{\partial t} + i n_{CA}\frac{\partial \varphi_A}{\partial t} = -\frac{i}{\hbar}\left(n_{CA}E_A + K_{\text{Co}}\left(n_{CA}n_{CB}\right)^{1/2}e^{i\delta}\right) \tag{3.43}$$

$$\frac{1}{2}\frac{\partial n_{CB}}{\partial t} + i n_{CB}\frac{\partial \varphi_B}{\partial t} = -\frac{i}{\hbar}\left(n_{CB}E_B + K_{\text{Co}}\left(n_{CA}n_{CB}\right)^{1/2}e^{-i\delta}\right) \tag{3.44}$$

The following expressions are obtained by separation of Equations 3.43 and 3.44 into real and imaginary parts:

$$\frac{\partial n_{CA}}{\partial t} = \frac{2K_{\text{Co}}}{\hbar}(n_{CA}n_{CB})^{1/2}\sin\delta = -\frac{\partial n_{CB}}{\partial t} \tag{3.45}$$

$$\begin{aligned}\frac{\partial \varphi_A}{\partial t} &= -\frac{1}{\hbar}\left(E_A + K_{\text{Co}}\left(\frac{n_{CB}}{n_{CA}}\right)^{1/2}\cos\delta\right)\\ \frac{\partial \varphi_B}{\partial t} &= -\frac{1}{\hbar}\left(E_B + K_{\text{Co}}\left(\frac{n_{CA}}{n_{CB}}\right)^{1/2}\cos\delta\right)\end{aligned} \tag{3.46}$$

The resulting DC current I through the contact is

$$I = 2|e|\frac{\partial n_{CA}}{\partial t}V \tag{3.47}$$

where V is the volume of superconductor A. Using Equation 3.45, the following current through the contact is obtained for two identical superconductors:

$$I = \frac{4|e|K_{\text{Co}}}{\hbar}n_C V\sin\delta = I_c\sin\delta \tag{3.48}$$

where I_c is the critical current for the Josephson junction. The Josephson DC current through the contact is between $-I_c$ and I_c, depending on the phase difference δ of the two Cooper-pair wave functions. This phase difference is time independent for zero voltage across the insulating layer.

Applying a DC voltage U across the insulating layer of the Josephson junction causes a time-dependent phase difference between the two Cooper-pair wave functions. For two identical superconductors, from Equation 3.46, we obtain the following expression:

$$\frac{d}{dt}(\varphi_B - \varphi_A) = \frac{1}{\hbar}(E_A - E_B) \tag{3.49}$$

An applied voltage U corresponds to an energy difference $E_A - E_B = 2eU$. The resulting phase difference between the Cooper-pair wave functions is

$$\delta(t) = \delta(0) + \frac{2|e|U}{\hbar}t \tag{3.50}$$

where

$$\nu_J = \frac{2|e|U}{h} = \frac{U}{\phi_0} \tag{3.51}$$

is the characteristic Josephson frequency. Using Equation 3.48 and taking into consideration the time dependence of the phase difference $\delta(t)$ between the two Cooper-pair wave functions, the following expression for the Josephson AC current I_{AC} is obtained:

$$I_{AC} = I_c \sin\left(2\pi \nu_J t + \delta(0)\right) \tag{3.52}$$

The AC Josephson current I_{AC} oscillates with a frequency $\nu_J = 483.6$ MHz/μV. The tunneling of a Cooper pair through the barrier is related to the emission or absorption of a photon with an energy $E = h\nu_J = 2|e|U$. The AC Josephson current flows in addition to the current resulting from single-electron tunneling. A more detailed description of the AC Josephson effects can be found in reference [3].

Next, macroscopic quantum interference effects in Josephson junctions will be discussed. These interference effects are closely connected to the phase shifts of the Cooper-pair wave functions caused by an applied magnetic field. The situation considered in the following calculations is illustrated in Figure 3.15. The magnetic field $B_y(x)$ is applied along the y-direction. A possible representation of the vector potential $\mathbf{A}$ within the barrier is

$$A_z(x) = -xB_y(0) \quad \text{for} \quad |x| \le d/2 \tag{3.53}$$

Inside the two identical superconductors A and B, the magnetic field $B_y(x)$ decays exponentially. At a sufficiently large distance from the insulating layer, the magnetic field is zero

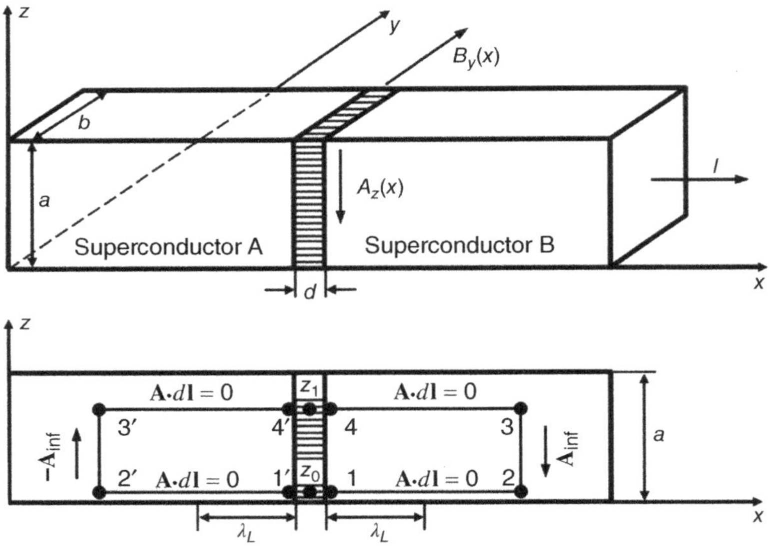

Figure 3.15 *S-I-S tunnel junction with a magnetic field applied along the y-direction (top) and integration path for the determination of the phase difference of the two Cooper-pair wave functions at an arbitrary point z_1 (Reprinted from "High-Temperature Superconductors: Materials, Properties, and Applications" by Rainer Wesche, Kluwer Academic Publishers, 1998, ISBN 0-7923-8386-9, Figure 2-32, p. 46 with kind permission from Springer Science+Business Media B.V.)*

and the vector potential reaches a constant value $\mathbf{A}_{\mathrm{inf}}$. Furthermore, the screening currents are zero deep inside the superconductor.

Combining Equations 3.16–3.18, the following condition for the phase shift is found:

$$\oint \nabla\varphi(\mathbf{l})\,d\mathbf{l} = \frac{2\pi}{\phi_0}\oint \mathbf{A}\cdot d\mathbf{l} \tag{3.54}$$

The phase difference between the two Cooper-pair wave functions at a reference point z_0 in the barrier (see Figure 3.15, bottom) is

$$\varphi_0 = \varphi(1) - \varphi(1') \tag{3.55}$$

The phase difference at an arbitrary point z_1 is

$$\varphi(z_1) = \varphi(4) - \varphi(4') \tag{3.56}$$

This phase difference is determined by integration of the vector potential $A_z(x)$ along the paths $1 \to 4$ and $1' \to 4'$ indicated in Figure 3.15,

$$\varphi(z_1) = \varphi_0 + \frac{2\pi}{\phi_0}\left\{\int_1^4 \mathbf{A}\cdot d\mathbf{l} - \int_{1'}^{4'} \mathbf{A}\cdot d\mathbf{l}\right\} \tag{3.57}$$

Because $\mathbf{A}\cdot d\mathbf{l} = 0$ on the horizontal paths, only the vertical paths $2 \to 3$ and $2' \to 3'$ contribute to the line integrals in Equation 3.57. The resulting phase difference of the Cooper-pair wave functions for two identical superconductors is

$$\varphi(z_1) = \varphi_0 + \frac{2\pi}{\phi_0}2zA_{\mathrm{inf}} \tag{3.58}$$

where $z = z_1 - z_0$. The total flux enclosed by the barrier is

$$\phi = \oint \mathbf{A}\cdot d\mathbf{l} = 2aA_{\mathrm{inf}} \tag{3.59}$$

The total flux is approximately given by

$$\phi = a(d + 2\lambda_L)B_y(0) \tag{3.60}$$

where d is the thickness of the insulating layer and λ_L the penetration depth. Thus, the phase difference of the two Cooper-pair wave functions can be written as

$$\varphi(z_1) = \varphi_0 + \frac{2\pi\phi}{\phi_0}\frac{z}{a} \tag{3.61}$$

The applied magnetic field gives rise to a spatial variation of the phase difference $\varphi(z)$ of the two Cooper-pair wave functions. Using Equation 3.61, the following expression for the Josephson DC current is obtained:

$$I = bJ_c\int \sin\varphi(z)\,dz \tag{3.62}$$

Integration of Equation 3.62 gives

$$I = I_c \sin \varphi_0 \frac{\sin \pi\phi/\phi_0}{\pi\phi/\phi_0} \tag{3.63}$$

where $I_c = J_c ab$. The phase shift φ_0 in Equation 3.63 is unknown. However, the maximum Josephson DC current I_{max} is reached for $\sin \varphi_0 = 1$, corresponding to a phase difference of $\pi/2$.

$$I_{max} = I_c \left| \frac{\sin \pi\phi/\phi_0}{\pi\phi/\phi_0} \right| \tag{3.64}$$

Figure 3.16 shows the resulting interference pattern for the Josephson DC current. This interference figure is well known from the diffraction of light at a single rectangular slit. The Josephson–Fraunhofer interference of the two Cooper-pair wave functions indicates that superconductivity is a macroscopic quantum phenomenon.

Finally, quantum interference effects occurring in a loop of two identical S-I-S junctions will be briefly described. We consider the double-junction loop schematically illustrated in Figure 3.17.

In the case of two identical Josephson contacts, the same currents I_1 and I_2 flow in both. It is furthermore assumed that the contribution of the currents to the magnetic field applied to the Josephson contacts can be neglected. Similar to the calculation of the phase difference in a single contact, the vector potential **A** has to be integrated along a path which includes the whole area of the double contact loop.

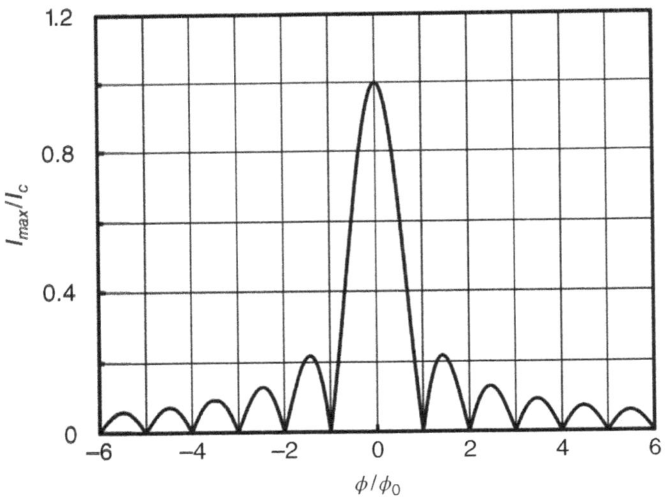

Figure 3.16 *Josephson–Fraunhofer diffraction pattern for the normalized maximum Josephson DC current I_{max}/I_c. Minima of the current occur at ϕ/ϕ_0 values of 1, 2, 3, etc. (Reprinted from "High-Temperature Superconductors: Materials, Properties, and Applications" by Rainer Wesche, Kluwer Academic Publishers, 1998, ISBN 0-7923-8386-9, Figure 2-33, p. 49 with kind permission from Springer Science+Business Media B.V.)*

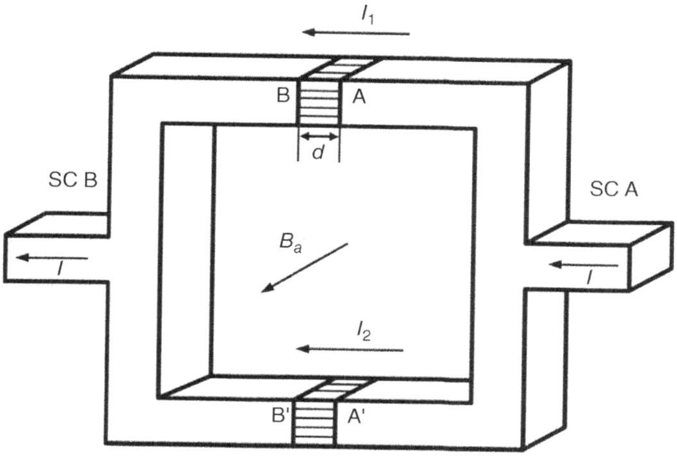

Figure 3.17 *Double contact junction consisting of two identical S-I-S junctions (Reprinted from "High-Temperature Superconductors: Materials, Properties, and Applications" by Rainer Wesche, Kluwer Academic Publishers, 1998, ISBN 0-7923-8386-9, Figure 2-34, p. 49 with kind permission from Springer Science+Business Media B.V.)*

The phase difference between the wave functions in the two junctions is, analogously to Equation 3.61, given by

$$\varphi_{A'B'} = \varphi_{AB} + 2\pi\phi_L/\phi_0 \tag{3.65}$$

where ϕ_L is the magnetic flux within the loop. Equation 3.48 is valid for each of the two contacts and the resulting total Josephson current is

$$I = I_c \left\{ \sin\varphi_{AB} + \sin\left(\varphi_{AB} + 2\pi\frac{\phi_L}{\phi_0}\right)\right\} \tag{3.66}$$

The phase difference within the sufficiently small individual contacts has been neglected in Equation 3.66. The undetermined phase difference $\varphi_{\rm AB}$ can be replaced by $\varphi_{\rm AB} = \varphi_0 + \pi\phi_L/\phi_0$ provided the following expression holds for the current flowing through the double contact

$$I = 2I_c \sin\varphi_0 \cos(\pi\phi_L/\phi_0) \tag{3.67}$$

The maximum possible value of $\sin\varphi_0$ is 1. Thus, the maximum Josephson current is

$$I_{\rm max} = 2I_c|\cos(\pi\phi_L/\phi_0)| \tag{3.68}$$

The maximum currents taking into account the phase difference within the individual loops can be obtained by superposition of the single contact equation 3.64 and the loop equation 3.68,

$$I_{\rm max} = 2I_c \left| \frac{\sin\pi\dfrac{\phi_J}{\phi_0}}{\pi\dfrac{\phi_J}{\phi_0}} \cos\frac{\pi\phi_L}{\phi_0} \right| \tag{3.69}$$

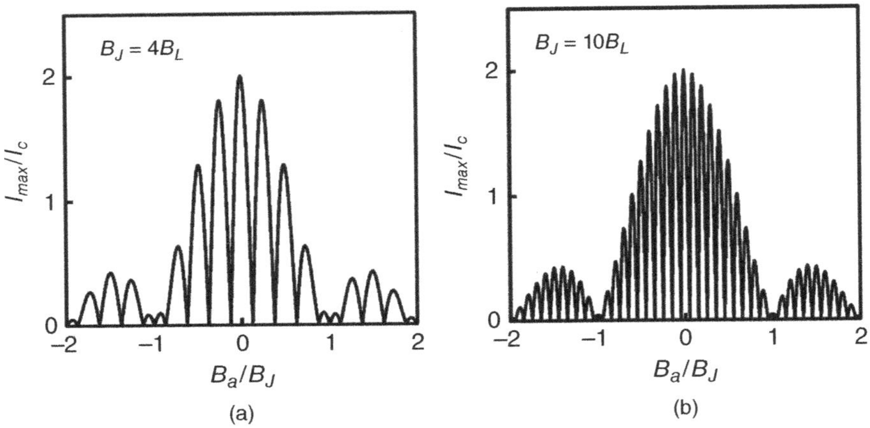

Figure 3.18 *Quantum interference effects in double-loop Josephson junctions (Reprinted from "High-Temperature Superconductors: Materials, Properties, and Applications" by Rainer Wesche, Kluwer Academic Publishers, 1998, ISBN 0-7923-8386-9, Figure 2-35, p. 51 with kind permission from Springer Science+Business Media B.V.)*

where ϕ_J is the flux in a single contact. The magnetic fluxes in the individual contacts and the loop are connected to the applied field B_a by

$$\phi_J = B_a A_J \qquad \phi_L = B_a A_L \tag{3.70}$$

where A_J and A_L are the areas of the single contacts and the loop, respectively. Defining the magnetic fields $B_J = \phi_0/A_J$ and $B_L = \phi_0/A_L$, Equation 3.69 can be written as

$$I_{\max} = 2I_c \left| \frac{\sin \pi \dfrac{B_a}{B_J}}{\pi \dfrac{B_a}{B_J}} \cos \frac{\pi B_a}{B_L} \right| \tag{3.71}$$

Using Equation 3.71 the diffraction patterns for B_J/B_L ratios of 4 and 10 have been calculated. Figure 3.18 shows the results. The maxima of the rapid loop oscillations are limited by the slower current variations within the single contacts. Experimental results for multiple Josephson junctions can be found in reference [17].

References

1. J. Bardeen, L.N. Cooper, and J.R. Schrieffer, Microscopic theory of superconductivity, *Phys. Rev.*, **106**, 162–164 (1957).
2. J. Bardeen, L.N. Cooper, and J.R. Schrieffer, Theory of superconductivity, *Phys. Rev.*, **108**, 1175–1204 (1957).
3. C.P. Poole, Jr.,, H.A. Farach, and R.J. Creswick, *Superconductivity*, Academic Press, San Diego, CA,1995.

4. H. Fröhlich, Theory of the superconducting ground state: I. The ground state at the absolute zero of temperature, *Phys. Rev.*, **79**, 845–856 (1950).
5. J. Bardeen, Wave functions of superconducting electrons, *Phys. Rev.*, **80**, 567–574 (1950).
6. C.A. Reynolds, B. Serin, W.H. Wright, and L.B. Nesbitt, Superconductivity of isotopes of mercury, *Phys. Rev.*, **78**, 487 (1950).
7. E. Maxwell, Isotope effect in the superconductivity of mercury, *Phys. Rev.*, **78**, 477 (1950).
8. J.W. Garland, Jr.,, Isotope effect in superconductivity, *Phys. Rev. Lett.*, **11**, 114–121 (1963).
9. B. Serin, C.A. Reynolds, and C. Lohman, The isotope effect in superconductivity: II. Tin and lead, *Phys. Rev.*, **86**, 162–164 (1952).
10. R. Doll and M. Näbauer, Experimental proof of magnetic flux quantization in a superconducting ring, *Phys. Rev. Lett.*, **7**, 51–52 (1961).
11. B.S. Deaver, Jr.,, and W.M. Fairbank, Experimental evidence for quantized flux in superconducting cylinders, *Phys. Rev. Lett.*, **7**, 43–46 (1961).
12. I. Giaever, Energy gap in superconductors measured by electron tunnelling, *Phys. Rev. Lett.*, **5**, 147–148 (1960).
13. I. Giaever, Electron tunnelling between two superconductors, *Phys. Rev. Lett.*, **5**, 464–466 (1960).
14. J. Nicol, S. Shapiro, and P.H. Smith, Direct measurement of the superconducting energy gap, *Phys. Rev. Lett.*, **5**, 461–464 (1960).
15. B.D. Josephson, Possible new effects in superconductive tunnelling, *Phys. Lett.*, **1**, 251–253 (1962).
16. B.D. Josephson, Coupled superconductors, *Rev. Mod. Phys.*, **36**, 216–220 (1964).
17. R.C. Jaklevic, J. Lambe, J.E. Mercereau, and A.H. Silver, Macroscopic quantum interference in superconductors, *Phys. Rev.*, **140A**, 1628–1637 (1965).

4

Type II Superconductors

4.1 Introduction

Soon after his discovery in 1911, Heike Kammerlingh Onnes had the vision to use superconducting magnets for the generation of high magnetic fields. However, the superconducting elements, such as mercury, tin, and lead, discovered in his laboratory at the University of Leiden, were all Type I superconductors, characterized by the full Meissner effect and low thermodynamic critical fields (B_c) of less than 100 mT (see Figure 2.15, Chapter 2). As a consequence, superconductivity in Type I superconductors can easily be destroyed by low magnetic fields. In Type I superconductors, both the vanishing of perfect diamagnetism and the re-establishment of the electrical resistance occur at the same magnetic field B_c.

The first experimental evidence for the existence of a different type of superconductors was found in the late 1920s [1]. In superconducting alloys, it was observed that perfect diamagnetism vanishes at considerably lower magnetic fields than resistance-free current transport. At the time this strange behavior of superconducting alloys was explained by in-homogeneities of the material leading to a matrix of thin superconducting filaments with diameters smaller than the London penetration depth (sponge hypothesis) [2, 3].

The experimental studies of Shubnikov and co-workers using single crystalline lead alloys revealed the true nature of Type II superconductivity [4]. To avoid any sponge-like behavior, the Pb–Tl and Pb–In single crystals were carefully annealed at temperatures close to their melting point. These experiments showed clearly that below a certain concentration of impurities, the material behaves like superconductive elements showing perfect diamagnetism (full Meissner effect). At higher concentrations of impurities, the magnetic behavior changes to that of Type II superconductors. They are characterized by the loss of perfect diamagnetism at the lower critical field B_{c1}. Above this field, magnetic flux starts to penetrate into the superconductor. Finally, at the upper critical field B_{c2}, the resistive normal-conducting state is re-established. An important aspect is the fact that the difference between B_{c1} and B_{c2} increases with increasing impurity content. This discovery of Shubnikov and co-workers opened the way towards the use of superconductors for the generation of high magnetic fields. In the next section, we will briefly describe the main

Physical Properties of High-Temperature Superconductors, First Edition. Rainer Wesche.

ideas of the Ginzburg–Landau (GL) theory [5], which perfectly describes the behavior of Type II superconductors.

4.2 The Ginzburg–Landau Theory[1]

The GL theory is a phenomenological theory based on the principles of thermodynamics. It is an application of the Landau theory of phase transitions to superconductivity [5]. In Chapter 2, the behavior of Type I superconductors was discussed. These superconductors are characterized by perfect diamagnetism. The screening currents flowing in the surface layer produce a magnetization which cancels the applied magnetic field in the interior of the superconductor. A second type of superconductors, called Type II superconductors, will be briefly described in this section. Only these Type II superconductors can be used for magnet and power applications. The GL theory [5–8] provides a quantitative description of their behavior.

A constant density n_C of the Cooper pairs within the whole superconductor is assumed in the London theory. The only characteristic length scale is the penetration depth λ_L describing the decay of a magnetic field inside a superconductor. By taking into consideration a variation of the Cooper pair density n_C, Ginzburg and Landau introduced a superconducting order parameter $\Phi(\mathbf{r})$, which is a complex quantity containing a phase factor $\exp(i\varphi)$:

$$\Phi(\mathbf{r}) = |\Phi(\mathbf{r})| \exp(i\varphi) \tag{4.1}$$

The square of the order parameter gives the Cooper pair density

$$n_C(\mathbf{r}) = |\Phi(\mathbf{r})|^2 \tag{4.2}$$

Above the critical temperature T_c, the order parameter is zero. In the vicinity of the transition temperature T_c, the Gibbs free energy can be expanded as a function of the order parameter $\Phi(\mathbf{r})$ [6–8]:

$$\begin{aligned} G_s(\Phi) = G_n + \int_V \Big\{ & \alpha|\Phi|^2 + \frac{1}{2}\beta|\Phi|^4 + \frac{1}{2m_C}|(-i\hbar\nabla - q_C\mathbf{A})\Phi|^2 \\ & + \frac{1}{2\mu_0}B^2(\mathbf{r}) - \mu_0\mathbf{H}(\mathbf{r})\cdot\mathbf{M}(\mathbf{r}) \Big\} dV \end{aligned} \tag{4.3}$$

where m_C is the mass, q_C is the charge of a Cooper pair, $B(\mathbf{r})$ is the magnetic flux density, $\mathbf{H}(\mathbf{r})$ is the magnetic field, $\mathbf{A}$ is the vector potential, and V is the volume of the superconductor. Furthermore, α and β are functions of temperature. The first two terms of the integrand in Equation 4.3 are the expansion of the energy difference of the normal and superconducting states at zero applied magnetic field close to the critical temperature. The third term describes the kinetic energy of the Cooper pairs. The last two terms of the integrand give the electromagnetic energy of the superconductor. Requiring a minimum of the Gibbs free

[1] Reprinted with minor changes from "*High-Temperature Superconductors: Materials, Properties, and Applications*" by Rainer Wesche, Kluwer Academic Publishers, 1998, ISBN 0-7923-8386-9, pp. 65–70 with kind permission from Springer Science+Business Media B.V.

energy with respect to the order parameter Φ and the vector potential $\mathbf{A}$ for the equilibrium state leads to the two GL equations, which describe the superconducting state [5–8]:

$$\frac{1}{2m_C}(-i\hbar\nabla - q_C\mathbf{A})^2\,\Phi(\mathbf{r}) + \alpha\,\Phi(\mathbf{r}) + \beta\,|\Phi(\mathbf{r})|^2\,\Phi(\mathbf{r}) = 0 \tag{4.4}$$

$$\mathbf{j} = \frac{i\hbar q_C}{2m_C}(\Phi(\mathbf{r})^*\nabla\Phi(\mathbf{r}) - \Phi(\mathbf{r})\nabla\Phi(\mathbf{r})^*) - \frac{q_C{}^2}{m_C}\mathbf{A}|\Phi(\mathbf{r})|^2 \tag{4.5}$$

where $\Phi(\mathbf{r})^*$ is the conjugate complex of the order parameter Φ and $\mathbf{j}$ is the current density of the supercurrents. At a superconductor–vacuum interface, we have the boundary condition

$$\mathbf{n}(-i\hbar\nabla - q_C\mathbf{A})\Phi = 0 \tag{4.6}$$

where $\mathbf{n}$ is a unit vector perpendicular to the interface, i.e., the current across the boundary is zero.

First, the zero field case $\mathbf{A} = 0$ will be considered, assuming in addition that the density n_C of the Cooper pairs is constant. This situation is found deep inside the superconductor far away from the phase boundaries, where Equation 4.3 reduces to

$$G_s = G_n + V\left(\alpha\,|\Phi_i|^2 + \frac{1}{2}\beta\,|\Phi_i|^4\right) \tag{4.7}$$

The equilibrium state corresponds to the minimum of the Gibbs free energy G_s. Thus, $dG_s/d|\Phi_i|^2 = 0$ and

$$\alpha + \beta|\Phi_i|^2 = 0 \tag{4.8}$$

Inserting $|\Phi_i|^2 = -\alpha/\beta$ into Equation 4.7 leads to the following expression for the condensation energy of the Cooper pairs:

$$G_n - G_s = \frac{1}{2}\frac{\alpha^2}{\beta}V = \frac{B_c{}^2 V}{2\mu_0} \tag{4.9}$$

where B_c is the thermodynamic critical field (see Figure 4.5) and V is the volume of the superconductor.

Next, the interface of normal and superconducting semi-infinite spaces will be considered. The following variation of the superconducting order parameter Φ results from the GL theory [6]:

$$\Phi(x) = \Phi_i \tanh\frac{x}{\sqrt{2}\xi} \tag{4.10}$$

where Φ_i is the order parameter deep inside the superconductor and ξ is the coherence length, the shortest length within which a significant change of the order parameter Φ and the Cooper pair density n_C can occur. Figure 4.1 shows the variation of the order parameter in the vicinity of a plane normal–superconductor interface. A more detailed introduction to the GL theory can be found in Refs. [6, 9]. In 1959, Gor'kov showed that the GL theory can be derived from the BCS theory [10].

In Section 2.5, it was shown that the difference between the Gibbs free energies of the normal and superconducting states is given by the condensation energy $E_c = VB_c{}^2/2\mu_0$ of the Cooper pairs (see Equation 2.62, Chapter 2). An applied magnetic field B is excluded from the interior of the superconductor.

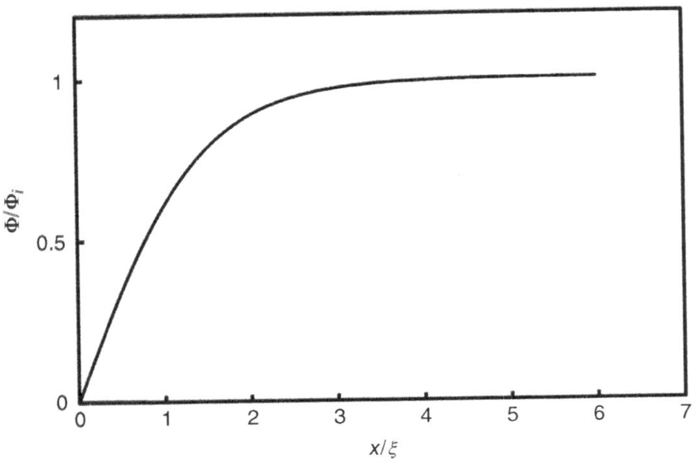

Figure 4.1 *Variation of the order parameter Φ in the superconducting phase in the vicinity of a plane phase boundary (Reprinted from "High-Temperature Superconductors: Materials, Properties, and Applications" by Rainer Wesche, Kluwer Academic Publishers, 1998, ISBN 0-7923-8386-9, Figure 2-45, p. 67 with kind permission from Springer Science+Business Media B.V.)*

The energy related to the exclusion of the magnetic field is given by

$$\Delta G = -\int_0^B \mathbf{m} \cdot d\mathbf{B} \tag{4.11}$$

where **m** is the magnetic moment of the superconductor and **B** is the applied magnetic field. This energy reduces the difference between the Gibbs free energies of the normal and superconducting states. In bulk specimens, the penetration depth λ_L is much smaller than the dimensions of the superconductor. For a slab with the magnetic field applied parallel to the faces, demagnetization effects can be neglected and $\mathbf{m} = -\mathbf{B}V/\mu_0$. However, the magnetic energy is considerably reduced for thin superconductors with dimensions comparable to the penetration depth λ_L. Figure 2.17 in Chapter 2 shows that an applied magnetic field can penetrate into the interior of a thin slab. As a consequence, magnetic fields well above the bulk critical field B_c are required to destroy the superconducting state in sufficiently thin specimens. Therefore, a subdivision into many superconducting and normal regions would allow the existence of superconductivity in the presence of large magnetic fields. However, experience shows that this does not occur in Type I superconductors. To understand this behavior, the surface energy associated with the boundaries between normal and superconducting regions has to be taken into consideration. The Cooper pair density n_C increases in the interfacial layer from zero to the value n_{Ci} deep inside the superconductor. This leads to a loss of condensation energy E_c. On the other hand, the magnetic field penetrates into a thin layer of the superconductor, and as a consequence the magnetic energy E_B required for field exclusion is reduced. For simplicity, the variation of the Cooper pair density n_C

and the magnetic field B near the phase boundaries is approximated by the following step functions:

$$\left.\begin{array}{l} n_c = 0 \\ n_c = n_{ci} \end{array}\right\} \quad \text{for} \quad \left\{\begin{array}{l} x \leq \xi \\ x > \xi \end{array}\right. \tag{4.12}$$

$$\left.\begin{array}{l} B = B_a \\ B = 0 \end{array}\right\} \quad \text{for} \quad \left\{\begin{array}{l} x \leq \lambda_L \\ x > \lambda_L \end{array}\right. \tag{4.13}$$

where B_a is the applied magnetic field, ξ is the coherence length, and λ_L is the penetration depth. The resulting surface energy per unit area of a plane for $B_a = B_c$ is

$$\Delta E_c - \Delta E_B \cong (\xi - \lambda_L)\frac{B_c^{\ 2}}{2\mu_0} \tag{4.14}$$

From this simple model, it follows that energy is required to form a normal–superconductor interface if $\xi > \lambda_L$. This behavior is the characteristic of Type I superconductors. The existence of a positive surface energy prevents a subdivision into many normal and superconducting regions. If, on the other hand, $\xi < \lambda_L$, energy is gained by the formation of a normal–superconductor interface. This situation is found in Type II superconductors, in which as a consequence of the negative surface energy, normal and superconducting regions can coexist. Figure 4.2 shows the variation of the magnetic field B and the Cooper pair density n_C in the vicinity of a plane normal–superconductor interface for both types of superconductors. The ratio of the coherence length ξ and the penetration depth λ_L is called the GL parameter κ:

$$\kappa = \frac{\lambda_L}{\xi} \tag{4.15}$$

In fact, the GL theory predicts that, depending on whether the parameter κ is smaller or larger than $1/\sqrt{2}$, the superconductor in question is of Type I or Type II, respectively. Table 4.1 gives for various superconductors the coherence length ξ and the GL parameter κ at $T = 0$. The coherence length is closely related to the mean free path of the electrons in the metal. Large values of the mean free path of the electrons lead to a large coherence length ξ. Most elements are therefore Type I superconductors.

Because of the reduced electron mean free paths in alloys and intermetallic compounds, these materials are generally Type II superconductors. The cuprate high-T_c superconductors are extreme Type II superconductors with κ-values of $\approx$100.

4.3 Magnetic Behavior of Type I and Type II Superconductors[2]

The magnetic induction **B**, the magnetic field **H**, and the magnetization **M** are related by the expression

$$\mathbf{B} = \mu_0(\mathbf{H} + \mathbf{M}) \tag{4.16}$$

[2] Partly reprinted from "*High-Temperature Superconductors: Materials, Properties, and Applications*" by Rainer Wesche, Kluwer Academic Publishers, 1998, ISBN 0-7923-8386-9, pp. 71–76 with kind permission from Springer Science+Business Media B.V.

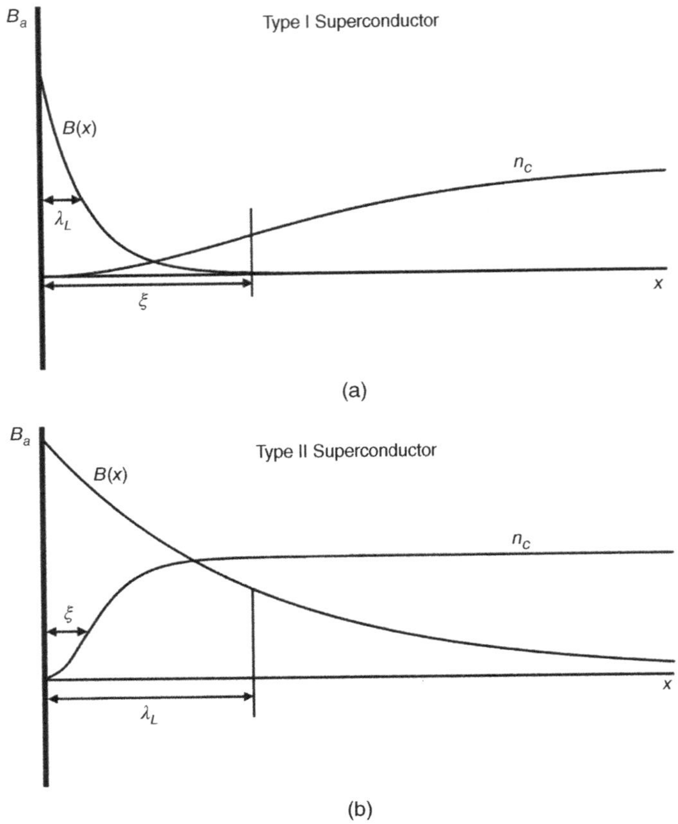

Figure 4.2 *Variation of the Cooper pair density n_C and the magnetic field B near a plane normal–superconductor interface for Type I (top) and Type II (bottom) superconductors (Reprinted from "High-Temperature Superconductors: Materials, Properties, and Applications" by Rainer Wesche, Kluwer Academic Publishers, 1998, ISBN 0-7923-8386-9, Figure 2-46, p. 69 with kind permission from Springer Science+Business Media B.V.)*

Table 4.1 *Coherence length ξ and GL parameter κ for selected superconductors at $T = 0$*

Material	T_c(K)	ξ(nm)	$\kappa = \lambda_L/\xi$	Reference
Al	1.18	1520	0.03	[6]
In	3.41	360	0.11	[6]
Cd	0.52	760	0.14	[6]
Sn	3.72	180	0.23	[6]
Ta	4.5	93	0.38	[7]
Pb	7.2	82	0.48	[6]
Nb	9.25	39	1.28	[6]
NbTi	9.5	4	75	[9]
Nb_3Sn	18	3	21.7	[9]
Nb_3Ge	23.2	3	30	[9]
Rb_3C_{60}	29.6	2	123.5	[11]
$YBa_2Cu_3O_{7-x}$	93	2	95	[6]

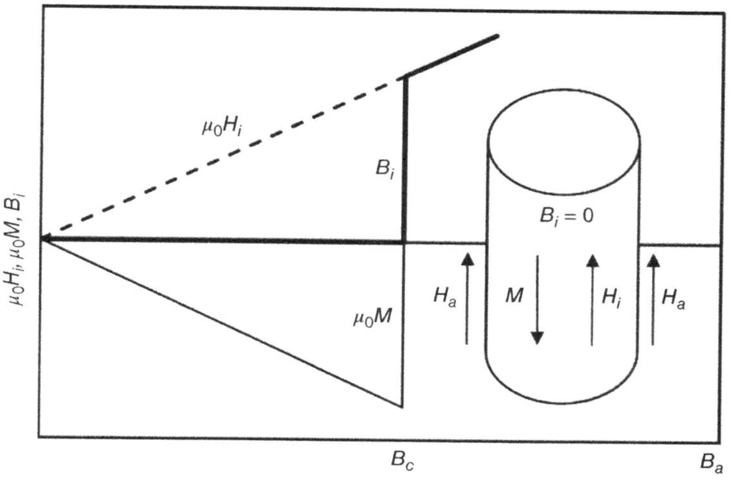

Figure 4.3 *Relations between the applied magnetic field H_a, the magnetization M, and the magnetic induction B_i inside a cylindrical Type I superconductor (see inset). In the superconducting state the magnetization M, generated by the screening currents, cancels the magnetic induction caused by the external field H_a, and hence $B_i = 0$. In the normal state above the thermodynamic critical field B_c the magnetization is zero, leading to $B_i = \mu_0 H_a$*

We consider a superconducting cylinder with the magnetic field applied parallel to the axis (see Figure 4.3). For this geometry, the magnetic fields on the two sides of the boundary are equal:

$$H_a = \frac{B_a}{\mu_0} = H_i \tag{4.17}$$

Inside a Type I superconductor, the magnetization M generated by the screening currents flowing in the surface layer cancels the magnetic induction B_a caused by the applied magnetic field H_a. As a consequence, the magnetic induction in the interior of the superconductor B_i is zero. At the critical field B_c, the screening currents vanish, the magnetization is negligibly small, and $B_i \approx B_a$. In Figure 4.3, the relations between the magnetization and the internal fields B_i and H_i inside a Type I superconductor are illustrated.

In an ideal Type II superconductor, no defects are present, which act as pinning centers (see the following and Section 4.4). In the absence of effective pinning, the magnetic flux is not hindered to move in and out of the Type II superconductor, and, hence, the magnetization is reversible. Figure 4.4 shows the magnetization curve of an ideal Type II superconductor. Below the lower critical field B_{c1}, an applied magnetic field is excluded from the interior of the superconductor just as in Type I superconductors. As soon as the applied magnetic field exceeds B_{c1}, magnetic flux can penetrate into the interior of Type II superconductors. In the resulting mixed state, the normal and superconducting regions coexist within the material. Finally, superconductivity is completely destroyed at the upper critical field B_{c2}. The thermodynamic critical field B_c, also indicated in Figure 4.4, is defined by the expression

$$\int_{B_{c1}}^{B_c} (B_a + \mu_0 M) dB_a = \mu_0 \int_{B_c}^{B_{c2}} -M\, dB_a \tag{4.18}$$

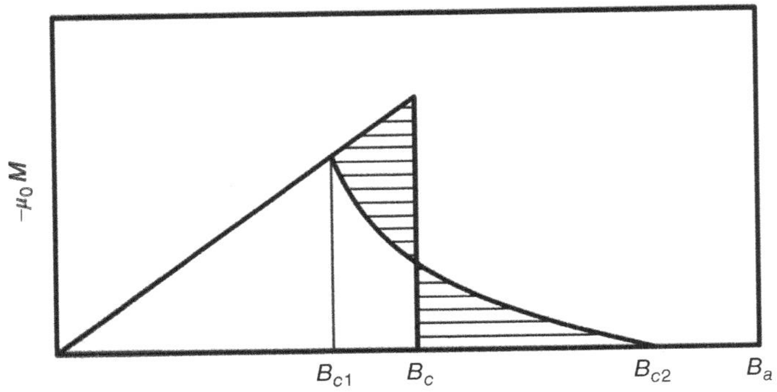

Figure 4.4 *Magnetization curve for an ideal Type II superconductor. The shaded areas above and below of the thermodynamic critical field B_c are equal (Reprinted from "High-Temperature Superconductors: Materials, Properties, and Applications" by Rainer Wesche, Kluwer Academic Publishers, 1998, ISBN 0-7923-8386-9, Figure 2-47, p. 71 with kind permission from Springer Science+Business Media B.V.)*

where B_a is the applied magnetic field and M is the magnetization of the superconductor. Equation 4.18 corresponds to the condition that the shaded areas left and right of B_c are equal. Using the GL parameter κ, the lower and upper critical fields can be related to the thermodynamic critical field B_c as follows [6]:

$$B_{c1} = \frac{B_c \ln \kappa}{\sqrt{2}\kappa} \tag{4.19}$$

$$B_{c2} = \sqrt{2}\kappa B_c \tag{4.20}$$

The thermodynamic critical field is given by [6]

$$B_c = \frac{\phi_0}{2\sqrt{2}\pi \lambda_L \xi} \tag{4.21}$$

Table 4.2 *Upper critical fields B_{c2} of selected high-field superconductors*

Material	T_c(K)	B_{c2}(T)($T = 0$)	Reference
NbTi	9.5	13	[9]
NbN	15	15	[9]
V_3Ga	15	23	[9]
V_3Si	17.1	23	[12]
Nb_3Sn	18	23	[12]
Nb_3Al	19.1	33	[12]
Nb_3Ge	23.2	38	[9]
$PbMo_6S_8$	15	60	[9]
Rb_3C_{60}	29.6	57	[11, 13, 14]

where ϕ_0 is the flux quantum. Values for the upper critical field B_{c2} of selected superconductors are listed in Table 4.2. Because of the low thermodynamic critical fields B_c of Type I superconductors, the effect of an applied magnetic field on the Gibbs free energy of the normal state was neglected in Section 2.5.

However, the paramagnetism of the conduction electrons has to be taken into consideration for the large upper critical fields of the Type II superconductors.

The magnetic energies corresponding to electron spins parallel and antiparallel to the applied magnetic field are E_+ and E_-, respectively:

$$\begin{aligned} E_+ &= \mu B_a \qquad S = +\tfrac{1}{2}\hbar \\ E_- &= -\mu B_a \qquad S = -\tfrac{1}{2}\hbar \end{aligned} \tag{4.22}$$

with $\hbar = 1.0546 \times 10^{-34}$ J s. The magnetic moment of a free electron is

$$\mu = \frac{1}{2} g \mu_B \tag{4.23}$$

where $g = 2.023$ and $\mu_B = 9.274\,02 \times 10^{-24}\,\mathrm{A\,m^2}$ is the Bohr magneton. The resulting energy difference for the two spin directions is

$$\Delta E = E_+ - E_- \cong 2\mu_B B_a \tag{4.24}$$

In the presence of an applied magnetic field, the number of electrons with the magnetic moment μ parallel to the applied magnetic field is enhanced at the expense of electrons with opposite μ. The resulting paramagnetic susceptibility χ_P is

$$\chi_P = \mu_0 {\mu_B}^2 D(E_F) \tag{4.25}$$

where $D(E_F)$ is the density of states at the Fermi energy and μ_0 is the permeability of free space. The Pauli spin susceptibility is independent of temperature. As a consequence of the paramagnetism of the conduction electrons, the Gibbs free energy of the normal state is reduced, whereas the free energy of the superconducting state is unchanged because the Cooper pairs have zero spin. The upper critical field B_{c2} is reduced by this effect. As soon as the energy $\Delta E = 2\mu_B B_a$ exceeds the binding energy of the Cooper pairs, these can be broken up into single electrons. A more detailed calculation gives for the Pauli limiting field B_{Pauli} [6]

$$B_{\mathrm{Pauli}} = \frac{\Delta}{\sqrt{2}\mu_B} \tag{4.26}$$

Figure 4.5 shows the magnetic fields for both Type I and Type II superconductors in the interior as a function of the applied field. The resulting phase diagrams for both types of superconductors are compared in Figure 4.6. In Type I superconductors, there exist two different phases, namely the superconducting and the normal, which cannot coexist. On the other hand, three different phases can exist in Type II superconductors.

Below the lower critical field $B_{c1}(T)$, the Type II superconductor in the superconducting phase is characterized by perfect diamagnetism, in which it behaves like a Type I superconductor. Above the upper critical field $B_{c2}(T)$, the material is in the normal state. In a Type

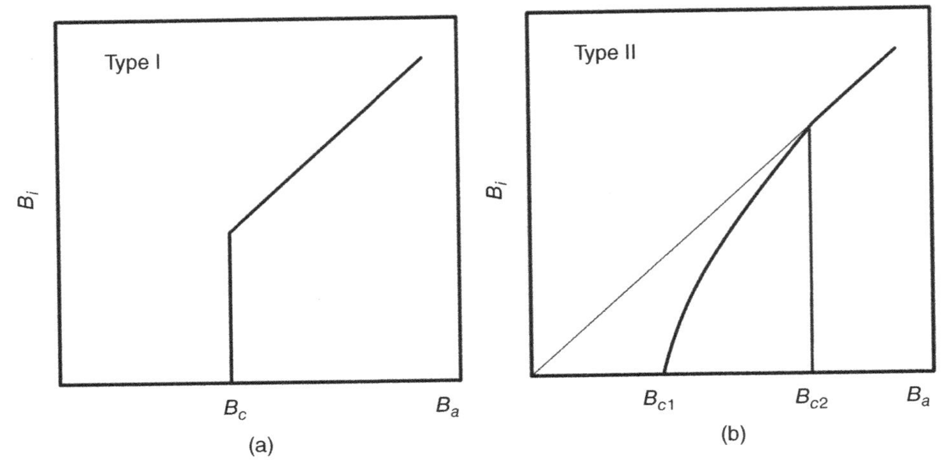

Figure 4.5 *Internal magnetic field B_i versus applied magnetic field B_a for both Type I (left) and Type II (right) superconductors. In Type I, superconductors the internal magnetic field is zero below B_c, while it is equal to B_a in the normal state above B_c. In Type II superconductors, magnetic flux is excluded from the interior up the lower critical field B_{c1}. Between the lower (B_{c1}) and the upper critical fields (B_{c2}), magnetic flux starts to enter into the interior of the superconductor. At B_{c2}, the normal state is re-established and $B_i \cong \mu_0 H_a = B_a$ (Reprinted from "High-Temperature Superconductors: Materials, Properties, and Applications" by Rainer Wesche, Kluwer Academic Publishers, 1998, ISBN 0-7923-8386-9, Figure 2-48, p. 73 with kind permission from Springer Science+Business Media B.V.)*

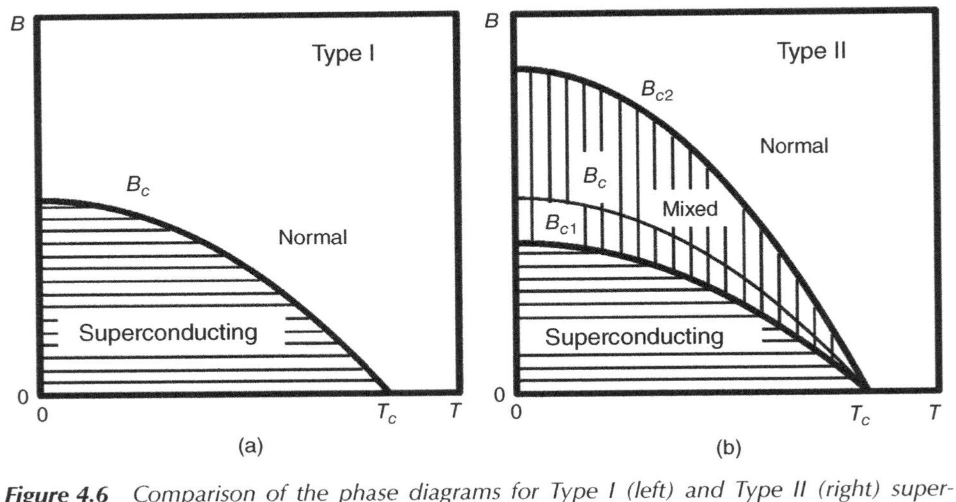

Figure 4.6 *Comparison of the phase diagrams for Type I (left) and Type II (right) superconductors (Reprinted from "High-Temperature Superconductors: Materials, Properties, and Applications" by Rainer Wesche, Kluwer Academic Publishers, 1998, ISBN 0-7923-8386-9, Figure 2-49, p. 74 with kind permission from Springer Science+Business Media B.V.)*

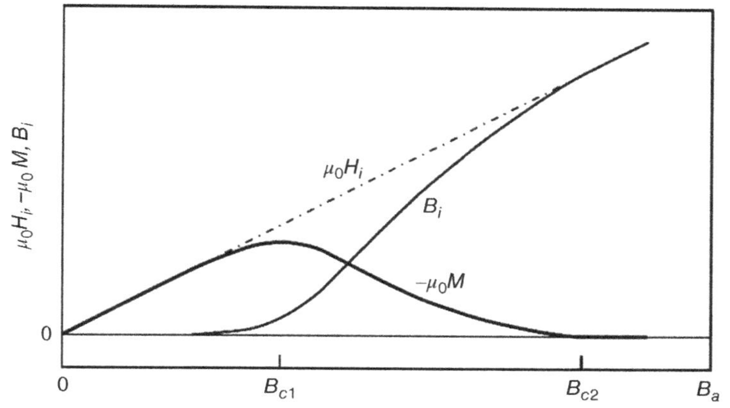

Figure 4.7 *Relations between the internal magnetic field H_i, the magnetization M and the magnetic induction B_i inside a real Type II superconductor. In the vicinity of B_{c1} the curves of the magnetization M and the magnetic induction B_i are rounded. In the normal state above B_{c2}, we find again $B_i \cong \mu_0 H_a = B_a$*

II superconductor between $B_{c1}(T)$ and $B_{c2}(T)$, the superconducting and normal regions can coexist. This mixed state is called the Shubnikov phase.

In a nonideal Type II superconductor, the defects present in the material act as pinning centers. The pinned flux lines (see in the following) are hindered to move in and out of the interior of the Type II superconductor, which leads to a hysteresis of the magnetization for increasing and decreasing the applied magnetic fields. The relations between flux pinning and critical current in a Type II superconductor are described in Section 4.4. Figure 4.7 shows the magnetic field in the interior, and the magnetization curve of a nonideal Type II superconductor. Because of flux pinning, the curves of the internal field and magnetization are rounded in the vicinity of the lower critical field.

In 1957, Abrikosov found solutions of the GL equations for particular situations. Based on his theoretical work, Abrikosov was able to explain the existence of Type II superconductors [15]. Figure 4.8 shows the resulting mixed state for Type II superconductors.

The mixed state is characterized by an array of normal cores within the superconductor, which allow magnetic flux to enter into the superconductor. Supercurrents encircle the normal cores in the mixed state, which are therefore called vortices. Figure 4.9 shows the variation of the Cooper pair density n_C and the magnetic flux density B in the interior of a Type II superconductor in the mixed state. The radius of each normal core is the coherence length ξ. The variation of the magnetic flux density B is determined by the penetration depth λ_L. The supercurrents encircling the normal cores cause a repulsive vortex–vortex interaction.

The exact theory of Type II superconductors predicts that the minimum of the free energy is reached for a hexagonal lattice of vortices. The flux within each of the normal cores is quantized. Calculations show that each vortex contains just one fluxoid ϕ_0. The average magnetic field B_i within a Type II superconductor is

$$B_i = N_v \phi_0 \tag{4.27}$$

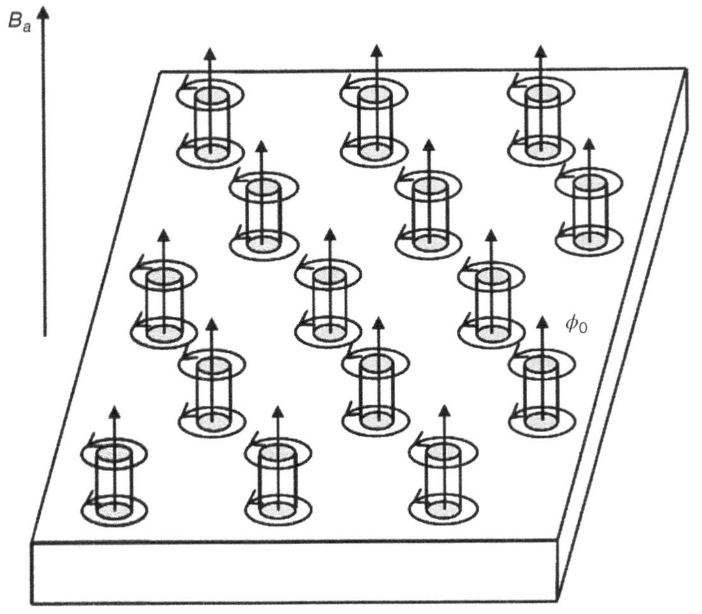

Figure 4.8 Type II superconductor in the mixed state. Each vortex contains just a single fluxoid (Reprinted from "High-Temperature Superconductors: Materials, Properties, and Applications" by Rainer Wesche, Kluwer Academic Publishers, 1998, ISBN 0-7923-8386-9, Figure 2-50, p. 74 with kind permission from Springer Science+Business Media B.V.)

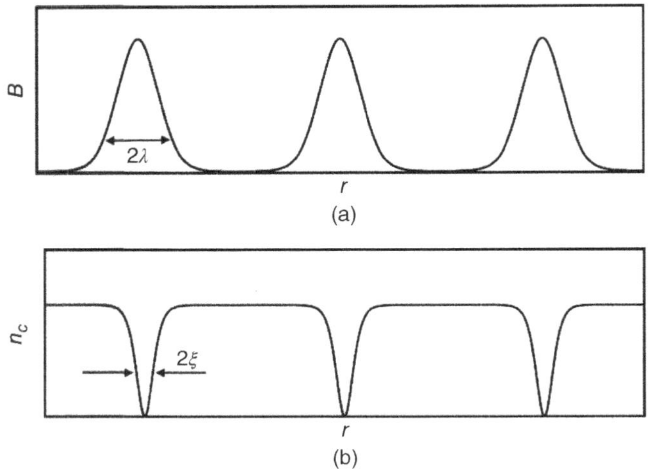

Figure 4.9 Variation of the magnetic flux density B (top) and the Cooper pair density n_C (bottom) in the mixed state of a Type II superconductor (Reprinted from "High-Temperature Superconductors: Materials, Properties, and Applications" by Rainer Wesche, Kluwer Academic Publishers, 1998, ISBN 0-7923-8386-9, Figure 2-51, p. 75 with kind permission from Springer Science+Business Media B.V.)

where N_v is the number of vortices per unit area and ϕ_0 is the flux quantum. The area occupied by each vortex in a hexagonal lattice is

$$A_v = \frac{1}{2}\sqrt{3}d^2 \tag{4.28}$$

where d is the average separation of the vortices. Using Equation 4.28, the following expression results for the average magnetic field B_i in the interior of a Type II superconductor

$$B_i = \frac{\phi_0}{\frac{1}{2}\sqrt{3}d^2} \tag{4.29}$$

A more detailed description of the flux line lattice in cuprate superconductors will be given in Chapter 10.

4.4 Critical Current Densities of Type I and Type II Superconductors[3]

Next, the critical current densities in both Type I and Type II superconductors will be considered. The critical currents in Type I superconductors are limited by the very low critical fields B_c of these materials. The current I flowing in a round superconducting wire produces a magnetic field at the surface

$$B = \frac{\mu_0 I}{2\pi r} \tag{4.30}$$

where r is the radius of the wire and μ_0 is the permeability of free space. The critical current I_c of a thick Type I superconductor is reached as soon as the magnetic field reaches the critical field B_c of the superconductor. Assuming that the current density is j_c within a surface layer of thickness λ_L and zero elsewhere, a critical current density $j_c = 7.9 \times 10^7 \mathrm{Acm}^{-2}$ is obtained for tin at $T = 0$ [7]. Thus, the critical current in Type I superconductors is limited mainly by the condition that the supercurrent flows only in a thin surface layer with a thickness given by the penetration depth λ_L for the magnetic field.

The situation is different for Type II superconductors. In the mixed state transport currents can flow in the whole conductor cross-section. Figure 4.10 shows the current–voltage characteristics for a Type II superconductor with different numbers of defects. The linear part corresponds to the flux flow resistivity ρ_{ff}:

$$\rho_{\mathrm{ff}} = \frac{|\mathbf{E_e}|}{j} \tag{4.31}$$

where $\mathbf{E_e}$ is the electric field and j is the transport current density. This flux flow resistivity is independent of the number of defects. To understand the occurrence of flux flow resistivity in Type II superconductors, the Lorentz force F_L acting on the vortices in the presence of an applied magnetic field has to be taken into account. Figure 4.11 shows that

[3] Reprinted with minor changes from "*High-Temperature Superconductors: Materials, Properties, and Applications*" by Rainer Wesche, Kluwer Academic Publishers, 1998, ISBN 0-7923-8386-9, pp. 76–78 with kind permission from Springer Science+Business Media B.V.

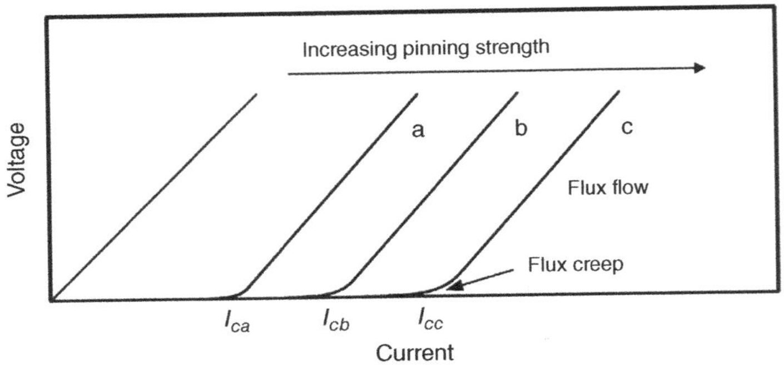

Figure 4.10 *Current–voltage characteristic for a Type II superconductor with different numbers of defects. Increasing pinning strength arises from an increasing number of defects acting as pinning centers (Reprinted from "High-Temperature Superconductors: Materials, Properties, and Applications" by Rainer Wesche, Kluwer Academic Publishers, 1998, ISBN 0-7923-8386-9, Figure 2-52, p. 77 with kind permission from Springer Science+Business Media B.V.)*

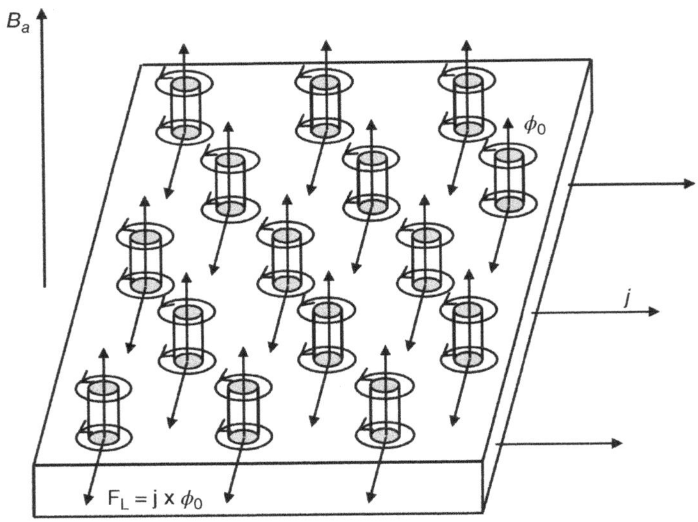

Figure 4.11 *The Lorentz force* $\mathbf{F}_L$ *causes flux flow perpendicular to the direction of the current and to the applied magnetic field (Reprinted from "High-Temperature Superconductors: Materials, Properties, and Applications" by Rainer Wesche, Kluwer Academic Publishers, 1998, ISBN 0-7923-8386-9, Figure 2-53, p. 78 with kind permission from Springer Science+Business Media B.V.)*

the Lorentz force $\mathbf{F_L} = \mathbf{j} \times \mathbf{\Phi_0}$ leads to a motion of the vortices perpendicular to the current and to the field direction. The resulting variation of the magnetic flux causes flux flow resistance. In an ideal Type II superconductor, the magnetic flux can move freely. Thus, the critical current density would be zero in the presence of a magnetic field perpendicular to the current direction. However, a small number of defects are always present in a

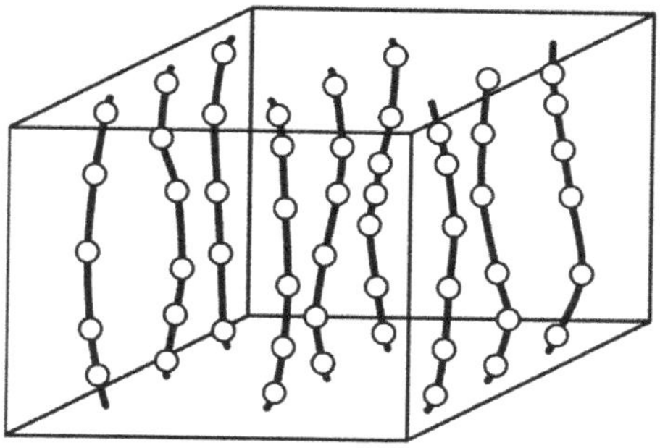

Figure 4.12 *Flux lines pinned at defects in a Type II superconductor. As a consequence of the interaction of flux lines and pinning centers, the hexagonal lattice of the vortices is slightly distorted (Reprinted from "High-Temperature Superconductors: Materials, Properties, and Applications" by Rainer Wesche, Kluwer Academic Publishers, 1998, ISBN 0-7923-8386-9, Figure 2-54, p. 78 with kind permission from Springer Science+Business Media B.V.)*

real Type II superconductor. These defects can act as pinning centers for the flux lines. For example, the energy of flux lines traversing normal precipitates is reduced. In the presence of a large number of normal precipitates, energetically favorable positions for the flux lines in the superconductor can exist. As a consequence, flux motion is suppressed as long as the Lorentz force is smaller than the pinning force. Figure 4.12 illustrates flux pinning at defects. The hexagonal flux line lattice may be slightly distorted due to the interaction of vortex lines and pinning centers. In addition to normal precipitates, grain boundaries and dislocations can act as pinning centers. Because of the extremely short coherence length ξ in high-T_c superconductors, even point defects are weak pinning centers in these materials. Flux pinning leads to irreversible magnetization curves. Thermally activated depinning can occur for Lorentz forces well below the pinning force. This effect is similar to diffusion in solids. Thermally activated flux creep is especially important at elevated temperatures in the high-T_c superconductors. Flux pinning and creep effects will be discussed in more detail in Chapter 10.

4.5 Anisotropic Superconductors[4]

The anisotropic GL theory can be used to describe the magnetic properties of layered superconductors. Examples of superconductors with layered crystal structures are MgB_2, the cuprates, and the iron pnictides. As a consequence of their layered structures, the penetration depth depends on the direction of flow of the screening currents. It is further assumed that the screening currents in the crystallographic a and b directions are equal. Thus, two different penetration depths λ_{ab} and λ_c result, where the indices ab and c refer to the directions of flow of the screening currents. Because of this anisotropy, two different

[4] Reprinted with minor changes from "*High-Temperature Superconductors: Materials, Properties, and Applications*" by Rainer Wesche, Kluwer Academic Publishers, 1998, ISBN 0-7923-8386-9, pp. 134–135 with kind permission from Springer Science+Business Media B.V.

GL parameters κ_{ab} and κ_c can be defined [6]:

$$\kappa_{ab} = \left(\frac{\lambda_{ab}\lambda_c}{\xi_{ab}\xi_c}\right)^{1/2}$$
$$\kappa_c = \frac{\lambda_{ab}}{\xi_{ab}} \tag{4.32}$$

where ξ_{ab} and ξ_c are the coherence lengths in the ab plane and along the c direction, respectively. The indices of the GL parameters give the direction of the applied magnetic field. The dimensionless anisotropy parameter γ_a can be expressed by the ratios of the following material properties [16]:

$$\gamma_a = \left(\frac{m_c}{m_{ab}}\right)^{1/2} = \frac{\lambda_c}{\lambda_{ab}} = \frac{\xi_{ab}}{\xi_c} = \frac{B_{c2,ab}}{B_{c2,c}} = \frac{B_{c1,c}}{B_{c1,ab}} \tag{4.33}$$

where m_{ab} and m_c are the effective masses of the charge carriers for currents in the ab plane and along the c direction, respectively. Furthermore, B_{c1} and B_{c2} are the lower and upper critical fields where the indices ab and c correspond to the field directions in the ab plane and along the c direction, respectively. The thermodynamic critical field B_c is given by [6, 9, 16]

$$B_c = \frac{\phi_0}{2\sqrt{2}\pi\lambda_{ab}\xi_{ab}} \tag{4.34}$$

where ϕ_0 is the flux quantum. The resulting expressions for the upper critical fields are [6, 9, 16]

$$B_{c2,c} = \frac{\phi_0}{2\pi{\xi_{ab}}^2} \tag{4.35}$$

$$B_{c2,ab} = \frac{\phi_0}{2\pi\xi_c\xi_{ab}} \tag{4.36}$$

The coherence length and the GL parameter can be derived from the penetration depth and critical field data.

References

1. A. Shepelev and D. Larbalestier, Die vergessene Entdeckung, *Phys. J.*, **10**(6), 51–53 (2011).
2. K. Mendelssohn and J.R. Moore, Supra-conducting alloys, *Nature*, **135**, 826–827 (1935).
3. C.J. Gorter, Note on the supraconductivity of alloys, *Physica*, **2**, 449–452 (1935).
4. L.V. Shubnikov, V.I. Khotkevich, Yu.D. Shepelev, and Yu.N. Ryabinin, Magnetic properties of superconducting metals and alloys, *Ukr. J. Phys.*, **53**, 42–52 (2008) (special issue).
5. V.L. Ginzburg, Nobel lecture: On superconductivity and superfluidity (what I have and have not managed to do) as well as on the "physical minimum" at the beginning of the XXI century, *Rev. Mod. Phys.*, **76**, 981–998 (2004).

6. C.P. Poole, Jr.,, H.A, Farach, and R.J. Creswick, *Superconductivity*, Academic Press, San Diego, CA, 1995.
7. W. Buckel, *Supraleitung, Grundlagen und Anwendungen*, Wiley-VCH, Weinheim, Germany, 1994.
8. M.V. Milošević and R. Geurts, The Ginzburg–Landau theory in application, *Physica C*, **470**, 791–795 (2010).
9. T.P. Orlando and K.A. Delin, *Foundations of Applied Superconductivity*, Addison-Wesley, Reading, MA, 1991.
10. L.P. Gor'kov, Microscopic derivation of the Ginzburg–Landau equations in the theory of superconductivity, *Sov. Phys. JETP*, **9**, 1364–1367 (1959).
11. G. Sparn, J.D. Thompson, R.L. Whetten S.-M. Huang, R.B. Kaner, F. Diederich, G. Grüner, and K. Holczer, Pressure and field dependence of superconductivity in Rb_3C_{60}, *Phys. Rev. Lett.*, **68**, 1228–1231 (1992).
12. J. Muller, A15-type superconductors, *Rep. Prog. Phys.*, **43**, 641–687 (1980).
13. S. Foner, E.J. McNiff, Jr.,, D. Heimann, S.-M. Huang, and R.B. Kaner, Measurements of upper critical field of K_3C_{60} and Rb_3C_{60} powders, *Phys. Rev. B*, **46**, 14936–14939 (1992).
14. C.E. Johnson, H.W. Jiang, K. Holczer R.B. Kaner, R.L. Whetten, and F. Diederich, Upper-critical-field – temperature phase diagram of alkali-metal-intercalated C_{60} superconductors, *Phys. Rev. B*, **46**, 5880–5882 (1992).
15. A.A. Abrikosov, On the magnetic properties of superconductors of the second group, *Sov. Phys.– JETP*, **5**, 1174–1182 (1957).
16. M. Tinkham, Introduction to the phenomenology of high temperature superconductors, *Physica C*, **235–240**, 3–8 (1994).

5
Cuprate Superconductors: An Overview

5.1 Introduction

The discovery by Bednorz and Müller [1] that superconductivity occurs in the La–Ba–Cu–O system at a temperature as high as 30 K was the beginning of a new era in the field of superconductivity. Soon after their discovery it was clarified that the superconducting phase in the La–Ba–Cu–O system is $La_{2-x}Ba_xCuO_4$, a complex nonstoichiometric cuprate, which is characterized by the presence of CuO_2 planes in the crystal structure. The discovery initiated an intense effort to find other cuprate superconductors with even higher critical temperatures. Within a year $YBa_2Cu_3O_{7-x}$ ($T_c = 93$ K), the first cuprate superconductor with a transition temperature well above the boiling point of nitrogen, was found [2]. Characteristic features of cuprate high-temperature superconductors are their chemical and structural complexity. In this chapter, we will provide an overview on the major families of superconducting cuprates. Some of their common structural features will be briefly described. An important aspect of these complex nonstoichiometric oxides is the possibility to tune the charge carrier density by means of changes in the chemical composition. This strongly affects the physical properties of the cuprates. The undoped parent compounds are Mott insulators. In spite of the fact that band theory would predict metallic behavior, the parent compounds (e.g., La_2CuO_4) show insulating behavior. Due to the strong electron–electron interaction, the electrons in the conduction band are prevented from moving freely. Because of the strong Coulomb repulsion, each electron is quasibound to a single copper ion. Additional charge carriers, which are introduced by doping, are mobile, leading to electrical conductivity and superconductivity. This change in electrical conductivity is accompanied by the vanishing of the antiferromagnetism of the parent compound.

Physical Properties of High-Temperature Superconductors, First Edition. Rainer Wesche.

5.2 Families of Superconductive Cuprates

Today, well over a hundred cuprate high-temperature superconductors are known. First, let us consider the layering schemes in the major families of cuprate superconductors. In the crystal structures of all cuprate superconductors, CuO_2 planes are present, which are responsible for the high-temperature superconductivity of these materials. The electrical current in the normal as well as in the superconducting state flows mainly along the CuO_2 planes, which are therefore called conduction planes. In the major families of cuprate superconductors, the CuO_2 planes are embedded between multilayers of LaO, BaO, TlO, or SrO. These layers of insulating character are required for the stabilization of the crystal structure, and to ensure charge neutrality. Because of the layered crystal structures, the physical properties of the cuprates in the normal as well as in the superconducting state are highly anisotropic.

In Figure 5.1, the layering schemes of La_2CuO_4 and $YBa_2Cu_3O_7$ are compared. In La_2CuO_4, each of the copper oxide blocks contains only a single CuO_2 plane. The copper oxide blocks are separated by double layers of LaO. In contrast to the conducting CuO_2 planes, the LaO layers show an insulating character. Impurity atoms added to them allow the introduction of additional charge carriers into the CuO_2 conduction planes. The double layers of LaO act as charge carrier reservoirs. This aspect will be discussed in more detail in Section 5.3. A common feature of $YBa_2Cu_3O_7$ and La_2CuO_4 is the CuO_2 conduction planes. However in the yttrium compound, each copper oxide block contains two CuO_2 planes, which are separated by a layer of yttrium ions. The yttrium ions can be replaced by rare earth elements (RE) (e.g., Er, Dy, Gd, Eu, Nd), leading to the chemical

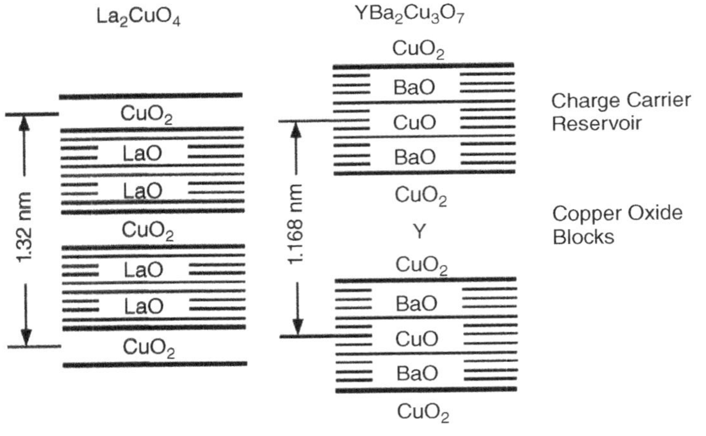

Figure 5.1 *Layering schemes of La_2CuO_4 and $YBa_2Cu_3O_7$. In the lanthanum compound, single CuO_2 sheets are embedded between double layers of LaO. Each copper oxide block of the yttrium compound consists of two CuO_2 sheets, separated by a layer of yttrium ions. The charge carrier reservoir is formed of two BaO layers with embedded CuO chains (Reprinted from "High-Temperature Superconductors: Materials, Properties, and Applications" by Rainer Wesche, Kluwer Academic Publishers, 1998, ISBN 0-7923-8386-9, Figure 3-1, p. 80 with kind permission from Springer Science+Business Media B.V.)*

formula $REBa_2Cu_3O_7$. A special feature of the $YBa_2Cu_3O_7$ and the related $REBa_2Cu_3O_7$ compounds is the presence of CuO chains along the crystallographic *b* direction in the center of the charge carrier reservoir formed of BaO double layers (see also Chapter 6). The *c* lattice parameters of La_2CuO_4 and $YBa_2Cu_3O_7$ are 1.32 and 1.168 nm, respectively. The crystallographic unit cell of the lanthanum compound contains two formula units, while that of the yttrium compound contains only one. Frequently, the chemical formulas are replaced by the abbreviations La-214 and RE-123, where the numbers refer to the atomic ratios. In most of the abbreviations, the number of oxygen atoms is omitted as in the case of RE-123.

The layering schemes of the first three members of the thallium-based cuprate superconductor family of chemical composition $Tl_2Ba_2Ca_{n-1}Cu_nO_{2n+4}$ (Tl-22($n-1$)n) are illustrated in Figure 5.2. The indicated unit cells contain two formula units of the thallium-based superconductor in question. In $Tl_2Ba_2CuO_6$ (Tl-2201), single layers of CuO_2 are separated by the charge carrier reservoirs. The sequence of the layers in the charge carrier reservoirs is BaO–TlO–TlO–BaO. We find the same sequence in the other Tl-22($n-1$)n compounds, as shown in Figure 5.2 for $Tl_2Ba_2CaCu_2O_8$ and $Tl_2Ba_2Ca_2Cu_3O_{10}$. In the thallium-based cuprates with $n \geq 2$, the adjacent CuO_2 planes within a single copper oxide block are separated by layers of Ca ions. In $Tl_2Ba_2Ca_2Cu_3O_{10}$, the copper oxide block consists of three CuO_2 layers. Again, layers of Ca ions are inserted between adjacent CuO_2 sheets. An interesting feature is the fact that the inner and outer CuO_2 layers in a single copper oxide block are structurally not identical. Because of these differences, it is possible that the charge carrier densities in the inner and outer CuO_2 planes are different.

The crystal structures of the members of the Bi-based high-temperature superconductor family $Bi_2Sr_2Ca_{n-1}Cu_nO_{2n+4}$ (Bi-22($n-1$)n) are closely related to their Tl-22($n-1$)n counterparts. Starting from Figure 5.2, we obtain the layering schemes of the Bi-22($n-1$)n cuprate superconductors simply by the replacement of the TlO layers by BiO layers, and of the BaO layers by SrO layers.

A further layering scheme is realized in the family of the mercury-based high-temperature superconductors. The chemical composition of the members of this family of cuprate superconductors is described by the formula $HgBa_2Ca_{n-1}Cu_nO_{2n+2}$ (Hg-12($n-1$)n). Figure 5.3 shows the layering schemes for the compounds with n up to 5. In $HgBa_2CuO_4$ (Hg-1201), the copper oxide blocks consist of a single CuO_2 layer, as in the case of the La-214, Tl-2201, and Bi-2201 compounds. The sequence of the layers in the charge carrier reservoirs is BaO–Hg–BaO. Most of the oxygen sites in the mercury plane are not occupied (see also Chapter 6).

The copper oxide blocks in the compounds with $n \geq 2$ are identical to the corresponding ones in the Bi-22($n-1$)n and Tl-22($n-1$)n compounds. An interesting feature of the mercury-based high-temperature superconductors is the possibility of synthesizing compounds with more than five CuO_2 planes in a single copper oxide block.

There exists a family of analogous $TlM_2Ca_{n-1}Cu_nO_{2n+3}$ (Tl-12($n-1$)n) compounds, where M stands for Ba or Sr. The Hg layer of the Hg-12($n-1$)n compounds is replaced by a TlO layer.

Using high-pressure synthesis, many additional families of cuprate superconductors can be synthesized. High-pressure synthesis allows the insertion of other layers of ions than bismuth, thallium, or mercury into the charge carrier reservoirs. For example, there exist families of $AlSr_2Ca_{n-1}Cu_nO_{2n+3}$ (Al-12($n-1$)n), $BSr_2Ca_{n-1}Cu_nO_{2n+3}$ (B-12($n-1$)n),

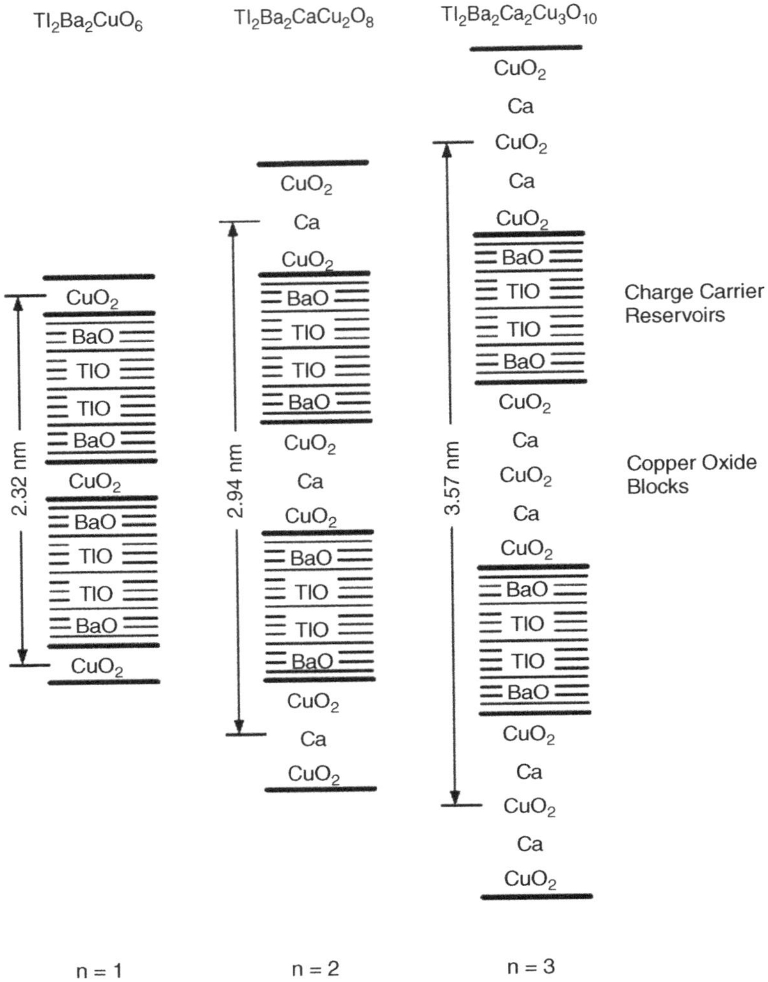

Figure 5.2 *Layering schemes of* $Tl_2Ba_2Ca_{n-1}Cu_nO_{2n+4}$ *up to n = 3. For n ≥ 2, the* CuO_2 *layers within a single copper oxide block are separated by layers of Ca ions. The sequence of the layers in the charge carrier reservoirs is BaO–TlO–TlO–BaO. Replacement of Tl by Bi and of Ba by Sr leads to the layering scheme of* $Bi_2Sr_2Ca_{n-1}Cu_nO_{2n+4}$ *(Reprinted from "High-Temperature Superconductors: Materials, Properties, and Applications" by Rainer Wesche, Kluwer Academic Publishers, 1998, ISBN 0-7923-8386-9, Figure 3-2, p. 81 with kind permission from Springer Science+Business Media B.V.)*

$GaSr_2Ca_{n-1}Cu_nO_{2n+3}$ (Ga-12($n-1$)n), $CuBa_2Ca_{n-1}Cu_nO_{2n+2}$ (Cu-12($n-1$)n), and $(Cu_{0.5}C_{0.5})Ba_2Ca_{n-1}Cu_nO_{2n+3}$ ((Cu,C)-12($n-1$)n) cuprate superconductors [3].

Next, let us consider the critical temperatures in the different families of the cuprate superconductors. In Table 5.1, the critical temperatures of various cuprate superconductors are listed [4–39]. In the $La_{2-x}M_xCuO_4$ family (M = Sr or Ba), a maximum critical temperature of 39 K has been reported with partial substitution of La by Sr. The critical temperature

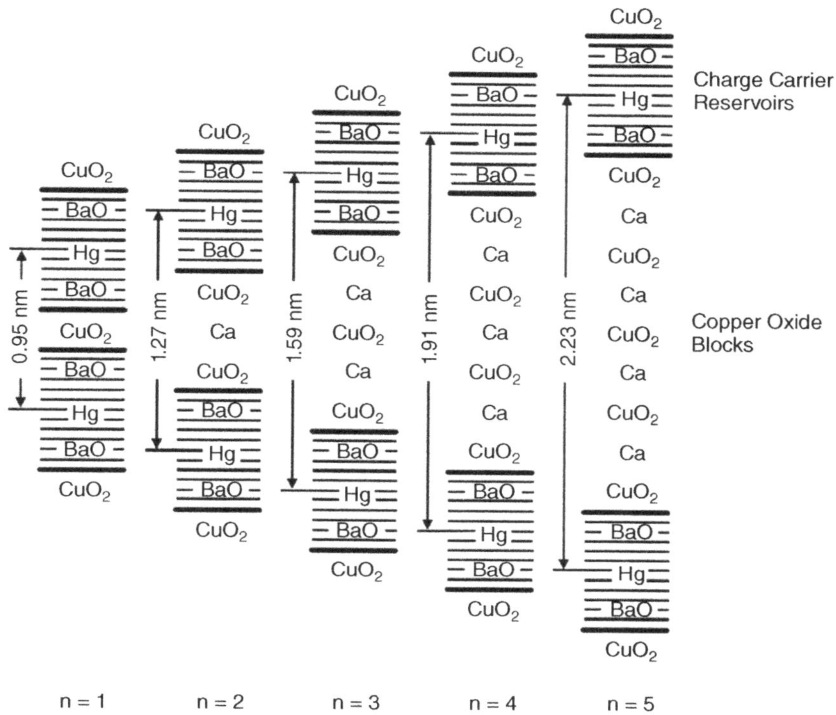

Figure 5.3 *Layering schemes of $HgBa_2Ca_{n-1}Cu_nO_{2n+2}$ up to n = 5. For n ≥ 2, the CuO_2 layers within a single copper oxide block are separated by layers of Ca ions. The sequence of the layers in the charge carrier reservoirs is BaO–Hg–BaO. Replacement of Hg by Tl leads to the layering scheme of $TlBa_2Ca_{n-1}Cu_nO_{2n+3}$ (Reprinted from "High-Temperature Superconductors: Materials, Properties, and Applications" by Rainer Wesche, Kluwer Academic Publishers, 1998, ISBN 0-7923-8386-9, Figure 3-3, p. 82 with kind permission from Springer Science+Business Media B.V.)*

of the RE-123 compounds varies between 93 and 98 K, depending on the rare earth element present in the compound. The critical temperature was found to increase with the ionic radius of the rare earth element in question [6]. The maximum critical temperature of $YBa_2Cu_4O_8$ is around 80 K [4]. This cuprate is structurally related to $YBa_2Cu_3O_{7-\delta}$. In the Bi-22$(n-1)n$ family, a maximum critical temperature of 122 K has been reported for Bi-2223 [3]. The formation and stability of the $n = 3$ phase is significantly improved by partial substitution of bismuth by lead. Even higher critical temperatures of 128 K have been achieved in the corresponding family of Tl-22$(n-1)n$ superconductors for the member with $n = 3$. In the Tl-12$(n-1)n$ family, compounds with thallium partially replaced by lead show slightly higher critical temperatures than pure compounds. For example, the critical temperature of Tl-1223 without lead substitution is around 110 K, whereas in (Tl, Pb)-1223, a T_c of 122 K has been achieved. A critical temperature as high as 135 K has been reached in the mercury compound $HgBa_2Ca_2Cu_3O_{8+\delta}$. A slightly higher T_c of 138 K

Table 5.1 *Critical temperatures of various cuprate superconductors*

Family	Compound	T_c(K)	Reference
La-214	$La_{1.85}Sr_{0.15}CuO_4$	39	[4]
	$La_{1.91}Ba_{0.09}CuO_{4-\delta}$	32.5	[5]
RE-123	$YBa_2Cu_3O_{7-\delta}$	93	[6]
	$GdBa_2Cu_3O_{7-\delta}$	94	[6]
	$EuBa_2Cu_3O_{7-\delta}$	95	[6]
	$NdBa_2Cu_3O_{7-\delta}$	96	[6]
	$LaBa_2Cu_3O_{7-\delta}$	98	[6]
Bi-22($n-1$)n	$Bi_{2.11}Sr_{1.89}CuO_{6+\delta}$	8.5	[7]
	$Bi_2(Sr_{1.6}La_{0.4})CuO_{6+\delta}$	23.5	[4]
	$Bi_2(Ca_{0.8}Sr_{0.8}La_{0.4})CuO_{6+\delta}$	80	[8]
	$Bi_{2.2}Sr_2Ca_{0.8}Cu_2O_{8+\delta}$	84	[9]
	$Bi_2Sr_2CaCu_2O_8$	87	[10]
	$Bi_{2.14}Sr_{1.77}Ca_{0.92}Cu_2O_{8+\delta}$	93	[11]
	$Bi_{1.8}Pb_{0.4}Sr_{1.72}Ca_2Cu_3O_{10+\delta}$	111	[12]
	$Bi_{1.7}Pb_{0.3}Sr_2Ca_2Cu_3O_{10}$	116	[13]
Tl-22($n-1$)n	$Tl_2Ba_2CuO_{6+\delta}$	93	[14]
	$Tl_{1.7}Ba_2Ca_{1.06}Cu_{2.32}O_{8+\delta}$	108	[15]
	$Tl_{1.64}Ba_2Ca_{1.87}Cu_{3.11}O_{10+\delta}$	125	[15]
	$Tl_2Ba_2Ca_2Cu_3O_{10-\delta}$	128	[16]
	$Tl_2Ba_2Ca_3Cu_4O_{12+\delta}$	114	[17]
Tl-12($n-1$)n	$Tl(Ba_{2-x}La_x)CuO_{5-\delta}$	45	[18]
	$Tl_{1.1}Ba_2Ca_{0.9}Cu_{2.1}O_{7.1}$	80	[19]
	$TlBa_2CaCu_2O_y$	110	[20]
	$Tl_{1.1}Ba_2Ca_{1.8}Cu_3O_{9.7}$	110	[19]
	$TlBa_2Ca_3Cu_4O_{11+\delta}$	114	[21]
	$(Tl_{0.5}Pb_{0.5})Sr_2CaCu_2O_7$	80	[4]
	$(Tl_{0.5}Pb_{0.5})Sr_2(Ca_{0.8}Y_{0.2})Cu_2O_7$	107	[4]
	$(Tl_{0.5}Pb_{0.5})Sr_2Ca_2Cu_3O_9$	118	[22]
	$(Tl_{0.5}Pb_{0.5})Sr_2Ca_2Cu_3O_9$	122	[4]
Hg-12($n-1$)n	$HgBa_2CuO_{4+\delta}$	94	[23]
	$HgBa_2CaCu_2O_{6+\delta}$	123	[24]
	$HgBa_2Ca_2C_3O_{8+\delta}$	133	[25]
	$HgBa_2Ca_2C_3O_{8+\delta}$	135	[26]
	$HgBa_2Ca_3C_4O_{10+\delta}$	126	[26]
	$HgBa_2Ca_4C_5O_{12+\delta}$	110	[26]
	$HgBa_2Ca_5C_6O_{14+\delta}$	107	[26]
	$Hg_{0.8}Tl_{0.2}Ba_2Ca_2Cu_3O_{8+x}$	138	[27]
	$Hg_{0.5}Pb_{0.5}Ba_2Ca_4Cu_5O_{12+\delta}$	115	[28]
Al-12($n-1$)n	$AlSr_2Ca_2Cu_3O_9$	78	[29]
	$AlSr_2Ca_3Cu_4O_{11}$	110	[30]
	$AlSr_2Ca_4Cu_5O_{13}$	83	[30]
B-12($n-1$)n	$BSr_2Ca_2Cu_3O_9$	75	[31]
	$BSr_2Ca_3Cu_4O_{11}$	110	[31, 32]
	$BSr_2Ca_4Cu_5O_{13}$	85	[31]

(continued overleaf)

Table 5.1 (continued)

Family	Compound	T_c(K)	Reference
Ga-12($n-1$)n	$GaSr_2Ca_2Cu_3O_9$	70	[33]
	$GaSr_2Ca_3Cu_4O_{11}$	107	[33]
Pb-12($n-1$)n	$PbSr_2CaCu_2O_x$	82	[34]
	$PbSr_2Ca_2Cu_3O_x$	122	[35]
Cu-12($n-1$)n	$CuBa_2Ca_2Cu_3O_{8+\delta}$	117	[36]
	$(Cu_{0.5}C_{0.5})Ba_2Ca_2Cu_3O_{9+\delta}$	67	[37, 38]
	$(Cu_{0.5}C_{0.5})Ba_2Ca_3Cu_4O_{11+\delta}$	117	[37, 38]
	$(Cu_{0.5}S_{0.5})Sr_2Ca_2Cu_3O_y$	≈100	[39]
	$(Cu_{0.5}S_{0.5})Sr_2Ca_6Cu_7O_y$	≈60	[39]

has been reported for $Hg_{0.8}Tl_{0.2}Ba_2Ca_2Cu_3O_{8+x}$. Critical temperatures well above 100 K have also been reached in Al-1234, B-1234, Ga-1234, and in (Cu, C)-1234 (see Table 5.1).

The critical temperature of the cuprate superconductors depends on the exact chemical composition, which determines the charge carrier density in the conductive CuO_2 layers. The variation of the charge carrier density by means of doping will be described in Section 5.3. Empirical rules for the critical temperatures of cuprate superconductors will be discussed in Chapter 7.

5.3 Variation of Charge Carrier Density (Doping)

In semiconductors, the charge carrier density can be increased by the addition of impurity atoms. Let us consider the example of silicon. Each silicon atom provides four valence electrons. If we replace one silicon atom by a trivalent boron atom, one valence electron is absent. The boron atoms are called acceptor atoms because they can accept one electron from the host material. Thus, the addition of boron leads to the introduction of defect electrons, i.e. holes, which behave like positively charged electrons. On the other hand, the addition of an arsenic atom with five valence electrons provides one excess electron to the host material. Arsenic is therefore called a donor atom. The energy levels of the impurity atoms are within the energy gap of the semiconductor. The energy levels of the acceptor atoms are slightly above the valence band, while those of the donor atoms are slightly below the conduction band. As a consequence, the acceptor or the donor levels can be easily activated to provide mobile holes or electrons to the semiconductor. The addition of impurity atoms to a semiconductor is called doping.

In a similar way the charge carrier density in the CuO_2 planes of cuprate high-temperature superconductors can be tuned. Let us consider the simplest case of La_2CuO_4. In this insulating parent compound, the oxidation state of the lanthanum ions is +3 and that of the copper ions +2. The eight electrons provided by the two lanthanum ions and the copper ion are taken up by the four O^{2-} ions. In principle, the oxidation state of copper ions can be +1, +2, and +3. Partial replacement of La^{3+} by Ba^{2+} leads to a reduced number of electrons, which can be transferred from the charge carrier reservoir to the CuO_2 planes. In order to ensure charge neutrality, one Cu^{3+} ion is formed in the CuO_2

planes for each Ba^{2+} added to the LaO double layers (see Figure 5.1). The Cu^{3+} defects act as mobile holes in the normal conducting state. For $La_{2-x}Ba_xCuO_4$, the fractions of Cu^{2+} and Cu^{3+} ions are $(1 - x)$ and x, respectively. We can define a formal valence $2 + p = 2 \cdot (1 - x) + 3x = 2 + x$ of the copper atoms.

So far, we have not taken into consideration that the oxygen content may deviate from stoichiometry. An additional way to introduce holes is by an excess of oxygen in the charge carrier reservoirs, which reduces the number of electrons that can be transferred from the cations of the charge carrier reservoirs to the CuO_2 planes. As a consequence, holes are introduced into the CuO_2 planes. The charge carrier density, and hence the critical temperature of $YBa_2Cu_3O_{7-\delta}$, is very sensitive to the exact oxygen content. The actual oxygen content of $YBa_2Cu_3O_{7-\delta}$ depends on the atmosphere (pure oxygen, air, or oxygen partial pressure of an Ar/O_2 mixture), the temperature, and the cooling rate. Typically, the exact oxygen content of cuprate superconductors is not known. The uncertainty of the oxygen content is the reason for introducing δ in the chemical formula (see Table 5.1).

In more complex cuprate superconductors, the cations may be disordered. For example, in $Bi_2Sr_2CaCu_2O_{8+\delta}$ the crystallographic sites of bismuth ions may be partly occupied by Sr ions and vice versa. In addition, calcium ions may be replaced by strontium ions. In principle, it would be more correct to write the chemical formula of Bi-2212 as $(Bi,Sr)_{2+x}(Sr,Ca)_{3+y}Cu_2O_{8+\delta}$. There exists a considerable range for the possible compositions of the Bi-2212 compound.

In addition to cation disorder, the ratios of Bi:Sr:Ca:Cu may deviate considerably from the ideal of 2:2:1:2. These effects lead to self-doping in the more complex cuprates (e.g., Bi-, Tl-, and Hg-based cuprate superconductors). This means that charge carriers can be introduced into the CuO_2 planes without the need to introduce impurity atoms.

Because of the close relation of chemical composition, charge carrier density and physical properties, knowledge of the exact chemical composition and the resulting charge carrier density is of crucial importance. In principle, the Hall effect can be used to determine the number density and the type of the charge carriers. In Figure 5.4, the generation of the Hall voltage is schematically illustrated. We consider a conductor with a magnetic field applied in the $+z$ direction and a current in the $-y$ direction. The Lorentz force

$$\mathbf{F_L} = q\mathbf{v} \times \mathbf{B} \tag{5.1}$$

acting on charge carriers with the charge q and the average velocity $\mathbf{v}$ leads to a charge separation. The direction of the electric field $\mathbf{E_e}$ depends on the type of the charge carriers. For electrons, an electric field in the $-x$ direction (Figure 5.4, top) is built up, while for holes the generated electric field points in the $+x$ direction (Figure 5.4, bottom). At equilibrium the electric force $q\mathbf{E_e}$ balances the Lorentz force $\mathbf{F_L}$,

$$q\mathbf{E_e} = q\mathbf{v} \times \mathbf{B}. \tag{5.2}$$

The average velocity $\mathbf{v}$ of the charge carriers and the current density $\mathbf{j}$ are related by the expression

$$\mathbf{j} = nq\mathbf{v} \tag{5.3}$$

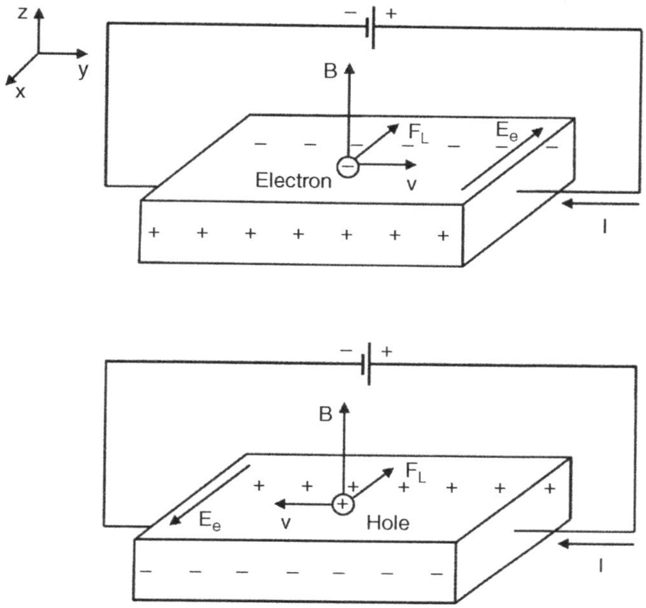

Figure 5.4 Schematic illustration of the Hall effect. For electrons (top) and holes (bottom), the resulting electric fields are in the opposite direction.

where n is the number density of electrons or holes. Combining Equations 5.2 and 5.3 leads to

$$\mathbf{E_e} = \frac{1}{nq}\mathbf{j} \times \mathbf{B} \tag{5.4}$$

where $q = +1.602 \times 10^{-19}$ A s for holes and $q = e = -1.602 \times 10^{-19}$ A s for electrons. The sign of the Hall coefficient

$$R_H = \pm\frac{1}{n|e|} \tag{5.5}$$

is positive for holes and negative for electrons. Measurements of the Hall voltage indicate that in the major families of the cuprate high-T_c superconductors, holes are the charge carriers.

It should be mentioned that there also exist electron-doped cuprate superconductors. The first electron-doped cuprate superconductor $Pr_{2-x}Ce_xCuO_{4-\delta}$ was discovered by Tokura *et al.* in 1989 [40]. The valence of Pr is +3, whereas that of Ce is +4. Substitution of Ce^{4+} for Pr^{3+} leads to electron doping of the CuO_2 planes. A critical temperature of 22.4 K [41] has been reported for $Pr_{2-x}Ce_xCuO_{4-\delta}$. Further electron-doped cuprate superconductors are $Nd_{2-x}Ce_xCuO_4$ ($T_c = 22.8$ K) and $Sm_{2-x}Ce_xCuO_4$ ($T_c = 17.6$ K) [41].

In the case when more than one band contributes to electrical conductivity, Equation 5.4 is not valid, and the evaluation of the measured Hall voltages is more complicated than described above.

In $(La,Sr)_2CuO_{4+\delta}$ and $YBa_2Cu_3O_{7-\delta}$, the oxidation state of La and Y is +3 and that of Sr and Ba +2, while the valence of copper depends on the exact chemical composition. In all cuprate superconductors, the oxygen content is nonstoichiometric. The formal valence 2+p of copper can be determined by iodometric titration [3, 42–47]. The cuprate sample in question is dissolved under an inert atmosphere in an acidic solution with an excess of KI. In this process, the following reactions occur:

$$Cu^{+} + I^{-} \rightarrow CuI \downarrow \tag{5.6}$$

$$Cu^{2+} + 2I^{-} \rightarrow CuI \downarrow + 0.5I_2 \tag{5.7}$$

$$Cu^{3+} + 3I^{-} \rightarrow CuI + I_2 \tag{5.8}$$

The reaction of $[CuO]^{p+}$ is summarized by the formula

$$[CuO]^{p+} + (2+p)\,I^{-} \rightarrow CuI \downarrow + 0.5\,(p+1)\,I_2 \tag{5.9}$$

Any Cu^+ present in the sample precipitates out as CuI. Only copper oxidized beyond Cu^+ contributes to the liberation of neutral iodine I_2. The amount of neutral iodine generated in this reaction can be determined by titration with sodium thiosulfate $Na_2S_2O_3$,

$$I_2 + 2Na_2S_2O_3 \rightarrow 2NaI + Na_2S_4O_6 \tag{5.10}$$

Assuming that the investigated sample is single phase, and that the ratio of the metallic cations is exactly known, then we have [44]

$$p = \frac{M_{sc} C_{ST} V_1}{W_1} - 1 \tag{5.11}$$

where M_{sc} is the molar weight of the cuprate in g/mol, W_1 the weight of the sample in grams, C_{ST} the concentration of the $Na_2S_2O_3$ solution in mol/l, and V_1 its volume in liters. Reduced systematic errors can be achieved by a double titration method. A more detailed description of iodometric titration can be found in references [3, 42–47].

In general, the maximum critical temperature in cuprate superconductors is typically found to be in the range of 2.15–2.20 for a formal valence $2 + p$. This means that the maximum critical temperature for the cuprate superconductor in question is reached for a carrier concentration of 0.15–0.2 holes per copper atom in the CuO_2 planes (see Chapter 7).

5.4 Summary

The high-temperature superconductors, discovered first by Bednorz and Müller, are complex copper oxides (cuprates). A common feature of all cuprate superconductors is the layered crystal structure with conducting CuO_2 layers sandwiched between other metal oxide layers of insulating character. These other metal oxide layers act as charge carrier reservoirs. In general, the cuprate high-temperature superconductors are nonstoichiometric oxides with varying oxygen content. The physical properties depend on the charge carrier concentration in the CuO_2 planes. In the insulating parent compounds of the cuprate superconductors, the oxidation state of the copper atoms in the CuO_2 planes is +2.

For example, excess oxygen in the charge carrier reservoirs, or substitution of trivalent lanthanum by divalent strontium, leads to the presence of holes in the conductive CuO_2 planes. As a consequence, the formal valence of copper in the CuO_2 planes is changed from +2 to $(2+p)$. In most of the cuprates, the maximum value of the critical temperature is reached for p values in the range of 0.15–0.2. At high carrier concentrations ($p > 0.3$), a transition to a nonsuperconducting metal has been observed. The phase diagram of the cuprate superconductors will be discussed in Chapter 8.

The most important families of cuprate superconductors are $(La,M)_2CuO_4$ (M: Sr or Ba), $REBa_2Cu_3O_{7-\delta}$ (RE: rare earth element or yttrium), $Bi_2Sr_2Ca_{n-1}Cu_nO_{2n+4}$, $Tl_2Ba_2Ca_{n-1}Cu_nO_{2n+4}$, $TlM_2Ca_{n-1}Cu_nO_{2n+3}$ (M: Sr or Ba), and $HgBa_2Ca_{n-1}Cu_nO_{2n+2}$. The maximum critical temperatures in the different families of cuprate superconductors vary from 39 K for the La-214 family to 135 K in the Hg-12 $(n-1)n$ family. For a further discussion of the empirical rules for the critical temperatures of cuprate superconductors see also Chapter 7.

References

1. J.G. Bednorz and K.A. Müller, Possible high T_c superconductivity in the Ba–La–Cu–O system, *Z. Phys. B*, **64**, 189–193 (1986).
2. M.K. Wu, J.R. Ashburn, C.J. Torng, P.H. Hor, R.L. Meng, L. Gao, Z.J. Huang, Y.Q. Wang, and C.W. Chu, Superconductivity at 93 K in a new mixed-phase, Y–Ba–Cu–O compound system at ambient pressure, *Phys. Rev. Lett.*, **58**, 908–910 (1987).
3. R. Wesche, *High-Temperature Superconductors: Materials, Properties and Applications*, Kluwer Academic Publishers, Norwell, MA, USA, 1998.
4. D.R. Harshman and A.P. Millis, Concerning the nature of high-T_c superconductivity: Survey of experimental properties and implications for interlayer coupling, *Phys. Rev. B*, **45**, 10684–10712 (1992).
5. M.K.R. Khan, H. Tanabe, I. Tanaka, and H. Kojima, Anisotropic electric resistivity of superconducting $La_{2-x}Ba_xCuO_{4-\delta}$ single crystal, *Physica C*, **258**, 315–320 (1996).
6. G.V.M. Williams and J.L. Tallon, Ion size effects on T_c and interplanar coupling in $RBa_2Cu_3O_{7-\delta}$, *Physica C*, **258**, 41–46 (1996).
7. J. Hejtmánek, M. Nevřiva, E. Pollert, D. Sedmidubský, and P. Vaŝek, Superconductivity and transport of $Bi_{2+x}Sr_{2-y}CuO_{6+\delta}$ single crystals, *Physica C*, **264**, 220–226 (1996).
8. T. Nakajima, M. Tange, T. Kizuka, H. Ikeda, R. Yoshizaki, and S. Sasaki, 80 K superconductivity in Bi-2201 phase, *Physica C*, **463–465**, 93–95 (2007).
9. S.A. Sunshine, T. Siegrist, L.F. Schneemeyer, D.W. Murphy, R.J. Cava, B. Batlogg, R.B. van Dover, R.M. Fleming, S.H. Glarum, S. Nakahara, R. Farrow, J.J. Krajewski, S.M. Zahurak, J.V. Waszczak, J.H. Marshall, P. Marsh, L.W. Rupp, Jr., and W.F. Peck, Structure and physical properties of single crystals of the 84-K superconductor $Bi_{2.2}Sr_2Ca_{0.8}Cu_2O_{8+\delta}$, *Phys. Rev. B*, **38**, 893–896 (1988).
10. S. Martin, A.T. Fiory, R.M. Fleming, G.P. Espinosa, and A.S. Cooper, Anisotropic critical current density in superconducting $Bi_2Sr_2CaCu_2O_8$ crystals, *Appl. Phys. Lett.*, **54**, 72–74 (1989).

11. Y. Kotaka, T. Kimura, H. Ikuta, J. Shimoyama, K. Kitazawa, K. Yamafuji, K. Kishio, and D. Pooke, Doping state and transport anisotropy in Bi-2212 single crystals, *Physica C*, **235–240**, 1529–1530 (1994).
12. W. Carrillo-Cabrera and W. Göpel, Influence of high temperature annealing on the $(Bi,Pb)_2Sr_2Ca_2Cu_3O_{10}$ phase and determination by its crystal structure by X-ray powder diffraction, *Physica C*, **161**, 373–389 (1989).
13. S. Bansal, T.K. Bansal, A.K. Jha, and R.G. Mendiratta, Effect of addition of lead on the superconductivity and structure of $Bi_2Sr_2Ca_2Cu_3O_{10}$ ceramics, *Physica C*, **173**, 260–266 (1991).
14. J.L. Wagner, O. Chmaissem, J.D. Jorgensen, D.G. Hinks, P.G. Radaelli, B.A. Hunter, and W.R. Jensen, Multiple defects in overdoped $Tl_2Ba_2CuO_{6+\delta}$, *Physica C*, **277**, 170–182 (1997).
15. S.S.P. Parkin, V.Y. Lee, E.M. Engler, A.I. Nazzal, T.C. Huang, G. Gorman, R. Savoy, and R. Beyers, Bulk superconductivity at 125 K in $Tl_2Ca_2Ba_2Cu_3O_x$, *Phys. Rev. Lett.*, **60**, 2539–2542 (1988).
16. A. Maignan, C. Martin, V. Hardy, Ch. Simon, M. Hervieu, and B. Raveau, A crystal study of the 128 K superconductor $Tl_2Ba_2Ca_2Cu_3O_{10-\delta}$: Synthesis and anisotropic magnetic properties, *Physica C*, **219**, 407–422 (1994).
17. D.M. Ogborne and M.T. Weller, Structure and oxygen stoichiometry in $Tl_2Ba_2Ca_3Cu_4O_{12-\delta}$: A high-resolution powder neutron-diffraction study, *Physica C*, **223**, 283–290 (1994).
18. H.C. Ku, M.F. Tai, J.B. Shi, M.J. Shieh, S.W. Hsu, G.H. Hwang, D.C. Ling, T.J. Watson-Yang, and T.Y. Lin, The occurrence of superconductivity in the $TlBa_2CuO_{5-\delta}$ type (1021) system, *Japanese J. Appl. Phys.*, **28**, L923–L925 (1989).
19. R. Beyers, S.S.P. Parkin, V.Y. Lee, A.I. Nazzal, R. Savoy, G. Gorman, and T.C. Huang, and S. Placa, Crystallography and microstructure of Tl–Ca–Ba–Cu–O superconducting oxides, *Appl. Phys. Lett.*, **53**, 432–434 (1988).
20. S. Nakajima, M. Kikuchi, Y. Syono, K. Nagase, T. Oku, N. Kobayashi, D. Shindo, and K. Hiraga, Improvement in superconductivity of $TlBa_2CaCu_2O_y$ system, *Physica C*, **170**, 443–447 (1990).
21. D.M. Ogborne and M.T. Weller, The structure of $TlBa_2Ca_3Cu_4O_{11}$, *Physica C*, **230**, 153–158 (1994).
22. M.D. Marcos and J.P. Attfield, Crystal structure of $Tl_{0.5}Pb_{0.5}Sr_2Ca_2Cu_3O_9$ at 300 K and around T_c (118 K), *Physica C*, **270**, 267–273 (1996).
23. S.N. Putilin, E.V. Antipov, O. Chmaissem, and M. Marezio, Superconductivity at 94 K in $HgBa_2CuO_{4+\delta}$, *Nature*, **362**, 226–228 (1993).
24. E.V. Antipov, J.J. Capponi, C. Chaillout, O. Chmaissem, S.M. Loureiro, M. Marezio, S.N. Putilin, A. Santoro, and J.L. Tholence, Synthesis and neutron powder diffraction study of the superconductor $HgBa_2CaCu_2O_{6+\delta}$, *Physica C*, **218**, 348–355 (1993).
25. E.V. Antipov, S.M. Loureiro, C. Chaillout, J.J. Capponi, P. Bordet, J.L. Tholence, S.N. Putilin, and M. Marezio, The synthesis and characterization of the $HgBa_2Ca_2Cu_3O_{8+\delta}$ and $HgBa_2Ca_3Cu_4O_{10+\delta}$ phases, *Physica C*, **215**, 1–10 (1993).
26. J.J. Capponi, E.M. Kopnin, S.M. Loureiro, E.V. Antipov, E. Gautier, C. Chaillout, B. Souletie, M. Brunner, J.L. Tholence, and M. Marezio, High pressure synthesis and heat treatments of the $HgBa_2Ca_4Cu_5O_{12+\delta}$ and $HgBa_2Ca_5Cu_6O_{14+\delta}$, *Physica C*, **256**, 1–7 (1996).

27. G.F. Sun, K.W. Wong, B.R. Xu, Y. Xin, and D.F. Lu, T_c enhancement of $HgBa_2Ca_2Cu_3O_{8+x}$ by Tl substitution, *Phys. Lett. A*, **192**, 122–124 (1994).
28. H. Schwer, J. Karpinski, L. Lesne, C. Rossel, A. Morawski, T. Lada, and A. Paszewin, X-ray single crystal structure analysis of the 1245 type superconductor $Hg_{0.5}Pb_{0.5}Ba_2Ca_4Cu_5O_{12+\delta}$, *Physica C*, **254**, 7–14 (1995).
29. A.T. Matveev and E. Takayama-Muromachi, High-pressure synthesis of a new superconductor, $AlSr_2Ca_{2-x}Y_xCu_3O_9$. The $n = 3$ member of the $AlSr_2Ca_{n-1}Cu_nO_{2n+3}$ family, *Physica C*, **254**, 26–32 (1995).
30. M. Isobe, T. Kawashima, K. Kosuda, Y. Matsui, and E. Takayama-Muromachi, A new series of high-T_c superconductors $AlSr_2Ca_{n-1}Cu_nO_{2n+3}$ ($n = 4$, $T_c = 110$ K; $n = 5$, $T_c = 83$ K) prepared at high pressure, *Physica C*, **234**, 120–126 (1994).
31. T. Kawashima, Y. Matsui, and E. Takayama-Muromachi, New series of oxide superconductors, $BSr_2Ca_{n-1}Cu_nO_{2n+3}$ ($n = 3$–5), prepared at high pressure, *Physica C*, **254**, 131–136 (1995).
32. E. Takayama-Muromachi, Y. Matsui, and K. Kosuda, New oxyborate superconductor, $BSr_2Ca_3Cu_4O_{11}$ ($T_c = 110$ K) prepared at high pressure, *Physica C*, **241**, 137–141 (1995).
33. J. Ramirez-Castellanos, Y. Matsui, E. Takayama-Muromachi, and M. Isobe, Microstructural characterization of $GaSr_2Ca_2Cu_3O_{9+\delta}$, $n = 3$ member of the homologous series of superconductors $GaSr_2Ca_{n-1}Cu_nO_{2n+3}$, *Physica C*, **251**, 279–284 (1995).
34. X.-J. Wu, T. Tamura, S. Adachi, C.-Q. Jin, T. Tatsuki, and H. Yamauchi, Structural study on high-pressure synthesized Pb–Sr–Ca–Cu–O system, *Physica C*, **247**, 96–104 (1995).
35. T. Tamura, S. Adachi, X.-J. Wu, T. Tatsuki, and K. Tanabe, Pb-1223 cuprate superconductor with T_c above 120 K, *Physica C*, **277**, 1–6 (1997).
36. C.-Q. Jin, S. Adachi, X.-J. Wu, H. Yamauchi, and S. Tanaka, 117 K superconductivity in the Ba–Ca–Cu–O system, *Physica C*, **223**, 238–242 (1994).
37. T. Kawashima, Y. Matsui, and E. Takayama-Muromachi, New oxycarbonate superconductors $(Cu_{0.5}C_{0.5})Ba_2Ca_{n-1}Cu_nO_{2n+3}$ ($n = 3$, 4) prepared at high pressure, *Physica C*, **224**, 69–74 (1994).
38. H. Kumakura, K. Togano, T. Kawashima, and E. Takayama-Muromachi, Critical current densities and irreversibility lines of new oxycarbonate superconductors $(Cu_{0.5}C_{0.5})Ba_2Ca_{n-1}Cu_nO_{2n+3}$ ($n = 3$, 4), *Physica C*, **226**, 222–226 (1994).
39. E. Takayama-Muromachi, Y. Matsui, and J. Ramirez-Castellanos, New series of oxysulphate superconductors $(Cu_{0.5}S_{0.5})Sr_2Ca_{n-1}Cu_nO_y$ ($n = 3$–7), prepared at high pressure, *Physica C*, **252**, 221–228 (1995).
40. Y. Tokura, H. Takagi, and S. Uchida, A superconducting copper oxide compound with electrons as the charge carriers, *Nature*, **337**, 345–347 (1989).
41. M.A. Crusellas, J. Fontcuberta, S. Piñol, J. Beille, and T. Grenet, Transport properties of $Ln_{2-x}Ce_xCuO_{4-y}$ single crystals under high pressure, *Physica C*, **209**, 537–548 (1993).
42. E.H. Appelman, L.R. Morss, A.M. Kini, U. Geiser, A. Umezawa, G.W. Crabtree, and K. Douglas Carlson, Oxygen content of superconducting perovskites, $La_{2-x}Sr_xCuO_y$ and $YBa_2Cu_3O_y$, *Inorg. Chem.*, **26**, 3237–3239 (1987).

43. A.I. Nazzal, V.Y. Lee, E.M. Engler, R.D. Jacowitz, Y. Tokura, and J.B. Torrance, New procedure for determination of $[Cu–O]^{+p}$ charge and oxygen content in high-T_c copper oxides, *Physica C*, **153–155**, 1367–1368 (1988).
44. Y. Maeno, H. Teraoka, K. Matsukuma, K. Yoshida, K. Sugiyama, F. Nakamura, and T. Fujita, Simple and reliable method for determination of oxygen content in high-T_c copper oxides, *Physica C*, **185–189**, 587–588 (1991).
45. W.M. Chen, W. Hong, J.F. Geng, X.S. Wu, W. Ji, L.Y. Li, L. Qui, and X. Jin, Iodometric titration for determining the oxygen content of samples doped with Fe and Co, *Physica C*, **270**, 349–353 (1996).
46. W.M. Chen, C.C. Lam, J.F. Geng, L.Y. Li, K.C. Hung, and X. Jin, A new calculating formula of iodometric titration for high T_c superconductors, *Physica C*, **270**, 155–158 (1996).
47. A. Fukuoka, A. Tokiwa-Yamamoto, M. Itoh, R. Usami, S. Adachi, H. Yamauchi, and K. Tanabe, Dependence of superconducting properties on Cu-valence determined by iodometry, *Physica C*, **265**, 13–18 (1996).

6
Crystal Structures of Cuprate Superconductors

6.1 Introduction

We have already noticed that layered crystal structures are a characteristic property of all cuprate superconductors. Electrically conducting copper oxide layers are sandwiched between insulating layers, which act as charge carrier reservoirs. In this chapter, we will consider the crystal structures of the cuprate superconductors in more detail.

The crystal structure of solids can be studied by different diffraction methods. In Section 6.2, X-ray and neutron diffraction will be briefly described. The wavelength of the X-rays or the particle waves must be comparable to or smaller than the lattice parameters of the investigated solids. Deceleration of 40 kV electrons by a copper target generates bremsstrahlung with wavelengths in the region of interest. The wavelength λ of neutrons is given by the de Broglie relation $\lambda = h/p$, where h is Planck's constant and p the momentum of the particle. X-rays, mainly interacting with the electrons of atoms, are only weakly scattered by light elements. It is therefore difficult to study the occupation of lattice sites by oxygen in the cuprate superconductors by means of X-ray diffraction. More information on the positions of oxygen atoms in high-temperature superconductors can be obtained by neutron diffraction, which is more sensitive to the detection of light atoms. Due to their magnetic moment, neutrons can be used to study the magnetic structure of a solid. Neutron scattering performed below the Néel temperature (T_N) responds to the magnetic unit cell of the antiferromagnetic state, whereas diffraction patterns taken above T_N provide information on the atomic positions, which can be compared with that derived from X-ray diffraction.

In addition, X-ray powder diffraction can be used to identify the different solid phases present in a sample. The X-ray powder diffraction pattern of a single-phase polycrystalline sample is like a fingerprint characteristic of the solid phase in question. In multiphase samples, the ratios of the intensities of the strongest reflections of the different phases

Physical Properties of High-Temperature Superconductors, First Edition. Rainer Wesche.

present can be used to estimate the fractions of these phases present in the sample. X-ray powder diffraction is therefore a powerful tool to find the optimum processing conditions to synthesize single-phase cuprate superconductors.

In Section 6.3, the crystal structures of various cuprate superconductors are described. In addition, typical X-ray powder diffraction patterns of various cuprate superconductors will be presented.

6.2 Diffraction Methods

6.2.1 Bragg Condition

The Bragg condition is illustrated in Figure 6.1. A small fraction of the incident beam is reflected by each of the parallel lattice planes. Similar to the reflection of light at a mirror, the angles of incidence and reflection at each of the lattice planes are equal. Constructive interference of the beams, reflected from adjacent lattice planes, occurs for path differences of $n \cdot \lambda$, where n is an integer number. Thus, the propagation directions of scattered waves satisfy the Bragg condition

$$2d \sin\theta = n\lambda \tag{6.1}$$

In this elastic scattering process, the energy of the X-ray photons is unchanged.

6.2.2 Miller Indices

The different planes in the crystal lattice are described by the Miller indices. They can be found by the following procedure. First, the intercepts of the lattice plane with the crystallographic axes are determined. The Miller indices, typically called (hkl), are the smallest possible integer number ratios of the inverse values of these intercepts. In Figure 6.2, several lattice planes and the corresponding Miller indices are shown.

In the case of the trigonal and hexagonal crystal systems, the Miller indices are frequently replaced by the Miller–Bravais indices ($hkil$). The values of h, k, and l are identical to those

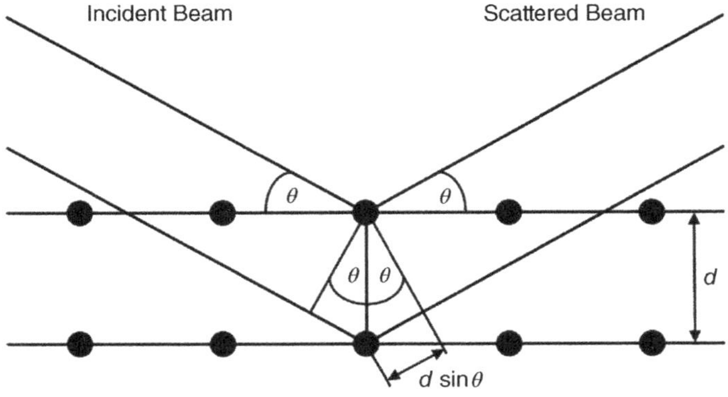

Figure 6.1 Sketch of Bragg scattering. Path differences equal to an integer number of the wavelength of the X-rays lead to constructive interference

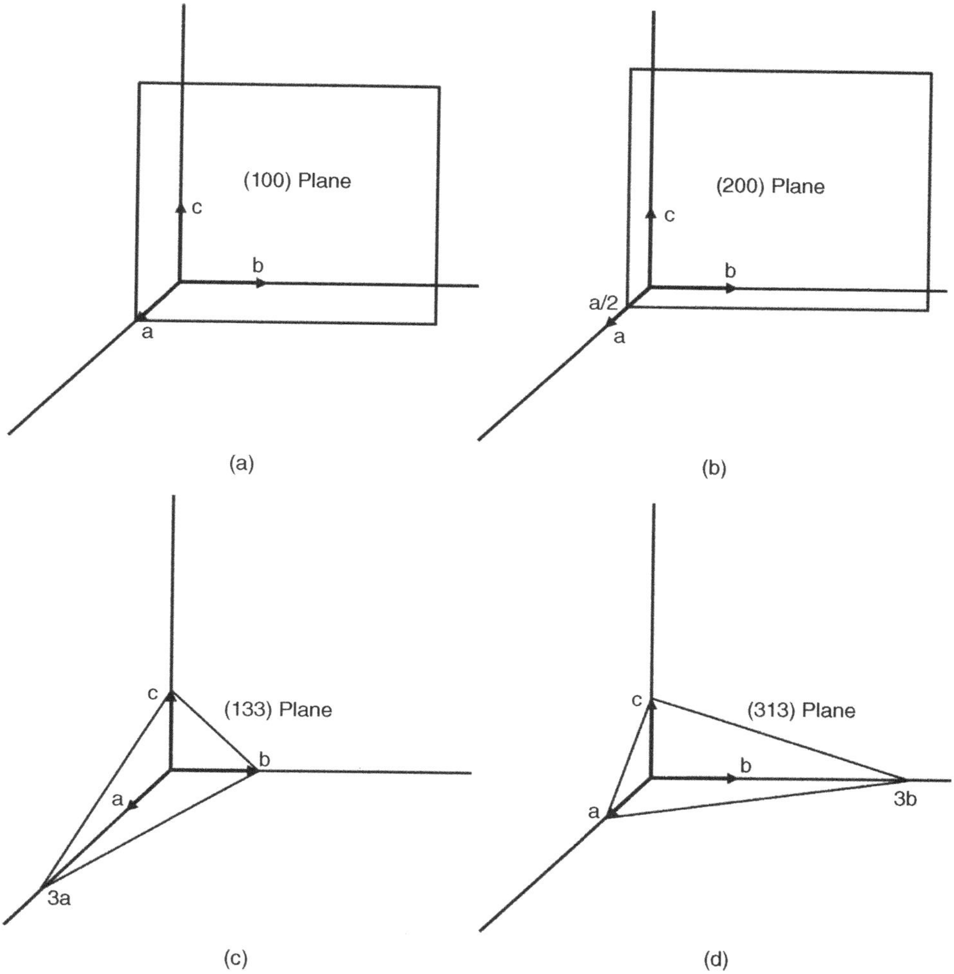

Figure 6.2 *Selected lattice planes and corresponding Miller indices (hkl)*

of the corresponding Miller indices. The extra index (i) is defined for convenience to be $i = -(h + k)$.

6.2.3 Classification of Crystal Structures

Crystal structures are classified according to their translational symmetry. Fourteen Bravais lattices exist in 3D space, and only five on a 2D plane. The Bravais lattices in two and three dimensions are listed in Table 6.1. The Bravais lattice symbols indicate both the crystal system and the centering type. The centering types in three dimensions and the corresponding symbols are primitive (P), face-centered ($C\,(A, B)$), body-centered (I), and all face-centered (F). In two dimensions, there exist only two centering types, namely primitive (p) and centered (c). For example, the lattice symbol for a face-centered cubic crystal structure

Table 6.1 *Bravais nets and lattices*

System	Lattice symbol	Nature of the vector set
Two dimensions		
Square (tetragonal)	*tp*	$a = b,\ \gamma = 90°$
Rectangular (orthorhombic)	*op*	$a \neq b,\ \gamma = 90°$
	oc	
Hexagonal	*hp*	$a = b,\ \gamma = 120°$
Oblique (monoclinic)	*mp*	$a \neq b,\ \gamma \neq 90°,\ \gamma \neq 120°$
Three dimensions		
Cubic	*cP*	$a = b = c,\ \alpha = \beta = \gamma = 90°$
	cI	
	cF	
Tetragonal	*tP*	$a = b \neq c,\ \alpha = \beta = \gamma = 90°$
	tI	
Orthorhombic	*oP*	$a \neq b \neq c,\ \alpha = \beta = \gamma = 90°$
	oC	
	oI	
	oF	
Trigonal (rhombohedral)	*hR*	$a = b = c,\ \alpha = \beta = \gamma < 120°,\ \neq 90°$
Hexagonal	*hP*	$a = b \neq c,\ \alpha = \beta = 90°,\ \gamma = 120°$
Monoclinic	*mP*	$a \neq b \neq c,\ \alpha = \gamma = 90°,\ \beta \neq 90°$
	mC	
Triclinic (anorthic)	*aP*	$a \neq b \neq c,\ \alpha \neq \beta \neq \gamma$

is *cF*. The angles between the different crystallographic axes are defined in Figure 6.3. Based on their symmetry properties, all crystal structures can be classified into 230 space groups [1, 2], described by the Mauguin–Hermann symbols. For example, in the space group symbol *I4/mmm*, the letter *I* stands for body centering, the number *4* indicates a fourfold rotational symmetry, and the three letters *m* represent three mirror planes. We shall see later that many cuprate superconductors crystallize in the symmetry of the space group *I4/mmm*.

6.2.4 X-ray Diffraction

For the determination of the crystal structure of a solid by means of X-ray diffraction, single crystals as well as powder samples can be used. The Laue method, which is schematically illustrated in Figure 6.4, uses a single crystal and needs a continuous X-ray spectrum. The incident beam is parallel to one of the principal crystal directions. The positions of the diffraction spots obey the Bragg condition. For each crystallographic plane, the wavelength appropriate to fulfill the Bragg condition is selected out of the continuous X-ray spectrum. The pattern of the diffraction spots reveals the symmetry of the crystal lattice in question. The main disadvantage of the Laue method is the fact that single crystals of sufficient size and quality are required for the determination of the crystal structure. Frequently, suitable crystals are not available immediately after the discovery of a new superconductor.

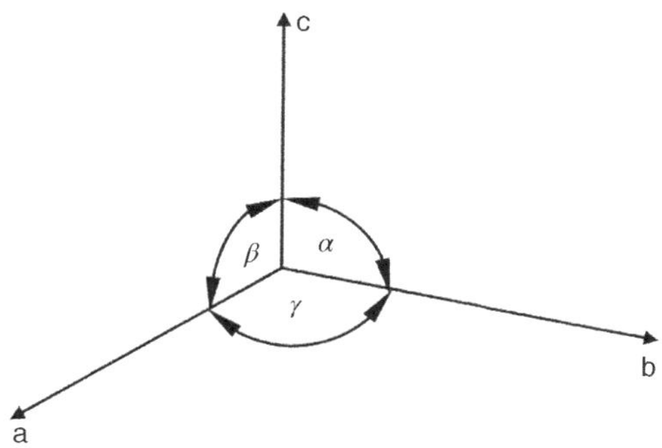

Figure 6.3 Definition of the angles between the crystallographic axes

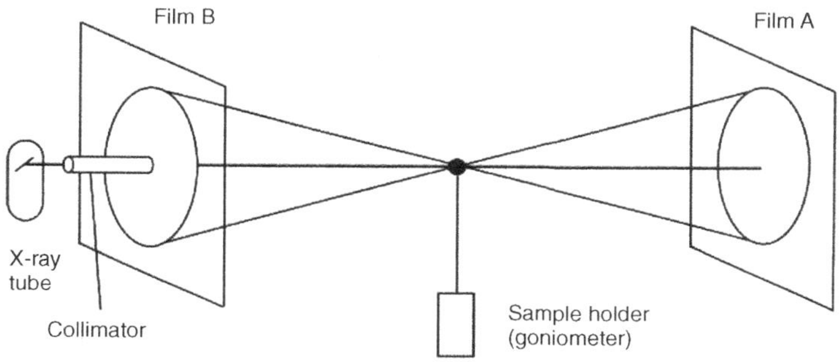

Figure 6.4 Sketch of the Laue method

As an alternative to the Laue method, X-ray powder diffraction can be used to determine the crystal structure. In Figure 6.5, the experimental setup of X-ray powder diffraction is sketched. A monochromatic X-ray beam is required for powder diffraction experiments. The characteristic X-rays produced by the excitation of inner electrons may be used as a source of monochromatic X-rays. Cu Kα radiation with a wavelength of 0.15418 nm is frequently used for X-ray powder diffraction; nickel filters are then used to absorb most of the Kβ line and the undesirable bremsstrahlung. Alternatively, a monochromator crystal may be used, which reflects only the Cu Kα radiation. The position of the X-ray tube is fixed. The width of the incident X-ray beam is limited by the collimators C1 and C2. The sample holder is rotated around its axis, starting from the angle θ_i with a step width $\Delta\theta$, up to the final Bragg angle θ_f. Simultaneously, the angle between the detector and the incident beam is varied, starting from $2\theta_i$ with a step width of $2\Delta\theta$, up to the final scattering angle of $2\theta_f$. The width of the reflected beam is limited by the collimators C3 and C4. The nickel filter, used to absorb most of the Kβ line and the undesired bremsstrahlung background, is positioned between the collimators C3 and C4. The positions of the diffracted X-ray beams satisfy the Bragg condition. The intensities of the scattered X-rays are plotted as a function of the scattering angle 2θ. In addition to the determination of the crystal structure of a newly discovered superconductor, impurity phases present can be detected by X-ray powder diffraction.

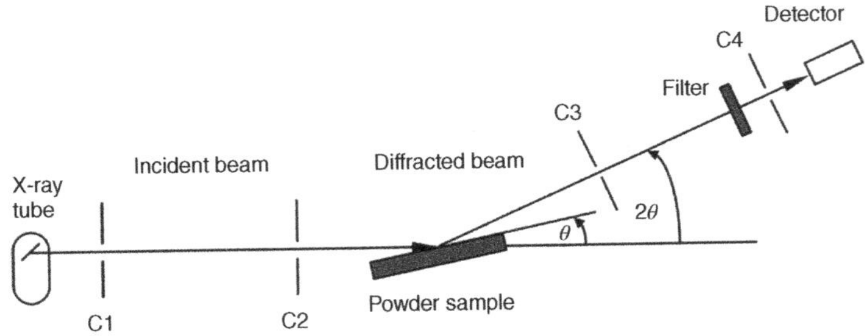

Figure 6.5 *Sketch of X-ray powder diffraction*

The geometrical theory of diffraction described so far is sufficient to determine the lattice constants of the sample. The observed intensities of the scattered beams provide additional information about the occupation of the lattice sites. This information can be extracted by a comparison of calculated and measured intensities. The intensity of the diffracted X-rays is proportional to the structure factor $F(hkl)$ [3]

$$F(hkl) = \sum_j f_j \exp(-i2\pi(x_j h + y_j k + z_j l)) \tag{6.2}$$

where the summation is over all atoms in the unit cell, f_j is the atomic form factor, $i^2 = -1$, h, k, l are the Miller indices, and the coordinates x_j, y_j, and z_j give the position of the jth atom in the unit cell. The atomic form factor is an atomic property mainly determined by the local electron concentration.

6.2.5 Neutron Diffraction

The neutrons and the atomic nuclei interact via the strong interaction. The scattering of neutrons at the atomic nuclei in a solid provides in principle the same information as X-ray diffraction. The X-rays interact with the electrons of the atoms, and hence their scattering depends strongly on the atomic number (proton number). X-rays are only weakly scattered by light atoms, whereas they are strongly scattered by heavy atoms with a large atomic number. In contrast, neutron scattering depends only weakly on the atomic mass, and hence can provide information on the positions of light atoms in the lattice. Furthermore, the interaction of the magnetic moment of the neutron with the ordered magnetic moments, for example in an antiferromagnetic solid, provides information on the magnetic unit cell of the solid. In antiferromagnetic solids, the magnetic structure is different from the crystal structure. Frequently, the magnetic moment of adjacent atoms changes its sign, and hence the distance of magnetically identical atoms is larger than that of crystallographic identical atoms.

The wavelength of neutrons is given by the formula $\lambda = 0.0286/E^{0.5}$, where λ is the wavelength of the neutrons in nanometers and E their kinetic energy in electron-volt (eV). An energy of 80 meV leads to a wavelength of 0.1 nm, which is comparable to the interatomic distances in a solid. Because of the low energy of the neutrons, inelastic scattering can be used to study excitations (e.g., phonons, magnons) in a solid. As an example, let us consider the generation or destruction of a phonon in an inelastic scattering process. For inelastic scattering, the Bragg condition can be written as

$$\mathbf{k} = \mathbf{k}' + \mathbf{G}_{\mathbf{hkl}} \pm \mathbf{q} \tag{6.3}$$

where $\mathbf{k}$ is the wave vector of the incident neutron beam, $\mathbf{k}'$ the wave vector of the scattered beam, $\mathbf{G}_{\mathbf{hkl}}$ a reciprocal lattice vector, and $\mathbf{q}$ the wave vector of the phonon. The plus sign describes the generation of a phonon, while the minus sign corresponds to the destruction of a phonon. In addition, energy must be conserved in the scattering process

$$\frac{\hbar^2 k^2}{2m_n} = \frac{\hbar^2 k'^2}{2m_n} \pm \hbar\omega(q) \tag{6.4}$$

Here ω is the angular frequency of the phonon. A more detailed description of elastic and inelastic neutron scattering can be found in Ref. [4].

6.3 Crystal Structures of the Cuprate High-Temperature Superconductors

6.3.1 The Crystal Structure of La_2CuO_4

The crystal structures of many cuprate high-temperature superconductors have been determined by means of X-ray powder diffraction. In addition, the technique provides information on the phase purity of the considered sample, and can be used to optimize the preparation conditions for the synthesis of the cuprate superconductor in question. It therefore seems useful not only to describe the crystal structure, but also to show typical X-ray powder diffraction patterns. In Figure 6.6, the X-ray powder diffraction pattern of $La_{2-x}Sr_xCuO_4$ ($x = 0.15$) is shown. The sketch is based on measurements using Ni-filtered Cu Kα radiation [5]. All peaks can be assigned to the La-214 phase, and most of them are labeled with the corresponding Miller indices.

The tetragonal unit cell of $La_{2-x}Sr_xCuO_4$ is shown in Figure 6.7. The lattice parameters are $a = 0.3779$ nm and $c = 1.3200$ nm for $x = 0.15$ [6]. Characteristic features of La_2CuO_4

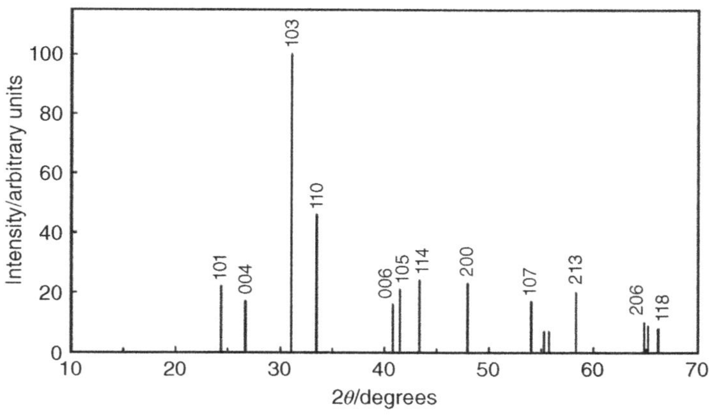

Figure 6.6 Sketch of the typical Cu Kα X-ray diffraction pattern of $La_{2-x}Sr_xCuO_4$ (data from [5]). For most of the peaks, the Miller indices are indicated

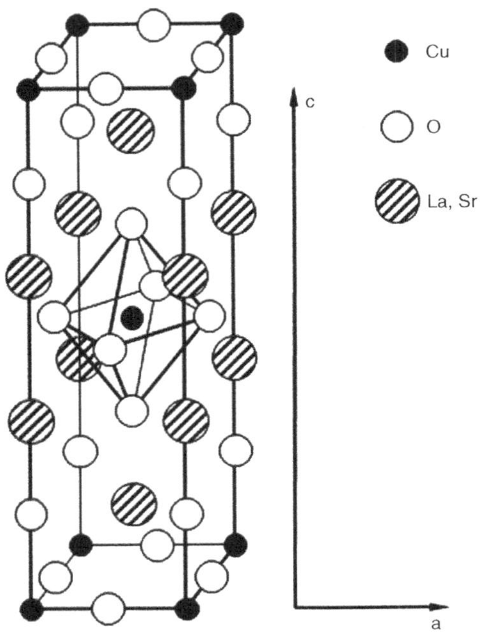

Figure 6.7 *Body-centered tetragonal unit cell of the crystal structure of* $La_{2-x}Sr_xCuO_4$ *(adapted from [6])*

are the CuO_2 planes at the top, the bottom, and in the center of the unit cell. This square coordinated CuO_2 layers are a common feature of all cuprate high-T_c superconductors. The current in the normal as well as in the superconducting state flows along the CuO_2 planes, which are therefore called conduction planes. In La_2CuO_4, the CuO_2 sheets are separated by insulating double layers of LaO. As shown for the copper ion in the center of the unit cell, each of the copper ions is in the center of an extremely elongated oxygen octahedron. However, the results of band structure calculations indicate that the copper–oxygen interactions along the c axis are small [7]. The normalized positions of all atoms in the unit cell are listed in Table 6.2. The unit cell contains two formula units.

The crystal structure of La_2CuO_4 belongs to the space group *I4/mmm*. Omitting the positions of equivalent atoms, and taking into account the symmetry properties, it is sufficient to provide the normalized positions of only four atoms. This minimum set of atomic positions is presented in Table 6.3.

For each of the atomic positions provided in Table 6.3, there exists an equivalent site at the position $z' = 1 - z$. From the symmetry operation, called body centering, additional atom positions at $x' = x \pm 1/2$, $y' = y \pm 1/2$; $z' = z \pm 1/2$ are resulted. Applying this symmetry operation exchanges edge and centered atoms. Face atoms are moved to another face site.

6.3.2 *The Crystal Structure of $YBa_2Cu_3O_{7-\delta}$*

Next, let us consider the crystal structure of $YBa_2Cu_3O_{7-\delta}$, the currently most important high-temperature superconductor for magnet and power applications. A sketch of the

Table 6.2 *Normalized atom positions in the tetragonal unit cell of* $La_{1.85}Sr_{0.15}CuO_4$ *(the lattice constants are a = 0.3779 mm and c = 1.3200 nm (data from [6]))*

Atoms	x	y	z	Atoms	x	y	z
La or Sr	0	0	−0.361	O	0	−0.5	−0.5
La or Sr	−0.5	0.5	−0.139	O	−0.5	0	−0.5
La or Sr	0.5	0.5	−0.139	O	0	0.5	−0.5
La or Sr	0.5	−0.5	−0.139	O	0.5	0	−0.5
La or Sr	−0.5	−0.5	−0.139	O	−0.5	0.5	−0.316
La or Sr	−0.5	0.5	0.139	O	0.5	0.5	−0.316
La or Sr	0.5	0.5	0.139	O	0.5	−0.5	−0.316
La or Sr	0.5	−0.5	0.139	O	−0.5	−0.5	−0.316
La or Sr	−0.5	−0.5	0.139	O	0	0	−0.184
La or Sr	0	0	0.361	O	0	−0.5	0
Cu	−0.5	0.5	−0.5	O	−0.5	0	0
Cu	0.5	0.5	−0.5	O	0	0.5	0
Cu	0.5	−0.5	−0.5	O	−0.5	0	0
Cu	−0.5	−0.5	−0.5	O	0	0	0.184
Cu	0	0	0	O	−0.5	0.5	0.316
Cu	−0.5	0.5	0.5	O	0.5	0.5	0.316
Cu	0.5	0.5	0.5	O	0.5	−0.5	0.316
Cu	0.5	−0.5	0.5	O	−0.5	−0.5	0.316
Cu	−0.5	−0.5	0.5	O	0	−0.5	0.5
				O	−0.5	0	0.5
				O	0	0.5	0.5
				O	0.5	0	0.5

Table 6.3 *Minimum set of normalized atom positions required to describe the unit cell of* $La_{2-x}Sr_xCuO_4$ *(data from [6])*

Atoms	x	y	z	Comments
La or Sr	0	0	0.361	Edge atoms
Cu	0	0	0	Edge atoms
O1	0	0.5	0	Face-atoms
O2	0	0	0184	Edge atoms

typical X-ray powder diffraction is shown in Figure 6.8. The indicated Miller indices are based on an orthorhombic unit cell with the lattice parameters $a = 0.38218$ nm, $b = 0.38913$ nm, and $c = 1.1677$ nm [8]. X-ray and neutron powder diffraction data can also be found in references [9–11]. A space group *Pmmm* is consistent with both X-ray and neutron diffraction data. The crystal structure of the closely related nonsuperconducting phase $YBa_2Cu_3O_6$ can be described by a tetragonal unit cell (space group *P4/mmm*) of nearly identical cell dimensions [7].

Figure 6.9 shows the orthorhombic unit cell of $YBa_2Cu_3O_{7-\delta}$. The unit cell contains the atoms corresponding to one formula unit. The presence of CuO chains along the crystallographic b direction is a special feature of the crystal structure of $YBa_2Cu_3O_{7-\delta}$. Each copper atom in the CuO chains is surrounded by four oxygen atoms, as shown in Figure 6.9.

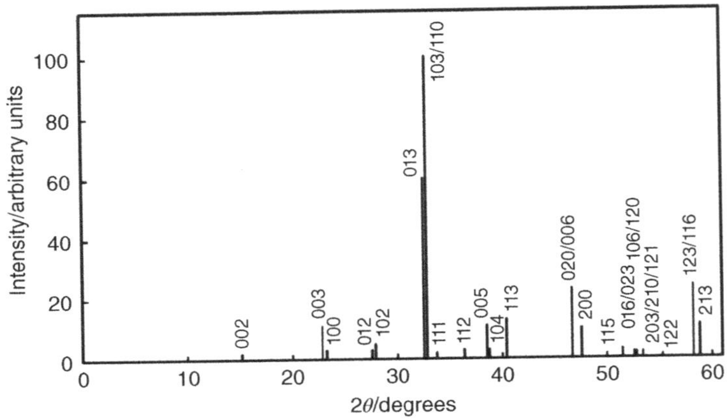

Figure 6.8 *Sketch of the typical Cu Kα X-ray diffraction pattern of $YBa_2Cu_3O_{7-x}$ (data from [8]). For most of the peaks, the Miller indices are indicated*

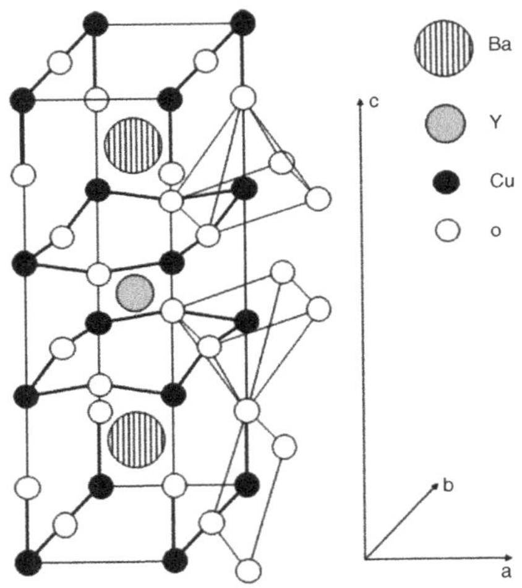

Figure 6.9 *Orthorhombic unit cell of $YBa_2Cu_3O_{7-\delta}$ (adapted from [12]). The atomic positions are listed in Table 6.4*

As in other cuprate superconductors, CuO_2 planes are present in the crystal structure. Two CuO_2 planes, separated by a layer of Y atoms, form the CuO_2 blocks of Y-123. The slightly different z positions of the copper and oxygen atoms lead to a buckling of the CuO_2 planes. Each of the copper atoms in the CuO_2 layers is in the center of the basis of a pyramid formed by five oxygen atoms (see Figure 6.9). Adjacent CuO_2 blocks are separated by double layers of BaO with embedded CuO chains.

Table 6.4 provides the normalized positions of the atoms in the unit cell of Y-123. The Cu1 atoms together with the O4 atoms form the CuO chains. The lattice sites of the Cu2, O2, and O3 atoms are all in the slightly buckled CuO_2 planes. The oxygen atoms occupy perovskite-like anion positions halfway between the copper ions. The O1 atoms are in

Table 6.4 *Normalized atom positions in the unit cell of Y-123 (data from [13])*

Atoms	x	y	z	Comments
Y	0.5	0.5	0.5	Centered atoms
Ba	0.5	0.5	±0.183	Centered atoms
Cu1	0	0	0	Edge atoms
Cu2	0	0	±0.355	Edge atoms
O1	0	0	±0.159	Edge atoms
O2	0.5	0	±0.378	Face atoms
O3	0	0.5	±0.378	Face atoms
O4	0	0.5	0	Edge atoms

the BaO layers, with no oxygen atoms present in the Y layers. Frequently, the formula $Ba_2YCu_3O_{7-\delta}$ is used instead of $YBa_2Cu_3O_{7-x}$ [8]. This alternative formula takes into account that the Y layers separate the two CuO_2 layers in the CuO_2 blocks, like the Ca layers in the Bi-, Tl-, and Hg-based cuprate superconductors.

6.3.3 The Crystal Structures of Bi-22(n − 1)n High-Temperature Superconductors

In this section, we will consider the crystal structures of the bismuth-based cuprate high-temperature superconductors. There exists a series of compounds described by the chemical formula $Bi_2Sr_2Ca_{n-1}Cu_nO_{2n+4}$. In single samples, the 2201, 2212, and 2223 phases may coexist. It is therefore of importance to study the phase purity by X-ray powder diffraction. In case of the Bi-2223 compound, it proved to be very difficult to synthesize single-phase material without partial substitution of bismuth by lead.

We will discuss here only the crystal structures of Bi-2212 and Bi-2223, the two most important bismuth-based cuprate superconductors. These two compounds are used in the first generation of high-T_c superconducting wires and tapes.

X-ray powder diffraction data can be found for example in [14–16]. The X-ray powder diffraction pattern of Bi-2212 is sketched in Figure 6.10. The Miller indices provided in the figure are based on a pseudotetragonal unit cell with the lattice parameters $a = b =$ 0.3830 nm and $c = 3.071$ nm [14]. The true symmetry of Bi-2212 is *Fmmm* or lower [7, 17]. The dimensions of the orthorhombic unit cell based on the space group *Fmmm* are $a = 0.5414$, $b = 0.5418$, and $c = 3.089$ nm [17]. As a special feature, an incommensurate superlattice structure with a period of 4.76 has been observed along the crystallographic b direction [17].

Figure 6.11 shows a sketch of the X-ray powder diffraction pattern of (Bi,Pb)-2223 measured with Ni-filtered Cu Kα radiation. The Miller indices are based on a tetragonal unit cell with the lattice parameters $a = 0.5411$ nm and $c = 3.707$ nm [16]. In principle, it would have been possible to use a pseudotetragonal unit cell with a lattice parameter $a' = (1/\sqrt{2})a = 0.3826$ nm, as in the case of Bi-2212, and an unchanged c lattice parameter. It should be noted that, due to the use of different unit cells, the Miller indices provided in Figures 6.10 and 6.11 are systematically different. For example, the Miller indices (1 1 3) ((Bi,Pb)-2223) and (1 0 3) (Bi-2212) correspond to analogous lattice planes.

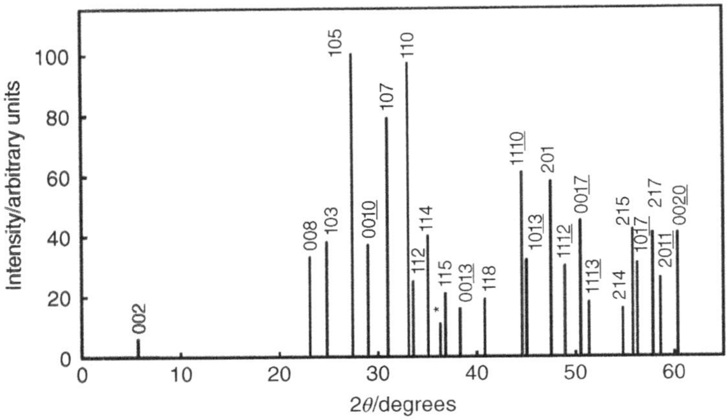

Figure 6.10 *Sketch of the typical Cu Kα X-ray diffraction pattern of Bi-2212 (data from [14]). The Miller indices are based on a pseudotetragonal unit cell with the lattice parameters a = b = 0.3830 nm and c = 3.071 nm*

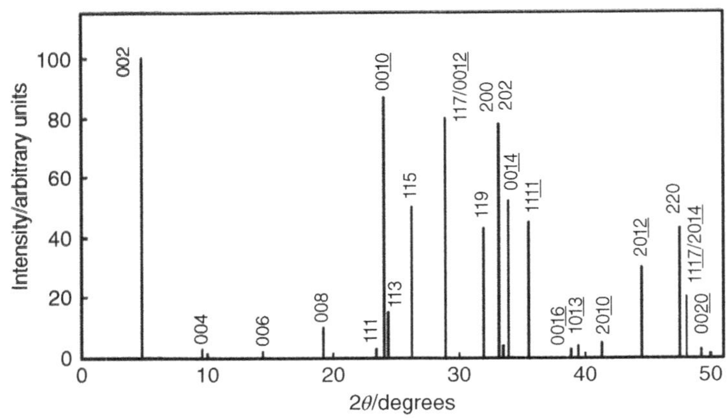

Figure 6.11 *Sketch of the typical Cu Kα X-ray diffraction pattern of (Bi,Pb)-2223 (data from [16]). The Miller indices are based on a pseudotetragonal unit cell with the lattice parameters a = b = 0.5411 nm and c = 3.707 nm*

In Figure 6.12, the pseudotetragonal unit cells of Bi-2212 and (Bi,Pb)-2223 are compared. The CuO_2 blocks of Bi-2212 are formed of two CuO_2 sheets, separated by a layer of Ca ions. Neighboring CuO_2 blocks are sandwiched between charge carrier reservoirs with the layer sequence SrO–BiO–BiO–SrO. The copper atoms in the CuO_2 sheets are at the center of the basal plane of a pyramid formed of five oxygen atoms. The Bi-2223 structure contains three CuO_2 sheets in a single CuO_2 block, which are separated by layers of Ca ions. In the central CuO_2 sheet, each copper atom is surrounded by four oxygen atoms. The copper atoms in the outer CuO_2 sheets are in the center of the basal plane of an oxygen pyramid. Adjacent CuO_2 blocks are separated by multilayers of SrO–BiO–BiO–SrO, as in the case of Bi-2212. The insertion of an additional CuO_2–Ca unit transforms the structure of Bi-2212 to that of Bi-2223.

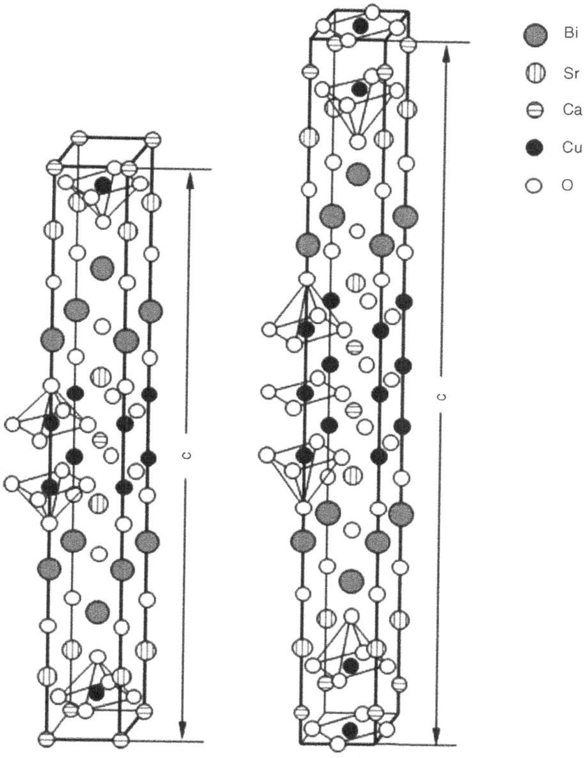

Figure 6.12 Comparison of the pseudotetragonal unit cells of Bi-2212 ($a \cong 0.38\,nm$, $c \cong 3.09\,nm$) and Bi-2223 ($a \cong 0.38\,nm$ and $c = 3.82\,nm$) (adapted from [12])

The normalized atom positions in the crystal structures of Bi-2212 and Bi-2223 are gathered in Tables 6.5 and 6.6. The positions are based on pseudotetragonal unit cells (*I4/mmm*). The application of the body-centering operation provides additional atom positions, as has been already discussed for the crystal structure of $La_{2-x}Sr_xCuO_4$ in Table 6.3. In the unit cells of Bi-2212 and Bi-2223, shown in Figure 6.12, the position $z = 0$ is in the central plane of the unit cell.

6.3.4 The Crystal Structures of Tl-based High-Temperature Superconductors

There exist two families of Tl-based cuprate high-temperature superconductors. The chemical formula of the members of the first family is $Tl_2Ba_2Ca_{n-1}Cu_nO_{2n+4}$ (Tl-22($n-1$)n). The Tl-22($n-1$)n compounds are the counterparts of the Bi-22($n-1$)n cuprate superconductors. The second family of Tl-based cuprate high-temperature superconductors is described by the chemical formula $TlM_2Ca_{n-1}Cu_nO_{n+3}$ (Tl-12($n-1$)n), where the M atoms are Sr or Ba.

As in the Bi-based superconductors, different Tl-based cuprate superconductors may coexist in a single sample. X-ray powder diffraction is therefore a useful tool to ensure

Table 6.5 *Normalized atom positions for the pseudotetragonal unit cell (I4/mmm) of Bi-2212 (the lattice parameters are $a \cong 0.38$ nm and $c = 3.09$ nm (data from [17]))*

Atoms	x	y	z	Comments
Bi	0	0	0.199	Edge atoms[a]
Sr	0.5	0.5	0.109	Centered atoms[a]
Cu	0	0	0.054	Edge atoms[a]
Ca	0.5	0.5	0	Centered atoms[a]
O1	0	0.5	0.054	Face atoms
O2	0	0	0.12	Edge atoms[a]
O3	0.5	0.5	0.198	Centered atoms[a]

[a] Atoms at centered (edge) positions in z planes move to edge (centered) positions in $z' = z \pm 0.5$ planes.

Table 6.6 *Normalized atom positions for the pseudotetragonal unit cell (I4/mmm) of Bi-2223 (the lattice parameters $a \cong 0.38$ nm and $c = 3.82$ nm (data from [18]))*

Atoms	x	y	z	Comments
Bi	0	0	0.211	Edge atoms[a]
Sr	0.5	0.5	0.135	Centered atoms[a]
Cu1	0	0	0	Edge atoms[a]
Cu2	0	0	0.091	Edge atoms[a]
Ca	0.5	0.5	0.046	Centered atoms[a]
O1	0	0.5	0	Face atoms
O2	0	0.5	0.091	Face atoms
O3	0	0	0.161	Edge atoms[a]
O4	0.5	0.5	0.213	Centered atoms[a]

[a] Atoms at centered (edge) positions in z planes move to edge (centered) positions in $z' = z \pm 0.5$ planes.

that the sample under investigation contains only a single phase. Let us first consider the X-ray powder diffraction patterns and the crystal structures of Tl-22($n-1$)n high-T_c superconductors.

Figure 6.13 shows a sketch of the X-ray powder diffraction pattern of $Tl_2Ba_2CuO_6$, which is based on measurements using Cu Kα radiation [19]. All reflections can be indexed to an orthorhombic unit cell of space group *Fmmm* with the lattice parameters $a = 0.5473$ nm, $b = 0.5483$ nm, and $c = 2.328$ nm.

In principle, the data are also consistent with a tetragonal crystal structure (space group *I4/mmm*). In Figure 6.14, the tetragonal unit cell of $Tl_2Ba_2CuO_6$ (Tl-2201) is shown. Each of the conducting CuO_2 blocks of Tl-2201 contains only a single layer of CuO_2. Neighboring CuO_2 blocks are separated by BaO–TlO–TlO–BaO layers, which form the charge carrier reservoirs of insulating character. Each copper atom is surrounded by six oxygen atoms, forming an octahedron.

The positions of the atoms in the tetragonal unit cell are listed in Table 6.7. It should be noted that, as in the case of the Bi-22($n-1$)n high-T_c superconductors, additional atom

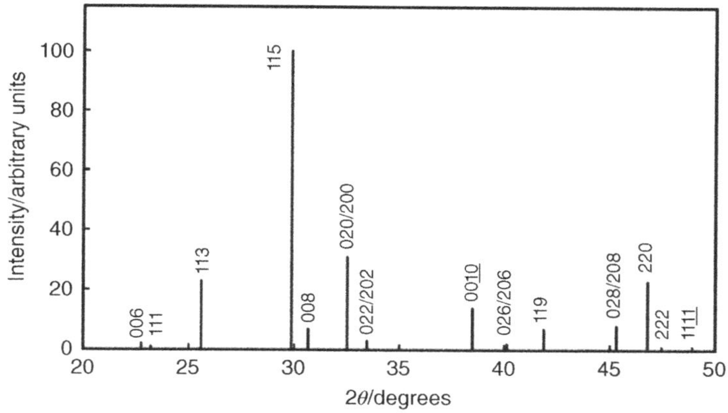

Figure 6.13 *Sketch of the typical Cu Kα X-ray diffraction pattern of Tl-2201 (data from [19]). The Miller indices are based on an orthorhombic unit cell (a = 0.5473 nm, b = 0.5483 nm, c = 2.328 nm)*

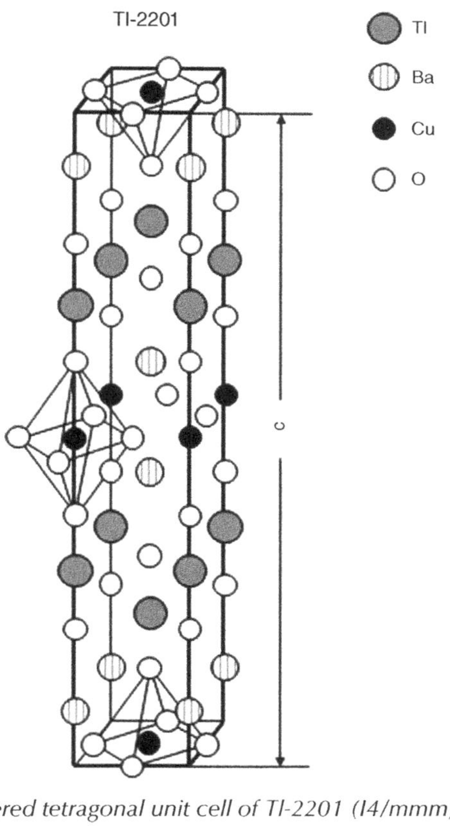

Figure 6.14 *Body-centered tetragonal unit cell of Tl-2201 (I4/mmm) with the lattice parameters a = 0.387 nm and c = 2.328 nm (adapted from [12])*

Table 6.7 Normalized atom positions in the tetragonal unit cell (I4/mmm) of Tl-2201 (the lattice parameters are $a = 0.387$ nm and $c = 2.328$ nm (data from [7, 19]))

Atoms	x	y	z	Comments
Tl	0.5	0.5	0.203	Centered atoms[a]
Ba	0	0	0.083	Edge atoms[a]
Cu	0.5	0.5	0	Centered atoms[a]
O1	0	0.5	0	Face atoms
O2	0.5	0.5	0.117	Centered atoms[a]
O3	0	0	0.211	Edge atoms[a]

[a]Atoms at centered (edge) positions in z planes move to edge (centered) positions in $z' = z \pm 0.5$ planes.

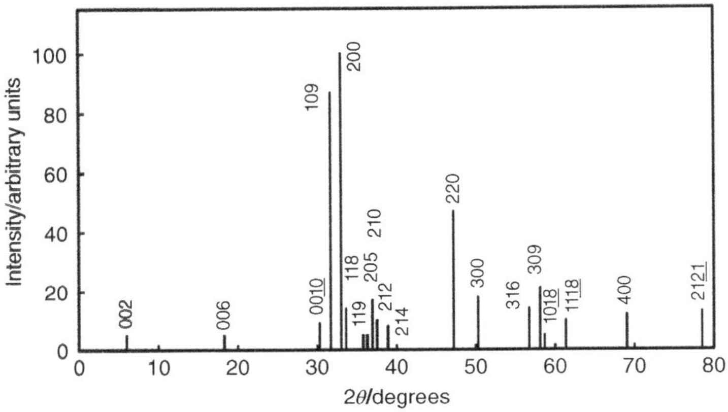

Figure 6.15 *Sketch of the typical Cu Kα X-ray diffraction pattern of Tl-2212 (data from [20]). The Miller indices are based on a tetragonal unit cell ($a = b = 0.544$ nm, $c = 2.955$ nm)*

positions are provided by body centering. The $z = 0$ plane is at the bottom of the tetragonal unit cell shown in Figure 6.14.

A sketch of a typical X-ray powder diffraction pattern of Tl-2212 made with Cu Kα radiation is shown in Figure 6.15. The Miller indices are based on a pseudotetragonal unit cell ($a = b = 0.544$ nm, $c = 2.955$ nm) [20].

Figure 6.16 shows the tetragonal unit cell of Tl-2212 (space group *I4/mmm*). Note that for the selected unit cell the lattice parameter a is only $(1/\sqrt{2}) \times 0.544$ nm $= 0.385$ nm. In the unit cell of Tl-2212, the BiO double layers of their Bi-2212 counterparts are replaced by TlO double layers, while the SrO layers present in Bi-2212 are replaced by BaO layers. As in their Bi-2212 counterpart, the copper oxide blocks consist of two CuO_2 sheets, separated by a layer of Ca ions. Each Cu atom in the CuO_2 planes is surrounded by a pyramid of five oxygen atoms. The normalized positions of the atoms in the unit cell of Tl-2212 are listed in Table 6.8.

The typical Cu Kα X-ray powder diffraction pattern of Tl-2223 is sketched in Figure 6.17. The Miller indices are based on a pseudotetragonal unit cell with the lattice parameters $a = b = 0.540$ nm, $c = 3.625$ nm [20]. Figure 6.18 shows an X-ray

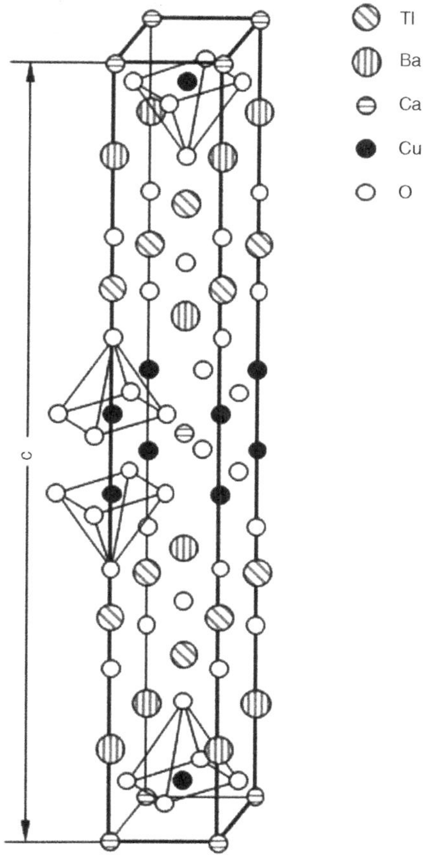

Figure 6.16 Body-centered tetragonal unit cell of Tl-2212 (I4/mmm) with cell dimensions of $a = 0.385$ nm and $c = 2.955$ nm (adapted from [21])

Table 6.8 Normalized atom positions in the tetragonal unit cell (I4/mmm) of Tl-2212 (the lattice parameters are $a \cong 0.39$ nm and $c \cong 2.94$ nm (data from [7]))

Atoms	x	y	Z	Comments
Tl	0.5	0.5	0.21	Centered atoms[a]
Ba	0	0	0.12	Edge atoms[a]
Cu	0.5	0.5	0.05	Centered atoms[a]
Ca	0	0	0	Edge atoms[a]
O1	0	0.5	0.05	Face atoms
O2	0.5	0.5	0.15	Centered atoms[a]
O3	0	0	0.22	Edge atoms[a]

[a] Atoms at centered (edge) positions in z planes move to edge (centered) positions in $z' = z \pm 0.5$ planes.

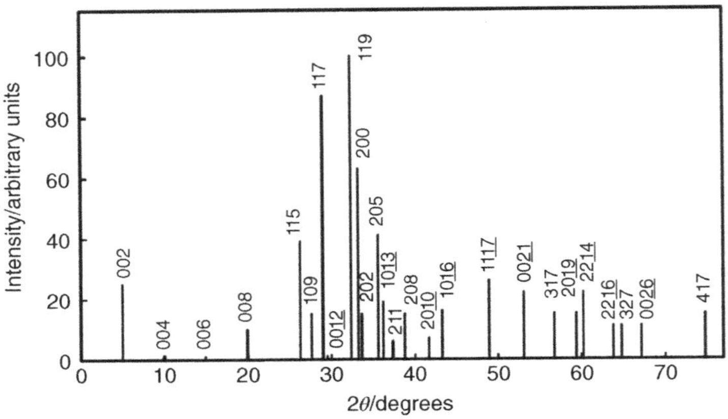

Figure 6.17 Sketch of the typical Cu Kα X-ray diffraction pattern of Tl-2223 (data from [20]). The Miller indices are based on a tetragonal unit cell (a = b = 0.540 nm, c = 3.625 nm)

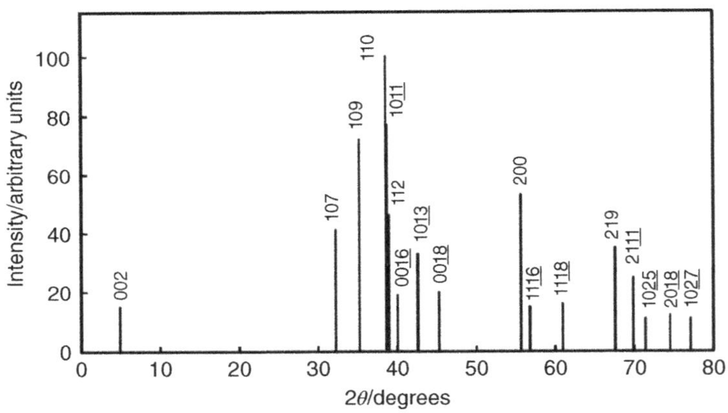

Figure 6.18 Sketch of the typical Co Kα X-ray diffraction pattern of Tl-2234 (data from [21]). The Miller indices are based on a tetragonal unit cell (a = 0.3851 nm and c = 4.201 nm)

powder diffraction pattern of Tl-2234. The measurements have been performed using Co Kα radiation ($\lambda = 0.1798$ nm). The Miller indices are based on a tetragonal body-centered unit cell (space group *I4/mmm*) with the lattice parameters $a = 0.3851$ nm and $c = 4.201$ nm [21].

In Figure 6.19, the tetragonal unit cells of Tl-2223 ($a = b = 0.382$ nm, $c = 3.625$ nm) and Tl-2234 ($a = b = 0.385$ nm and $c = 4.201$ nm) are compared. The sequence of the layers in the charge carrier reservoirs of Tl-2223 and Tl-2234 is BaO–TlO–TlO–BaO, which is identical in all Tl-22($n-1$)n compounds (see Figures 6.14 and 6.16). However, the number n of CuO_2 sheets per copper oxide block increases from 2 in the Tl-2212 to 3 in Tl-2223 and 4 in Tl-2234. In all Tl-22($n-1$)n compounds with $n \geq 2$, the CuO_2 sheets in an individual copper oxide block are separated by a layer of Ca ions. In principle, the insertion of an additional (CuO_2–Ca) block leads from Tl-22($n-1$)n to the next member Tl-22n($n+1$) of the Tl cuprate family.

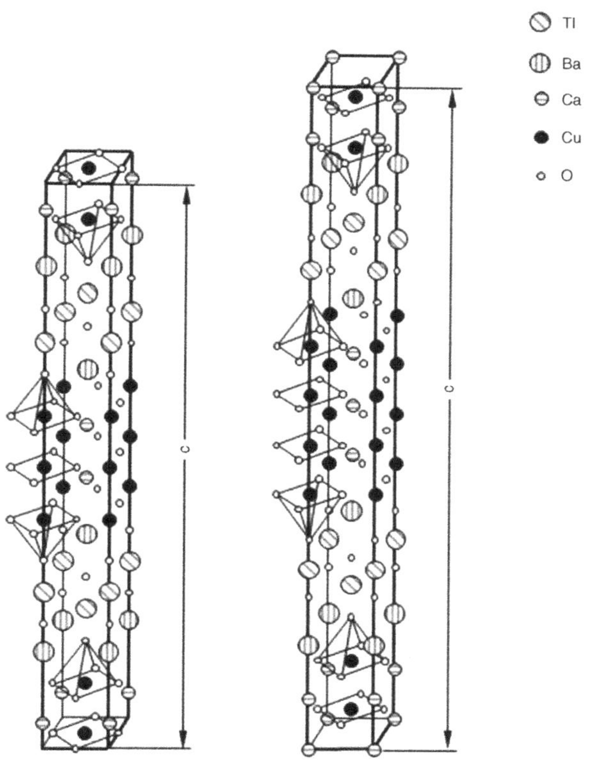

Figure 6.19 *Body-centered tetragonal unit cells (I4/mmm) of Tl-2223 (a ≅ 0.38 nm, c ≅ 3.63 nm) and Tl-2234 (a ≅ 0.39 nm, c ≅ 4.20 nm) (adapted from [22])*

The normalized atom positions in the unit cells of Tl-2223 and Tl-2234 are listed in Tables 6.9 and 6.10 [23–24]. For both compounds, the atom positions are based on a body-centered tetragonal unit cell (*I4/mmm*).

As an example of the second family of thallium-based cuprate superconductors, we consider $TlBa_2Ca_2Cu_3O_9$. A sketch of the X-ray powder diffraction pattern of Tl-1223 is shown in Figure 6.20 [25]. The crystal structure of Tl-1223 belongs to the space group *P4/mmm* [22]. The primitive tetragonal unit cell of Tl-1223 is shown in Figure 6.21. Note that in the primitive tetragonal unit cell, there is no exchange of edge and center atoms by changing the z position from z to $z' = z \pm 0.5$. As in the Tl-2223 counterpart, the copper oxide blocks consist of three CuO_2 sheets. Neighboring CuO_2 planes are separated by a layer of calcium ions. In the central CuO_2 sheet, each of the copper atoms is surrounded by four oxygen atoms, while in the outer CuO_2 planes, the copper atoms are in the center of the basal plane of an oxygen tetrahedron. A common feature of the Tl-12($n-1$)n compounds is a charge carrier reservoir containing a single TlO layer. The sequence of the layers in the charge carrier reservoirs is BaO–TlO–BaO. The Tl-1212 compound contains one (CuO_2–Ca) unit less than Tl-1223. The normalized atom positions in the primitive tetragonal unit cell are listed in Table 6.11.

Table 6.9 *Normalized atom positions in the tetragonal unit cell (I4/mmm) of Tl-2223 (the lattice parameters are a = 0.3845 nm and c ≅ 3.567 nm (data from [23]))*

Atoms	x	y	z	Comments
Tl	0	0	0.22	Edge atoms[a]
Ba	0.5	0.5	0.144	Centered atoms[a]
Cu1	0	0	0	Edge atoms[a]
Cu2	0	0	0.089	Edge atoms
Ca	0.5	0.5	0.05	Centered atoms[a]
O1	0	0.5	0	Face atoms
O2	0	0.5	0.088	Face atoms
O3	0	0	0.166	Edge atoms[a]
O4	0.5	0.5	0.22	Centered atoms[a]

[a]Atoms at centered (edge) positions in z planes move to edge (centered) positions in $z' = z \pm 0.5$ planes.

Table 6.10 *Normalized atom positions in the tetragonal unit cell (I4/mmm) of Tl-2234 (the lattice parameters are a = 0.3850 nm and c = 4.1984 nm (data from [24]))*

Atoms	x	y	z	Comments
Tl	0.5	0.5	0.226	Centered atoms[a]
Ba	0	0	0.16	Edge atoms[a]
Cu1	0.5	0.5	0.038	Centered atoms[a]
Cu2	0.5	0.5	0.113	Edge atoms[a]
Ca1	0	0	0	Edge atoms[a]
Ca2	0	0	0.077	Edge atoms[a]
O1	0.5	0	0.038	Face atoms
O2	0.5	0	0.113	Face atoms
O3	0.5	0.5	0.178	Centered atoms[a]
O4	0.5	0.5	0.268	Centered atoms[a]

[a]Atoms at centered (edge) positions in z planes move to edge (centered) positions in $z' = z \pm 0.5$ planes.

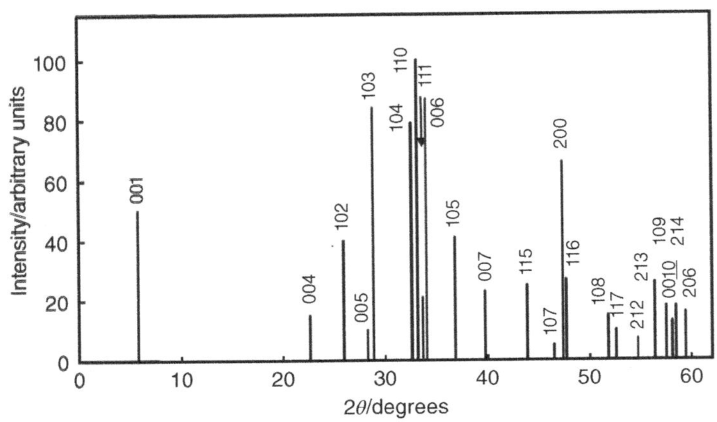

Figure 6.20 *Sketch of the X-ray powder diffraction pattern of Tl-1223 (data from [25])*

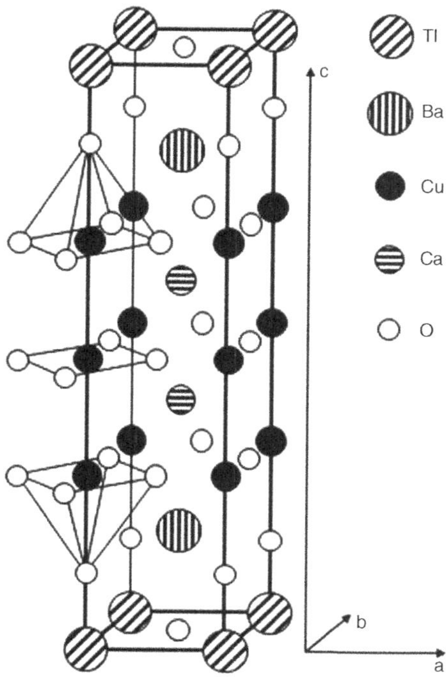

Figure 6.21 Primitive tetragonal unit cell of Tl-1223 (adapted from [12])

Table 6.11 Normalized atom positions in the primitive tetragonal unit cell (P4/mmm) of Tl-1223 (in the primitive unit cell edge and center, atoms do not exchange their positions in different planes along the c direction. The lattice parameters are a = b = 0.384 nm and c = 1.587 nm (data from [22]))

Atoms	x	y	z	Comments
Tl	0	0	0	Edge atoms
Ba	0.5	0.5	0.176	Centered atoms
Cu1	0	0	0.5	Edge atoms
Cu2	0	0	0.302	Edge atoms
Ca	0.5	0.5	0.397	Centered atoms
O1	0.5	0	0.5	Face atoms
O2	0.5	0	0.304	Face atoms
O3	0	0	0.132	Edge atoms
O4	0.5	0.5	0	Centered atoms

6.3.5 The Crystal Structures of Hg-based High-Temperature Superconductors

The mercury-based cuprates $HgBa_2Ca_{n-1}Cu_nO_{2n+2}$ are structurally closely related to the Tl-12(n – 1)n compounds. They are distinguished by the replacement of the TlO layers by HgO layers. A sketch of the X-ray powder diffraction pattern of $HgBa_2CuO_4$ is shown in Figure 6.22. The Miller indices are based on a primitive tetragonal unit cell with the lattice

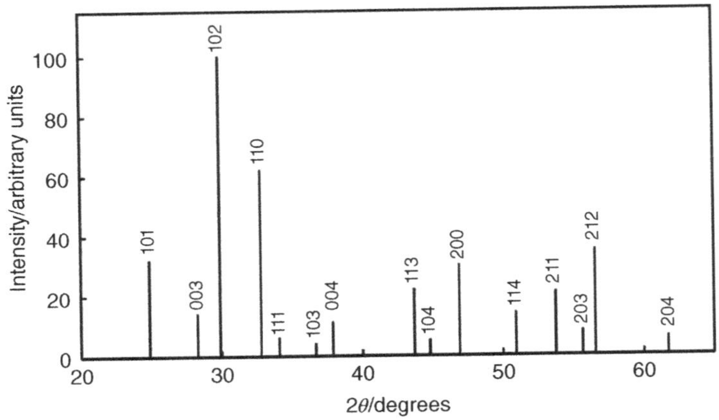

Figure 6.22 Sketch of the X-ray powder diffraction pattern of Hg-1201. The Miller indices are based on a primitive tetragonal unit cell with the lattice parameters a = 0.388 nm and c = 0.953 nm (data from [26])

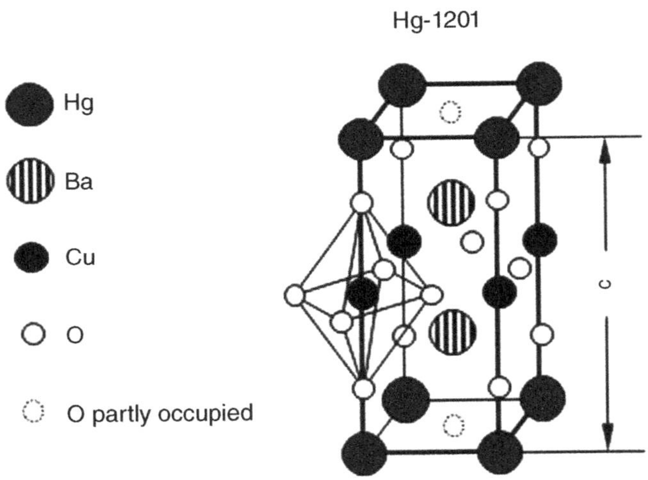

Figure 6.23 Primitive tetragonal unit cell of Hg-1201 (adapted from [12])

parameters $a = 0.388$ nm and $c = 0.953$ nm [26]. Hg-1201 crystallizes with the symmetry of the space group *P4/mmm* [27–29].

The primitive tetragonal unit cell of Hg-1201 [27] is illustrated in Figure 6.23. In the crystal structure of $HgBa_2CuO_4$, single CuO_2 layers are sandwiched between charge carrier reservoirs formed of a BaO–HgO–BaO block. Only 6.3% of the oxygen sites in the HgO layers are occupied [28]. The copper atoms sit in the center of an oxygen octahedron. The normalized atom positions in the primitive tetragonal unit cell are listed in Table 6.12.

Sketches of the X-ray powder diffraction patterns of Hg-1212 [30, 31] and Hg-1223 [32] are presented in Figures 6.24 and 6.25. All compounds of the mercury-based high-T_c superconductor family Hg-12$(n-1)n$ crystallize with the symmetry of the space group

Table 6.12 Normalized atom positions in the primitive tetragonal unit cell (P4/mmm) of Hg-1201 (the lattice parameters are $a=b=0.388$ nm and $c=0.951$ nm (data from [28]))

Atoms	x	y	z	Comments
Hg	0	0	0	Edge atoms
Ba	0.5	0.5	0.299	Centered atoms
Cu	0	0	0.5	Edge atoms
O1	0.5	0	0.5	Face atoms
O2	0	0	0.207	Edge atoms
O3	0.5	0.5	0	Centered atoms

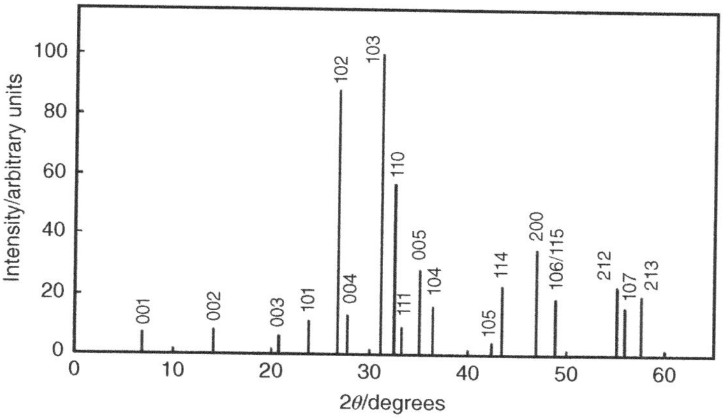

Figure 6.24 Sketch of the X-ray powder diffraction pattern of Hg-1212 (Cu Kα radiation). The Miller indices are based on a primitive tetragonal unit cell with the lattice parameters $a = 0.3862$ nm and $c = 1.2707$ nm (data from [30])

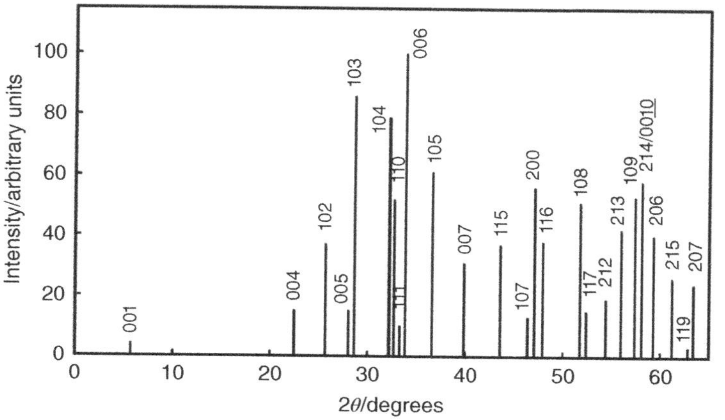

Figure 6.25 Sketch of the X-ray powder diffraction pattern of Hg-1223 (Cu Kα radiation). The Miller indices are based on a primitive tetragonal unit cell with the lattice parameters $a = 0.3852$ nm and $c = 1.583$ nm (data from [32])

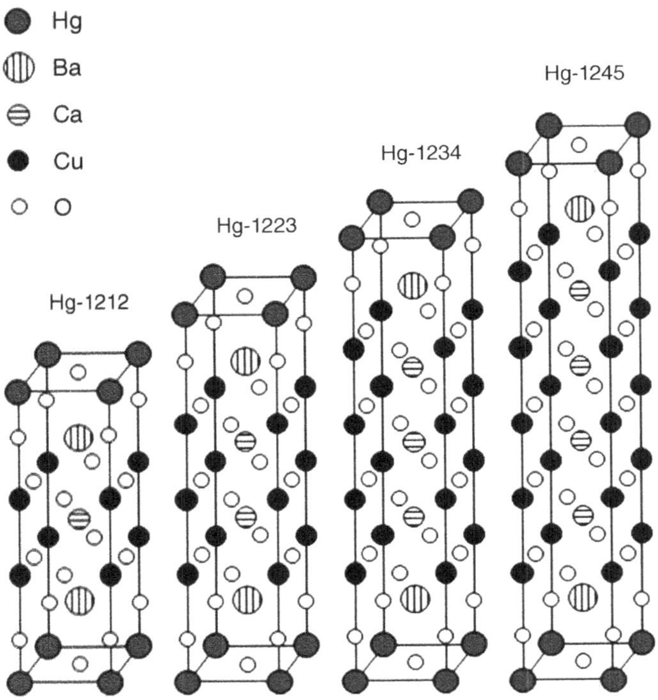

Figure 6.26 *Comparison of the primitive tetragonal unit cells of the Hg-12(n − 1)n cuprate superconductors with n = 2, 3, 4, and 5 (adapted from [12, 33])*

P4/mmm. In this cuprate family, compounds with more than five CuO_2 sheets per copper oxide block have been successfully synthesized.

The crystal structures of the Hg-12($n-1$)n compounds for n in the range of 2–5 are compared in Figure 6.26. In each of these compounds, neighboring CuO_2 sheets within a single copper oxide block are separated by layers of Ca ions. The sequence of the layers in the charge carrier reservoirs is BaO–HgO–BaO, as in the single-layer compound Hg-1201. The normalized atom positions in the primitive tetragonal unit cells of Hg-1212 [33], Hg-1223 [33], and Hg-1234 [34] are provided in Tables 6.13–6.15, respectively.

6.3.6 Lattice Parameters of Cuprate Superconductors

Most of the cuprate high-T_c superconductors discussed here crystallize with an orthorhombic or tetragonal crystal structure [35–45]. In the former case, the difference between the lattice parameters a and b is very small. Because of the common structural feature of the CuO_2 planes, the lattice parameters a and b are both close to 0.38 nm. In most of the cuprate high-T_c superconductor families, the number of CuO_2 sheets within the CuO_2 blocks can be varied from 1 to n, leading to a systematic increase of the c lattice parameter with increasing n. In general, the lattice parameters of cuprate superconductors are intimately connected with the exact chemical composition. The lattice parameters of a particular cuprate superconductor therefore vary depending on the different synthesis conditions.

Table 6.13 *Normalized atom positions in the primitive tetragonal unit cell (P4/mmm) of Hg-1212 (the lattice parameters are a = b = 0.385 nm and c = 1.275 nm (data from [33]))*

Atoms	*x*	*y*	*z*	Comments
Hg	0	0	0	Edge atoms
Ba	0.5	0.5	0.22	Centered atoms
Cu	0	0	0.375	Edge atoms
Ca	0.5	0.5	0.5	Centered atom
O1	0	0.5	0.38	Face atoms
O2	0	0	0.154	Edge atoms
O3[a]	0.5	0.5	0	Centered atoms

[a]Only 6% of O3 sites are occupied.

Table 6.14 *Normalized atom positions in the primitive tetragonal unit cell (P4/mmm) of Hg-1223 (the lattice parameters are a = b = 0.385 nm and c = 1.595 nm (data from [33]))*

Atoms	*x*	*y*	*z*	Comments
Hg	0	0	0	Edge atoms
Ba	0.5	0.5	0.18	Centered atoms
Cu1	0	0	0.5	Edge atoms
Cu2	0	0	0.302	Edge atoms
Ca	0.5	0.5	0.4	Centered atom
O1	0	0.5	0.5	Face atoms
O2	0	0.5	0.304	Face atoms
O3	0.5	0.5	0	Centered atoms
O4[a]	0	0	0.124	Edge atoms

[a]Only 6% of O4 sites are occupied.

Table 6.15 *Normalized atom positions in the primitive tetragonal unit cell (P4/mmm) of Hg-1234 (the lattice parameters are a = 0.385 nm and c = 1.894 nm (data from [34]))*

Atoms	*x*	*y*	*z*	Comments
Hg[a]	0	0	0	Edge atoms
Ba	0.5	0.5	0.144	Centered atoms
Cu1	0	0	0.248	Edge atoms
Cu2	0	0	0.415	Edge atoms
Ca1	0.5	0.5	0.329	Centered atoms
Ca2	0.5	0.5	0.5	Centered atoms
O1	0	0.5	0.251	Face atoms
O2	0	0.5	0.414	Face atoms
O3	0	0	0.091	Edge atoms
O4[b]	0.5	0.5	0	Centered atoms

[a]Only 79% of the Hg sites are occupied.
[b]Only 60% of O4 sites are occupied.

Table 6.16 *Lattice parameters of various cuprate high-T_c superconductors*

Compound	Symmetry	*a* (nm)	*b* (nm)	*c* (nm)	T_c (K)
$La_{1.9}Sr_{0.1}CuO_4$	Tetragonal	0.3784	0.3784	1.3211	33 [36]
$La_{1.875}Sr_{0.125}CuO_4$	Tetragonal	0.3778	0.3778	1.3216	36 [36]
$La_{1.85}Sr_{0.15}CuO_4$	Tetragonal	0.3779	0.3779	1.3200	36 [6]
$La_{1.85}Sr_{0.15}CuO_4$	Tetragonal	0.3779	0.3779	1.323	39 [36]
$La_{1.775}Sr_{0.225}CuO_4$	Tetragonal	0.3771	0.3771	1.3247	29 [36]
$La_{1.725}Sr_{0.275}CuO_4$	Tetragonal	0.3767	0.3767	1.3225	22 [36]
$Nd_{1.85}Ce_{0.15}CuO_{3.93}$	Tetragonal	0.395	0.395	1.207	24 [12]
$YBa_2Cu_3O_{6.67}$	Orthorhombic	0.3831	0.3889	1.1736	60 [36]
$YBa_2Cu_3O_{6.9}$	Orthorhombic	0.3822	0.3891	1.1677	91 [8]
$YBa_2(Cu_{0.975}Zn_{0.025})_3O_7$	Orthorhombic	0.3820	0.3890	1.1673	62 [36]
$YBa_2(Cu_{0.95}Zn_{0.05})_3O_7$	Orthorhombic	0.3820	0.3885	1.1671	36 [36]
$HoBa_2Cu_3O_7$	Orthorhombic	0.3846	0.3881	1.1640	92 [36]
$YBa_2Cu_4O_8$	Orthorhombic	0.3839	0.3869	2.7243	80 [12]
$HoBa_2Cu_4O_8$	Orthorhombic	0.3855	0.3874	2.7295	80 [36]
$Bi_2Sr_2CuO_{6+\delta}$	Orthorhombic	0.5361	0.5370	2.4369	8.5 [36]
$Bi_2(Sr_{1.6}La_{0.4})CuO_{6+\delta}$	Orthorhombic	0.5370	0.5400	2.450	23.5 [36]
$Bi_{2.2}Sr_2Ca_{0.8}Cu_2O_{8+\delta}$	Orthorhombic	0.5414	0.5418	3.089	84 [17]
$Bi_2Sr_2CaCu_2O_8$	Orthorhombic	0.5413	0.5411	3.091	89 [36]
$Bi_2Sr_2CaCu_2O_{8+\delta}$	Orthorhombic	0.5408	0.5413	3.081	75 [36]
$Bi_2Sr_2CaCu_2O_{8+\delta}$	Orthorhombic	0.5407	0.5415	3.0874	83 [37]
$Bi_2Sr_{1.5}Ca_{1.5}Cu_2O_{8.19}$	Orthorhombic	0.5402	0.5408	3.0699	83 [37]
$Bi_2Sr_{1.5}Ca_{1.5}Cu_2O_{8.22}$	Orthorhombic	0.5396	0.5400	3.0635	73 [37]
$Bi_2Sr_{1.5}Ca_{1.5}Cu_2O_{8.22}$	Orthorhombic	0.5396	0.5404	3.0618	67 [37]
$Bi_2Sr_2Ca_{0.9}Y_{0.1}Cu_2O_{8.23}$	Orthorhombic	0.5408	0.5416	3.0807	90 [37]
$Bi_2Sr_2Ca_{0.7}Y_{0.3}Cu_2O_{8.30}$	Orthorhombic	0.5410	0.5425	3.0619	86 [37]
$Bi_2Sr_2Ca_2Cu_3O_{10+\delta}$	Orthorhombic	0.541	0.539	3.82	115 [18]
$Bi_{1.7}Pb_{0.25}Sr_{1.47}Ca_{2.54}Cu_{3.1}O_{10+\delta}$	Orthorhombic	0.5411	0.5411	3.707	105 [16]
$(Bi,Pb)_2Sr_{1.72}Ca_2Cu_3O_{10+\delta}$	Orthorhombic	0.5392	0.5395	3.6985	111 [38]
$(Bi_{1.6}Pb_{0.4})Sr_2Ca_2Cu_3O_{10}$	Orthorhombic	0.5413	0.5413	3.7100	107 [36]
$Tl_2Ba_2CuO_6$	Tetragonal	0.3866	0.3866	2.3239	85 [36]
$Tl_2Ba_2CuO_{6+\delta}$	Orthorhombic	0.5473	0.5483	2.3277	93 [19]
$Tl_{1.7}Ba_2Ca_{1.06}Cu_{2.32}O_{8+\delta}$	Tetragonal	0.3857	0.3857	2.939	108 [12]
$Tl_2Ba_2CaCu_2O_8$	Tetragonal	0.3855	0.3855	2.9318	99 [36]
$Tl_2Ba_2Ca_2Cu_3O_{10}$	Tetragonal	0.3850	0.3850	3.588	125 [36]
$Tl_{1.64}Ba_2Ca_{1.87}Cu_{3.11}O_{10+\delta}$	Tetragonal	0.3822	0.3822	3.626	125 [12]
$Tl_2Ba_2Ca_3Cu_4O_{12+\delta}$	Tetragonal	0.3850	0.3850	4.1984	114 [24]
$Tl_{1.1}Ba_2Ca_{0.9}Cu_{2.1}O_{7.1}$	Tetragonal	0.3851	0.3851	1.2728	80 [39]
$Tl_{1.1}Ba_2Ca_{1.8}Cu_{3.0}O_{9.7}$	Tetragonal	0.3843	0.3843	1.5871	110 [39]
$TlBa_2Ca_3Cu_4O_{12+\delta}$	Tetragonal	0.3848	0.3848	1.9001	114 [40]
$(Tl,Cr)Sr_2(Ca,Tl)Cu_2O_7$	Tetragonal	0.3816	0.3816	1.2022	100 [41]
$(Tl_{0.5}Pb_{0.5})Sr_2CaCu_2O_7$	Tetragonal	0.3802	0.3802	1.2107	80 [36]
$(Tl_{0.5}Pb_{0.5})Sr_2Ca_2Cu_3O_9$	Tetragonal	0.3821	0.3821	1.5294	122 [36]
$HgBa_2CuO_{4+\delta}$	Tetragonal	0.380	0.380	0.9509	94 [42]
$HgBa_2CaCu_2O_{6+\delta}$	Tetragonal	0.3859	0.3859	1.2657	123 [43]
$HgBa_2CaCu_2O_{6+\delta}$	Tetragonal	0.3856	0.3856	1.2652	121 [35]
$HgBa_2Ca_2Cu_3O_{8+\delta}$	Tetragonal	0.3853	0.3853	1.5818	133 [35]

Table 6.16 *(continued)*

Compound	Symmetry	*a* (nm)	*b* (nm)	*c* (nm)	T_c (K)
$HgBa_2Ca_2Cu_3O_{8+\delta}$	Tetragonal	0.3855	0.3855	1.5842	110 [44]
$HgBa_2Ca_3Cu_4O_{10+\delta}$	Tetragonal	0.3854	0.3854	1.9006	126 [35]
$HgBa_2Ca_4Cu_5O_{12+\delta}$	Tetragonal	0.3852	0.3852	2.2141	110 [45]
$HgBa_2Ca_5Cu_6O_{14+\delta}$	Tetragonal	0.3852	0.3852	2.526	107 [45]
$(Hg_{0.9}Pb_{0.1})Ba_2Ca_2Cu_3O_{8+\delta}$	Tetragonal	0.3850	0.3850	1.5836	118 [44]
$(Hg_{0.8}Pb_{0.2})Ba_2Ca_2Cu_3O_{8+\delta}$	Tetragonal	0.3848	0.3848	1.5831	125 [44]
$(Hg_{0.7}Pb_{0.3})Ba_2Ca_2Cu_3O_{8+\delta}$	Tetragonal	0.3843	0.3843	1.5824	130 [44]
$(Hg_{0.6}Pb_{0.4})Ba_2Ca_2Cu_3O_{8+\delta}$	Tetragonal	0.3842	0.3842	1.5818	120 [44]
$(Hg_{0.5}Pb_{0.5})Ba_2Ca_2Cu_3O_{8+\delta}$	Tetragonal	0.3840	0.3840	1.5822	103 [44]
$(Hg_{0.5}Pb_{0.5})Ba_2Ca_4Cu_5O_{12+\delta}$	Tetragonal	0.3853	0.3853	2.2172	115 [12]
$(Tl_{0.5}Pb_{0.5})Sr_2Ca_2Cu_3O_9$	Tetragonal	0.3815	0.3815	1.5280	118 [12]
$BSr_2Ca_2Cu_3O_{9+\delta}$	Tetragonal	0.3821	0.3821	1.3854	75 [12]
$BSr_2Ca_3Cu_4O_{11+\delta}$	Tetragonal	0.3836	0.3836	1.7082	110 [12]
$BSr_2Ca_4Cu_5O_{13+\delta}$	Tetragonal	0.3837	0.3837	2.022	85 [12]
$AlSr_2Ca_{1.5}Y_{0.5}Cu_3O_{9+\delta}$	Tetragonal	0.3836	0.3836	1.4405	78 [12]
$AlSr_2Ca_3Cu_4O_{11+\delta}$	Tetragonal	0.3839	0.3839	1.772	110 [12]
$AlSr_2Ca_4Cu_5O_{13+\delta}$	Tetragonal	0.3845	0.3845	2.087	83 [12]

The c lattice parameters of the different members of the Tl-22($n-1$)n family are approximately given by the formula $c \cong (1.7 + 0.62\,n)$ nm [7]. The crystal structures of all Tl-22($n-1$)n cuprate high-T_c superconductors are close to an ideal tetragonal structure (symmetry group *I4/mmm*). The structures of the Tl-22($n-1$)n compounds are closely related to those of their Bi-22($n-1$)n counterparts. However, the structural modulations generally found in the bismuth-based compounds are missing.

The crystal structures of Tl-12($n-1$)n compounds are primitive tetragonal (space group *P4/mmm*). Their lattice parameters are close to $a \cong 0.38$ nm and $c \cong (0.63 + 0.32\,n)$ nm. Their mercury-based counterparts Hg-12($n-1$)n also crystallize with the symmetry of the space group *P4/mmm*. The lattice parameters of the mercury-based compounds are $a \cong$ 0.39 nm and $c \cong (0.95 + 0.32(n-1))$ nm [35].

A further aspect of importance is the systematically different distance of adjacent CuO_2 blocks. In the Bi-22($n-1$)n and Tl-22($n-1$)n compounds, this distance is relatively large as compared to Tl-12($n-1$)n and Hg-12($n-1$)n high-T_c superconductors. For example, the distance between the CuO_2 planes in Tl-2201 is 1.16 nm, but only 0.95 nm in Hg-1201. The large distance between adjacent CuO_2 blocks in the Bi-22($n-1$)n and Tl-22($n-1$)n compounds seems to cause a weak superconducting coupling of adjacent CuO_2 blocks, which leads to a pronounced anisotropy of the superconducting properties.

The lattice parameters of various cuprate high-T_c superconductors are listed in Table 6.16. The lattice parameters depend, like many other physical properties, on the exact oxygen content, and hence on the synthesis conditions.

References

1. W.H. Zachariasen, *Theory of X-ray Diffraction in Crystals*, Dover Publications, Inc., New York, 1994.

2. T. Hahn (ed.), *International Tables for Crystallography – Brief Teaching Edition of Volume A: Space-group Symmetry*, D. Reidel Publishing Company, Dordrecht, Boston, Lancaster, Tokyo, 1988.
3. Ch. Kittel, *Introduction to Solid State Physics*, John Wiley & Sons, Inc., New York, 1976.
4. G. Schatz and A. Weidinger, *Nuclear condensed matter physics*, John Wiley & Sons Ltd, Chichester, England, 1996.
5. G. Rietveld, M. Glastra, and D. van de Marel, Doping dependence of the chemical potential in cuprate high-T_c superconductors I. $La_{2-x}Sr_xCuO_4$, *Physica C*, **241**, 257–272 (1995).
6. M. Decroux, A. Junod, A. Bezinge, D. Cattani, J. Cors, J.L. Jorda, A. Stettler, M. François, K. Yvon, Ø. Fischer, and J. Muller, Structure, resistivity, critical field, specific heat jump at T_c, Meissner effect, a.c. and d.c. susceptibility of the high-temperature superconductor $La_{2-x}Sr_xCuO_4$, *Europhys. Lett.*, **3**, 1035–1040 (1987).
7. R.M. Hazen, Crystal structures of high-temperature superconductors, in *Physical Properties of High Temperature Superconductors II*, D.M. Ginsberg (ed.), World Scientific, Singapore, 1990.
8. R.J. Cava, B. Batlogg, R.B. van Dover, D.W. Murphy, S. Sunshine, T. Siegrist, J.P. Remeika, E.A. Rietman, S. Zahurak, and G.P. Espinosa, Bulk superconductivity at 91 K in single-phase oxygen-deficient perovskite $Ba_2YCu_3O_{9-\delta}$, *Phys. Rev. Lett.*, **58**, 1676–1679 (1987).
9. R. Beyers, G. Lim, E.M. Engler, R.J. Savoy, T.M. Shaw, T.R. Dinger, W.J. Gallagher, and R.L. Sandstrom, Crystallography and microstructure of $Y_1Ba_2Cu_3O_{9-x}$, a perovskite-based superconducting oxide, *Appl. Phys. Lett.*, **50**, 1918–1920 (1987).
10. M.A. Beno, L. Soderholm, D.W. Capone, D.G. Hinks, J.D. Jorgensen, J.D. Grace, I.K. Schuller, C.U. Segre, and K. Zhang, Structure of the single-phase high-temperature superconductor $YBa_2Cu_3O_{7-\delta}$, *Appl. Phys. Lett.*, **51**, 57–59 (1987).
11. J.J. Capponi, C. Chaillout, A.W. Hewat, P. Lejay, M. Marezio, N. Nguyen, B. Raveau, J.L. Soubeyroux, J.L. Tholence, and R. Tournier, Structure of the 100 K superconductor $Ba_2YCu_3O_7$ between (5–300) K by neutron powder diffraction, *Europhys. Lett.*, **3**, 1301–1307 (1987).
12. R. Wesche, *High-temperature Superconductors: Materials, Properties and Applications*, Kluwer Academic Publishers, Norwell, MA, USA,1998.
13. A. Junod, A. Bezinge, T. Graf, J.L. Jorda, J. Muller, L. Antognazza, D. Cattani, J. Cors, M. Decroux, Ø. Fischer, M. Banovski, P. Genoud, L. Hoffmann, A.A. Manuel, M. Peter, E. Walker, M. François, and K. Yvon, Structure, resistivity, critical field, specific-heat jump at T_c, ac and dc susceptibility of high-T_c superconductor $YBa_2Cu_3O_7$, *Europhys. Lett.*, **4**, 247–252 (1987).
14. K.M. Stephens, D.A. Robinson, A. Alvanipour, W.S. Hinton, and A. Morrobel-Sosa, A triphenyl bismuth-based solution route to Bi:Sr:Ca:Cu:O superconductors, *Physica C*, **168**, 351–358 (1990).
15. S.C. Bhargava, J.S. Chakrabarty, R. Sharma, C.V. Tomy, and S.K. Malik, Preparation of 2212 phase of Bi superconductor with T_c of 92 K, *Solid St. Commun.*, **78**, 397–401 (1991).

16. G. Calestani, C. Rizzoli, G.D. Andreetti, E. Buluggiu, D.C. Giori, A. Valenti, A. Vera, and G. Amoretti, Composition effects on the formation and superconducting character of $c \approx 31$ Å and $c \approx 37$ Å phases in the Bi–Sr–Ca–Cu–O and Bi–Pb–Sr–Ca–Cu–O systems. X-ray and ESR analysis, *Physica C*, **158**, 217–224 (1989).
17. S.A. Sunshine, T. Siegrist, L.F. Schneemayer, D.W. Murphy, R.J. Cava, B. Batlogg, R.B. van Dover, R.M. Fleming, S.H. Glarun, S. Nakahara, R. Farrow, J.J. Krajewski, S.M. Zahurak, J.V. Wasczczak, J.H. Marshall, P. Marsh, L.W. Rupp, Jr., and F.W. Peck, Structure and physical properties of single crystals of the 84-K superconductor $Bi_{2.2}Sr_2Ca_{0.8}Cu_2O_{8+\delta}$, *Phys. Rev. B*, **38**, 893–896 (1988).
18. C.J.D. Hetherington, R. Ramesh, M.A. O'Keefe, R. Kilaas, G. Thomas, S.M. Green, and H.L. Luo, *Appl. Phys. Lett.*, **53**, 1016–1018 (1988).
19. J.L. Wagner, O. Chmaissem, J.D. Jorgensen, D.G. Hinks, P.G. Radaelli, B.A. Hunter, and W.R. Jensen, Multiple defects in overdoped $Tl_2Ba_2CuO_{6+\delta}$: Effects on structure and superconductivity, *Physica C*, **277**, 170–182 (1997).
20. R.M. Hazen, L.W. Finger, R.J. Angel, C.T. Prewitt, N.L. Ross, C.G. Hadidiacos, P.J. Heaney, D.R. Veblen, Z.Z. Sheng, A. El Ali, and A.M. Hermann, 100-K superconducting phases in the Tl–Ca–Ba–Cu–O system, *Phys. Rev. Lett.*, **60**, 1657–1660 (1988).
21. M.R. Presland, J.L. Tallon, P.W. Gilberd, and R.S. Liu, Bulk single-superconducting-phase thallium "2234" superconductor – $Tl_{2-x}Ba_2Ca_{3+x}Cu_4O_{12-\delta}$, *Physica C*, **191**, 307–315 (1992).
22. S.S.P. Parkin, V.Y. Lee, A.I. Nazzal, R. Savoy, R. Beyers, and S.J. La Placa, $Tl_1Ca_{n-1}Ba_2Cu_nO_{2n+3}$ ($n = 1, 2, 3$): A new class of crystal structures exhibiting volume superconductivity at up to ≅110 K, *Phys. Rev. Lett.*, **61**, 750–753 (1988).
23. T. Kajitani, K. Hiraga, S. Nakajima, M. Kikuchi, Y. Syono, and C. Kabuto, X-ray diffraction analysis on 2223 Tl-oxide single crystals, *Physica C*, **161**, 483–492 (1989).
24. D.M. Ogborne and M.T. Weller, Structure and oxygen stoichiometry in $Tl_2Ba_2Ca_3Cu_4O_{12-\delta}$ – A high-resolution powder neutron-diffraction study, *Physica C*, **223**, 283–290 (1994).
25. S. Mikusu, N. Urita, Y. Hashinaka, K. Tokiwa, A. Iyo, Y. Tanaka, and T. Watanabe, Transport properties of $TlBa_2Ca_2Cu_3O_y$ in an over-doped state, *Physica C*, **442**, 91–96 (2006).
26. M. Itoh, A. Tokiwa-Yamamoto, S. Adachi, and H. Yamauchi, Increase of T_c in $HgBa_2CuO_y$ by reducing treatment, *Physica C*, **212**, 271–273 (1993).
27. S.M. Loureiro, E.T. Alexandre, E.V. Antipov, J.J. Capponi, S. de Brion, B. Souletie, J.L. Tholence, M. Marezio, Q. Huang, and A. Santoro, Suppression of superconductivity and the overdoped region in $HgBa_2CuO_{4+\delta}$, *Physica C*, **243**, 1–9 (1995).
28. H.R. Khan, O. Loebich, P. Fabbricatore, A. Sciutti, and B. Zhang, Synthesis, X-ray diffraction and AC magnetic susceptibility investigations of $HgBa_2CuO_{4+\delta}$, *Physica C*, **229**, 165–168 (1994).
29. O. Chmaissem, Q. Huang, S.N. Putilin, M. Marezio, and A. Santoro, Neutron powder diffraction study of the crystal structures of $HgBa_2CuO_{4+\delta}$ and $HgBaO_2$, *Physica C*, **212**, 259–265 (1993).
30. R.L. Meng, L. Beauvais, X.N. Zhang, Z.J. Huang, Y.Y. Sun, Y.Y. Xue, and C.W. Chu, Synthesis of the high-temperature superconductors $HgBa_2CaCu_2CuO_{6+\delta}$ and $HgBa_2Ca_2Cu_3O_{8+\delta}$, *Physica C*, **216**, 21–28 (1993).

31. R.L. Meng, Y.Y. Sun, J. Kulik, Z.J. Huang, F. Chen, Y.Y. Xue, and C.W. Chu, Superconductivity at 112–117 K in $HgBa_2CaCu_2O_{6+\delta}$, *Physica C*, **214**, 307–312 (1993).
32. M. Paranthaman, Single-step synthesis of bulk $HgBa_2Ca_2Cu_3O_{8+\delta}$, *Physica C*, **222**, 7–12 (1994).
33. M. Cantoni, A. Schilling, H.-U. Nissen, and H.R. Ott, Characterisation of superconducting Hg–Ba–Ca–Cu oxides – Structural and physical aspects, *Physica C*, **215**, 11–18 (1993).
34. S.M. Loureiro, E.V. Antipov, E.M. Kopnin, M. Brunner, J.J. Capponi, and M. Marezio, Structure and superconductivity of the $HgBa_2Ca_3Cu_4O_{10+\delta}$ phase, *Physica C*, **257**, 117–124 (1996).
35. E.V. Antipov, S.M. Loureiro, C. Chaillout, J.J. Capponi, P. Bordet, J.L. Tholence, S.N. Putilin, and M. Marezio, The synthesis and characterization of the $HgBa_2Ca_2Cu_3O_{8+\delta}$ and $HgBa_2Ca_3Cu_4O_{10+\delta}$, *Physica C*, **215**, 1–10 (1993).
36. D.R. Harshman and A.P. Mills, Jr., Concerning the nature of high-T_c superconductivity: Survey of experimental properties and implications for interlayer coupling, *Phys. Rev. B*, **45**, 10684–10712 (1992).
37. W.A. Groen, D.M. de Leeuw, and L.F. Feiner, Hole concentration and T_c in $Bi_2Sr_2CaCu_2O_{8+\delta}$, *Physica C*, **165**, 55–61 (1990).
38. W. Carrilo-Cabrera and W. Göpel, Influence of high-temperature annealing on the $(Bi,Pb)_2Sr_2Ca_2Cu_3O_{10}$ phase and determination of its crystal structure by X-ray powder diffractometry, *Physica C*, **161**, 373–389 (1989).
39. R. Beyers, S.S.P. Parkin, V.Y. Lee, A.I. Nazzal, R. Savoy, G. Gorman, T.C. Huang, and S. La Placa, Crystallography and microstructure of Tl–Ca–Ba–Cu–O superconducting oxides, *Appl. Phys. Lett.*, **53**, 432–434 (1988).
40. D.M. Ogborne and M.T. Weller, The structure of $TlBa_2Ca_3Cu_4O_{11}$, *Physica C*, **230**, 153–158 (1994).
41. Y.F. Li, O. Chmaissem, and Z.Z. Sheng, Crystal structure and T_c of 1212-type cuprate $(Tl,Cr)Sr_2(Ca,Tl)Cu_2O_7$, *Physica C*, **248**, 42–48 (1995).
42. S.N. Putilin, E.V. Antipov, O. Chmaissem, and M. Marezio, Superconductivity at 94 K in $HgBa_2CuO_{4+\delta}$, *Nature*, **362**, 226–228 (1993).
43. E.V. Antipov, J.J. Capponi, C. Chaillout, O. Chmaissem, S.M. Loureiro, M. Marezio, S.N. Putilin, A. Santoro, and J.L. Tholence, Synthesis and neutron powder diffraction study of the superconductor $HgBa_2CaCu_2O_{6+\delta}$ before and after heat treatment, *Physica C*, **218**, 348–355 (1993).
44. X.S. Wu, H.M. Shao, S.S. Jiang, C. Gou, D.F. Chen, D.W. Wang, and Z.H. Wu, Syntheses and neutron powder diffraction studies of the superconductor $HgBa_2Ca_2Cu_3O_{8+\delta}$ by Pb substitution, *Physica C*, **261**, 189–195 (1996).
45. J.J. Capponi, E.M. Kopnin, S.M. Loureiro, E.V. Antipov, E. Gautier, C. Chaillout, B. Souletie, M. Brunner, J.L. Tholence, and M. Marezio, High-pressure synthesis and heat treatments of the $HgBa_2Ca_4Cu_5O_{12+\delta}$ and $HgBa_2Ca_5Cu_6O_{14+\delta}$ phases, *Physica C*, **256**, 1–7 (1996).

7
Empirical Rules for the Critical Temperature

7.1 Introduction

In Chapter 5, the cuprate high-T_c superconductors were described as complex nonstoichiometric oxides, whose actual oxygen content depends on the heat treatment conditions during synthesis. In addition, the charge carrier density in the conducting CuO_2 planes can be adjusted by a partial substitution of the host atoms in the charge carrier reservoirs by atoms with a different valence. In this chapter, we will discuss the relations between the charge carrier density in the CuO_2 planes and the critical temperature in various cuprate high-T_c superconductors. We will restrict our considerations to hole-doped cuprate superconductors.

In Chapter 6, we considered the crystal structures of various families of cuprate high-T_c superconductors. The different members of a homologous series of cuprate superconductors (e.g., Hg-12($n-1$)n) are distinguished by the number of CuO_2 sheets within the copper oxide blocks, which are separated by the layers of the charge carrier reservoirs. In contrast to the copper oxide blocks, the number of layers in the charge carrier reservoirs is unchanged for the different members of a homologous series of cuprate superconductors. In principle, it would be expected that an increasing number of CuO_2 sheets within the copper oxide blocks would be beneficial for superconductivity. In fact, the critical temperature was found to increase with the number n of CuO_2 sheets within the single copper oxide blocks, for a certain family of cuprate superconductors, up to a maximum of T_c which is typically reached for the compounds with $n = 3$. In this chapter, we will describe the dependence of the critical temperature on the number of CuO_2 sheets in the copper oxide blocks.

As a direct consequence of the fact that the charge carrier density in the conducting CuO_2 planes is intimately connected to the exact chemical composition, the critical temperature

Physical Properties of High-Temperature Superconductors, First Edition. Rainer Wesche.

is also related to the lattice parameters. In addition, metal properties like the density of states at the Fermi level depend on the lattice parameters. Hence, the critical temperature depends on the lattice parameters even if the charge carrier density is not affected. The easiest way to vary the lattice parameters is to apply pressure. We will, therefore, present data on the pressure dependence of the critical temperature. An alternative to the application of pressure is the replacement of larger ions by smaller ions of the same valence leading to a chemical pressure.

7.2 Relations between Charge Carrier Density and Critical Temperature

The superconductivity of the cuprate high-temperature superconductors occurs in the CuO_2 layers of the copper oxide blocks. The critical temperature is closely related to the formal valence $2 + p$ of the copper atoms, where p is the number of mobile charge carriers. For the insulating parent compounds, the formal valence of copper is 2, and, hence, p is zero. By means of doping and oxygen nonstoichiometry, the value of p can be varied. For example, a formal valence of 2.2 means that the oxidation state of 20% of the copper atoms in the CuO_2 planes is +3. These Cu^{3+} defects can move freely in the CuO_2 planes and, therefore, have to be considered as mobile holes. The critical temperature of the high-T_c superconductors shows a parabolic dependence on the number N_h of holes per copper atom in the conduction planes [1–6]. Empirically, it was found that the dependence of the critical temperature on N_h can be well described by the expression [3, 4]

$$T_c = T_{c,\mathrm{opt}}(1 - A(N_h - N_{\mathrm{opt}})^2) \tag{7.1}$$

where T_c is the critical temperature, $T_{c,\mathrm{opt}}$ is the maximum value of T_c for the cuprate in question, A is a constant, N_h is the number of holes per CuO_2, and N_{opt} is the number of holes leading to $T_{c,\mathrm{opt}}$. The maximum value of the critical temperature is typically reached for N_h close to 0.16, and, hence, a formal copper valence of $\approx$2.16.

In Figure 7.1, the parabolic dependence of the critical temperature on the number of holes per CuO_2 unit is presented for $YBa_2Cu_3O_{7-\delta}$ [7], $Y_{1-x}Ca_xBa_2Cu_3O_{7-\delta}$ [7], $Bi_2Sr_2CaCu_2O_{8+\delta}$ [8], and $La_{2-x}Sr_xCuO_{4+\delta}$ [5]. The maximum critical temperature of $YBa_2Cu_3O_{7-\delta}$ is 92 K at $N_{\mathrm{opt}} = 0.16$ holes per CuO_2. The solid line is a fit to the T_c data of $YBa_2Cu_3O_{7-\delta}$ and $Y_{1-x}Ca_xBa_2Cu_3O_{7-\delta}$ using Equation 7.1. The fit provides $A = 68.37$ using fixed values of $T_{c,\mathrm{opt}} = 92$ K and $N_{\mathrm{opt}} = 0.16$.

The Bi-2212 data show again a parabolic dependence of T_c on the number of holes per CuO_2. The double-point dash line is a fit using Equation 7.1, which provides $T_{c,\mathrm{opt}} =$ 90.49 K, $N_{\mathrm{opt}} = 0.169$, and $A = 61.725$.

The T_c data of $La_{2-x}Sr_xCuO_{4+\delta}$ [5] can be well represented by the dashed line. The parameters obtained from the fit are $T_{c,\mathrm{opt}} = 37.67$ K, $N_{\mathrm{opt}} = 0.1845$, and $A = 47.33$. An interesting aspect is the fact that $La_{2-x}Sr_xCuO_{4+\delta}$ is nonsuperconducting when the number of the charge carriers per CuO_2 unit is less than 0.05. The vanishing of superconductivity in the case of an extremely small electrical conductivity in the normal state seems to be plausible. An unexpected phenomenon is the transition to a nonsuperconducting metal for $N_h \geq 0.32$ holes per CuO_2. Bi-2212 and Y-123 show a similar behavior.

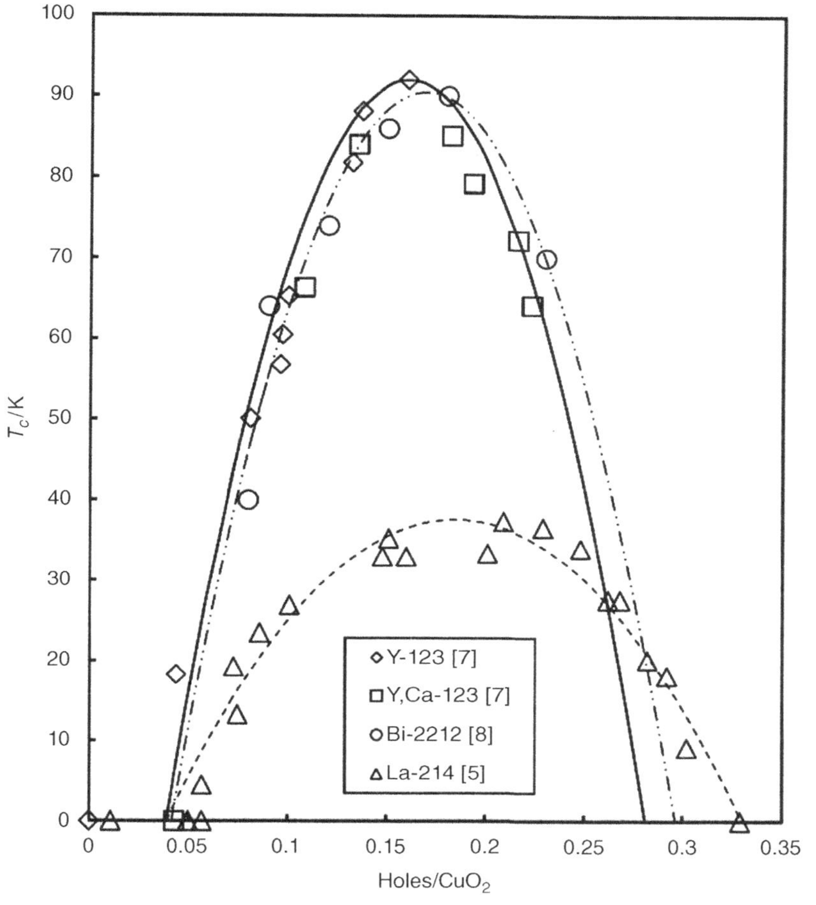

Figure 7.1 *Critical temperatures of $YBa_2Cu_3O_{7-\delta}$ (Y-123), $Y_{1-x}Ca_xBa_2Cu_3O_{7-\delta}$ (Y, Ca-123), $Bi_2Sr_2CaCu_2O_{8+\delta}$ (Bi-2212), and $La_{2-x}Sr_xCuO_{4+\delta}$ (La-214) vs. the actual number of holes per CuO_2. The inverted parabolas are fits based on Equation 7.1 (results from [5, 7, 8])*

In Figure 7.2, the T_c versus N_h curves of the first three members of the Hg-12($n-1$)n family are compared. As in the case of Y-123, Bi-2212, and La-214, we find a parabolic dependence of the critical temperature on the number of holes per CuO_2 unit. The fit parameters, which provide the $T_c(N_h)$ curves shown in Figure 7.2, are collected in Table 7.1. The maximum values of T_c occur for N_{opt} values close to 0.16. The data suggest that N_{opt} decreases slightly with increasing number n of CuO_2 planes in the copper oxide blocks. The factor A and $T_{c,opt}$ increase with increasing number n of CuO_2 planes in the copper oxide blocks. Above a carrier concentration of $\approx$0.3 holes per CuO_2, the Hg-12($n-1$)n compounds do not superconduct. For very low carrier concentrations ($N_h < 0.05$), the Hg-12($n-1$)n compounds are nonsuperconducting, like other cuprates.

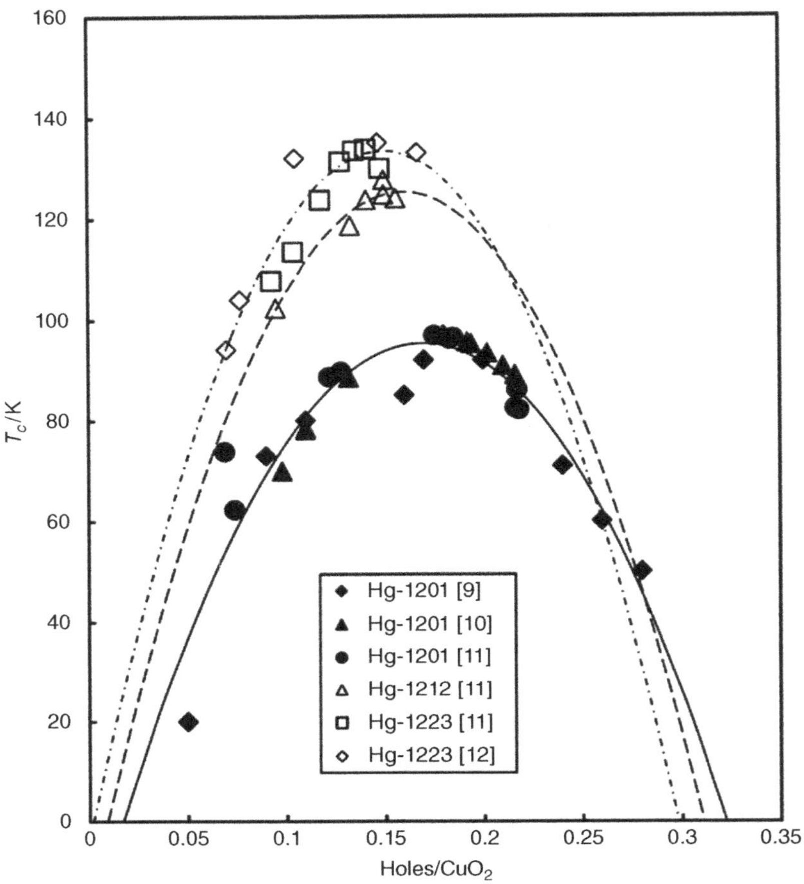

Figure 7.2 *Critical temperatures of* $HgBa_2CuO_{4+\delta}$ *(Hg-1201),* $HgBa_2CaCu_2O_{6+\delta}$ *(Hg-1212), and* $HgBa_2Ca_2Cu_3O_{8+\delta}$ *(Hg-1223) vs. the actual number of holes per* CuO_2*. The inverted parabolas are fits based on Equation 7.1 (results from [9–12])*

Table 7.1 *Fit parameters describing the* $T_c(N_h)$ *curves of Hg-1201, Hg-1212, and Hg-1223*

Compound	$T_{c,opt}$(K)	N_{opt}	A
Hg-1201	95.38	0.170	42.86
Hg-1212	125.23	0.16[a]	43.90
Hg-1223	133.37	0.15[a]	45.61

[a] Fixed value.

7.3 Effect of the Number of CuO_2 Planes in the Copper Oxide Blocks

Typically, the maximum possible critical temperature in the different families of cuprate superconductors is reached for the member with three CuO_2 planes in a single copper oxide block. In Figure 7.3, this behavior is illustrated for Bi-22($n-1$)n [13–15], Tl-22($n-1$)n [15–17] and (Hg,Tl)-22($n-1$)n [18]. The critical temperature of Bi-2201 is well below 20 K, while (Hg,Tl)-2201 is nonsuperconducting. In the compounds Bi-2212 and (Hg,Tl)-2212 with two CuO_2 planes in the copper oxide blocks, the critical temperature is much higher and exceeds 80 K. The critical temperature of 92 K of $YBa_2Cu_3O_{7-\delta}$ is comparable to the T_c-values of Bi-2212 and (Hg,Tl)-2212. In all cuprate families considered, the highest T_c is reached for the compound with three CuO_2 planes in the copper oxide blocks. For $n \geq 4$, the T_c-values are well below the corresponding maximum values, and tend to decrease further for (Hg,Tl)-2245. The Tl-22($n-1$)n compounds show higher critical temperatures than the members of the other two cuprate families. In particular, the T_c of Tl-2201 is as high as 92 K. It can be seen that T_c increases with number n of CuO_2 planes up to $n = 3$. However, for $n \geq 4$, the number of charge carriers provided by the charge carrier reservoirs may no longer be sufficient to reach the optimum hole concentration in the CuO_2 planes. Furthermore, the inner and outer CuO_2 planes in compounds with $n \geq 3$ are structurally not equivalent. The outer planes are in direct contact with the layers of the charge carrier reservoirs, while the inner planes are not.

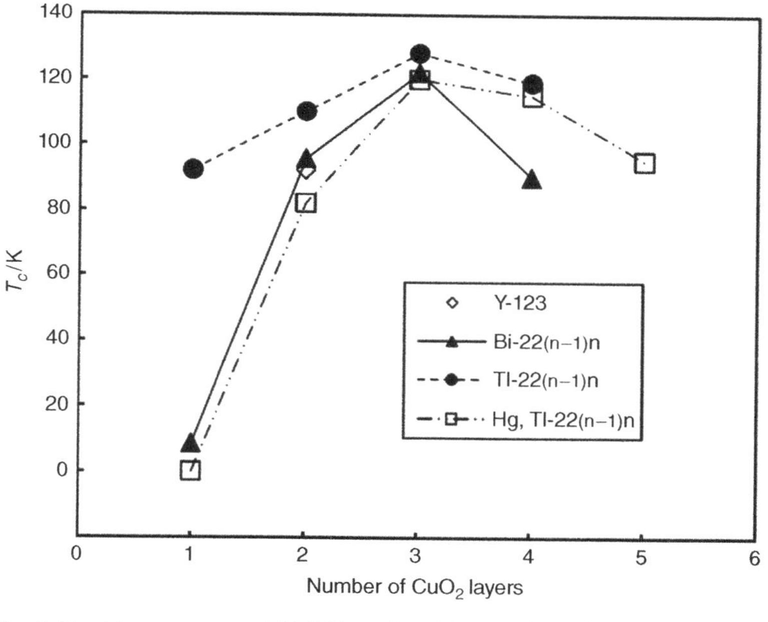

Figure 7.3 *Critical temperature of M-22(n − 1)n (M = Bi, Tl, and Hg & Tl) vs. the number of CuO_2 planes in a single copper oxide block (data from [13–18]). Typically, the maximum of T_c has been achieved in the compound with three CuO_2 planes per copper oxide block. For comparison, the critical temperature of $YBa_2Cu_3O_{7-x}$ (two CuO_2 layers per copper oxide block) is also shown*

In several $MBa_2Ca_{n-1}Cu_nO_x$ (M = Hg, Tl, (Cu,C)) and $MSr_2Ca_{n-1}Cu_nO_x$ (M = (Hg,Cr), (Cu,Cr)) cuprate superconductor families it is possible to synthesize compounds with five or even more CuO_2 planes in a single copper oxide block. Of special importance is the dependence of the critical temperature on n for $n \geq 5$. In Figure 7.4, the critical temperatures of various M-12($n-1$)n compounds are presented. In the (Cu,Cr)-12($n-1$)n cuprate family, the critical temperature decreases with increasing n for $n > 3$, and the compounds do not superconduct for $n = 8$ and 9. (Hg,Cr)-12($n-1$)n shows for $n = 4$ and 5 a similarly pronounced decrease in T_c as the corresponding members of the (Cu,Cr)-12($n-1$)n family. However, no T_c data for larger n are available. A very interesting behavior was found for the Hg-12($n-1$)n family. As usual T_c reaches a maximum for $n = 3$ decreases up to $n = 6$, while for n, from 7 to 9, the critical temperature is constant and as high as 100 K to the first approximation. For n in the range of 4–6, the reductions of the critical temperatures in the Tl-12($n-1$)n and (Cu,C)-12($n-1$)n families are similar to those found in the Hg-12($n-1$)n system.

The observed behavior of the critical temperature in the (Cu,Cr)-12($n-1$)n family would be in line with the assumption that the charge carriers are more or less uniformly distributed among the CuO_2 planes in the copper oxide blocks. For $n \geq 8$, the number of charge carriers provided by the charge carrier reservoirs is no longer sufficient to reach the minimum number of around 0.05 holes per CuO_2 necessary to establish superconductivity. On the other hand, the fact that in the Hg-12($n-1$)n family the critical temperature of the compounds with n in the range of 6–9 are all close to 100 K suggests that the charge carriers are nonuniformly distributed among the CuO_2 planes in the copper oxide blocks. The observed behavior would be consistent with a number of holes per CuO_2 in the outer planes, which

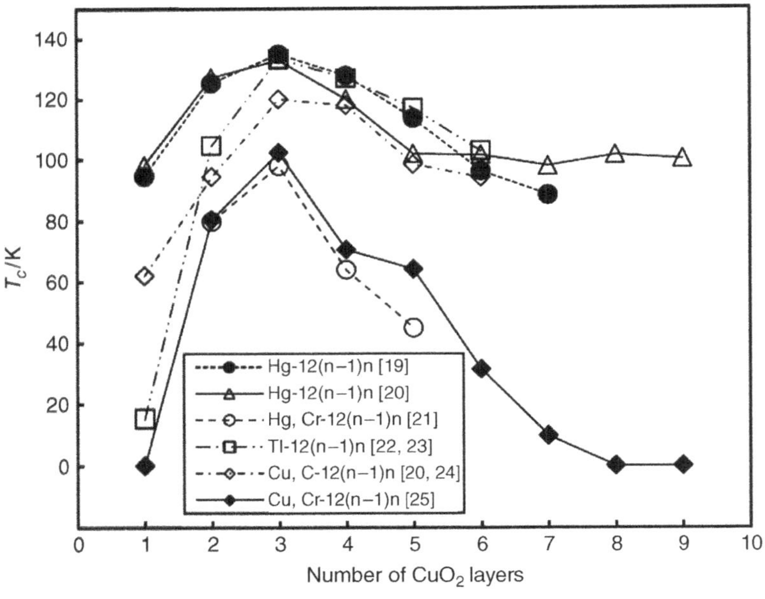

Figure 7.4 *Critical temperature of M-12(n – 1)n (M = Tl, Hg, Hg & Cr, Cu & C, and Cu & Cr) vs. the number of CuO_2 planes in a single copper oxide block (results from [19–25])*

is sufficient to generate superconductivity in the outer CuO_2 planes closest to the charge carrier reservoirs, whereas the number of holes per CuO_2 in the inner CuO_2 planes is too small for superconductivity.

Experimental evidence for this scenario has been found from ^{63}Cu and ^{65}Cu nuclear magnetic resonance (NMR) [26, 27] and muon spin rotation (μSR) studies [28, 29]. Kotegawa *et al.* performed systematic studies of the ^{63}Cu Knight shift in cuprate high-temperature superconductors with three and more CuO_2 planes in a single copper oxide block [26]. Before presenting the results of their investigations, let us briefly consider NMR which is based on the interaction of a nonzero nuclear spin with an applied magnetic field. Due to the applied magnetic field (B_a), each energy level splits into $2I + 1$ sublevels with the energies

$$E_M = -\gamma \hbar B_a M \tag{7.2}$$

where γ is the gyromagnetic ratio and the value of the quantum number M varies between $-I$ and I. Depending on whether the value of the nuclear spin is an integer or half-integer, M is also integer or half-integer. For $I = 3/2$, the possible M values are $-3/2, -1/2, +1/2$, and $+3/2$. The NMR lines originate from transitions between these sublevels. NMR provides information on the microscopic environment of the atomic nucleus being considered. The applied magnetic field interacts with the electrons around the nucleus. In a diamagnetic system, the applied magnetic field is lowered, while in a paramagnetic system, it is enhanced. The shift of the NMR lines is related to the fact that the applied magnetic field B_a is modified $B_n = (1 - \sigma_c)B_a$ at the nucleus, where σ_c is called the chemical shift. Typical values of the chemical shift σ_c are between 10^{-3} and 10^{-6}. The Knight shift is caused by the interaction of the nucleus with the polarized conduction electrons, which also causes the Pauli spin susceptibility discussed in Chapter 4. For *s*-wave electrons, which have a finite probability to be found at the position of the nucleus, the Knight shift is [30]

$$K = \frac{\Delta\omega}{\omega} = \frac{2}{3}\chi_P \frac{|\psi(0)|^2}{N} \tag{7.3}$$

where χ_P is the Pauli spin susceptibility, $|\Psi(0)|^2$ the probability to find the electron at the nucleus, and N is the average electron density. The Knight shift is typically of the order of a few tenth of a percentage. An introduction to NMR can be found in [30]. The results of Kotegawa *et al.* [26] indicate that the holes provided by the charge carrier reservoirs are distributed in a nonuniform way among the CuO_2 planes in cuprates with three or more CuO_2 planes in the copper oxide blocks. In general, the number of holes doped into the outer planes adjacent to the charge carrier reservoirs is larger than that introduced to the inner layers. The apical oxygen atoms above the copper atoms of the outer CuO_2 planes seem to be responsible for the preferred insertion of holes into these layers. The fractions of holes in a single inner or outer CuO_2 layer versus the average number of holes per CuO_2 ($N_{h,\text{total}}/n$) are presented in Figure 7.5. The data are based on the assumption that $2f_{\text{op}} + (n-2)f_{\text{ip}} = 1$. This means that in the systems with $n = 5$, where the central CuO_2 plane (ip*) is not equivalent to the other two inner planes (ip), a possible difference of N_h(ip) and N_h(ip*) has been neglected. The distribution of the holes among the inner and outer layers becomes increasingly nonuniform with increasing average number of holes per CuO_2 unit. For compounds with three CuO_2 layers per copper oxide block, the difference of the fractions f_{ip} and f_{op} is less pronounced than in

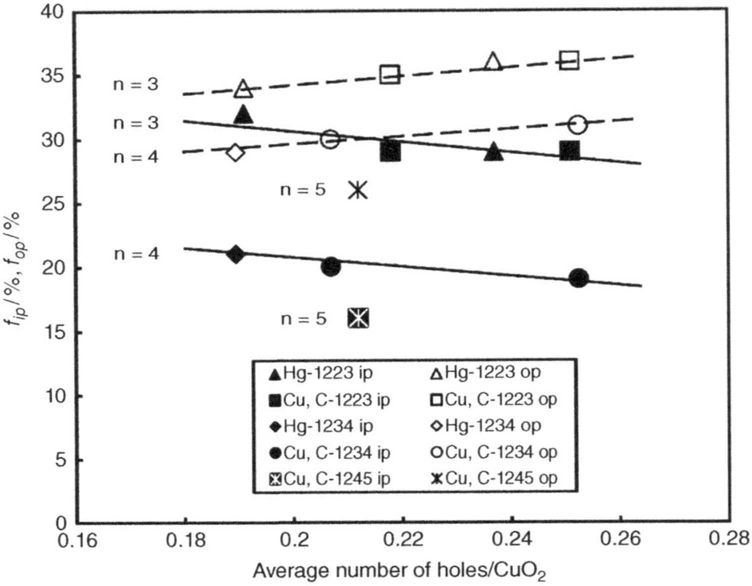

Figure 7.5 *Fraction of the holes in the inner and outer layers of cuprate superconductors with three to five CuO_2 planes in the copper oxide blocks (data from [26]). The nonuniformity of the distribution of the charge carriers on the inner and outer layers increases with increasing average number of holes per CuO_2*

compounds with four or five CuO_2 planes in the copper oxide blocks, as illustrated in Figure 7.5.

Investigations of $HgBa_2Ca_4Cu_5O_y$ by means of μSR are in line with a nonuniform distribution of the holes among the inner and outer CuO_2 planes. More interestingly, these results [28, 29] suggest that, in this compound, superconductivity and antiferromagetism coexist on a microscopic level. The nearly optimally doped outer planes are responsible for bulk superconductivity below 108 K, while the strongly underdoped inner planes undergo a transition to antiferromagnetism below 60 K.

7.4 Effect of Pressure on the Critical Temperature

The investigation of the dependence of the critical temperature on pressure is useful in the search for new superconductors with further enhanced T_c-values. Very soon after the discovery of high-temperature superconductivity in $La_{2-x}Ba_xCuO_4$ (La-214), it was found that the critical temperature of this cuprate increases rapidly with the applied pressure ($dT_c/dp = +8$ K/GPa). The large effect of pressure on T_c prompted Chu and co-workers to replace the La^{3+} ions with smaller Y^{3+} ions, leading to the discovery of $YBa_2Cu_3O_{7-\delta}$

[31], the first superconductor with a T_c well above the boiling point of liquid nitrogen. Another highlight is the fact that the critical temperature of $HgBa_2Ca_2Cu_3O_{8+\delta}$, which is as high as 135 K at ambient pressure, can be increased to ≈160 K under high pressure [11]. No substitution has yet been found to reach a similarly high critical temperature at ambient pressure.

As a starting point for the further considerations, we may assume that the pressure dependence of the critical temperature is caused only by the pressure dependence of the number of holes (N_h) per CuO_2 [32]. Using Equation 7.1, we find

$$\frac{dT_c}{dp} = \left(\frac{dT_c}{dN_h}\right)\left(\frac{dN_h}{dp}\right) = -2AT_{c,\text{opt}}(N_h - N_{\text{opt}})\frac{dN_h}{dp} \tag{7.4}$$

Here, N_h is the actual number of holes per CuO_2, N_{opt} is the number of holes per CuO_2 leading to the maximum critical temperature ($T_{c,\text{opt}}$). This charge transfer model can be improved by taking into account an additional intrinsic strain effect, which describes the change of the critical temperature originating from the change in the lattice parameters [32]. With this modification, we find

$$\frac{dT_c}{dp} = \left(\frac{dT_c}{dp}\right)_{\text{in}} + \left(\frac{dT_c}{dN_h}\right)\left(\frac{dN_h}{dp}\right) \tag{7.5}$$

The use of Equation 7.1 and the assumption that A and N_{opt} are independent of pressure lead to

$$\frac{dT_c}{dp} = \frac{dT_{c,\text{opt}}}{dp}(1 - A(N_h - N_{\text{opt}})^2) - \frac{dN_h}{dp}(2AT_{c,\text{opt}}(N_h - N_{\text{opt}})) \tag{7.6}$$

For nearly optimally doped samples, $N_h \cong N_{\text{opt}}$ the quadratic term in $(N_h - N_{\text{opt}})$ can be neglected, and we obtain

$$\frac{dT_c}{dp} = \frac{dT_{c,\text{opt}}}{dp} - \frac{dN_h}{dp}(2AT_{c,\text{opt}}(N_h - N_{\text{opt}})) \tag{7.7}$$

This modified charge transfer model is a useful starting point, but does not include all the complex phenomena which have been observed in the cuprate superconductors. In $La_{2-x}Ba_xCuO_4$, the pressure dependence of T_c is affected by a phase transition to a low temperature tetragonal phase below 60 K, which is detrimental for the critical temperature. The application of high pressures as large as 12 GPa seems to suppress this phase transition, leading to large values of dT_c/dp [32]. Another aspect of importance is the ordering of oxygen defects. For example in underdoped $YBa_2Cu_3O_{7-x}$, the critical temperature could be sharply reduced by quenching of the sample in liquid nitrogen. The reduction of T_c is caused by a lowered carrier concentration in the CuO_2 planes caused by the reduced local order of oxygen defects in the CuO chains of the Y-123 compound [32].

In Figure 7.6, the critical temperature of Bi-based high-T_c superconductors is shown as a function of hydrostatic pressure. In Bi-2212 as well as Bi,Pb-2223, the critical temperature increases with increasing pressure. The pressure derivatives of the critical temperature are in the range of 0.93–1.92 K/GPa.

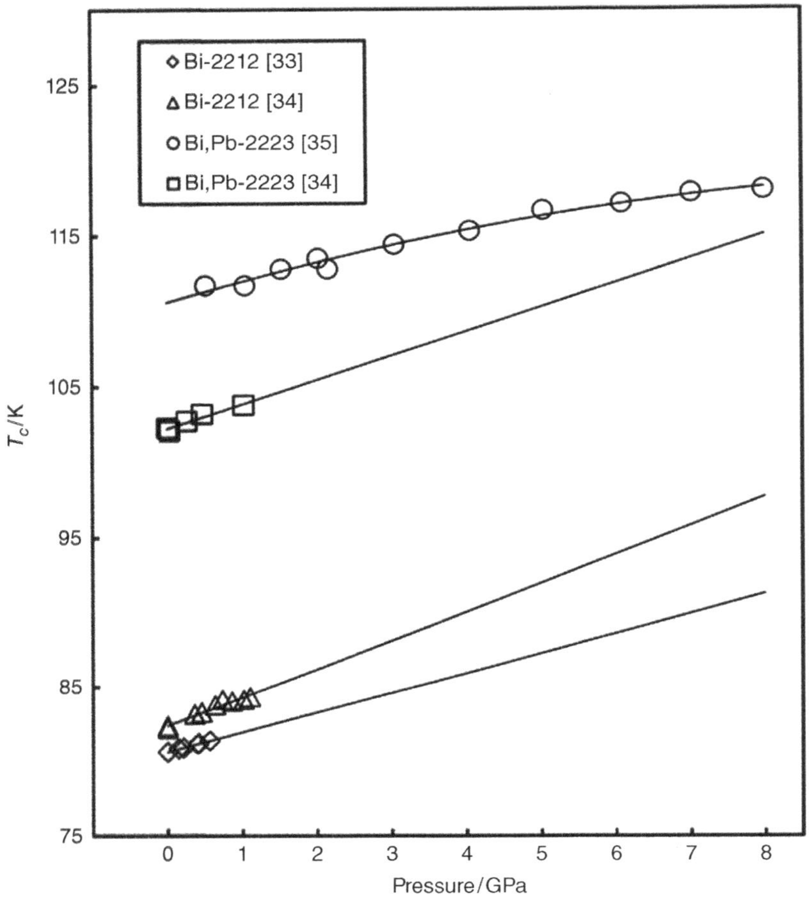

Figure 7.6 *Critical temperature of Bi-2212 and Bi,Pb-2223 vs. hydrostatic pressure. In both bismuth-based high-T_c superconductors, the critical temperature increases moderately with applied pressure (results from 33–35])*

The pressure dependences of the critical temperatures of Tl-2201, Tl-2212, and Tl-2223 are compared in Figure 7.7. The critical temperature of the overdoped Tl-2201 decreases with increasing hydrostatic pressure. In Tl-2201, oxygen ordering effects are of importance, and the critical temperature depends on the full temperature and pressure history. The intrinsic pressure dependence $(dT_c/dp)_{\mathrm{in}}$ is positive [32]. The other Tl-based high-T_c superconductors show an initial increase of T_c with increasing pressure. For the optimally doped Tl-2223, characterized by a T_c of $\approx$129 K, the pressure derivative dT_c/dp is smaller than that for Tl-2223 compounds with initial T_c of approximately 100 and 116 K. For these Tl-2223 compounds with nonoptimized concentration of holes in the CuO_2 planes, the change of the carrier density with pressure increases T_c.

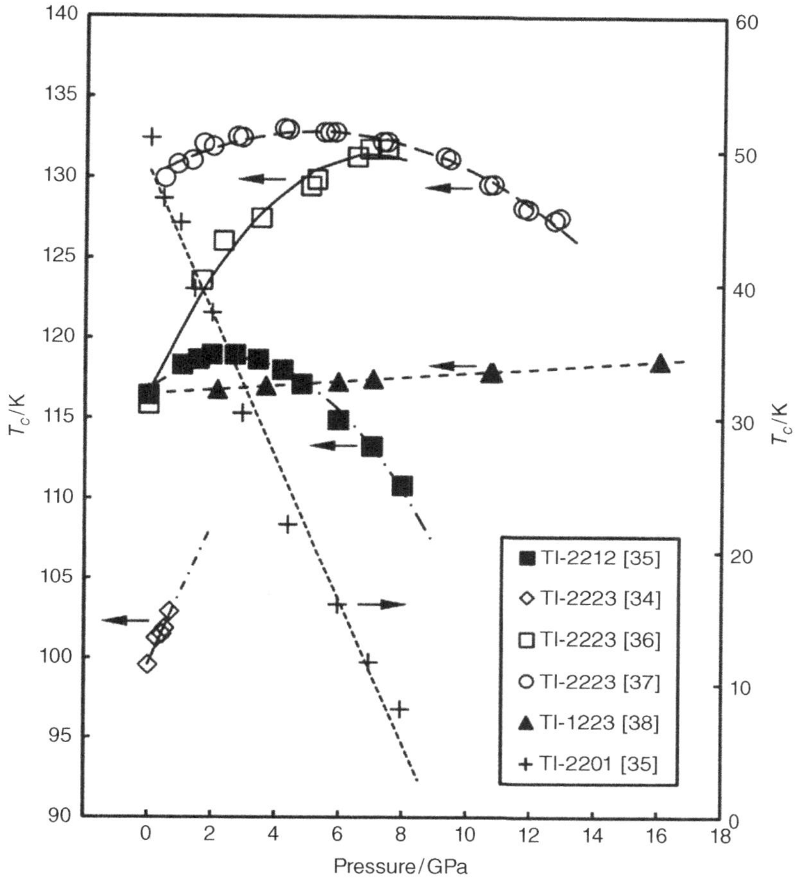

Figure 7.7 *Critical temperature of thallium-based high-T_c superconductors vs. hydrostatic pressure. In contrast to other cuprate superconductors, the T_c of Tl-2201 decreases with increasing pressure. This behavior has been attributed to charge re-distribution effects in this overdoped cuprate superconductor. Note the different scale for Tl-2201 (results from [34–38])*

Figure 7.8 illustrates the pronounced increase of the critical temperature of Hg-1201, Hg-1212, and Hg-1223 with increasing hydrostatic pressure. The data of different experiments show a considerable variation of the T_c of Hg-1223 at high pressure. The extremely high critical temperatures above 160 K, reported by Gao *et al.* [11], have been attributed to the presence of shear stress in this quasi-hydrostatic pressure experiment. The maximum values of T_c in the different compounds are 117 K at 23 GPa in Hg-1201, 152 K at 29 GPa in Hg-1212, and 164 K at 31 GPa in Hg-1223.

The next aspect to be considered is the initial slope of T_c versus pressure for nearly optimally doped Hg-based high-T_c superconductors. The data for hydrostatic pressures

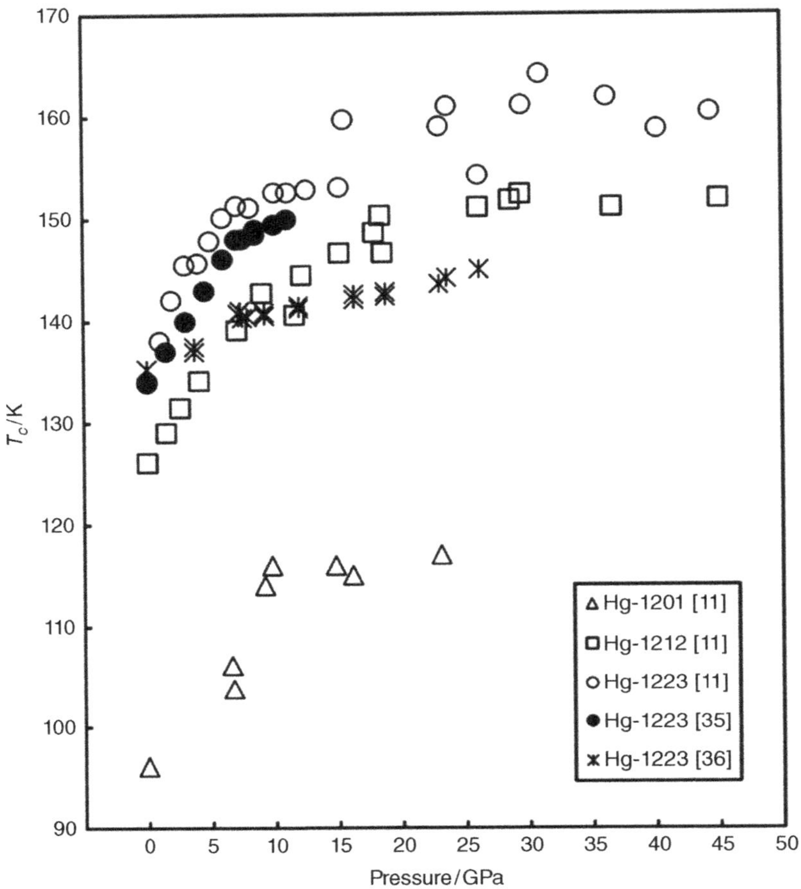

Figure 7.8 *Critical temperature of mercury-based high-T_c superconductors vs. hydrostatic or quasi-hydrostatic pressure. The critical temperature of Hg-1223 increases up to 164 K at a pressure of 31 GPa (results from [11, 35, 36])*

up to 1 GPa are presented in Figure 7.9. An interesting aspect is the observation that dT_c/dp for the Hg-12($n-1$)n compounds with n between 1 and 3 are nearly identical. The measurements of Klehe *et al.* [39] are consistent with a value of +1.75 K/GPa in all three Hg-based compounds. Their relative volume derivatives $d\ln T_c/d\ln V$ are close to −1.2 [32]. For optimally doped Y-123, Tl- and Bi-based superconductors, intrinsic relative volume derivatives $d\ln T_c/d\ln V$ close to −1.2 have also been found, leading to the following volume dependence [32]:

$$T_c \propto V^{-1.2} \tag{7.8}$$

The measured initial values of the pressure derivative of various high-T_c superconductors are shown in Table 7.2 [11, 33–44]. These values are for hydrostatic or quasi-hydrostatic pressure experiments on optimally and nonoptimally doped cuprate

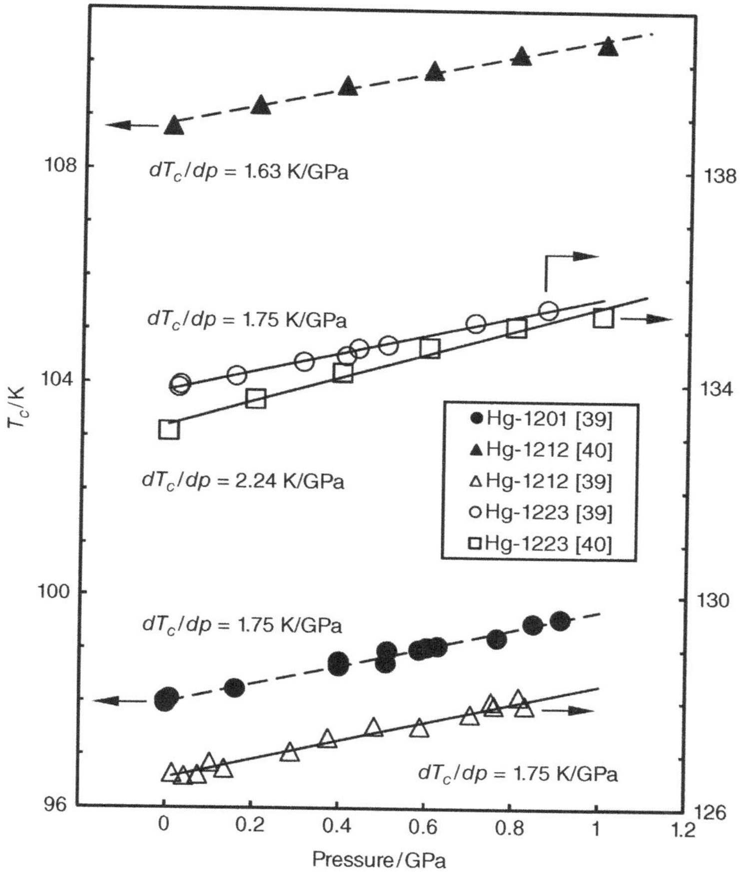

Figure 7.9 *Critical temperature of mercury-based high-T_c superconductors vs. hydrostatic pressure up to 1 GPa. The initial pressure derivatives of the first three members of the Hg-12(n − 1)n family are nearly identical (results from [39, 40])*

high-T_c superconductors. The measured pressure derivatives dT_c/dp depend strongly on the doping level of the investigated sample. Typically, the pressure derivatives of T_c of optimally doped samples are considerably smaller than those of nonoptimally doped samples. For example, the pressure derivative of Tl-2223 varies from 1.7 K/GPa ($T_c \cong 129$ K) for the nearly optimally doped sample to 5 K/GPa for a sample with a critical temperature of only 99.5 K.

In hydrostatic pressure experiments, the unit cell is compressed along all three principal crystallographic axes. The observed intrinsic pressure effect, leading to an increase of the critical temperature with decreasing volume, may be caused either by the reduction of the area of the CuO_2 planes or by the reduction of the distance of adjacent CuO_2 planes along the c direction. To distinguish between these two possibilities, uniaxial pressure experiments are required. For the determination of the pressure derivatives dT_c/dp_a, dT_c/dp_b, and dT_c/dp_c (indices a, b, c identify the crystallographic axes), single crystals of high

Table 7.2 *Initial slope of the T_c vs. p curves of various cuprate high-T_c superconductors*

Compound	dT_c/dp(K/GPa)	Comments	Reference
$YBa_2Cu_3O_{6+x}$	0.4	$x = 1$	[42]
	4.3	$0.4 \leq x \leq 0.8$	[42]
	8.0	$x \approx 0.35$	[42]
$YBa_2Cu_3O_{6+x}$	13.7	$x = 0.63$	[43]
	7.0	$x = 0.73$	[43]
	2.2	$x = 0.93$	[43]
$YBa_2Cu_4O_8$	5.5		[44]
$GdBa_2Cu_4O_8$	4.5		[44]
$Bi_2Sr_2CaCu_2O_{8+x}$	1.4	$x = 0.11$	[33]
	1.5	$x = 0.15$	[33]
	1.6	$x = 0.24$	[33]
$Bi_2Sr_2CaCu_2O_8$	1.6	Polycrystalline	[34]
	1.9	Single crystal	[34]
$Bi_{1.6}Pb_{0.4}Sr_2Ca_2Cu_3O_{10}$	1.7		[34]
$Bi_{1.68}Pb_{0.32}Sr_{1.75}Ca_{1.85}Cu_{2.65}O_{10+x}$	1.1	$p \leq 3$ GPa	[35]
$Tl_2Ba_2CuO_{6+x}$	−5.4	$p \leq 8$ GPa	[35]
$Tl_2Ba_2CaCu_2O_{8+x}$	1.8		[35]
$Tl_2Ba_2Ca_2Cu_3O_{10+x}$	5.0	$T_c(0) = 99.5$ K	[34]
$Tl_2Ba_2Ca_2Cu_3O_{10+x}$	4.4	$T_c(0) \cong 116$ K	[36]
$Tl_2Ba_2Ca_2Cu_3O_{10+x}$	1.7	$T_c(0) \cong 129$ K	[37]
$(Tl,Pb,Bi)_1(Sr,Ba)_2Ca_2Cu_3O_9$	0.13	$T_c(0) \cong 116$ K	[38]
$HgBa_2CuO_{4+x}$	1.95	$T_c(0) \cong 96$ K	[11]
$HgBa_2CuO_{4+x}$	1.72	$T_c(0) \cong 98$ K	[39]
$HgBa_2CaCu_2O_{6+x}$	1.80	$T_c(0) = 126.6$ K	[39]
$HgBa_2CaCu_2O_{6+x}$	1.9	$T_c(0) = 126$ K	[11]
$HgBa_2CaCu_2O_{6+x}$	1.63	$T_c(0) \cong 109$ K	[40]
$HgBa_2Ca_2Cu_3O_{8+x}$	3.3	$T_c(0) \cong 134$ K	[11]
$HgBa_2Ca_2Cu_3O_{8+x}$	2.0	$T_c(0) \cong 134$ K	[35]
$HgBa_2Ca_2Cu_3O_{8+x}$	≈0.8	$T_c(0) \cong 135$ K	[41]
$HgBa_2Ca_2Cu_3O_{8+x}$	1.71	$T_c(0) \cong 134$ K	[39]
$HgBa_2Ca_2Cu_3O_{8+x}$	2.24	$T_c(0) \cong 133$ K	[40]

quality and sufficient size are required. Welp *et al.* [45] have reported a direct measurement of the changes of T_c caused by uniaxial pressure along the main crystallographic axes of an untwinned single crystal of $YBa_2Cu_3O_{7-x}$. They found $dT_c/dp_a = -2.0 \pm 0.2$ K/GPa, $dT_c/dp_b = +1.9 \pm 0.2$ K/GPa and $dT_c/dp_c = -0.3 \pm 0.1$ K/GPa. The results show a pronounced in-plane pressure effect of opposite sign along the a and b axes, which nearly cancel. On the other hand, the uniaxial pressure effect along the c direction is very small.

Another way to determine the uniaxial pressure dependence of $YBa_2Cu_3O_{7-x}$ was used by Meingast *et al.* [46]. They performed ultra-high-resolution measurements of thermal expansion along the main crystallographic axes. Using the Ehrenfest relationship

$$\frac{dT_c}{dp_i} = \frac{\Delta\alpha_i V_m T_c}{\Delta C_p} \tag{7.9}$$

Table 7.3 *Critical temperature vs. uniaxial pressure data of various cuprate high-T_c superconductors*

Compound	dT_c/dp_i(K/GPa)	Comments	Reference
$YBa_2Cu_3O_{7-x}$	-1.9	$dT_c/d\varepsilon_a = +217$ K, $p \parallel a$	[46]
	$+2.2$	$dT_c/d\varepsilon_b = -316$ K, $p \parallel b$	
	≈ 0	$dT_c/d\varepsilon_c = +30$ K, $p \parallel c$	
		pressure $\Rightarrow \varepsilon$ negative	
$YBa_2Cu_3O_{7-x}$	-2.0 ± 0.2	$dT_c/d\varepsilon_a = +230$ K, $p \parallel a$	[45]
	$+1.9 \pm 0.2$	$dT_c/d\varepsilon_b = -220$ K, $p \parallel b$	
	-0.3 ± 0.1	$dT_c/d\varepsilon_c = +18$ K, $p \parallel c$	
		pressure $\Rightarrow \varepsilon$ negative	
		$C_{11}/C_{22}/C_{33} = 231/268/186$ GPa	
		$C_{12}/C_{13}/C_{23} = 132/71/95$ GPa	
$YBa_2Cu_4O_8$	$+3.7 - 5.0$	$p \parallel a$	[47]
	$+0.3 - 0.4$	$p \parallel b$	
	0 ± 0.7	$p \parallel c$	
$Bi_2Sr_2CaCu_2O_{8+x}$	0.9 ± 0.1	$p \parallel a, b$	[48]
	0.0 ± 0.4	$p \parallel c$	
$Bi_2Sr_2CaCu_2O_{8+x}$	$+1.5$	$p \parallel a, b$	[49]
	-4.5	$p \parallel c$	
$Bi_2Sr_2CaCu_2O_{8+x}$	$+1.6 \pm 0.2$	$p \parallel a$	[50]
	$+2.0 \pm 0.2$	$p \parallel b$	
	-2.8 ± 0.9	$p \parallel c$	
$HgBa_2CuO_{4+x}$	$+2.3 \pm 0.2$	$p \parallel a, b$	[51]
	-3.6 ± 0.3	$p \parallel c$	

where $\Delta\alpha_i$ is the mean-field jump in the thermal expansion coefficient along the i direction ($i = a, b, c$) at the superconducting transition, ΔC_p is the jump in the specific heat, and V_m is the molar volume. Using for $\Delta C_p T_c^{-1}$, a value of 50 mJ mol^{-1}K^{-2}, they found $dT_c/dp_a = -1.9$ K/GPa, $dT_c/dp_b = +2.2$ K/GPa and $dT_c/dp_c \approx 0$ K/GPa. These data are in reasonably good agreement with the results published by Welp *et al.* [45].

Using the measured uniaxial pressure dependence, the strain dependences of the critical temperature can be calculated

$$\frac{dT_c}{d\varepsilon_i} = -\sum_j C_{ji} \frac{dT_c}{dp_j} \tag{7.10}$$

Here, the C_{ji}s are the components of the elastic tensor and ε_i is the strain along the direction i. The measured values of dT_c/dp_i of various cuprate superconductors are listed in Table 7.3.

Very recently, the dependence of the critical temperature of $HgBa_2CuO_{4+x}$ on uniaxial pressure has been measured by Hardy *et al.* [51]. Their results, also presented in Table 7.3, indicate that the T_c of $HgBa_2CuO_{4+x}$ can be enhanced by a reduction of the area of the CuO_2 planes. The T_c of this compound can also be increased by enlarging the distance between the CuO_2 planes. Hence, the most efficient way to reach higher critical temperatures is to increase the ratio c/a. In Hg-1201, the copper oxide blocks are made up of a single CuO_2 plane, and, hence, the *c* lattice parameter can be enlarged only by thickening the BaO–Hg–BaO layers. The hydrostatic pressure derivative dT_c/dp measured in the multilayer compounds Hg-1212 and Hg-1223 is nearly identical to the value obtained in the single-layer compound. This suggests that the positive effect of an enlarged *c* lattice parameter is related to a larger distance between adjacent copper oxide blocks, and not a larger distance between adjacent CuO_2 planes within a single copper oxide block [51].

7.5 Summary

The critical temperature of cuprate superconductors depends on the actual number of holes per CuO_2. The maximum critical temperature is typically reached for a carrier concentration around 0.16 holes per CuO_2. In other words, the highest critical temperature occurs for a formal valence of +2.16 of the copper atoms in the CuO_2 planes. The maximum critical temperature in different families of cuprate superconductors is typically reached in the member with three CuO_2 planes in a single copper oxide block. For more than three such planes per block, the critical temperatures decrease with increasing number of CuO_2 planes. The reasons for this are too small number of holes provided by the charge carrier reservoirs, and a nonuniform distribution of the holes among the inner and outer CuO_2 layers within the individual copper oxide blocks. The maximum critical temperatures which can be achieved in different families of cuprate superconductors vary considerably. The buckling of the copper oxide planes in $YBa_2Cu_3O_{7-x}$ or the structural modulations in Bi-based superconductors seem to hinder very high T_c-values. A tetragonal structure and plane CuO_2 sheets, as found in the Hg-12$(n-1)n$ compounds, seem to favor high critical temperatures. The results of hydrostatic and uniaxial pressure experiments indicate that the critical temperature can be enhanced by reducing the area of the CuO_2 planes. Very recent results obtained for Hg-1201 suggest that a large c/a ratio is beneficial for achieving high T_c-values.

References

1. M.R. Presland, J.L. Tallon, R.G. Buckley, R.S. Liu, and N.E. Flower, General trends in oxygen stoichiometry effects on T_c in Bi and Tl superconductors, *Physica C*, **176**, 95–105 (1991).
2. J.L. Tallon, C. Bernhard, H. Shaked, R.L. Hitterman, and J.D. Jorgensen, Generic superconducting phase behavior in high-T_c cuprates: T_c variation with hole concentration in $YBa_2Cu_3O_{7-\delta}$, *Phys. Rev. B*, **51**, 12911–12914 (1995).
3. J.T. Kucera and J.C. Bravman, Transport characterization of calcium-doped $YBa_2Cu_3O_{7-\delta}$ thin films, *Phys. Rev. B*, **51**, 8582–8590 (1995).
4. A. Schmehl, B. Goetz, R.R. Schulz, C.W. Schneider, H. Bielefeldt, H. Hilgenkamp, and J. Mannhart, Doping-induced enhancement of the critical currents of grain boundaries in $YBa_2Cu_3O_{7-\delta}$, *Europhys. Lett.*, **47**, 110–115 (1999).
5. J.B. Torrance, Y. Tokura, A.I. Nazzal, A. Bezinge, T.C. Huang, and S.S.P. Parkin, Anomalous disappearance of high-T_c superconductivity at high hole concentration in metallic $La_{2-x}Sr_xCuO_4$, *Phys. Rev. Lett.*, **61**, 1127–1130 (1988).

6. Y. Kubo, Y. Shimakawa, T. Manako, and H. Igarashi, Transport and magnetic properties of $Tl_2Ba_2CuO_{6+\delta}$ showing a δ-dependent gradual transition from an 85-K superconductor to a non-superconducting metal, *Phys. Rev. B*, **43**, 7875–7882 (1991).
7. K. Widder, D. Berner, J. Münzel, H.P. Geserich, M. Kläser, G. Müller-Vogt, and Th. Wolf, Charge carrier distribution in $Y_{1-z}Ca_zBa_2Cu_3O_x$, *Physica C*, **267**, 254–260 (1996).
8. W.A. Groen, D.M. de Leeuw, and L.F. Feiner, Hole concentration and T_c in $Bi_2Sr_2CaCu_2O_{8+\delta}$, *Physica C*, **165**, 55–61 (1990).
9. A. Fukuoka, A. Tokiwa-Yamamoto, M. Itoh, R. Usami, S. Adachi, H. Yamauchi, and K. Tanabe, Dependence of superconducting properties on the Cu-valence determined by iodometry in $HgBa_2CuO_{4+\delta}$, *Physica C*, **265**, 13–18 (1996).
10. J. Hofer, J. Karpinski, M. Willemin, G.I. Meijer, E.M. Kopnin, R. Molinski, H. Schwer, C. Rossel, and H. Keller, Doping dependence of superconducting parameters in $HgBa_2CuO_{4+\delta}$ single crystals, *Physica C*, **297**, 103–110 (1998).
11. L. Gao, Y.Y. Xue, F. Chen, Q. Xiong, R.L. Meng, D. Ramirez, C.W. Chu, J.H. Eggert, and H.K. Mao, Superconductivity up to 164 K in $HgBa_2Ca_{m-1}Cu_mO_{2m+2+\delta}$ (m = 1, 2, and 3) under quasihydrostatic pressures, *Phys. Rev. B*, **50**, 4260–4263 (1994).
12. S. Mikusu, N. Urita, Y. Hashinaka, K. Tokiwa, A. Iyo, Y. Tanaka, and T. Watanabe, Transport properties of $TlBa_2Ca_2Cu_3O_y$ in an over-doped state, *Physica C*, **442**, 91–96 (2006).
13. J. Hejtmánek, M. Nevřiva, E. Pollert, D. Sedmidubský, and P. Vašek, Superconductivity and transport of $Bi_{2+x}Sr_{2-y}CuO_{6+\delta}$ single crystals, *Physica C*, **264**, 220–226 (1996).
14. Th. Schweizer, R. Müller, and L.J. Gauckler, A wet chemistry method to determine the Bi and Cu valencies in Bi–Sr–Ca–Cu-O (2212) high-temperature superconductors, *Physica C*, **225**, 143–148 (1994).
15. R. Hott, H. Rietschel, and M. Sander, Hoch-T_c-Supraleiter: Wie steht es mit den Anwendungen?, *Phys. Bl.*, **48**, 355–358 (1992).
16. R.S. Liu, J.L. Tallon, and P.P. Edwards, An efficient and reproducible approach for attaining superconductivity at 128 K in $Tl_2Ba_2Ca_2Cu_3O_{10-\delta}$, *Physica C*, **182**, 119–122 (1991).
17. A. Maignan, C. Martin, M. Huve, J. Provost, M. Hervieu, C. Michel, and B. Raveau, The "2201" thallium cuprate: T_c's up to 92 K can be achieved by "hydrogen" annealing, *Physica C*, **170**, 350–360 (1990).
18. T. Tatsuki, A. Tokiwa-Yamamoto, T. Tamura, S. Adachi, and K. Tanabe, Annealing study on $(Hg,Tl)_2Ba_2Ca_{n-1}Cu_nO_y$ ($n = 1$–5) superconductors, *Physica C*, **278**, 160–168 (1997).
19. B.A. Scott, E.Y. Suard, C.C. Tsuei, D.B. Mitzi, T.R. McGuire, B.-H. Chen, and D. Walker, Layer dependence of the superconducting transition temperature of $HgBa_2Ca_{n-1}Cu_nO_{2n+2+\delta}$, *Physica C*, **230**, 239–245 (1994).
20. A. Iyo, Y. Tanaka, Y. Kodama, H. Kito, K. Tokiwa, and T. Watanabe, Synthesis and physical properties of multilayered cuprates, *Physica C*, **445–448**, 17–22 (2006).
21. E. Kandyel, X.-J. Wu, S. Adachi, and S. Tajima, Synthesis of superconducting $(Hg_{0.7}Cr_{0.3})Sr_2Ca_{n-1}Cu_nO_y$ ($n = 2$–5) using high-pressure technique, *Physica C*, **322**, 9–18 (1999).

22. H. Kusuhara, T. Kotani, H. Takei, and K. Tada, Superconducting phases with single Tl-O layer structure preparation: (Tl, Pb)$Ca_{n-1}Ba_2Cu_nO_{2n+3}$ ($n = 2, 3, 4, 5, 6$), *Japan. J. Appl. Phys.*, **28**, L1772–L1774 (1989).
23. A. Iyo, Y. Aizawa, Y. Tanaka, M. Tokumoto, K. Tokiwa, T. Watanabe, and H. Ihara, High-pressure synthesis of $TlBa_2Ca_{n-1}Cu_nO_y$ ($n = 3$ and 4) with $T_c = 133.5$ K ($n = 3$) and 127 K ($n = 4$), *Physica C*, **357–360**, 324–328 (2001).
24. H. Wakamatsu, T. Ogata, N. Yamaguchi, S. Mikusu, K. Tokiwa, T. Watanabe, K. Ohki, K. Kikunaga, N. Terada, N. Kikuchi, Y. Tanaka, and A. Iyo, Fabrication of (Cu,C)Ba_2CuO_y superconducting thin film by RF magnetron sputtering, *J. Phys.: Conf. Ser.*, **43**, 289–292 (2006).
25. S.M. Loureiro, Y. Matsui, and E. Takayama-Muromachi, New series of high-T_c Cr-based superconductors, *Physica C*, **302**, 244–256 (1998).
26. H. Kotegawa, Y. Tokunaga, K. Ishida, G.-Q Zheng, Y. Kitaoka, K. Asayama, H. Kito, A. Iyo, H. Ihara, K. Tanaka, K. Tokiwa, and T. Watanabe, NMR study of carrier distribution and superconductivity in multilayered high-T_c cuprates, *J. Phys. Chem. Solids*, **62**, 171–175 (2001).
27. H. Kotegawa, Y. Tokunaga, Y. Araki, G.-q. Zheng, Y. Kitaoka, K. Tokiwa, T. Watanabe, A. Iyo, H. Kito, Y. Tanaka, and H. Ihara, Microscopic coexistence of antiferromagnetism and superconductivity in $HgBa_2Ca_4Cu_5O_y$: Cu NMR-study, *Physica C*, **388–389**, 237–238 (2003).
28. K. Tokiwa, S. Mikusu, W. Higemoto, K. Nishiyama, A. Iyo, Y. Tanaka, H. Kotegawa, H. Mukuda, Y. Kitaoka, and T. Watanabe, Muon spin rotation study of magnetism in multilayer $HgBa_2Ca_4Cu_5O_y$ superconductor, *Physica C*, **460–462**, 892–895 (2007).
29. K. Tokiwa, S. Ito, H. Okumoto, W. Higemoto, K. Nishiyama, A. Iyo, Y. Tanaka, and T. Watanabe, μSR study on multi-layered $HgBa_2Ca_4Cu_5O_y$ (Hg-1245) superconductor, *Physica C*, **388–389**, 243–244 (2003).
30. G. Schatz and A. Weidinger, *Nuclear Condensed Matter Physics*, John Wiley & Sons, Ltd, Chichester, England, 1996.
31. M.K. Wu, J.R. Ashburn, C.J. Torng, P.H. Hor, R.L. Meng, L. Gao, Z.J. Huang, Y.Q. Wang, and C.W. Chu, Superconductivity at 93 K in a new mixed-phase Y–Ba–Cu–O compound system, *Phys. Rev. Lett.*, **58**, 908–910 (1987).
32. J.S. Schilling, High-pressure effects, in *Handbook of High-Temperature Superconductivity: Theory and Experiment*, J.R. Schrieffer and J.S. Brooks (eds.), Springer, New York, 2007.
33. R. Sieburger, P. Müller, and J.S. Schilling, Pressure dependence of the superconducting transition temperature in $Bi_2Sr_2CaCu_2O_{8+y}$ as a function of oxygen content, *Physica C*, **181**, 335–339 (1991).
34. R. Kubiak, K. Westerholt, G. Pelka, and H. Bach, Pressure dependence of the superconducting transition temperature of Bi- and Tl-based high-T_c superconductors, *Physica C*, **166**, 523–529 (1990).
35. H. Takahashi, A. Tokiwa-Yamamoto, N. Môri, S. Adachi, H. Yamauchi, and S. Tanaka, Large enhancement of T_c in the 134 K superconductor $HgBa_2Ca_2Cu_3O_y$ under high pressure, *Physica C*, **218**, 1–4 (1993).
36. D.D. Berkley, E.F. Skelton, N.E. Moulton, M.S. Osofsky, W.T. Lechter, V.M. Browning, and D.H. Liebenberg, Pressure dependence of the superconducting transition temperature in single crystals of $Tl_2Ba_2Ca_2Cu_3O_{10-x}$, *Phys. Rev. B*, **47**, 5524–5527 (1993).

37. D. Tristan Jover, R.J. Wijngaarden, R.S. Liu, J.L. Tallon, and R. Griessen, Superconductivity at 133 K in $Tl_2Ba_2Ca_2Cu_3O_{10+x}$, *Physica C*, **218**, 24–28 (1993).
38. G. Triscone, R.E. Gladyshevskii, S.H. Han, J. Herrmann, and M.B. Maple, Effect of pressure on the electrical resistivity of a 116 K Tl-1223 superconducting ceramic, *Physica C*, **272**, 21–25 (1996).
39. A.-K. Klehe, J.S. Schilling, J.L. Wagner, and D.G. Hinks, Hydrostatic pressure dependence of the superconducting transition temperature of $HgBa_2CaCu_2O_{6+\delta}$ and $HgBa_2Ca_2Cu_3O_{8+\delta}$, *Physica C*, **223**, 313–320 (1994).
40. Y.S. Yao, Y.J. Su, W. Liu, Z.X. Liu, H. Gong, J.W. Li, G.H. Cao, F. Wu, and Z.X. Zhao, Superconductivity, microstructure and pressure effect on T_c of Hg-based compounds prepared under high pressure, *Physica C*, **224**, 91–98 (1994).
41. D. Tristan Jover, R.J. Wijngaarden, A. Schilling, H.R. Ott, and R. Griessen, Pressure dependence of the superconducting critical temperature T_c of $HgBa_2Ca_2Cu_3O_{8+x}$ up to 26 GPa, *Physica C*, **235–240**, 893–894 (1994).
42. C.C. Almasan, S.H. Han, B.W. Lee, L.M. Paulius, M.B. Maple, B.W. Veal, J.W. Downey, A.P. Paulikas, Z. Fisk, and J.E. Schirber, Pressure dependence of T_c and charge transfer in $YBa_2Cu_3O_x$ ($6.35 \leq x \leq 7$), *Phys. Rev. Lett.*, **69**, 680–683 (1992).
43. M. Baran and I. Fita, High effect of pressure on the critical temperature in $GdBa_{1.5}Sr_{0.5}Cu_3O_x$, *Physica C*, **261**, 125–130 (1996).
44. M. Baran, V. Dyakonov, L. Gladczuk, G. Levchenko, S. Piechota, and H. Szymczak, Comparative study of the pressure effect on the critical parameters of $GdBa_2Cu_4O_8$ and $YBa_2Cu_4O_8$, *Physica C*, **241**, 383–388 (1995).
45. U. Welp, M. Grimsditch, S. Fleshler, W. Nessler, J. Downey, G.W. Crabtree, and J. Guimpel, Effect of uniaxial stress on the superconducting transition in $YBa_2Cu_3O_7$, *Phys. Rev. Lett.*, **69**, 2130–2133 (1992).
46. C. Meingast, O. Kraut, T. Wolf, H. Wühl, A. Erb, and G. Müller-Vogt, Large *a-b* anisotropy of the expansivity anomaly at T_c in untwinned $YBa_2Cu_3O_{7-\delta}$, *Phys. Rev. Lett.*, **67**, 1634–1637 (1991).
47. C. Meingast, J. Karpinski, E. Jilek, and E. Kaldis, Uniaxial pressure dependence of T_c of single crystalline$YBa_2Cu_4O_8$ via thermal expansion, *Physica C*, **209**, 591–596 (1993).
48. H. Kierspel, H. Winkelmann, T. Auweiler, W. Schlabitz, B. Büchner, V.H.M. Duijn, N.T. Hien, A.A. Menovsky, and J.J.M. Franse, Thermal expansion, specific heat, and uniaxial pressure dependences of T_c in $Bi_2Sr_2CaCu_2O_{8+\delta}$, *Physica C*, **262**, 177–186 (1996).
49. N. Watanabe, K. Fukamachi, Y. Ueda, K. Tsushima, A.M. Balbashov, T. Nakanishi, and N. Môri, Effect of hydrostatic and uniaxial stress on T_c for single crystals of $Bi_2Sr_2CaCu_2O_x$, *Physica C*, **235–240**, 1309–1310 (1994).
50. C. Meingast, A. Junod, and E. Walker, Superconducting fluctuations and uniaxial-pressure dependence of T_c of a $Bi_2Sr_2CaCu_2O_{8+x}$ single crystal from high-resolution thermal expansion, *Physica C*, **272**, 106–114 (1996).
51. F. Hardy, N.J. Hillier, C. Meingast, D. Colson, Y. Li, N. Barišić, G. Yu, X. Zhao, M. Greven, and J.S. Schilling, Enhancement of the critical temperature of $HgBa_2CuO_{4+\delta}$ by applying uniaxial and hydrostatic pressure: Implications for a universal trend in cuprate superconductors, *Phys. Rev. Lett.*, **105**, 167002 (2010).

8

Generic Phase Diagram of Cuprate Superconductors

8.1 Introduction

In this chapter, we will discuss the generic phase diagram of hole-doped cuprate high-T_c superconductors. In Chapter 7, we have already discussed the bell-shaped dependence of the critical temperature on the concentration of holes in the CuO_2 planes. We will see that the cuprate high-T_c superconductors are complex materials showing a wealth of physical phenomena depending on the actual carrier concentration. The most prominent of these phenomena is superconductivity at high temperatures. The symmetry of the superconducting order parameter has been found to be of *d*-wave and not of *s*-wave symmetry as in the conventional metallic superconductors. An interesting aspect is the fact that the parent compounds of cuprate high-T_c superconductors are Mott insulators. The normal-state properties of optimally doped cuprate superconductors have been found to be unusual (linear dependence of the resistivity over a large range in temperature, strong dependence of the Hall coefficient on temperature, divergence of the resistivity anisotropy, violation of the Wiedemann–Franz law close to zero temperature). In the overdoped region, the normal-state properties are closer to that of a normal metal, which is well described by the Fermi liquid theory. A further outstanding feature of the cuprate high-T_c superconductors is a pseudogap existing at temperatures well above the critical temperature, especially in the underdoped region.

8.2 Generic Phase Diagram of Hole-Doped Cuprate Superconductors

8.2.1 Generic Phase Diagram: An Overview

In Figure 8.1, a simplified sketch of the generic phase diagram [1–15] of hole-doped cuprate high-T_c superconductors is shown. The phase diagram indicates that the cuprate

Physical Properties of High-Temperature Superconductors, First Edition. Rainer Wesche.

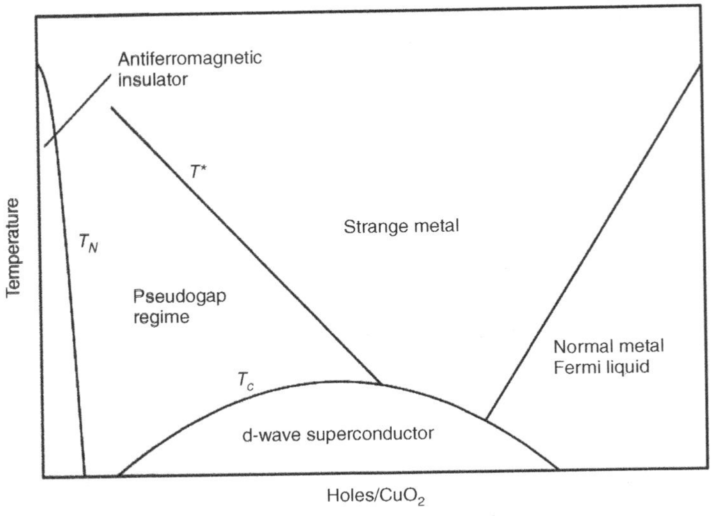

Figure 8.1 *Simplified generic phase diagram of hole-doped cuprate high-T_c superconductors, illustrating that these materials are close to a metal–insulator transition. The physical properties of the cuprates show considerable changes with increasing concentration of holes in the CuO_2 planes*

superconductors are close to a metal–insulator transition. The insulating parent compounds of the high-T_c cuprate superconductors are Mott insulators. For example, La_2CuO_4 and $YBa_2Cu_3O_6$ are the insulating parent compounds of $La_{2-x}Sr_xCuO_{4+\delta}$ and $YBa_2Cu_3O_{7-\delta}$, respectively. Figure 8.2 shows the antiferromagnetic ordering of the Cu moments in La_2CuO_4 [1, 16]. In the parent compounds, magnetic ordering sets in at Néel temperatures T_N which can reach values in the 250–400 K range. The presence of holes in the CuO_2 planes, needed to make the material superconduct, reduces the Néel temperature to zero.

At carrier concentrations close to 0.05 holes per CuO_2 unit, a superconducting phase starts to evolve. As already discussed in Chapter 7, the critical temperature shows an inverted parabola dependence on the number of holes per CuO_2 unit. The maximum critical temperature is typically reached at carrier concentrations between 0.15 and 0.18 holes per CuO_2 unit. For further increasing carrier concentrations, the critical temperature decreases again, and above a doping level of 0.3 holes per CuO_2 unit, the cuprates do not superconduct.

A special feature of the superconducting state in the cuprates is the d-wave symmetry of the superconducting order parameter. In the conventional metallic superconductors, the s-wave symmetry of the order parameter leads to an energy gap which is independent of the direction of the wave vector **k** of the quasi-particles. On the other hand, the width of the energy gap of a d-wave superconductor depends on the direction of the wave vector **k**. In the nodal directions (see Figure 8.9), the energy gap of a d-wave superconductor vanishes. A more detailed description of d-wave symmetry will be presented in Section 8.2.2.

In tunneling experiments, the temperature dependence of the superconducting energy gap has been studied [17]. A striking feature of these experiments is the observation of an

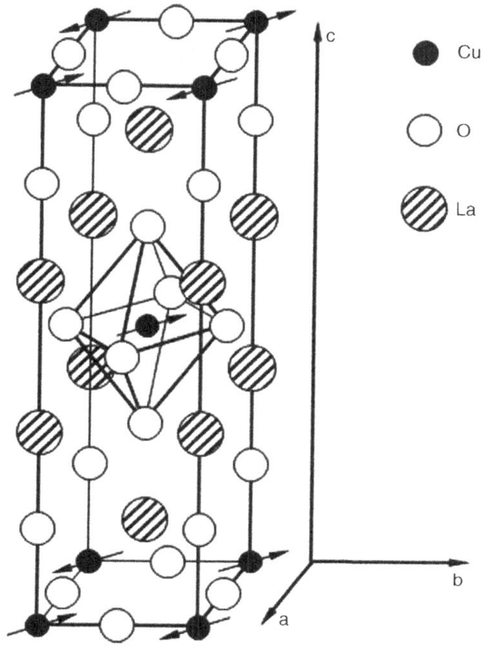

Figure 8.2 *Crystal structure of* La_2CuO_4 *and the possible alignment of the copper spins in the antiferromagnetic ground state (adapted from [1, 16])*

amazing gap-like feature at temperatures well above the critical temperature. This so-called pseudogap is most prominent in the underdoped region of the phase diagram. It can be observed up to a temperature known as T^* (see Figure 8.1). It is absent in highly overdoped cuprate superconductors. Possible explanations for the occurrence of the pseudogap will be discussed in Section 8.2.3.

8.2.2 Symmetry of the Superconducting Order Parameter

The superconducting order parameter is the collective wave function which describes the Cooper pairs in a superconductor (see Chapter 4). Its symmetry is closely related to the pairing mechanism of the Cooper pairs. In the previous section, we have seen that the parent compounds of the cuprate high-T_c superconductors are antiferromagnetic Mott insulators. In principle, it could be imagined that the formation of Cooper pairs in the cuprates is mediated by spin fluctuations. Such a mechanism for the pairing of Cooper pairs would require $dx^2 - dy^2$ symmetry of the order parameter. In conventional metallic superconductors, the order parameter has the s-wave symmetry. In an s-wave superconductor, the superconducting energy gap is independent of, or only weakly dependent on, the direction of the wave vector $\mathbf{k}$ of the single electrons. In contrast, the energy gap of a d-wave superconductor depends on the direction of the wave vector $\mathbf{k}$. The directional dependence of the energy gap is described by [18]

$$\Delta(\mathbf{k}) \sim \Delta_{\max}\left[\cos\left(k_x a\right) - \cos\left(k_y a\right)\right] \tag{8.1}$$

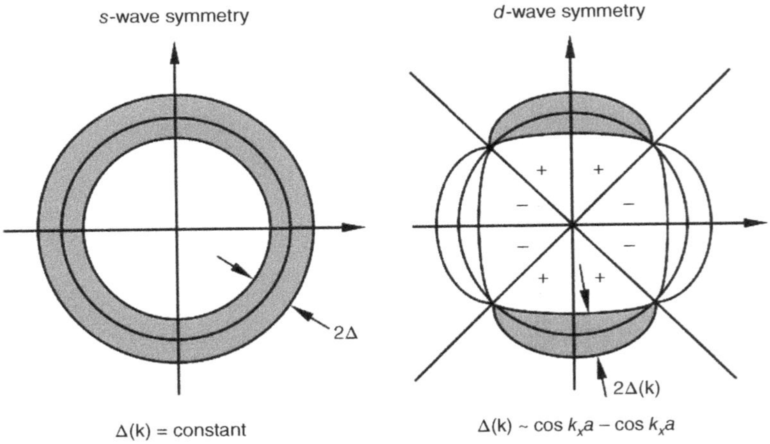

Figure 8.3 *Comparison of the energy gaps for single electrons (quasi-particles) of s- (left) and d-wave (right) superconductors. For s-wave symmetry, the energy gap is independent of the direction of the wave vector* ***k****, whereas it depends on* ***k*** *in a d-wave superconductor. In the nodal directions the order parameter changes its sign (adapted from [5])*

where Δ_{max} is the maximum gap value and a is the lattice parameter in the CuO_2 planes. In Figure 8.3, the dependence of the superconducting energy gap on the wave vector **k** is illustrated for the s- and d-wave symmetries. The energy gap of d-wave superconductors is strongly anisotropic with nodes along the [1 1 0] direction in k space. Between the lobes in the k_x and k_y directions, the order parameter changes its sign. This means that, for Cooper pairs with orthogonal relative momenta, there exists a relative phase shift of π in the superconducting wave function. In the cuprates, Equation 8.1 is expected to describe the behavior of the energy gap within the CuO_2 planes. In fact, experimental evidence for the symmetry of the superconducting order parameter of the cuprates has been found, as will be described below.

Experimental evidence for the $d_x2_{-y}2$ symmetry of the order parameter has been found from the flux quantization in tri-crystal superconducting rings of $YBa_2Cu_3O_{7-\delta}$ [19, 20]. In a junction between two d-wave superconductors, the direction of the tunneling current depends on the relative orientation of the lobes of the wave functions of the Cooper pairs with respect to the junction interface. Siegrist and Rice [21] have shown that the supercurrent I_s^{ij} across the junction can be expressed by

$$I_s^{ij} = (A^{ij} \cos 2\varphi_i \cos 2\varphi_j) \sin \Delta\varphi_{ij} \tag{8.2}$$

Here, A^{ij} is a constant characteristic of the considered junction ij, and φ_i and φ_j are the angles between the crystallographic axes and the junction interface between superconductors i and j. For a superconducting ring with a single π junction ($I_s^{ij} < 0$), Siegrist and Rice [21] have found a spontaneous magnetization of $\phi_0/2$. In a multiple junction ring with an odd number of π junctions, the spontaneous magnetization is again $\phi_0/2$ [19, 21]. A half-integer spontaneous magnetization of a π-ring (see Figure 8.4) has been measured by means of a scanning SQUID microscope. To ensure that the results are not affected by systematic experimental errors (e.g., the roughness of the interface) the flux through

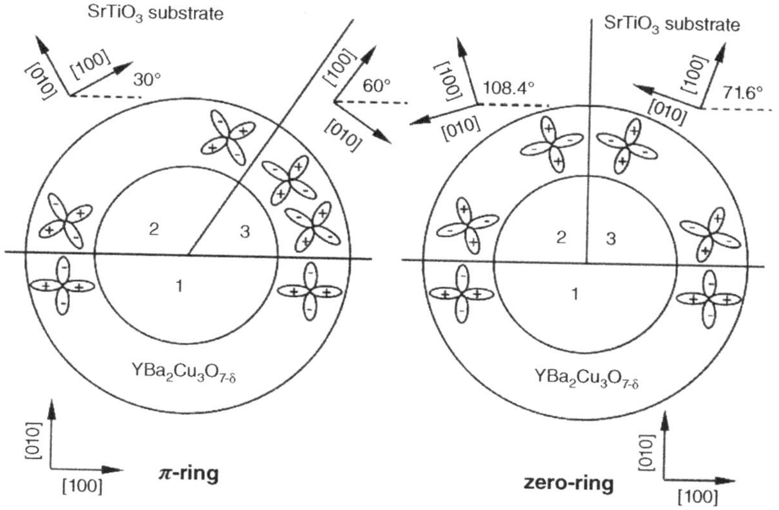

Figure 8.4 *Crystallography of zero- (right) and π-rings (left) of epitaxial* $YBa_2Cu_3O_{7-\delta}$ *films grown on a tri-crystal (1 0 0)* $SrTiO_3$ *substrate. The rings are patterned into the films by means of photolithography. The grain boundaries act as junction barriers. In the π-ring the order parameter changes its sign at only one (1–3) of the three interfaces, i.e. one of the interfaces is a π junction. In contrast, the zero-ring contains two π junctions (1–2 and 1–3) with phase shifts (adapted from [20])*

a zero-ring (see Figure 8.4) has also been measured. For the zero ring, there exists no spontaneous magnetization in the ground state; this has been confirmed experimentally [19, 20]. For these investigations, π- and zero-rings of $YBa_2Cu_3O_{7-\delta}$ have been deposited on a (1 0 0) $SrTiO_3$ tri-crystal substrate. The $YBa_2Cu_3O_{7-\delta}$ rings, of 24 μm inner diameter and 10 μm width, have been etched by photo-lithography. The crystal structure in the single-crystalline substrate is transmitted to the $YBa_2Cu_3O_{7-\delta}$ film. The grain boundaries act as the junction interfaces. In the π-ring, the sign of the superconducting order parameter changes at the 1–3 junction, while it is unchanged at the other two junctions; i.e., the ring contains a single π junction. In contrast, in the zero-ring we have two junctions (1–2, 1–3) with and one without a sign change, as illustrated in Figure 8.4.

Other ways to determine the symmetry of the superconducting order parameter have been presented in [18]. In one type of experiment, the magnetic flux through a corner SQUID magnetometer is measured; another is the measurement of the Fraunhofer diffraction pattern of a single Josephson junction. A sketch of a Pb-$YBa_2Cu_3O_{7-\delta}$ corner junction is presented in Figure 8.5. A magnetic field is applied perpendicular to the *ab* plane of $YBa_2Cu_3O_{7-\delta}$ single crystal. In the corner junction under consideration, the tunneling current flows in the *ac*- and *bc*-faces. In small junctions, the magnetic field generated by the tunneling currents can be neglected. For a rectangular tunneling junction (see Chapter 3, Equations 3.63 and 3.64) with a tunneling area A, the maximum Josephson current is

$$I_{\max}(\phi) = I_c \left| \frac{\sin\left(\pi\phi/\phi_0\right)}{(\pi\phi/\phi_0)} \right| \tag{8.3}$$

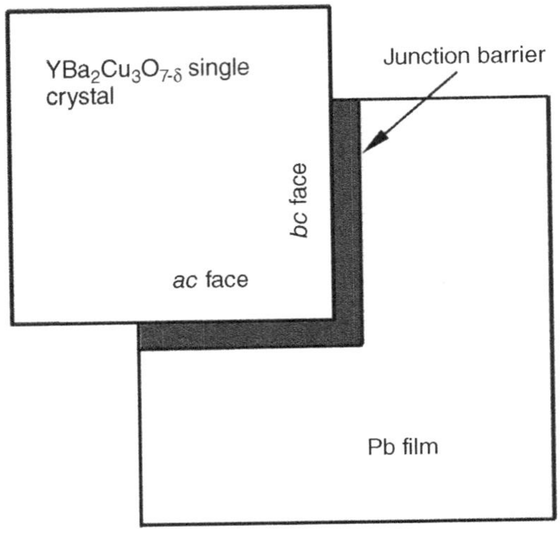

Figure 8.5 *Schematic illustration of a symmetric Y-123-Pb corner junction with equally large ac- and bc-faces and uniform critical current density. Depending on the symmetry of the superconducting order parameter of the Y-123, we would obtain the usual Fraunhofer diffraction pattern for the s-wave symmetry (see Figure 8.6). In the case of the d-wave symmetry, a modified modulation with $I_c(\phi = 0) = 0$ according to Equation 8.4 would be expected (see Figure 8.6)*

Here, $\phi_0 = h/2|e|$ is the flux quantum (see Chapter 3), $I_c = J_c A$, and $\phi = B\,w\,t$ (w: width of the junction, t: magnetic barrier thickness, B: applied magnetic field) the total magnetic flux through the junction. In the case of an s-wave superconductor, the phase in the two crystal faces of a symmetric corner junction would be the same, and, hence, Equation 8.3 would describe the critical current. For a symmetric corner junction of a d-wave superconductor, the sign of the order parameter would be opposite in the a- and b-directions (see Figure 8.5) leading to a modified current modulation [18]:

$$I_{max}(\phi) = I_c \left| \frac{\sin^2(\pi\phi/2\phi_0)}{(\pi\phi/2\phi_0)} \right| \tag{8.4}$$

The two different diffraction patterns are compared in Figure 8.6. For the zero applied field, i.e., $\phi = 0$, the tunneling currents into the ac- and bc-planes of a symmetric d-wave corner junction would cancel each other. The maximum current would be reached for a flux of $\pm 0.74\phi_0$. In a symmetric d-wave corner junction, the current is zero when the flux through each half of the junction is ϕ_0.

Experiments on symmetric Y-123 corner junctions are in line with the d-wave symmetry of the superconducting order parameter [18, 22]. The observed current shows a pronounced dip at $\phi = 0$, as is expected for the d-wave symmetry. The fact that the superconducting order parameter of the cuprates is of $d_x2_{-y}2$ symmetry indicates that pairing by spin fluctuations is possible, but does not rule out other pairing mechanisms which are also compatible with the d-wave symmetry.

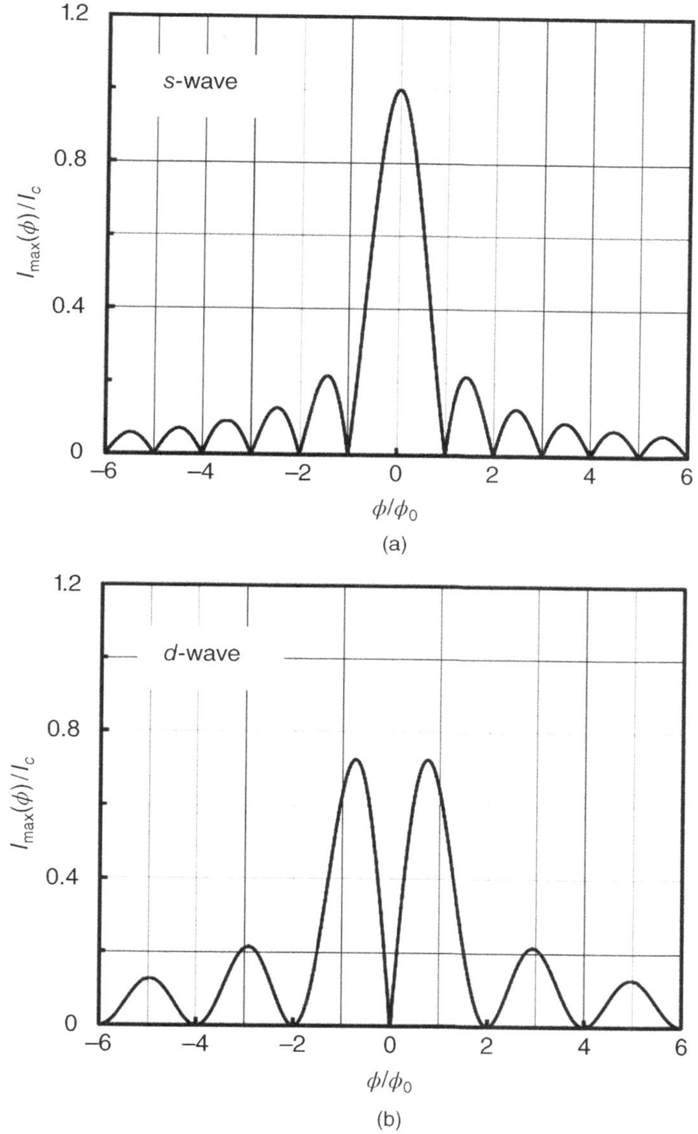

Figure 8.6 *Expected modulation of the current as a function of the flux ϕ through a symmetric corner junction depending on whether the superconducting order parameter of Y-123 has s-wave (Equation 8.3) or d-wave symmetry (Equation 8.4)*

8.2.3 The Pseudogap

It is beyond the scope of the present book to describe in detail the physics of the pseudogap in cuprate superconductors. There exists a large body of experimental results using a variety of experimental methods (nuclear magnetic resonance (NMR), muon spin rotation (μSR), scanning tunneling spectroscopy (STS), angle-resolved photoemission spectroscopy (ARPES), neutron scattering, Nernst effect). Unfortunately, the experimental results, obtained with different methods, seem to favor different descriptions of the pseudogap. There exist also various theoretical models to describe the pseudogap of the cuprate superconductors. We will not attempt to describe all the available experimental data, or to compare them with the different theoretical models. Instead, only a few ideas on the pseudogap will be presented.

The superconducting energy gap and the pseudogap of Bi-2212 have been studied by Renner *et al.* using STS [17]. Their tunneling spectra measured on an underdoped $Bi_2Sr_2CaCu_2O_{8+\delta}$ single crystal with a T_c of 83.0 K is shown in Figure 8.7. The conductance curves, obtained from STS, provide the density of states around the Fermi energy. For comparison, the density of states of an *s*-wave and that of a *d*-wave superconductor are shown in Figure 8.8. Due to the vanishing energy gap at the nodes (see Figure 8.9)

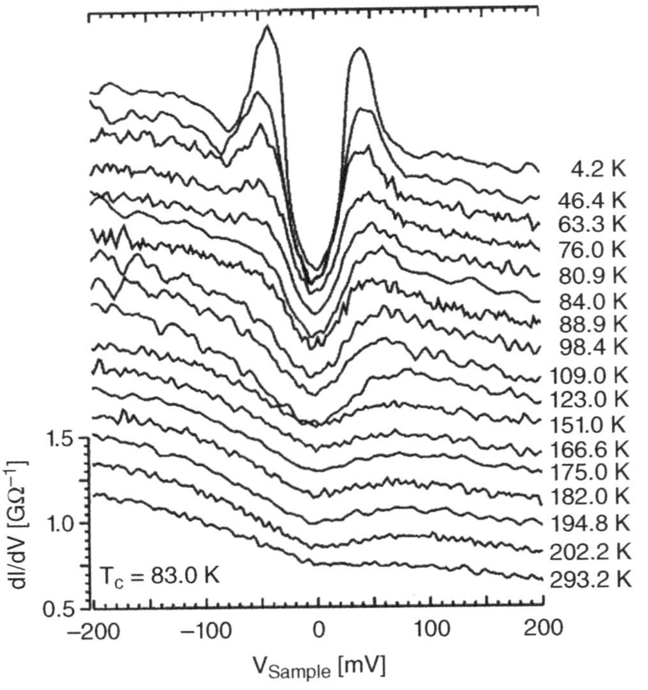

Figure 8.7 Tunneling conductance of underdoped Bi-2212. A gap-like feature (pseudogap) can be observed at temperatures much higher than the critical temperature of 83 K (Reprinted with permission from Renner et al. [17, Figure 2], Copyright (1997) by the American Physical Society, http://journals.aps.org/prl/abstract/10.1103/PhysRevLett.80.149, [17])

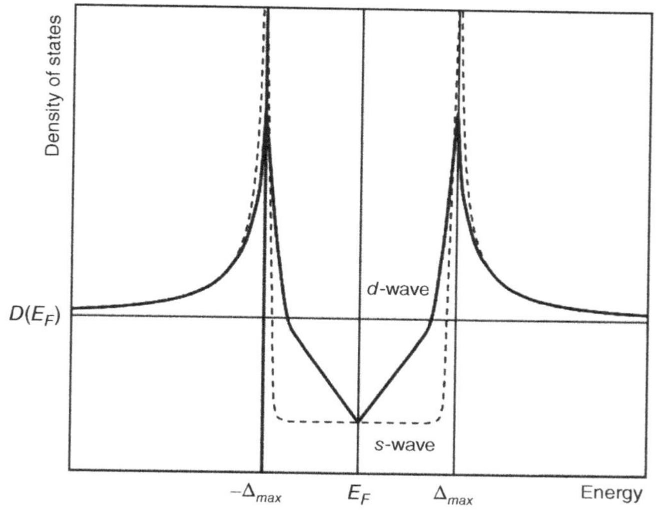

Figure 8.8 *Sketch of the density of states of the s- and d-wave superconductors close to the Fermi energy. Because of the vanishing of the gap in the nodal directions (see Figure 8.9), the density of states increases linearly within the gap (adapted from [5])*

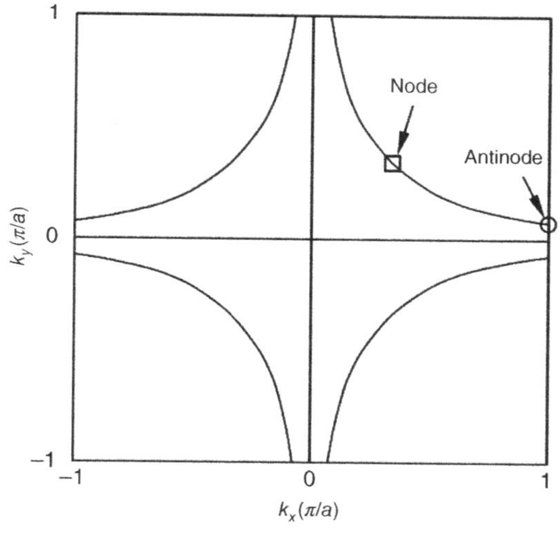

Figure 8.9 *Sketch of the Fermi surface of the CuO_2 planes. The nodal (open square) and anti-nodal (open circle) directions are indicated*

of a *d*-wave superconductor, there is a nonvanishing density of states within the energy gap of Bi-2212 at temperatures well below $T_c = 83$ K. A striking feature is the fact that at temperatures well above T_c, a gap-like feature is present in the conductance curves. This is the signature of the pseudogap. The data obtained from STS suggest that there is a continuous transition from the superconducting to the pseudogap state.

One hypothesis to explain the existence of the pseudogap proposes that preformed Cooper pairs exist at temperatures well above the critical temperature. However, no phase coherence can be established because of their extremely small number density. Evidence for pairing above the critical temperature has been found, for example, from ARPES [23–25], angular magnetoresistivity (AMR) [26], and Nernst effect measurements [27, 28]. Due to a thermal gradient along the sample in the presence of a transverse magnetic field, a transverse voltage is generated. This effect is called the Nernst effect (see Chapter 12). In the presence of vortices, the Nernst effect is strongly enhanced. If the thermal gradient is in the x direction and the magnetic field along the z direction, an electric field is generated by the movement of the vortices, leading to an electric field $\mathbf{E_e} = \mathbf{B} \times \mathbf{v}$ ($\mathbf{E_e}$ is the electric field, $\mathbf{B}$ is the magnetic field, and $\mathbf{v}$ is the velocity of the vortices). The Nernst coefficient is defined as $\nu = E_{ey}/(B|\nabla T|)$. Measurements of the Nernst effect indicate that, due to fluctuations, short-lived vortices (superconducting fluctuations) can exist well above T_c in underdoped cuprate superconductors. On the other hand, the maximum temperature for the existence of vortices seems to be well below the pseudogap temperature T^* [28], and, hence, it is not clear if noncoherent Cooper pairs are responsible for the pseudogap.

An alternative hypothesis is based on the assumption that the pseudogap phenomenon competes with superconductivity. Ma *et al.* [29] performed a combined scanning tunneling microscopy and ARPES study of $Bi_2Sr_{2-x}La_xCuO_{6+\delta}$. Below the critical temperature, they found two distinct gaps, a small one around the nodal direction and a large one in the antinodal direction. The small gap near the nodal direction disappears at the critical temperature, whereas the large antinodal gap has been found to exist below and above T_c. The small and the large gaps seem to coexist in the antinodal region. A similar behavior has been reported for underdoped Bi-2212 [30, 31]. For the gap in the nodal direction, a BCS-like temperature dependence of the energy gap has been found [31]. The nodal gap seems to be related to superconductivity, while the gap in the antinodal region persisting above T_c seems to be related to the pseudogap.

Next, we have to consider possible competing orders within the pseudogap regime. As a first example, let us consider charge stripes, which are sketched in Figure 8.10. In this scenario, charge stripes and antiferromagnetic domains coexist. The copper spins are ordered within the antiferromagnetic domains, while the doped holes are arranged in charge stripes. The charge stripes also act as the domain boundaries. Evidence for charge stripes has been reported for $La_{2-x}Ba_xCuO_4$ ($x = 0.125$) and other 214 compounds [32–35]. In principle, the charge stripes may be fluctuating instead of static. Fluctuating charge stripes have been observed by scanning tunneling microscopy in $Bi_2Sr_2CaCu_2O_{8+x}$ [36].

An alternative ordering in the pseudogap regime is based on loop currents. In Figure 8.11, planar orbital loop currents are shown. The loop currents flow in a triangle made of two oxygen and one copper atom [12]. More complex current paths involving the apical oxygen atoms have been proposed for Hg-1201 [11]. In a very recent review article by Zaanen [37], it is stated that in the pseudogap a competing order is at work. Close to optimal doping ($\approx$0.19 holes per CuO_2), the pseudogap vanishes. This behavior has been attributed to a quantum phase transition. One puzzling aspect is the fact that it seems to be possible that several different exotic ordering phenomena can exist in the pseudogap. It is beyond the scope of the present book to discuss all these ordering phenomena. A more detailed description can be found in [37].

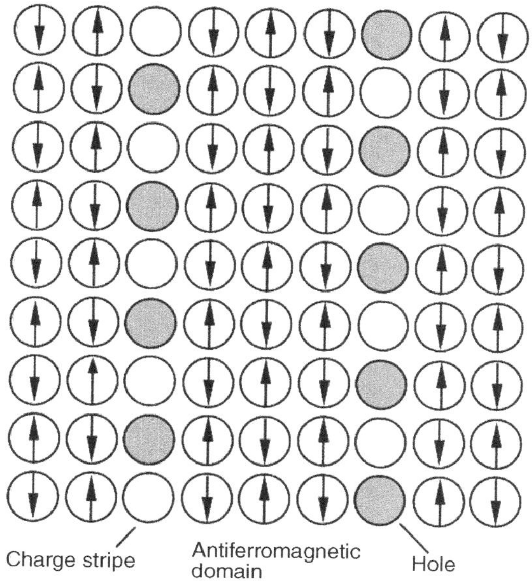

Figure 8.10 Sketch of vertical charge stripes acting as antiferromagnetic domain boundaries (adapted from [1])

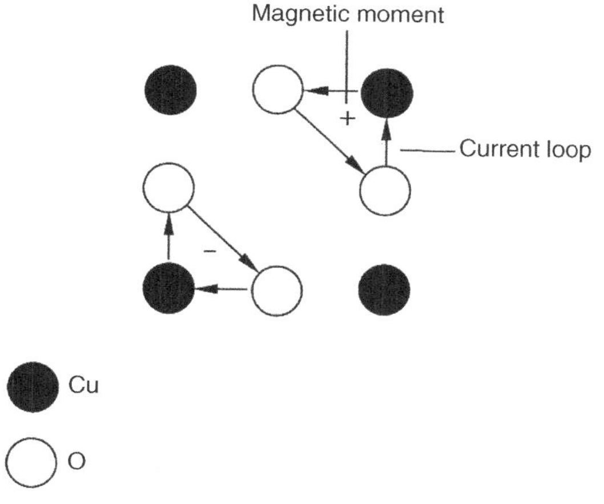

Figure 8.11 Sketch of possible orbital currents in the pseudogap state (adapted from [11, 12])

8.3 Summary

The generic phase diagram of hole-doped cuprate high-temperature superconductors indicates that the parent compounds are Mott insulators. Doping of charge carriers into the CuO_2 planes rapidly destroys the antiferromagnetic state and establishes superconductivity. A carrier concentration of slightly less than 0.2 holes per CuO_2 unit leads to the optimum critical temperature. Above this optimum doping, the critical temperature decreases with increasing charge carrier density, and finally at high doping levels (>0.3 holes per CuO_2 unit) the cuprates do not superconduct. In the high charge carrier density region (overdoped or nonsuperconducting metal), the cuprates show a Fermi liquid behavior. Optimally doped cuprate superconductors behave differently, and are often referred to as strange metals. In contrast to a T^2 dependence of the electrical resistivity expected from Fermi liquid theory, their resistivity increases linearly with temperature.

The superconducting order parameter of the cuprate high-temperature superconductors has a *d*-wave symmetry. In a *d*-wave superconductor, the width of the energy gap depends on the wave vector **k** of the quasiparticles. In the nodal directions, the width of the energy gap is zero, whereas the maximum gap is found in the antinodal directions.

Complex behavior has been found in the underdoped region close to the antiferromagnetic state. Even well above the critical temperature, a pseudogap has been observed. As a first hypothesis, the occurrence of the pseudogap has been attributed to pairs which lack phase coherence. Evidence for the existence of Cooper pairs above the critical temperature has been found using the Nernst effect. However, the maximum temperature for such superconducting fluctuations seems to be well below T^*. In a second approach, the pseudogap is considered to compete with superconductivity, involving such phenomena as fluctuating charge stripes and unusual orbital currents. The pseudogap of the underdoped cuprates is still subject to debate. The complexity of the physics of the cuprates is closely related to the evolution of the Fermi surface, which is fully developed in the overdoped region, whereas it vanishes in a Mott insulator. On the way from the Mott insulator to the overdoped region, Fermi arcs and pockets of increasing size have been observed byARPES.

References

1. J. Orenstein and A. Millis, Advances in the physics of high-temperature superconductivity, *Science*, **288**, 468–474 (2000).
2. M. Buchanan, Mind the pseudogap, *Nature*, **409**, 8–11 (2001).
3. L. Alff, Y. Krockenberger, B. Welter, M. Schonecke, R. Gross, D. Manske, and M. Naito, A hidden pseudogap under the "dome" of superconductivity in electron-doped high-temperature superconductors, *Nature*, **422**, 698–701 (2003).
4. A.J. Millis, Gaps and our understanding, *Science*, **314**, 1888–1889 (2006).
5. R. Wesche, High-temperature superconductors, in *Springer Handbook of Electronic and Photonic Materials*, S. Kasap and P. Capper (eds.), Springer Science & Business Media, Inc., New York, 2006.
6. N. Doiron-Leyraud, C. Proust, D. LeBoeuf, J. Levallois, J.-P. Bonnemaison, R. Liang, D.A. Bonn, W.N. Hardy, and L. Taillefer, Quantum oscillations and the Fermi surface in an underdoped high-T_c superconductor, *Nature*, **447**, 565–568 (2007).
7. R.A. Fisher, J.E. Gordon, and N.E. Phillips, High-T_c superconductors: Thermodynamic properties, in *Handbook of High-Temperature Superconductivity: Theory and Experiment*, J.R. Schrieffer and J.S. Brooks (eds.), Springer, New York, 2007.
8. P.A. Lee, *t-J* model and the gauge theory description of underdoped cuprates, In *Handbook of High-Temperature Superconductivity: Theory and Experiment*, J.R. Schrieffer and J.S. Brooks (eds.), Springer, New York, 2007.

9. S.H. Naqib and R.S. Islam, Extraction of the pseudogap energy scale from the static magnetic susceptibility of single and double CuO_2 plane high-T_c cuprates, *Supercond. Sci. Technol.*, **21**, 105017 (2008).
10. Y. Kohsaka, C. Taylor, P. Wahl, A. Schmidt, Jhinhwan Lee, K. Fujita, J.W. Alldredge, K. McElroy, Jinho Lee, H. Eisaki, S. Uchida, D.-H. Lee, and J.C. Davis, How Cooper pairs vanish approaching the Mott insulator in $Bi_2Sr_2CaCu_2O_{8+\delta}$, *Nature*, **454**, 1072–1078 (2008).
11. Y. Li, V. Balédent, N. Barišić, Y. Cho, B. Fauqué, Y. Sidis, G. Yu, X. Zhao, P. Bourges, and M. Greven, Unusual magnetic order in the pseudogap region of the superconductor $HgBa_2CuO_{4+\delta}$, *Nature*, **455**, 372–375 (2008).
12. C. Varma, Mind the pseudogap, *Nature*, **468**, 184–185 (2010).
13. M.J. Lawler, K. Fujita, J. Lee, A.R. Schmidt, Y. Kohsaka, C.K. Kim, H. Eisaki, S. Uchida, J.C. Davis, J.P. Sethna, and E.-A. Kim, Intra-unit-cell electronic nematicity of the high-T_c copper-oxide pseudogap states, *Nature*, **466**, 347–351 (2010).
14. M.R. Norman, Fermi surface reconstruction and the origin of high-temperature superconductivity, *Physics*, **3**, 86 (2010).
15. I.M. Vishik, W.S. Lee, R.-H. He, M. Hashimoto, Z. Hussain, T.P. Devereaux, and Z.-X. Shen, ARPES studies of cuprate Fermiology: Superconductivity, pseudogap and quasiparticle dynamics, *New J. Phys.*, **12**, 105008 (2010).
16. J.M. Tranquada, Neutron scattering studies of antiferromagnetic correlations in cuprates, in *Handbook of High-Temperature Superconductivity: Theory and Experiment*, J.R. Schrieffer and J.S. Brooks (eds.), Springer, New York, 2007.
17. Ch. Renner, B. Revaz, J.-Y. Genoud, K. Kadowaki, and Ø. Fischer, Pseudogap precursor of the superconducting gap in under- and overdoped $Bi_2Sr_2CaCu_2O_{8+\delta}$, *Phys. Rev. Lett.*, **80**, 149–152 (1998).
18. D.J. Van Harlingen, Phase-sensitive tests of the symmetry of the pairing state in the high-temperature superconductors – Evidence for $d_x2_{-y}2$ symmetry, *Rev. Mod. Phys.*, **67**, 515–535 (1995).
19. C.C. Tsuei, J.R. Kirtley, C.C. Chi, Lock See Yu-Jahnes, A. Gupta, T. Shaw, J.Z. Sun, and M.B. Ketchen, Pairing symmetry and flux quantization in a tricrystal superconducting ring of $YBa_2Cu_3O_{7-\delta}$, *Phys. Rev. Lett.*, **73**, 593–596 (1994).
20. J.R. Kirtley, C.C. Tsuei, J.Z. Sun, C.C. Chi, Lock See Yu-Jahnes, A. Gupta, M. Rupp, and M.B. Ketchen, Symmetry of the order parameter in the high-T_c superconductor $YBa_2Cu_3O_{7-\delta}$, *Nature*, **373**, 225–228 (1995).
21. M. Sigrist and T.M. Rice, Paramagnetic effect in high T_c superconductors – A hint for *d*-wave superconductivity, *J. Phys. Soc. Japan*, **61**, 4283–4286 (1992).
22. D.A. Wollman, D.J. Van Harlingen, J. Giapintzakis, and D.M. Ginsberg, Evidence for d_{x2-y2} pairing from the magnetic field modulation of $YBa_2Cu_3O_{7-\delta}$-Pb Josephson junctions, *Phys. Rev. Lett.*, **74**, 797–800 (1995).
23. A. Kanigel, U. Chatterjee, M.R. Norman, G. Koren, K. Kadowaki, and J.C. Campuzano, Evidence for pairing above the transition temperature of cuprate superconductors from the electronic dispersion in the pseudogap phase, *Phys. Rev. Lett.*, **101**, 137002 (2008).
24. H.-B. Yang, J.D. Rameau, P.D. Johnson, T. Valla, A. Tsvelik, and G.D. Gu, Emergence of preformed Cooper pairs from the doped Mott insulating state in $Bi_2Sr_2CaCu_2O_{8+\delta}$, *Nature*, **456**, 77–80 (2008).

25. K. Nakayama, T. Sato, Y. Sekiba, K. Terashima, P. Richard, T. Takahashi, K. Kudo, N. Okumura, T. Sasaki, and N. Kobayashi, Evolution of a pairing-induced pseudogap from the superconducting gap of $(Bi,Pb)_2Sr_2CuO_6$, *Phys. Rev. Lett.*, **102**, 227006 (2009).
26. V. Sandu, E. Cimpoiasu, T. Katuwal, Shi Li, M.B. Maple, and C.C. Almasan, Evidence for vortices in the pseudogap region of $Y_{1-x}Pr_xBa_2Cu_3O_7$, *Phys. Rev. Lett.*, **93**, 177005 (2004).
27. Z.A. Xu, N.P. Ong, Y. Wang, T. Takeshita, and S. Uchida, Vortex-like excitations and the onset of superconducting phase fluctuation in underdoped $La_{2-x}Sr_xCuO_4$, *Nature*, **406**, 486–488 (2000).
28. N.E. Hussey, Normal state transport properties, in *Handbook of High-Temperature Superconductivity: Theory and Experiment*, J.R. Schrieffer and J.S. Brooks (eds.), Springer, New York, 2007.
29. J.-H. Ma, Z.-H. Pan, F.C. Niestemski, M. Neupane, Y.-M. Xu, P. Richard, K. Nakayama, T. Sato, T. Takahashi, H.-Q. Luo, L. Fang, H.-H. Wen, Z. Wang, H. Ding, and V. Madhavan, Coexistence of competing orders with two energy gaps in real and momentum space in the high temperature superconductor $Bi_2Sr_{2-x}La_xCuO_{6+\delta}$, *Phys. Rev. Lett.*, **101**, 207002 (2008).
30. K. Tanaka, W.S. Lee, D.H. Lu, A. Fujimori, T. Fujii, Risdiana, I. Terasaki, D.J. Scalapino, T.P. Devereaux, Z. Hussain, and Z.X. Shen, Distinct Fermi-momentum-dependent energy gaps in deeply underdoped Bi2212, *Science*, **314**, 1910–1913 (2006).
31. W.S. Lee, I.M. Vishik, K. Tanaka, D.H. Lu, T. Sasagawa, N. Nagaosa, T.P. Devereaux, Z. Hussain, and Z.-X. Shen, Abrupt onset of a second energy gap at the superconducting transition of underdoped Bi2212, *Nature*, **450**, 81–84 (2007).
32. J.M. Tranquada, H. Woo, T.G. Perring, H. Goka, G.D. Gu, G. Xu, M. Fujita, and K. Yamada, Quantum magnetic excitations from stripes in copper oxide superconductors, *Nature*, **429**, 534–538 (2004).
33. L. Tassini, F. Venturini, Q.-M. Zhang, R. Hackl, N. Kikugawa, and T. Fujita, Dynamical properties of charged stripes in $La_{2-x}Sr_xCuO_4$, *Phys. Rev. Lett.*, **95**, 117002 (2005).
34. C.C. Homes, S.V. Dordevic, G.D. Gu, Q. Li, T. Valla, and J.M. Tranquada, Charge order, metallic behaviour, and superconductivity in $La_{2-x}Ba_xCuO_4$ with $x = 1/8$, *Phys. Rev. Lett.*, **96**, 257002 (2006).
35. E. Berg, E. Fradkin, S.A. Kivelson, and J.M. Tranquada, Striped superconductors: Howspin, charge and superconducting orders intertwine in the cuprates, *New J. Phys.*, **11**, 115004 (2009).
36. C.V. Parker, P. Aynajian, E.H. da Silva Neto, A. Pushp, S. Ono, J. Wen, Z. Xu, G. Gu, and A. Yazdani, Fluctuating stripes at the onset of the pseudogap in the high-T_c superconductor $Bi_2Sr_2CaCu_2O_{8+x}$, *Nature*, **468**, 677–680 (2010).
37. J. Zaanen, A modern, but way to short history of the theory of superconductivity at high temperature, in *100 Years of Superconductivity*, H. Rogalla and P.H. Kes (eds.), CRC Press, Boca Raton, FL, 2012.

9

Superconducting Properties of Cuprate High-T_c Superconductors

9.1 Introduction

The description of the generic phase diagram of hole-doped cuprate high-T_c superconductors in Chapter 8 revealed the complexity of the physics of these materials. Next, let us consider the superconducting properties of the cuprate high-T_c superconductors. First, the values of the characteristic length scales and the Ginzburg–Landau parameter are presented in Section 9.2. In the layered crystal structures, CuO_2 blocks are separated by insulating layers which act as charge carrier reservoirs, leading to highly anisotropic superconducting properties. As a further consequence, the critical current densities are high along the copper oxide planes, but very low perpendicular to them (i.e., along the c direction). This leads to two different penetration depths λ_{ab} and λ_c, where the indices give the direction of the screening currents. There also exist two different coherence lengths, namely ξ_{ab} in the CuO_2 planes and ξ_c perpendicular to them. Relations between different material properties and the definition of the Ginzburg–Landau parameters κ_{ab} and κ_c have already been presented in Section 4.5 of Chapter 4.

In Section 9.3, the values of the maximum width of the d-wave gaps will be presented. The magnetic phase diagram of cuprate superconductors will be briefly described in Section 9.4. The lower and upper critical fields depend on whether the magnetic field is applied parallel or perpendicular to the CuO_2 planes. In the symbols for the upper ($B_{c2,ab}$, $B_{c2,c}$) and lower critical fields ($B_{c1,ab}$, $B_{c1,c}$), the indices ab and c indicate the direction of the applied magnetic field.

For most applications of superconductivity, high critical current densities are of crucial importance. In Section 9.5, the critical currents in polycrystalline samples, bulk material, and cuprate films are compared. Evidence for considerably different critical currents within grains and across grain boundaries will be discussed. Because of the importance of current

Physical Properties of High-Temperature Superconductors, First Edition. Rainer Wesche.

flow across grain boundaries for the development of long superconducting wires, we will discuss this aspect in more detail in Section 9.6.

9.2 Characteristic Length Scales

For the characteristic length scales in cuprate superconductors we have [1]

$$\xi_c < \xi_{ab} << \lambda_{ab} < \lambda_c \tag{9.1}$$

The fact that the penetration depth λ_{ab} for the flow of the screening currents in the CuO_2 planes is considerably smaller than λ_c for screening currents perpendicular to the CuO_2 planes simply indicates that the in-plane critical currents are much larger than the out-of-plane ones. The coherence length is the shortest length within which a significant change in the Cooper pair density is possible. Alternatively, we may consider the coherence length as the size of a Cooper pair. The coherence length ξ_{ab} along the CuO_2 planes is typically much larger than ξ_c perpendicular to them. The characteristic length scales of various cuprate superconductors at zero temperature are listed in Table 9.1 [2–36]. The values in round brackets indicate the error bars. Like other physical properties, both the penetration depth and the coherence length depend on the actual carrier concentration, which is typically in the range of 3×10^{21} to 17×10^{21} cm^{-3} in cuprate superconductors [1]. For comparison, the carrier concentrations in copper and aluminum are 84.5×10^{21} and 180.6×10^{21} cm^{-3}, considerably larger than in cuprate superconductors. In Table 9.1, the values of the Ginzburg–Landau parameter $\kappa_c = \lambda_{ab}/\xi_{ab}$ are quoted.

The values of the Ginzburg–Landau parameter κ_c are between 22 and 120, and hence the cuprates are extreme Type II superconductors. For screening currents circulating in the CuO_2 planes, the penetration depth (λ_{ab}) is typically in the range of 100–300 nm. In general, λ_c is much larger than λ_{ab}, reaching values of the order of 1000 nm. The extreme shortness of the coherence lengths, especially perpendicular to the CuO_2 planes, is the reason that grain boundaries can act as weak links for the supercurrents in cuprate superconductors. The coherence length ξ_c is of the order of several tenths of a nanometer, which is comparable to interatomic distances. For comparison, the coherence length of the metallic superconductors NbTi, Nb_3Sn, and Nb_3Ge is between 3 and 4 nm.

In Figure 9.1, the temperature dependences of the penetration depths λ_{ab} of various cuprate high-T_c superconductors are shown. The penetration depth is closely related to the number density of the Cooper pairs (see Equation 2.31 in Chapter 2). In low-temperature superconductors, the dependence of the penetration depth on temperature is well described by the empirical relation $\lambda(T) = \lambda_0(1-(T/T_c)^4)^{-0.5}$ (Equation 2.24, Chapter 2). As illustrated in Figure 9.1, the $\lambda_{ab}(T)$ data for Y-123 [15] can be well represented by this relation, but not for Bi-2201 [20], (Hg,Cu)-1201 [35] or Tl-2223 [29]. The $\lambda_{ab}(T)$ data of these high-T_c superconductors can be fitted by the alternative expression $\lambda_{ab}(T) = \lambda_{ab}(0)$ $(1-(T/T_c))^a$ (where $a \approx -0.3$) [29] (see Figure 9.1). At higher temperatures, the empirical relation (Equation 2.24) typically provides a good approximation, as illustrated for (Hg,Cu)-1201 [34] in Figure 9.1. The temperature dependence of the coherence length ξ_{ab} of Tl-2223 can be described up to $T/T_c = 0.95$ by the relation $\xi_{ab} = \xi_{ab}(0)(1 - (T/T_c))^b$ ($b \approx -0.52$) [29].

Table 9.1 *Characteristic length scales and the Ginzburg–Landau parameters of various cuprate superconductors*

Compound	T_c (K)	λ_{ab} (nm)	λ_c (nm)	ξ_{ab} (nm)	ξ_c (nm)	κ_c	Reference
$La_{2-x}Sr_xCuO_4$	36	258					[2]
$La_{1.8}Sr_{0.2}CuO_4$	36.5	250		2.1		120	[3]
$La_{1.5}Nd_{0.4}Sr_{0.1}CuO_4$	10			6.3	0.21		[4]
$La_{1.48}Nd_{0.4}Sr_{0.12}CuO_4$[a]	6.7			9.1	0.27		
$La_{1.45}Nd_{0.4}Sr_{0.15}CuO_{4+\delta}$	24			4.8	0.19		
$Nd_{1.85}Ce_{0.15}CuO_{4-\delta}$[b]	26.2			5.33	0.71		[5]
$Sr_{0.9}La_{0.1}CuO_2$[b]	43	116(2)					[6]
$YBa_2Cu_3O_{6+x}$	56	175(6)				50	[7]
$YBa_2Cu_3O_{6+x}$	20.1		27,800				[8]
	18.1		31,100				
	16.4		35,700				
	12		50,800				
	7.4		89,800				
	4		206,000				
$YBa_2Cu_3O_{6.9}$	92.5	140[c]		2.2[c]		62	[9]
$YBa_2Cu_3O_{7-x}$	88.8	26	125	3.4	0.7	37	[10]
$YBa_2Cu_3O_{7-\delta}$	92.5			1.6	0.3	≈56	[11]
$YBa_2Cu_3O_{7-\delta}$	91	130	450	1.3	0.2	100	[12]
$YBa_2Cu_3O_{7-\delta}$	89.7	140					[13]
$YBa_2Cu_3O_{7-\delta}$	≈92	200(40)	1420(150)				[14]
$YBa_2Cu_3O_{7-\delta}$	≈82	141.5(3.0)	>700				[15]
Y-123 thin film	35.4				0.33		[16]
	78.1				0.30		
	81.9				0.29		
	88.5				0.25		
	90				0.21		
	87.4				0.19		
	81.2				0.14		
	75.2				0.11		
$YBa_2Cu_3O_{7-\delta}$	92	152					[17]
	80	169					
	70	179					
	57	189					
	41	196					
$Y_{0.95}Ca_{0.05}Ba_2Cu_3O_y$	84.2			1.6(2)			[18]
$Y_{0.95}Ca_{0.05}Ba_2Cu_{2.94}Zn_{0.06}O_y$	65.1			1.9(2)			
$Y_{0.95}Ca_{0.05}Ba_2Cu_{2.88}Zn_{0.12}O_y$	33.3			2.6(2)			
$Y_{0.9}Ca_{0.1}Ba_2Cu_3O_y$	80.6			3.4(2)			
$YBa_2Cu_3O_{7-\delta}$	92.6	96.2	697				[19]
	90.1	86.3	619				
$Y_{0.992}Pr_{0.008}Ba_2Cu_3O_{7-\delta}$	91.2	105.4	829				

(continued overleaf)

Table 9.1 *(continued)*

Compound	T_c (K)	λ_{ab} (nm)	λ_c (nm)	ξ_{ab} (nm)	ξ_c (nm)	κ_c	Reference
$Y_{0.987}Pr_{0.013}Ba_2Cu_3O_{7-\delta}$	89.9	84.3	622				[19]
$Y_{0.976}Pr_{0.024}Ba_2Cu_3O_{7-\delta}$	90.2	84.0	620				
$Bi_2(Sr,La)_2CuO_y$	28.3	438		4.0		≈110	[20]
$Bi_2Sr_{1.6}La_{0.4}CuO_{4+\delta}$	22.1(2)	290(8)		13(3)		≈22	[21]
$(Bi,Pb)_2(Sr,La)_2CuO_{6+\delta}$	35	358					[2]
$Bi_2Sr_{1.5}Ca_{1.5}Cu_2O_{8+\delta}$	82.4	200(6)		6(1)		≈33	[21]
$Bi_{2.2}Sr_2Ca_{0.8}Cu_2O_{8+\delta}$	85			3.8	0.16		[22]
Bi-2212	≈84			2.7	0.18		[23]
$(Bi,Pb)_2Sr_2CaCu_2O_8$	94	178(8)		2.0(1)		92	[24]
$Bi_2Sr_2Ca_2Cu_3O_x$	106.5			2.9	0.093		[25]
$(Bi,Pb)_2Sr_2Ca_2Cu_3O_x$	109	245		2.9		84	[26]
$Tl_2Ba_2CaCu_2O_x$	≈100			2	0.03		[27]
$Tl_2Ba_2Ca_2Cu_3O_{10+\delta}$	123	117[d]		1.1[d]		≈73	[28]
		124[e]		1.4[e]			
$Tl_2Ba_2Ca_2Cu_3O_{10-\delta}$	126	240	1100				[29]
$Tl_2Ba_2Ca_2Cu_3O_{10}$	121.5	163		1.36		≈100	[30]
$(Tl,Pb,Bi)(Sr,Ba)_2Ca_2Cu_3O_9$	116	137		2.1		65.2	[31]
$HgBa_2CuO_{4+\delta}$	95	117		2.1		≈48	[28]
$HgBa_2CuO_{4+\delta}$	93.2	140		2		70	[32]
$HgBa_2CuO_{4+\delta}$	70(1)	216(17)		3.3(7)			[33]
$\delta = 4.9\%$							
$\delta = 5.5\%$	78.2(4)	161(17)		3.4(7)			
$\delta = 5.5\%$	78.5(5)	200(28)		3.0(6)			
$\delta = 6.6\%$	88.5(9)	153(23)		2.3(5)			
$\delta = 9.6\%$	95.6(6)	145(9)		2.0(4)			
$\delta = 9.7\%$	95.3(6)	165(9)		2.0(4)			
$\delta = 10.0\%$	94.1(2)	158(12)		2.2(4)			
$\delta = 10.1\%$	93.4(2)	156(9)		1.8(4)		≈100	
$\delta = 10.3\%$	92.5(5)	139(17)		2.0(4)			
$\delta = 10.5\%$	90.9(2)	156(15)		1.7(3)			
$\delta = 10.8\%$	89.1(5)	177(20)		1.8(4)			
$(Hg,Cu)Ba_2CuO_{4+\delta}$	96	247					[34]
$Hg_{0.8}Cu_{0.2}Ba_2CuO_{4+\delta}$	96	162(5)		1.43(8)		113	[35]
	92	196(5)		1.70(8)		115	
	85	213(5)		2.2(2)		97	
	75	235(5)					
	62	283(6)					
$HgBa_2Ca_2Cu_3O_{8+\delta}$	132	150					[36]
$Ca_{2-x}Na_xCuO_2Cl_2$	28	316					[2]

[a] Charge stripe ordering;
[b] Electron-doped cuprate superconductor;
[c] $\lambda(0)$ and $\xi(0)$ indices not specified.
[d] Aligned;
[e] Random

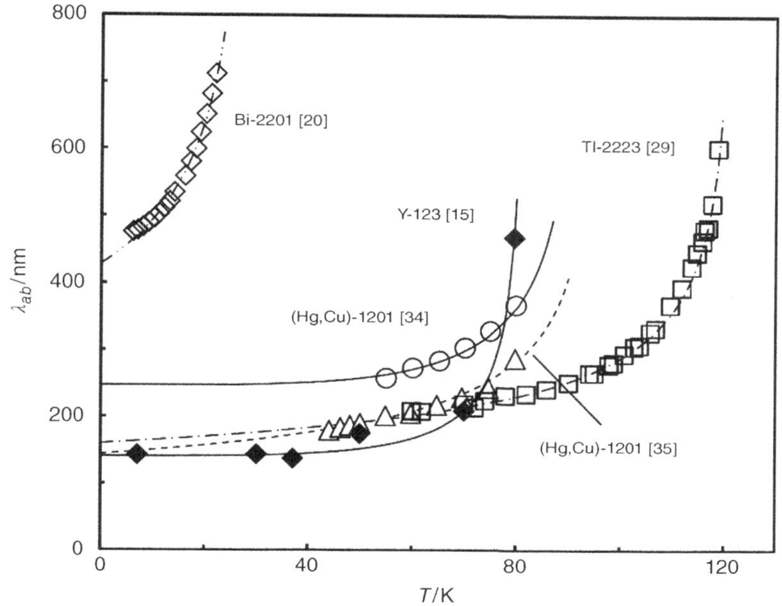

Figure 9.1 *Penetration depth λ_{ab} versus temperature of various high-T_c superconductors. The solid lines are fits to the empirical relation $\lambda_{ab}(T) = \lambda_{ab}(0)(1-(T/T_c)^4)^{-0.5}$ (Equation 2.24, Chapter 2). The dotted lines are fits to the alternative expression $\lambda_{ab}(T) = \lambda_{ab}(0)(1-(T/T_c))^a$ [29]. The values of the fit parameter a are in the range of 0.33–0.38 (data from [15, 20, 29, 34, 35])*

9.3 Superconducting Energy Gap

We have already discussed in Chapter 8 the fact that the superconducting order parameter of the cuprate high-T_c superconductors is of d-wave symmetry, as a consequence of which the energy gap is zero in the nodal directions and reaches its maximum value in the antinodal direction. In underdoped cuprate superconductors, there exists a pseudogap at temperatures well above T_c. The pseudogap seems to be in competition with superconductivity. Here we will only discuss the maximum width of the superconducting energy gap. The values of the half-width Δ of the energy gap of various cuprate superconductors are listed in Table 9.2 [37–53].

The relation between the maximum width of the superconducting energy gap and the critical temperature is illustrated in Figure 9.2. In optimally doped and overdoped cuprate superconductors, the ratio $2\Delta_{max}/k_BT_c$ is typically between 5 and 10. Only for the underdoped region which have higher values of this ratio been reported. In conventional metallic superconductors, the ratio of $2\Delta/k_BT_c$ is only 3.5 (see Figure 3.4, Chapter 3). The relatively large value $2\Delta_{max}/k_BT_c \approx 8$ has been considered as evidence for strong coupling in the cuprate superconductors. A few investigations of the energy gap using muon spin rotation (μSR) suggest that in La-214 [39] and Y-123 [45] two superconducting energy gaps are present. The μSR data indicate that in La-214 as well as in Y-123, a larger gap of d-wave symmetry coexists with a much smaller one of s-wave symmetry. Another interesting aspect

Table 9.2 *Energy gap of various cuprate superconductors*

Compound	T_c (K)	Δ_{max} (meV)	$2\Delta_{max}/k_BT_c$	Reference
$La_{1.8}Sr_{0.2}CuO_4$	38	11.5	7.1	[37]
$La_{1.85}Sr_{0.15}CuO_4$	37	13	7.9	[38]
$La_{1.83}Sr_{0.17}CuO_4$ [a]	35.5	8.0(1) [b]	5.24(7)	[39]
		1.54(8) [c]	1.00(5)	
$La_{1.855}Sr_{0.145}CuO_4$	36	14	9.0	[40]
$La_{1.895}Sr_{0.105}CuO_4$	30	26	20.1	
$YBa_2Cu_3O_7$	89.5	35	9.1	[41]
$YBa_2Cu_3O_{7-x}$ [d]	90	18	4.7(1.2)	[42]
$YBa_2Cu_3O_7$	92	≈32	8	[43]
$YBa_2Cu_3O_{7-x}$	90	≈31	8	[44]
$YBa_2Cu_3O_{7-x}$ [a]	91.2	22.9(1) [b]	5.83	[45]
		0.71(1) [c]	0.18	
$Bi_{2.07}Sr_{1.8}CuO_{6+x}$	7	1.85	6.1	[46]
$Bi_2Sr_2CaCu_2O_{8+x}$	≈92	35	≈6.5	[41]
$Bi_2Sr_2CaCu_2O_{8+x}$	80	≈29.5	≈8.6	[47]
$Bi_{2.1}Sr_{1.9}CaCu_2O_{8+x}$	85	23	6.2	[46]
$Bi_2Sr_2CaCu_2O_8$	83	30(4)	≈8	[48]
$Bi_2Sr_2CaCu_2O_{8+x}$		40 [e]		[49]
		55 [e]		
$Bi_2Y_{0.1}Sr_{1.9}CaCu_2O_{8+x}$	94	27	6.6	[46]
$Bi_2Sr_2Ca_2Cu_3O_{10+x}$	109.5(5)	41–58 [f]		[50]
$Bi_2Sr_2Ca_2Cu_3O_{10+x}$		30 [g]		[51]
		21.5 [h]		
$(Bi,Pb)_2Sr_2Ca_2Cu_3O_{10+x}$	108	30	6.4	[46]
$Pr_{2-x}Ce_xCuO_{4-\delta}$ [i] $x=0.13$	17(1)	2.5	3.5(5)	[52]
$x=0.15$	19(1)	3.25	4.0(4)	
$x=0.16$	16(1)	2.6	3.8(5)	
$x=0.17$	13(1)	1.3	2.3(1.3)	
$x=0.18$	11(1)	1.0	2.1(1.5)	
$x=0.19$	8.0(4)	0.9	2.6(9)	
$Tl_2Ba_2CaCu_2O_{8+x}$	94.5	25(3)	6.1(7)	[53]
$Tl_2Ba_2Ca_2Cu_3O_{10+x}$	114	30(5)	6.1(1.0)	[53]

[a] Evidence for two gaps;
[b] *d*-wave gap;
[c] *s*-wave gap;
[d] Film of 1 μm thickness;
[e] Underdoped samples;
[f] Spatial variation of the energy gap;
[g] Inner CuO_2 planes;
[h] Outer CuO_2 planes;
[i] Electron doped.

is a spatial variation of the energy gap of Bi-2212 [49] and Bi-2223 [50], which has been detected by scanning tunneling microscopy (STM). The studies revealed the existence of regions with a smaller energy gap and large sharp coherence peaks, and other regions with a larger gap and small broad peaks. Both regions are embedded in a background with an intermediate gap. A large value of the energy gap seems to be correlated with a larger local

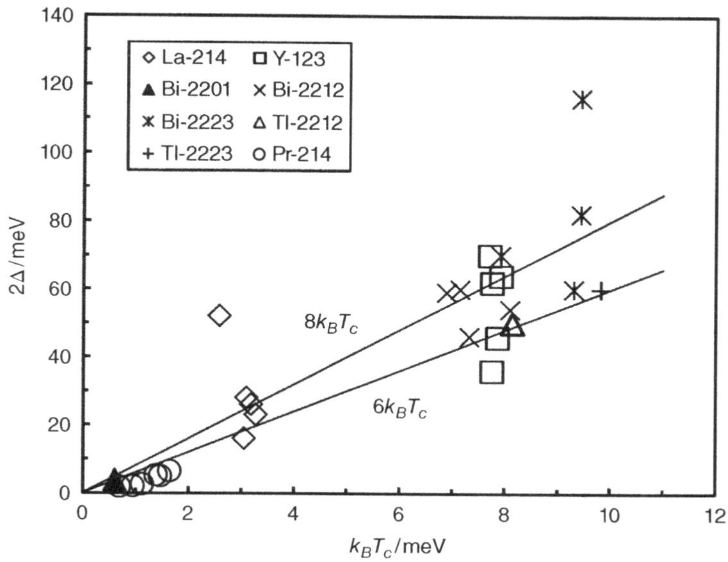

Figure 9.2 *Maximum values of $2\Delta_{max}$ versus k_BT_c of various cuprate superconductors. The lines indicate ratios $2\Delta_{max}/k_BT_c$ of 6 and 8 (data from Table 9.2)*

concentration of interstitial oxygen atoms [54]. In underdoped Bi-2212, the fraction of the area with a large gap has been observed to increase with decreasing number of holes per CuO_2 unit. Thus, good superconductivity has been attributed to regions with small gaps, whereas the large gap seems to be characteristic of underdoped cuprates.

9.4 Magnetic Phase Diagram and Irreversibility Line

In Figure 9.3, the simplified magnetic phase diagram of cuprate high-T_c superconductors is presented. The inset shows for comparison the magnetic phase diagram of a metallic Type II superconductor. In both materials, magnetic flux is excluded from the interior below the lower critical field B_{c1}. In this low field region, the cuprate superconductor is in the perfectly diamagnetic Meissner phase. A new phenomenon is the existence of an irreversibility field (B_{irr}), which is well below the upper critical field (B_{c2}). Below the irreversibility field high-T_c superconductors behave like metallic Type II superconductors in the mixed state (Shubnikov phase). Below B_{irr}, there exists a flux-line lattice with pinned vortices, and hence current flow without resistance is possible. Above the irreversibility line, the critical current is zero because of the absence of flux pinning. Typically, the temperature dependence of the irreversibility field can be described by a scaling relation of the form

$$B_{irr}(T) = B_0\left(1 - \frac{T}{T_c}\right)^{\alpha} \tag{9.2}$$

Here B_0 and α are the fitting parameters characteristic of the cuprate superconductor in question. The absence of pinning above the irreversibility line has been attributed to melting

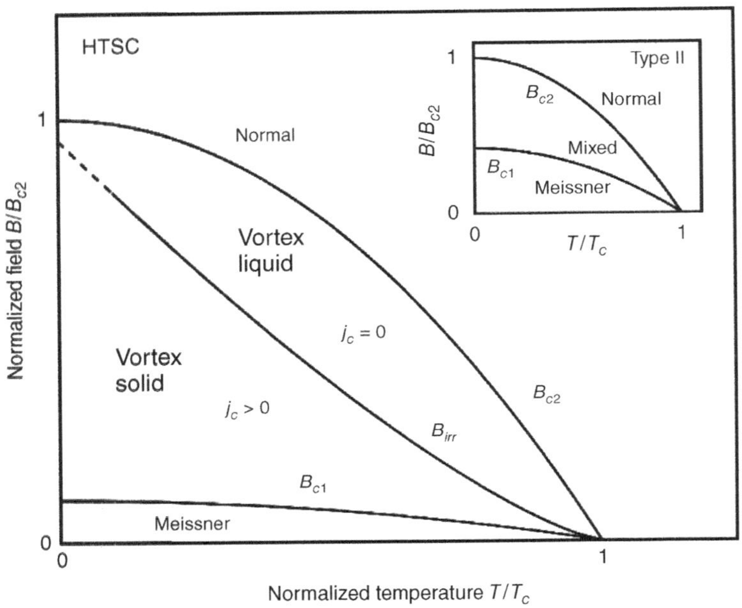

Figure 9.3 *Sketch of the simplified magnetic phase diagram of cuprate high-T_c superconductors. For comparison, the magnetic phase diagram of metallic Type II superconductors is shown in the inset. An outstanding feature of the magnetic phase diagram of cuprate superconductors is the existence of the irreversibility line B_{irr}, above which the flux lines are not pinned, and hence the critical current density is zero*

of the flux-line lattice. It has been predicted that in cuprate superconductors with a low degree of disorder, the transition from a vortex solid to a vortex liquid should be of first order, and hence accompanied by a latent heat [55]. A latent heat of 0.6 $k_B T_m$ per vortex and layer (T_m melting temperature) has been found in specific heat measurements of $YBa_2Cu_3O_7$ single crystals [56]. Similar latent heat of 0.45 $k_B T_m$/vortex/layer has been measured in untwinned Y-123 single crystals [57]. These experimental results suggest that in less disordered cuprate superconductors it should be possible to observe a first-order melting transition. A more conventional explanation for the occurrence of the irreversibility line is thermally activated depinning. Independent of the cause of the existence of the irreversibility line, current transport without resistance is limited to temperatures and fields well below it.

The existence of an irreversibility line was first observed by Müller *et al.* in measurements of the susceptibility of $(La,Ba)_2CuO_4$ [58]. The irreversibility temperature can be deduced from susceptibility versus temperature curves (see Figure 9.4). The magnetic susceptibility (χ) is defined as

$$\chi = \mu_0 \frac{M}{B} \tag{9.3}$$

where M is the magnetization and B the magnetic induction. Figure 9.4 shows a sketch of the susceptibility of $(La,Ba)_2CuO_4$ for field cooling and zero-field cooling. Due to the trapped

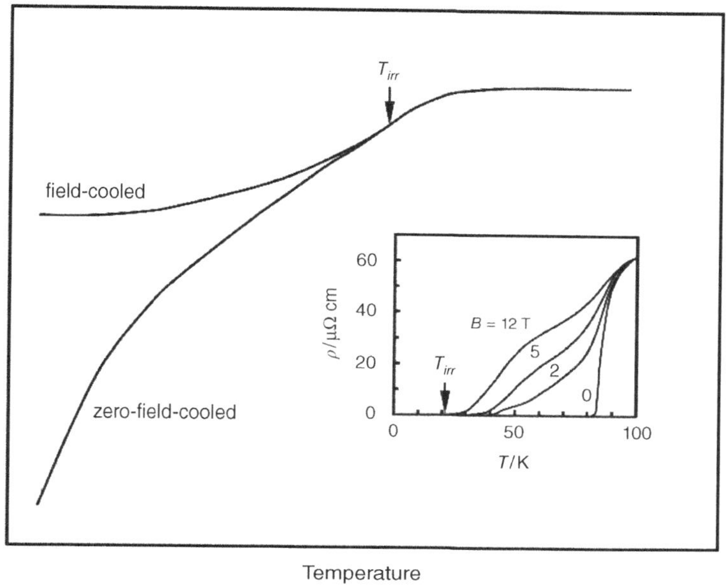

Figure 9.4 *Sketch of susceptibility of (La,Ba)$_2$CuO$_4$ obtained by field cooling and zero-field cooling (adapted from [58]). Above the irreversibility temperature (T_{irr}), the magnetic behavior is reversible. The inset shows the resistivity of a Bi-2212 single crystal measured with magnetic fields applied along the crystallographic c direction (adapted from [59]). The broadening of the resistive transition is a consequence of the fact that B_{irr} is much smaller than B_{c2}*

flux, the susceptibility measured for field cooling is much smaller than that obtained for zero-field cooling. Above the irreversibility temperature the susceptibilities for field cooling and zero-field cooling are identical, i.e., the magnetic behavior is reversible. Another way to determine the irreversibility temperature is the measurement of electrical resistivity in a magnetic field. In the presence of a magnetic field, the resistivity versus temperature curves are considerably broadened as compared to measurements without applied field (see, e.g., [59]). This behavior is illustrated in the inset of Figure 9.4.

We will discuss the irreversibility fields in more detail in Chapter 10. At the upper critical field B_{c2} superconductivity is completely destroyed and the normal state is reached, as in metallic Type II superconductors.

Next, let us consider the typical values of the lower and upper critical fields in cuprate superconductors. The upper critical fields are extremely high and can reach values well above 100 T at low temperatures. Consequently in most high-T_c superconductors the upper critical field cannot be measured directly at 4.2 K. The upper critical field at zero temperature can be estimated from the value of the derivative dB_{c2}/dT at $T = T_c$ using the Werthamer–Helfand–Hohenberg (WHH) theory [60]

$$B_{c2}(0) = -0.69 T_c \left(\frac{dB_{c2}}{dT} \right)_{T=T_c} \tag{9.4}$$

In principle, the lower critical field can be directly determined from the measurement of magnetic hysteresis loops. The upper and lower critical fields of various cuprate high-T_c superconductors are listed in Table 9.3. In general, we find that

$$B_{c2,ab} > B_{c2,c} >> B_{c1,c} > B_{c1,ab} \tag{9.5}$$

As other physical properties, the lower and upper critical fields depend on the actual carrier concentration in the CuO_2 planes. In strongly underdoped $YBa_2Cu_3O_{6+x}$, the lower critical field ($B_{c1,c}$) decreases with decreasing charge carrier concentration [7] (see Table 9.3). Reduced carrier concentrations are reflected in lowered critical temperatures.

Figure 9.5 shows the lower critical fields of four Y-123 samples as a function of temperature. For all considered samples, the temperature dependence of the lower critical field can be approximated by the expression $B_{c1}(T) = B_{c1}(0)\,(1-(T/T_c)^a)$, where $B_{c1}(0)$ is the lower critical field at zero temperature. In the case of Y-123, the lower critical field data can be well represented with an exponent $a = 2.5$. For two Y-123 samples, the lower critical field has been measured for magnetic fields applied parallel and perpendicular to the CuO_2 planes. In both samples, the lower critical field $B_{c1,c}$ for B parallel to the crystallographic c axis is much larger than $B_{c1,ab}$ (B parallel to the CuO_2 planes). A comparison of the values of $B_{c2,ab}$ and $B_{c2,c}$, presented in Table 9.3 indicates that the anisotropy of the superconducting properties of Bi-22(n − 1)n and Tl-22(n − 1)n is much more pronounced than that of Y-123. The values of $B_{c2,ab}$ are extremely high and may reach values of several hundred tesla. Even for the unfavorable field direction, the upper critical field $B_{c2,c}$ may reach values as high as 100 T.

9.5 Critical Current Densities in Cuprate Superconductors

9.5.1 Definitions of the Critical Current

In contrast to the critical temperature and the critical fields, the critical current density is not an intrinsic property of the considered superconductor. A current without resistance can flow in a Type II superconductor only when the flux lines are pinned and cannot move. The ability to pin flux lines depends on the microstructure of the material. For example, flux pinning can be provided by nonsuperconducting precipitates, point defects, or grain boundaries. In Section 9.6, we show that in cuprate superconductors, grain boundaries can act as barriers to the supercurrents.

As a basis for later discussion, we will briefly describe different definitions of the critical current. In principle, it is not possible to measure a resistance that is in fact zero. Consequently, it is necessary to use a criterion which defines the critical current. A first option to define the critical current is an electric field criterion E_{ec}. In the field of high-temperature superconductivity, a criterion of 1 μV/cm is frequently used, whereas the stricter criterion of 0.1 μV/cm is typically applied to metallic superconductors. In the transition region, the electric field and the current are related by the power law

$$E_e = E_{ec}\left(\frac{I}{I_c}\right)^n \tag{9.6}$$

Table 9.3 *Upper and lower critical fields of various cuprate superconductors*

Compound	T_c (K)	$B_{c1,ab}$ (mT)	$B_{c1,c}$ (mT)	$B_{c2,ab}$ (T)	$B_{c2,c}$ (T)	Reference
$La_{1.85}Sr_{0.15}CuO_4$	35			50[a]		[61]
$La_{1.85}Ba_{0.15}CuO_4$	30			36[a]		[61]
$La_{1.85}Sr_{0.15}CuO_4$	36.5			64[a]		[62]
$La_{1.8}Sr_{0.2}CuO_4$	36.5			50[a]		[3]
$La_{1.5}Nd_{0.4}Sr_{0.1}CuO_4$	10			242	8.5	[4]
$La_{1.48}Nd_{0.4}Sr_{0.12}CuO_4$	6.7			132	4	[4]
$La_{1.45}Nd_{0.4}Sr_{0.15}CuO_4$	24			360	14.2	[4]
$YBa_2Cu_3O_{6+x}$	56		23.8(1.5)			[7]
	22		5.8			
	17.6		4.1			
	15		3.3			
	13.6		2.7			
	8.9		1.4			
$YBa_2Cu_3O_7$	92.2			300[a]		[63]
$YBa_2Cu_3O_7$	91	18	53			[12]
$YBa_2Cu_3O_7$	88.2	25(2)	85(4)			[64]
$YBa_2Cu_3O_7$	88.8			140	29	[10]
$YBa_2Cu_3O_{7-x}$	92.5			674	122	[11]
$YBa_2Cu_3O_{7-x}$	92.6			974	134.4	[19]
	90.1			858	119.6	
$Y_{0.992}Pr_{0.008}Ba_2Cu_3O_{7-x}$	91.2			1005	127.7	[19]
$Y_{0.987}Pr_{0.013}Ba_2Cu_3O_{7-x}$				963	130.5	[19]
$Y_{0.976}Pr_{0.024}Ba_2Cu_3O_{7-x}$				1010	136.8	[19]
$Nd_{1.85}Ce_{0.15}CuO_{4-x}$[c]	26.2			87	11.6	[5]
$MoSr_2YCuO_8$	25			21.8		[65]
	27.3			27.5		
$Bi_2(Sr,La)_2CuO_y$	28.3				20.2	[20]
$Bi_{2.2}Sr_2Ca_{0.8}CuO_{8+x}$	85			533	22	[22]
$Bi_2Sr_2CaCu_2O_{8+x}$	84			500	33	[23]
$Bi_{2.1}Sr_{2.1}Ca_{0.9}Cu_2O_{8+x}$	≈85			542	71.5	[66]
$(Bi,Pb)_2Sr_2CaCu_2O_8$	94				89(6)	[24]
$Bi_2Sr_2Ca_2Cu_3O_{10+x}$	110			1210	39	[25]
$Bi_{1.7}Pb_{0.3}Sr_2Ca_2Cu_3O_{10+x}$		90[a]		184[a]		[67]
$Tl_2Ba_2CuO_{6+x}$	86			297	21	[25]
$Tl_2Ba_2CuO_{6+x}$	9.7			0.85[a]		[68]
$Tl_2Ba_2CuO_{6+x}$	15.7			1.9[a]		[68]
$TlBa_2CaCu_2O_{8+x}$	97			120	27	[25]
$Tl_2Ba_2Ca_2Cu_3O_{10+x}$	123				170	[28]
$Tl_2Ba_2Ca_2Cu_3O_{10+x}$	121.5				115	[30]
$Tl_2Sr_2Ca_3Cu_4O_{12+x}$	93			71	16	[25]
$(Tl,Pb,Bi)(Sr,Ba)_2Ca_2Cu_3O_9$	116		37		74.9	[31]
$HgBa_2CuO_{4+x}$	95				66	[28]
$HgBa_2CuO_{4+x}$	93.4				≈100	[33]
$Hg_{0.8}Cu_{0.2}Ba_2CuO_{4+x}$	96				160(20)	[35]
	92				114(11)	
	85				67(15)	
$HgBa_2Ca_2Cu_3O_{8+x}$	132				≈100	[36]

[a]Field direction not specified;
[b]Charge stripe ordering.
[c]Electron doped.

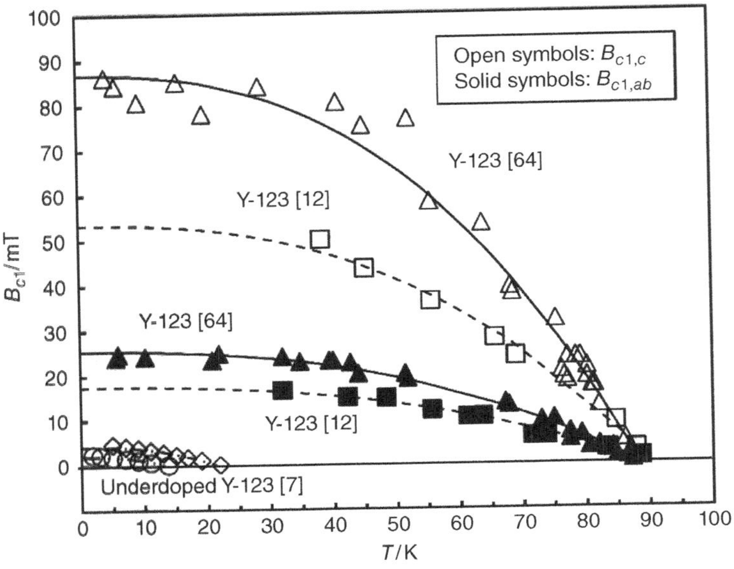

Figure 9.5 *Lower critical fields of four Y-123 samples as functions of temperature. Data indicate a considerable anisotropy of the lower critical field depending on whether the applied field is parallel (open symbols) or perpendicular (solid symbols) to the c direction. The lines are fits to the function $B_{c1}(T) = B_{c1}(0)\,(1-(T/T_c)^a)$, where $B_{c1}(0)$, T_c and a are the fitting parameters (data from [7, 12, 64])*

Here, E_e is the electric field, E_{ec} the electric field criterion used to define the critical current I_c and I the transport current. The parameter n, which is defined by Equation 9.6, is a measure of the sharpness of the transition. Using Equation 9.6, we can compare the values of the critical currents I_{c1} and I_{c2} defined by the criteria of 0.1 and 1 μV/cm, respectively.

$$I_{c2} = I_{c1}\left(\frac{E_{ec2}}{E_{ec1}}\right)^{1/n} = I_{c1} \times 10^{1/n} \tag{9.7}$$

The dependence of the critical current on the electric field criterion is more pronounced when the n factor is low. If we assume to have a critical current of 100 A at $E_{c1} = 0.1$ μV/cm, then the corresponding I_{c2} at 1 μV/cm would be ≈139 A for an n factor of 7, while it would be only ≈112 A for $n = 20$ (see Figure 9.6).

An alternative definition of the critical current is the offset criterion [69, 70], which takes into account different n factors. The offset criterion $I_{c,\text{offset}}$ is defined by taking the tangent to the $E_e - I$ curve at a selected field value E_{ec} and extrapolating to zero electric field (see Figure 9.7). The critical current $I_{c,\text{field}}$ defined by an electric field criterion, and the offset critical current $I_{c,\text{offset}}$, are related by the expression

$$I_{c,\text{offset}} = I_{c,\text{field}}\left(1 - \frac{1}{n}\right) \tag{9.8}$$

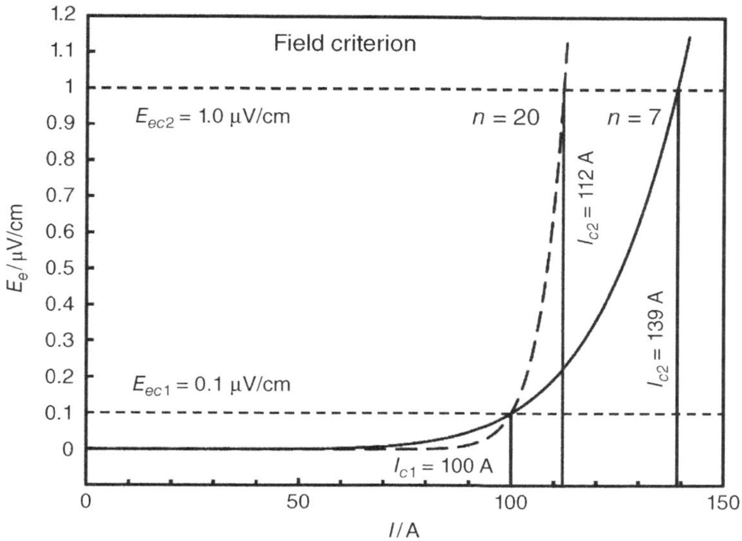

Figure 9.6 *Definition of the critical current by means of electric field criteria of 0.1 and 1 μV/cm. In both considered superconductors, the critical current defined by an electric field criterion of 0.1 μV/cm is 100 A. Nevertheless the E_e–I curves characterized by n factors of 7 and 20 are considerably different. Using an electric field criterion of 1 μV/cm, we would find significantly different critical currents of 112 A (n = 20) and 139 A (n = 7)*

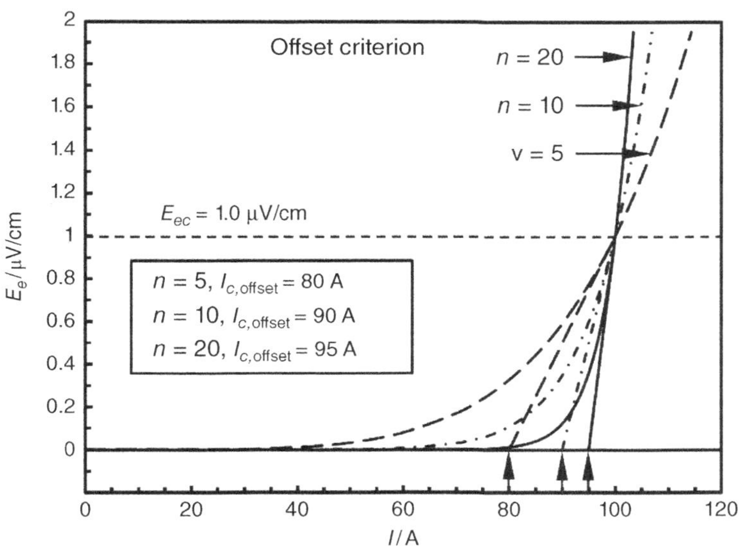

Figure 9.7 *Illustration of the definition of the critical current by an offset criterion of 1 μV/cm. Taking the tangent to the E_e–I curve at E_{ec} = 1 μV/cm, and extrapolating to zero electric field, provide the offset critical current. The offset critical currents of 80, 90, and 95 A reflect the different n factors of 5, 10, and 20, respectively. Using an electric field criterion of 1 μV/cm to define the critical current would lead to identical I_c values of 100 A*

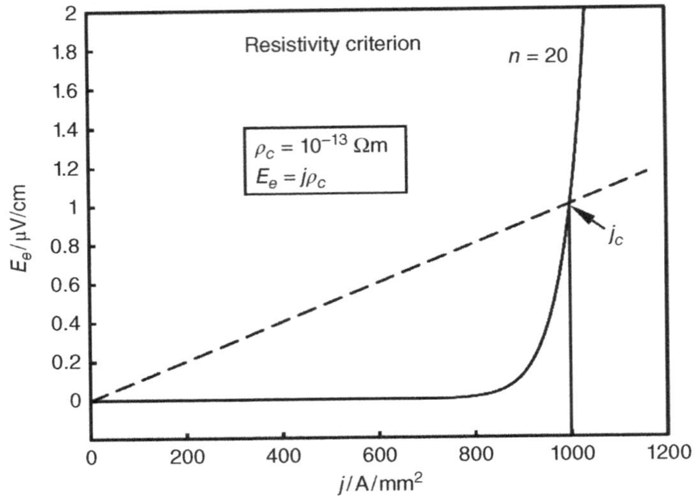

Figure 9.8 *Definition of the critical current with a resistivity criterion. In the considered example a resistivity criterion of* 10^{-13} *Ω m is used*

In Figure 9.7, we use an electric field of 1 μV/cm in the definition of the offset critical current. Three superconductors with the same $I_{c,\text{field}}$ of 100 A ($E_{ec} = 1$ μV/cm) but different n factors of 5, 10, and 20 are compared. Due to different n factors, we obtain different offset critical currents of 80 A ($n = 5$), 90 A ($n = 10$), and 95 A ($n = 20$). The offset criterion takes into account the different broadening of the transitions. In spite of identical $I_{c,\text{field}}$ values for $E_{ec} = 1$ μV/cm, the critical currents $I_{c,\text{field}}$ at an electric field criterion of 0.1 μV/cm would be 63 A ($n = 5$), 79 A ($n = 10$), and 89 A ($n = 20$). For the same value of E_{ec}, the use of an offset criterion provides a better estimation of the current-carrying capacity of the superconductor at lower electric fields than an electric-field criterion.

A third way to define the critical current is based on a predefined resistivity ρ_c (see Figure 9.8). A resistivity criterion of 10^{-13} Ωm is typically used to define the critical current. This value of ρ_c has to be compared with the resistivity of copper of 1.67×10^{-8} Ωm at room temperature. Electric field E_e, critical current density j, and resistivity ρ of a normal conductor are related by Ohm's law.

$$E_e = \rho j \tag{9.9}$$

As illustrated in Figure 9.8, the critical current is defined by the intersection of the straight line corresponding to Ohm's law and the E_e–j curve of the superconductor.

9.5.2 Critical Currents in Polycrystalline Cuprate Superconductors

Soon after the discovery that the critical temperature of $YBa_2Cu_3O_{7-x}$ exceeds 77 K, the boiling point of liquid nitrogen, it was observed that the transport critical current density in sintered Y-123 samples is disappointingly low, typically well below 1000 A/cm² (see, e.g., [71–74]). The dependence of the transport critical current on the applied magnetic field is illustrated in Figure 9.9 (top) for fields of less than 50 mT. The critical current of

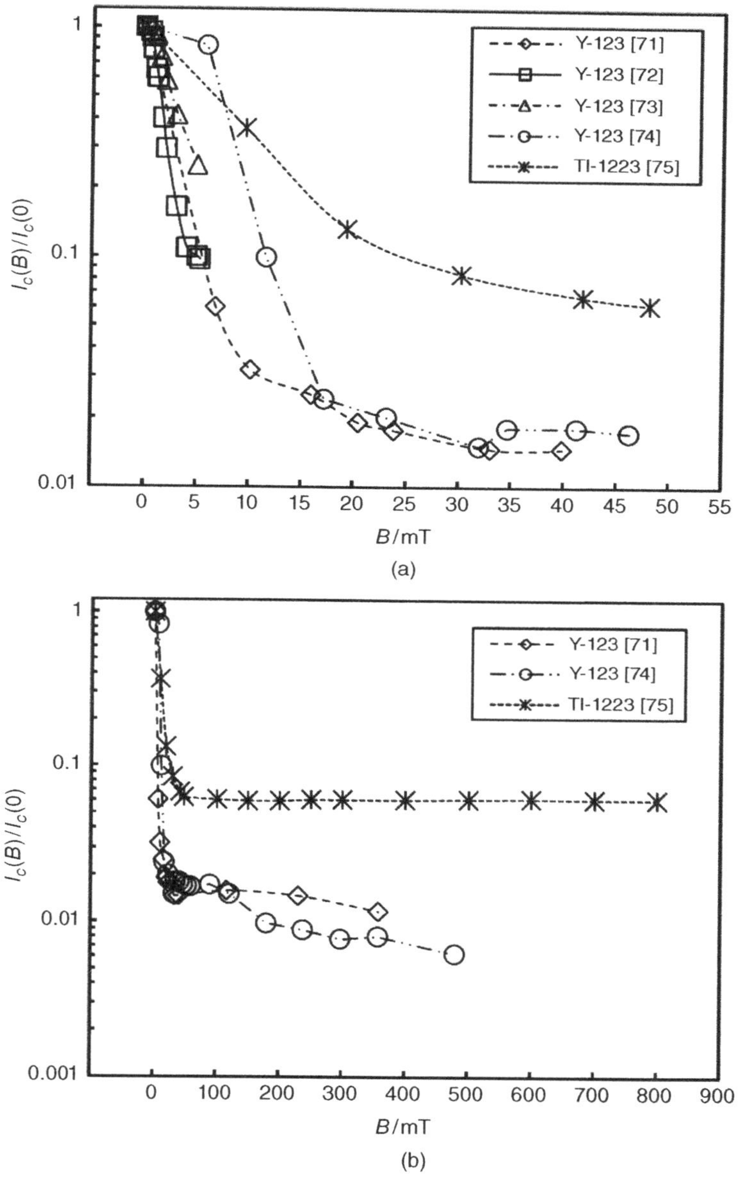

Figure 9.9 *Critical currents of various Y-123 [71–74] and Tl-1223 [75] samples versus applied magnetic field. All values are normalized to the corresponding critical current at 77 K and zero applied field. The low field behavior shown in the top panel indicates that magnetic fields of less than 20 mT can reduce the critical current by an order of magnitude. On the other hand, the small critical currents surviving at higher fields above 100 mT depend only weakly on the applied magnetic field (bottom panel)*

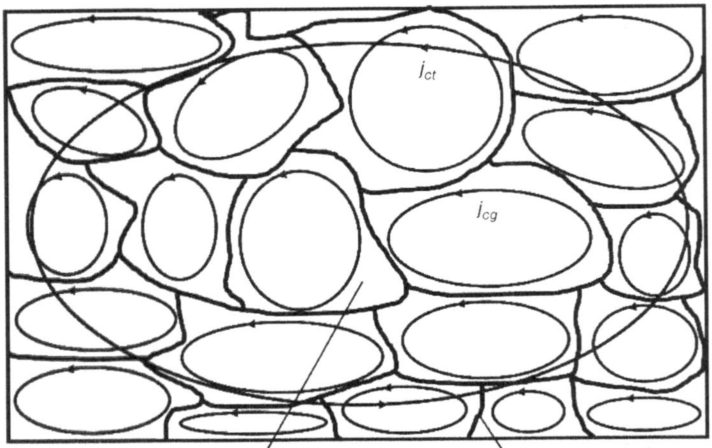

Figure 9.10 *Sketch of a granular cuprate superconductor. Within the individual grains the critical current density (j_{cg}) is high, whereas the transport critical current density $j_{ct} \approx j_{cgb}$ crossing the grain boundaries is much lower. The critical current density j_{cg} within the individual grains is limited by flux pinning. The grain boundaries, which act as barriers for the supercurrents, have to be considered as SIS or SIN Josephson junctions*

sintered Y-123 samples is reduced by more than an order of magnitude due to an applied magnetic field as low as 10 mT. The reduction of the critical current in Tl-1223 tape with a silver–gold sheath [75] is less pronounced. As illustrated in Figure 9.9b, the critical current only depends weakly on the applied magnetic field in the field range of 50–800 mT. The low transport critical current densities in sintered cuprate superconductors have been attributed to granularity. Measurements of magnetization indicate that in sintered cuprate superconductors the critical current density j_{cg} within the individual grains is orders of magnitude larger than the transport critical current j_{ct}, which is limited by the critical current j_{cgb} across the grain boundaries (see Figure 9.10). The grain boundaries act as weak links and have to be considered as Josephson junctions. The small values of j_{cgb} limit the transport critical current to relatively low values. The rapid drop of the critical current caused by extremely low magnetic fields is in line with the hypothesis that the grain-boundary–grain system behaves like a Josephson junction. The small critical currents still present at higher fields may be the consequence of a few grain boundaries acting as strong links. We will discuss the grain-boundary weak links in more detail in Section 9.6.

A further consequence of granularity is the dependence of the transport critical current on history. Figure 9.11 shows the history effect for Y-123/Ag [72], Tl-1223/Ag [76], and Bi-2223/Ag [77] composite conductors. The samples have been cooled to 77 K (Y-123, Tl-1223) or to 4.2 K (Bi-2223) in zero field. The transport critical current was measured in step-wise increasing and then decreasing magnetic fields. The data clearly show that the critical currents found for increasing field are lower than those measured in decreasing fields. At the zero applied field, the initial critical current is higher than the value found after reducing the field from its maximum value to zero.

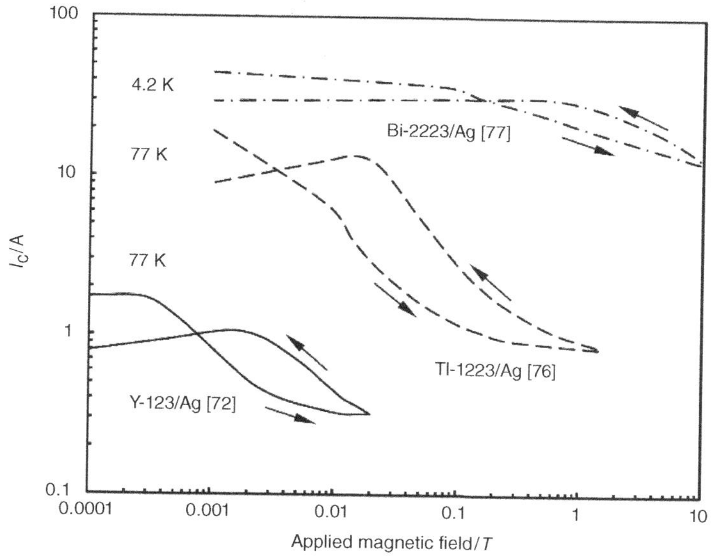

Figure 9.11 *Hysteresis of the transport critical current of Y-123/Ag, Tl-1223/Ag, and Bi-2223/Ag composites. Critical currents measured after zero-field-cooling in increasing magnetic fields are lower than those found for decreasing fields. At very low fields a crossover is observed and the critical currents for decreasing field are lower than those for increasing field (data from [72, 76, 77])*

Evetts and Glowacki proposed a simple model based on the superposition of the magnetic field generated in single grains or colonies of strongly coupled grains, and the applied magnetic field at the grain boundaries, which provides a qualitative explanation of the history effect [72]. At very low magnetic fields, the magnetic flux enters the samples only along the network of weakly coupled grain boundaries, whereas the field does not penetrate into the grains or strongly coupled colonies of grains. At higher fields, the magnetic flux starts to enter the grains. The flow of screening currents close to the surface of the individual grains reduces the magnetic field inside the grains while they enhance the magnetic field at the grain boundaries. The enhanced magnetic field at the grain boundaries is reflected by the compression of the field lines in this region, as illustrated in Figure 9.12 (left). On the other hand, the direction of the screening currents is reversed for decreasing applied magnetic fields, and hence the magnetic field at the grain boundaries is reduced (see Figure 9.12, center). As a consequence of these effects, the critical current measured in increasing field is lower than that found for the decreasing field. Finally, when the applied magnetic field is again zero, there is still some trapped flux in the grains or in colonies of strongly coupled grains which causes a small magnetic field at the grain boundaries (see Figure 9.12, right). The transport critical current is therefore lower than the initial value measured immediately after cool down. A quantitative description of the history effect (see [76, 77]), which would require a more detailed knowledge of the weak link network, the demagnetization factors of grains or grain colonies, and flux pinning within the individual grains, is beyond the scope of this book.

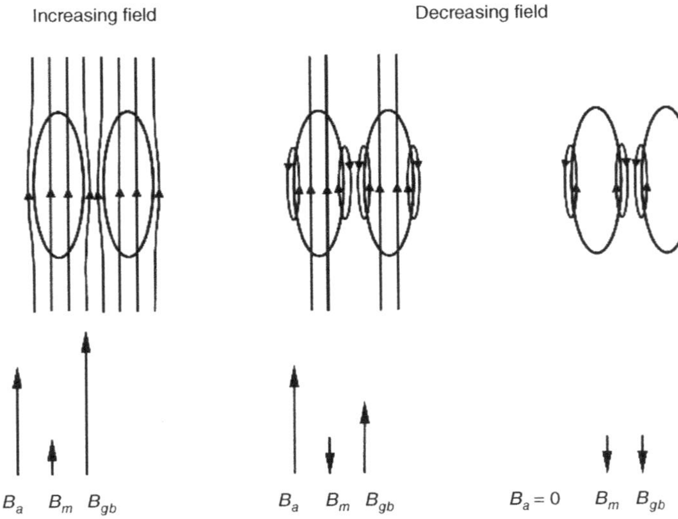

Figure 9.12 The magnetic fields at the grain boundaries, which limit the transport critical current, are the superposition of the applied magnetic field B_a and the magnetic field B_m generated by the screening currents in the individual grains or colonies of strongly coupled grains. For increasing field (left) the field at the grain boundary is enhanced, while for decreasing field it is reduced (center). Coming back to zero applied field the individual grains are still magnetized, and hence a small magnetic field is present at the grain boundaries, leading to a critical current lower than the initial value measured immediately after zero-field cooling (right)

9.5.3 Critical Currents in Bulk Cuprate Superconductors

Partial melting and melt texture growth have been found to be efficient ways to improve the current flow across the grain boundaries in bulk cuprate superconductors. The processing of these materials will be described in Chapter 14. In this section, we will consider only the critical current densities in bulk cuprate superconductors. Figure 9.13 shows the critical current densities of various RE-123 (RE = rare earth element or *Y*) bulk superconductors as a function of the applied magnetic field. The critical current densities at 77 K and zero applied field are in the range of 40 000–310 000 A/cm^2. These values are several orders of magnitude larger than those found for sintered cuprate superconductors. The much higher critical current densities indicate that the critical current is not limited by grain boundary weak links. Nevertheless, the critical current drops rapidly in relatively low magnetic fields in most of the considered samples. For well-textured bulk material, the critical current density depends not only on the absolute value of the applied field but also on its direction. Fields parallel to the crystallographic *c* direction lead to a much more pronounced drop of the critical current density than fields applied along the *ab* planes (see Figure 9.13). The current-carrying capacity of bulk material is dominated by the intragrain critical current density j_{cg} within the single grains, which is limited by flux pinning (see Chapter 10). It has been found that secondary phases (e.g., Y_2BaCuO_5) and additions (e.g., ZnO_2, TiO_2) can improve the flux pinning. The in-field critical current densities seem to depend also on the rare earth elements present in the RE-123 compounds.

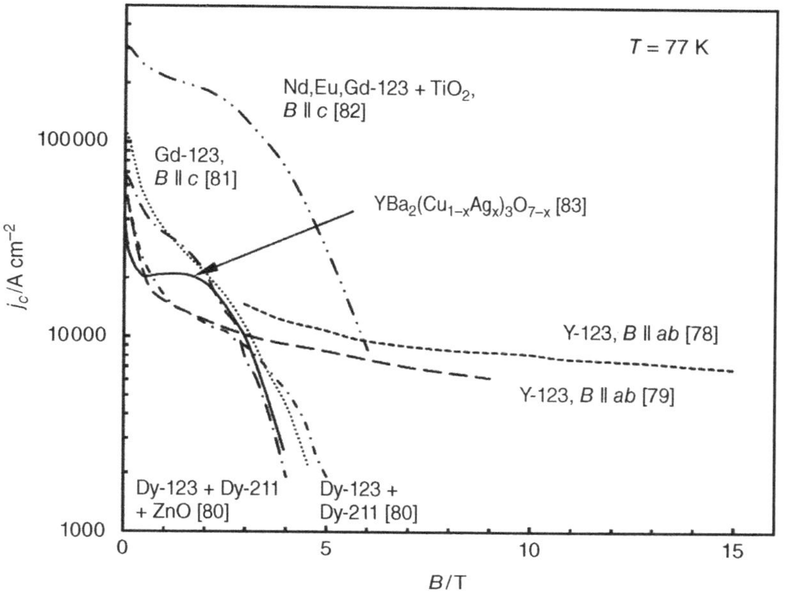

Figure 9.13 *Critical current densities versus field of various bulk cuprate high-T_c superconductors at 77 K. The drop of the critical current density is much less pronounced for fields applied along the CuO_2 (ab) planes than for fields parallel to the c direction (data from [78–83])*

For $(Nd_{0.33}Eu_{0.33}Gd_{0.33})_1Ba_2Cu_3O_{7-x}$ with TiO_2 additions [82], extremely high critical current densities have been reported even for the unfavorable field direction $B \parallel c$ (see Figure 9.13).

Bi-2212 and Bi-2223 bulk hollow cylinders have been successfully manufactured [84, 85]. The critical current density of the hollow cylinders can be well represented by a scaling relation already used to describe the j_c of Bi-2212/Ag wires without an applied field [86]

$$j_c(T) = j_c(0)\left(1 - \frac{T}{T_c}\right)^{\gamma} \tag{9.10}$$

Here T_c is the critical temperature, $j_c(0)$ is the critical current density at zero temperature, and γ is a scaling exponent. For the melt-cast processed Bi-2212 tubes, the values of the scaling parameters are $T_c = 92$ K, $j_c(0) = 25{,}000$ A/cm^2, and $\gamma = 2.5$ [85]. The corresponding values for the Bi-2223 tubes are $T_c = 108$ K, $j_c(0) = 5000$ A/cm^2, and $\gamma = 1.5$ [85]. The fact that, especially at 77 K, the critical current densities in Bi-2212 and Bi-2223 bulk superconductors are much lower than those achieved in Y-123 is a consequence of the less effective pinning in the two Bi-based compounds.

9.5.4 Critical Currents in Superconducting Films

The highest critical current densities have been achieved in textured films, the preparation of which will be described in Chapter 16. Figure 9.14 shows the critical current densities of

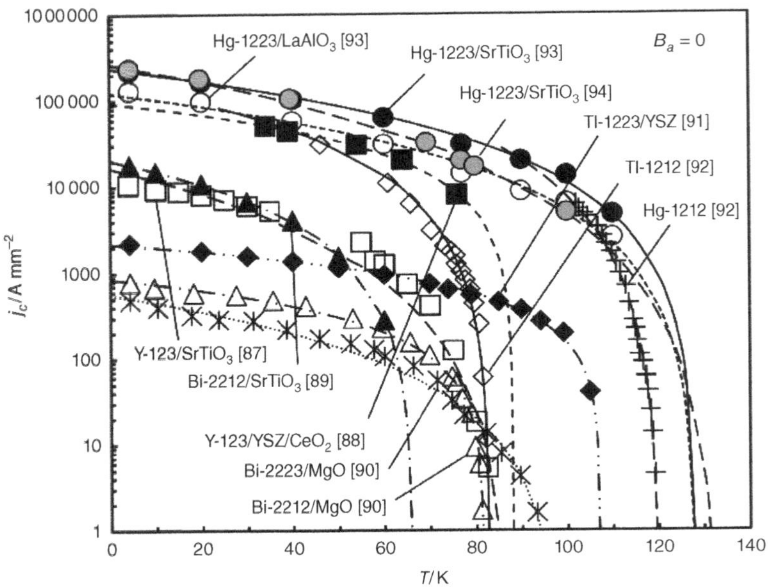

Figure 9.14 *Critical current densities of various cuprate high-T_c superconductor films versus temperature for zero applied field. Because the critical current densities in textured superconducting films are much larger than in bulk material, j_c is plotted in A/mm^2 (10^5 A/cm^2 = 1000 A/mm^2). The lines are fits based on Equation 9.10 (data from [87–94])*

various cuprate high-T_c films as a function of temperature without an applied field. The critical current densities of the films depend on the substrate, the texture, and the flux pinning. The critical current density of Bi-2212 films on $SrTiO_3$ single crystal substrates exceeds 10 000 A/mm^2 at 4 K, whereas the corresponding values for films deposited on MgO substrates are only around 800 A/mm^2 (see Figure 9.14). At low temperatures, the critical current densities of Bi-2212/MgO and Bi-2223/MgO are comparable. The critical current density at 77 K depends strongly on the transition temperature of the superconductor in question. Thus, the j_c of Bi-2223 at 77 K is significantly larger than that of Bi-2212 in spite of similar critical current densities at 4 K. Because of extremely high transition temperatures, j_c values of more than 5000 A/mm^2 have been achieved in films of Hg-1212 and Hg-1223 even at temperatures of 100 K. Well textured Y-123 films (see Chapter 15) are the basis of second-generation high-temperature superconductors (coated conductors) for magnet and power applications. In the Y-123 film deposited on a biaxially textured nickel tape with CeO_2 and yttria-stabilized zirconia (YSZ) buffer layers, the critical current density reaches nearly 8000 A/mm^2 (0.8 MA/cm^2) even at 77 K (see Figure 9.14). Nowadays the critical current densities in the Y-123 or RE-123 films of coated conductors exceed 1 MA/cm^2 at 77 K without an applied magnetic field. For most of the high-temperature superconductor films, the temperature dependence of the critical current density can be well represented by the scaling law $j_c(T) = j_c(0)\,(1 - T/T_c)^{\gamma}$ (Equation 9.10), as illustrated by the lines in Figure 9.14.

The physical properties of cuprate high-T_c superconductors are highly anisotropic because of their layered crystal structures. The supercurrents flow preferentially along the CuO_2 (*ab*) planes, which are separated by the insulating charge carrier reservoirs. As a consequence, flux lines related to a magnetic field applied along the *ab* planes of a textured cuprate film are pinned by the layers forming the charge carrier reservoirs. It is energetically favorable that the positions of the normal conducting vortex cores coincide with the regions of weakened superconductivity in the charge carrier reservoirs. This intrinsic pinning (see also Chapter 10) hinders the movement of flux lines along the crystallographic *c* direction. A Lorentz force parallel to the *c* direction results for currents within the CuO_2 planes in the presence of a transverse magnetic field applied along the *ab* planes. On the other hand, the Lorentz force is parallel to the *ab* planes for current flow along the CuO_2 planes and a magnetic field applied along the *c* direction. For this field–current orientation, the intrinsic pinning is not active, and hence the flux lines can easily be moved along the *ab* planes (see also Chapter 10). Figure 9.15 shows the dependence of the critical current density at 77 K on the magnetic field for Sm-123, Y-123, and Tl-1223 films. Only the j_c data of the Bi-2212 film have been measured at a temperature of 50 K. As expected from the intrinsic pinning model, the critical current densities for fields parallel to the *c* direction are, in all considered high-T_c superconductor films, considerably smaller than those for fields along the *ab* planes. The current always flows along the CuO_2 planes. The Sm-123 and Y-123 films show a moderate anisotropy

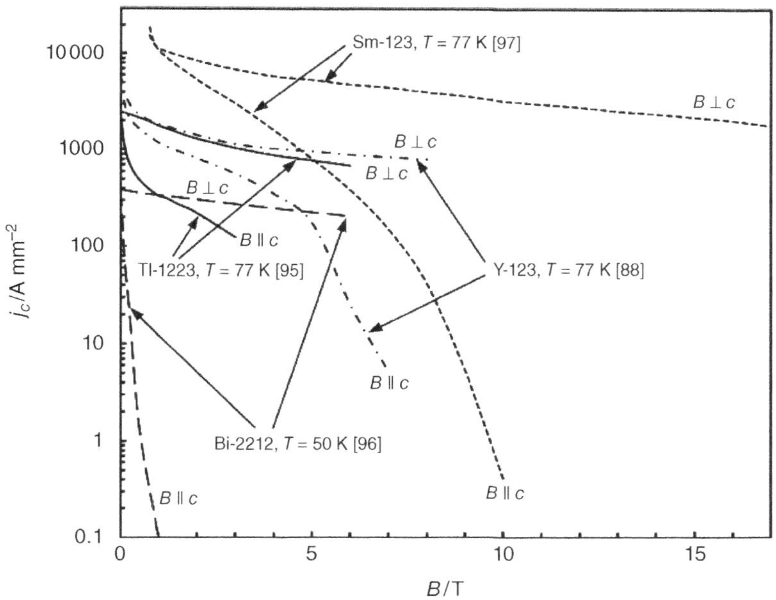

Figure 9.15 *Critical current densities of selected high-T_c superconductor films versus field for $B \perp c$ and $B \parallel c$. For all samples except the Bi-2212 film ($T = 50$ K), data have been measured at 77 K. In general, the critical currents for $B \perp c$ are considerably larger than those for $B \parallel c$ (data from [88, 95–97])*

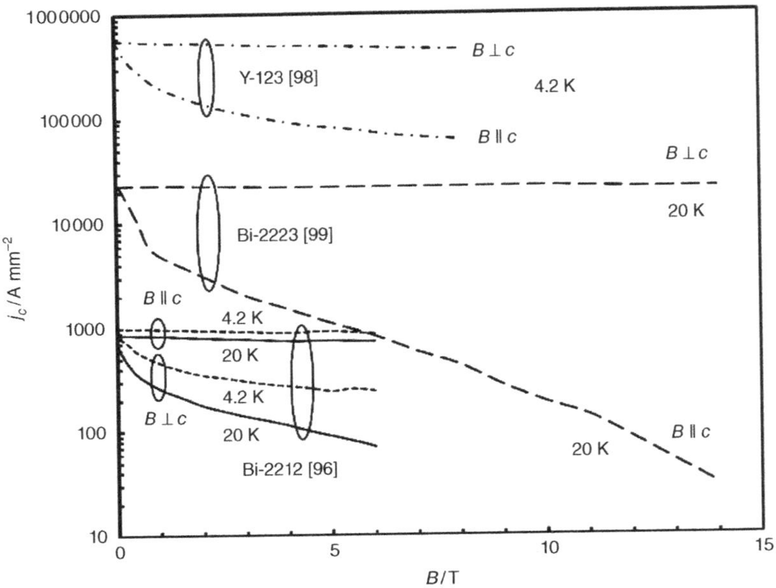

Figure 9.16 Low temperature critical current densities of selected high-T_c superconductor films versus field for $B \perp c$ and $B \parallel c$. A comparison with Figure 9.15 indicates that the anisotropy of the critical current of the Bi-2212 film with respect to the field direction is much smaller at 20 and 4.2 K than at 50 K (data from [96, 98, 99])

of the critical current density with respect to the direction of the applied magnetic field even at a temperature of 77 K. A slightly more pronounced anisotropy has been found for the Tl-1223 film. The effect is most pronounced for the Bi-2212 film in spite of the fact that a temperature of only 50 K is considered.

The critical current densities of Bi-2212, Bi-2223, and Y-123 films at low temperatures are presented in Figure 9.16. A comparison of the j_c data of the Bi-2212 film presented in Figures 9.15 and 9.16 indicates that the anisotropy of the critical current density decreases rapidly with reduced temperature. At 4.2 K and 6 T the ratio of $j_c(B \perp c)$ to $j_c(B \parallel c)$ is ≈3.5 for the Bi-2212 film, while it is around a factor of 6.5 for the Y-123 film. For the Bi-2223 film, the ratio $j_c(B \perp c)/j_c(B \parallel c)$ is ≈27 at 20 K and 6 T compared to a value of ≈10 for the Bi-2212 film (20 K, 6 T). It should be noted that the anisotropy of j_c increases with an improved degree of grain alignment. In the case of a less perfect texture there exists, even for B nominally parallel to the *ab* planes, a field component parallel to the *c* direction, which tends to reduce the measured anisotropy of j_c.

In the discussion of the anisotropy of the critical current with respect to the field direction, we have restricted ourselves to the extreme cases that the applied field is along the crystallographic *c* direction or along the *ab* planes. In general, the critical current density of a well-textured high-T_c superconductor film is a function of temperature, field, and the angle θ between the crystallographic *c* direction and the field direction (see Figure 9.17). The magnetic field is always applied transverse to the current, which flows along the CuO_2

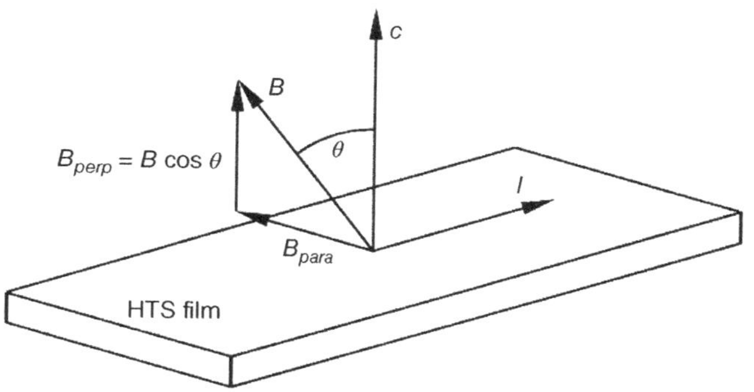

Figure 9.17 *Definition of the angle θ*

planes. For Bi-2223 films [99], the critical current density is determined mainly by the perpendicular field component, leading to the following relation:

$$j_c(B,\theta) = j_c^{\mathrm{perp}}(B_{sc}) \tag{9.11}$$

Here j_c^{perp} is the critical current density for magnetic fields perpendicular to the *ab* planes, and the scaling field $B_{sc} = B\cos\theta$ is simply the component of the total field B perpendicular to the CuO_2 planes of the textured film. This scaling relation is also expected to provide a good approximation for the angular dependence of the critical current density of textured Bi-2212 films. For RE-123 films, the situation is more complex. A theory proposed by Blatter *et al.* [100] leads to the following scaling field for uncorrelated point defects:

$$B_{sc} = \varepsilon(\theta)B = \left(\cos^2\theta + \frac{\sin^2\theta}{\gamma_a^2}\right)^{1/2} B \tag{9.12}$$

where $\varepsilon(\theta)$ is an angle dependent anisotropy parameter and $\gamma_a = (m_c/m_{ab})^{0.5}$ the anisotropy of the effective electron mass, which arises from the interaction with the complex lattices under consideration. For Y-123 γ_a is in the range 5–7. The resulting critical current density is proportional to B_{sc}^{-1} [101]. In Figure 9.18, the angle dependence of the inverse anisotropy parameter $1/\varepsilon$ is presented for $\gamma_a = 5$. In the angle range of 69° to 83°, the j_c of the Y-123 films studied by Yamasaki *et al.* [101] has been found to be proportional to B_{sc}^{-1}. Intrinsic pinning and stacking faults parallel to the *ab* planes, which are not taken into account, lead to a more pronounced peak for $\theta = 90°$ (i.e., $B \parallel ab$ planes). On the other hand, the introduction of artificial pinning centers (e.g., nano-dots, nano-rods) can cause an additional maximum for $\theta = 0°$ ($B \parallel c$) [102]. The deviations of typical j_c versus angle curves from that expected for random point defects are shown in Figure 9.18.

So far, we have considered only the critical current density for supercurrents, which flow along the CuO_2 (*ab*) planes. Based on the fact that the CuO_2 planes are conductive, whereas the charge carrier reservoirs separating adjacent CuO_2 blocks are insulating, we expect that the critical current density perpendicular to the CuO_2 planes is much smaller than that along them. The critical currents along the crystallographic *c* direction have to be considered as tunneling currents through the insulating layers of the charge carrier reservoirs.

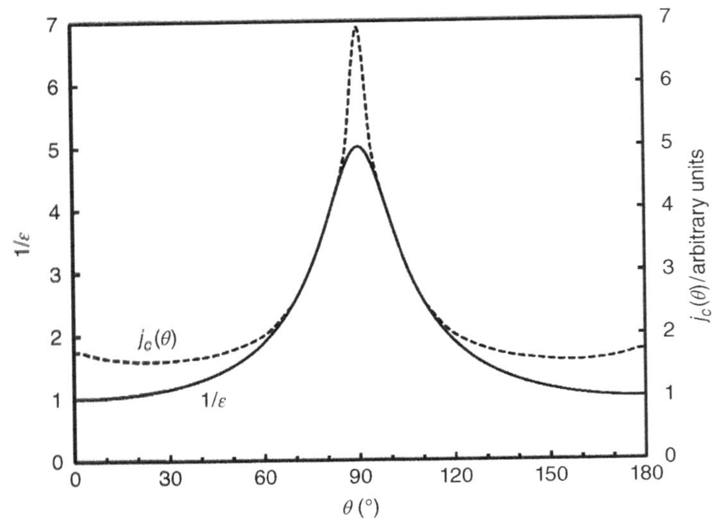

Figure 9.18 Comparison of the inverse of the angle-dependent anisotropy parameter ε with a typical j_c versus angle curve of a Y-123 film at 77 K and low field. The values of ε are based on an anisotropy $\gamma_a = 5$ of the effective mass of the electrons (adapted from [101])

In Hg-1212 films, critical current densities as high as 10^8 A/cm^2 have been measured along the CuO_2 planes, whereas the j_c along the c direction is only 5000 A/cm^2 [103]. For Hg-1223 films with critical current densities of 10^7 A/cm^2 at 10 K, a giant anisotropy of the effective masses m_c and m_{ab} has been reported. The value of the ratio m_c/m_{ab} is as high as 32 000 [104].

9.6 Grain-Boundary Weak Links

In the previous sections, we have presented data on the critical current densities in sintered polycrystalline cuprate superconductors, in bulk material, and in textured films. We have already seen (see Section 9.5.2) that the transport critical current densities in sintered polycrystalline cuprate high-T_c superconductors are disappointingly low – apparently due to granularity. Magnetization measurements show (see, e.g., [1, 105, 106]) that the j_c values within individual grains are much larger than the transport critical current densities. The observed behavior is caused by grain boundaries acting as weak links for the supercurrents. Taking into account that the coherence length along the crystallographic c direction is extremely short in cuprate superconductors (of the order of interatomic distances) it seems plausible that grain boundaries can act as barriers for the supercurrents. The fact that the critical current densities in textured thin films are extremely high indicates that not all grain boundaries act as weak links. In the textured films, there exist many low-angle grain boundaries, which seem to be not detrimental to the transport critical current density.

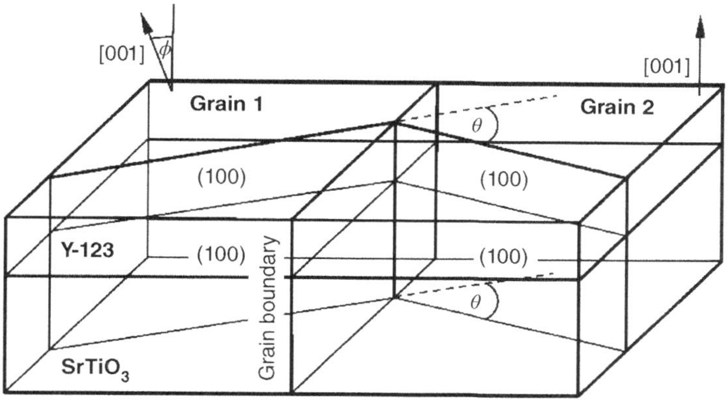

Figure 9.19 *Crystallography of Y-123 films epitaxially grown on $SrTiO_3$ bicrystals. The two Y-123 grains are separated by an artificial [100] tilt boundary. The right grain is rotated by an angle θ around the common c axis (adapted from [107])*

In order to clarify the effect of the misalignment of adjacent grains on the critical current density across the grain boundary between them, Dimos and coworkers studied the orientation dependence of the j_c values of artificial grain boundaries [107, 108]. In their excellent work, they used $SrTiO_3$ single crystal substrates on which thin Y-123 films were grown epitaxially. The use of $SrTiO_3$ bicrystals allows the preparation of two Y-123 grains with a well-defined misorientation angle θ, which are connected by an artificial grain boundary. The bicrystals were obtained by hot pressing of two $SrTiO_3$ single crystals at a temperature of 1450°C. The deposition of the Y-123 films was performed by electron-beam evaporation of yttrium, barium, and copper or laser ablation from a sintered Y-123 pellet. To reach the hole concentration, leading to the highest critical temperature, the films were slowly cooled in an oxygen atmosphere. A sketch of the crystallography of the $SrTiO_3$ bicrystals and the epitaxially grown Y-123 films is presented in Figure 9.19. The right grain is rotated with respect to the left grain by an angle θ around the common c axis ([0 0 1] tilt boundary). The c axes of the two grains were misaligned by the small angle ϕ. The misalignment angles were determined by Laue X-ray diffraction with an accuracy of 0.5°. Figure 9.20 shows a sketch of the three microbridges of 10 μm width, patterned into the two grains and the grain boundary region by means of an excimer laser. The critical currents of the three microbridges were measured by the four-point method.

The results of the Dimos experiment [107] clearly indicate that the critical current density across the grain boundaries decreases rapidly with the increasing misalignment angle θ. For a grain boundary characterized by $\theta = 0°$ and $\phi = 3°$, a critical current density as high as 4 MA/cm^2 was measured at ≈5 K. This value has to be compared with critical current densities of 7.14 and 8.0 MA/cm^2 in the two grains. The ratio of the grain boundary j_c to the average value of the two grains is 0.53. This ratio drops to 0.05 at θ values of 10°. For misalignment angles above 20°, the j_c across the grain boundary is around 2% of the average j_c in the grains. In Figure 9.21, the critical current densities across several artificial bicrystal

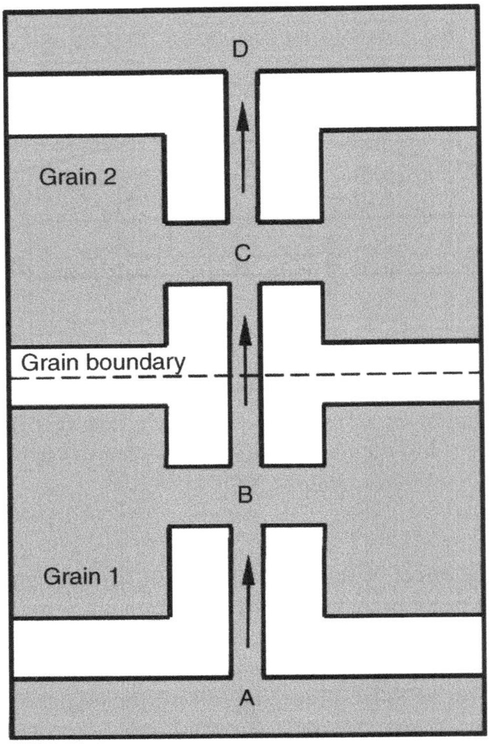

Figure 9.20 *Sketch of the microbridges patterned into the two Y-123 grains and the grain boundary region for the four-point measurement of the critical current (after [107])*

grain boundaries are shown [108–112]. In general, the j_c across different types of grain boundaries ([1 0 0], [0 0 1] tilt boundaries and [1 0 0] twist boundaries – for more details see [108]) decreases with increasing misalignment angle. Figure 9.21 suggests that the critical current density across the grain boundaries drops exponentially with the increasing misalignment angle:

$$j_c = j_{c0} \exp\left(-\frac{\theta}{\theta_0}\right) \tag{9.13}$$

Here j_{c0} is the critical current density across a grain boundary with vanishing misalignment of the adjacent grains, and θ_0 the angle leading to a 63% reduction of the grain boundary j_c. For [0 0 1] tilt grain boundaries in Y-123 films, values $j_{c0} \cong 2 \times 10^7$ A/cm^2 and $\theta_0 \cong 6.3°$ have been reported for measurements at 4.2 K [110]. The corresponding values for Bi-2223 [0 0 1] tilt boundaries are $j_{c0} \cong 6 \times 10^6$ A/cm^2 and $\theta_0 \cong 5°$ at 26 K [111].

The study of the properties of bulk bicrystal grain boundaries of Y-123 [113] showed a rapid drop of the critical current densities across the grain boundaries with increasing misalignment angle of the adjacent grains, as in the case of artificial grain boundaries.

The experimental results confirm that high-angle grain boundaries can act as barriers to the flow of the supercurrents. The fact that the j_c across low-angle grain boundaries is not significantly reduced, as compared to the j_c values in the individual grains, is in line with

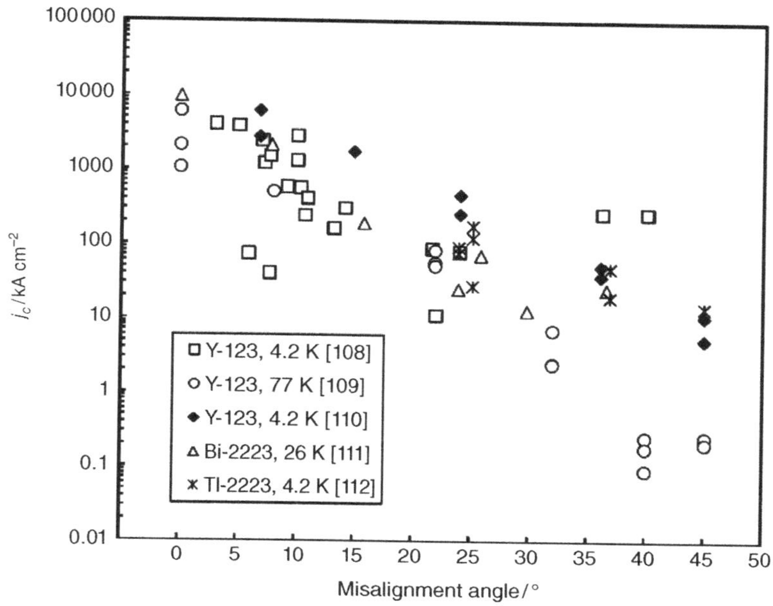

Figure 9.21 *Critical current density across artificial grain boundaries versus misalignment angle. For all considered grain boundaries the critical current density drops nearly exponentially with increasing misalignment angle (data from [108–112])*

the high transport critical current densities observed in textured cuprate high-T_c films. Possible reasons for the weak-link behavior of high-angle grain boundaries are the structural distortions in this region, which may be accompanied by the segregation of impurities and variation of the oxygen content at grain boundaries. Evidence for the variation of the oxygen content is a considerably reduced critical temperature found at Bi-2212 bulk bicrystal grain boundaries [114].

One reason for the weak-link behavior of the grain boundaries is the extremely short coherence length in connection with the relatively low carrier concentration of 5×10^{21} cm^{-3} in optimally doped cuprate superconductors, which has to be compared to a conduction electron concentration of 85×10^{21} cm^{-3} in copper. Hilgenkamp and Mannhart have pointed out similarities between semiconductor-metal interfaces and the grain boundaries in high-T_c superconductors [115, 116]. Figure 9.22 (top) shows the band bending at the interface between a p-type semiconductor and a metal with a lower work function. Electrons flow from the metal into the semiconductor, leading to an excess of negatively charged acceptor ions and a depletion of holes in the valence band. Currents from the metal to the semiconductor would lead to a further decrease in the number of holes in the interfacial region, and hence would enlarge the space-charge region. Currents from the metal to the semiconductor are therefore negligible (Schottky contact). Due to large dielectric constants ($\varepsilon \approx 10$) and low carrier concentrations, band bending effects can occur at the grain boundaries of cuprate high-T_c superconductors. As a consequence the grain boundaries can act as double Schottky barriers, as illustrated in Figure 9.22 (bottom). The estimated Thomas–Fermi screening length λ_{TF} is between 0.5 and 1 nm in

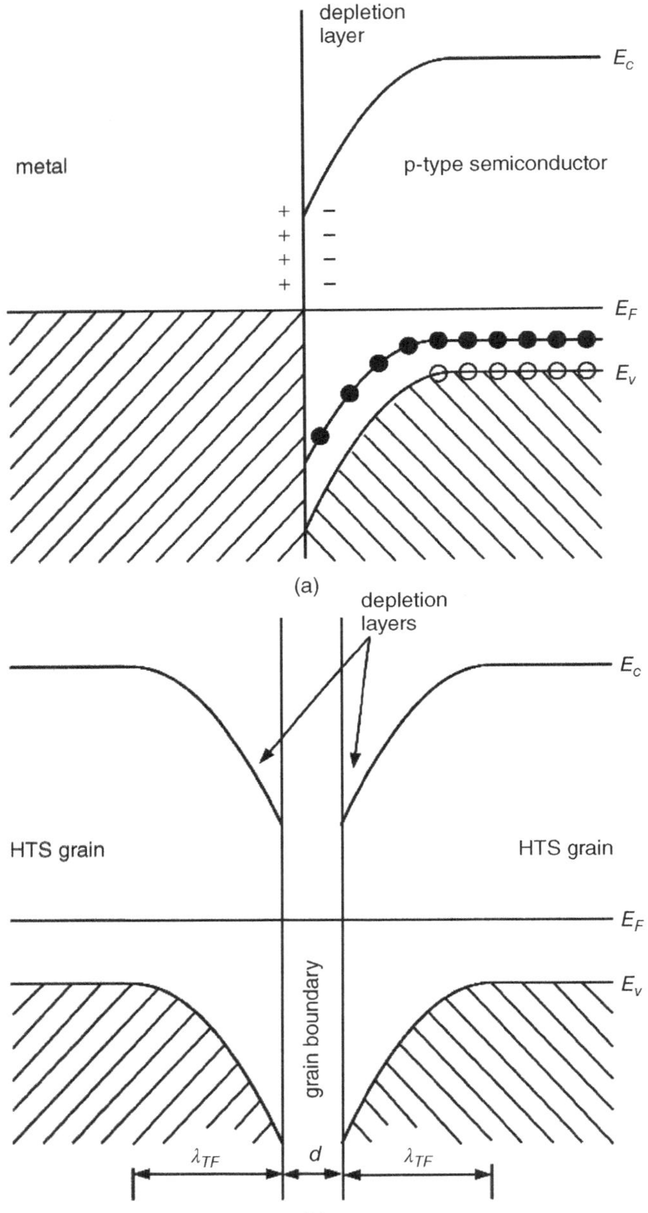

Figure 9.22 *Space-charge effects leading to band bending at a semiconductor–metal interface (top) and at the grain boundaries of cuprate high-T_c superconductors (bottom) (adapted from [115])*

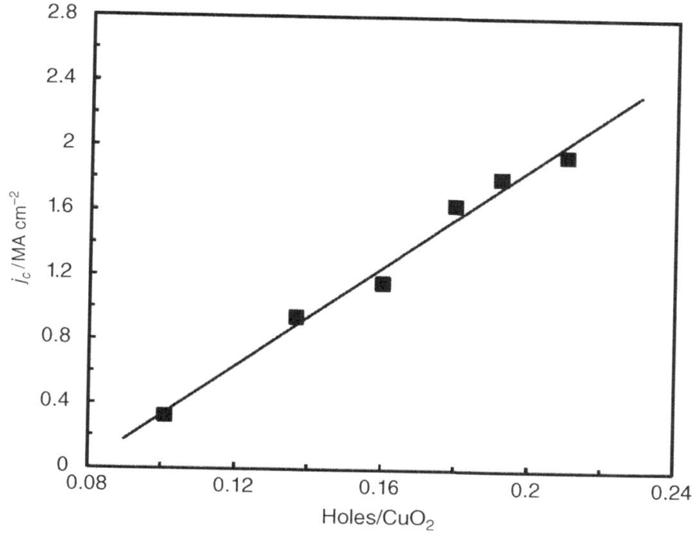

Figure 9.23 *Critical current density of a symmetric 24° [0 0 1] tilt boundary versus number of holes per CuO_2 unit. The hole concentration in the bicrystalline $Y_{0.7}Ca_{0.3}Ba_2Cu_3O_{7-\delta}$ films has been varied by means of annealing at different oxygen pressures (data from [110])*

cuprate superconductors [115], comparable to the coherence length. The total width of the insulating layer at the grain boundary is $t = d + 2\lambda_{TF}$, where d is the typical structural width of the grain boundary, which is of the order of 0.4 nm. According to [115] the thickness of the depletion layer is 0.16 nm for $\varepsilon = 10$, an in-built voltage of 0.1 V, and a carrier concentration of 4.5×10^{21} cm^{-3}. Thus, the total thickness of the insulating layer is significantly enhanced by the space-charge region.

Within the double Schottky barrier scenario, it should be possible to improve the electronic properties of the interface by increasing the charge carrier concentration, for example by means of Ca doping in Y-123 films. Experimental evidence for an increase of the critical current densities across symmetric 24° [0 0 1] tilt boundaries is presented in Figure 9.23. As expected, the critical current density at 4.2 K across the grain boundary increases with increasing number of holes per CuO_2. This is even true in the overdoped region, where T_c starts to decrease.

9.7 Summary

Due to the layered crystal structures both the penetration depth and the coherence length are strongly anisotropic. The coherence length ξ_{ab} along the CuO_2 planes is of the order of a few nanometers, whereas ξ_c along the crystallographic c direction is as small as 0.2 nm in many cuprate high-T_c superconductors. The values of the penetration depth λ_{ab} related to screening currents flowing in the CuO_2 planes are of the order of 100–200 nm. The screening currents in the c direction have to cross the insulating layers of the charge carrier reservoirs, and have to be considered as tunneling currents. As a consequence, λ_c is much larger than λ_{ab} and can reach values above 1000 nm even at zero temperature. In many high-T_c superconductors the Ginzburg–Landau parameter $\kappa_c = \lambda_{ab}/\xi_{ab}$ is as high as $\approx$100 indicating that the cuprates are extreme Type II superconductors.

As in metallic Type II superconductors, flux is excluded from the interior of a cuprate superconductor below the lower critical field B_{c1}. Between the lower and upper critical fields, magnetic flux can enter the interior of the superconductor within the normal conducting cores of the vortices, which coexist with superconducting regions in the mixed state. Above the upper critical field B_{c2}, the material is in the normal state. In contrast to metallic Type II superconductors, there exists an irreversibility line well below B_{c2}. Above the irreversibility line the critical current density is zero because of vanishing pinning of the flux lines. This phenomenon has been attributed to melting of the flux-line lattice, or thermally activated depinning. Below the irreversibility line the vortex matter is in a solid glass-like phase, and hence the flux lines are pinned.

The critical currents in sintered cuprate high-T_c superconductors are typically well below 1000 A/cm^2. Moreover, they decrease strongly even in extremely low fields of the order of 10 mT. This behavior has been attributed to grain boundaries, which act as barriers to the supercurrents. On the other hand, extremely high critical current densities of the order of 1 MA/cm^2 have been achieved in textured films. This considerably different behavior suggests that not all grain boundaries act as barriers to the transport currents. Experiments on artificial bicrystal grain boundaries show that the critical current density across the grain boundary decreases exponentially with increasing misalignment angle of the adjacent grains. These experimental results suggest that a texture is required to reach high transport critical current densities.

References

1. R. Wesche, *High-Temperature Superconductors: Materials, Properties and Applications*, Kluwer Academic Publishers, Norwell, MA, 1998.
2. R. Khasanov, T. Kondo, S. Strässle, D.O.G. Heron, A. Kaminski, H. Keller, S.L. Lee, and T. Takeuchi, Evidence for a competition between the superconducting state and the pseudogap state of $(BiPb)_2(SrLa)_2CuO_{6+\delta}$ from muon spin rotation experiments, *Phys. Rev. Lett.*, **101,** 227002 (2008).
3. A.J. Panson, G.R. Wagner, A.I. Braginski, J.R. Gavaler, M.A. Janocko, H.C. Pohl, and J. Talvacchio, Properties of $La_{1.8}Sr_{0.2}CuO_4$ superconductors, *Appl. Phys. Lett.*, **50**, 1104–1106 (1987).
4. X.Q. Xiang, Y.Q. Zhang, J.F. Ding, and X.G. Li, Effect of stripe ordering on the anisotropic superconductivity in $La_{1.6-x}Nd_{0.4}Sr_xCuO_4$ single crystals, *Physica C*, **468**, 2336–2340 (2008).
5. Y. Wang and H. Gao, Anisotropic magnetotransport of superconducting and normal state in an electron-doped $Nd_{1.85}Ce_{0.15}CuO_{4-\delta}$ single crystal, *Physica C*, **470**, 689–692 (2010).
6. A. Shengelaya, R. Khasanov, D.G. Eshchenko, D. Di Castro, I.M. Savić, M.S. Park, K.H. Kim, S.-I. Lee, K.A. Müller, and H. Keller, Muon-spin-rotation measurements of the penetration depth of the infinite layer electron-doped $Sr_{0.9}La_{0.1}CuO_2$ cuprate superconductor, *Phys. Rev. Lett.*, **94**, 127001 (2005).

7. R. Liang, D.A. Bonn, W.N. Hardy, and D. Broun, Lower critical field and superfluid density of highly underdoped $YBa_2Cu_3O_{6+x}$ single crystals, *Phys. Rev. Lett.*, **94**, 117001 (2005).
8. A. Hosseini, D.M. Broun, D.E. Sheehy, Survival of the *d*-wave superconducting state near the edge of antiferromagnetism in the cuprate phase diagram, *Phys. Rev. Lett.*, **93**, 107003 (2004).
9. R.J. Cava, B. Batlogg, R.B. van Dover, D.W. Murphy, S. Sunshine, T. Siegrist, J.P. Remeika, E.A. Rietman, S. Zahurak, and G.P. Espinosa, Bulk superconductivity at 91 K in single-phase oxygen-deficient perovskite $Ba_2YCu_3O_{9-\delta}$, *Phys. Rev. Lett.*, **58**, 1676–1679 (1987).
10. T.K. Worthington, W.J. Gallagher, and T.R. Dinger, Anisotropic nature of high-temperature superconductivity in single-crystal $Y_1Ba_2Cu_3O_{7-x}$, *Phys. Rev. Lett.*, **59**, 1160–1163 (1987).
11. U. Welp, W.K. Kwok, G.W. Crabtree, K.G. Vandervoort, and J.Z. Liu, Magnetic measurements of the upper critical field of $YBa_2Cu_3O_{7-\delta}$ single crystals, *Phys. Rev. Lett.*, **62**, 1908–1911 (1989).
12. L. Krusin-Elbaum, A.P. Malozemoff, Y. Yeshurun, D.C. Cronemeyer, and F. Holtzberg, Temperature dependence of lower critical fields in Y–Ba–Cu–O crystals, *Phys. Rev. B*, **39**, 2936–2939 (1989).
13. L. Krusin-Elbaum, R.L. Greene, F. Holtzberg, A.P. Malozemoff, and Y. Yeshurun, Direct measurement of the temperature-dependent magnetic penetration depth in Y–Ba–Cu–O crystals, *Phys. Rev. Lett.*, **62**, 217–220 (1989).
14. N. Athanassopoulou and J.R. Cooper, Temperature-dependent anisotropic magnetic penetration depth of oxygen-depleted $YBa_2Cu_3O_{7-\delta}$, *Physica C*, **259**, 326–336 (1996).
15. D.R. Harshman, L.F. Schneemeyer, J.V. Waszczak, G. Aeppli, R.J. Cava, B. Batlogg, L.W. Rupp, E.J. Ansaldo, and D.Ll . Williams, Magnetic penetration depth in single-crystal $YBa_2Cu_3O_{7-\delta}$, *Phys. Rev. B*, **39**, 851–854 (1989).
16. C. Camerlingo and G. Jung, Coherence length in deoxygenated (103)/(013) oriented YBCO superconductor films, *Physica C*, **460–462**, 805–806 (2007).
17. M.R. Trunin, Yu.A. Nefyodov, and A.F. Shevchun, Superfluid density in the underdoped $YBa_2Cu_3O_{7-x}$: Evidence for *d*-density-wave order of the pseudogap, *Phys. Rev. Lett.*, **92**, 067006 (2004).
18. I. Kokanović, J.R. Cooper, and M. Matusiak, Nernst effect measurements of epitaxial $Y_{0.95}Ca_{0.05}Ba_2(Cu_{1-x}Zn_x)_3O_y$ and $Y_{0.9}Ca_{0.1}Ba_2Cu_3O_y$ superconducting films, *Phys. Rev. Lett.*, **102**, 187002 (2009).
19. A. Kortyka, R. Puzniak, A. Wisniewski, M. Zehetmayer, H.W. Weber, Y.Q. Cai, and X. Yao, Anisotropy of the superconducting state parameters and intrinsic pinning in low-level Pr-doped $YBa_2Cu_3O_{7-\delta}$ single crystals, *Supercond. Sci. Technol.*, **23**, 065001 (2010).
20. M. Akamatsu, L.X. Chen, H. Ikeda, and R. Yoshizaki, Magnetic properties of the optimally doped Bi-2201 single crystals, *Physica C*, **235–240**, 1619–1620 (1994).
21. P.J. Baker, T. Lancaster, S.J. Blundell, F.L. Pratt, M.L. Brooks, and S.-J. Kwon, Tuning the interlayer spacing of high-T_c Bi-based superconductors by intercalation: Measuring the penetration depth and the two-dimensional superfluid density, *Phys. Rev. Lett.*, **102**, 087002 (2009).

22. T.T.M. Palstra, B. Batlogg, L.F. Schneemeyer, R.B. van Dover, and J.V. Waszczak, Angular dependence of the upper critical field of $Bi_{2.2}Sr_2Ca_{0.8}Cu_2O_{8+\delta}$, *Phys. Rev. B*, **38**, 5102–5105 (1988).
23. J.H. Kang, R.T. Kampwirth, and K.E. Gray, High critical field anisotropy of superconducting Bi–Sr–Ca–Cu oxide from highly oriented thin films, *Appl. Phys. Lett.*, **52**, 2080–2082 (1988).
24. Lu Zhang, J.Z. Liu, and R.N. Shelton, Upper critical field of $(Bi,Pb)_2Sr_2CaCu_2O_8$ single crystals, *Phys. Rev. B*, **45**, 4978–4982 (1992).
25. I. Matsubara, H. Tanigawa, T. Ogura, H. Yamashita, and M. Kinoshita, Upper critical field and anisotropy of the high-T_c $Bi_2Sr_2Ca_2Cu_3O_x$ phase, *Phys. Rev. B*, **45**, 7414–7417 (1992).
26. I. Matsubara, R. Funahashi, K. Ueno, H. Yamashita, and T. Kawai, Lower critical field and reversible magnetization of $(Bi,Pb)_2Sr_2Ca_2Cu_3O_x$ superconducting whiskers, *Physica C*, **256**, 33–38 (1996).
27. J.H. Kang, K.E. Gray, R.T. Kampwirth, and D.W. Day, Large anisotropy in the upper critical field of sputtered thin films of superconducting Tl–Ba–Ca–Cu–O, *Appl. Phys. Lett.*, **53**, 2560–2562 (1988).
28. J.R. Thompson, J.G. Ossandon, D.K. Christen, B.C. Chakoumakos, Y.R. Sun, M. Paranthaman, and J. Brynestad, Vortex fluctuations, magnetic penetration depth, and H_{c2} in Hg- and Tl-based high-T_c superconductors, *Phys. Rev. B*, **48**, 14 031–14 034 (1993).
29. A. Maignan, C. Martin, V. Hardy, Ch. Simon, M. Hervieu, and B. Raveau, A crystal study of the 128 K superconductor $Tl_2Ba_2Ca_2Cu_3O_{10-\delta}$ Synthesis and anisotropic magnetic properties, *Physica C*, **219**, 407–412 (1994).
30. G. Brandstätter, F.M. Sauerzopf, H.W. Weber, F. Ladenberger, and E. Schwarzmann, Upper critical field, penetration depth, and GL parameter of Tl-2223 single crystals, *Physica C*, **235–240**, 1845–1846 (1994).
31. G. Triscone, A. Junod, and R.E. Gladyshevskii, Magnetic and thermal properties of the 116 K superconductor Tl-1223, *Physica C*, **264**, 233–249 (1996).
32. G. Le Bras, L. Fruchter, V. Vulcanescu, V. Viallet, A. Bertinotti, A. Forget, J. Hammann, J.-F. Marucco, and D. Colson, Superconducting properties and anisotropy of a $HgBa_2CuO_{4+\delta}$, *Physica C*, **271**, 205–213 (1996).
33. J. Hofer, J. Karpinski, M. Willemin, G.I. Meijer, E.M. Kopnin, R. Molinski, H. Schwer, C. Rossel, and H. Keller, Doping dependence of superconducting parameters in $HgBa_2CuO_{4+\delta}$ single crystals, *Physica C*, **297,** 103–110 (1998).
34. D. Pelloquin, V. Hardy, A. Maignan, and B. Raveau, Single crystals of the 96 K superconductor $(Hg,Cu)Ba_2CuO_{4+\delta}$: Growth, structure and magnetism, *Physica C*, **273**, 205–212 (1997).
35. G. Villard, A. Daignere, D. Pelloquin, and A. Maignan, Effect of underdoping on the superconductivity of $(Hg,Cu)Ba_2CuO_{4+\delta}$ '1201' single crystals, *Physica C*, **314,** 196–204 (1999).
36. V. Vulcanescu, L. Fruchter, A. Bertinotti, D. Colson, G. Le Bras, and J.-F. Marucco, Magnetization and anisotropy of a $HgBa_2Ca_2Cu_3O_{8+\delta}$ single crystal, *Physica C*, **259**, 131–134 (1996).

37. H. Seidel, F. Hentsch, M. Mehring, J.G. Bednorz, and K.A. Müller, Nuclear spin-lattice relaxation of ^{139}La in the superconducting phase of $La_{1.8}Sr_{0.2}CuO_{4-\delta}$, *Europhys. Lett.*, **5**, 647–650 (1988).
38. K. Terashima, H. Matsui, T. Sato,T. Takahashi, M. Kofu, and K. Hirota, Anomalous momentum dependence of the superconducting coherence peak and its relation to the pseudogap of $La_{1.85}Sr_{0.15}CuO_4$, *Phys. Rev. Lett.*, **99**, 017003 (2007).
39. R. Khasanov, A. Shengelaya, A. Maisuradze, F. La Mattina, A. Bussmann-Holder, H. Keller, and K.A. Müller, Experimental evidence for two gaps in the high-temperature $La_{1.83}Sr_{0.17}CuO_4$ superconductor, *Phys. Rev. Lett.*, **98**, 057007 (2007).
40. M. Shi, J. Chang, S. Pailhés, M.R. Norman, J.C. Campuzano, M. Månsson, T. Claesson, O. Tjernberg, A. Bendounan, L. Patthey, N. Momono, M. Oda, M. Ido, C. Mudry, and J. Mesot, Coherent *d*-wave superconducting gap in underdoped $La_{2-x}Sr_xCuO_4$ by angle-resolved photoemission spectroscopy, *Phys. Rev. Lett.*, **101**, 047002
(2008).
41. T. Ekino, T. Minami, H. Fujii, and J. Akimitsu, Superconducting energy gaps in $YBa_2Cu_3O_7$ and $Bi_2Sr_2CaCu_2O_8$, *Physica C*, **235–240**, 1899–1900 (1994).
42. R.T. Collins, Z. Schlesinger, R.H. Koch, R.B. Laibowitz, T.S. Plaskett, P. Freitas, W.J. Gallagher, R.L. Sandstrom, and T.R. Dinger, Comparative study of superconducting energy gaps in oriented films and polycrystalline bulk samples of Y–Ba–Cu–O, *Phys. Rev. Lett.*, **59**, 704–707 (1987).
43. R.T. Collins, Z. Schlesinger, F. Holtzberg, and C. Feild, Infrared evidence for gap anisotropy, *Phys. Rev. Lett.*, **63**, 422–425 (1989).
44. Z. Schlesinger, R.T. Collins, D.L. Kaiser, and F. Holtzberg, Superconducting energy gap and normal-state reflectivity of single crystal Y–Ba–Cu–O, *Phys. Rev. Lett.*, **59**, 1958–1961 (1987).
45. R. Khasanov, S. Strässle, D. Di Castro, T. Masui, S. Miyasaka, S. Tajima, A. Bussmann-Holder, and H. Keller, Multiple gap symmetries for the order parameter of cuprate superconductors from penetration depth measurements, *Phys. Rev. Lett.*, **99**, 237601 (2007).
46. N. Hudákova, P. Samuely, P. Szabó, V. Plecháček, K. Knížek, and D. Sedmidubský, Scaling of the superconducting order parameter in Bi cuprates with T_c, *Physica C*, **246**, 163–168 (1995).
47. R. Escudero, E. Guarner, and F. Morales, Superconducting energy gap of single-crystal $Bi_2Sr_2Ca_1Cu_2O_{8+\delta}$ by electron tunnelling, *Physica C*, **166**, 15–18 (1990).
48. R. Manzke, T. Buslaps, R. Claessen, and J. Fink, On the superconducting energy gap in $Bi_2Sr_2CaCu_2O_8$ investigated by high-resolution angle-resolved photoemission, *Europhys. Lett.*, **9**, 477–482 (1989).
49. A.C. Fang, L. Capriotti, D.J. Scalapino, S.A. Kivelson, N. Kaneko, M. Greven, and A. Kapitulnik, Gap-inhomogeneity-induced electronic states in superconducting $Bi_2Sr_2CaCu_2O_{8+\delta}$, *Phys. Rev. Lett.*, **96**, 017007 (2006).
50. N. Jenkins, Y. Fasano, C. Berthod, I. Maggio-Aprile, A. Piriou, E. Giannini, B.W. Hoogenboom, C. Hess, T. Cren, and Ø Fischer, Imaging the essential role of spin fluctuations in high-T_c superconductivity, *Phys. Rev. Lett.*, **103**, 227001 (2009).

51. S. Ideta, K. Takashima, M. Hashimoto, T. Yoshida, A. Fujimori, H. Anzai, T. Fujita, Y. Nakashima, A. Ino, M. Arita, H. Namatame, M. Taniguchi, K. Ono, M. Kubota, D.H. Lu, Z.-X. Shen, K.M. Kojima, and S. Uchida, Enhanced superconducting gaps in the trilayer high-temperature $Bi_2Sr_2Ca_2Cu_3O_{10+\delta}$ cuprate superconductor, *Phys. Rev. Lett.*, **104**, 227001 (2010).
52. Y. Dagan, R. Beck, and R.L. Greene, Dirty superconductivity in the electron-doped cuprate $Pr_{2-x}Ce_xCuO_{4-\delta}$: Tunneling study, *Phys. Rev. Lett.*, **99**, 147004 (2007).
53. I. Takeuchi, J.S. Tsai, Y. Shimakawa, T. Manako, and Y. Kubo, Energy gap of Tl–Ba–Ca–Cu–O compounds by tunneling, *Physica C*, **158**, 83–87 (1989).
54. K. McEllroy, D.H. Lee, J.E. Hoffman, K.M. Lang, J. Lee, E.W. Hudson, H. Eisaki, S. Uchida, and J.C. Davis, Coincidence of checkerboard charge order and antinodal state decoherence in strongly underdoped superconducting $Bi_2Sr_2CaCu_2O_{8+\delta}$, *Phys. Rev. Lett.*, **94**, 197005 (2005).
55. I.F. Herbut and Z. Tešanović, First-order melting of vortex lattice in strongly type-II three-dimensional superconductors, *Physica C*, **255**, 324–328 (1995).
56. A. Junod, M. Roulin, J.-Y. Genoud, B. Revaz, A. Erb, and E. Walker, Specific heat peaks observed up to 16 T on the melting line of the vortex lattice in $YBa_2Cu_3O_7$, *Physica C*, **275**, 245–258 (1997).
57. A. Schilling, R.A. Fisher, N.E. Phillips, U. Welp, D. Dasgupta, W.K. Kwok, and G.W. Crabtree, Calorimetric measurement of the latent heat of vortex-lattice melting in untwined $YBa_2Cu_3O_{7-\delta}$, *Nature*, **382**, 791–793 (1996).
58. K.A. Müller, M. Takashige, and J.G. Bednorz, Flux trapping and superconductive glass state in La_2CuO_{4-y}:Ba, *Phys. Rev. Lett.*, **58**, 1143–1146 (1987).
59. T.T.M. Palstra, B. Batlogg, L.F. Schneemeyer, and J.V. Waszczak, Thermally activated dissipation in $Bi_{2.2}Sr_2Ca_{0.8}Cu_2O_{8+\delta}$, *Phys. Rev. Lett.*, **61**, 1662–1665 (1988).
60. N.R. Werthamer, E. Helfand, and P.C. Hohenberg, Temperature and purity dependence of the superconducting critical field, H_{c2}. III. Electron spin and spin-orbit effects, *Phys. Rev.*, **147**, 295–302 (1966).
61. D.W. Capone, D.G. Hinks, J.D. Jorgensen, and K. Zhang, Upper critical fields and high superconducting transition temperatures of $La_{1.85}Sr_{0.15}CuO_4$ and $La_{1.85}Ba_{0.15}CuO_4$, *Appl. Phys. Lett.*, **50**, 543–544 (1987).
62. M. Decroux, A. Junod, A. Bezinge, D. Cattani, J. Cors, J.L. Jorda, A. Stettler, M. François, K. Yvon, Ø. Fischer, and J. Muller, Structure, resistivity, critical field, specific-heat jump at T_c, Meissner effect, a.c. and d.c. susceptibility of the high-temperature superconductor $La_{2-x}Sr_xCuO_4$, *Europhys. Lett.*, **3**, 1035–1040 (1987).
63. A. Junod, A. Bezinge, T. Graf, J.L. Jorda, J. Muller, L. Antognazza, D. Cattani, J. Cors, M. Decroux, Ø. Fischer, M. Banovski, P. Genoud, L. Hoffmann, A.A. Manuel, M. Peter, E. Walker, M. François, and K. Yvon, Structure, resistivity, critical field, specific-heat jump at T_c, Meissner effect, a.c. and d.c. susceptibility of high-temperature superconductor $YBa_2Cu_3O_7$, *Europhys. Lett.*, **4**, 247–252 (1987).
64. D.-H. Wu and S. Sridhar, Pinning forces and lower critical fields in $YBa_2Cu_3O_y$ crystals: Temperature dependence and anisotropy, *Phys. Rev. Lett.*, **65**, 2074–2077 (1990).
65. N. Balchev, K. Nenkov, V. Antonov, J. Pirov, and B. Kunev, Superconductivity and critical fields in undoped and Sn-doped $MoSr_2YCu_2O_{8-\delta}$, *Physica C*, **470**, 2040–2046 (2010).

66. J.N. Li, K. Kadowaki, M.J.V. Menken, Y.K. Huang, K. Bakker, A.A. Menovsky, and J.J.M. Franse, Upper critical fields of single-crystalline and polycrystalline Ca–Sr–Bi–Cu–O compounds, *Appl. Phys. A,* **47**, 209–211 (1988).
67. F. Shi, T.S. Rong, S.Z. Zhou, X.F. Wu, J. Du, Z.H. Shi, C.G. Cui, R.Y. Jin, J.L. Zhang, Q.Z. Ran, and N.C. Shi, Preparation process, crystal structure, and physical properties of the 110-K single-phase Pb–Bi–Sr–Ca–Cu–O superconductor, *Phys. Rev. B,* **41**, 6541–6546 (1990).
68. M. Hasegawa, K. Izawa, A. Shibata, and Y. Matsuda, Growth and upper critical field of heavily overdoped $Tl_2Ba_2CuO_x$ single crystals, *Physica C,* **377**, 459–465 (2002).
69. J.W. Ekin, Offset criterion for determining superconductor critical current, *Appl. Phys. Lett.,* **55**, 905–907 (1989).
70. J.W. Ekin, D.K. Finnemore, Quiang Li, J. Tenbrink, and W. Carter, Effect of axial strain on the critical current of Ag-sheathed Bi-based superconductors in magnetic fields up to 25 T, *Appl. Phys. Lett.,* **61**, 858–860 (1992).
71. K. Kwasnitza and Ch. Widmer, New programmable switching and storage effects in superconducting polycrystalline $YBa_2Cu_3O_7$ due to hysteretic critical intergrain transport current, *Cryogenics,* **29**, 1035–1040 (1989).
72. J. Evetts and B.A. Glowacki, Relation of critical current irreversibility to trapped flux and microstructure in polycrystalline $YBa_2Cu_3O_7$, *Cryogenics,* **28**, 641–649 (1988).
73. R.B. Stephens, Critical current limitations in ceramic oxide superconductors, *Cryogenics,* **29**, 399–404 (1989).
74. H. Küpfer, I. Apfelstedt, R. Flükiger, C. Keller, R. Meier-Hirmer, B. Runtsch, A. Turowski, U. Wiech, and T. Wolf, Investigation of inter- and intragrain critical currents in high T_c ceramic superconductors, *Cryogenics,* **28**, 650–660 (1988).
75. R.E. Gladyshevskii, A. Perin, B. Hensel, R. Flükiger, R. Abraham, K. Lebbou, M.Th. Cohen-Adad, and J.-L. Jordan, Preparation of *in-situ* reaction and physical characterization of Ag(Au) and Ag(Pd) sheathed $(Tl,Pb,Bi)(Sr,Ba)_2Ca_2Cu_3O_{9-\delta}$ tapes, *Physica C,* **255**, 113–123 (1995).
76. F.A. List, D.M. Kroeger, and V. Selvamanickam, Hysteresis of the transport critical current of (Tl,Pb)-(Sr,Ba)-Ca-Cu-O and Bi–Sr–Ca–Cu–O conductors: Effects of temperature and magnetic field, *Physica C,* **275**, 220–230 (1997).
77. L. Cesnak, T. Melišek, P. Kováč, and I. Hušek, Magnetic field hysteresis of critical currents in Bi(2223)/Ag tapes and a way to overcome it, *Cryogenics,* **37**, 823–827 (1997).
78. K. Kimura, M. Morita, M. Tanaka, S. Takebayashi, L. Trouilleux, K. Miyamoto, M. Hashimoto, K. Watanabe, S. Awaji, and N. Kobayashi, Transport critical current density of melt-processed (QMG) $YBa_2Cu_3O_x$ bulk superconductors, *Cryogenics,* **33**, 506–509 (1993).
79. M. Lepropre, I. Monot, M.P. Delamare, M. Hervieu, Ch. Simon, J. Provost, G. Desgardin, B. Raveau, J.M. Barbut, D. Bourgault, and D. Braithwaite, Critical currents up to 71000 $A\,cm^{-2}$ at 77 K in melt textured YBCO doped with $BaSnO_3$, *Cryogenics,* **34**, 63–67 (1994).
80. H. Ikuta, T. Yamada, M. Yoshikawa, Y. Yanagi, Y. Ytoh, B. Latha, and U. Mizutani, Melt processing of (rare earth)-Ba-Cu-O bulk superconductors, *Supercond. Sci. Technol.,* **18**, S119–S125 (2005).

81. K. Iida, N.H. Babu, Y.H. Shi, D.A. Cardwell, and M. Murakami, Gd-Ba-Cu-O bulk superconductors by a seeded infiltration growth technique under reduced oxygen partial pressure, *Supercond. Sci. Technol.*, **19**, 641–647 (2006).
82. M. Muralidhar, M. Jirsa, and M. Tomita, Flux pinning and superconducting properties of melt-textured NEG-123 superconductor with TiO_2 addition, *Physica C*, **470**, 592–597 (2010).
83. P. Diko, V. Antal, K. Zmorayová, M. Šefčiková, X. Chaud, J. Kováč, X. Yao, I. Chen, M. Eisterer, and H.W. Weber, The influence of post-growth thermal treatments on the critical current density of TSMG YBCO bulk superconductors, *Supercond. Sci. Technol.*, **23**, 124002 (2010).
84. S. Elschner, J. Bock, and S. Gauss, Development of HTSC-rings for inductive fault current limiters, in *Advances in Superconductivity VII*, K. Yamafuji and T. Morishita (eds.), Springer-Verlag, Tokyo, 1995.
85. J.-F. Fagnard, S. Elschner, J. Bock, M. Dirickx, B. Vanderheyden, and P. Vanderbemden, Shielding efficiency and $E(J)$ characteristics measured on large melt cast Bi-2212 hollow cylinders in axial magnetic fields, *Supercond. Sci. Technol.*, **23**, 095012 (2010).
86. R. Wesche, Temperature dependence of critical currents in superconducting Bi-2212/Ag wires, *Physica C*, **246**, 186–194 (1995).
87. L.H. Allen, P.R. Broussard, J.H. Claassen, and S.A. Wolf, Temperature and field dependence of the critical current densities of Y–Ba–Cu–O films, *Appl. Phys. Lett.*, **53**, 1338–1340 (1988).
88. D.P. Norton, A. Goyal, J.D. Budai, D.K. Christen, D.M. Kroeger, E.D. Specht, Q. He, B. Saffian, M. Paranthaman, C.E. Klabunde, D.F. Lee, B.C. Sales, and F.A. List, Epitaxial $YBa_2Cu_3O_7$ on biaxially textured nickel (001): An approach to superconducting tapes with high critical current density, *Science*, **274**, 755–757 (1996).
89. P. Schmitt, L. Schultz, and G. Saemann-Ischenko, Electrical properties of $Bi_2Sr_2CaCu_2O_x$ thin films prepared in situ by pulsed laser deposition, *Physica C*, **168**, 475–478 (1990).
90. J.T. Kucera, D.G. Steel, D.W. Face, J.M. Graybeal, T.P. Orlando, and D.A. Rudman, Critical current densities in textured thin films of Bi–Sr–Ca–Cu-oxide, *Physica C*, **162–164**, 671–672 (1989).
91. J.E. Tkaczyk, J.A. DeLuca, P.L. Karas, P.J. Bednarczyk, D.K. Christen, C.E. Klabunde, and H.R. Kerchner, Enhanced transport critical current at high fields after heavy ion irradiation of textured $TlBa_2Ca_2Cu_3O_z$ thick films, *Appl. Phys. Lett.*, **62**, 3031–3033 (1993).
92. A.A. Gapud, J.Z. Wu, L. Fang, S.L. Yan, Y.Y. Xie, M.P. Siegal, and D.L. Overmyer, Supercurrents in $HgBa_2CaCu_2O_{6+\delta}$ and $TlBa_2CaCu_2O_7$ epitaxial thin films, *Appl. Phys. Lett.*, **74**, 3878–3880 (1999).
93. S.H. Jun, J.Z. Wu, S.C. Tidrow, and D.W. Eckart, Growth of $HgBa_2Ca_2Cu_3O_{8+\delta}$ thin films on $LaAlO_3$ substrates using fast temperature ramping Hg-vapor annealing, *Appl. Phys. Lett.*, **68**, 2565–2567 (1996).
94. S.H. Jun and J.Z. Wu, Superconductivity above 130 K in high-quality mercury-based thin films, *Appl. Phys. Lett.*, **68**, 862–864 (1996).

95. S. Tönies, H.W. Weber, G. Gritzner, O. Heiml, and M.H. Eder, Tl-1223 thick films – A competitor for Y-123 coated conductors?, *IEEE Trans. Appl. Supercond.*, **13**, 2618–2621 (2003).
96. C. Stölzel, M. Huth, and H. Adrian, *c*-axis oriented thin $Bi_2Sr_2CaCu_2O_{8+\delta}$ films prepared by flash-evaporation, *Physica C*, **204**, 15–20 (1992).
97. S. Awaji, M. Namba, K. Watanabe, M. Miura, Y. Ichino, Y. Yoshida, Y. Takai, and K. Matsumoto, J_c and B_i properties of $Sm_{1+x}Ba_{2-x}Cu_3O_y$ films with nano-particles, *Physica C*, **463–465**, 669–673 (2007).
98. L. Schultz, B. Roas, P. Schmitt, P. Kummeth, and G. Saemann-Ischenko, Comparison of the critical current anisotropy in YBaCuO and BiSrCaCuO films, *IEEE Trans. Magn.*, **27**, 990–996 (1991).
99. H. Yamasaki, K. Endo, S. Kosaka, M. Umeda, S. Misawa, S. Yoshida, and K. Kajimura, Magnetic-field angle dependence of the critical current density in high-quality $Bi_2Sr_2Ca_2Cu_3O_x$ thin films, *IEEE Trans. Appl. Supercond.*, **3**, 1536–1539 (1993).
100. G. Blatter, V.B. Geshkenbein, and A.I. Larkin, From isotropic to anisotropic superconductors: A scaling approach, *Phys. Rev. Lett.*, **68**, 875–878 (1992).
101. H. Yamasaki, K. Ohki, I. Yamaguchi, M. Sohma, W. Kondo, H. Matsui, T. Manabe, and T. Kumagai, Strong flux pinning due to dislocations associated with stacking faults in $YBa_2Cu_3O_{7-\delta}$ thin films prepared by fluorine-free metal organic deposition, *Supercond. Sci. Technol.*, **23**, 105004 (2010).
102. V.S. Dang, P. Mikheenko, A. Sarkar, M.M. Awang Kechik, J.S. Abell, and A. Crisan, Increased critical current density and pinning in thick Ag/$YBa_2Cu_3O_{7-x}$ multilayers, *Physica C*, **470**, 1238–1241 (2010).
103. L. Krusin-Elbaum, C.C. Tsuei, and A. Gupta, High current densities above 100 K in the high-temperature superconductor $HgBa_2CaCu_2O_{6+\delta}$, *Nature*, **373**, 679–681 (1995).
104. Y. Tsabba and S. Reich, Giant mass anisotropy and high critical current in Hg-1223 superconducting films, *Physica C*, **269**, 1–4 (1996).
105. S. Senoussi, M. Oussena, and S. Hadjoudj, On the critical fields and current densities of $YBa_2Cu_3O_7$ and $La_{1.85}Sr_{0.15}CuO_4$ superconductors, *J. Appl. Phys.*, **63**, 4176–4178 (1988).
106. M. Oussena, S. Senoussi, and G. Collin, Magnetic hysteresis in $La_{1.85}Sr_{0.15}CuO_4$, *Europhys. Lett.*, **4**, 625–629 (1987).
107. D. Dimos, P. Chaudhari, J. Mannhart, and F.K. LeGoues, Orientation dependence of grain-boundary critical currents in $YBa_2Cu_3O_{7-\delta}$ bicrystals, *Phys. Rev. Lett.*, **61**, 219–222 (1988).
108. D. Dimos, P. Chaudhari, and J. Mannhart, Superconducting transport properties of grain boundaries in $YBa_2Cu_3O_7$ bicrystals, *Phys. Rev. B*, **41**, 4038–4049 (1990).
109. J.A. Alarco, E.Olsson, Z.G. Ivanov, D. Winkler, E.A. Stepantsov, O.I. Lebedev, A.L. Vasilev, A.Ya. Tzalenchuk, and N.A. Kiselev, Microstructure and properties of artificial grain boundaries in epitaxial $YBa_2Cu_3O_{7-\delta}$ thin films grown on [001] tilt Y–ZrO_2 bicrystals, *Physica C*, **247**, 263–269 (1995).

110. A. Schmehl, B. Goetz, R.R. Schulz, C.W. Schneider, H. Bielefeldt, H. Hilgenkamp, and J. Mannhart, Doping-induced enhancement of the critical currents of grain boundaries in $YBa_2Cu_3O_{7-\delta}$, *Europhys. Lett.*, **47**, 110–115 (1999).
111. A. Attenberger, J. Hänisch, B. Holzapfel, and L. Schultz, Electrical transport properties of Bi(2223) [0 0 1] tilt grain boundary junctions, *Physica C,* **372–376**, 649–651 (2002).
112. E. Sarnelli, P. Chaudhari, W.Y. Lee, and E. Esposito, Transport properties in Tl-Ba-Ca-Cu-O grain boundary junctions on $SrTiO_3$ bicrystal substrates, *Appl. Phys. Lett.,* **65**, 362–364 (1994).
113. V.R. Todt, X.F. Zhang, D.J. Miller, M. St. Louis-Weber, and V.P. Dravid, Controlled growth of bulk bicrystals and the investigation of microstructure-property relations of $YBa_2Cu_3O_x$ grain boundaries, *Appl. Phys. Lett.,* **69**, 3746–3748 (1996).
114. Q. Li, Y.N. Tsay, Y. Zhu, M. Suenaga, G.D. Gu, and N. Koshizuka, Large T_c depression at low angle [1 0 0] tilt grain boundaries, *Appl. Phys. Lett.,* **70**, 1164–1166 (1997).
115. J. Mannhart and H. Hilgenkamp, Interfaces involving complex superconductors, *Physica C,* **317–318**, 383–391 (1999).
116. H. Hilgenkamp and J. Mannhart, Superconducting and normal-state properties of $YBa_2Cu_3O_{7-\delta}$-bicrystal grain boundary junctions in thin films, *Appl. Phys. Lett.,* **73**, 265–267 (1998).

10

Flux Pinning in Cuprate High-T_c Superconductors

10.1 Introduction

In this chapter, we will discuss flux pinning in cuprate high-T_c superconductors. It is beyond the scope of this book to discuss pinning theories applicable to cuprate superconductors. Instead, we will describe the consequences of strong anisotropy on the flux pinning in high-T_c superconductors. In Section 4.4, we have already mentioned that current flow without loss is possible in a Type II superconductor only when the flux lines are pinned. Without pinning, the flux lines would move in the superconductor due to the Lorentz force, leading to flux flow resistance. The pinning behavior of cuprate high-T_c superconductors is much more complex than that of low-temperature superconductors because of the much larger thermal energy at elevated operating temperatures, and the consequences of the layered crystal structures. After presentation of experimental evidence for the existence of flux lines in cuprate high-T_c superconductors (see Section 10.2), we will briefly consider the possibility of intrinsic pinning by the insulating charge carrier reservoirs (see Section 10.3). As a consequence of weak pinning and high transition temperatures, thermally activated flux creep is possible in the high-T_c superconductors, which is reflected by giant flux creep (see Section 10.4). In the discussion of the magnetic phase diagram of cuprate high-T_c superconductors, we have already introduced the irreversibility line. In Section 10.5, we will consider the interplay of structural features, artificial pinning centers and irradiation, and their effect on the position of the irreversibility line in the magnetic phase diagram. Finally, we will briefly summarize the aspects of pinning behavior in cuprate high-T_c superconductors in Section 10.6.

Physical Properties of High-Temperature Superconductors, First Edition. Rainer Wesche.

10.2 Vortex Lattice

Above the lower critical field, magnetic flux can enter the interior of a Type II superconductor via the flux lines. In this mixed state, superconducting regions and normal conducting vortex cores coexist. Each vortex is encircled by screening currents, which lead to mutual repulsion of the vortex cores. As a consequence, a hexagonal flux-line lattice is formed in isotropic metallic Type II superconductors. As soon as a sufficiently large fraction of the flux lines is pinned, the vortex–vortex interaction prevents flux motion. The effectiveness of collective pinning is determined by the stiffness of the flux-line lattice.

According to a proposal by Buckel [1], the energy required for the creation of a flux line can be estimated from the lower critical field B_{c1}. At the lower critical field, the loss of condensation energy and the gain of magnetic expulsion energy become equal (Equation 2.60 in Chapter 2), and hence flux lines start to enter the interior of the superconductor. The formation energy for a single flux line is

$$E_{\text{fl}} = \int_0^{B_{c1}} m dB \tag{10.1}$$

where $m = (\phi_0 / \mu_0) l_{\text{fl}}$ is the magnetic moment connected with a single flux quantum in a flux line of length l_{fl}. The required energy per unit length is

$$\frac{E_{\text{fl}}}{l_{\text{fl}}} = \frac{\phi_0}{\mu_0} B_{c1} \tag{10.2}$$

Using the definition of the Ginzburg–Landau parameter $\kappa = \lambda_L/\xi$ and Equations 4.19 and 4.21 in Chapter 4, we find

$$\frac{E_{\text{fl}}}{l_{\text{fl}}} = \frac{1}{4\pi\mu_0} \left(\frac{\phi_0}{\lambda_{\text{L}}}\right)^2 \ln\left(\frac{\lambda_{\text{L}}}{\xi}\right) \tag{10.3}$$

Here, λ_L is the penetration depth, ξ the coherence length, and ϕ_0 the flux quantum. For example, the flux line can be pinned by a normal conducting precipitate. When the flux line is at the position of a normal region, no energy is required to form the normal vortex core, and the resulting pinning energy can be estimated with Equation 10.3. In general, regions with defects leading to a reduced Cooper pair density can act as pinning centers.

In single crystals of $YBa_2Cu_3O_7$, the flux-line lattice has been observed by means of a high-resolution Bitter technique [2, 3]. The single crystals were cooled to 4.2 K in the presence of a small magnetic field. After field cooling, the surfaces of the $YBa_2Cu_3O_7$ samples were decorated with small magnetic nickel particles. Nickel was evaporated in the presence of ≈0.2 mbar helium gas. A baffle was used to hinder direct evaporation onto the surface of the $YBa_2Cu_3O_7$ sample. The fine nickel particles were preferentially deposited on those parts of the sample surface where a magnetic field is present in the vortex cores of the flux lines. After warming up to room temperature, the resulting images of the flux-line lattice were examined by scanning electron microscopy. The results indicated that each vortex in $YBa_2Cu_3O_7$ contains just one single flux quantum ϕ_0 as in metallic Type II superconductors. Magnetic decoration was also successfully used to study the flux-line lattice in $Bi_2Sr_2CaCu_2O_{8+x}$ single crystals before and after irradiation with 340 MeV Xe ions [4].

Another way to study the flux-line lattice of superconductors is the use of scanning tunneling microscopy [5, 6]. Scanning tunneling microscopy was first applied to visualize the hexagonal flux-line lattice in the low-temperature superconductor $NbSe_2$ [7, 8]. The same experimental technique was applied by Maggio-Aprile *et al.* to study the flux-line lattice of $YBa_2Cu_3O_7$ [9, 10]. They found a distorted quadratic lattice with vortices of elliptical shape. The elliptic shape of the vortices is a consequence of the anisotropic coherence lengths within the CuO_2 planes.

10.3 Consequences of Anisotropy and Intrinsic Pinning

In Chapter 9, we have seen that the critical current density in textured films depends not only on the absolute value but also on the direction of the applied magnetic field. The critical current density declines rapidly for magnetic fields perpendicular to the CuO_2 (*ab*) planes, whereas the field dependence is much less pronounced for fields parallel to them. Figure 10.1 shows the position of the vortices in the layered crystal structure for these two field directions. For a magnetic field applied perpendicular to the CuO_2 planes and a current flow along the planes, the Lorentz force acting on the vortices is along the CuO_2 planes and transverse to the directions of the applied field and the current (see Figure 10.1a). The rapid decline of the critical current density for this field direction indicates that the flux pinning is not very efficient. The cuprate superconductors can be considered as a sequence of alternating superconducting CuO_2 blocks and insulating charge carrier reservoirs along the crystallographic *c* direction. The Cooper-pair density in the charge carrier reservoirs is considerably lower than that in the CuO_2 blocks. The strong variation of the Cooper-pair density is possible because of the extremely short coherence length along the *c* direction, which is comparable to interatomic distances. There is a tendency for the flux lines along the *c* direction to be cut within the weakly superconducting charge carrier reservoirs, leading to short flux-line segments or even – in extreme cases – pancake vortices. The latter exist only within the copper oxide blocks, and are therefore easily movable. Due to this segmentation of the flux lines, pinning by point defects is not very efficient for magnetic fields perpendicular to the CuO_2 planes. The degree of the anisotropy of the critical current density with respect to the direction of the applied field is closely related to the thickness of the charge carrier reservoirs and to the number of CuO_2 planes in the copper oxide blocks (see also Section 10.5). In Tl-22($n-1$)n and Bi-22($n-1$)n compounds the anisotropy is very pronounced because of the relatively thick BaO–TlO–TlO–BaO and SrO–BiO–BiO–SrO charge carrier reservoirs. In Tl-12($n-1$)n and Hg-12($n-1$)n compounds, the distance between adjacent CuO_2 blocks is smaller than in the 22($n-1$)n cuprates. The charge carrier reservoirs consist of only three layers, namely BaO–TlO–BaO or BaO–Hg–BaO. As a consequence, the anisotropy of the 12($n-1$)n compounds is much less pronounced than that of the 22($n-1$)n compounds. Bi-2212 and Tl-2212 show a slightly larger anisotropy than the corresponding Bi-2223 and Tl-2223 compounds.

For fields parallel to the CuO_2 planes, the Lorentz force on the vortices is perpendicular to these planes (see Figure 10.1b). The much less pronounced decrease in the critical current density suggests that the flux lines are much more efficiently pinned for this field direction than for magnetic fields perpendicular to the CuO_2 planes. It may be expected that the vortices are efficiently pinned by charge carrier reservoirs with reduced Cooper-pair

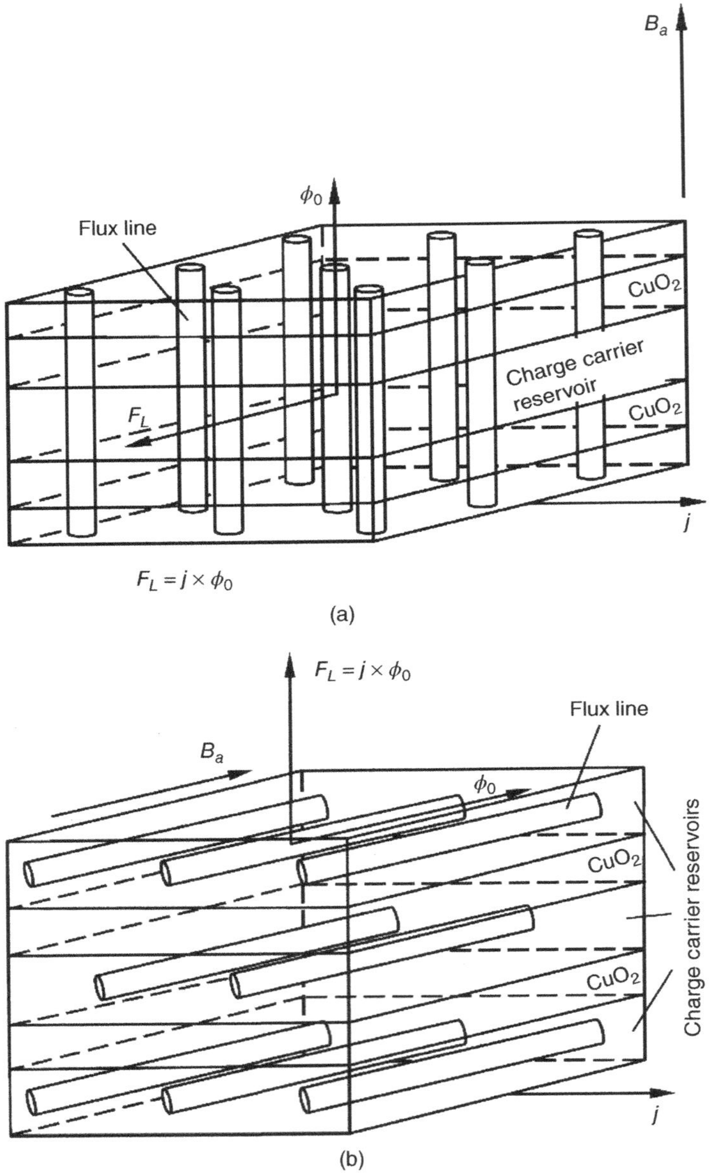

Figure 10.1 *(a) For magnetic fields applied perpendicular to the CuO_2 planes, the Lorentz force is parallel to the CuO_2 planes. In the charge carrier reservoirs, the density of the Cooper pairs is significantly reduced, and hence the flux lines are frequently cut within the charge carrier reservoirs, leading in the worst case to pancake vortices which are restricted to single CuO_2 blocks. These pancake vortices can easily be moved along the ab planes. (b) For magnetic fields applied parallel to the CuO_2 plains, the Lorentz force is parallel to the c direction. The charge carrier reservoirs with their reduced Cooper pair density provide an efficient pinning of the flux lines (intrinsic pinning), hindering their movement parallel to the c direction*

density. The layered crystal structures of the cuprates provide intrinsic pinning for fields along the CuO_2 planes. The effect of intrinsic pinning on the critical current densities in cuprate superconductors has been discussed in more detail by Tachiki and Takahashi [11].

10.4 Thermally Activated Flux Creep

In cuprate high-T_c superconductors, magnetic relaxation effects are much more pronounced than in low-temperature superconductors. In magnetic relaxation experiments, the superconductor under investigation is cooled in zero field to the envisaged temperature, which is well below the critical temperature. In the case of studies of the relaxation of the remanent magnetization, a magnetic field is applied and then removed. After field removal, the relaxation of the magnetization with time is measured. Alternatively the relaxation of the magnetization in the presence of an applied magnetic field can be studied. In general, a considerable reduction in the magnetization has been observed over long periods in cuprate superconductors. For magnetic fields applied perpendicular to the CuO_2 planes, the relaxation of the magnetization is much faster than for fields parallel to these planes. In $YBa_2Cu_3O_7$ single crystals, relaxation of the magnetization has been observed after zero-field-cooling as well as field cooling [12]. The results suggest that the magnetic relaxation effects are a consequence of pronounced thermally activated flux creep. Because of the relatively weak pinning and the elevated temperatures, thermally activated flux creep is of much more importance in cuprate high-T_c superconductors than in low-T_c superconductors.

In the discussion of thermally activated flux creep, we follow the description given in Ref. [13]. Diffusion in solids is similar to thermally activated flux creep, as illustrated in Figure 10.2. Even for thermal energies k_BT considerably smaller than the pinning energy V_p, there is a finite probability for the flux lines to hop to adjacent pinning centers. The Lorentz forces acting on the flux lines reduce the pinning potential, and hence favor thermally activated flux creep.

Conventional flux creep theory assumes that the pinning potential V_p decreases linearly with the current density

$$V_p(j) = V_{p0}\left(1 - \frac{j}{j_c}\right) \tag{10.4}$$

Here V_{p0} is the height of the barrier at zero current and j_c the critical current density in the absence of thermal activation. The hopping rate ν of the flux lines is given by an Arrhenius law

$$\nu = \nu_0 \exp\left(-\frac{V_p(j)}{k_BT}\right) \tag{10.5}$$

where the attempt frequency ν_0 is typically in the range of 10^{10}–10^{12} Hz. For $k_BT << V_{p0}$, a logarithmic relaxation results

$$M(t) = M_0\left(1 - \frac{k_BT}{V_{p0}} \ln\left(\frac{t}{\tau_0}\right)\right) \tag{10.6}$$

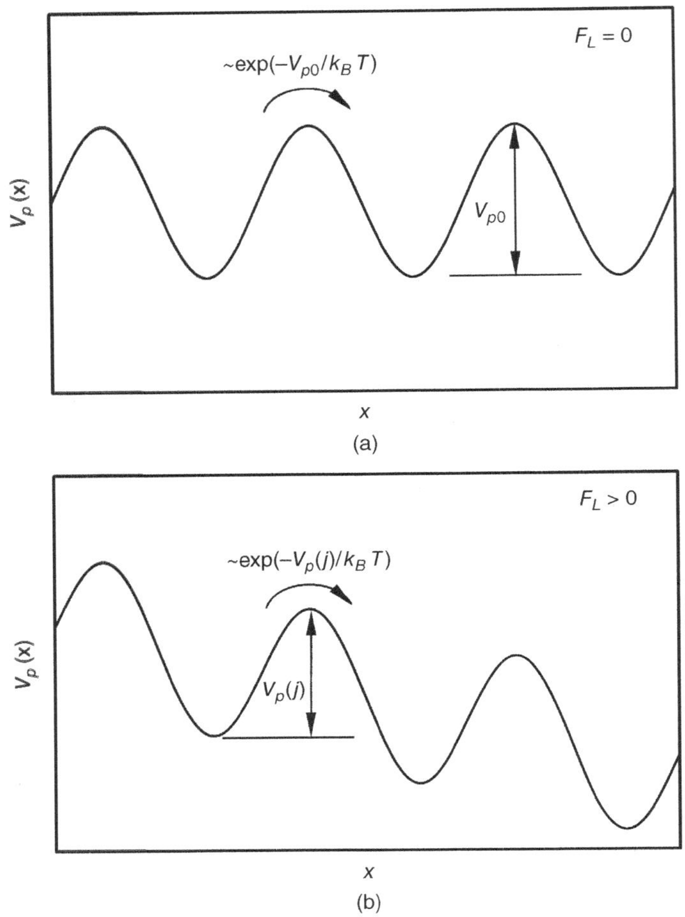

Figure 10.2 *(a) In the absence of a transport current, thermally activated flux creep is similar to diffusion in a solid. There exists a finite probability of flux-line hopping to the neighboring potential well. (b) The Lorentz force caused by the transport current, which acts on the flux lines, leads to a reduction in the potential well, and hence supports flux-line hopping (adapted from [13]).*

Here M_0 is the initial value of the magnetization and the time constant τ_0 is closely related to the attempt frequency ν_0. In experimental studies of the timely evolution of the magnetization, logarithmic as well as nonlogarithmic relaxations have been observed in cuprate superconductors [12, 14–20]. At elevated temperatures, a nonlogarithmic behavior was found, while a logarithmic relaxation was observed at sufficiently low temperatures.

Another way to get information on the pinning potential is the study of the resistive transition in the presence of an applied magnetic field. Measurements in high magnetic fields have shown that the pinning potential $V_p(j)$ depends logarithmically on the current density [21, 22]

$$V_p(j) = V_{p0} \ln\left(\frac{j_0}{j}\right) \tag{10.7}$$

At the critical current density j_0, the barrier height is zero. Vinokur *et al.* [23] have shown that a logarithmic dependence of the pinning potential on the current density leads to a power law for the magnetic relaxation

$$M(t) = M_0\left(\frac{t}{\tau_0}\right)^{-\alpha} \tag{10.8}$$

where $\alpha = k_B T / V_{p0}$ and τ_0 is again a time constant characteristic of the magnetic relaxation process.

Finally, let us briefly consider the results of collective pinning theories [24–26]. They predict a nonlogarithmic relaxation of the magnetization, which is described by [24]

$$M(t) = M_0\left[1 + \mu\frac{k_B T}{V_{p0}} \ln\left(\frac{t}{\tau_0}\right)\right]^{-\frac{1}{\mu}} \tag{10.9}$$

The effective pinning potential related to collective pinning is [24]

$$V_p(j) = \frac{V_{p0}}{\mu}\left[\left(\frac{j_0}{j}\right)^{\mu} - 1\right] \tag{10.10}$$

Feigel'man *et al.* [25] showed that the value of μ depends on the flux bundle size. For individual vortex hopping, they found $\mu = 1/7$, while the values of μ for small and large bundles are 3/2 and 7/9, respectively. The experimental investigations by Ren and de Groot [24] indicate that magnetic relaxation in an $YBa_2Cu_3O_{7-\delta}$ single crystal can be well described by collective pinning. They used μ as a fitting parameter. Their magnetic relaxation data, measured at 50 K, provide nearly constant μ values of $\approx$0.2 between 1 and 4 T. For fields above 4 T, μ increases continuously and reaches a value above 2 at 8 T. Measurements at a magnetic field of 2 T along the crystallographic c direction provide values of μ between 0.2 and 0.6 in the temperature range of 10–65 K. A significantly higher value of $\approx$1.5 was obtained at 70 K and 2 T.

Figure 10.3 shows the relaxation of the remanent magnetization of a Bi-2212 ceramic [18], a Bi-2212 single crystal [19], and a (Bi,Pb)-2223 ceramic [16]. In the case of the Bi-2212 ceramic, the remanent state was prepared by increasing the magnetic field to 1 T within 16 minutes. After waiting for at least 1 h, the magnetic field was removed. The relaxation data, shown in Figure 10.3, are normalized to the remanent magnetization measured 20 s after the end of the field sweep. The best fit to the measured data was found for a logarithmic dependence of the pinning potential on the current density (Equation 10.7), leading to a power law dependence of $M_{rem}(t)$ (Equation 10.8). Using Equation 10.8, a pinning potential $V_{p0} = 22 \pm 2$ meV was found. The initial magnetization of the Bi-2212 ceramic cylinder was 247 000 $\pm$ 8000 A/m.

The remanent magnetization of the Bi-2212 single crystal was measured after a field of 0.1 T was applied along the c direction and removed after 10–100 ms. The observed relaxation of the remanent magnetization can be well described by the scaling relation [19]

$$M(t) = M_0 \left(1 - \left[\frac{k_B T}{V_{p0}} \ln\left(1 + \frac{t}{\tau_0}\right)\right]^{1/\alpha}\right) \tag{10.11}$$

where M_0 is the remanent magnetization at $t = 0$ and α a scaling exponent. The activation energy, related to the above expression, is $V_p(j) = V_{p0}(1 - j/j_c)^\alpha$.

The remanent magnetization of the (Bi,Pb)-2223 ceramic samples was measured after cooling to the desired temperature in zero field, and switching a magnetic field of 1 T on and off. The field sweep was performed within about 20 s. The remanent magnetization was found to decay logarithmically.

The data, presented in Figure 10.3, indicate that the observed relaxation of the remanent magnetization depends on temperature, the time window considered, and the sample type. For the three different Bi-based cuprate superconductors, the relaxation was found to be described by different $M(t)$ relations in the time windows considered. For the Bi-2212 single crystal, nonlogarithmic relaxation was observed at times of less than 10 s after removal of the applied magnetic field. In the case of the (Bi,Pb)-2223 and the Bi-2212 ceramic, the measurements of magnetic relaxation start at times t_0 of 10 and 20 s, respectively. The observed relaxation is proportional to ln t for times larger than t_0. The different relaxation

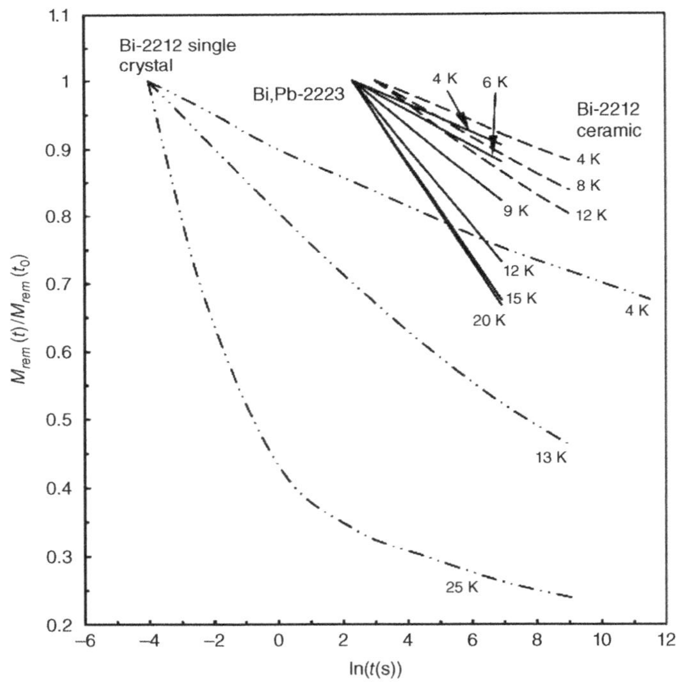

Figure 10.3 *Relaxation of the remanent magnetization in a Bi-2212 single crystal, a Bi-2212 ceramic, and a (Bi,Pb)-2223 sample (data from [16, 18, 19]). For the Bi-2212 single crystal, the magnetic relaxation is clearly nonlogarithmic*

found for a Bi-2212 single crystal, a Bi-2212 ceramic, and a (Bi,Pb)-2223 ceramic sample reflects differences in the pinning potential. The pinning in the Bi-2212 ceramic is more effective than that in the Bi-2212 single crystal.

In general, the relaxation rate S of the magnetization can be used to derive an effective activation energy V_{p0}. From the single barrier expression (Equation 10.6), we get

$$S = \frac{1}{M_0}\frac{dM}{d\ln\left(\frac{t_b}{\tau_0}\right)} = -\frac{k_B T}{V_{p0}} \tag{10.12}$$

Here t_b is the time at which the measurement begins. For a nonlogarithmic relaxation, the effective pinning potential V_{p0} can be obtained from a fit of the measured data using the scaling relation $M(t)$ in question. The experimentally determined values of $dM(t)/d\ \ln(t/\tau_0)$ reach a maximum at intermediate temperatures. Hagen and Griessen proposed that the occurrence of the maximum of the relaxation rate at intermediate temperatures can be explained by a distribution of activation energies [27].

Figure 10.4 shows the temperature dependence of $dM/d\ln(t)$ of Y-123 and Bi-2212 single crystals. After zero-field-cooling, the magnetization data were collected in magnetic fields of 0.06 T (Y-123) and 0.1 T (Bi-2212) applied parallel to the c direction. For Bi-2212, a maximum of $dM/d\ln(t)$ was found at a temperature of 15 K, while for Y-123

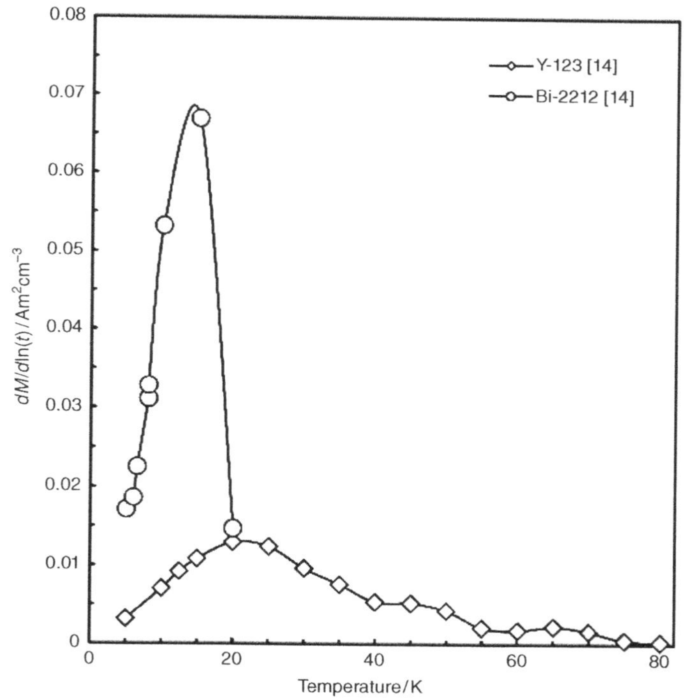

Figure 10.4 *Comparison of the magnetic relaxation rates in Y-123 and Bi-2212 crystals (data from [14]). The relaxation rates in the Y-123 crystal are much smaller than those obtained for the Bi-2212 crystal, which suggests that the pinning in the Y-123 is much better than in Bi-2212*

the maximum value is reached at 20 K. The value of $V_{p0} = 0.015$ eV found for Y-123 is around a factor of 2 larger than the value of 0.0008 eV obtained for Bi-2212 [14].

Figure 10.5 shows the magnetic relaxation of various cuprate superconductors. The normalized $M(t)/M(t_0)$ data of the above mentioned Y-123 and Bi-2212 single crystals are also shown [14]. In the case of Y-123, the value of t_0 is 200 s. The relaxation of the magnetization of Tl-2223 and Tl-1223 [28], both measured at 30 K and 1 T, can be directly compared. In Tl-2223, the decrease in the magnetization is much more pronounced than in Tl-1223. The magnetic relaxation of (Bi,Pb)-2223 at 10 K and 0.5 T [29] is comparable to that of Tl-1223 at 30 K and 1 T. The measurements performed on a textured Y-123 tape, before and after irradiation with 180 MeV Cu^{11+} ions, clearly indicate that the magnetic relaxation after irradiation is significantly reduced [30]. This result reflects the fact that the defects generated by the irradiation provide a more efficient pinning of the flux lines.

Another way to get information on thermally activated flux creep is the study of the electrical resistivity at the transition to the normal state in the presence of a magnetic field. The flux creep resistance is given by

$$\rho = \rho_{\text{ff}} \exp\left(-\frac{V_{p0}(B)}{k_B T}\right) \tag{10.13}$$

where V_{p0} is the activation energy and ρ_{ff} the resistivity for $1/T \to 0$.

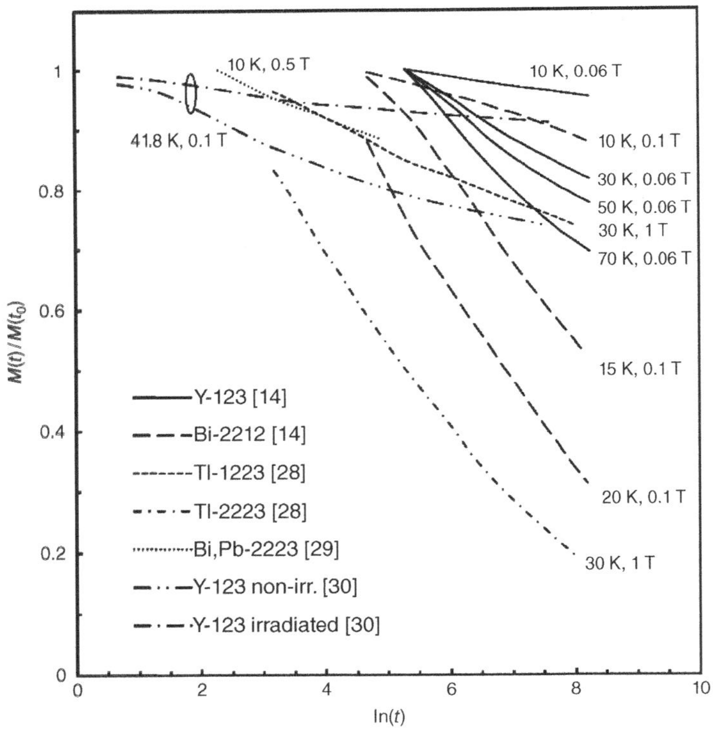

Figure 10.5 Relaxation of the in-field magnetization in various cuprate superconductors (data from [14, 28–30])

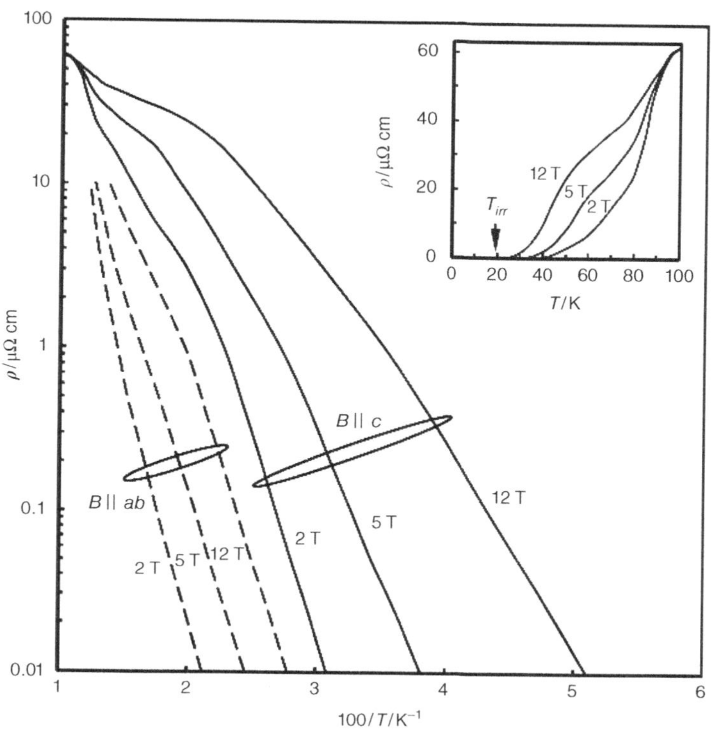

Figure 10.6 *Arrhenius plots of the resistance of a Bi-2212 single crystal for magnetic fields of 2, 5, and 12 T applied parallel or perpendicular the ab (CuO_2) planes. Below 1 μΩ cm, the lines are straight, as expected for thermally activated flux creep. The inset shows the resistivity versus temperature curves for fields parallel to the c direction (data from [31])*

Figure 10.6 shows Arrhenius plots of the resistance versus temperature curves of a Bi-2212 single crystal [31] for fields parallel and perpendicular to the crystallographic *c* direction. For fields of 2, 5, and 12 T parallel to the *c* direction, the corresponding resistivity versus temperature curves are shown in the inset. The width of the transition increases with increasing applied field. Well below a resistivity of 1 μΩ cm, the resistivity in the Arrhenius plot depends linearly on $1/T$, as expected for thermally activated flux creep. The slopes of the resistivity lines in this low resistivity region provide the activation energy V_{p0}. The data, presented in Figure 10.6, clearly indicate that the resistivity slopes for fields parallel to the *ab* planes are steeper than those for fields parallel to the *c* direction. This result shows that the pinning energy for fields along the CuO_2 (*ab*) planes is significantly larger than that for fields perpendicular to these planes.

Figure 10.7 shows Arrhenius plots of the resistivity of textured Y-123 films with the addition of 4 wt% $BaZrO_3$ [32]. The magnetic fields in the range of 1–8 T are applied parallel to the *c* direction. Well below a resistivity of 10 μΩ cm, the dependence of the resistivity on the inverse temperature is linear, as expected for thermally activated flux creep. The slopes of the resistivity lines increase with decreasing magnetic fields, and hence V_{p0} is enhanced at lower fields (see also Figure 10.10).

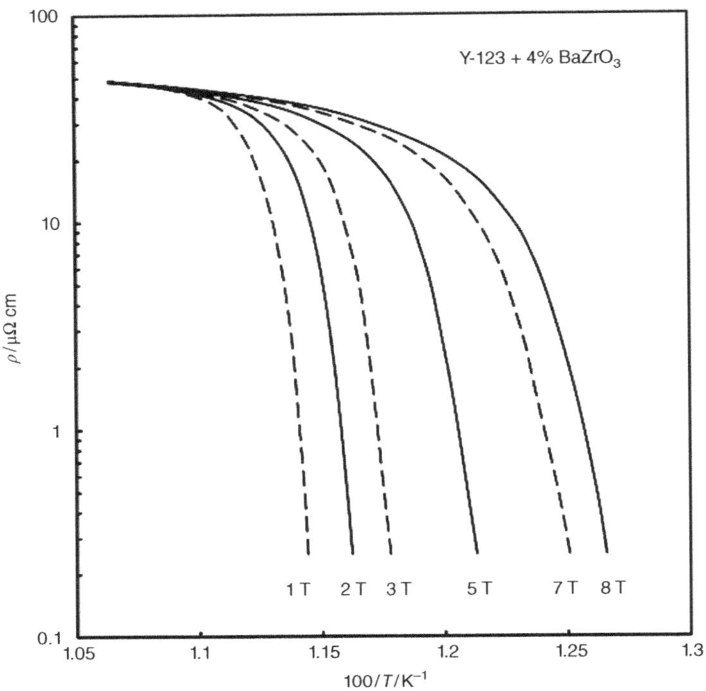

Figure 10.7 *Arrhenius plots of the resistance of a textured Y-123 film with 4 wt% $BaZrO_3$ for magnetic fields of 1–8 T parallel to the c direction. Below 5 μΩ cm, the lines are straight, as expected for thermally activated flux creep. The slopes become steeper at low fields, reflecting an increase in the activation barrier at lower fields (data from [32])*

In Figure 10.8, Arrhenius plots of the resistivity of (Y,Gd)-123 textured films [33] with zirconium additions are shown for fields in the range of 1–17 T. Well below $\rho = 10$ μΩ cm, the resistivity again depends linearly on $1/T$, which is consistent with thermally activated flux creep.

Values of the barrier height V_{p0}, found from the resistivity at the transition to the normal state in the presence of a magnetic field, are presented in Figures 10.9 and 10.10. In Figure 10.9, the activation energies for Hg-1212 [34] and Gd-123 [32] are shown. In Hg-1212, the activation energy reaches a value of nearly 10 eV at a field of 0.5 T parallel to the *ab* planes. The activation energy drops to less than 0.2 eV at a field of 9 T. In magnetic fields perpendicular to the *ab* planes, significantly lower activation energies were found. For example, the barrier height is only slightly above 1 eV at a field of 0.5 T, and drops to 70 meV at 9 T.

In the case of Gd-123, the values of V_{p0} depend less strongly on the applied magnetic field. The activation energy reaches nearly 7 eV at a field of 0.5 T parallel to the *ab* planes. Even at a field of 8 T, a barrier height of around 1.6 eV has been found. For fields applied perpendicular to the *ab* planes, the activation energy drops from 3.7 to 0.7 eV in the field range of 0.5–8 T.

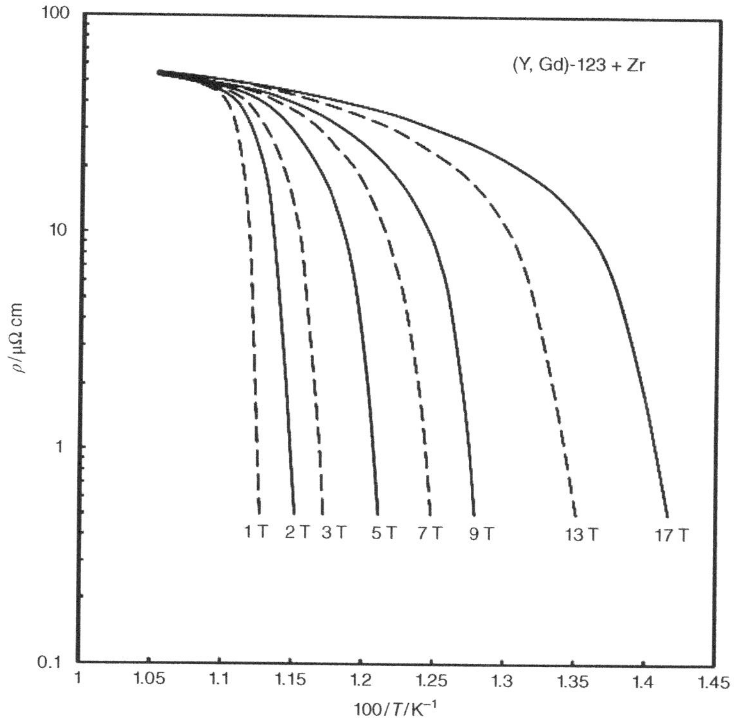

Figure 10.8 *Arrhenius plots of the resistance of a textured (Y,Gd)-123 film, with 1 wt% Zr doping, for magnetic fields of 1–17 T parallel to the c direction. Below 5 μΩ cm, the lines are straight as expected for thermally activated flux creep. The slopes become steeper at low fields, reflecting an increase in the activation barrier at lower fields (data from [33])*

In Figure 10.10, the barrier heights of textured Y-123 [35] and Gd-123 films [32] with various contents of $BaZrO_3$ (BZO) are compared. In all measurements, the magnetic field was applied parallel to the crystallographic *c* direction, i.e., perpendicular to the CuO_2 planes. Comparable activation energies have been found in Y-123 and Gd-123 films without BZO additions. In the investigated range with up to 4% BZO addition, the barrier height increases with the weight fraction of BZO added to Y-123. The improved pinning in the Y-123 films with BZO additions is caused by the self-assembly of $BaZrO_3$ rods along the *c* direction during film growth. The improved pinning for fields parallel to the *c* direction is reflected by the enhanced barrier heights. For Gd-123 films, the addition of 4 wt% of $BaZrO_3$ leads to a significantly less pronounced dependence of the activation energy on applied field. However, the activation energy is lower than for the undoped compound below a field of 3 T. Up to a field of 8 T, the V_{p0} values found for Gd-123 with 4 wt% BZO are lower than those achieved in Y-123 with the same amount of BZO. The data in Figure 10.10 suggest that for fields above 8 T, the barrier height would be highest in the Gd-123 film with 4 wt% BZO.

Finally, it should be mentioned that measurements of the current–voltage characteristics provide information on the dependence of the activation energy V_p on the current density.

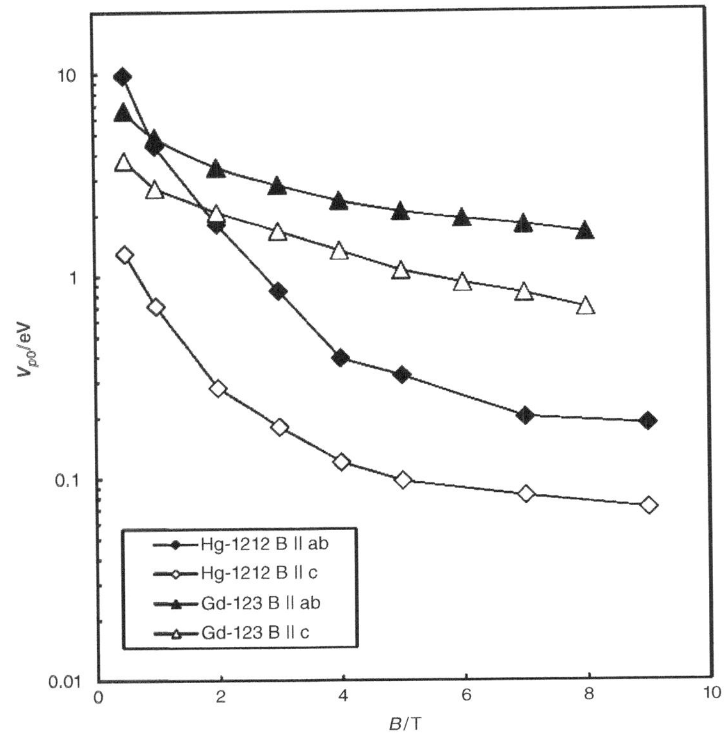

Figure 10.9 *Comparison of the activation energies in Hg-1212 and Gd-123 films in fields parallel and perpendicular to the ab planes (data from [32, 34])*

The electric field generated by thermally activated flux creep can be written as

$$E_e = j\rho_{ff} \exp\left(-\frac{V_p\left(j/j_0, B, T\right)}{k_B T}\right) \tag{10.14}$$

Here ρ_{ff} is the flux flow resistivity for $1/T \to 0$. Experimental studies of $E_e - j$ characteristics of Y-123 films [36] suggest that $V_p(j) \propto (j_0/j)^\mu$, as expected for a vortex glass state (see Section 10.5).

10.5 Irreversibility Lines

In Figure 9.3 of Chapter 9, we presented a simplified magnetic phase diagram of cuprate high-temperature superconductors. An outstanding feature is the existence of an irreversibility line well below the upper critical field. Above the irreversibility line the flux lines are not pinned, and hence the critical current is zero. High-T_c superconductors can be used in superconducting magnets or power applications only well below the irreversibility line. We have already mentioned that one possible explanation for the occurrence of the irreversibility temperature is the melting of the vortex lattice. For the vortex solid below

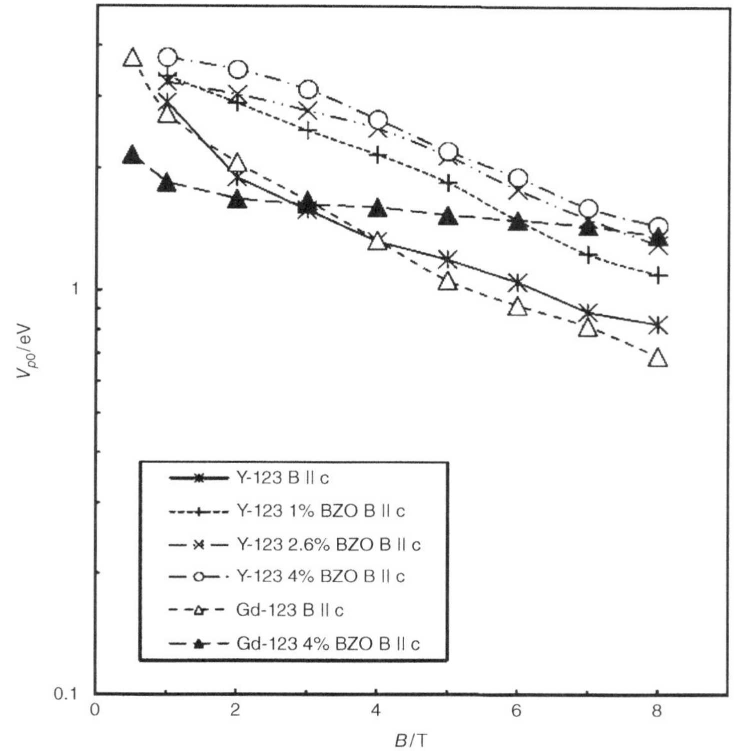

Figure 10.10 *Comparison of the activation energies in Y-123 and Gd-123 films with and without $BaZrO_3$ doping in fields parallel to the c direction (data from [32, 35])*

the irreversibility line, a glass-like behavior was reported soon after the discovery of high-temperature superconductivity. In such a scenario, the irreversibility temperature has to be considered as the glass temperature above which a vortex liquid exists. According to the theory of Fisher [37], the vortex glass state is a real superconductor below the glass temperature. Figure 10.11 shows a sketch of the voltage versus temperature curves for the vortex glass scenario. The dependence of the voltage on current can be described by the power law $V \propto I^n$ at the glass temperature T_g. At temperatures above T_g, the voltage does not vanish even in the limit of $I \to 0$. The lg V versus lg I curves bend upward. On the other hand, the resistance vanishes for $I \to 0$ at temperatures below the glass temperature, i.e., the lg V versus lg I curves bend downward. This means that in the vortex glass picture, there exists a real superconducting state below the glass temperature. In investigations of the current–voltage characteristics of high-T_c superconductors, evidence for the vortex glass scenario was found [38–40].

As an alternative to melting of the flux-line lattice, the vanishing critical current above the irreversibility line may be attributed to thermally activated depinning. Instead of discussing the reasons for the occurrence of the irreversibility line in more detail, we will focus on trends for the position of the irreversibility line in the magnetic phase diagram of various

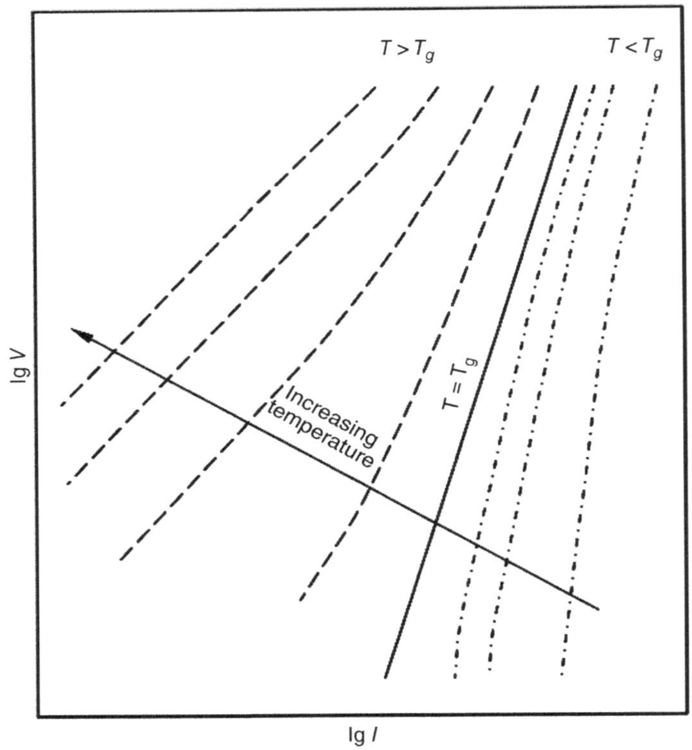

Figure 10.11 *Sketch of the voltage versus current curves expected for a vortex-glass state. At the glass temperature T_g, the voltage can be described by a power law $V \propto I^n$. Below the glass temperature, the V versus I curves have a downward curvature, leading to vanishing voltage, in the limit of small I, while they have an upward curvature above T_g, i.e., V does not vanish in the same limit*

cuprate superconductors. Interesting aspects are the shift of the irreversibility line to higher fields caused by irradiation, or the introduction of artificial pinning centers.

Figure 10.12 shows the irreversibility lines of Bi-based cuprate superconductors. Due to the relatively low value of 20.2 T for the upper critical field of the Bi-2201 single crystal, the irreversibility field could be measured down to low temperatures. For example, the irreversibility field reaches a value of 12.8 T below a temperature of 2 K [41]. Although the shapes of the irreversibility lines of the four Bi-based high-T_c superconductors are similar, their positions are shifted according to the very different values of the critical temperature (Bi-2201: 28.3 K, Bi-2212 single crystal: 77.8 K [42], melt-processed Bi-2212: 92.8 K [42], (Bi,Pb)-2223 tape: 107.7 K [42]).

In Figure 10.13, the irreversibility lines of Tl-based cuprate superconductors are compared [28, 43, 44]. It should be mentioned that the measured irreversibility field depends not only on the sample properties, but also on the measurement technique and the criterion used to define the onset of irreversibility. A straightforward comparison is, therefore, only possible for data published by the same authors. The data taken from Ref. [28] clearly

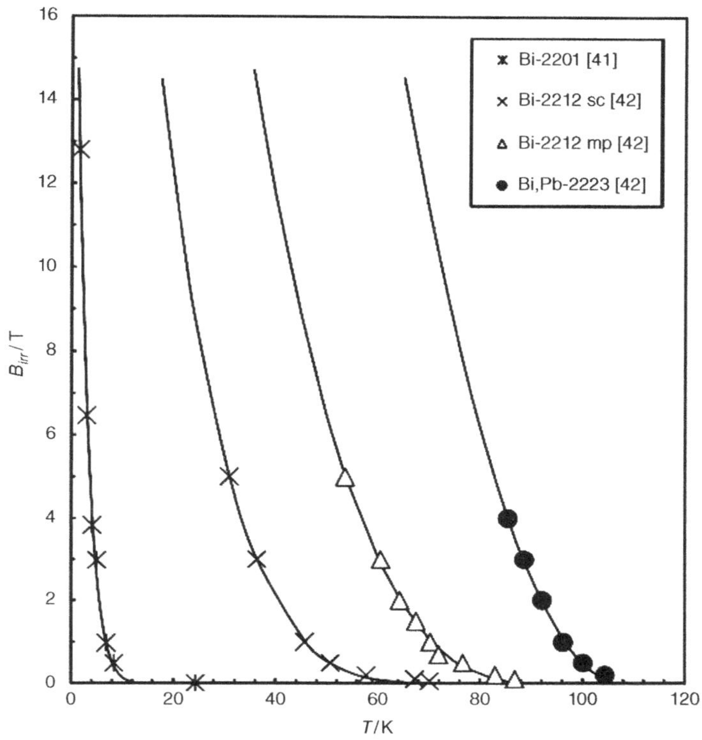

Figure 10.12 *Irreversibility lines of a $Bi_2(Sr,La)_2CuO_y$ single crystal [41], a Bi-2212 single crystal [42], a melt-processed Bi-2212 sample [42], and a Bi,Pb-2223 tape [42]. The solid lines are fits based on the relation $B_{irr} = B_0\,(1-T/T_c)^{\alpha}$ (Equation 9.2, Chapter 9)*

indicate that the irreversibility fields of Tl-1223 are much higher than those of Tl-2223. The two compounds are distinguished mainly by the thicknesses of the charge carrier reservoirs, which consist of BaO-TlO-TlO-BaO layers in Tl-2223, but contain only a single layer of thallium oxide (BaO-TlO-BaO) in Tl-1223. Due to their insulating character, the thicker charge carrier reservoirs in Tl-2223 lead to a less good coupling of adjacent superconducting CuO_2 blocks. In consequence, the anisotropy of the superconducting properties is more pronounced in this material than in Tl-1223. The position of the irreversibility line of Tl-2223 [43] is only slightly higher than that of Tl-2234 [43].

It has already been shown (see Section 9.4, Table 9.3 in Chapter 9) that the upper critical fields B_{c2} of cuprate high-T_c superconductors depend on the direction of the applied magnetic field with respect to the superconducting CuO_2 planes. For fields parallel to the CuO_2 planes (crystallographic *ab* planes), B_{c2} is much higher than for fields perpendicular to the CuO_2 planes (parallel to *c*). In general, the irreversibility fields are much lower than the upper critical field. Consequently, the irreversibility fields for a field perpendicular to the CuO_2 planes are much lower than those for parallel fields. For Y-123 [45] and Hg-1212 [46], this anisotropy of the irreversibility field is illustrated in Figure 10.14. The anisotropy of B_{irr} for Y-123 is much less pronounced than that found for Hg-1212. It will be shown

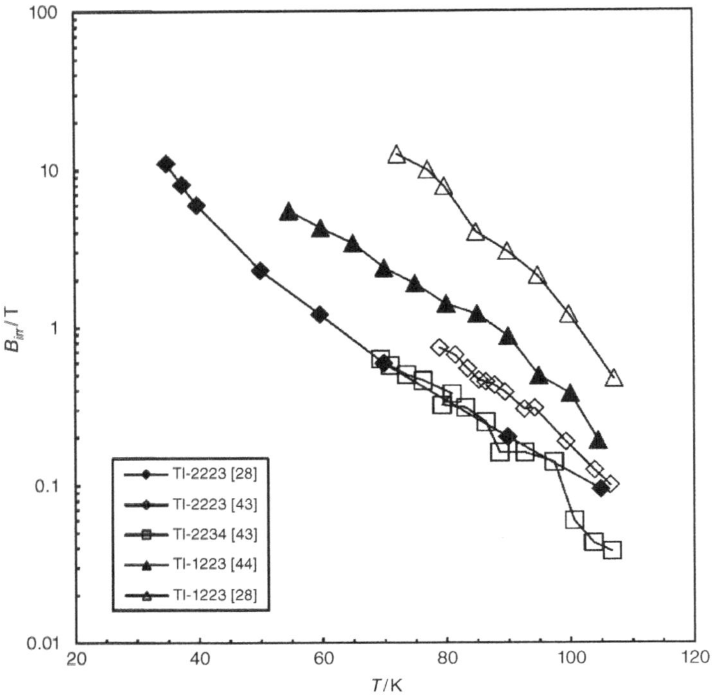

Figure 10.13 *Comparison of the irreversibility lines of Tl-1223 [28, 44], Tl-2223 [28, 43], and Tl-2234 [43]*

later (Figures 10.17 and 10.19) that the slope of the irreversibility line of Y-123 is larger than those of other cuprate superconductors.

Figure 10.15 shows the irreversibility lines of various Hg-12($n-1$)n high-T_c superconductors [47–50]. The data taken from Ref. [47] indicate that the positions of the irreversibility lines of Hg-1201, Hg-1212, and Hg-1223 mainly reflect the different critical temperatures of the mercury-based high-T_c superconductors. In the series of Hg-12($n-1$)n compounds, the behavior of the irreversibility line for $n > 4$ was studied. The irreversibility field of Hg-1234 [49] is significantly lower than that of Hg-1223 [49]. It has already been mentioned that in compounds with more than three CuO_2 planes in the copper oxide blocks, the critical temperature starts to decrease as compared to the maximum value typically reached for $n = 3$. This behavior has been attributed to different carrier concentrations in nonequivalent CuO_2 layers. In compounds with three or more CuO_2 layers in a single copper oxide block, there exist structurally different CuO_2 planes. For example, in Hg-1234 each copper oxide block consists of four CuO_2 layers. The outer two are adjacent to the layers of the charge carrier reservoirs, whereas each of the inner layers is separated from the charge carrier reservoirs by one of the outer CuO_2 layers. A comparison of the irreversibility fields of Hg-1245 [50] and Hg-12910 [50] indicates a further reduction in the irreversibility field with increasing number n of CuO_2 planes per copper oxide block.

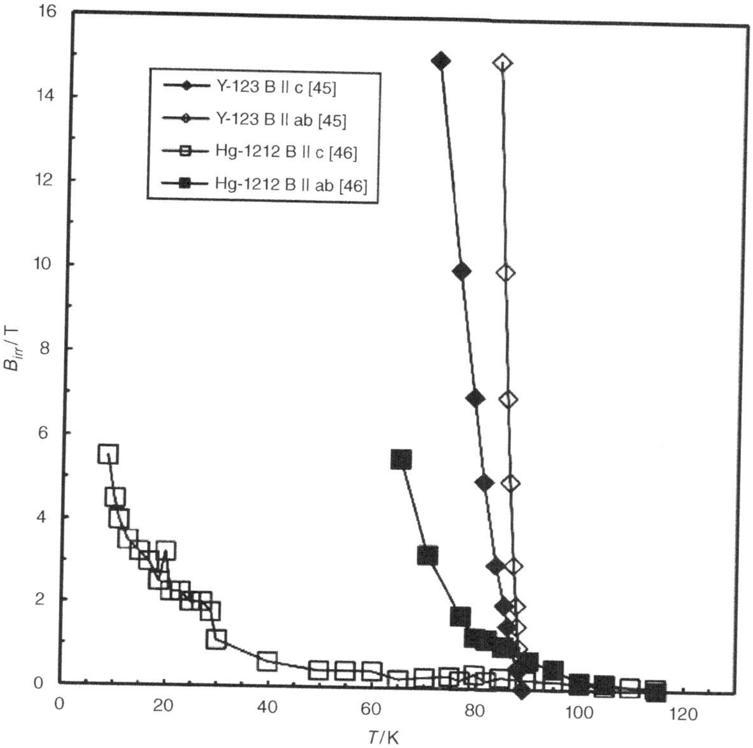

Figure 10.14 *Irreversibility fields of Y-123 [45] and Hg-1212 [46] for fields parallel and perpendicular to the CuO_2 (ab) planes. The anisotropy of the irreversibility field with respect to the direction of the applied magnetic field is much more pronounced for Hg-1212*

In Figure 10.16, the irreversibility lines of several RE-123 (RE: Y or another rare earth element) samples are shown [33, 42, 45, 51]. All considered RE-123 samples show a steep slope of B_{irr} close to T_c. A comparison of the data of Y-123 [45] and Sm-123 [45] indicates that the properties of Sm-123 are superior to those of Y-123. The fact that the irreversibility field of RE-123 superconductors is closer to the upper critical field than in other high-T_c superconductors seems to be related to the unique structural feature that conducting CuO chains are embedded in the layers of the charge carrier reservoirs (BaO-CuO-BaO). A large anisotropy and an accompanied strong reduction in the irreversibility field is a characteristic of Bi-2212 and Tl-2212. The irreversibility lines of Tl-1223 and Hg-1223 are typically higher than those of Bi- or Tl-22($n-1$)n, but well below that of Y-123 or other RE-123, when the data are plotted as a function of the reduced temperature T/T_c (see Figure 10.17, data from [34, 52, 53]).

Figure 10.18 shows the irreversibility lines of Hg-1212 [48] before and after oxygenation. Due to the increase in the critical temperature caused by the oxygenation, the irreversibility line is also shifted to higher temperatures. Next, the effect of irradiation on the position of the irreversibility line is considered. A more detailed discussion on irradiation effects can be found in Ref. [13]. The low values of the irreversibility field

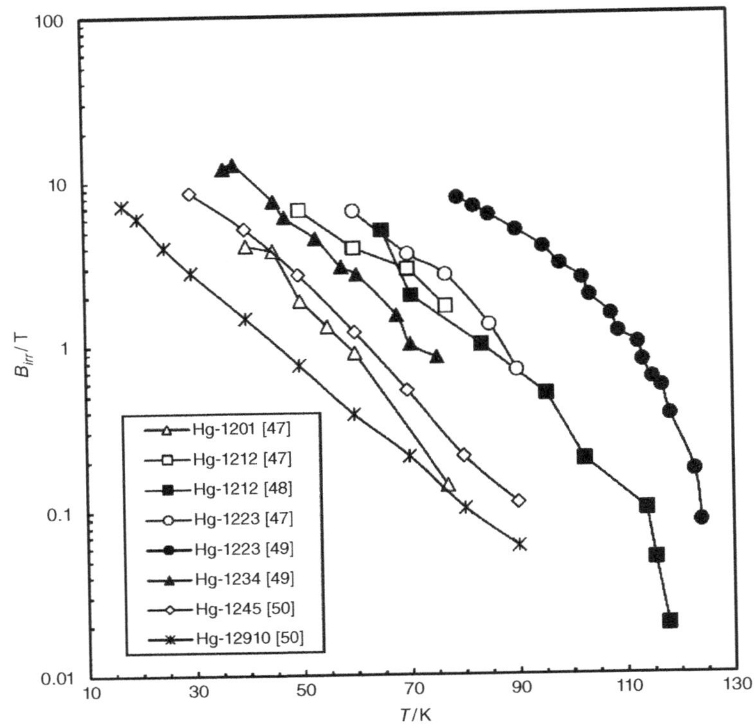

Figure 10.15 *Irreversibility lines of various mercury-based cuprate superconductors (results from [47–50])*

for magnetic fields applied along the *c* direction seem to be closely related to the weak pinning in this case. In principle, the amorphous ion tracks generated by heavy ion irradiation along the *c* direction can provide significantly improved pinning. Figure 10.18 shows the irreversibility lines of (Bi,Pb)-2223 [54] and Bi-2212 tapes [55] before and after irradiation. The (Bi,Pb)-2223 tapes were irradiated by 1.08 GeV Au^{23+} ions up to a fluence of 1.7×10^{11} Au ions/cm^2. This fluence leads to a large number of columnar defects parallel to the *c* direction, which corresponds to a matching field of 3.5 T. The formation of continuous amorphous tracks was visualized by means of transmission electron microscopy (TEM). The Bi-2212 tapes were irradiated by 230 MeV Au^{14+} ions with a fluence of up to 1.6×10^{11} Au ions/cm^2. The data presented in Figure 10.18 clearly indicate that in both high-T_c superconductors, the irreversibility field is significantly enhanced due to the improved pinning provided by the columnar defects parallel to the *c* direction.

In Figure 10.19, the irreversibility lines of Y-123 [30], Tl-2212 [52], and Bi-2212 films [56] and single crystals [52] obtained before and after irradiation are compared. At higher fluences, the critical temperature of the cuprate superconductor is reduced by heavy ion

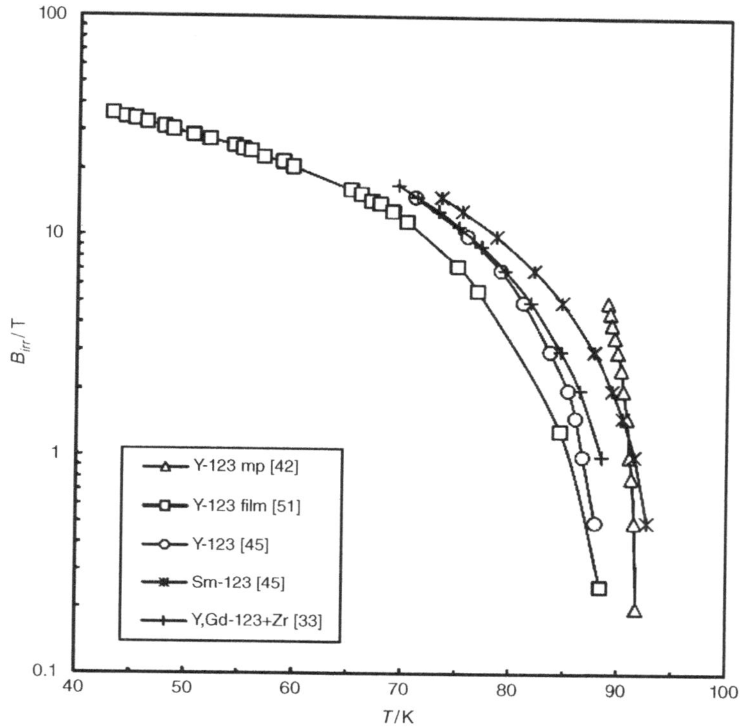

Figure 10.16 *Irreversibility lines of various RE-123 cuprate superconductors (results from [33, 42, 45, 51])*

irradiation. The irreversibility field is, therefore, plotted as a function of the normalized temperature T/T_c. The textured Y-123 thick film was irradiated with 180 MeV Cu^{11+} ions at 100 K along the average c direction up to a fluence of 10^{12} ions/cm^2. After irradiation, the steepness of the irreversibility line was even more pronounced than before. The Bi- and Tl-2212 single crystals [52] were irradiated with 6 GeV Pb ions at 300 K. The ion flux was again parallel to the c direction. For both highly anisotropic compounds, the irreversibility line is shifted after irradiation to considerably higher reduced temperatures. The Bi-2212 films [56] were irradiated by 1.162 GeV Au ions up to a dose of 10^{11} ions/cm^2. Again, the irreversibility line is shifted to higher reduced temperatures.

The results clearly indicate that improved pinning introduced by heavy ion irradiation leads to significantly higher values of the irreversibility fields and temperatures. However, heavy ion irradiation is not a practicable way to improve the properties of high-T_c wires or tapes of kilometer length. To improve the pinning properties of RE-123-coated conductors, artificial pinning centers have been successfully introduced (nanodots, nanorods of secondary phases). The introduction of artificial pinning centers in RE-123-coated conductors will be discussed in Chapter 15.

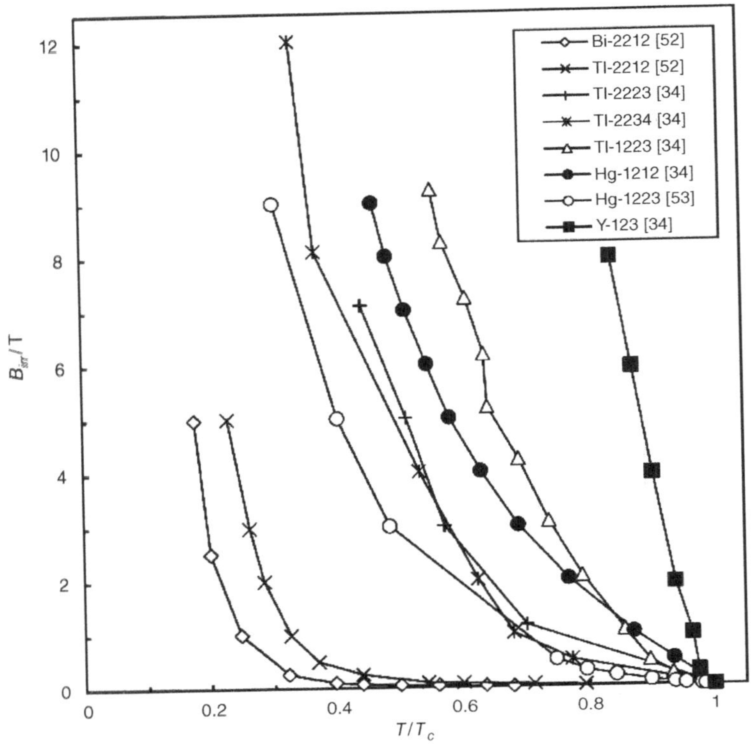

Figure 10.17 *Irreversibility fields of various cuprate high-T_c superconductors as functions of normalized temperature T/T_c (data from [34, 52, 53])*

10.6 Summary

The flux-line lattice in high-T_c superconductors has been made visible by means of magnetic decoration techniques and the use of scanning tunneling microscopy. As in metallic low-T_c superconductors, each vortex contains just one flux quantum.

The anisotropy of the critical current density and irreversibility are closely related to the electronic anisotropy. We may consider the structure of the cuprate superconductors as alternating layers of charge carrier reservoirs and CuO_2 blocks along the crystallographic *c* direction. Adjacent CuO_2 blocks and the charge carrier reservoir separating them may be considered as a built-in Josephson junction. The thicker the layers of the insulating charge carrier reservoirs, the more weakly the adjacent superconducting CuO_2 blocks are coupled. Hence the electronic anisotropy is expected to be more pronounced in cuprates with thicker insulating charge carrier reservoirs. The more weakly the latter are coupled, the more pronounced is the electronic anisotropy. It seems to be reasonable that the anisotropy of Bi- and Tl-22($n-1$)n superconductors with thicker charge carrier reservoirs is more pronounced than that of Tl- and Hg-12($n-1$)n cuprates with thinner charge carrier reservoirs. The RE-123 compounds show much less anisotropy than Bi-, Tl-, and Hg-based cuprate superconductors, especially close to T_c. This behavior of the RE-123 compounds

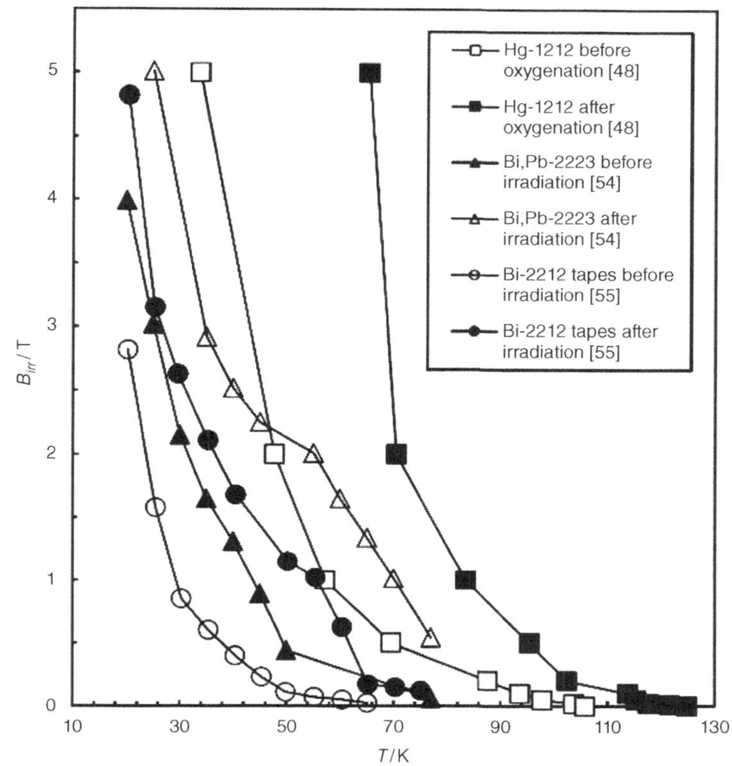

Figure 10.18 *The effects of oxygenation and of heavy ion irradiation on the positions of the irreversibility lines of Hg-1212 [48], (Bi,Pb-2223) [54], and Bi-2212 [55]*

seems to arise because only in these materials conducting CuO chains are embedded in the charge carrier reservoirs. In general, this picture is supported by the observed relaxation of the magnetization, and by the position of the irreversibility line.

An interesting aspect is the fact that the formation of amorphous tracks along the *c* direction, due to heavy ion irradiation, shifts the irreversibility line to higher reduced temperatures T/T_c. In highly anisotropic Bi- and Tl-2212 superconductors, the effect is very pronounced. For bismuth-based high-T_c superconductors, irradiations were performed at angles of 60° ((Bi,Pb)-2223 tapes) and 45° (Bi-2212 single crystal) with respect to the *ab* planes [13, 57, 58]. Measurements of the critical current density of the (Bi,Pb)-2223 tapes at 60 K indicated a peak of j_c for a magnetic field of 1 T parallel to the ion tracks. This means that at 60 K and 1 T, not only pancake vortices but also real flux lines exist, and hence the ion tracks can act as pinning centers. However, at a lower temperature of 40 K this peak vanishes, implying that only two-dimensional pinning is being observed [57].

In the case of the Bi-2212 single crystal, the resistance versus angle curves, measured at 70 K, showed a minimum for an applied field of 0.3 T parallel to the ion tracks. At a higher field of 2 T, the minimum of the resistance parallel to the ion tracks disappeared [58]. In a wider range of field and temperatures [13, 57, 58], there exists in general no significant

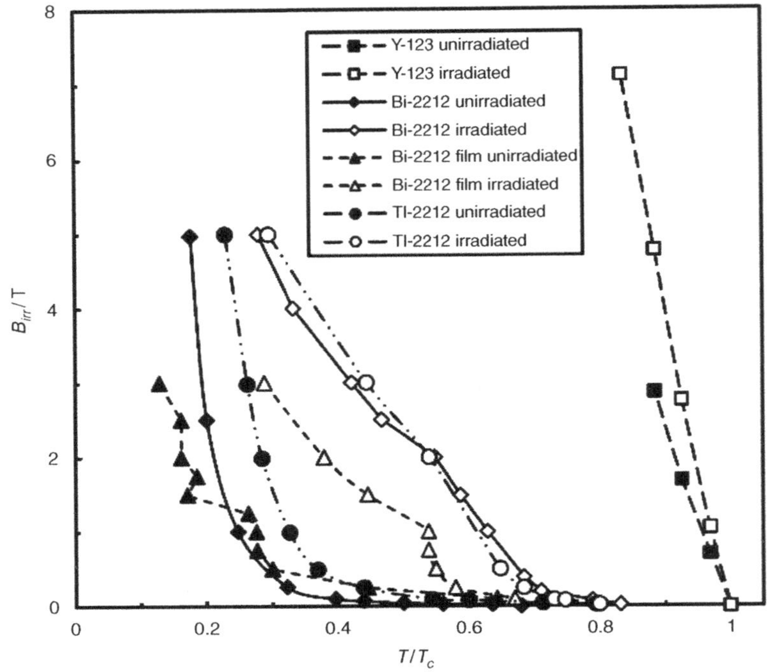

Figure 10.19 *Effect of heavy ion irradiation on the irreversibility lines of Y-123 [30], Bi-2212 and Tl-2212 single crystals [52], and Bi-2212 films [56]. The shift of the irreversibility line to higher normalized temperatures T/T_c is most pronounced for the Bi-2212 and Tl-2212 single crystals. In the case of Y-123, the slope of the irreversibility line is already very steep before irradiation. After radiation, the slope is even steeper. Nevertheless, the shift of B_{irr} is less remarkable than for Bi- and Tl-2212*

enhancement of the critical current density for magnetic fields parallel to ion tracks which are inclined to the *ab* planes. This seems to be the consequence of the decay of the flux lines into pancake vortices. For less anisotropic superconductors (e.g., Y-123), enlarged pinning is not expected to be restricted to columnar defects perpendicular to the *ab* planes.

If we suppose that the occurrence of the irreversibility lines is due to depinning, the observed shifts to higher reduced temperatures simply reflect the better pinning. Assuming the irreversibility line to be caused by the melting of the flux-line lattice, it would necessarily follow that the melting temperature is influenced by the number density and the strength of the present pinning centers, as proposed by Obaidat *et al.* [59].

References

1. W. Buckel, *Supraleitung, Grundlagen und Anwendungen*, Wiley-VCH, Weinheim, Germany, 1994.

2. P.L. Gammel, D.J. Bishop, G.J. Dolan, J.R. Kwo, C.A. Murray, L.F. Schneemeyer, and J.V. Waszczak, Observation of hexagonally correlated flux quanta in $YBa_2Cu_3O_7$, *Phys. Rev. Lett.*, **59**, 2592–2595 (1987).
3. G.J. Dolan, G.V. Chandrashekhar, T.R. Dinger, C. Feild, and F. Holtzberg, Vortex structure in $YBa_2Cu_3O_7$ and evidence for intrinsic pinning, *Phys. Rev. Lett.*, **62**, 827–830 (1989).
4. M. Kraus, M. Leghissa, and G. Saemann-Ischenko, Wechselspiel linienhafter Objekte in der Nanowelt – Flusslinien und Kernspuren in Hochtemperatur-Supraleitern, *Phys. Bl.*, **50**, 333–338 (1994).
5. S. Behler, Flusslinien-Pinning – sichtbar gemacht mit dem Rastertunnelmikroskop, *Phys. Bl.*, **50**, 428 (1994).
6. Ch. Renner, I. Maggio-Aprile, and Ø. Fischer, Rastertunnelspektroskopie auf Hochtemperatursupraleitern – Ortsaufgelöste Tunnelspektroskopie und Abbildung des Flusswirbelgitters, *Phys. Bl.*, **54**, 427–430 (1998).
7. H.F. Hess, R.B. Robinson, R.C. Dynes, J.M. Valles, Jr., and J.V. Wasczcak, Scanning tunneling microscope observation of the Abrikosov flux line lattice and the density of states near and inside a fluxoid, *Phys. Rev. Lett.*, **62**, 214–216 (1989).
8. S. Behler, S.H. Pan, P. Jess, A. Baratoff, H.-J. Güntherodt, F. Lévy, G. Wirth, and J. Wiesner, Vortex pinning in ion-irradiated $NbSe_2$ studied by scanning tunneling microscopy, *Phys. Rev. Lett.*, **72**, 1750–1753 (1994).
9. I. Maggio-Aprile, Ch. Renner, A. Erb, E. Walker, and Ø. Fischer, Direct vortex lattice imaging and tunneling spectroscopy of flux lines on $YBa_2Cu_3O_{7-\delta}$, *Phys. Rev. Lett.*, **75**, 2754–2757 (1995).
10. I. Maggio-Aprile, Ch. Renner, A. Erb, E. Walker, and Ø. Fischer, Critical currents approaching the depairing limit at a twin boundary in $YBa_2Cu_3O_{7-\delta}$, *Nature*, **390**, 487–490 (1997).
11. M. Tachiki and S. Takahashi, Effect of intrinsic pinning on critical current in cuprate superconductors, *Cryogenics*, **32**, 923–929 (1992).
12. Y. Yeshurun and A.P. Malozemoff, Giant flux creep and irreversibility in an Y–Ba–Cu–O crystal: An alternative to the superconducting glass model, *Phys. Rev. Lett.*, **60**, 2202–2205 (1988).
13. R. Wesche, *High-Temperature Superconductors: Materials, Properties and Applications*, Kluwer Academic Publishers, Norwell, MA, 1998.
14. Y. Yeshurun, A.P. Malozemoff, T.K. Worthington, R.M. Yandrofski, L. Krusin-Elbaum, F.H. Holtzberg, T.R. Dinger, and G.V. Chandrashekhar, Magnetic properties of YBaCuO and BiSrCaCuO crystals: A comparative study of flux creep and irreversibility, *Cryogenics*, **29**, 258–262 (1989).
15. L. Civale, A.D. Marwick, M.W. McElfresh, T.K. Worthington, A.P. Malozemoff, F.H. Holtzberg, J.R. Thompson, and M.A. Kirk, Defect independence of the irreversibility line in proton-irradiated Y–Ba–Cu–O crystals, *Phys. Rev. Lett.*, **65**, 1164–1167 (1990).
16. M. Mittag, R. Job, and M. Rosenberg, Magnetic relaxation and flux creep in ceramic (Bi, Pb)-2223 HTSC, *Physica C*, **174**, 101–108 (1991).
17. P. Svedlindh, C. Rossel, K. Niskanen, P. Norling, P. Nordblad, L. Lundgren, and G.V. Chandrashekhar, Non-logarithmic magnetic relaxation in $Bi_{2.2}Sr_{1.7}CaCu_2O_8$ single crystals; evidence for collective flux pinning, *Physica C*, **176**, 336–346 (1991).

18. A. Spirgatis, R. Trox, J. Kötzler, and J. Bock, Non-logarithmic flux creep in Bi-2212, *Cryogenics*, **33**, 138–141 (1993).
19. M. Nideröst, T. Teruzzi, R. Frassanito, A. Amann, P. Visani, and A.C. Mota, Short-time flux dynamics in Bi(2212) single crystal, *Physica C*, **235–240**, 2891–2892 (1994).
20. A. Gupta, E.D. Tuset, M.G. Karkut, and K. Fossheim, Short time magnetic relaxation in $Bi_2Sr_2Ca_2Cu_3O_{10}$ tapes: A study of thermally activated flux motion and vortex glass transition, *Physica C*, **272**, 33–42 (1996).
21. E. Zeldov, N.M. Amer, G. Koren, A. Gupta, M.W. McElfresh, and R.J. Gambino, Flux creep characteristics in high-temperature superconductors, *Appl. Phys. Lett.*, **56**, 680–682 (1990).
22. E. Zeldov, N.M. Amer, G. Koren, and A. Gupta, Flux creep in $Bi_2Sr_2CaCu_2O_8$ epitaxial films, *Appl. Phys. Lett.*, **56**, 1700–1702 (1990).
23. V.M. Vinokur, M.V. Feigel'man, and V.B. Geshkenbein, Exact solution for flux creep with logarithmic $U(j)$ dependence: Self-organized critical state in high-T_c superconductors, *Phys. Rev. Lett.*, **67**, 915–918 (1991).
24. Y. Ren and P.A.J. de Groot, Experimental studies on magnetic relaxations in $YBa_2Cu_3O_{7-\delta}$: Logarithmic, power law or collective pinning?, *Cryogenics*, **33**, 357–361 (1993).
25. M.V. Feigel'man, V.B. Geshkenbein, A.I. Larkin, and V.M. Vinokur, Theory of collective flux creep, *Phys. Rev. Lett.*, **63**, 2303–2306 (1989).
26. M.V. Feigel'man, V.B. Geshkenbein, and V.M. Vinokur, Flux creep and current relaxation in high-T_c superconductors, *Phys. Rev. B*, **43**, 6263–6265 (1991).
27. C.W. Hagen and R. Griessen, Distribution of activation energies for thermally activated flux motion in high-T_c superconductors: An inversion scheme, *Phys. Rev. Lett.*, **62**, 2857–2860 (1989).
28. D.N. Zheng, A.M. Campbell, R.S. Liu, and P.P. Edwards, Critical current, magnetic irreversibility line and relaxation in a single Tl–O layer 1223 superconductor, *Cryogenics*, **33**, 46–49 (1993).
29. S. Chu and M.E. McHenry, Irreversibility lines and pinning force density of aligned $(Bi,Pb)_2Sr_2Ca_2Cu_3O_x$ single crystals, *IEEE Trans. Appl. Supercond.*, **7**, 1150–1153 (1997).
30. H. Kumakura, K. Togano, N. Tomita, E. Yanagisawa, S. Okayasu, and Y. Kazumata, Comparative study on 180 MeV Cu^{11+} irradiation effect on textured $YBa_2Cu_3O_x$ and $Bi_2Sr_2CaCu_2O_y$, *Physica C*, **251**, 231–237 (1995).
31. T.T.M. Palstra, B. Batlogg, L.F. Schneemeyer, and J.V. Waszczak, Thermally activated dissipation in $Bi_{2.2}Sr_2Ca_{0.8}Cu_2O_{8+\delta}$, *Phys. Rev. Lett.*, **61**, 1662–1665 (1988).
32. K. Schlesier, H. Huhtinen, and P. Paturi, Reduced intrinsic and strengthened columnar pinning of undoped and 4 wt% $BaZrO_3$-doped $GdBa_2Cu_3O_{7-\delta}$ thin films: A comparative resistivity study near T_c, *Supercond. Sci. Technol.*, **23**, 055010 (2010).
33. S. Awaji, M. Namba, K. Watanabe, M. Miura, M. Yoshizumi, T. Izumi, and Y. Shiohara, Flux pinning properties of TFA-MOD (Y, Gd)$Ba_2Cu_3O_x$ tapes with $BaZrO_3$ nanoparticles, *Supercond. Sci. Technol.*, **23**, 014006 (2010).
34. M. Rupp, A. Gupta, and C.C. Tsuei, Magnetic field induced broadening of the resistive transition in epitaxial *c*-axis-oriented $HgBa_2CaCu_2O_{6+\delta}$ films, *Appl. Phys. Lett.*, **67**, 291–293 (1995).

35. M. Safonchik, K. Traito, S. Tuominen, P. Paturi, H. Huhtinen, and R. Laiho, Magnetic field dependence of the optimal $BaZrO_3$ concentration in nanostructured $YBa_2Cu_3O_{7-\delta}$ films, *Supercond. Sci. Technol.*, **22**, 065006 (2009).
36. P. Berghuis, R. Herzog, R.E. Somekh, J.E. Evetts, R.A. Doyle, F. Baudenbacher, and A.M. Campbell, Current–voltage measurements as a probe of the activation barriers for flux creep in thin films of $YBa_2Cu_3O_{7-\delta}$, *Physica C*, **256**, 13–32 (1996).
37. M.P.A. Fisher, Vortex-glass superconductivity: A possible new phase in bulk high-T_c oxides, *Phys. Rev. Lett.*, **62**, 1415–1418 (1989).
38. R.H. Koch, V. Foglietti, W.J. Gallagher, G. Koren, A. Gupta, and M.P.A. Fisher, Experimental evidence for vortex-glass superconductivity in Y–Ba–Cu–O, *Phys. Rev. Lett.*, **63**, 1511–1514 (1989).
39. H. Yamasaki, K. Endo, S. Kosaka, M. Umeda, S. Yoshida, and K. Kajimura, Current-voltage characteristics and quasi-two-dimensional vortex glass transition in epitaxial $Bi_2Sr_2Ca_2Cu_3O_x$ films, *Cryogenics*, **35**, 263–269 (1995).
40. Y. Mawatari, H. Yamasaki, S. Kosaka, and M. Umeda, Critical current properties and vortex-glass – liquid transition in Ag-sheathed Bi-2223 tapes, *Cryogenics*, **35**, 161–167 (1995).
41. M. Akamatsu, L.X. Chen, H. Ikeda, and R. Yoshizaki, Magnetic properties of the optimally doped Bi-2201 single crystals, *Physica C*, **235–240**, 1619–1620 (1994).
42. N. Ihara and T. Matsushita, Effect of flux creep on irreversibility lines in superconductors, *Physica C*, **257**, 223–231 (1996).
43. M.R. Presland, J.L. Tallon, P.W. Gilberd, and R.S. Liu, Bulk single-superconducting-phase thallium "2234" superconductor – $Tl_{2-x}Ba_2Ca_{3+x}Cu_4O_{12-\delta}$, *Physica C*, **191**, 307–315 (1992).
44. G. Triscone, A. Junod, and R.E. Gladyshevskii, Magnetic and thermal properties of the 116 K superconductor Tl-1223, *Physica C*, **264**, 233–249 (1996).
45. R. Fuger, M. Eisterer, S.S. Oh, and H.W. Weber, Superior properties of SmBCO coated conductors at high magnetic fields and elevated temperatures, *Physica C*, **470**, 323–325 (2010).
46. L. Krusin-Elbaum, C.C. Tsuei, and A. Gupta, High current densities above 100 K in the high-temperature superconductor $HgBa_2CaCu_2O_{6+\delta}$, *Nature*, **373**, 679–681 (1995).
47. U. Welp, G.W. Crabtree, J.L. Wagner, and D.G. Hinks, Flux pinning and the irreversibility lines in the $HgBa_2CuO_{4+\delta}$, $HgBa_2CaCu_2O_{6+\delta}$ and $HgBa_2Ca_2Cu_3O_{8+\delta}$ compounds, *Physica C*, **218**, 373–378 (1993).
48. Z.J. Huang, Y.Y. Xue, R.L. Meng, and C.W. Chu, Irreversibility line of the $HgBa_2CaCu_2O_{6+\delta}$ high-temperature superconductors, *Phys. Rev. B*, **49**, 4218–4221 (1994).
49. O. Laborde, J.L. Tholence, J. Voiron, M. Saint-Paul, B. Souletie, and J.-J. Capponi, Irreversibility line and critical current of superconducting $HgBa_2Ca_{n-1}Cu_nO_{2n+2+\delta}$ ($n = 3$, 4 and 5), *Physica C*, **235–240**, 2717–2718 (1994).
50. P.M. Shirage, A. Iyo, D.D. Shivagan, A. Crisan, Y. Tanaka, Y. Kodama, and H. Kito, Critical current densities and irreversibility fields of a $HgBa_2Ca_{n-1}Cu_nO_{2n+2+\delta}$ sample containing $n = 6$–15 phases, *Physica C*, **468**, 1287–1290 (2008).
51. J. Hänisch, N. Kozlova, C. Cai, K. Nenkov, G. Fuchs, and B. Holzapfel, Determination of the irreversibility field of YBCO thin films from pulsed high-field measurements, *Supercond. Sci. Technol.*, **20**, 228–231 (2007).

52. V. Hardy, J. Provost, D. Groult, M. Hervieu, B. Raveau, S. Durčok, E. Pollert, J.C. Frison, J.P. Chaminade, and M. Pouchard, Strong shift of the irreversibility line in bismuth and thallium based 2212 single crystals irradiated by 6.0 GeV Pb ions, *Physica C*, **191**, 85–96 (1992).
53. P. Schilbe, I. Didschuns, K. Lüders, G. Fuchs, M. Baenitz, K.A. Lokshin, D.A. Pavlov, E.V. Antipov, and H.R. Khan, Temperature behavior of irreversibility fields of various Hg-based superconducting ceramics, *Physica C*, **372–376**, 1865–1867 (2002).
54. L. Civale, A.D. Marwick, R. Wheeler IV, M.A. Kirk, W.L. Carter, G.N. Riley, Jr., and A.P. Malozemoff, Superconducting current density enhancement by heavy ion irradiation of Bi-2223 silver-clad tapes, *Physica C*, **208**, 137–142 (1993).
55. X. Gao, Y. Kazumata, H. Kumakura, and K. Togano, Effects of 230 MeV Au^{14+} irradiation on $Bi_2Sr_2CaCu_2O_x$, *Physica C*, **250**, 325–330 (1995).
56. H. Frank, J. Lethen, L. Buschmann, B. Decker, J. Wiesener, G. Wirth, P. Wagner, H. Adrian, P. Lemmens, and G. Güntherodt, Dimensional crossover in the pinning of heavy-ion irradiated Bi-2212 films, *Physica C*, **259**, 142–150 (1996).
57. P. Kummeth, C. Struller, H.-W. Neumüller, and G. Saemann-Ischenko, Dimensionality of flux pinning in $(Bi,Pb)_2Sr_2Ca_2Cu_3O_{10+\delta}$ Ag tapes with columnar defects – Crossover from two-dimensional to three-dimensional behavior, *Appl. Phys. Lett.*, **65**, 1302–1304 (1994).
58. F. Warmont, V. Hardy, Ch. Goupil, Ch. Simon, J. Provost, and A. Ruyter, Angle-resolved transport measurements in a Bi-2212 crystal with inclined columnar defects: Study of the directional effects, *Physica C*, **277**, 61–69 (1997).
59. I.M. Obaidat, U. Al Khawaja, and M. Benkraouda, Roles of pinning strength and density in vortex melting, *Supercond. Sci. Technol.*, **21**, 085004 (2008).

11

Transport Properties

11.1 Introduction

In the previous chapters, the superconducting properties of the cuprate high-T_c superconductors have been described. The further aspects to be considered in this chapter are the resistivity in the normal conducting state and the thermal conductivity, both of which are strongly anisotropic because of the layered crystal structures of the cuprate superconductors. The transport properties vary considerably for different types of samples. The degree of disorder in a sintered cuprate superconductor is much more pronounced than in single crystals or textured films. In sintered polycrystalline material, only averaged values of the electrical and thermal conductivity can be measured. On the other hand, the anisotropy of the transport properties can be investigated in textured bulk material, in well-textured films, and in single crystals. Because of the complex chemistry of cuprate high-T_c superconductors, nonsuperconducting minor phases may be present, affecting the electrical and thermal conductivities. In general, the transport properties depend on the exact chemical composition. The resistivity in the normally conducting state depends strongly on the actual number density of the mobile charge carriers, which also determines the critical temperature.

In a comparison with the transport properties of good normal conductors such as copper and aluminum, the considerably different number densities of the mobile charge carriers need to be taken into account. For example, the number density of the conduction electrons in copper and aluminum is 84.5×10^{21} cm^{-3} and 180.6×10^{21} cm^{-3}, respectively [1]. In cuprate superconductors, the typical number density of mobile holes is in the range of 3×10^{21} to 17×10^{21}cm^{-3} [2, 3], roughly an order of magnitude less than that in good normal conductors. A further important difference is the fact that the crystal structures of aluminum and copper are face-centered cubic. As a consequence, their thermal and electrical conductivities are isotropic and both quantities can be described by a single value. In conductors having a crystal structure of lower symmetry, the transport properties depend on the crystallographic direction. Many cuprate high-T_c superconductors crystallize in tetragonal or orthorhombic symmetry.

Knowledge of the electrical resistivity in the normal state and the thermal conductivity is frequently needed for the design of superconducting devices. Moreover, theories of high-temperature superconductivity in cuprates must be able to explain both the superconducting and the normal-state properties.

11.2 Normal-State Resistivity

As a first example, the resistivity of sintered $La_{2-x}Sr_xCuO_4$ will be considered. The resistivity of sintered samples is an average over all crystallographic orientations. Figure 11.1 shows the resistivity of $La_{2-x}Sr_xCuO_4$ with x in the range 0.06–0.25. With increasing strontium content, the formal valence $2+p$ of copper increases (i.e., the number of mobile holes per copper atom increases). As we have already discussed in Chapter 7, the critical temperature of cuprate superconductors shows an inverse parabolic dependence on the number of holes per copper atom. Figure 11.1 indicates that the critical temperature reaches values of ≈36 [4] and ≈38 K [5] for two samples with $x = 0.15$. The critical temperatures for the other x values of 0.06, 0.076, 0.1, 0.2, and 0.25 are ≈11 K, ≈24 K, ≈33 K, ≈32 K, and ≈17 K, respectively [5]. In this range of x, the resistivity decreases with increasing x.

Now increasing strontium content x is accompanied by an increasing number density of holes. The electrical conductivity is given by $\sigma = n_h e^2 \tau / m$, where n_h is the number density of holes, e the electron charge and τ the relaxation time (Equation 2.18 in Chapter 2). Consequently, it seems to be reasonable that the electrical resistivity of sintered $La_{2-x}Sr_xCuO_4$ samples decreases with increasing x. The fact that the two samples with $x = 0.15$ show

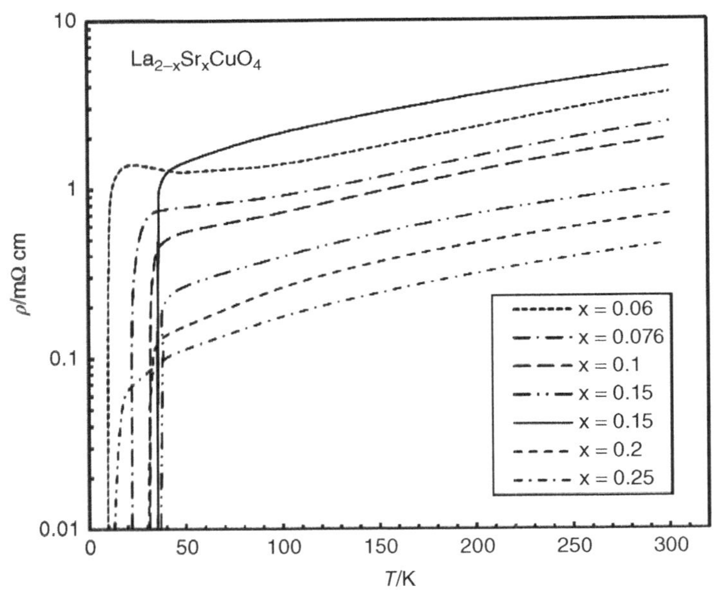

Figure 11.1 Resistivity versus temperature of sintered $La_{2-x}Sr_xCuO_4$ samples. The values of x range from 0.06 to 0.25 (data from [4, 5])

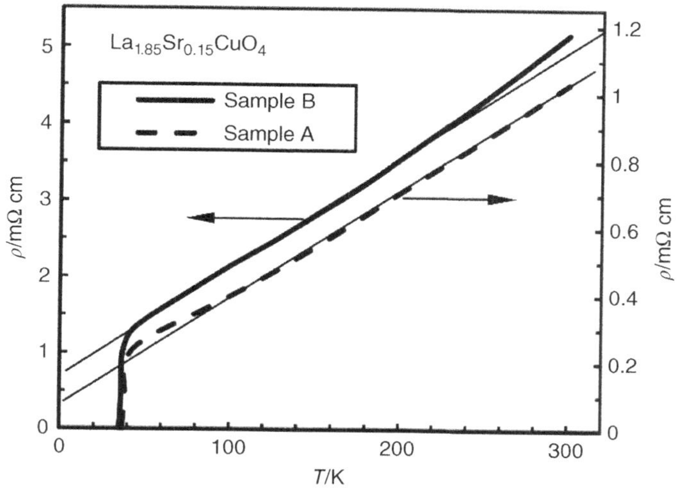

Figure 11.2 *Resistivity versus temperature curves of two nearly optimally doped* $La_{1.85}Sr_{0.15}CuO_4$ *samples. Over a wide range of temperatures, the resistivity depends linearly on temperature (data from [4, 5])*

considerably different normal-state resistivities of around 5 [4] and 1 mΩcm [5] clearly indicates that the electrical conductivity is also strongly dependent on the sample quality.

Figure 11.2 shows that the normal-state electrical resistivity of optimally doped $La_{1.85}Sr_{0.15}CuO_4$ between 50 and 300 K depends nearly linearly on temperature. This linear dependence of the resistivity on temperature is a characteristic feature of the resistivity of optimally doped high-T_c superconductors for current flow along the CuO_2 planes (see also single crystal data below) [6]. This behavior needs to be explained by a theory of high-T_c superconductivity.

Figure 11.3 shows the temperature dependence of the resistivity of various bismuth-based cuprate superconductors (data from [7–14]). The data also indicate that the resistivity varies strongly for different samples. The resistivity along the CuO_2 planes shows metallic behavior, i.e., the resistivity increases with increasing temperature. This behavior is in fact observed for the in-plane resistivity ρ_{ab} of the $Bi_2Sr_2CuO_6$ single crystal [7]. The value of ρ_{ab} is around 0.28 mΩcm at 100 K. In contrast, the resistivity of the $Bi_{2.15}Sr_{1.85}CuO_{6+x}$ single crystal [8] increases with decreasing temperatures at the lowest temperatures, as in semiconductors. Furthermore, the resistivity of ≈1.2 mΩcm at 100 K is significantly larger than that of the $Bi_2Sr_2CuO_6$ single crystal. For the Bi-2212 single crystals A [10] and B [11], the temperature dependence of the resistivity shows metallic behavior. At a temperature of 200 K, the in-plane resistivity of Bi-2212 A is more than a factor of 20 larger than that of Bi-2212 B. The lowest resistivity of all considered bismuth-based high-T_c superconductors shows the single crystal Bi-2212 B with a value as low as ≈0.11 mΩcm at 300 K. As we will see later, the anisotropy of the resistivity in the CuO_2 planes (ρ_a/ρ_b) is of the order of 2. The resistivities of the two (Bi,Pb)-2223 polycrystalline samples [12, 14] and the (Bi,Pb)-2223 fiber [13] are all around 1 mΩcm at a temperature of 290 K. The highest resistivity of 4.2 mΩcm was found for the Bi-2212 film [9].

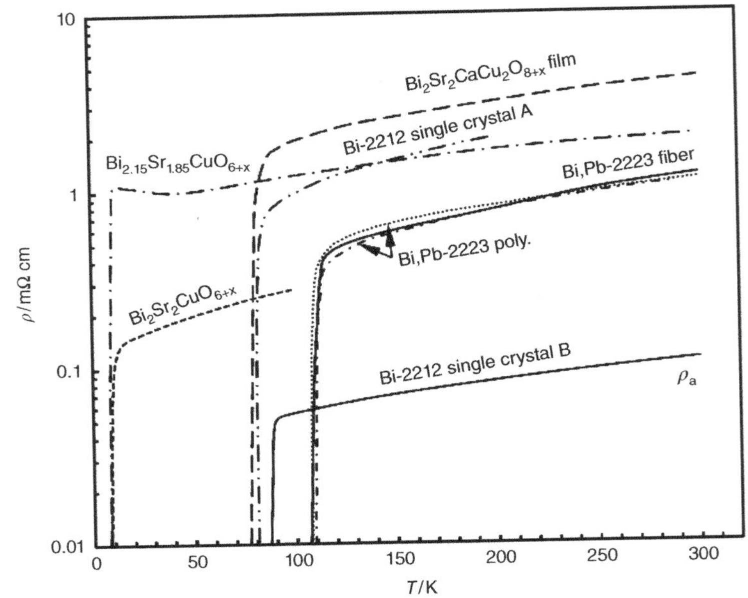

Figure 11.3 *Resistivity versus temperature of various bismuth-based cuprate high-T_c superconductors (data from [7–14])*

Figure 11.4 shows examples of resistivity versus temperature curves of various thallium-based cuprate superconductors [15–19]. All considered samples show a metallic behavior, i.e., increasing resistance with increasing temperature. The values of the resistivity at room temperature are in the range of 1 to 40 mΩcm. Resistivity data of mercury-based high-temperature superconductors [20–24] are presented in Figure 11.5. As in the case of the thallium-based cuprate superconductors, all the samples show metallic behavior. In the sintered samples, the resistivity at room temperature is in the range of 8 to 47 mΩcm. Only the Hg-1223 film shows a significantly lower resistivity of ≈0.09 mΩ cm at 200 K.

Figure 11.6 shows the resistivity of different types of RE-123 (RE = Y or another rare earth element) samples [25–28]. The Y-123 film exhibits the lowest resistivity, while the highest resistivity was found in the sintered Y-123 and La-123 samples. In the textured Er-123 bulk material, an intermediate value of the resistivity was measured. All considered samples show metallic behavior.

In Figure 11.7, the effect of different heat treatments on the critical temperature and the normal-state resistivity is illustrated for various cuprate superconductors. The $Bi_{2.1}Sr_2Ca_{0.95}Y_{0.05}Cu_2O_{8+\delta}$ samples [29] were annealed at temperatures of 500°C, 700°C, 820°C, and 840°C, followed by rapid cool-down. The heat treatments were performed in air. The results indicate that the critical temperature increases with increasing annealing temperature and reaches a maximum for an annealing temperature of 820°C. Further enhanced annealing temperatures lead to a reduction of T_c. This behavior suggests that the as-prepared Bi-2212 is overdoped. Annealing at higher temperatures leads to a loss

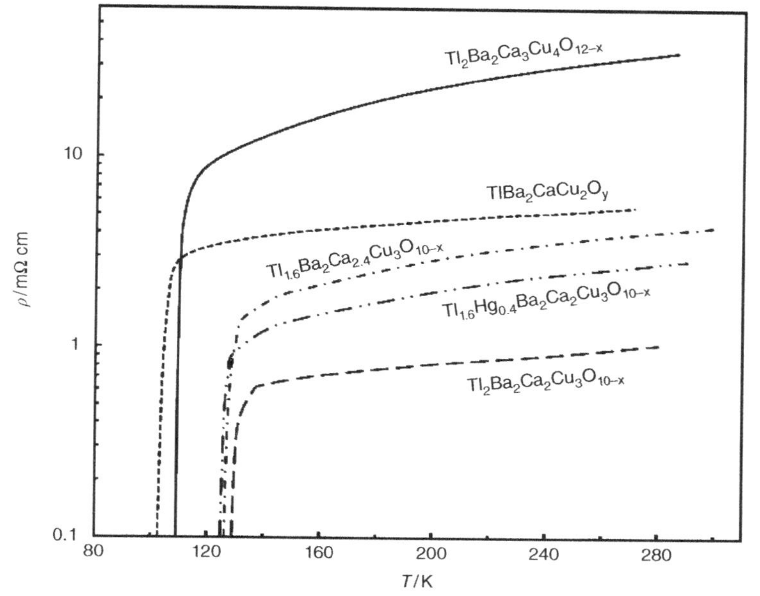

Figure 11.4 *Resistivity versus temperature of various thallium-based cuprate high-T_c superconductors (data from [15–19])*

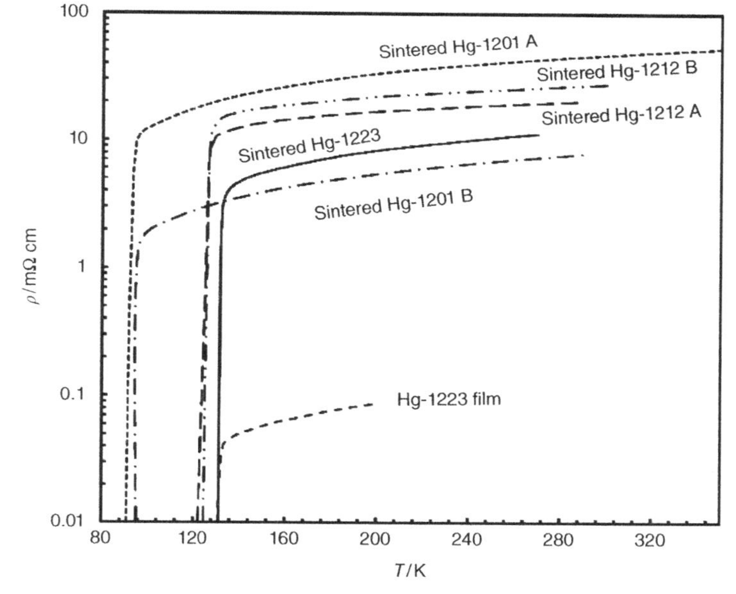

Figure 11.5 *Resistivity versus temperature of various mercury-based cuprate high-T_c superconductors (data from [20–24])*

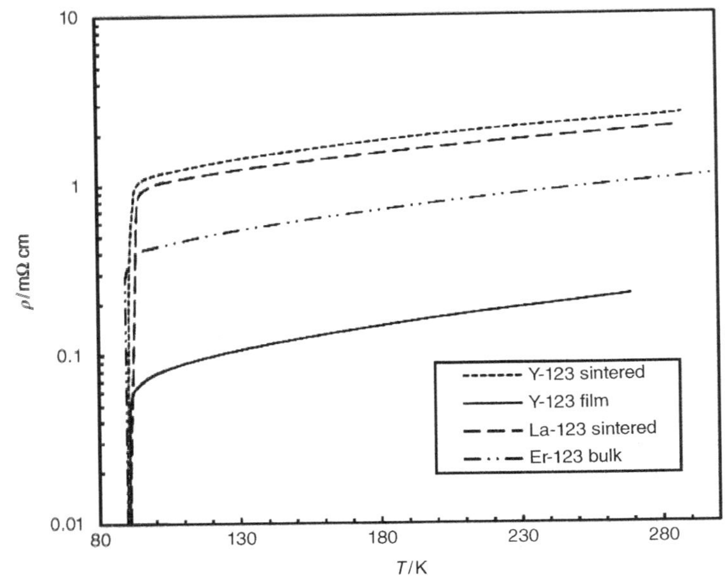

Figure 11.6 *Resistivity versus temperature of RE-123 high-T_c superconductors (data from [25–28])*

of oxygen, which leads to a reduction of the number of mobile holes per Cu atom in the CuO_2 planes, and to a more resistive normal state.

In the case of $Tl_{1.6}Hg_{0.4}Ba_2Ca_2Cu_3O_{10-x}$ [18], annealing in oxygen does not strongly affect the critical temperature. Nevertheless, the normal-state resistance is significantly reduced, perhaps because of the ordering effects of the oxygen.

For the textured Er-123 bulk material [28], annealing at 500°C in 1 bar flowing oxygen leads to a higher normal-state resistivity than heat treatment at only 400°C. However, the critical temperature is not significantly affected.

The critical temperatures of the two as-prepared Hg-1223 samples [23, 24] are well below 130 K. Heat treatment in an oxygen atmosphere increases the critical temperature in both samples to above 130 K. The increase of T_c is accompanied by a significantly reduced normal-state resistivity. This behavior suggests that the as-prepared Hg-1223 samples are underdoped. In spite of similar critical temperatures, which suggest that the number densities of mobile holes in both samples are comparable, their normal-state resistivities after annealing in oxygen differ by a factor of ≈100. The in-plane resistivity of the epitaxial film is as low as 0.087 mΩ cm at 200 K [24]. For the Hg-1212 sample, oxygen annealing again leads to an increased critical temperature and a reduced normal-state resistivity. The room temperature resistivity of the as-prepared Hg-1212 sample is as high as 130 mΩ cm [22]. Even after two heat treatments in an oxygen atmosphere, the resistivity at 300 K exceeds 27 mΩ cm.

As a further example, we consider the effect of heat treatments under different oxygen partial pressures on the T_c and the resistivity of thin Bi-2212 single crystals [30]. The single crystals were annealed for 1 h at 745°C in atmospheres with oxygen partial pressures of

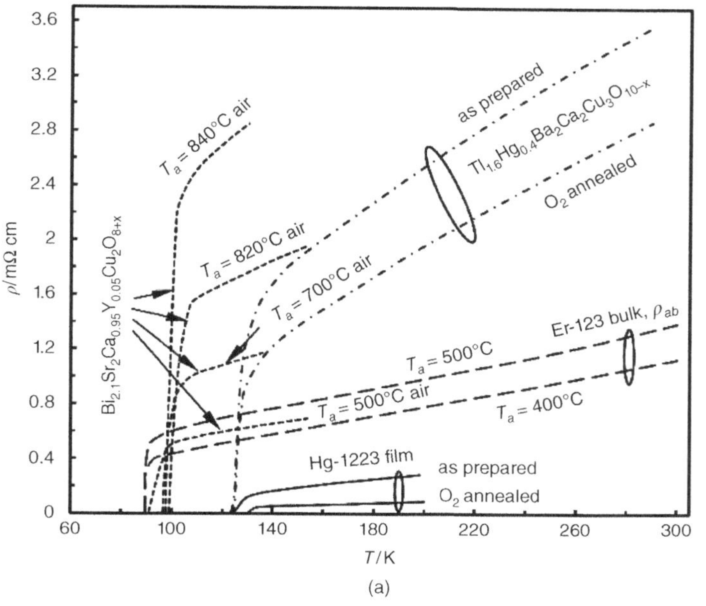

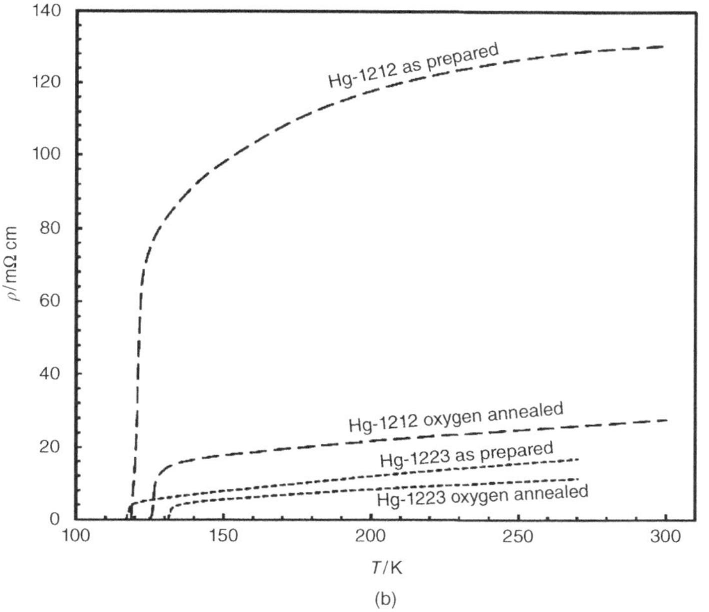

Figure 11.7 *Effect of different heat treatments on the T_c and the normal-state resistivity of various cuprate superconductors (data from [18, 22–24, 28, 29])*

0.04, 0.3, 10 and 1000 mbar, followed by rapid cooling to room temperature. The heat treatments led to a reversible change in the critical temperature [30]. The reduction of T_c with decreasing oxygen partial pressure is accompanied by an increased normal-state resistivity. The values of the critical temperature based on the midpoint of the transition, and the corresponding normal-state resistivities, are listed in Table 11.1.

The results in Table 11.1 suggest that the introduction of oxygen by heat treatment at high oxygen partial pressure increases the number of mobile holes, which is reflected in the increasing T_c. As a consequence, the normal-state resistivity decreases. Annealing at 1 bar oxygen pressure re-establishes the original T_c value [30].

In Figure 11.8, the effects of irradiation on the normal-state resistivity of Bi-2201 and (Bi,Pb)-2223 are presented. The Bi-2201 sample [7] was irradiated by electrons. The

Table 11.1 *Variation of the critical temperature and normal-state resistivity of $Bi_2Sr_2CaCu_2O_8$ single crystals with heat treatment at different oxygen pressures [30]*

pO_2 (mbar)	T_c(K)	ρ(300 K) ($\mu\Omega$cm)	ρ(100 K) ($\mu\Omega$cm)
0.04	≈47	518	454[a]
0.3	≈59	394	189
10	≈75	310	106
1000	≈87	178	54

[a] $\rho_{min} \cong 415\ \mu\Omega$cm at 150 K, $d\rho/dT < 0$ for $T < 150$ K.

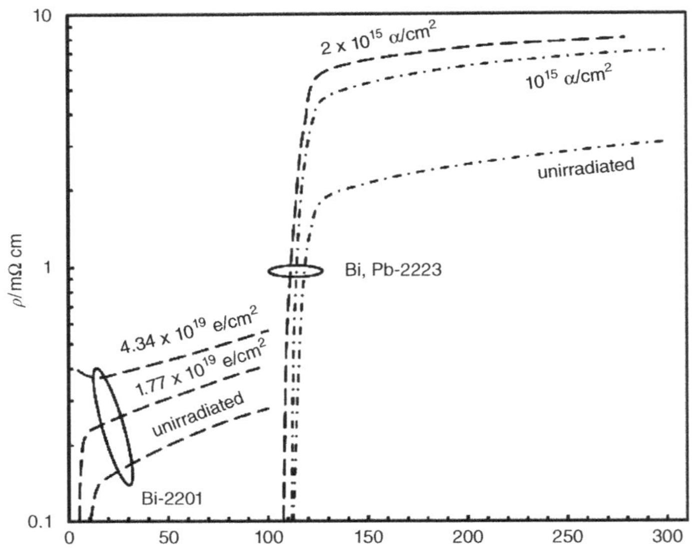

Figure 11.8 *Effect of irradiation with electrons and α particles on the T_c and the resistivity of Bi-2201 single crystals and of polycrystalline (Bi,Pb)-2223, respectively (data from [7, 31])*

fluence of $1.77 \times 10^{19}\,\mathrm{e/cm^2}$ led to a significant reduction of the critical temperature. Moreover, the normal-state resistivity is significantly enhanced. A higher fluence of $4.34 \times 10^{19}\,\mathrm{e/cm^2}$ destroyed superconductivity and led to a further increase of the resistivity. As a second example, the resistivity versus temperature data of a (Bi,Pb)-2223 sample before and after irradiation with α particles are presented in Figure 11.8 [31]. Again the irradiation led to a reduced T_c and an enhanced resistivity. As in metals like copper, the irradiation damage increases the normal-state resistivity. The decreasing critical temperature suggests that increasing disorder, caused by irradiation, weakens superconductivity.

Last but not least, the anisotropy of the resistivity with respect to the different crystallographic directions will be considered. For the investigation of this anisotropy, high-quality single crystals or well-textured films are required. In general, the resistivity along the crystallographic c-axis is much larger than the resistivity for current flow along the CuO_2 planes. The in-plane resistivity of cuprates crystalizing in tetragonal symmetry can be described by a single value because $\rho_a = \rho_b = \rho_{ab}$. On the other hand, the values of ρ_a and ρ_b are different in Bi-2212 or Y-123 having orthorhombic crystal structures.

In Figure 11.9 (top), the in-plane resistivities of various cuprate high-T_c superconductors are presented [32–38]. For the Y-123 single crystal, only the in-plane resistivity ρ_{ab} averaged over the a- and b-directions is available. The separate measurement of ρ_a and ρ_b would require the use of single crystals free of twinning. The in-plane resistivity of most of the considered single crystals shows metallic behavior (i.e., the resistivity increases with increasing temperature). The resistivity of the different samples at room temperature varies between $26\,\mu\Omega\mathrm{cm}$ and $6\,\mathrm{m}\Omega\mathrm{cm}$.

For the two Bi-2212 single crystals A and B, the values of ρ_a and ρ_b were measured. In both single crystals, the resistivity along the a-axis (ρ_a) was found to be larger than that along the b-axis (ρ_b). The ratio of ρ_a to ρ_b in both cases is above 2.

In Figure 11.9 (bottom), the values of the resistivity ρ_c for currents perpendicular to the CuO_2 planes are presented [32–37, 39]. In general, the resistivity ρ_c for currents along the c direction is much larger than the in-plane resistivity ρ_{ab}. A further important aspect is that, except for Bi-2201 and (Bi,Pb)-2223, the resistivity perpendicular to the CuO_2 planes increases with decreasing temperature. This behavior is similar to that of semiconductors. The values of the resistivity at room temperature vary between $9\,\mathrm{m}\Omega\mathrm{cm}$ and $3.6\,\Omega\mathrm{cm}$.

Using the data shown in Figure 11.9, the anisotropy ratios ρ_c to ρ_{ab} or ρ_c to ρ_b have been calculated and are presented in Figure 11.10. They vary from 30 (Y-123) to more than 100 000 (Bi-2212 B) at room temperature. Typically the ratios increase with decreasing temperature.

Resistivity data of various cuprate superconductors are listed in Table 11.2 [4, 5, 7–10, 13–28, 30–32, 35–38, 40–80]. The quoted values of the critical temperature have been taken from the cited articles. In many cases they have been estimated from the resistivity-versus-temperature curves. In polycrystalline samples, the current percolates through the sample preferring paths of lower resistivity. The resistivity of sintered cuprate superconductors typically shows a metallic behavior, like the in-plane resistivity.

The resistivity data listed in Table 11.2 illustrate the effects of different sample types, of changes in composition, and of different heat treatments. For some of the single crystals, the values of the out-of-plane resistivity ρ_c are also quoted.

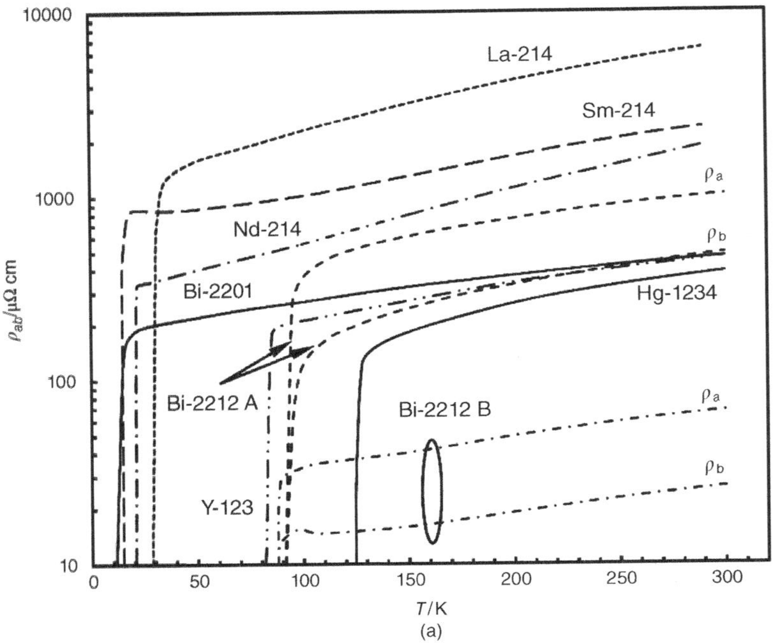

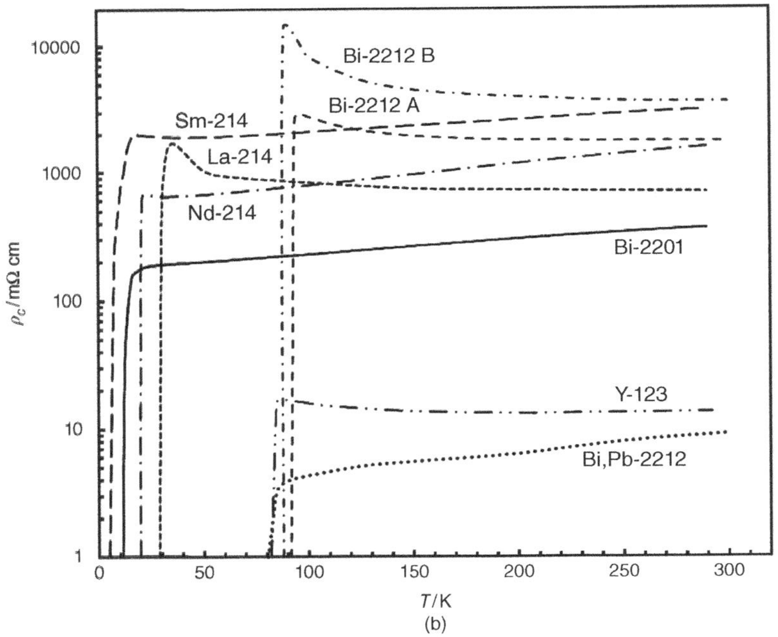

Figure 11.9 *Top panel: in-plane resistivity of various high-T_c superconductors. For the Bi-2212 single crystals A and B, resistivity data are presented for current flow along the crystallographic a- and b-axes. The in-plane resistivity shows metallic behavior (data from [32–38]). Bottom panel: out-of-plane resistivity of cuprate superconductors, frequently showing semiconducting behavior (data from [32–37, 39])*

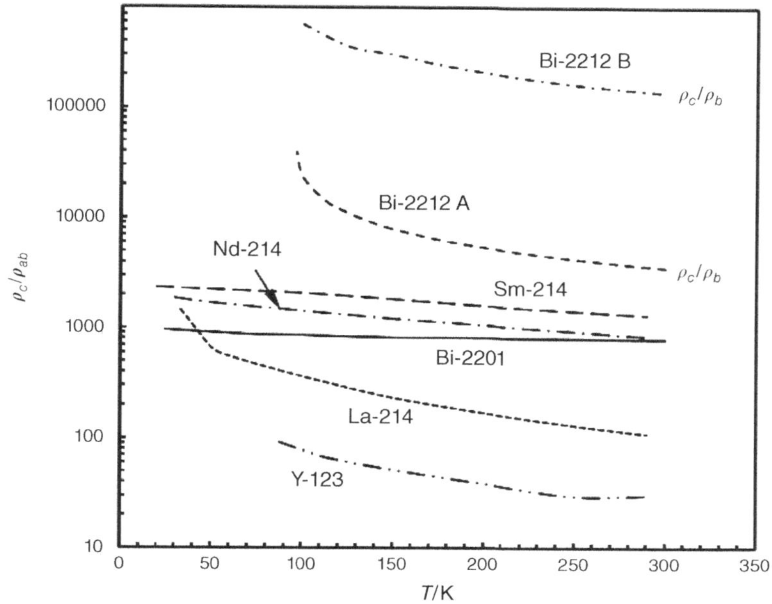

Figure 11.10 *Anisotropy of the normal-state resistivity (ρ_c/ρ_{ab} or ρ_c/ρ_b) versus temperature (data from Figure 11.9)*

An interesting aspect is the dependence of T_c and resistivity on the film thickness. The in-plane resistivity data of ultrathin $La_{1.85}Sr_{0.15}CuO_4$ (0 0 1) films, presented in Table 11.2, indicate that the critical temperature of a film of only 3 nm thickness reaches a value of 32 K, while the resistivity is 1.5 mΩcm at 250 K. The critical temperature increases with increasing film thickness and a T_c value of 44 K was found at a film thickness of 90 nm. On the other hand, the resistivity decreases with increasing film thickness and reaches a value of 0.3 mΩcm at 250 K for the 90 nm thick film. The preparation and the properties of superconducting films will be discussed in more detail in Chapter 16.

Very recently, it has been pointed out that some features in the dependence of normal-state resistivity on temperature seem to be closely related to the occurrence of the pseudogap at temperatures well above T_c (see Chapter 8). A characteristic feature of optimally doped cuprate superconductors is the fact that the resistivity depends linearly on temperature over a broad range of temperature from slightly above T_c up to several hundred kelvin. Ito *et al.* investigated the in-plane resistivity of $YBa_2Cu_3O_{7-y}$ single crystals [81]. They established a correlation between the deviation from a linear temperature dependence of the resistivity and the change in the spectrum of the spin fluctuations. The temperature at which the resistivity starts to deviate from a linear dependence can therefore be used to estimate the temperature T^* below which the pseudogap opens (see the sketch in Figure 11.11) [60, 62]. Raffy *et al.* showed that the out-of-plane resistivity versus temperature data of $Bi_2Sr_2CaCu_2O_y$ thin films with various oxygen doping collapse onto a single curve down to the onset of superconducting fluctuations when the temperature is normalized to T_{min}, where ρ_c reaches a minimum (see the sketch in Figure 11.11) [62].

Table 11.2 Electrical resistivity of various cuprate superconductors

Compound	T_c(K)	T(K)	ρ(mΩcm)	Comments	Reference
$La_{1.85}Sr_{0.15}CuO_4$				Ultrathin (0 0 1) films	[40]
$d = 3$ nm	32	50	0.69[a]		
		250	1.5[a]		
$d = 6$ nm	40	50	0.16[a]		
		250	0.55[a]		
$d = 10$ nm	40	50	0.099[a]		
		250	0.37[a]		
$d = 23$ nm	44	50	0.067[a]		
		250	0.32[a]		
$d = 45$ nm	44	50	0.065[a]		
		250	0.32[a]		
$d = 90$ nm	44	50	0.057[a]		
		250	0.3[a]		
$La_{1.94}Sr_{0.06}CuO_4$	11	50	1.3	Polycrystalline	[5]
		250	2.9		
$La_{1.924}Sr_{0.076}CuO_4$	25	50	0.78	Polycrystalline	[5]
		250	2.0		
$La_{1.9}Sr_{0.1}CuO_4$	33	50	0.56	Polycrystalline	[5]
		250	1.6		
$La_{1.85}Sr_{0.15}CuO_4$	38	50	0.27	Polycrystalline	[5]
		250	0.87		
$La_{1.85}Sr_{0.15}CuO_4$	36	50	1.4	Polycrystalline	[4]
		250	4.3		
$La_{1.8}Sr_{0.2}CuO_4$	≈32	50	0.16	Polycrystalline	[5]
		250	0.59		
$La_{1.8}Sr_{0.2}CuO_4$	≈36	50	2.5	Polycrystalline	[41]
		250	4.9		
$La_{1.75}Sr_{0.25}CuO_4$	≈16	50	0.12	Polycrystalline	[5]
		250	0.39		
$La_{1.94}Ba_{0.06}CuO_{4-\delta}$	≈15	50	2.8[a]	Single crystals	[32]
		50	2300[b]		
		250	7.9[a]		
		250	1600[b]		
$La_{1.92}Ba_{0.08}CuO_{4-\delta}$	≈28	50	2.4[a]		
		50	1700[b]		
		250	7.0[a]		
		250	900[b]		
$La_{1.91}Ba_{0.09}CuO_{4-\delta}$	≈32	50	1.6[a]		
		50	1050[b]		
		250	5.4[a]		
		250	700[b]		
$YBa_2Cu_3O_7$	90.7	100	0.078[a]	Epitaxial film	[26]
		270	0.224[a]		
$YBa_2Cu_3O_{7-\delta}$	≈93	100	0.057	Thin film	[42]
		250	0.133		

(continued overleaf)

Table 11.2 *(continued)*

Compound	T_c(K)	T(K)	ρ(mΩcm)	Comments	Reference
$YBa_2Cu_3O_{7-\delta}$	≈90	100	0.53[a]	Melt textured	[43]
		250	1.38[a]		
		100	4.7[b]		
		250	9.2[b]		
$YBa_2Cu_3O_{7-\delta}$	≈92	100	2.6[a]	Melt textured	[44]
		250	5.8[a]		
		100	28.5[b]		
		250	34[b]		
$YBa_2Cu_3O_{7-\delta}$	≈91	100	0.15[a]	Melt textured	[44]
		250	0.47[a]		
		100	13[b]		
		250	28[b]		
$YBa_2Cu_3O_{7-\delta}$	≈88	100	0.29	Polycrystalline	[45]
		250	0.56		
$YBa_2Cu_3O_{7-\delta}$	≈92	100	1.15	Polycrystalline	[25]
		250	2.3		
$YBa_2Cu_3O_{7-\delta}$	≈92	100	4.2	Polycrystalline	[46]
		250	6.9		
$YBa_2Cu_3O_{7-\delta}$	91	100	0.3	Polycrystalline	[47]
		250	0.7		
$YBa_2Cu_3O_{7-\delta}$	≈91	100	0.32	(0 0 1) film	[48]
		250	0.74		
$YBa_2Cu_3O_{7-\delta}$	≈92	100	1.0[c]	(1 0 0) film	[49]
		100	16.5[b]		
		250	1.4[c]		
		250	22.2[b]		
$YBa_2Cu_3O_{7-\delta}$	90.5	100	1.8[c]	(1 0 0) film	[50]
		100	32[b]		
		250	2.15[c]		
		250	27[b]		
$YBa_2Cu_3O_{7-\delta} + BaZrO_3$				Polycrystalline	[51]
0 wt%	92.52	100	0.68		
		300	1.23		
1.0 wt%	92.66	100	1.01		
		300	1.97		
2.5 wt%	92.15	100	0.93		
		300	1.46		
5.0 wt%	92.32	100	1.21		
		300	1.69		
10.0 wt%	92.02	100	1.97		
		300	2.18		
$YBa_2Cu_{2.97}Ni_{0.03}O_{7-\delta}$	85.7	100	1.77	Polycrystalline	[52]
		250	2.53		
$YBa_2Cu_{2.94}Ni_{0.06}O_{7-\delta}$	81.4	100	0.69	Polycrystalline	[52]
		250	1.0		
$YBa_2Cu_{2.91}Ni_{0.09}O_{7-\delta}$	76.5	100	3.61	Polycrystalline	[52]
		250	4.6		

(continued overleaf)

Table 11.2 *(continued)*

Compound	T_c(K)	T(K)	ρ(mΩ cm)	Comments	Reference
$Y_{0.88}Ca_{0.12}Ba_2Cu_3O_{6.785}$	83	100	0.083[d]	Single crystals	[53]
		100	3.74[b]		
		250	0.22[d]		
		250	6.34[b]		
$Y_{0.88}Ca_{0.12}Ba_2Cu_3O_{6.793}$	82	100	0.068[d]	Single crystals	[53]
		100	2.34[b]		
		250	0.19[d]		
		250	4.63[b]		
$Y_{0.86}Ca_{0.14}Ba_2Cu_3O_{6.775}$	80	100	0.054[d]	Single crystals	[53]
		250	0.15[d]		
$Y_{0.86}Ca_{0.14}Ba_2Cu_3O_{6.788}$	78	100	0.045[d]	Single crystals	[53]
		250	0.13[d]		
$ErBa_2Cu_3O_{7-\delta}$	≈90	100	0.43[a]	Bulk material	[28]
		250	0.95[a]		
$NdBa_2Cu_3O_{7-\delta}$	≈92	290	0.84	Polycrystalline	[54]
$NdBa_{1.9}La_{0.1}Cu_3O_{7-\delta}$	≈83	290	1.3		
$NdBa_{1.8}La_{0.2}Cu_3O_{7-\delta}$	≈58	290	2.1		
$NdBa_{1.7}La_{0.3}Cu_3O_{7-\delta}$	≈40	290	3.8		
$LaBa_2Cu_3O_{7-\delta}$	≈93.5	100	1.0	Polycrystalline	[27]
		250	1.95		
$YBa_2Cu_4O_8$	≈78	100	6.2	Polycrystalline	[55]
		250	10		
$YBa_2Cu_4O_8$	74	100	0.31	Polycrystalline	[56]
		250	0.73		
$Y_{0.95}Ca_{0.05}Ba_2Cu_4O_8$	80.5	100	0.40	Polycrystalline	[56]
		250	0.82		
$Y_{0.9}Ca_{0.1}Ba_2Cu_4O_8$	82	100	0.39	Polycrystalline	[56]
		250	0.76		
$Bi_2Sr_2CuO_{6+\delta}$	≈22	293	25	Polycrystalline	[57]
$Bi_2Sr_2CuO_{6+\delta}$	≈9	20	0.15[a]	Single crystal	[7]
		100	0.28[a]		
$Bi_{2.11}Sr_{1.89}CuO_{6+\delta}$	6.8	130	0.27[a]	Single crystal	[8]
		300	0.48[a]	Oxygen annealed	
$Bi_{2.11}Sr_{1.89}CuO_{6+\delta}$	8.5	130	1.3[a]	Single crystal	[8]
		300	1.9[a]	Argon annealed	
$Bi_{2.15}Sr_{1.85}CuO_{6+\delta}$	6.5	130	0.4[a]	Single crystal	[8]
		300	0.65[a]	Oxygen annealed	
$Bi_{1.95}Sr_{1.65}La_{0.4}CuO_{6+\delta}$	≈14	100	0.27[a]	Single crystal	[35]
		300	0.47[a]		
$Bi_{1.7}Pb_{0.3}Sr_{1.9}La_{0.1}CuO_{6+\delta}$	40.8	100	0.07[a]	Single crystal	[58]
		300	0.17[a]	As grown	
$Bi_{2.2}Sr_2Ca_{0.8}Cu_2O_{8+\delta}$	84	100	0.04[a]	Single crystal	[59]
		300	0.13[a]		

(continued overleaf)

Table 11.2 *(continued)*

Compound	T_c(K)	T(K)	ρ(mΩcm)	Comments	Reference
$Bi_2Sr_2CaCu_2O_{8+\delta}$	≈88	100	0.033[d]	Single crystal	[36]
		100	0.015[c]		
		100	8400[b]		
		300	0.068[d]		
		300	0.026[c]		
		300	3570[b]		
$Bi_2Sr_2CaCu_2O_{8+\delta}$	83	100	0.96[a]	Single crystal	[10]
		200	1.97[a]		
$Bi_2Sr_2CaCu_2O_{8+\delta}$	≈47	100	0.45[a]	Single crystal	[30]
		300	0.52[a]		
	≈59	100	0.19[a]		
		300	0.39[a]		
	≈75	100	0.11[a]		
		300	0.31[a]		
	≈87	100	0.054[a]		
		300	0.18[a]		
$Bi_2Sr_2CaCu_2O_{8+\delta}$	≈93	125	0.103[a]	Single crystal	[60]
		200	0.177[a]		
$Bi_{2.14}Sr_{1.77}Ca_{0.92}Cu_2O_{8+\delta}$	93	125	0.52[d]	Single crystal	[37]
		125	0.20[c]		
		125	2170[b]		
		300	1.02[d]		
		300	0.49[c]		
		300	1740[b]		
$Bi_2Sr_2CaCu_2O_{8+\delta}$	83	120	0.13[d]	Single crystal	[61]
		120	0.16[c]		
		300	0.31[d]		
		300	0.35[c]		
$Bi_2Sr_2CaCu_2O_{8+\delta}$	≈85	100	0.24[a]	Thin films	[62]
		100	4650[b]		
		300	0.49[a]		
		300	4630[b]		
$Bi_2Sr_2CaCu_2O_{8+\delta}$	≈80	125	2.30[a]	Thin films	[9]
		300	4.23[a]		
$Bi_{2.2}Sr_{1.8}Ca_{1.05}(Cu,Li)_{2.15}$ $O_{8+\delta}$ single crystals	89	125	0.25[a]	Li/(Li + Cu) = 1.4%	[63]
		300	0.40[a]		
	92	125	0.12[a]	Li/(Li + Cu) = 3.0%	
		300	0.20[a]		
	93	125	0.135[a]	Li/(Li + Cu) = 13%	
		300	0.30[a]		
$Bi_{1.8}Pb_{0.35}Sr_{1.87}Ca_2Cu_3$ $O_{10+\delta}$	111	125	0.5	Fibers	[13]
		300	1.2		
$Bi_{1.8}Pb_{0.4}Sr_2Ca_2CuO_{10+\delta}$	110	125	0.44	Polycrystalline	[14]
		290	1.08		

(continued overleaf)

Table 11.2 *(continued)*

Compound	T_c(K)	T(K)	ρ(mΩcm)	Comments	Reference
$Bi_{1.4}Pb_{0.6}Sr_2Ca_2Cu_3O_{10+\delta}$	105	125	3.1	Polycrystalline	[64]
		290	5.4		
$Bi_{1.9}Pb_{0.3}Sr_{1.6}Ca_{1.9}Cu_{3.4}$	≈110	125	3.9	Polycrystalline	[65]
$O_{10+\delta}$		200	5.0		
$Bi_{1.84}Pb_{0.34}Sr_{1.91}Ca_{2.03}$	112	150	2.1	Polycrystalline	[31]
$Cu_{3.06}O_{10+\delta}$		300	3.1		
$Bi_{1.8}Pb_{0.26}Sr_2Ca_2Cu_3O_{10+\delta}$	108	150	0.65	Polycrystalline	[12]
		300	1.13		
$TlBa_2CaCu_2O_{7+\delta}$	≈105	125	3.5	Polycrystalline	[15]
		270	5.5		
$Tl_2Ba_2CaCu_2O_{8+\delta}$	97	125	0.54	Polycrystalline film	[66]
		300	1.24		
$Tl_2Ba_2CaCu_2O_{8+\delta}$	102	150	0.58	Film *c*-axis texture	[67]
		290	1.04		
$Tl_2Ba_2Ca_2Cu_3O_{10+\delta}$	116	150	0.088	Film *c*-axis texture	[67]
		290	0.164		
$Tl_{1.6}Ba_2Ca_{2.4}Cu_3O_{10+\delta}$	128	150	1.95	Polycrystalline	[16]
		300	4.35		
$Tl_{1.7}Ba_2Ca_{2.3}Cu_3O_{10+\delta}$	125	150	≈1.6	Polycrystalline	[68]
		250	≈3		
$Tl_2Ba_2Ca_2Cu_3O_{10+\delta}$	≈129	140	0.63	Polycrystalline	[17]
		280	1.04		
$Tl_{1.6}Hg_{0.4}Ba_2Ca_2Cu_3O_{10-\delta}$	126	140	1.21	Polycrystalline	[18]
		290	2.87		
$Tl_2Ba_2Ca_3Cu_4O_{12+\delta}$	110	150	14.5	Polycrystalline	[19]
		290	36		
$(Tl,Pb)Sr_2Y_{0.4}Ca_{0.6}Cu_2$	≈89	120		Polycrystalline	[69]
$O_{7+\delta}$, Tl:Pb = 1:1		250	19.2		
$(Tl,Pb)Sr_2Y_{0.25}Ca_{0.75}Cu_2$	≈95	120		Polycrystalline	[69]
$O_{7+\delta}$, Tl:Pb = 1:1		250	14.1		
$(Tl,Pb)Sr_2Y_{0.1}Ca_{0.9}Cu_2$	≈96	120		Polycrystalline	[69]
$O_{7+\delta}$, Tl:Pb = 1:1		250	5.4		
$(Tl,Pb)Sr_2CaCu_2O_{7+\delta}$	≈81	120		Polycrystalline	[69]
Tl:Pb = 1:1		250	2.65		
$HgBa_2CuO_{4+\delta}$	≈93	150	24	Polycrystalline	[20]
		300	47		
$HgBa_2CuO_{4+\delta}$	≈95	150	3.9	Polycrystalline	[21]
		290	7.85		
$HgBa_2CaCu_2O_{6+\delta}$	≈126	150	13.6	Polycrystalline	[21]
		290	20.3		
$HgBa_2CaCu_2O_{6+\delta}$	126	150	14	Polycrystalline	[70]
		300	20		

(continued overleaf)

Table 11.2 *(continued)*

Compound	T_c(K)	T(K)	ρ(mΩcm)	Comments	Reference
$HgBa_2CaCu_2O_{6+\delta}$	126	150	17.7	Polycrystalline	[22]
		300	27.6		
$HgBa_2CaCu_2O_{6+\delta}$	121	150	0.14[a]	Film *c*-axis texture	[71]
		300	0.4[a]		
$HgBa_2CaCu_2O_{6+\delta}$	124	150	0.68	Thin film	[72]
		250	1.39	Cl doping	
	122	150	0.27	Thin film	[72]
		250	0.53	F doping	
$HgBa_2Ca_2Cu_3O_{8+\delta}$	134	200	18	Polycrystalline	[70]
		300	27		
$HgBa_2Ca_2Cu_3O_{8+\delta}$	134	200	44.5	Polycrystalline	[73]
		300	66.7		
$HgBa_2Ca_2Cu_3O_{8+\delta}$	132	180	7.3	Polycrystalline	[23]
		270	11.3		
$HgBa_2Ca_2Cu_3O_{8+\delta}$	132	160	0.064[a]	Film *c*-axis texture	[24]
		200	0.087[a]		
$HgBa_2Ca_2Cu_3O_{8+\delta}$	≈131	180	7.3	Polycrystalline	[74]
		290	10.6		
$HgBa_2Ca_2Cu_3O_{8+\delta}$	≈135	180	15.2	Polycrystalline	[75]
		290	27.1		
$HgBa_2Ca_2Cu_3O_{8+\delta}$	≈132	180	7.3	Polycrystalline	[76]
		270	11.2		
$HgBa_2Ca_3Cu_4O_{10+\delta}$	126	150	0.18[a]	Single crystal	[38]
		300	0.39[a]		
$(Hg_{0.7}Tl_{0.3})_2Ba_2Y_{0.7}\ Ca_{0.3}$	≈57	100	1.75	Polycrystalline	[77]
$Cu_2O_{8-\delta}$		290	2.55		
$(Cu_{0.5}C_{0.5})Ba_2Ca_2Cu_3O_{9+\delta}$	67	100	0.16	Polycrystalline	[78]
		150	0.29		
$(Cu_{0.5}C_{0.5})Ba_2Ca_3Cu_4$	117	150	0.32	Polycrystalline	[78]
$O_{11+\delta}$		190	0.46		
$BSr_2Ca_3Cu_4O_{11+\delta}$	≈109	150	0.42	Polycrystalline	[79]
		180	0.52		
$(B_{0.6}C_{0.4})Sr_2Ca_2Cu_3O_{9+\delta}$	≈110	150	1.25	Polycrystalline	[80]
		300	2.88		
$(B_{0.6}C_{0.4})(Sr_{0.75}Ba_{0.25})_2Ca_2$	≈110	150	2.23	Polycrystalline	[80]
$Cu_3O_{9+\delta}$		300	4.37		
$(B_{0.6}C_{0.4})(Sr_{0.5}Ba_{0.5})_2Ca_2$	≈120	150	2.46	Polycrystalline	[80]
$Cu_3O_{9+\delta}$		300	4.93		
$(B_{0.6}C_{0.4})(Sr_{0.25}Ba_{0.75})_2Ca_2$	≈120	150	1.20	Polycrystalline	[80]
$Cu_3O_{9+\delta}$		300	2.54		

[a]In-plane resistivity ρ_{ab}.
[b]Out-of-plane resistivity ρ_c,
[c]ρ_b,
[d]ρ_a.

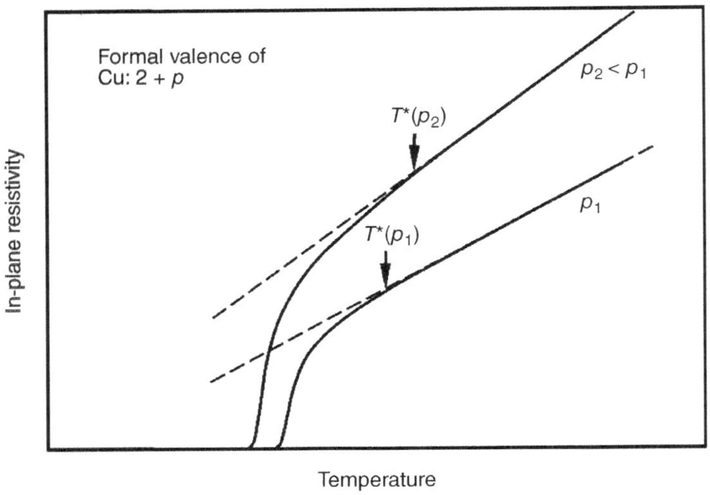

(a)

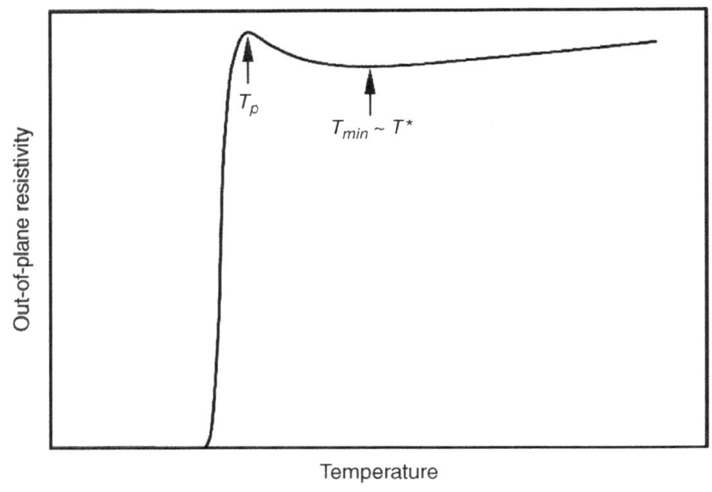

(b)

Figure 11.11 *Illustration of the determination of the pseudogap temperature T^* from resistivity versus temperature curves. At higher temperatures the in-plane resistivity (top) depends linearly on temperature. The first deviation from linearity well above the transition temperature has been attributed to the opening of the pseudogap [60, 62, 81]. The pseudogap temperature T^* depends on the number of mobile holes per Cu atom in the CuO_2 planes. Alternatively the minimum of the out-of-plane resistivity may be used to estimate T^* [62] . The temperature T_p, where ρ_c reaches its peak value immediately above the superconducting transition, has been attributed to the onset of superconducting fluctuations [62]*

The temperature T_{min} is used to estimate the temperature T^* separating the pseudogap region from the normal state (see Chapter 8). The use of the minimum of the out-of-plane resistivity tends to provide higher values of T^* than the first deviation from linearity of the in-plane resistivity.

11.3 Thermal Conductivity

In this section, the thermal conductivity of various cuprate superconductors will be considered. Thermal conductivity data are frequently needed for engineering purposes. For example, current leads (see, e.g., [82]) are required to connect low-temperature superconducting magnets, operated at temperatures close to 4 K, with power supplies at room temperature. To reduce the heat load to the 4 K level, the low temperature part ($T \leq 77\,\mathrm{K}$) of the current leads can be made of a bulk high-T_c superconductor. Due to the absence of Joule heating, it is possible to cool the superconducting part of the current lead only by heat conduction from the cold end, while the copper part is actively cooled by liquid nitrogen and the boiled-off gas. The heat flux to the cold end is

$$\frac{dQ}{dt} = \frac{A}{L}\int_{4\,\mathrm{K}}^{77\,\mathrm{K}} k(T)dT \tag{11.1}$$

Here, A is the cross-section of the bulk superconductor, L its length, and k its thermal conductivity, which depends on temperature T.

An introduction to the principles which govern the thermal conductivity of solids can be found in [1, 83]. Here, we will describe only some basic ideas of heat conduction in solids as in [2]. The steady-state thermal current density $j_Q = (1/A)(dQ/dt)$ can be used to define the thermal conductivity k:

$$j_Q = -k\frac{dT}{dx} \tag{11.2}$$

In metals, the conduction electrons and the phonons contribute to thermal conductivity. The two contributions behave like two resistors in parallel in the analog electrical circuit [83]

$$k = k_{\mathrm{el}} + k_{\mathrm{ph}} \tag{11.3}$$

where k_{el} is the electron and k_{ph} the phonon contribution. In analogy to heat conduction in a gas, the phonon contribution can be written as

$$k_{\mathrm{ph}} = \frac{1}{3}v_{\mathrm{ph}}L_{\mathrm{ph}}C_{\mathrm{ph}} \tag{11.4}$$

where v_{ph} is the velocity, L_{ph} the mean free path, and C_{ph} the specific heat of the phonons per unit volume. The mean free path of the phonons is limited by phonon–phonon, phonon–imperfection, and phonon–electron scattering. The corresponding thermal resistances add as the resistances in a series connection [83]. The phonon-specific heat is proportional to T^3 at low temperatures (see Section 2.5, Equation 2.47), while it reaches a constant value of $3R$ (R is the gas constant) well above the Debye temperature θ_D. Furthermore, the mean free path depends on temperature. At high temperatures the number of activated phonons is proportional to temperature. Omitting the phonon–imperfection and the phonon–electron scattering, this leads to a $1/T$ dependence of the phonon mean free path (see, e.g., [1]).

The electronic contribution to the thermal conductivity is

$$k_{\text{el}} = \frac{1}{3} v_F L_{\text{el}} C_{\text{el}} \tag{11.5}$$

where v_F is the Fermi velocity, $L_{\text{el}} = v_F \tau$ the mean free path of the electrons, and $C_{\text{el}} = (1/3)\pi^2 D(E_F) k_B{}^2 T$ the contribution of the electrons to the specific heat per unit volume. The density of states per unit volume of a free electron gas at the Fermi energy is $D(E_F) = 3n_e/2E_F$, where n_e is the number density of the conduction electrons. Using C_{el} and $D(E_F)$, the electronic contribution to thermal conductivity can be written as

$$k_{\text{el}} = \frac{\pi^2 n_e k_B{}^2 T \tau}{3m} \tag{11.6}$$

where m is the electron mass and τ the relaxation time. The electron mean free path is limited by electron–phonon and electron–imperfection scattering. The two thermal resistances related to these scattering mechanisms add like resistances in series [83]. The use of the expression $C_{\text{el}} = \gamma T$ (see Section 2.5) leads to

$$k_{\text{el}} = \frac{1}{3} \gamma v_F{}^2 T \tau \tag{11.7}$$

In the case in which the relaxation times for electrical and thermal processes are identical, the ratio of the thermal and electrical conductivity of the electronic contribution becomes

$$\frac{k_{\text{el}}}{\sigma} = \frac{\pi^2 n_e k_B{}^2 T \tau / 3m}{n_e e^2 \tau / m} = \frac{\pi^2}{3} \left(\frac{k_B}{e} \right)^2 T \tag{11.8}$$

Equation 11.8 is the well-known Wiedemann–Franz law $(k_{\text{el}}/\sigma) = LT$, where L is the Lorenz number. The value of the Lorenz number is $2.45 \times 10^{-8} \text{W}\Omega/\text{K}^2$. The relaxation time determining the electrical resistivity is known to be proportional to T^{-3} for $T << \theta_D$ and to T^{-1} for $T >> \theta_D$ [84]. Assuming, as in the Wiedemann–Franz law, that the relaxation times for electrical and thermal processes are equal, the resulting electronic contribution to the thermal conductivity is constant at high temperatures, while it is proportional to T^{-2} at temperatures much lower than the Debye temperature. Electron-impurity and electron-defect scattering causes a constant residual resistivity at temperatures close to absolute zero. As a consequence the thermal conductivity is proportional to T at temperatures close to $T = 0$. The behavior of pure metals at not too low a temperature is typically well described by the Wiedemann–Franz law [83].

The heat conductivity of pure metals is dominated by the electronic contribution even at room temperature. In disordered alloys, the electronic and the phonon contributions may be comparable. Below the critical temperature the conduction electrons in a superconductor start to form Cooper pairs. Because Cooper pairs do not exchange energy with the lattice, they do not contribute to heat conduction. As a consequence, the thermal conductivity of conventional superconductors decreases rapidly below the critical temperature.

Thermal conductivity can be measured by a steady-state heat flux method [83, 85]. A sketch of the experimental setup is presented in Figure 11.12. By means of resistive heaters and cooling with liquid helium, the temperatures of the sample and the copper block at the

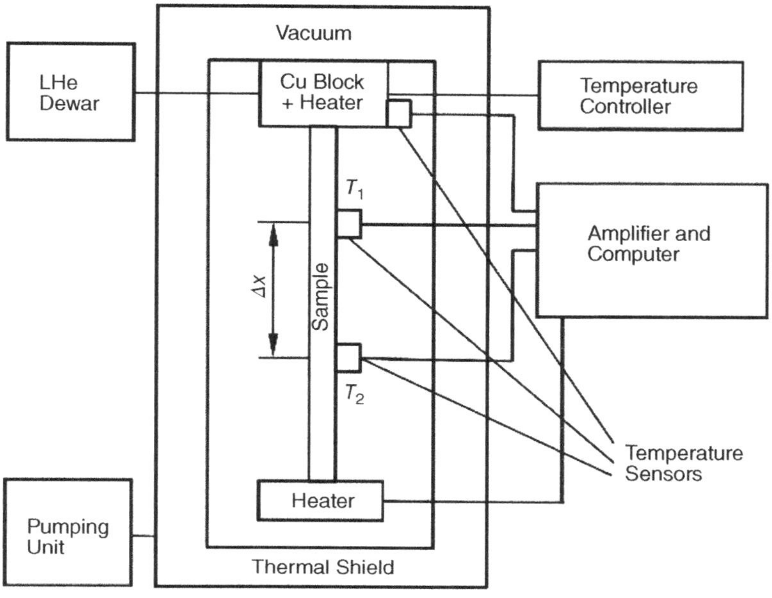

Figure 11.12 *Sketch of the setup for the measurement of the thermal conductivity by means of a steady-state heat flux method (adapted from [83, 85])*

top can be adjusted to values in the range of 4.2–300 K. The use of calibrated silicon diodes allows the measurement of temperature with an accuracy of ±50 mK. The heat flux dQ/dt can be obtained from the latent heat of evaporation of the liquid helium and the measured heater power. The heat flux and the thermal conductivity are related by the expression

$$\left|\frac{dQ}{dt}\right| = kA\left|\frac{\Delta T}{\Delta x}\right| \tag{11.9}$$

Here, A is the cross-sectional area of the sample, k the thermal conductivity, Δx the distance of the temperature sensors, and ΔT the difference of temperatures T_1 and T_2.

Figure 11.13 shows the thermal conductivity data of various polycrystalline cuprate superconductors [47, 64, 86–89]. The thermal conductivity of cuprate high-T_c superconductors is typically between 1 and 10 $\mathrm{Wm^{-1}K^{-1}}$ at temperatures close to room temperature. Measured resistivity data and the use of the Wiedemann–Franz law allow the estimation of the electronic contribution to the thermal conductivity. These estimations suggest that the heat transport in the cuprate superconductors is dominated by the phonon contribution at elevated temperatures [43, 46, 47, 64, 90–94]. Below the critical temperature, a broad maximum of the thermal conductivity has frequently been observed. This behavior has been attributed to reduced phonon–electron scattering in the superconducting state, caused by the condensation of the mobile charge carriers into Cooper pairs [95].

Figure 11.14 shows the thermal conductivity of three c-axis aligned $(Bi,Pb)_2Sr_2Ca_2Cu_3O_{10+x}$ cuprate superconductors [93, 94, 96]. As in the case of the polycrystalline samples, there exists a maximum in the thermal conductivity below T_c. For samples A and

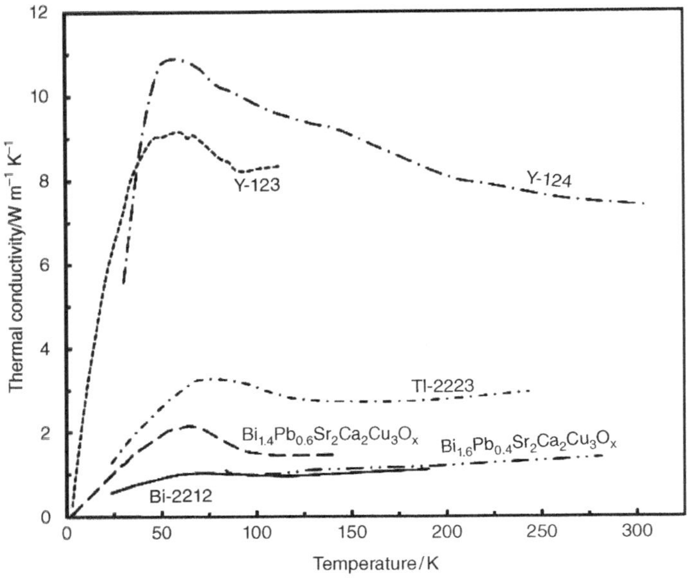

Figure 11.13 *Thermal conductivity of various polycrystalline cuprate high-T_c superconductors (data from [47, 64, 86–89])*

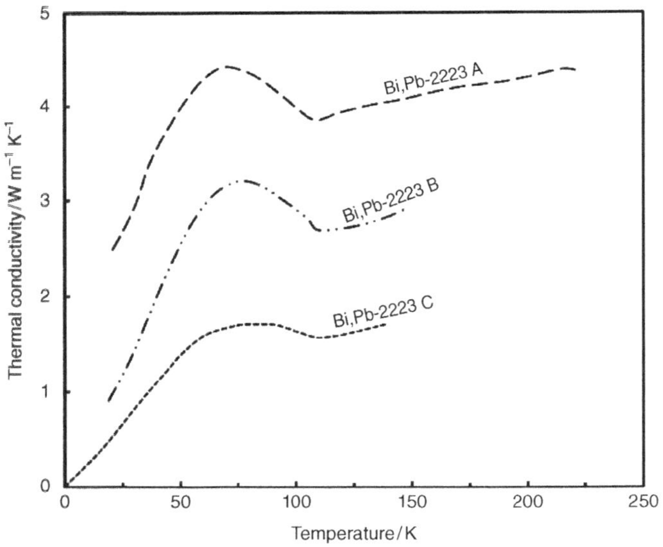

Figure 11.14 *Thermal conductivity of three c-axis aligned $(Bi,Pb)_2Sr_2Ca_2Cu_3O_{10+x}$ cuprate superconductors. As in polycrystalline samples the thermal conductivity reaches a broad maximum below T_c (data for samples A, B, C from [96, 94, 93], respectively)*

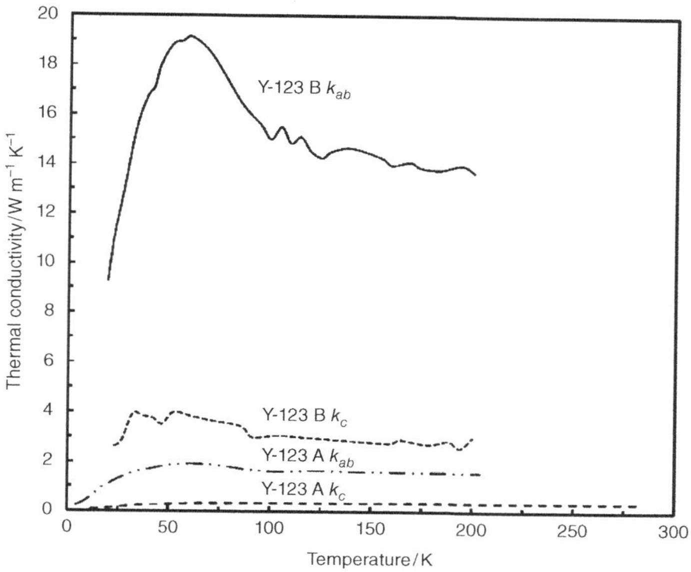

Figure 11.15 *In- and out-of-plane thermal conductivity of two melt-textured Y-123 samples. In spite of a difference of nearly one order of magnitude in the thermal conductivities of Y-123 A [43] and Y-123 B [44], the ratio of k_{ab} to k_c in both samples is close to 5*

B, the thermal conductivity is significantly larger than that of the polycrystalline samples presented in Figure 11.13.

The thermal conductivity of melt-textured Y-123, measured along the CuO_2 planes and perpendicular to them, provides first evidence for anisotropic heat transport (see Figure 11.15). For both samples, the in-plane thermal conductivity is nearly a factor of 5 larger than that along the *c* direction. It is interesting to notice that the thermal conductivity of the sample Y-123 B [44] is around an order of magnitude larger than that of the sample Y-123 A [43].

Figure 11.16 shows the thermal conductivities of various bulk cuprate superconductors [97–102]. The values of the thermal conductivity in the considered temperature range are between 1 and 10 $Wm^{-1}K^{-1}$. Extremely low values of the thermal conductivity have been found for Bi-2212 bulk material without added silver [97, 98]. The results are in line with the data for polycrystalline Bi-based superconductors presented in Figure 11.13. The thermal conductivity of Bi-2212 can be significantly enhanced by the addition of 10 wt% silver. The thermal conductivities of RE-123 superconductors are typically larger than those of Bi-based cuprates. The textured Er-123 bulk material shows a moderate anisotropy of the in- and out-of-plane thermal conductivities.

Measurement of the thermal conductivity of textured bulk material indicates that the heat transport in the cuprates is anisotropic. A more detailed investigation of this aspect requires the use of single crystals. The results of the measurement of the thermal conductivity of various single crystals are presented in Figure 11.17 [90–92, 103–104]. The in-plane thermal conductivity k_{ab} shows, similarly to polycrystalline material, a pronounced maximum below the critical temperature. On the other hand, no maximum in the out-of-plane thermal

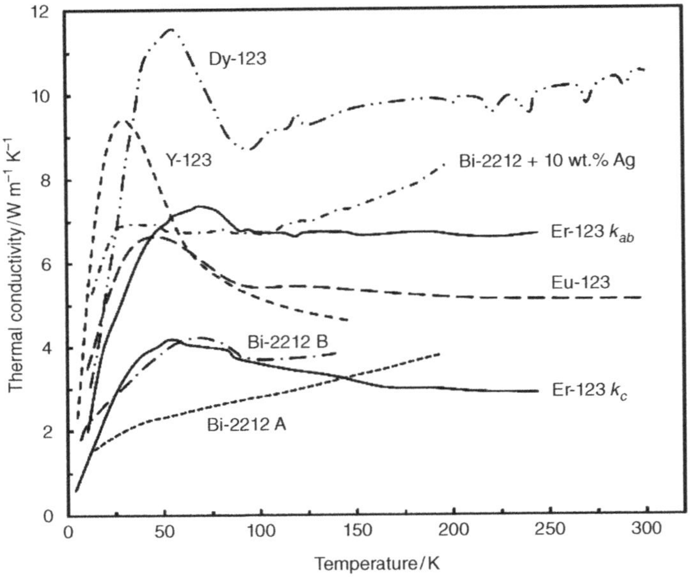

Figure 11.16 *Thermal conductivities of various cuprate bulk superconductors (data from [97–102])*

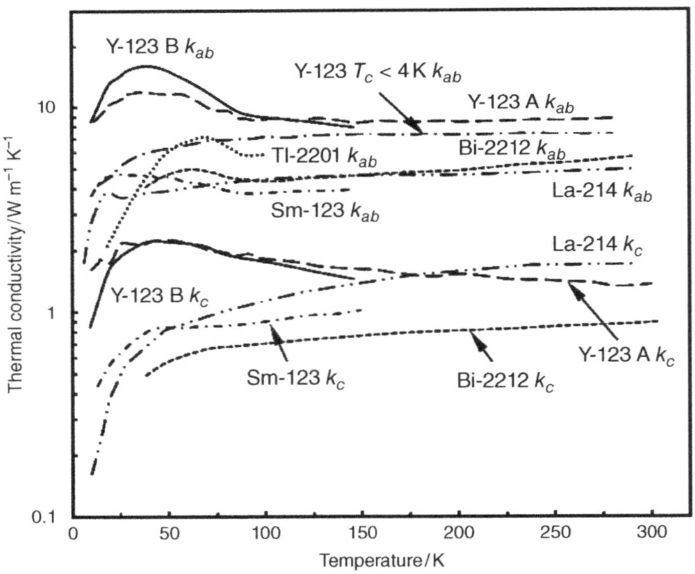

Figure 11.17 *In- and out-of-plane thermal conductivities of various cuprate superconductor single crystals (data from [90–92, 103, 104])*

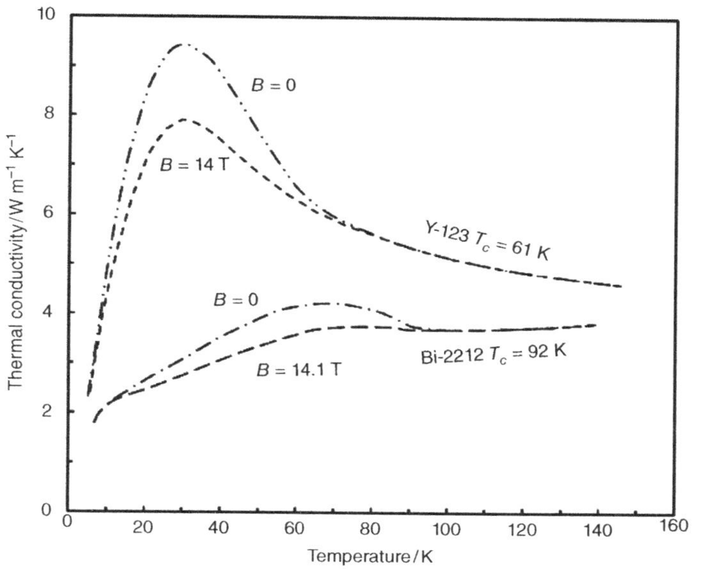

Figure 11.18 *Comparison of the thermal conductivities of Y-123 and Bi-2212 at* B = 0 *and* B≈14 T *(data from [98, 99])*

conductivity was found for Bi-2212 [91] and La-214 [92] single crystals. For the RE-123 single crystals, a maximum of the thermal conductivity below T_c was found in the in- and out-of-plane heat transport. The ratio of k_{ab} to k_c is around 3 in the La-214 single crystal [92] close to room temperature. A more pronounced anisotropy was observed for the Bi-2212 single crystal. At temperatures well above T_c the in-plane thermal conductivity is a factor of 6 larger than the out-of-plane thermal conductivity [91]. A similar ratio of k_{ab} to k_c was found for the two Y-123 single crystals A [103] and B [104]. An interesting observation is the fact that for a nonsuperconducting Y-123 single crystal ($T_c < 4$ K) no maximum in the in-plane thermal conductivity was found below 100 K [103]. This is in line with the hypothesis that the reduced scattering of the phonons by the reduced number of single electrons present in the superconducting state is responsible for the maximum in the thermal conductivity below T_c. For the Sm-123 single crystal, the value of the ratio of k_{ab}/k_c is of the order of 4 well above T_c.

The last aspect to be considered is the effect of an applied magnetic field on the thermal conductivity. In Figure 11.18, the thermal conductivities of Bi-2212 ($T_c = 92$ K) [98] and of underdoped Y-123 ($T_c = 61$ K) [99], measured without applied field and at ≈14 T, are compared. In both superconductors, the application of a magnetic field reduces the thermal conductivity in the superconducting state. The effect is most pronounced in the region of the maximum below T_c. The reduction of the thermal conductivity in the superconducting state has been attributed to scattering of the phonons and the quasiparticles (unpaired electrons) with the vortices present above the lower critical field B_{c1}. The Cooper pairs do not contribute to heat conduction.

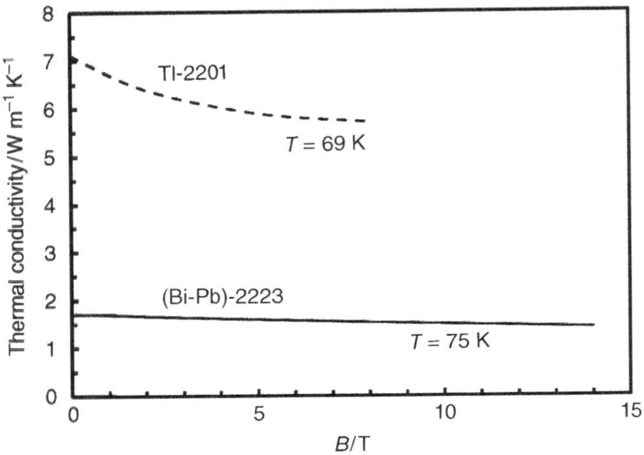

Figure 11.19 *Dependence of the thermal conductivity on the applied magnetic field. The temperatures 69 K (Tl-2201 [90]) and 75 K ((Bi,Pb)-2223 [93]) are close to the maximum of the thermal conductivity in the superconducting state (results from [90, 93])*

Figure 11.19 shows the dependence of the thermal conductivity of (Bi,Pb)-2223 [94] and Tl-2201 [90] on the applied magnetic field in the superconducting state. At the temperatures under consideration, the maximum of the thermal conductivity is reached at zero field. In both cuprates the thermal conductivity decreases with increasing magnetic field. The effect is more pronounced in Tl-2201.

11.4 Summary

The in-plane resistivity of optimally doped cuprate superconductors shows a linear dependence on temperature over a wide range of temperatures. In general, the in-plane resistivity shows metallic behavior, which means that the resistivity decreases with decreasing temperature. On the other hand, the out-of-plane resistivity at low temperatures frequently shows an increase of the resistivity with decreasing temperature, as in semiconductors. The electrical resistivity of single crystals, textured bulk material, or epitaxial films shows a pronounced anisotropy, with the out-of-plane resistivity being much larger than the in-plane resistivity. In cuprates with an orthorhombic crystal structure, there also exists a small anisotropy of the two in-plane resistivities ρ_a and ρ_b. For Bi-2212 single crystals, it was found that the resistivity ρ_a, measured along the crystallographic a direction, is a factor of ≈ 2 larger than ρ_b. The resistivity depends on the exact chemical composition, which determines the number of mobile holes per CuO_2. The oxygen content depends on the synthesis conditions (annealing temperature and oxygen partial pressure). As in normal metals an increased resistivity was found after irradiation. The in-plane resistivity of single crystals is typically in the range of 0.1 to 1 mΩcm at room temperature.

The thermal conductivity of cuprate high-T_c superconductors is orders of magnitude lower than that of pure copper. Due to the relatively low concentration of mobile charge carriers, heat is conducted mainly by phonons. The values of the in-plane thermal conductivity are significantly larger than those for out-of-plane heat transport. The in-plane thermal conductivity shows a broad maximum below the critical temperature. The increase of thermal conductivity in the superconducting state has been attributed to reduced scattering of phonons by electrons, because of the condensation of single electrons into Cooper pairs. An alternative to this hypothesis is an increased electronic contribution to the thermal conductivity due to the reduced electron–electron scattering in the superconducting state [94]. Typical values of the in-plane thermal conductivity are in the range of 1 to $10\,\mathrm{W\,m^{-1}\,K^{-1}}$ at room temperature. The application of a magnetic field leads to reduced values of the thermal conductivity in the superconducting state. Similarly to electrical resistivity, the values of thermal conductivity show a considerable variation for different samples.

References

1. Ch. Kittel, *Introduction to Solid State Physics*, John Wiley & Sons, Inc., New York, 1976.
2. R. Wesche, *High-Temperature Superconductors: Materials, Properties and Applications*, Kluwer Academic Publishers, Norwell, MA 1998.
3. D.R. Harshman and A.P. Millis, Concerning the nature of high-T_c superconductivity: Survey of experimental properties and implications for interlayer coupling, *Phys. Rev. B,* **45**, 10684–10712 (1992).
4. M. Decroux, A. Junod, A. Bezinge, D. Cattani, J. Cors, J.L. Jorda, A. Stettler, M. François, K. Yvon, Ø. Fischer, and J. Muller, Structure, resistivity, critical field, specific heat jump at T_c, Meissner effect, a.c. and d.c. susceptibility of the high-temperature superconductor $La_{2-x}Sr_xCuO_4$, *Europhys. Lett.,* **3**, 1035–1040 (1987).
5. H. Takagi, T. Ido, S. Ishibashi, M. Uota, S. Uchida, and Y. Tokura, Superconductor-to-nonsuperconductor transition in $(La_{1-x}Sr_x)_2CuO_4$ as investigated by transport and magnetic measurements, *Phys. Rev. B,* **40**, 2254–2261 (1989).
6. M.A. Mojumder, Temperature-dependence of cuprate resistivity: An old problem revisited, *Physica C,* **466**, 148–156 (2007).
7. F. Rullier-Albenque, A. Legris, H. Berger, and L. Forro, Effect of electron irradiation on $Bi_2Sr_2CaCu_2O_8$ and $Bi_2Sr_2CuO_6$ superconductors, *Physica C,* **254**, 88–92 (1995).
8. J. Hejtmánek, M. Nevřiva, E. Pollert, D. Sedmidubský, and P. Vašek, Superconductivity and electric transport of $Bi_{2+x}Sr_{2-y}CuO_{6+\delta}$ single crystals, *Physica C,* **264**, 220–226 (1996).
9. C. Stölzel, M. Huth, and H. Adrian, *c*-axis oriented thin $Bi_2Sr_2CaCu_2O_{8+\delta}$-films prepared by flash evaporation, *Physica C,* **204**, 15–20 (1992).
10. D. Shi, M. Tang, Y.C. Chang, P.Z. Jiang, K. Vandervoort, B. Malecki, and D.J. Lam, Transport critical current density in $Bi_2Sr_2CaCu_2O_x$ single crystals, *Appl. Phys. Lett.,* **54**, 2358–2360 (1989).
11. D.-S. Jeon, M. Akamatsu, H. Ikeda, and R. Yoshizaki, Zn substitution effect for Cu on the anisotropic transport properties in $Bi_2Sr_2CaCu_2O_{8+y}$ single crystals, *Physica C,* **253**, 102–108 (1995).

12. V. Plecháček, H. Hejdová, and Ž. Trejbalová, Preparation of single high T_c phase samples of Bi–Pb–Sr–Ca–Cu–O superconductors, *Cryogenics,* **30**, 11–13 (1990).
13. W. Carillo-Cabrera, W. Göpel, G.F. de la Fuente, and H.R. Verdún, Preparation and growth of (Bi,Pb)-Sr-Ca-Cu-O superconductor fibers, *Appl. Phys. Lett.,* **55**, 1032–1034 (1989).
14. W. Carillo-Cabrera and W. Göpel, Influence of high-temperature annealing on the $(Bi,Pb)_2Sr_2Ca_2Cu_3O_{10}$ phase and determination of its crystal structure by X-ray powder diffractometry, *Physica C,* **161**, 373–389 (1989).
15. S. Nakajima, M. Kikuchi, Y. Syono, K. Nagase, T. Oku, N. Kobayashi, D. Shindo, and K. Hiraga, Improvement in superconductivity of $TlBa_2CaCu_2O_y$ system by introduction of oxygen loss, *Physica C,* **170**, 443–447 (1990).
16. R.S. Liu, J.L. Tallon, and P.P. Edwards, An efficient and reproducible approach for attaining superconductivity at 128 K in $Tl_2Ba_2Ca_2Cu_3O_{10-\delta}$, *Physica C,* **182**, 119–122 (1991).
17. B.T. Ahn, W.Y. Lee, and R. Beyers, Effect of oxygen pressure on $Tl_2Ca_2Ba_2Cu_3O_{10\pm\delta}$ formation, *Appl. Phys. Lett.,* **60**, 2150–2152 (1992).
18. Y.X. Jia, C.S. Lee, and A. Zettl, Stabilization of the $Tl_2Ba_2Ca_2Cu_3O_{10}$ superconductor by Hg doping, *Physica C,* **234**, 24–28 (1994).
19. M. Kikuchi, S. Nakajima, Y. Syono, K. Hiraga, T. Oku, D. Shindo, N. Kobayashi, H. Iwasaki, and Y. Muto, Preparation of the bulk superconductor $Tl_2Ba_2Ca_3Cu_4O_{12}$, *Physica C,* **158**, 79–82 (1989).
20. I.C. Chang, M.D. Lan, T.J. Goodwin, J.Z. Liu, P. Klavins, and R.N. Shelton, Magnetic and transport properties of $HgBa_2CuO_{4+\delta}$, *Physica C,* **223**, 207–212 (1994).
21. L. Gao, Y.Y. Xue, F. Chen, Q. Xiong, R.L. Meng, D. Ramirez, C.W. Chu, J.H. Eggert, and H.K. Mao, Superconductivity up to 164 K in $HgBa_2Ca_{m-1}Cu_mO_{2m+2+\delta}$ under quasihydrostatic pressures, *Phys. Rev. B,* **50**, 4260–4263 (1994).
22. Y.T. Ren, J. Clayhold, F. Chen, Z.J. Huang, X.D. Qiu, Y.Y. Sun, R.L. Meng, Y.Y. Xue, and C.W. Chu, Thermopower and resistivity measurements on oxygen-annealed $HgBa_2CaCu_2O_{6+\delta}$, *Physica C,* **217**, 6–10 (1993).
23. H.M. Shao, L.J. Shen, J.C. Shen, X.Y. Hua, P.F. Yuan, and X.X. Yao, Synthesis of single-phase $HgBa_2Ca_2Cu_3O_{8+\delta}$ superconductor, *Physica C,* **232**, 5–9 (1994).
24. Y. Moriwaka, T. Sugano, C. Gasser, A. Fukuoka, K. Nakanishi, S. Adachi, and K. Tanabe, *Appl. Phys. Lett.,* **69**, 3423–3425 (1996).
25. I. Schildermans, M. Van Bael, E. Knaepen, J. Yperman, J. Mullens, and L.C. Van Poucke, Synthesis of the high temperature superconductor $YBa_2Cu_3O_{7-\delta}$ by the hydroxide co-precipitation method, *Physica C,* **278**, 55–61 (1997).
26. R. Hopfengärtner, B. Hensel, and G. Saemann-Ischenko, Analysis of the fluctuation-induced excess dc conductivity of epitaxial $YBa_2Cu_3O_7$ films: Influence of a short-wavelength cutoff in the fluctuation spectrum, *Phys. Rev. B,* **44**, 741–749 (1991).
27. T. Wada, N. Suzuki, T. Maeda, A. Maeda, S. Uchida, K. Uchinokura, and S. Tanaka, High transition temperature superconductor $LaBa_2Cu_3O_{7-y}$ with zero resistance at 92 K, *Appl. Phys. Lett.,* **52**, 1989–1991 (1988).
28. T. Naito, H. Fujishiro, K.Iida, and M. Murakami, Anisotropic thermal conductivity of Er–Ba–Cu–O bulk superconductors, *Physica C,* **468**, 1428–1430 (2008).

29. J.L. Tallon, R.G. Buckley, M.P. Staines, M.R. Presland, and P.W. Gilberd, Enhancing T_c in $Bi_{2.1}CaSr_2Cu_2O_{8+\delta}$ from 90 to 101 K by yttrium substitution, *Appl. Phys. Lett.*, **54**, 1591–1593 (1989).
30. L. Forró, J.R. Cooper, B. Leontić, and B. Keszei, Reversible changes in the superconducting transition temperature of $Bi_2Sr_2CaCu_2O_8$ single crystals by varying the oxygen concentration, *Europhys. Lett.*, **10**, 371–374 (1989).
31. S.K. Bandyopadhyay, P. Barat, P. Sen, U. De, A. De, P.K. Mukhopadhyay, S.K. Kar, and C.K. Majumdar, Effect of alpha irradiation on polycrystalline Bi-2223 superconductors, *Physica C*, **228**, 109–114 (1994).
32. M.K.R. Khan, H. Tanabe, I. Tanaka, and H. Kojima, Anisotropic electric resistivity of superconducting $La_{2-x}Ba_xCuO_{4-\delta}$ single crystals, *Physica C*, **258**, 315–320 (1996).
33. M.A. Crusellas, J. Fontcuberta, S. Piñol, J. Beille, and T. Grenet, Transport properties of $Ln_{2-x}Ce_xCuO_{4-y}$ single crystals under high pressure, *Physica C*, **209**, 537–548 (1993).
34. T. Penney, S. von Molnár, D. Kaiser, F. Holtzberg, and A.W. Kleinsasser, Strongly anisotropic electrical properties of single-crystal $YBa_2Cu_3O_{7-x}$, *Phys. Rev. B*, **38**, 2918–2921 (1988).
35. R. Jin, H.R. Ott, and D.P. Grindatto, Normal-state resistivity of superconducting $Bi_{1.95}Sr_{1.65}La_{0.4}CuO_{6+\delta}$ single crystals, *Physica C*, **250**, 395–402 (1995).
36. S. Martin, A.T. Fiory, R.M. Fleming, G.P. Espinosa, and A.S. Cooper, Anisotropic critical current density in superconducting $Bi_2Sr_2CaCu_2O_8$ crystals, *Appl. Phys. Lett.*, **54**, 72–74 (1989).
37. Y. Kotaka, T. Kimura, H. Ikuta, J. Shimoyama, K. Kitazawa, K. Yamafuji, K. Kishio, and D. Pooke, Doping state and transport anisotropy in Bi-2212 single crystals, *Physica C*, **235–240**, 1529–1530 (1994).
38. J. Löhle, J. Karpinski, A. Morawski, and P. Wachter, Electrical resistivity and Hall effect of single crystal $HgBa_2Ca_3Cu_4O_{10+\delta}$, *Physica C*, **266**, 104–108 (1996).
39. L. Winkeler, S. Sadewasser, B. Beschoten, H. Frank, F. Nouvertné, and G. Güntherodt, Interlayer coupling in Pb-substituted $Bi_2Sr_2CaCu_2O_{8+\delta}$ single crystals, *Physica C*, **265**, 194–200 (1996).
40. H. Sato, Thickness dependence of superconductivity and resistivity in $La_{1.85}Sr_{0.15}CuO_4$ films, *Physica C*, **468**, 991–995 (2008).
41. R.J. Cava, R.B. van Dover, B. Batllog, and E.A. Rietman, Bulk superconductivity at 36 K in $La_{1.8}Sr_{0.2}CuO_4$, *Phys. Rev. Lett.*, **58**, 408–410, (1987).
42. P.C. McIntyre, M.J. Cima, and Man Fai Ng, Metalorganic deposition of high-J_c $Ba_2YCu_3O_{7-x}$ thin films from trifluoroacetate precursors onto (1 0 0) $SrTiO_3$, *J. Appl. Phys.*, **68**, 4183–4187 (1990).
43. N. Wendling, J.M. Barbut, D. Bourgault, J. Chaussy, and J. Mazuer, Electrical and thermal conductivities of magnetically melt textured $YBa_2Cu_3O_{7-\delta}$, *Physica C*, **235–240**, 1517–1518 (1994).
44. M. Ikebe, H. Fujishiro, T. Naito, K. Noto, S. Kohayashi, and S. Yoshizawa, Thermal conductivity of YBCO(123) and YBCO(211) mixed crystals prepared by MMTG, *Cryogenics*, **34**, 57–61 (1994).
45. J. van den Berg, C.J. van der Beek, P.H. Kes, G.J. Nieuwenhuys, J.A. Mydosh, H.W. Zandbergen, F.P.F. van Berkel, R. Steens, and D.J.W. Ijdo, Interrelation of structure,

dimensionality and superconductivity in the high-temperature superconductor $YBa_2Cu_3O_{7-\delta}$, *Europhys. Lett.*, **4**, 737–742 (1987).

46. B.M. Terzijska, R. Wawryk, D.A. Dimitrov, Cz. Marucha, V.T. Kovachev, and J. Rafalowicz, Thermal conductivity of YBCO and thermal conductance at YBCO/ruby boundary Part 1: Joint experimental set-up for simultaneous measurements and experimental study in the temperature range 10–260 K, *Cryogenics*, **32**, 53–59 (1992).
47. U. Gottwick, R. Held, G. Sparn, F. Steglich, H. Rietschel, D. Ewert, B. Renker, W. Bauhofer, S. von Molnar, M. Wilhelm, and H.E. Hoenig, Transport properties of $YBa_2Cu_3O_7$: Resistivity, thermal conductivity, thermopower and Hall effect, *Europhys. Lett.*, **4**, 1183–1188 (1987).
48. P. Zhao, A. Ito, R. Tu, and T. Goto, High-speed growth of $YBa_2Cu_3O_{7-\delta}$ film with high critical temperature on MgO single crystal substrate by laser chemical vapor deposition, *Supercond. Sci. Technol.*, **23**, 125010 (2010).
49. G.Y. Sung and J.D. Suh, Superconducting and structural properties of in-plane aligned *a*-axis oriented $YBa_2Cu_3O_{7-x}$ thin films, *Appl. Phys. Lett.*, **67**, 1145–1147 (1995).
50. D. Fuchs, E. Brecht, P. Schweiss, I. Loa, C. Thomsen, and R. Schneider, Growth and characterization of *a*-axis oriented $YBa_2Cu_3O_{7-x}$ thin films on (1 0 0) $LaSrGaO_4$ substrates, *Physica C*, **280**, 167–177 (1997).
51. A. Mohanta and D. Behera, Fluctuation induced magneto-conductivity studies in $YBa_2Cu_3O_{7-\delta} + xBaZrO_3$, *Physica C*, **470**, 295–303 (2010).
52. G.K. Bichile, D.G. Kuberkar, K.M. Jadhav, S.S. Shah, and R.G. Kulkarni, Influence of nickel substitution on flux pinning and critical currents in $YBa_2Cu_3O_{7-\delta}$, *Cryogenics*, **31**, 833–838 (1991).
53. K. Nagasao, T. Matsui, and S. Tajima, Rapid change of electronic anisotropy in overdoped $(Y, Ca)Ba_2Cu_3O_{7-\delta}$, *Physica C*, **468**, 1188–1191 (2008).
54. S.R. Ghorbani and E. Rostamabadi, The normal state transport properties of $NdBa_{2-x}La_xCu_3O_{7-\delta}$: Evidence of localization hole by La, *Physica C*, **468**, 60–65 (2008).
55. S. Adachi, T. Sugano, A. Fukuoka, N. Seiji, M. Itoh, P. Laffez, and H. Yamauchi, A simple route to single-phase $YBa_2Cu_4O_8$ at ambient pressure, *Physica C*, **233**, 149–154 (1994).
56. X.G. Zheng, M. Suzuki, C. Xu, H. Kuriyaki, and K. Hirakawa, Synthesis of Ca-substituted $Y_{1-x}Ca_xBa_2Cu_4O_8$ at ambient pressure using CuI, *Physica C*, **271**, 272–276 (1996).
57. C. Michel, M. Hervieu, M.M. Borel, A. Grandin, F. Deslandes, J. Provost, and B. Raveau, Superconductivity in the Bi–Sr–Cu–O system, *Z. Phys. B*, **68**, 421–423 (1987).
58. Y. Arao, M. Tange, M. Yokoshima, H. Ikeda, and R. Yoshizaki, Optimization of the Bi-2201 superconductor with Pb and La co-doping, *Physica C*, **426–431**, 351–354 (2005).
59. S.A. Sunshine, T. Siegrist, L.F. Schneemeyer, D.W. Murphy, R.J. Cava, B. Batlogg, R.B. van Dover, R.M. Fleming, S.H. Glarum, S. Nakahara, R. Farrow, J.J. Krajewski, S.M. Zahurak, J.V. Waszczak, J.H. Marshall, P. Marsh, L.W. Rupp, Jr., and W.F. Peck, Structure and physical properties of single crystals of the 84-K superconductor $Bi_{2.2}Sr_2Ca_{0.8}Cu_2O_{8+\delta}$, *Phys. Rev. B*, **38**, 893–896 (1988).

60. Z.A. Xu, J.Q. Shen, S. Ooi, T. Mochiku, and K. Hirata, Pseudogap opening, superconducting fluctuation and vortex-like excitation above T_c in slightly overdoped $Bi_2Sr_2CaCu_2O_{8+\delta}$, *Physica C,* **421**, 61–66 (2005).
61. M.A. Quijada, D.B. Tanner, R.J. Kelley, and M. Onellion, *ab*-plane anisotropy in single-domain $Bi_2Sr_2CaCu_2O_8$ high-temperature superconductors, *Physica C,* **235–240**, 1123–1124 (1994).
62. H. Raffy, V. Toma, C. Murrills, and Z.Z. Li, *c*-axis resistivity of $Bi_2Sr_2CaCu_2O_y$ thin films at various oxygen doping: Phase diagram and scaling law, *Physica C,* **460–462**, 851–853 (2007).
63. T. Horiuchi, K. Kitahama, and T. Kawai, Preparation of Li-containing $Bi_2Sr_2CaCu_2O_8$ single crystals and their Li solubility limit, *Physica C,* **221**, 143–148 (1994).
64. K. Mori, M. Sasakawa, T. Igarashi, Y. Isikawa, K. Sato, K. Noto, and Y. Muto, Thermal conductivity, electrical resistivity and specific heat of high-T_c $Bi_{1.4}Pb_{0.6}Sr_2Ca_2Cu_3O_x$ superconductor, *Physica C,* **162–164**, 512–513 (1989).
65. T. Uzumaki, K. Yamanaka, N. Kamehara, and K. Niwa, Preparation and magnetic properties of Bi–Pb–Sr–Ca–Cu–O superconducting ceramics, *Appl. Phys. Lett.,* **54**, 2253–2255 (1989).
66. D.S. Ginley, J.F. Kwak, R.P. Hellmer, R.J. Baughman, E.L. Venturini, and B. Morosin, Sequential electron beam evaporated films of $Tl_2CaBa_2Cu_2O_y$ with zero resistance at 97 K, *Appl. Phys. Lett.,* **53**, 406–408 (1988).
67. M. Hong, S.H. Liou, D.D. Bacon, G.S. Grader, J. Kwo, A.R. Kortan, and B.A. Davidson, Superconducting Tl–Ba–Ca–Cu–O films by sputtering, *Appl. Phys. Lett.,* **53**, 2102–2104 (1988).
68. T. Kaneko, H. Yamauchi, and S. Tanaka, Zero-resistance temperature of Tl-based "2223" superconductor increased to 127 K, *Physica C,* **178**, 377–382 (1991).
69. R. Vijayaraghavan, N. Rangavittal, G.U. Kalkurni, E. Grantscharova, T.N. Guru Row, and C.N.R. Rao, Structure and superconducting properties of $Tl_{1-y}Pb_yY_{1-x}Ca_xSr_2Cu_2O_7$, *Physica C,* **179**, 183–190 (1991).
70. R.L. Meng, L. Beauvais, X.N. Zhang, Z.J. Huang, Y.Y. Sun, Y.Y. Xue, and C.W. Chu, Synthesis of the high-temperature superconductors $HgBa_2CaCu_2O_{6+\delta}$ and $HgBa_2Ca_2Cu_3O_{8+\delta}$, *Physica C,* **216**, 21–28 (1993).
71. M. Rupp, A. Gupta, and C.C. Tsuei, Magnetic field induced broadening of resistive transition in epitaxial *c*-axis-oriented $HgBa_2CaCu_2O_{6+\delta}$ films, *Appl. Phys. Lett.,* **67**, 291–293 (1995).
72. B.W. Kang, A.A. Gapud, X. Fei, T. Aytug, and J.Z. Wu, Minimization of detrimental effect of air in $HgBa_2CaCu_2O_{6+\delta}$ thin film processing, *Appl. Phys. Lett.,* **72**, 1766–1768 (1998).
73. K. Isawa, A. Tokiwa-Yamamoto, M. Itoh, S. Adachi, and H. Yamauchi, Encapsulation method for the synthesis of nearly single-phase superconducting $HgBa_2Ca_2Cu_3O_{8+\delta}$ with $T_c \geq 135$ K, *Physica C,* **222**, 33–37 (1994).
74. Y.-Y. Xie, J.Z. Wu, A.A. Gapud, Y. Yu, and Y. Xin, Synthesis of Hg-1223 superconductors using a cation-exchange process, *Physica C,* **322**, 19–24 (1999).
75. Z.J. Huang, R.L. Meng, X.D. Qiu, Y.Y. Sun, J. Kulik, Y.Y. Xue, and C.W. Chu, Superconductivity, structure and resistivity in $HgBa_2Ca_2Cu_3O_{8+\delta}$, *Physica C,* **217**, 1–5 (1993).

76. C.C. Lam, S.D. Wang, H.M. Shao, G.J. Shen, J.C.L. Chow, P.C.W. Fung, S.L. Ho, H.Y. Lai, and K.C. Hung, TEM observation of intergrowth structures in the $HgBa_2Ca_2Cu_3O_y$ ceramic superconductor, *Physica C,* **247**, 43–54 (1995).
77. A. Tokiwa-Yamamoto, T. Tatsuki, X.-J. Wu, S. Adachi, and K. Tanabe, Effect of Tl doping on the phase stability of $Hg_2Ba_2(Y,Ca)_1Cu_2O_{8-\delta}$ superconductors prepared by a high-pressure technique, *Physica C,* **259**, 36–42 (1996).
78. T. Kawashima, Y. Matsui, and E. Takayama-Muromachi, New oxycarbonate superconductors $(Cu_{0.5}C_{0.5})Ba_2Ca_{n-1}Cu_nO_{2n+3}$ ($n = 3, 4$) prepared at high pressure, *Physica C,* **224**, 69–74 (1994).
79. E. Takayama-Muromachi, Y. Matsui, and K. Kosuda, New oxyborate superconductor, $BSr_2Ca_3Cu_4O_{11}$ ($T_c = 110$ K) prepared at high pressure, *Physica C,* **241**, 137–141 (1995).
80. A. Iyo and M. Tokumoto, Preparation and superconductivity of $(B_{1-x}C_x)(Sr_{1-y}Ba_y)_2Ca_2Cu_3O_z$, *Physica C,* **311**, 35–41 (1999).
81. T. Ito, K. Takenaka, and S. Uchida, Systematic deviation from *T*-linear behaviour in the in-plane resistivity of $YBa_2Cu_3O_{7-y}$: Evidence for dominant spin scattering, *Phys. Rev. Lett.,* **70**, 3995–3998 (1993).
82. R. Wesche and A.M. Fuchs, Design of superconducting current leads, *Cryogenics,* **34**, 145–154 (1994).
83. J.G. Hust, Thermal conductivity and thermal diffusivity, in *Materials at Low Temperatures*, R.P. Reed and A.F. Clark (eds.), American Society for Metals, Metals Park, OH, USA, 1983.
84. C.P. Poole, Jr., H.A, Farach, and R.J. Creswick, *Superconductivity*, Academic Press, San Diego, 1995.
85. S. Yang, B. Chen, E.E. Hellstrom, E. Stiers, and J.M. Pfotenhauer, Thermal conductivity and contact conductance, *IEEE Trans. Appl. Supercond.,* **5**, 1471–1474 (1995).
86. I.H. Gul, M. Anis-ur-Rehman, and A. Maqsood, Temperature dependence of thermal and electrical conductivity of Bi-based high-T_c (2223) superconductor, *Physica C,* **450**, 83–87 (2006).
87. S.D. Peacor and C. Uher, Thermal conductivity of Bi-Sr-Ca-Cu-O superconductors: Correlation with the low-temperature specific-heat behavior, *Phys. Rev. B,* **39**, 11559–11562 (1989).
88. C. Uher, S.D. Peacor, and J. Shewchun, Thermal conductivity of $Tl_2Ba_2Ca_2Cu_3O_{10}$ ceramics from 300 K down to 0.1 K, *Physica C,* **177**, 23–26 (1991).
89. B. Sundqvist and B.M. Andersson, Thermal conductivity of polycrystalline $YBa_2Cu_4O_8$ and the phonon transport model, *Physica C,* **235–240**, 1377–1378 (1994).
90. F. Yu, V.N. Kopylov, M.B. Salamon, N.N. Kolesnikov, M.A. Hubbard, H.M. Duan, and A.M. Hermann, Field dependent thermal conductivity of $Tl_2Ba_2CuO_6$: Evidence for Andreev scattering of quasiparticles, *Physica C,* **267**, 308–312 (1996).
91. M.F. Crommie and A. Zettl, Thermal-conductivity anisotropy of single-crystal $Bi_2Sr_2CaCu_2O_8$, *Phys. Rev. B,* **43**, 408–412 (1991).
92. D.T. Morelli, G.L. Doll, J. Heremans, M.S. Dresselhaus, A. Cassanho, D.R. Gabbe, and H.P. Jenssen, Anisotropic thermal conductivity of superconducting lanthanum cuprate, *Phys. Rev. B,* **41**, 2520–2523 (1990).

93. K. Mori, A. Tanaka, K. Nishimura, J. Sakurai, T. Sasaki, N. Kobayashi, Y. Tanaka, and M. Mimura, Thermal conductivity and electrical resistivity of *c*-axis oriented polycrystalline $(Bi,Pb)_2Sr_2Ca_2Cu_3O_x$ in a magnetic field, *Physica C,* **235–240**, 1509–1510 (1994).
94. M. Matsukawa, K. Iwasaki, K. Noto, T. Sasaki, N. Kobayashi, K. Yoshida, K. Zikihara, and M. Ishihara, Thermal conductivity of *c*-axis aligned $(Bi,Pb)_2Sr_2Ca_2Cu_3O_y$ in the superconducting and mixed states, *Cryogenics,* **37**, 255–262 (1997).
95. L. Tewordt and Th. Wölkhausen, Theory of thermal conductivity of the lattice for high-T_c superconductors, *Solid State Commun.*, **70**, 839–844 (1989).
96. S. Castellazzi, M.R. Cimberle, C. Ferdeghini, E. Giannini, G. Grasso, D. Marrè, M. Putti, and A.S. Siri, Thermal conductivity of a BSCCO(2223) *c*-oriented tape: A discussion on the origin of the peak, *Physica C,* **273**, 314–322 (1997).
97. M. Matsukawa, F. Tatezaki, K. Noto, H. Fujishiro, K. Michishita, and Y. Kubo, Thermal conductivity of Ag-doped Bi-2212 superconducting materials prepared by the floating zone method, *Cryogenics,* **34**, 685–688 (1994).
98. K. Krishana, N.P. Ong, Q. Li, G.D. Gu, and N. Koshizuka, Plateaus observed in the field profile of thermal conductivity in the superconductor $Bi_2Sr_2CaCu_2O_8$, *Science,* **277**, 83–85 (1997).
99. A. Waske, C. Hess, B. Büchner, V. Hinkov, and C.T. Lin, Thermal conductivity of underdoped $YBa_2Cu_3O_y$, *Physica C,* **460–462**, 746–747 (2007).
100. T. Chakraborty, B. Gahtori, A. Sony, G.S. Okram, S.K. Agarwal, Y.-S. Chen, Y.-K. Kuo, Geetha , A. Rao, and C.K. Sarkar, Electrical and thermal properties of bulk superconductors $Eu_{0.95}Pr_{0.05}Ba_2(Cu_{1-x}M_x)_3O_{7-\delta}$ (M = Fe, Ni, Zn and Mn), *Physica C,* **470**, 244–249 (2010).
101. H. Fujishiro, S. Nariki, and M. Murakami, Thermal conductivity and thermoelectric power of DyBaCuO bulk superconductors, *Supercond. Sci. Technol.*, **19**, S447–S450 (2006).
102. T. Naito, H. Fujishiro, K. Iida, S Nariki, and M. Murakami, Thermal conductivity of Er–Ba–Cu–O and Ho–Ba–Cu–O superconducting bulks, *Supercond. Sci. Technol.*, **21**, 085001 (2008).
103. S.J. Hagen, Z.Z. Wang, and N.P. Ong, Anisotropy of the thermal conductivity of $YBa_2Cu_3O_{7-y}$, *Phys. Rev. B,* **40**, 9389–9392 (1989).
104. M. Matsukawa, H. Noto, A. Tamura, H. Furusawa, X. Yao, S. Nimori, N. Kobayashi, and Y. Shiohara, In-plane thermal conductivity of large single crystals of Sm-substituted $(Y_{1-x}Sm_x)Ba_2Cu_3O_{7-\delta}$, *Supercond. Sci. Technol.*, **19**, 777–782 (2006).

12

Thermoelectric and Thermomagnetic Effects

12.1 Introduction

In normal conducting metals and semiconductors the transport of charge is always accompanied by heat transport, which leads to the Seebeck, Peltier, and Thomson effects. The presence of a magnetic field leads to the Nernst effect, the thermal analog of the Hall effect. In this chapter, these thermoelectric effects will be briefly described. In the absence of a magnetic field the current density $\mathbf{j}$ in an isotropic conductor can be written as [1]

$$\frac{1}{\rho}\mathbf{j} = -S\nabla T - \nabla\left(\varphi - \frac{\mu}{e}\right) \tag{12.1}$$

Here ρ is the electrical resistivity, S is the Seebeck coefficient (thermopower or thermoelectric power), φ is the electrical potential, μ is the chemical potential and e is the absolute value of the electron charge. At absolute zero the chemical potential is equal to the Fermi energy. Independent of temperature, the value of the Fermi–Dirac distribution function is $f(E) = \{\exp[(E - \mu)/k_BT] + 1\}^{-1} = 0.5$ for $E = \mu$. In an open circuit the current density is zero and the electric field generated by a temperature gradient along the conductor is

$$\mathbf{E}_\mathrm{e} = S\nabla T \tag{12.2}$$

where $\mathbf{E}_\mathrm{e} = \nabla(\varphi - \mu/e)$. The thermopower originates from three different physical causes. The first is that the mobile charge carriers preferentially diffuse from the warm to the cold end of the conductor. The second is the dependence of the chemical potential of the electrons on temperature. The third cause is that electrons are carried toward the cold end by the phonons (phonon drag effect).

Physical Properties of High-Temperature Superconductors, First Edition. Rainer Wesche.

In a thermocouple, two metal junctions 1 and 2 are held at two different temperatures T_1 and T_2 (see Figure 12.1, top). The voltage U generated in the circuit is

$$U = \int_{T_1}^{T_2} (S_B(T) - S_A(T))\, dT \tag{12.3}$$

where S_A and S_B are the Seebeck coefficients of metals A and B, respectively. In the case of a small temperature difference, the temperature dependences of the Seebeck coefficients can be neglected, and the thermovoltage can be approximately written as $U \approx (S_B - S_A)(T_2 - T_1)$. On closing the circuit, the thermocurrent at the hot junction would flow from the metal with lower Seebeck coefficient to that with the higher one. Figure 12.1 shows the polarity that results for $T_2 > T_1$ and $S_A > S_B$. When one of the junctions is held at $T_1 = 0°C$, the thermocouple can be used to measure the temperature T_2. It should be noted that the thermopower vanishes in the superconducting state because the Cooper pairs do not transport heat.

In the absence of a magnetic field, the thermal current density $\mathbf{j_Q}$, the temperature gradient along the conductor, and the current density $\mathbf{j}$ are related by [1]

$$\mathbf{j_Q} = -k\nabla T + \Pi\mathbf{j} + \left(\varphi - \frac{\mu}{e}\right)\mathbf{j} \tag{12.4}$$

where Π is the Peltier coefficient and k is the thermal conductivity. From Equation 12.4, we obtain for the heat flux

$$\frac{dQ}{dt} = -kA\frac{dT}{dx} + \Pi I + UI \tag{12.5}$$

Here A is the cross-section of the conductor and I is the current through it. In a conductor carrying the current I, the Peltier heat $Q_P = \Pi I t$ is generated in addition to the Joule heat $U I t$ (where t is the time). In contrast to the Joule heating, the sign of the Peltier heat depends on the current direction. We now consider the closed circuit shown in Figure 12.1 (bottom). It is supposed that $T_2 > T_1$ and $\Pi_A > \Pi_B$. At the junction with the lower temperature T_1 the current would flow from metal A with the higher Peltier coefficient to metal B with the lower one. The Peltier heat $Q_P = (\Pi_A - \Pi_B)\, It > 0$ is absorbed by the right junction at T_1. The absorbed power due to the Peltier effect is

$$\frac{dQ_P}{dt} = (\Pi_A - \Pi_B)I = \Pi_{BA} I \tag{12.6}$$

where Π_A and Π_B are the Peltier coefficients of conductors A and B, respectively. The Peltier heat $Q_P = (\Pi_B - \Pi_A)\, It = \Pi_{AB}\, I\, t < 0$ is emitted at the left junction at the higher temperature T_2.

A conductor with a temperature gradient along its length, which carries a current, emits, or absorbs heat. This phenomenon is called the Thomson effect. The heat production q per unit volume and unit time is

$$q = \rho j^2 - \mu_T j\, \frac{dT}{dx} \tag{12.7}$$

Here ρ is the resistivity of the conductor, j is the current density, and μ_T is the Thomson coefficient.

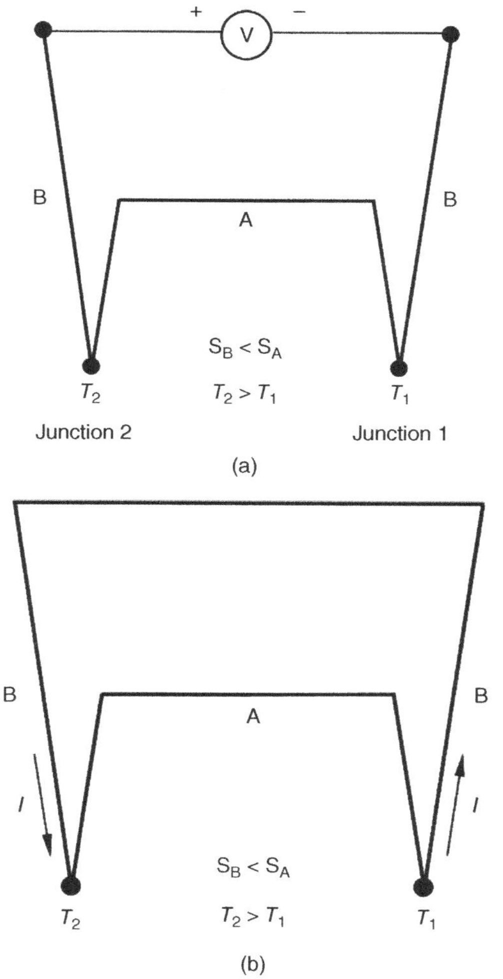

Figure 12.1 *Top: sketch of a thermocouple with the two junctions 1 and 2 of metals A and B at temperatures T_1 and T_2, respectively. At the metal interfaces 1 and 2, electrons move from the metal with the lower work function to that with the higher one. For $T_1 = T_2$, the thermovoltages at the two junctions cancel each other. In general, one junction (T_1) is kept at 0 °C and the thermovoltage can be used to measure T_2. Bottom: in a closed circuit with two junctions of metals A and B, there flows a thermocurrent I. In the junction with the higher temperature T_2 the current flows from metal B with the lower Seebeck coefficient to metal A with the higher one. Peltier heat is emitted at the left junction (T_2) and absorbed at the right junction (T_1)*

The Peltier, the Seebeck, and the Thomson coefficients are connected by the Thomson relations, the first of which is [1]

$$\mu_T = -T\frac{dS}{dT} \tag{12.8}$$

The second Thomson relation is [1]

$$\Pi = S \cdot T \tag{12.9}$$

Next we consider a nonmagnetic normal conductor with a magnetic field B_z applied transverse to a temperature gradient dT/dx (see Figure 12.2). For zero current, an electric field E_{ey} perpendicular to both the applied magnetic field and the temperature gradient is generated. The Nernst coefficient ν is defined as

$$\nu = \frac{E_{ey}}{B_z\left(-\frac{dT}{dx}\right)} \tag{12.10}$$

The Nernst effect is very small in normal metals but can be of importance in Type II superconductors because of the contribution of the vortices [2]. The vortex Nernst effect is illustrated in Figure 12.3. In the vortex liquid state, the vortices are mobile and diffuse from the warm to the cold end. Each of the vortex cores is encircled by screening currents. Due to the superposition of the thermal drift velocity v of the vortices and the velocity v_e of the electrons relative to the center of the vortex, they result in different velocities $v_t = v + v_e$ and $v_t = v - v_e$ on the two sides of the vortex (see Figure 12.3). As a consequence, there exists a net Lorentz force along the y direction, which causes an electric field $\mathbf{E_e} = \mathbf{v} \times \mathbf{B}$ [2].

The Nernst coefficient can be rewritten as [3, 4]

$$\nu = \frac{E_{ey}}{B_z\left(-\frac{dT}{dx}\right)} = \left[\frac{\alpha_{xy}}{\sigma} - S\tan\theta\right]\frac{1}{B_z} \tag{12.11}$$

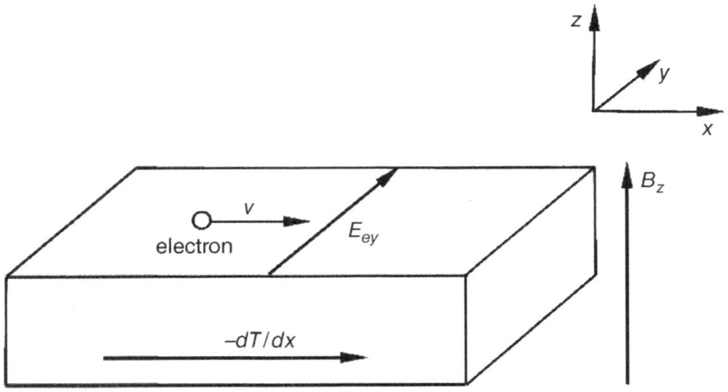

Figure 12.2 Illustration of the Nernst effect. For a magnetic field applied along the z axis and a thermal gradient along the –x axis, the conduction electrons diffuse from the warm to the cold end and are deflected by the Lorentz force. As a consequence, an electric field E_{ey} is generated

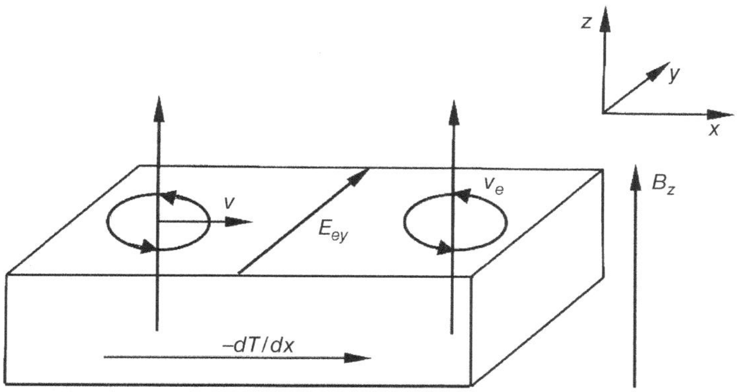

Figure 12.3 *Sketch of the vortice Nernst effect in a Type II superconductor. The vortices diffuse with velocity v from the warm to the cold end. Due to the superposition of the drift velocity v and the electron velocity v_e around the vortex core, there results for a magnetic field along z a net Lorentz force along y (adapted from [5])*

Here α_{xy} is the off-diagonal Peltier conductivity in the units (A $K^{-1}m^{-1}$) ($j_y = \alpha_{xy}(-dT/dx)$), $\theta = \sigma_{xy}/\sigma$ is the Hall angle (σ diagonal element, σ_{xy} off-diagonal element of the conductivity tensor), and S is the thermopower (Seebeck coefficient, units (μV K^{-1})). The off-diagonal term includes the normal state term α^n_{xy} and the vortex contribution α^s_{xy}. The investigation of the Nernst effect can provide information on the existence of vortices originating from superconducting fluctuations above the critical temperature in the pseudogap region. A more detailed description of the Nernst effect in cuprate high-T_c superconductors can be found in [5].

12.2 Thermoelectric Power of Cuprate Superconductors

The investigation of the thermopower provides a deeper insight into properties of the normal state. Figure 12.4 shows the Seebeck coefficients of Y-123 [6], Dy-123 [7], and neodymium-doped Bi-2234 [8]. For the Y-123 samples, a positive Hall number has been found, indicating that electrical conduction is dominated by mobile holes. The Hall number depends nearly linearly on temperature [6]. The Seebeck coefficients of the two Y-123 samples S1 and S2 are considerably different. For sample S2, the Seebeck coefficient S is always positive for temperatures between T_c and 300 K. As expected, the value of S is zero below the transition temperature. Sample S1 shows a negative value of S after 20 thermal cycles between ≈80 K and 300 K. After reannealing of sample S1 in an oxygen atmosphere at 500 °C for 12 h, S becomes very small, and positive above ≈190 °C. The results suggest that two bands with carriers of opposite charge contribute to the thermoelectric power in Y-123 [6]. At low temperatures, the sign of S is determined by the product of the charge sign and the sign of the energy derivative $dD(E)/dE$ of the density of states at the Fermi energy. The sign of $dD(E)/dE$ depends on the exact oxygen content of the sample [6].

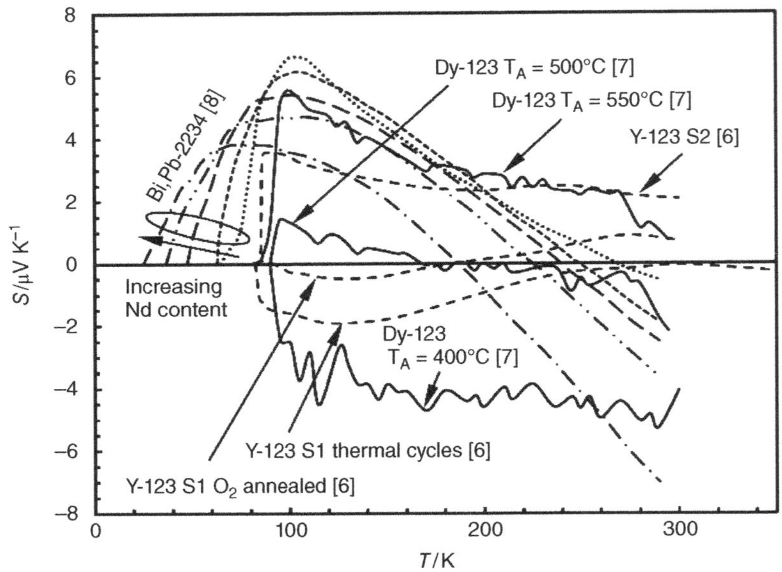

Figure 12.4 *Temperature dependence of the Seebeck coefficient of Y-123, Dy-123, and neodymium-doped (Bi,Pb)-2234 (data from [6–8])*

The Dy-123 bulk samples [7] are mixtures of $DyBa_2Cu_3O_{7-\delta}$ and Dy_2BaCuO_5 (Dy-211) with a molar ratio of Dy-123:Dy-211 = 100:5. The data shown in Figure 12.4 are the *ab* plane Seebeck coefficients. As in the case of the Y-123, the Seebeck coefficient is zero below the critical temperature and depends at higher temperatures on the heat treatment conditions. The samples were heat treated in flowing oxygen at temperatures of 400 °C, 500 °C, and 550 °C for 100 h. The sample heat treated at 400 °C is characterized by a negative Seebeck coefficient, whereas annealing at 550 °C leads to positive values. The Dy-123 sample annealed at 500 °C shows an intermediate behavior, with negative S below 170 K and positive above that. The samples heat treated at higher temperatures are expected to have a larger oxygen deficiency (δ). The sign change of the thermoelectric power has been attributed to variations in the oxygen deficiency [7].

The data obtained for $Bi_{1.7}Pb_{0.3-x}Nd_xSr_2Ca_3Cu_4O_{12+y}$ show the effect of different neodymium substitutions ($0 \leq x \leq 0.1$) on the thermoelectric power of (Bi,Pb)-2234 [8]. In the sample without Nd substitution, the highest peak value of the thermopower was found. The height of the peak decreases with increasing Nd content. In addition, the critical temperature decreases with increasing substitution of neodymium for lead. At high temperatures, the sign of the Seebeck coefficient changes from positive to negative. The thermopower data were successfully described within a two-band model [8]. The theoretical description is based on a narrow band close to the Fermi energy, which is superimposed on a broad band. The Seebeck coefficient can be well described above the peak position by [8]

$$S = \frac{AT}{B^2 + T^2} + \alpha_1 T \tag{12.12}$$

Here $A{\cdot}T$ and B/T are related to the contributions of mobile holes and semiconductor-like electrons, respectively, to electrical conduction. The second term $\alpha_1 \cdot T$ describes a normal band contribution to the thermopower. In Equations 12.13 and 12.14, E_0 is the center of the narrow band and Γ is its width

$$A = 2\frac{(E_0 - E_F)}{e} \tag{12.13}$$

$$B^2 = 3\frac{(E_0 - E_F)^2 + \Gamma^2}{\pi^2 {k_B}^2} \tag{12.14}$$

A, B, α_1, $(E_0 - E_F)$ and Γ were used as fitting parameters. The values obtained for the sample with $x = 0$ are [8]: $A = 1208.4\ \mu\text{V}$, $B = 65.6\ \text{K}$, $\alpha_1 = -0.016\ \mu\text{V/K}^2$, $E_0 - E_F = 7.03\ \text{K}$ and $\Gamma = 118.78\ \text{K}$. The corresponding values for the sample with $x = 0.1$ are [8]: $A = 3442.1\ \mu\text{V}$, $B = 166.7\ \text{K}$, $\alpha_1 = -0.0550\ \mu\text{V/K}^2$, $E_0 - E_F = 20.0\ \text{K}$ and $\Gamma = 301.70\ \text{K}$.

Figure 12.5 shows the Seebeck coefficients of Hg-1212 [9], Eu-123 [10], $NdBa_{2-x}La_xCu_3O_{7-\delta}$ [11] and $Bi_{2-x}Sr_xCuO_{6+\delta}$ [12]. Similar to the data presented in Figure 12.4 the thermoelectric power is very sensitive to the exact chemical composition and the heat treatment conditions. The thermopower of the $HgBa_2CaCu_2O_{6+\delta}$ sample as grown is positive and reaches a value of ~12 μV/K at 140 K. The Seebeck coefficient drops with increasing temperature and reaches a value of <7 μV/K at 300 K. After an additional annealing for 10 h at 200 °C in oxygen the critical temperature was shifted from 120 to 125 K. In addition, the peak value of the Seebeck coefficient was reduced from 12 to 9.5 μV/K and reached a value of 5 μV/K at 300 K. A second heat treatment for

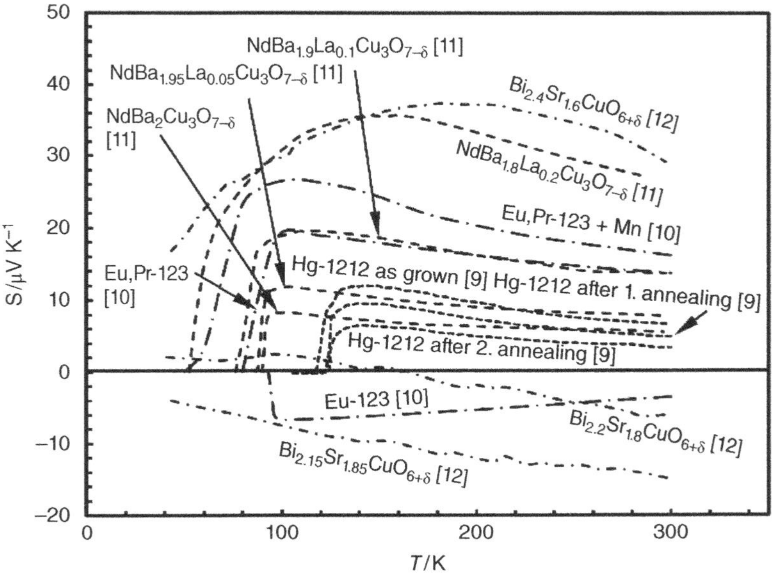

Figure 12.5 *Temperature- and doping-dependences of the Seebeck coefficient of Hg-1212, Eu-123, Nd-123, and Bi-2201 (data from [9–12])*

35 h at 200 °C in oxygen increased the critical temperature slightly to 126 K, whereas the thermoelectric power is further reduced. The peak value is reduced to ≈6.5 μV/K, while the value of S is < 3.5 μV/K at 300 K.

For the pure $EuBa_2Cu_3O_{7-\delta}$ (Eu-123) sample, the thermoelectric power is negative for temperatures between T_c and 300 K. Praseodymium doping $(Eu_{0.95}Pr_{0.05})Ba_2Cu_3O_{7-\delta}$ ((Eu,Pr)-123)) led to a reduction of T_c and a sign change of the thermoelectric power. Partial replacement of Cu by Mn ($(Eu_{0.95}Pr_{0.05})Ba_2(Cu_{1-0.02}Mn_{0.02})_3O_{7-\delta}$) reduces the critical temperature further. Moreover, the values of the Seebeck coefficient are shifted to even higher positive values than that in the case of (Eu,Pr)-123.

In the $NdBa_{2-x}La_xCu_3O_{7-\delta}$ system, the thermoelectric power is positive and increases with increasing x, while the critical temperature decreases with increasing lanthanum content. In addition, the position of the maximum of $S(T)$ is shifted to higher temperatures with increasing x. The thermoelectric power, measured at 290 K, can be used to determine the hole concentration [11]. The results are presented in Table 12.1.

The thermoelectric power of $Bi_{2+x}Sr_{2-x}CuO_{6+\delta}$ varies considerably with substitution of bismuth for strontium. For $x = 0.15$, the values of S are negative in the whole temperature range of 40–300 K. The thermopower drops from −4 μV/K at 40 K to − 15 μV/K at 300 K. On the other hand, the thermoelectric power is always positive for $x = 0.4$. The Seebeck coefficient reaches a maximum of ~37 μV/K at 180 K. The thermopower of $Bi_{2.2}Sr_{1.8}CuO_{6+\delta}$ is positive at low temperatures, but negative above ≈ 160 K.

As a last example, we consider the thermoelectric power of the electron-doped cuprate superconductor $Nd_{2-x}Ce_xCuO_{4-\delta}$ [13–15]. The temperature dependence of the thermoelectric power of several samples of $Nd_{2-x}Ce_xCuO_{4-\delta}$ is shown in Figure 12.6. For samples A, B, and C, characterized by $x = 0.15$, the thermoelectric power is positive. The maximum values of S exceed 2 μV/K for sample C after annealing at 900 °C in flowing nitrogen for 6 h. The peak value of S is only 1.7 μV/K for the as-grown sample C. The single-crystal sample A was deoxygenized by annealing for 1 h at 1080 °C in flowing nitrogen, leading to lower values of S than for the optimally annealed sample for which $T_c = 24$ K. In spite of the fact that the optimum annealing was applied to the two single crystals A and B, the T_c of B (24.4 K) is slightly higher than that of A. Furthermore, the thermoelectric power of single crystal B is lower than that of single crystal A. Single crystal B was oxygenated by an additional heat treatment at 640 °C in oxygen, reducing its thermoelectric power. For sample D, the thermopower is small and positive at temperatures slightly above T_c but

Table 12.1 *Variation of the Seebeck coefficient and the number of holes per copper atom p with lanthanum doping in the $NdBa_{2-x}La_xCu_3O_{7-\delta}$ system [11]*

Doping x	$S(290\ K)$ (μV/K)	Hole concentration p
0	5.5	0.136
0.05	7.9	0.127
0.10	13.7	0.113
0.15	18.6	0.105
0.20	27.4	0.095
0.30	35.1	0.088

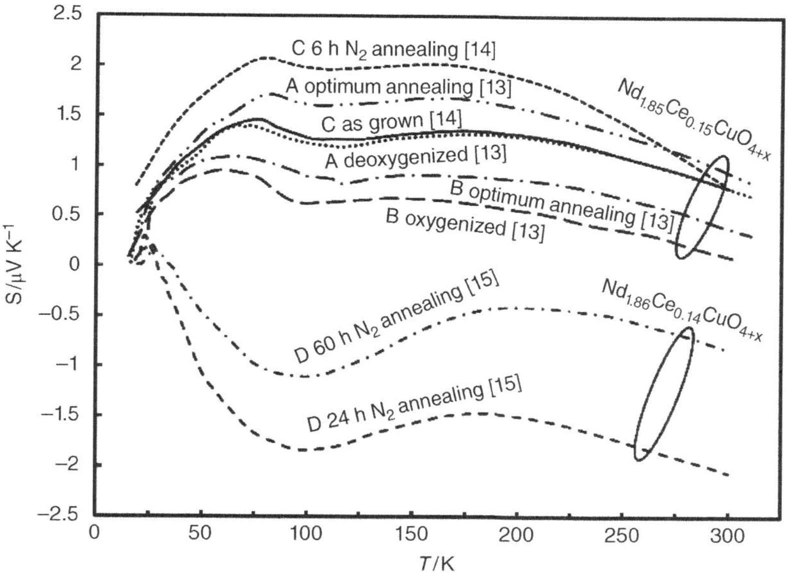

Figure 12.6 *Seebeck coefficient of the nominally electron-doped cuprate superconductor* $Nd_{2-x}Ce_xCuO_{4+x}$ *versus temperature (data from [13–15])*

negative at higher temperatures, reaching values of −0.8 (after 60 h annealing at 900 °C under flowing nitrogen) and −2 μV/K (24 h at 300 K in nitrogen).

The thermoelectric power of samples A and B at higher temperatures was analyzed within a two-band model similar to that used for the (Bi,Pb)-2234 samples (see above). The total thermopower is [13]

$$S = \frac{\sigma_h}{\sigma_h + \sigma_e} S_h + \frac{\sigma_e}{\sigma_h + \sigma_e} S_e \tag{12.15}$$

Here S_h and S_e are the Seebeck coefficients for holes and electrons, and σ_h and σ_e the electrical conductivities for the hole and the electron bands, respectively. The narrow hole-like band contribution can be written as $S_h = A \cdot T/(B^2 + T^2)$, where A and B are defined in Equations 12.13 and 12.14. The resulting expression for the total thermoelectric power is then [13]

$$S = \left(\frac{\sigma_h}{\sigma_h + \sigma_e}\right)\left(\frac{AT}{B^2 + T^2}\right) + \frac{\sigma_e}{\sigma_h + \sigma_e}\alpha_1 T \tag{12.16}$$

Well above 100 K the dependence of the thermoelectric power of $Nd_{1.85}Ce_{0.15}CuO_{4-\delta}$ on temperature can be well represented by Equation 12.16 (for more details, see [13]).

12.3 Nernst Effect

In Chapter 8, the generic phase diagram of cuprate high-T_c superconductors was discussed. Measurements of the tunneling conductance show the existence of a gap-like feature well above the critical temperature, especially in underdoped cuprates. One hypothesis to explain the occurrence of the pseudogap is based on the assumption that Cooper pairs are already formed at temperatures well above T_c at such small number density that phase coherence cannot be established. The superconducting state is characterized not only by the formation of Cooper pairs but also by the fact that all Cooper pairs occupy the same quantum state. The collective of the phase coherent Cooper pairs can be described by the wave function $\psi = n_C{}^{1/2} \exp(i\varphi(\mathbf{r}))$ (see also Chapter 3). The phase coherence of the Cooper pairs is effective over long distances, and hence the phase $\varphi(\mathbf{r})$ is a macroscopically observable quantity (flux quantization, Josephson effects). The Nernst effect allows us to search for Cooper pairs without phase coherence that are formed at temperatures above T_c. Their existence would be reflected in the presence of vortices, leading to a Nernst effect much larger than in normal conductors.

The Nernst effect has been studied in such materials as $La_{2-x}Sr_xCuO_4$ [3, 5, 16], $YBa_2Cu_3O_{7-\delta}$ [4, 5, 17, 18], $(Y,Ca)Ba_2(Cu,Zn)_3O_{7-\delta}$ [19], $Bi_2Sr_{2-y}La_yCuO_{6+\delta}$ [5], $Bi_2Sr_2CaCu_2O_{8+\delta}$ [5, 20, 21], $SmBa_2Cu_3O_{6+\delta}$ [22], and $Nd_{2-x}Ce_xCuO_{4-\delta}$ [5]. Figure 12.7 shows the Nernst signal, defined as $e_y = E_{ey}/(dT/dx)$, of various RE-123 (RE = Y or another rare earth element) superconductors in a magnetic field of 6 T. For all these materials, there exists a pronounced positive peak at temperatures around the transition temperature T_c, caused by the contribution of vortices to the Nernst signal. The vortex contribution depends strongly on the applied magnetic field (see, e.g., [5]). As soon as the temperature is low enough, the vortices are pinned, and hence their contribution to

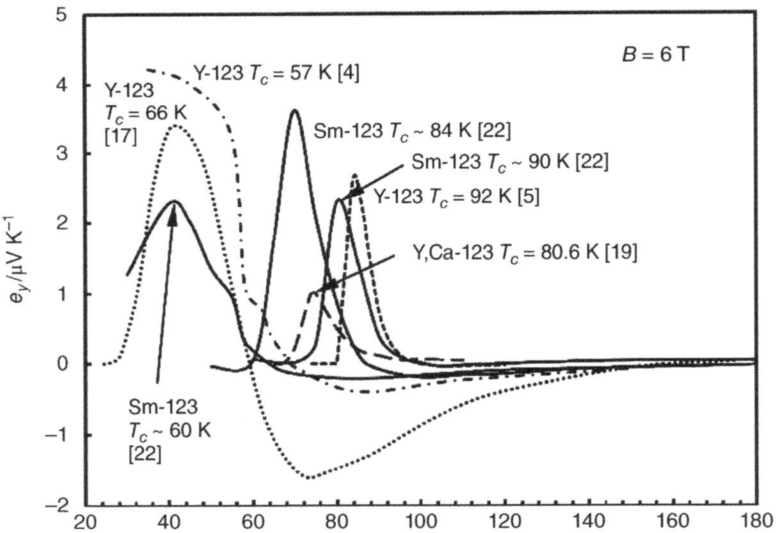

Figure 12.7 Nernst signal $e_y = E_{ey}/(dT/dx)$ of various RE-123 (RE = Y or another rare earth element) superconductors at $B = 6$ T versus temperature. The large positive peak is caused by vortices. At the lowest temperatures, the vortices are pinned and hence do not contribute to the Nernst signal. The typically small Nernst signal at higher temperatures originates from quasi-particles (unpaired electrons) (data from [4, 5, 17, 19, 22])

the Nernst signal vanishes at the lowest temperatures. The typically small Nernst signal at high temperatures is caused by quasi-particles (unpaired electrons). The peak values of the Nernst signal are reached slightly below T_c and reach values between ≈ 1 and ≈ 4 μV/K.

Figure 12.8 shows the Nernst signals of $La_{2-x}Sr_xCuO_4$ ($B = 6$ T), Bi-2212 ($B = 5$ T), and Bi-2223 ($B = 6$ T) superconductors. As in Figure 12.7, there exists a pronounced positive peak caused by the vortex contribution to the Nernst signal. The peak values at $B = 6$ T are between 4.8 and 6.2 μV/K for $La_{2-x}Sr_xCuO_4$, while they are in the range of 1.4–2.4 μV/K for Bi-2212 at $B = 5$ T. A maximum value of 2.6 μV/K was found for Bi-2223 at $B = 6$ T.

Interestingly, a significant vortex contribution to the Nernst signal was observed at temperatures well above T_c, especially in underdoped cuprate superconductors. This provides evidence that, due to superconducting fluctuations, there exist some non-phase-coherent (preformed) Cooper pairs well above T_c in the pseudogap region. However, the onset temperature for the formation of preformed Cooper pairs is well below the temperature T^*, where the pseudogap opens. These results suggest that the value of T_c of hole-doped cuprate superconductors is determined by the loss of long-range phase coherence of the Cooper pairs [5].

Very recent investigations of the Nernst effect have focused more on the quasi-particle contribution to the Nernst signal. Thus for nontwinned Y-123 single crystals ($T_c = 66$ K) (see Figure 12.7), Daou *et al.* found a pronounced negative quasi-particle contribution with an onset temperature well above T_c [17]. This quasi-particle contribution was found to be strongly anisotropic depending on whether the thermal gradient was parallel to the

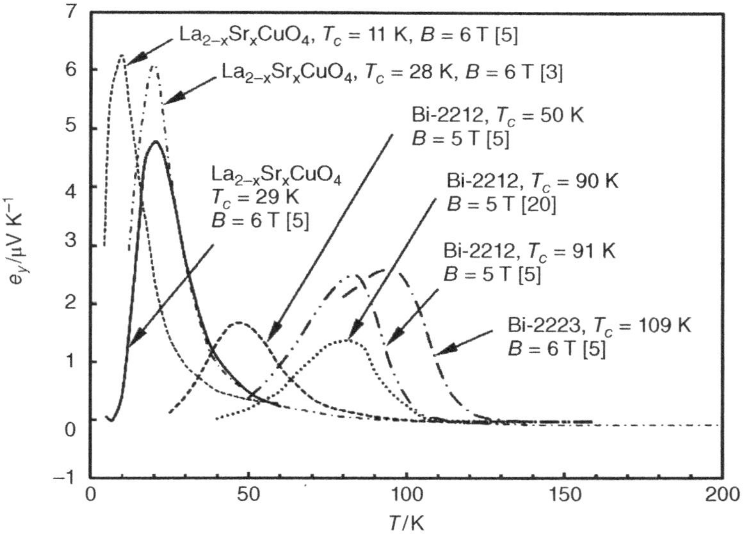

Figure 12.8 *Nernst signal $e_y = E_{ey}/(dT/dx)$ of $La_{2-x}Sr_xCuO_4$ ($B = 6$ T), Bi-2212 ($B = 5$ T), and Bi-2223 superconductors ($B = 6$ T) versus temperature. The large positive peak is caused by vortices. At the lowest temperatures the vortices are pinned and hence do not contribute to the Nernst signal. The typically small Nernst signal at higher temperatures originates from quasi-particles (unpaired electrons) (data from [3, 5, 20])*

crystallographic *a*- or *b*-axis. Daou et al. attributed this extremely anisotropic dip in the Nernst signal to the occurrence of charge stripes. The significantly negative Nernst signal sets in at a temperature equal to T^*, where the pseudogap opens.

12.4 Summary

The Seebeck effect can be used to study the normal-state properties of cuprate superconductors, whereas the Nernst effect provides deeper insight into the superconducting state. As expected, the thermoelectric power vanishes in the superconducting state. In general, the Seebeck coefficient depends strongly on the heat treatment, the exact chemical composition, and the oxygen content. Depending on these factors, the Seebeck coefficient may be positive or negative or even change sign with temperature. The results suggest that both hole-like and electron-like bands contribute to current transport. The high temperature values of the Seebeck coefficient can be typically explained by a two-band model.

The Nernst effect can be used to study the formation of preformed Cooper pairs well above the transition temperature T_c. Typically, a large vortex contribution to the Nernst signal was observed around the critical temperature T_c, which vanished only at temperatures well above T_c. These results indicate that Cooper pairs can exist above T_c due to superconducting fluctuations. The loss of superconductivity at T_c seems to be caused by the loss of long-range phase coherence. However, the onset temperature for the formation of preformed pairs is lower than the temperature T^* for the opening of the pseudogap.

Very recent investigations suggest that the quasi-particle (single electron) contribution to the Nernst effect is sensitive to charge ordering phenomena (e.g. charge stripes) in the pseudogap region.

References

1. B.M. Jaworski and A.A. Detlaf, Physik Griffbereit, Verlag Vieweg + Sohn GmbH, Braunschweig, Germany, 1972.
2. C.P. Poole, Jr., H.A, Farach, and R.J. Creswick, Superconductivity, Academic Press, San Diego, 1995.
3. Z.A. Xu, N.P. Ong, Y. Wang, T. Kakeshita, and S. Uchida, Vortex-like excitations and the onset of superconducting phase fluctuation in underdoped $La_{2-x}Sr_xCuO_4$, *Nature*, **406**, 486–488 (2000).
4. F. Rullier-Albenque, R. Tourbot, H. Alloul, P. Lejay, D. Colson, and A. Forget, Nernst effect and disorder in the normal state of high-T_c cuprates, *Phys. Rev. Lett.*, **96**, 067002 (2006).
5. Y. Wang, L. Li, and N.P. Ong, Nernst effect in high-T_c superconductors, *Phys. Rev. B*, **73**, 024510 (2006).
6. U. Gottwick, R. Held, G. Sparn, F. Steglich, H. Rietschel, D. Ewert, B. Renker, W. Bauhofer, S. von Molnar, M. Wilhelm, and H.E. Hoenig, Transport properties

of $YBa_2Cu_3O_7$: Resistivity, thermal conductivity, thermopower and Hall effect, *Europhys. Lett.*, **4**, 1183–1188 (1987).

7. H. Fujishiro, S. Nariki, and M. Murakami, Thermal conductivity and thermoelectric power of DyBaCuO bulk superconductors, *Supercond. Sci. Technol.*, **19**, S447–S450 (2006).
8. B. Özkurt, B. Özçelik, K. Kıymaç, M. Ali Aksan, and M. Eyyüphan Yakıncı, Thermoelectric power and thermal conduction studies on the Nd substituted BPSCCO (2234) superconductors, *Physica C*, **467**, 112–119 (2007).
9. Y.T. Ren, J. Clayhold, F. Chen, Z.J. Huang, X.D. Qiu, Y.Y. Sun, R.L. Meng, Y.Y. Xue, and C.W. Chu, Thermopower and resistivity measurements on oxygen-annealed $HgBa_2CaCu_2O_{6+\delta}$, *Physica C*, **217**, 6–10 (1993).
10. T. Chakraborty, B. Gahtori, A. Soni, G.S. Okram, S.K. Agarwal, Y.-S. Chen, Y.-K. Kuo, Geetha, A. Rao, and C.K. Sarkar, Electrical and thermal properties of bulk superconductors $Eu_{0.95}Pr_{0.05}Ba_2(Cu_{1-x}M_x)_3O_{7-\delta}$ (M = Fe, Ni, Zn and Mn), *Physica C*, **470**, 244–249 (2010).
11. S.R. Ghorbani and E. Rostamabadi, The normal state transport properties of $NdBa_{2-x}La_xCu_3O_{7-\delta}$: Evidence of localization hole by La, *Physica C*, **468**, 60–65 (2008).
12. J. Hejtmánek, M. Nevřiva, E. Pollert, D. Sedmidubský, and P. Vašek, Superconductivity and electric transport of $Bi_{2+x}Sr_{2-y}CuO_{6+\delta}$ single crystals, *Physica C*, **264**, 220–226 (1996).
13. L. Cheng, H.S. Yang, C.H. Sun, H.X. Gao, J.B. Wang, and X.D. Chen, Thermopower and resistivity for $Nd_{1.85}Ce_{0.15}CuO_{4-\delta}$ single crystals: Existence of a hole like band, *Physica C*, **468**, 1024–1027 (2008).
14. J. Liu, H.S. Yang, Y.S. Chai, C.H. Sun, H.X. Gao, L. Cheng, J.B. Wang, and L.Z. Cao, Transport properties in $Nd_{1.85}Ce_{0.15}CuO_4$ single crystals under different annealing conditions, *Supercond. Sci. Technol.*, **22**, 015013 (2009).
15. C.H. Sun, H.S. Yang, L. Cheng, J.B. Wang, and L.Z. Cao, A study of the transport properties and the annealing effect in $Nd_{1.86}Ce_{0.14}CuO_{4+\delta}$ single crystal, *Supercond. Sci. Technol.*, **21**, 085018 (2008).
16. O. Cyr-Choinière, R. Daou, F. Laliberté, D. LeBoeuf, N. Doiron-Leyraud, J. Chang, J.-Q. Jan, J.-G. Cheng, J.-S. Zhou, J.B. Goodenough, S. Pyon, T. Takayama, H. Tagaki, Y. Tanaka, and L. Taillefer, Enhancement of the Nernst effect by stripe order in a high-T_c superconductor, *Nature*, **458**, 743–745 (2009).
17. R. Daou, J. Chang, D. LeBoeuf, O. Cyr-Choinière, F. Laliberté, N. Doiron-Leyraud, B.J. Ramshaw, R. Liang, D.A. Bonn, W.N. Hardy, and L. Taillefer, Broken rotational symmetry in the pseudogap phase of a high-T_c superconductor, *Nature*, **463**, 519–522 (2010).
18. J. Chang, R. Daou, C. Proust, D. LeBoeuf, N. Doiron-Leyraud, F. Laliberté, B. Pingault, B.J. Ramshaw, R. Liang, D.A. Bonn, W.N. Hardy, H. Takagi, A.B. Antunes, I. Sheikin, K. Behnia, and L. Taillefer, Nernst and Seebeck coefficients of the cuprate superconductor $YBa_2Cu_3O_{6.67}$: A study of Fermi surface reconstruction, *Phys. Rev. Lett.*, **104**, 057005 (2010).

19. I. Kokanović, J.R. Cooper, and M. Matusiak, Nernst effect measurements of epitaxial $Y_{0.95}Ca_{0.05}Ba_2(Cu_{1-x}Zn_x)_3O_y$ and $Y_{0.9}Ca_{0.1}Ba_2Cu_3O_y$ superconducting films, *Phys. Rev. Lett.*, **102**, 187002 (2009).
20. Z.A. Xu, J.Q. Shen, S. Ooi, T. Mochiku, and K. Hirata, Pseudogap opening, superconducting fluctuation and vortex-like excitation above T_c in slightly overdoped $Bi_2Sr_2CaCu_2O_{8+\delta}$, *Physica C*, **421**, 61–66 (2005).
21. G. Bridoux, P. Pedrazzini, F. De la Cruz, and G. Nieva, Angular field dependence of the Nernst effect in $Bi_2Sr_2CaCu_2O_{8+\delta}$, *Physica C*, **460–462**, 841–842 (2007).
22. Z.A. Xu, E. Ahmed, Z.W. Zhu, J.Q. Shen, and X. Yao, Nernst effect in oxygen-depleted $SmBa_2Cu_3O_{6+x}$ single crystals, *Physica C*, **460-462**, 833–834 (2007).

13

Specific Heat

13.1 Introduction

In Section 2.5, thermodynamics of the superconducting state was briefly described. In the simplest case, the specific heat of a metal is the sum of the lattice specific heat (phonon contribution) and the contribution of the conduction electrons. As already discussed in Section 2.5, the phonon contribution is proportional to T^3 at low temperatures. In the Debye model, the crystal lattice is considered as a continuous isotropic medium. The Debye frequency $\omega_D = 2\pi\nu_D$ is the upper limit for the frequencies that can exist in a solid, corresponding to the low-wavelength limit. The Debye frequency can be used to define the Debye temperature $\theta_D = h\nu_D/k_B$, where h is Planck's constant and k_B is Boltzmann's constant. The expression $C_{\rm ph} = 234R(T/\theta_D)^3$ (R is the gas constant) is a good approximation for the phonon specific heat at temperatures $< \theta_D/10$. At temperatures well above the Debye temperature, the phonon specific heat approaches the classical value of $3R$. Because of Pauli's exclusion principle, thermal excitation is possible only for a small fraction of the conduction electrons close to the Fermi energy. As a consequence, the conduction electrons contribute typically $<1\%$ to the total specific heat at room temperature. The electronic specific heat, which depends linearly on temperature, is the dominating contribution to the specific heat at sufficiently low temperatures. In Section 2.5, we have already shown that there exists a jump in the specific heat at the critical temperature.

For materials capable of magnetization, there exists a magnetic contribution $M\,dH$ (M is the magnetization and H is the magnetic field strength) to the total energy of the system, and hence a magnetic contribution to the specific heat [1]. In paramagnetic salts, a Schottky anomaly has been observed. The Schottky specific heat at low temperatures may be significantly larger than the electronic and the lattice specific heats. The Schottky effect originates, e.g., from ions having two energy levels separated by a very small energy difference $k_B T_{\rm Sch}$. In a small temperature interval close to $T_{\rm Sch}$ there will occur many transitions from the lower to the higher energy level, which leads to a large contribution to the specific heat. At temperatures much higher than $T_{\rm Sch}$ the two levels are nearly equally occupied and the contribution of transitions between the two levels to the specific heat will be negligible [1].

Physical Properties of High-Temperature Superconductors, First Edition. Rainer Wesche.

It is beyond the scope of this chapter to give a detailed description of all physical phenomena related to the specific heat of solids. An introduction to the aspects related to specific heat can be found in [1]. In Section 13.2, low-temperature specific heat data of various cuprate superconductors will be presented. Section 13.3 is devoted to the change of the specific heat at the transition, including measurements with an applied magnetic field. Specific heat data for a larger temperature range, frequently required for engineering purposes, are presented in Section 13.4. Section 13.5 provides a short discussion of the specific heat of the cuprate high-T_c superconductors.

13.2 Specific Heat at Low Temperatures

In this section, low-temperature specific heat data of various cuprate superconductors are presented. In general, the measurement of the specific heat is performed at constant pressure, while in theoretical expressions typically the specific heat at constant volume is considered. For crystalline solids, the difference of C_p and C_v is <10% at room temperature, decreases with decreasing temperature, and finally vanishes at zero temperature [1]. In the case of aluminum, the difference of C_p and C_v is 4.55% at room temperature and drops to 0.48% at 80 K [1]. For simplicity, we will not distinguish between C_p and C_v in what follows. The specific heat data will be presented in the form C/T versus T^2. In the normal conducting state and in the absence of a Schottky effect, the specific heat is the sum of the electronic contribution $C_{\mathrm{el}} = \gamma T$ and the phonon specific heat $C_{\mathrm{ph}} = A_{\mathrm{ph}} T^3$, and hence in a plot of C/T versus T^2 there results a straight line. The slope provides $A_{\mathrm{ph}} = 234R/\theta_D{}^3$ (where R is the gas constant and θ_D is the Debye temperature), while γ is obtained from the intercept at $T = 0$.

Figure 13.1 shows low-temperature specific heat data of various Y-123 samples [2–7]. An upturn of the specific heat has been found at the lowest temperatures (Schottky anomaly). The Schottky contribution has been ascribed to the presence of $BaCuO_2$ as an impurity phase [2, 5]. Omitting the Schottky anomaly, the low-temperature specific heat can be approximated by $C = \gamma T + A_{\mathrm{ph}} T^3$. An interesting feature is the fact that linear extrapolation from above the Schottky anomaly to zero temperature leads to a nonzero γ value. For a metallic superconductor, it would be expected that the specific heat at $T << T_c$ is the sum of an exponential electronic term and a cubic lattice term [2]. The measured specific heat can be affected by the presence of small amounts of impurity phases (e.g., CuO, $BaCuO_2$). In Figure 13.1, the specific data for Y-123 pellets without and with the addition of 23 wt% of silver are compared. The measured specific heat of the Y-123 silver composite is slightly larger than the value obtained from the fractions of silver and Y-123 in the mixture and the specific heat data of both constituents [6]. The waviness of some C/T versus T^2 curves, presented in Figure 13.1, is an indicator of the scatter in the data. In Figure 13.1, the specific heat is provided in the unit mJ mole^{-1} K^{-1}. Another frequently used unit is mJ g at^{-1} K^{-1}. A gram atom of a substance includes N_A atoms, where $N_A = 6.02214 \times 10^{23}$ is the Avogadro number. One mole of $YBa_2Cu_3O_7$ contains 13 N_A atoms, and hence the molar mass is equal to 13 g atoms. The parameters used to describe the low-temperature specific heats of different Y-123 samples are gathered in Table 13.1. In general, the values of γ are strongly sample dependent. The obtained Debye temperatures of Y-123 are typically between 371 and 410 K.

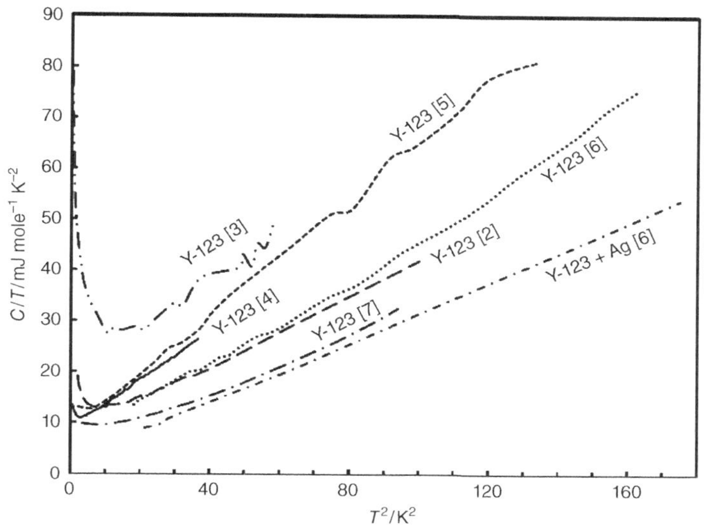

Figure 13.1 Low-temperature specific heat of various Y-123 samples presented as C/T versus T^2 (results from [2–7])

Table 13.1 Parameters used to describe the low-temperature specific heat of various Y-123 samples shown in Figure 13.1

Formula	γ(mJ mole^{-1}K^{-2})	θ_D(K)	Comments	Reference
$C/T = \gamma + A_{ph}T^2$ [a]	5.6	410	$40\,K^2 \le T^2 \le 100\,K^2$	[2]
$C/T = \gamma + A_{ph}T^2$ [a]	9.4	147	$0.09\,K^2 \le T^2 \le 64\,K^2$	[3]
$C/T = \gamma + A_{ph}T^2$ *	≈8.6[b]	≈374[b]	$0.16\,K^2 \le T^2 \le 36\,K^2$	[4]
$C/T = \gamma + A_{ph}T^2$ [a]	4.3 ± 0.2[c]	371 ± 3 [c]	$0.16\,K^2 \le T^2 \le 36\,K^2$	[4]
$C/T = \gamma + A_{ph}T^2$ [a]	7.3	410	$25\,K^2 \le T^2 \le 81\,K^2$	[6]

[a]Without Schottky contribution
[b]Single transition
[c]Double transition (data not shown in Figure 13.1).

Figure 13.2 shows the low-temperature specific heat data of various Bi-2212 and (Bi,Pb)-2223 samples [8–12]. The data are presented as C/T versus T^2. Again, the low-temperature specific heat varies considerably from sample to sample. Some of the samples show an upturn of the specific heat at the lowest temperatures. The low-temperature specific heat of a Bi-2212 single crystal taken from [8] can be well represented by the expression

$$C = \gamma T + A_{ph1}\, T^3 + A_{ph2}\, T^5 \tag{13.1}$$

The fit for the temperature range 1.25–4.5 K provides $\gamma = 2.07\,\mathrm{mJ\,mole^{-1}\,K^{-2}}$, $A_{ph1} = 1.29\,\mathrm{mJ\,mole^{-1}\,K^{-4}}$, and $A_{ph2} = 0.0228\,\mathrm{mJ\,mole^{-1}\,K^{-6}}$. The term $A_{ph2}\,T^5$ takes into account the deviation of the true phonon spectrum from the Debye model. Sasaki *et al.* [10] fitted the low-temperature specific heat data of Bi-2212 and (Bi,Pb)-2223 by

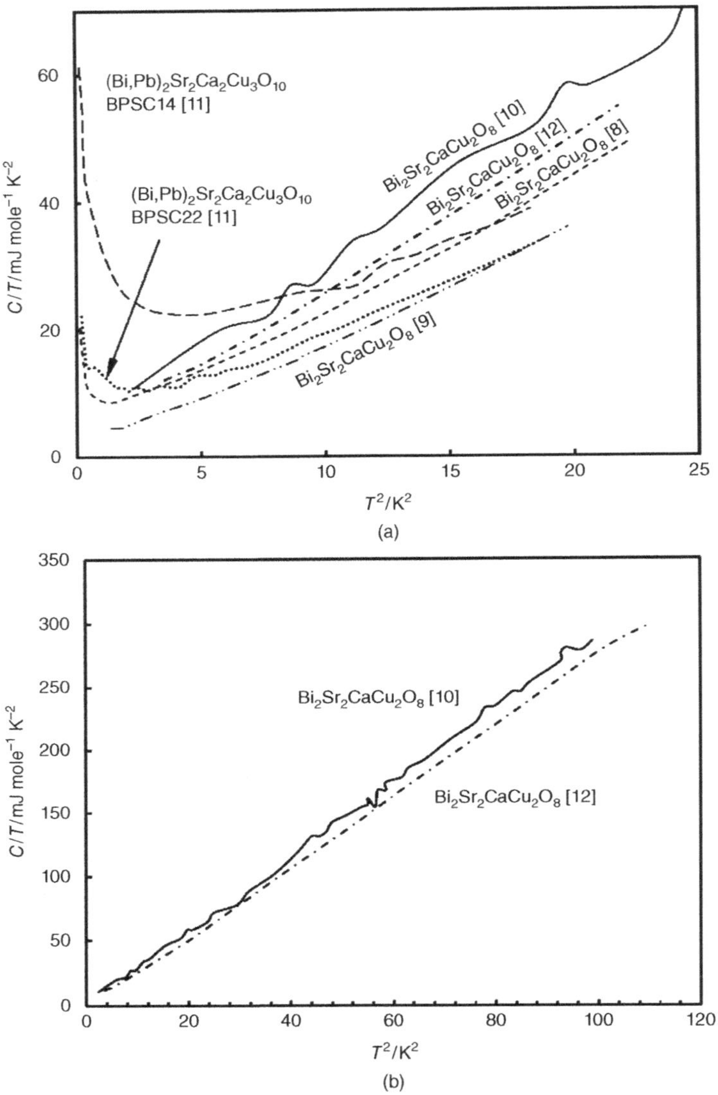

Figure 13.2 *Low-temperature specific heat of various bismuth-based superconductors presented as C/T versus T^2 (results from [8–12])*

the equation

$$C = \alpha T^{-2} + \gamma T + A_{\text{ph1}}\ T^3 + A_{\text{ph2}}\ T^5 \tag{13.2}$$

They found for Bi-2212 the fit parameters $\alpha = 22.2$ mJ K mole^{-1}, $\gamma = 0.1$ mJ mole^{-1} K^{-2}, $A_{\text{ph1}} = 2.83$ mJ mole^{-1}K^{-4}, and $A_{\text{ph2}} = 0.0016$ mJ mole^{-1}K^{-6}. The estimated Debye temperature is 217 K [10].

Table 13.2 Parameters used to represent the low-temperature specific heat of (Bi,Pb)-2223 based on Equation 13.3 in the temperature range of 2–14 K [11]

Sample	α(mJ mole^{-1}K)	γ(mJ mole^{-1}K^{-2})	θ_D(K)	T_E (K)	N_E(mJ mole^{-1}K^{-1})
BPSC14	50.86	10.44	292	54.07	5298
BPSC22	23.18	1.96	282	53.34	5701
BPSC22	38.02	0	276	54.43	5195

Collocott and Driver [11] described the specific heat data of Bi-2212 and (Bi,Pb)-2223 between 5 and 20 K by Equation 13.3

$$C = \alpha T^{-2} + \gamma T + A_{\text{ph}}\, T^3 + N_E \left(\frac{T}{T_E}\right)^2 \frac{\exp\left(\frac{T}{T_E}\right)}{\left[1 - \exp\left(\frac{T}{T_E}\right)\right]^2} \tag{13.3}$$

The fourth term represents an Einstein specific heat, which is characterized by the excitation of oscillators with a single frequency, where T_E is the Einstein temperature and N_E is the Einstein constant. The occurrence of an Einstein mode seems to be in line with the phonon dispersion curves of Bi-2212 with an optical mode in the zone center [11]. As in Equation 13.2, the first term approximates the Schottky contribution to the specific heat. The scaling parameters for the (Bi,Pb)-2223 samples BPSC14 and BPSC22 are listed in Table 13.2. In general, the parameters depend on the temperature range selected for the fit.

Figure 13.3 shows low-temperature specific heat data of Hg-1201 [13], Y-124 [5], Y-247 [14], La_2CuO_4 [15], and $La_{1.8}Sr_{0.2}CuO_4$ [16]. Again the data are represented as C/T versus T^2.

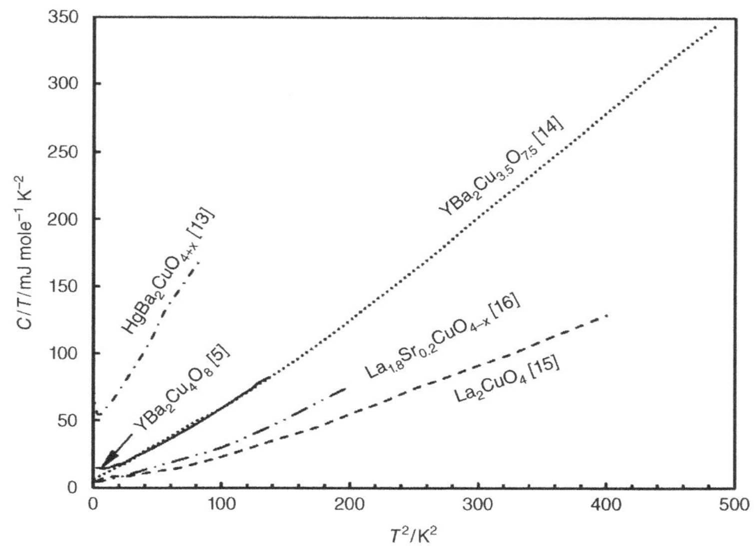

Figure 13.3 Low-temperature specific heat of various cuprate high-T_c superconductors presented as C/T versus T^2 (results from [5, 13–16])

13.3 Specific Heat Jump at the Transition to Superconductivity

In this section, the anomaly in the specific heat data close to the transition temperature T_c is considered. Figure 13.4 shows the specific heat of various cuprate superconductors close to their transition temperatures in the absence of an applied magnetic field [17–24]. In order to highlight the jump in the specific heat, the data are plotted as *C/T* versus *T*. The transition temperature of $YBa_2Cu_4O_8$ (Y-124) [17] is 81.4 K. A jump of $\Delta C/T_c$ of ~ 16 mJ mole^{-1} K^{-2} and a Debye temperature of 350 ± 10 K have been obtained. The Gd-123 [18], Dy-123 [19], and Y-123 A [19] samples show pronounced specific-heat anomalies at the transition temperatures. The values of $\Delta C/T_c$ are ≈44, ≈28, and ≈50 mJ mole^{-1} K^{-2} for Gd-123, Dy-123, and Y-123 A, respectively. For the sample Y-123 B, a smaller value of the specific heat jump normalized to T_c of only 17 mJ mole^{-1} K^{-2} was found. Especially for Bi-2212 the jump in the specific heat would be hardly visible in a plot of *C* versus *T* [23], while a small and broadened anomaly is found in the *C/T* versus *T* data. The sample (Bi,Pb)-2223 A is characterized by a critical temperature of 106.5 K, a Debye temperature of 245 K, and a specific heat anomaly $\Delta C/T_c$ of 33.6 mJ mole^{-1} K^{-2} [22]. In spite of the fact that T_c = 107 K of sample (Bi,Pb)-2223 B is nearly identical to that of sample A the value of $\Delta C/T_c$ = 24 mJ mole^{-1} K^{-2} [21] is significantly lower than that found for sample A. The jump in the specific heat of Tl-2223 normalized to the critical temperature of ≈108 K reaches a value of 52 mJ mole^{-1} K^{-2} [20].

To complete the picture, the results of specific heat measurements in the presence of a magnetic field are also presented. However, a detailed discussion of the in-field specific heat is beyond the scope of the present book. Figure 13.5 compares the specific heats of Bi-2212 single crystals measured with and without an applied magnetic field [9, 25]. For both Bi-2212 single crystals A [25] and B [9], a pronounced anomaly in the *C/T* versus *T* data is visible at the transition temperature in the absence of a magnetic field. A magnetic field of 14 T applied along the CuO_2 (*ab*) planes decreases the peak value of *C/T* only marginally. In addition, the hump in *C/T* around T_c is shifted to lower temperatures. This behavior has been attributed to superconducting fluctuations above T_c, which are of importance only in the absence of a magnetic field [25]. A magnetic field of 14 T applied along the crystallographic *c* direction strongly suppresses the specific heat anomaly around T_c. The in-field specific heat data illustrate the remarkable anisotropy of the superconducting properties in the cuprate high-T_c superconductors.

In Figure 13.6, the in-field and zero-field specific heat data of three Y-123 samples are compared [26–28]. In the absence of a magnetic field, all the three Y-123 samples exhibit a pronounced jump in the *C/T* versus *T* data. Applying a magnetic field of 20 T parallel to the crystallographic *c* direction suppresses the jump in *C/T* in the single crystal sample Y-123 A. For the single crystal sample B, measurements were performed at 14 T with the magnetic field parallel as well as perpendicular to the *ab* planes. The field parallel to the *ab* planes affects the jump in *C/T* only weakly, whereas the same field parallel to the *c* direction effectively suppresses the hump in *C/T*. In the case of the polycrystalline Y-123 sample, an intermediate magnetic field effect was observed. Such behavior can be expected from the different orientations of the grains in the polycrystalline sample with respect to the direction of the applied magnetic field of 14 T.

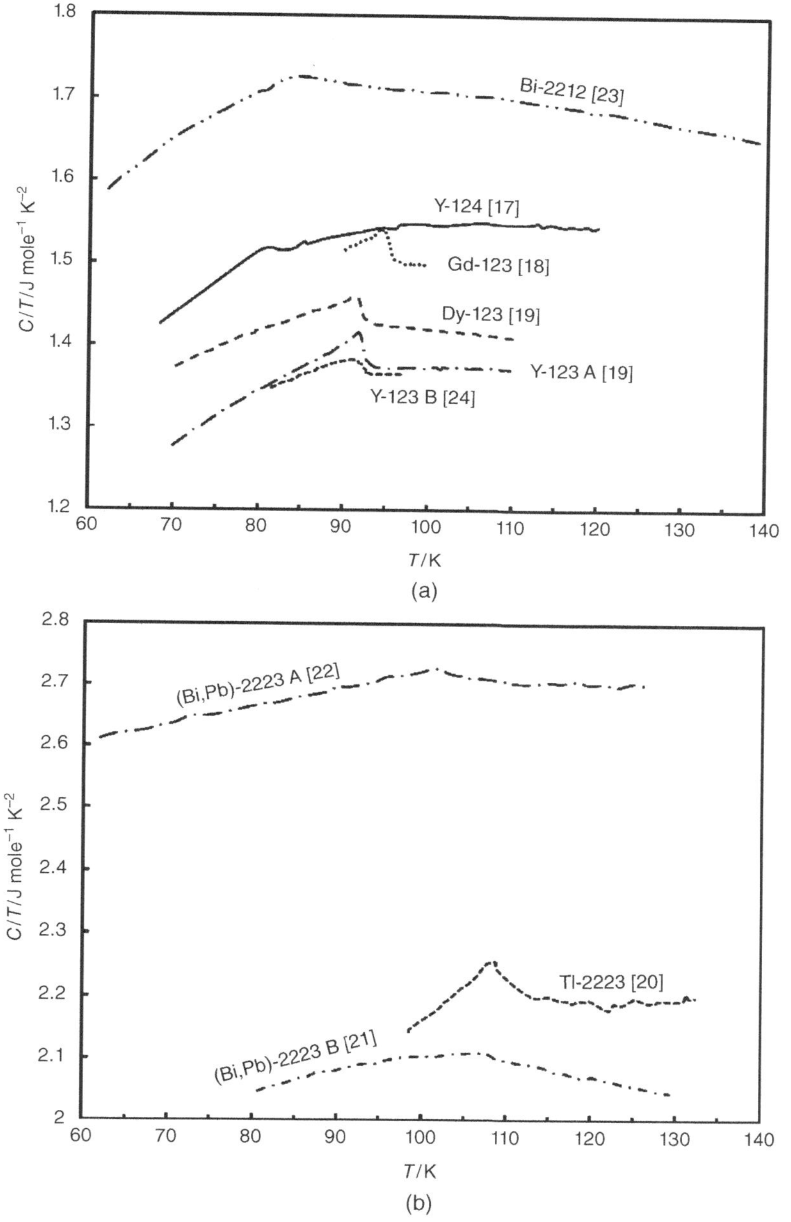

Figure 13.4 *Specific heat of various cuprate high-T_c superconductors close to the transition temperatures in question. The data (Bi-2212, RE-123 (RE = Y, Dy or Gd), Y-124 (top panel), Tl-2223, (Bi,Pb)-2223 bottom panel) are plotted as C/T versus T (results from [17–24])*

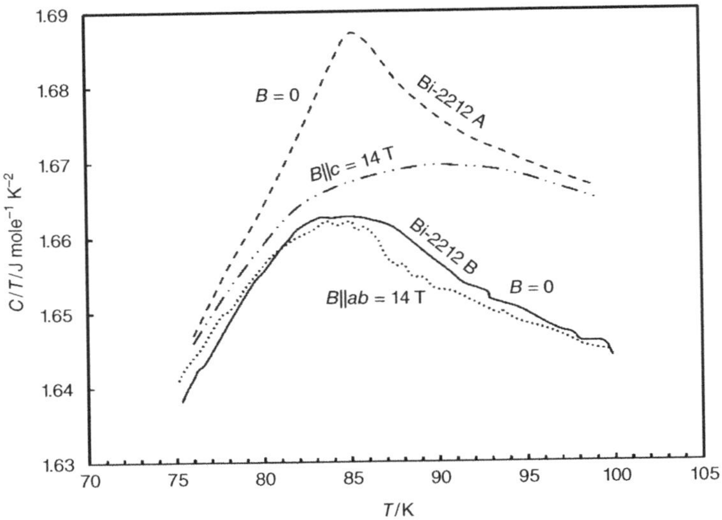

Figure 13.5 *Specific heats of two Bi-2212 single crystals in the presence and absence of an applied magnetic field in the vicinity of T_c. The data are presented as C/T versus T. A magnetic field of 14 T parallel to the c direction strongly suppresses the specific heat anomaly of Bi-2212 at T_c, whereas the same field parallel to the ab planes reduces the specific heat marginally and shifts the anomaly to slightly lower temperatures (results from [9, 25])*

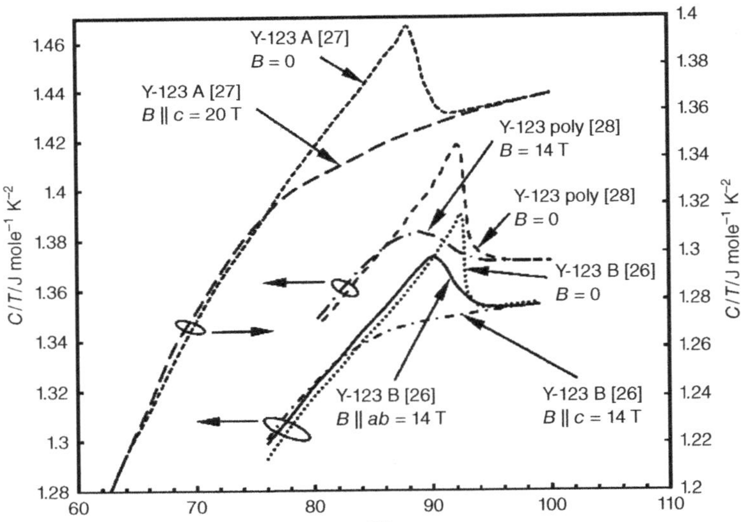

Figure 13.6 *Specific heats of various Y-123 samples in the presence and absence of an applied magnetic field in the vicinity of T_c. The data are plotted as C/T versus T. Magnetic fields parallel to the c direction strongly suppress the jump in the specific heat at T_c. The magnetic field effect is much less pronounced for fields parallel to the ab (CuO_2) planes (results from [26–28])*

13.4 Specific Heat Data up to Room Temperature

Finally, specific heat data for temperatures up to room temperature, frequently required for engineering purposes, will be presented. In contrast to the previous sections, the data will be presented as C versus T, and the specific heat will be in the unit $J\,g^{-1}K^{-1}$.

Figure 13.7 shows the specific heat data of various yttrium-based superconductors. The specific heats of all samples are very similar. For better readability, the specific heat data of Y-123 B [28], Y-123 A [6], Y-124 B [5], and Y-124 A [17] are shifted by 0.01, 0.02, 0.03, and $0.04\,J\,g^{-1}\,K^{-1}$, respectively. Only the data for Y-123 C are presented without an offset. The specific heats of the samples Y-123 B, Y-123-C, and Y-124 A all reach a value of $\approx 0.43\,J\,g^{-1}\,K^{-1}$ at room temperature.

Figure 13.8 shows the specific heat data of various bismuth-based cuprate high-T_c superconductors [9, 21–23, 29]. The specific heat of the sample Bi-2212 A reaches a value of $\sim 0.34\,J\,g^{-1}\,K^{-1}$ at 290 K. The specific heat of $0.42\,J\,g^{-1}\,K^{-1}(T \approx 290\,K)$ of the sample (Bi,Pb)-2223 A is comparable to the values found for the yttrium-based superconductors.

In Figure 13.9, the specific heat data of Tl-2201 [30], Tl-2223 [20], Hg-1201 [13], and Hg-1223 [31] are presented. At 200 K, the specific heat of these thallium- and mercury-based cuprate superconductors is between 0.26 and $0.37\,J\,g^{-1}\,K^{-1}$.

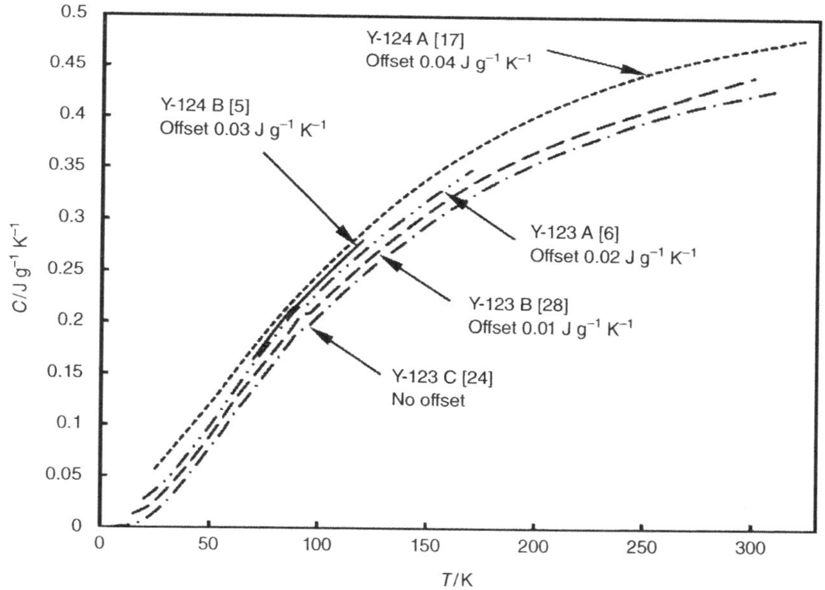

Figure 13.7 Specific heat versus temperature data of various yttrium-based cuprate superconductors (results from [5, 6, 17, 24, 28]). For better readability, the specific heat data of Y-123 B, Y-123 A, Y-124 B, and Y-124 A are shifted by 0.01, 0.02, 0.03, and $0.04\,J\,g^{-1}\,K^{-1}$, respectively

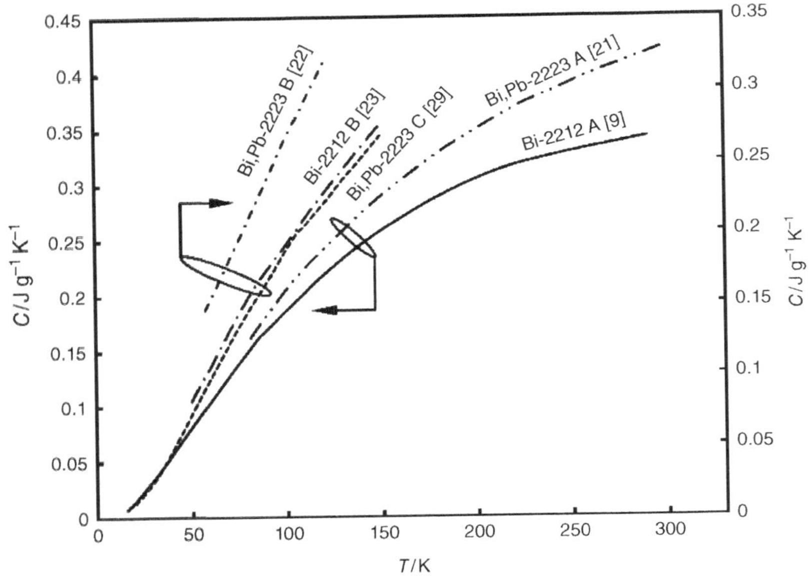

Figure 13.8 *Specific heat versus temperature of various bismuth-based cuprate superconductors (results from [9, 21–23, 29])*

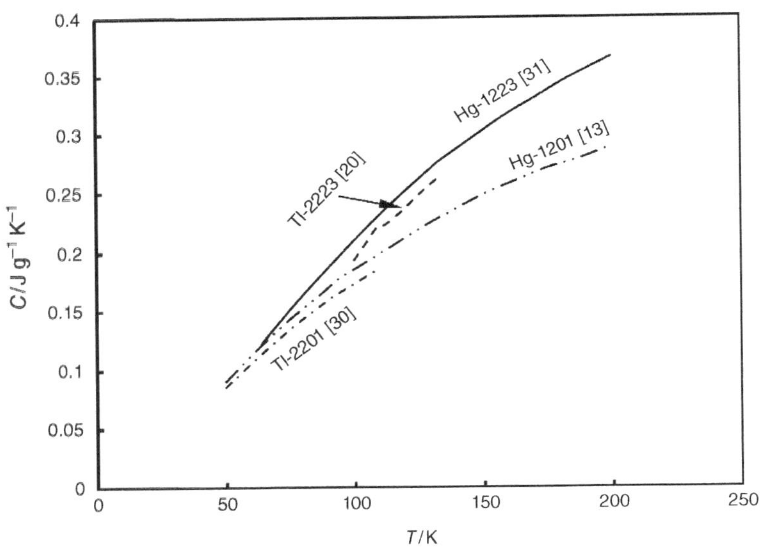

Figure 13.9 *Specific heat versus temperature of several thallium- and mercury-based cuprate superconductors (results from [13, 20, 30, 31])*

13.5 Summary

The specific heat of cuprate high-T_c superconductors shows a complex behavior at low temperatures. Frequently an upturn of the specific heat is observed at the lowest temperatures. This Schottky contribution depends strongly on minor phases present in the investigated samples. Another aspect is the occurrence of a linear term γT in the specific heat at low temperatures. The determination of the value of γ is difficult in the presence of a significant Schottky anomaly. In general, the Debye temperature, γ, and other fit parameters depend strongly on the temperature range.

RE-123 (RE is a rare earth element) cuprate high-T_c superconductors show a pronounced jump in the specific heat at the critical temperature T_c. On the other hand, only a broad hump was observed in the specific heat of Bi-2212 around T_c. This broadened specific heat anomaly has been attributed to superconducting fluctuations. A magnetic field applied along the crystallographic *c* direction strongly suppresses the specific heat anomaly at T_c, while the effect of magnetic fields applied parallel to the CuO_2 (*ab*) planes is much less pronounced.

The room temperature specific heat is typically in the range of 0.3–0.5 $J\,g^{-1}\,K^{-1}$. The Debye temperatures of cuprate high-T_c superconductors are typically in the range of 195–410 K [32].

References

1. L.L. Sparks, Specific heat, in *Materials at Low Temperatures*, R.P. Reed and A.F. Clark (eds.), American Society for Metals, Metals Park, 1983.
2. A. Junod, A. Bezinge, D. Cattani, M. Decroux, D. Eckert, M. François, A. Hewat, J. Muller, and K. Yvon, Structural, thermal and magnetic properties of the $YBa_2Cu_3O_{6.9}$ superconductor prepared by citrate pyrolysis, *Helvetica Physica Acta,* **61**, 460–469 (1988).
3. S. Vieira, S. Bourgeal, R. Villar, A. Aguilo, M.A. Ramos, C. Moure, and P. Duran, Low temperature thermal expansion and specific heat of a high T_c ceramic $Y_1Ba_2Cu_3O_{7-\delta}$, *Solid State Commun.,* **65,** 1555–1557 (1988).
4. M. Ishikawa, Y. Nakazawa, T. Takabatake, A. Kishi, R. Kato, and A. Maesono, Specific heat study on samples of $Ba_2YCu_3O_7$ with a double superconducting transition, *Solid State Commun.,* **66,** 201–204 (1988).
5. A. Schilling, A. Bernasconi, H.R. Ott, and F. Hulliger, Specific-heat, resistivity and magnetization study on polycrystalline $YBa_2Cu_4O_8$, *Physica C,* **169,** 237–244 (1990).
6. L. Ghivelder, P. Rodrigues Jr., P. Pureur, J.H. Souza, and S. Reich, Specific heat of superconducting YBaCuO–Ag nonrandom composite, *Physica C,* **194,** 97–104 (1992).
7. H. Gao, Y. Wang, L. Shan, X. Yao, and H.H. Wen, Low temperature specific heat of slightly overdoped $YBa_2Cu_3O_{7-\delta}$, *Physica C,* **432,** 293–298 (2005).
8. R. Caspary, A. Höhr, H. Spille, F. Steglich, H. Schmidt, R. Ahrens, G. Roth, H. Rietschel, Y.F. Yan, and Z.X. Zhao, Low-temperature specific heat in high magnetic field of $Bi_2Sr_2CaCu_2O_8$, *Europhys. Lett.,* **8,** 639–643 (1989).

9. A. Junod, K.-Q. Wang, T. Tsukamoto, G. Triscone, B. Revaz, E. Walker, and J. Muller, Specific heat up to 14 tesla and magnetization of a $Bi_2Sr_2CaCu_2O_8$ single crystal thermodynamics of a 2D superconductor, *Physica C,* **229,** 209–230 (1994).
10. Takato Sasaki, Y. Muto, T. Shishido, Takahiko Sasaki, T. Kajitani, M. Furuyama, N. Kobayashi, and T. Fukuda, Low temperature specific heat of single crystal $Bi_2Sr_2Ca_1Cu_2O_z$, *Physica C,* **162–164,** 504–505 (1989).
11. S.J. Collocott and R. Driver, Specific heat of the high-T_c superconductor $(Bi, Pb)_2Sr_2Ca_2Cu_3O_{10}$, and related phases Ca_2CuO_3 and $(Ca_{0.86}Sr_{0.14})CuO_2$ from 0.4 to 20 K, *Physica C,* **167,** 598–608 (1990).
12. J.S. Urbach, D.B. Mitzi, A. Kapitulnik, J.Y.T. Wei, and D.E. Morris, Low-temperature specific heat of single-crystal $Bi_2CaSr_2Cu_2O_8$ and $Tl_2Ca_2Ba_2Cu_3O_{10}$, *Phys. Rev. B,* **39,** 12 391–12 394 (1989).
13. B.F. Woodfield, C.W. Chu, R.A. Fisher, J.E. Gordon, S.B. Long, N.E. Philips, and Q. Xiong, The specific heat of $HgBa_2CuO_{4+\delta}$, *Physica C,* **235–240,** 1741–1742 (1994).
14. A. Junod, D. Sanchez, J.-Y. Genoud, T. Graf, G. Triscone, and J. Muller, Specific heat of the 95 K superconductor $YBa_2Cu_{3.5}O_{7.5}$ from 1 to 330 K, *Physica C,* **185–189,** 1399–1400 (1991).
15. M. Rosenberg, M. Mittag, R. Job, W. Chojetzki, R. Wernhardt, H. Sabrowsky, and R. Neubauer, A specific heat and magnetization study of high T_c ceramic superconductors, *Z. Phys. B,* **69,** 151–157 (1987).
16. J.M. Ferreira, B.W. Lee, Y. Dalichaouch, M.S. Torikachvili, K.N. Yang, and M.B. Maple, Low-temperature specific heat of the high-T_c superconductors $La_{1.8}Sr_{0.2}CuO_{4-\delta}$ and $RBa_2Cu_3O_{7-\delta}$ ($R = Y$, Eu, Ho, Tm, and Yb), *Phys. Rev. B,* **37,** 1580–1586 (1988).
17. A. Junod, D. Eckert, T. Graf, E. Kaldis, J. Karpinski, S. Rusiecki, D. Sanchez, G. Triscone, and J. Muller, Specific heat of the superconductor $YBa_2Cu_4O_8$ from 1.5 K to 330 K, *Physica C,* **168,** 47–56 (1990).
18. B. Gahtori, S.K. Agarwal, T. Chakraborty, A. Rao, and Y.-K. Kuo, Specific heat and correlation between resistivity and thermoelectric power of $GdBa_2(Cu_{1-x}Mn_x)_3O_{7-\delta}$ HTSC system for $x \leq 0.02$, *Physica C,* **469,** 27–29 (2009).
19. W. Schnelle, E. Braun, H. Broicher, R. Dömel, S. Ruppel, W. Braunisch, J. Harnischmacher, and D. Wohlleben, Fluctuation specific heat and thermal expansion of YBaCuO and DyBaCuO, *Physica C,* **168,** 465–474 (1990).
20. A.K. Bandyopadhyay, P. Maruthikumar, G.L. Bhalla, S.K. Agarwal, and A.V. Narlikar, Low temperature specific heat of single phase samples of bismuth and thallium based high-T_c cuprates, *Physica C,* **165,** 29–34 (1990).
21. J.E. Gordon, S. Prigge, S.J. Collocott, and R. Driver, Specific heat of $(Bi,Pb)_2Sr_2Ca_2 Cu_3O_{10-y}$ from 78 to 300 K, *Physica C,* **185–189,** 1351–1352 (1991).
22. M. Ausloos, M. Benhaddou, and R. Cloots, Specific heat of $Bi_{1.7}Pb_{0.3}Sr_2Ca_2Cu_3O_{10-\delta}$ by a new ac-differential method, *Physica C,* **235-240,** 1767–1768 (1994).
23. H. Kierspel, H. Winkelmann, T. Auweiler, W. Schlabitz, B. Büchner, V.H.M. Duijn, N.T. Hien, A.A. Menovsky, and J.J.M. Franse, Thermal expansion, specific heat, and uniaxial pressure dependences of T_c in $Bi_2Sr_2CaCu_2O_{8+\delta}$, *Physica C,* **262,** 177–186 (1996).
24. V.G. Bessergenev, Yu.A. Kovalevskaya, V.N. Naumov, and G.I. Frolova, Electron and phonon characteristics of $YBa_2Cu_3O_{7-\delta}$, *Physica C,* **245,** 36–40 (1995).

25. H. Kierspel, E. Kopanakis, B. Büchner, A. Freimuth, W. Schlabitz, V.H.M. Duijn, N.T. Hien, A.A. Menovsky, and J.J.M. Franse, Specific heat of $Bi_2Sr_2CaCu_2O_{8+x}$ in magnetic fields up to 16 Tesla, *Physica C,* **235–240,** 1765–1766 (1994).
26. E. Janod, A. Junod, K.-Q. Wang, G. Triscone, R. Calemczuk, and J.-Y. Henry, Specific heat up to 14 tesla of a $YBa_2Cu_3O_{6.92}$ single crystal, *Physica C,* **234,** 269–279 (1994).
27. A. Junod, E. Bonjour, R. Calemczuk, J.Y. Henry, J. Muller, G. Triscone, and J.C. Vallier, Specific heat of an $YBa_2Cu_3O_7$ single crystal in fields up to 20 T, *Physica C,* **211,** 304–318 (1993).
28. A. Mirmelstein, A. Junod, K.-Q. Wang, E. Janod, and J. Muller, Specific heat of $YBa_2Cu_3O_{7-\delta}$ ceramics with single and double superconducting transitions in magnetic fields up to 14 T, *Physica C,* **241,** 301–310 (1995).
29. N. Kobayashi, K. Egawa, K. Miyoshi, H. Iwasaki, H. Ikeda, and R. Yoshizaki, Fluctuation effects in specific heat and magnetization of *c*-axis aligned $(Bi, Pb)_2Sr_2Ca_2Cu_3O_x$, *Physica C,* **219,** 265–272 (1994).
30. A. Mirmelstein, A. Junod, G. Triscone, K.-Q. Wang, and J. Muller, Specific heat of $Tl_2Ba_2CuO_6$ ("2201") 90 K superconducting ceramics in magnetic fields up to 14 T, *Physica C,* **248,** 335–342 (1995).
31. O. Jeandupeux, A. Schilling, and H.R. Ott, Specific heat of superconducting $HgBa_2Ca_2Cu_3O_8$ between 65 and 200 K, *Physica C,* **216,** 17–20 (1993).
32. A. Junod, in *Physical Properties of High Temperature Superconductors II*, D.M. Gisberg (ed.), World Scientific, Singapore, 1990.

14

Powder Synthesis and Bulk Cuprate Superconductors

14.1 Introduction

In the previous chapters, the physical properties of cuprate superconductors have been described. So far, we have not considered their processing. The first step in the fabrication of superconducting wires and tapes by the powder-in-tube (PIT) method is the synthesis of the appropriate cuprate superconductor powders.

In this chapter, an overview of the synthesis of cuprate superconductor powders will be provided, focusing on Y-123, bismuth-, thallium- and mercury-based cuprate superconductors. In the Bi-22$(n-1)n$, Tl-22$(n-1)n$, Tl-12$(n-1)n$, and Hg-12$(n-1)n$ high-T_c superconductor families, there exist several compounds which are distinguished only by the number n of the CuO_2 layers in the copper oxide blocks. The CuO_2 layers are separated by $(n-1)$ layers of calcium ions. As a consequence the prepared powders are typically mixtures of different members of the same cuprate superconductor family. The synthesis of a single member of such a family may be possible only in a narrow window of processing conditions (starting materials, maximum reaction temperature, reaction time, and oxygen partial pressure). A more detailed description of the processing of cuprate superconductor powders can be found in [1].

A second aspect to be considered is the manufacture and the properties of bulk cuprate superconductors. Bulk high-T_c superconductors can be used for current leads, which are required to connect superconducting magnets operated at low temperatures with power supplies at room temperature. They can be used in fault current limiters, as magnetic bearings, and as magnetic shielding. When a bulk high-T_c superconductor is cooled to well below its transition temperature in the presence of a high magnetic field, a significant magnetic induction can be frozen into it. As long as the material is kept at low temperature it may be used as a permanent magnet.

Physical Properties of High-Temperature Superconductors, First Edition. Rainer Wesche.

14.2 Synthesis of Cuprate Superconductor Powders

14.2.1 Yttrium-based Superconductors

We follow the description of the processing of yttrium-based superconductors presented in [1]. The superconductor $YBa_2Cu_3O_{7-\delta}$ can easily be prepared by the solid-state reaction of appropriate amounts of high purity Y_2O_3, $BaCO_3$, and CuO [2–9]. In laboratory scale processing, the precursor powders are thoroughly mixed and ground with a mortar and pestle. The chemical reaction is

$$Y_2O_3 + 4BaCO_3 + 6CuO \rightarrow 2YBa_2Cu_3O_{6.5} + 4CO_2 \uparrow \quad (14.1)$$

In order to avoid contamination of the superconductor powder, the reaction is performed in platinum crucibles (see, e.g., [3]). The powders are calcined in air or flowing oxygen at temperatures in the range of 925–950 °C for 10 to 24 h. To further improve the homogeneity of the $YBa_2Cu_3O_{6.5}$ powder, the calcined powder is reground and heat-treated again at 925–950 °C. After a further grinding and mixing step, the powder is pressed into pellets and sintered at 900–950 °C under flowing oxygen. The maximum critical temperature of 92 K can be achieved only in the fully oxidized material. The pellets must therefore be annealed at 400–500 °C under flowing oxygen, followed by slow cooling to room temperature [3–10]:

$$2YBa_2Cu_3O_{6.5} + (0.5 - \delta)O_2 \rightarrow 2YBa_2Cu_3O_{7-\delta} \quad (14.2)$$

The preparation of $YBa_2Cu_3O_{7-\delta}$ pellets via solid-state reaction is shown in Figure 14.1. The effect of the use of CuO_2, BaO, and BaO_2 on the formation of $YBa_2Cu_3O_{7-\delta}$ was studied by Peacor *et al.* [11]. The results of their synthesis studies show that the use of BaO_2 is beneficial for the formation of $YBa_2Cu_3O_{7-\delta}$, whereas the use of BaO and Cu_2O as precursor materials seems to provide no real advantage in the processing of Y-123 powders. The substitution of $BaCO_3$ by BaO_2 leads to the elimination of the repeated calcination steps, and hence a reduced reaction time [11].

Disadvantages of the preparation of Y-123 powders by solid-state reaction are long reaction times, time-consuming intermediate grindings, and an unavoidable nonhomogeneity on the atomic scale. Alternative powder synthesis routes, which avoid these problems, are citrate pyrolysis [12, 13] and oxalate co-precipitation [14]. First, citrate pyrolysis will be briefly described. A sketch of the main steps of citrate pyrolysis is presented in Figure 14.2. Appropriate amounts of Y_2O_3, $BaCO_3$, and CuO are dissolved in hot nitric acid (HNO_3) and carefully mixed with the addition of citric acid. The resulting blue solution is neutralized with ammonia (NH_3 + H_2O) and the pH value is controlled. The spontaneous combustion of the yttrium, barium, and copper citrates leads to an expanded dark foam of submicron particles. By means of annealing at 400 °C under vacuum, organic contaminants are removed. The resulting powder contains the metal ions of yttrium, barium, and copper in the ratio 1:2:3. After heat treatment at 950 °C for 18 h under flowing oxygen, the precursor powder is crushed, thoroughly mixed, and pressed into pellets. Finally, the pellets are sintered for 18 h at 980 °C under flowing oxygen, followed by cool-down at a rate of 50 °C/h to room temperature.

A further synthesis route is based on the co-precipitation of yttrium, barium, and copper oxalates [14]. A sketch of this powder processing route is presented in Figure 14.3. The

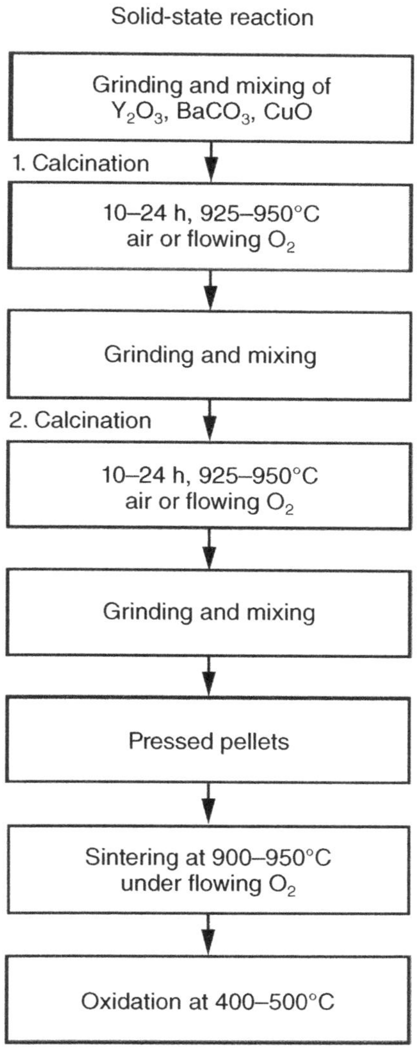

Figure 14.1 *Sketch of the synthesis of* $YBa_2Cu_3O_{7-\delta}$ *by solid-state reaction starting from* Y_2O_3, $BaCO_3$, *and CuO (adapted from [1])*

nitrate powders $Y(NO_3)_3 \cdot H_2O$, $Ba(NO_3)_2$, and $Cu(NO_3)_2 \cdot H_2O$ are used as starting materials for oxalate co-precipitation. Each is dissolved separately in a mixture of ethanol and distilled water. The three solutions are mixed in a such a way that a cation ratio of Y : Ba : Cu = 1 : 2 : 3 is obtained. The addition of oxalic acid initiates the co-precipitation of yttrium, barium, and copper oxalates. The precipitates, obtained by filtering of the solution, are washed with ethanol and then the powder is calcined for 3 h in air. The milled powder is pressed into pellets and sintered at 960 °C for 3 h in air. A further annealing at 960 °C in oxygen and cooling at a rate of 100 °C/h to room temperature are necessary to optimize the oxygen content in the Y-123 superconductor.

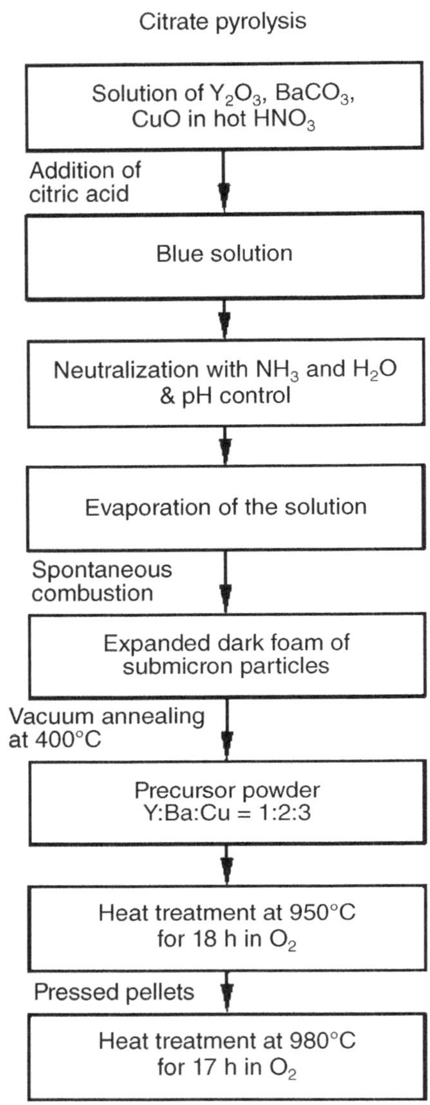

Figure 14.2 *Sketch of the synthesis of $YBa_2Cu_3O_{7-\delta}$ by citrate pyrolysis (adapted from [1])*

14.2.2 Bismuth-based Superconductors

The bismuth-based superconductors $Bi_2Sr_2CaCu_2O_{8+x}$ (Bi-2212) and $(Bi,Pb)_2Sr_2Ca_2Cu_3O_{10+x}$ (Bi-2223) were used in the first generation of high-temperature superconductor wires and tapes. Ag/Bi-2212 wires as well as Ag/Bi-2223 tapes are manufactured by the powder-in-tube (PIT) method (see Chapter 15). The preparation of the Bi-2212 and (Bi,Pb)-2223

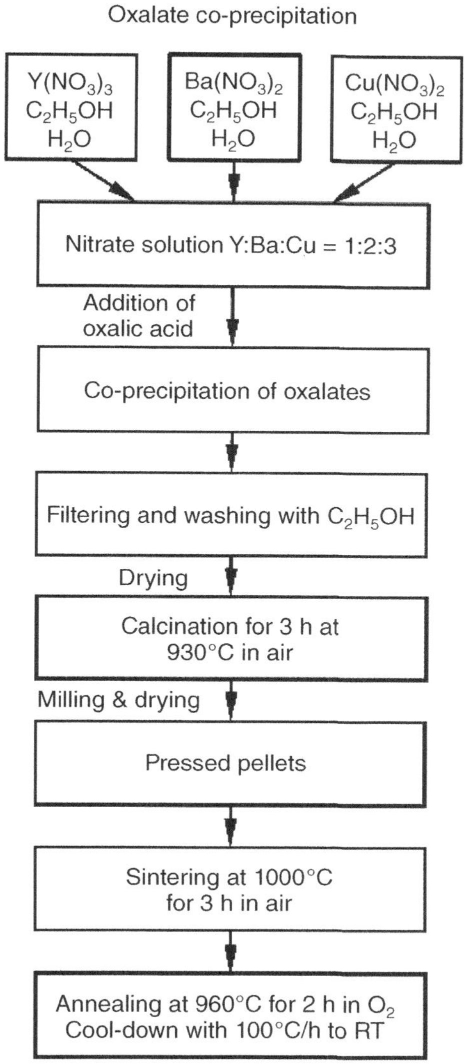

Figure 14.3 *Sketch of the synthesis of* $YBa_2Cu_3O_{7-\delta}$ *by oxalate co-precipitation (adapted from [1])*

powders is an important step in the wire and tape manufacture. We follow the description of the processing of bismuth-based superconductors presented in [1].

The three most important members of the Bi-22($n-1$)n cuprate superconductor family are $Bi_2Sr_2CuO_{6+x}$ ($T_c \approx 10\,K$), $Bi_2Sr_2CaCu_2O_{8+x}$ ($T_c \approx 90\,K$), and $Bi_2Sr_2Ca_2Cu_3O_{10+x}$ ($T_c \approx 110\,K$). In the synthesis of Bi-2212 powders, frequently the Bi-2201 and Bi-2223 phases are formed as impurity phases. With the use of appropriate preparation conditions, nearly phase-pure Bi-2212 powder can be obtained. The synthesis of phase-pure Bi-2223 powders without lead substitution for bismuth proved to be difficult.

Idemoto *et al.* studied the standard enthalpies of formation in the Bi–Sr–Ca–Cu–O system [15]. They considered the following reactions:

$$\begin{aligned}&(2.09/2)Bi_2O_3 + 1.89SrO + 2.10CaO + 2.99CuO + 0.29PbO + 0.20O_2\\&\quad\rightarrow Bi_{2.09}Sr_{1.88}Cu_{1.03}O_{6.16} + 0.01SrO + 2.10CaO + 1.96CuO + 0.29PbO\\&\quad+ 0.1425O_2 + 142.2\,kJ\end{aligned} \tag{14.3}$$

The energy released by the reaction of simple oxides to form Bi-2201 is 142.2 kJ. Equation 14.4 describes the reaction of Bi-2201 with CaO and CuO to form Bi-2212:

$$\begin{aligned}&Bi_{2.09}Sr_{1.88}Cu_{1.03}O_{6.16} + 0.01SrO + 2.10CaO + 1.96CuO + 0.29PbO + 0.1425O_2\\&\quad\rightarrow Bi_{2.07}Sr_{1.89}Ca_{0.96}Cu_{2.08}O_{8.23} + 0.01Bi_2O_3 + 1.14CaO + 0.91CuO\\&\quad+ 0.29PbO + 0.1025O_2 - 26.1\,kJ\end{aligned} \tag{14.4}$$

The energy of 26.1 kJ is consumed by the formation of Bi-2212.

Finally, the reaction of Bi-2212 with CaO and CuO to (Bi,Pb)-2223 is considered.

$$\begin{aligned}&Bi_{2.07}Sr_{1.89}Ca_{0.96}Cu_{2.08}O_{8.23} + 0.01Bi_2O_3 + 1.14CaO + 0.91CuO + 0.29PbO\\&\quad+ 0.1025O_2 \rightarrow Bi_{1.80}Pb_{0.29}Sr_{1.82}Ca_{2.10}Cu_{2.99}O_{10.30} + (0.29/2)Bi_2O_3\\&\qquad+ 0.07SrO - 39.0\,kJ\end{aligned} \tag{14.5}$$

The energy of 39 kJ is consumed by the formation of (Bi,Pb)-2223. In all reactions the number of gram atoms of metallic components was kept constant. Equations 14.3–14.5 indicate that the chemical stability increases from Bi-2223 to Bi-2212 and from Bi-2212 to Bi-2201, and hence Bi-2201 is the most stable of the three superconducting phases.

Single-phase Bi-2212 can be synthesized by the solid-state reaction of Bi_2O_3, $SrCO_3$, $CaCO_3$, and CuO. The reaction is summarized by the following equation:

$$Bi_2O_3 + 2SrCO_3 + CaCO_3 + 2CuO \rightarrow Bi_2Sr_2CaCu_2O_8 + 3CO_2\uparrow \tag{14.6}$$

Appropriate amounts of the initial oxide and carbonate powders are thoroughly mixed and ground. During calcination at a temperature above 750 °C in air, the carbonates are decomposed and the Bi-2212 phase starts to form. Typically, three to four calcination steps with intermittent grindings are performed to enhance the homogeneity of the powder and to reduce the residual carbon content. It has been demonstrated that, for a powder mixture of Bi_2O_3, $SrCO_3$, CaO, and CuO with an initial carbon content of 2.46% by weight, the residual carbon content decreases with the increasing number of calcinations. After the first, second, and third calcinations, the residual carbon content drops to ≈0.7%, ≈0.045%, and less than 0.02%, respectively [16]. The effect of the annealing time in pure oxygen or vacuum on the residual carbon content of Bi-2212 powders was investigated by Zhang and Hellstrom [17]. An initial carbon content of 0.16% was reduced to 0.033% after 48 h of annealing in vacuum. Annealing in pure oxygen at a temperature of 800 °C for 48 h led to an even smaller carbon content of 0.014%.

It may be noted that because in Bi-2212 bismuth ions can occupy strontium sites, and strontium and calcium can replace each other, the exact chemical composition of the Bi-2212 phase can vary. The exact chemical composition including the oxygen content determines the charge carrier concentration, the critical temperature, and the superconducting properties.

In Table 14.1, the conditions for the synthesis of Bi-2212 powders reported in the literature are presented. In addition to solid-state reaction of oxides and carbonates, Bi-2212 powders can be prepared by pyrolysis of organic salts [18] and oxalate co-precipitation [19].

Hasegawa *et al.* [18] dissolved bismuth octylate ($Bi(C_7H_{15}COO)_3$), strontium octylate ($Sr(C_7H_{15}COO)_2$), calcium octylate ($Ca(C_7H_{15}COO)_2$), and copper octylate ($Cu(C_7H_{15}COO)_2$) with the atomic ratio of Bi : Sr : Ca : Cu = 2 : 2 : 1 : 2 in an organic solvent. A silver tape of 2 mm width and 0.1 mm thickness was coated with the solvent and subjected to pyrolysis at 500 °C in air. A further heat treatment in air at 890 °C followed by slow cooling to 830 °C led to the formation of the Bi-2212 phase.

Kazin *et al.* [19] dissolved Bi_2O_3, $Sr(NO_3)_2$, $CaCO_3$, and CuO in nitric acid. To this solution oxalic acid was added, leading to the co-precipitation of the oxalates. These were filtered and washed with ethanol, dried at 100 °C, and finally decomposed by annealing at 810 °C for 10 h in alumina crucibles. After regrinding, the powder was heat treated at 835 °C for 10 h.

Due to the relatively low chemical stability of Bi-2223, single-phase material can be prepared only in a narrow window of heat treatment conditions and chemical compositions. The partial substitution of lead for bismuth accelerates the formation of the (Bi,Pb)-2223

Table 14.1 *Conditions for the synthesis of Bi-2212 powder (data taken from [1])*

Initial powders	Heat treatment conditions	Duration	Reference
Bi_2O_3, $SrCO_3$, $CaCO_3$, CuO	820 °C	3 × 12 h	[20]
Bi_2O_3, $SrCO_3$, CaO, CuO	800–850 °C, air	3×	[21, 22]
Bi_2O_3, $Sr(NO_3)_2$, CaO, CuO	800–850 °C, air	3 × 24 h	[23]
Bi_2O_3, $SrCO_3$, $CaCO_3$, CuO	840–860 °C, air	32–109 h	[24]
	840–860 °C, air	8 h	
Bi_2O_3, $SrCO_3$, $CaCO_3$, CuO	820 °C, 865 °C	2 × 18 h, 2 h	[25]
Bi_2O_3, $SrCO_3$, $CaCO_3$, CuO	750 °C, 800 °C, air	2 × 24 h	[26]
	820 °C, 850 °C, air	66 h, 174 h	
Bi_2O_3, $SrCO_3$, $CaCO_3$, CuO	780 °C, 810 °C, 820 °C, air	2 × 12 h, 70 h	[27]
Bi_2O_3, $SrCO_3$, $CaCO_3$, CuO	700–830 °C, air	≤ 24 h	[28]
	800–870 °C, air	≤ 7 × 48 h	
	830 °C, air	≤ 7 days	
Bi_2O_3, $SrCO_3$, CaO, CuO	840 °C–880 °C, air	4 × 20 h	[29]
$Bi(NO_3)_3$, $Sr(NO_3)_2$, $Ca(NO_3)_2$, $Cu(NO_3)_2$	125 °C + 450 °C, air	2 × 3 h	[30]
	750–850 °C, air	3 × 12 h	
	890 °C, 21% O_2 in N_2		
	Slow cooling to 870 °C	0.1 °C/min	
	850 °C, 21% O_2 in N_2	12 h	
Bi_2O_3, $SrCO_3$, $CaCO_3$, CuO	750 °C, 800 °C, air	24 h, 48 h	[31]
	820 °C, air	24 h, 60 h	
	670 °C, N_2	15 h	
$Bi(OH)_3$, $Ca(OH)_2$, $Cu(OH)_2$ $Sr_2Cu(OH)_6$	700–800 °C 2% O_2 in Ar	4–10 h	[32]

phase. Pierre *et al.* [33] studied the effect of partial lead substitution on the formation of (Bi,Pb)-2223. They determined the volume fractions of different phases formed from powder mixtures $Bi_{2-x}Pb_xSr_2Ca_2Cu_3O_y$, heat treated for 60 h at 880–885 °C, by X-ray powder diffraction. For $x \leq 0.3$, (Bi,Pb)-2212 is the major phase, while (Bi,Pb)-2223 is formed as a minor phase. At $x = 0.6$, the volume fraction of (Bi,Pb)-2223 reaches a maximum volume fraction of about 60%. The volume fraction of (Bi,Pb)-2212 is around 30%. Above $x = 0.4$ $CaPbO_4$ is formed as an impurity phase. For $x = 0.6$ the volume fraction of $CaPbO_4$ is between 5% and 10%. For a mixture with the composition $Bi_{1.4}Pb_{0.6}Sr_2Ca_2Cu_3O_y$, the amount of the 2223 phase increased at the expense of 2212 and $CaPbO_4$ up to sintering times of 120 h. Sintering times above 120 h do not significantly increase the volume fraction of 2223, which is around 80% after 120 h. Volume fractions above 80% of the 2223 phase can only be achieved by repeated sintering with intermediate grinding. Four to five sintering steps of ≈40 h duration are required to get nearly single-phase material. Pierre *et al.* [33] conclude that the optimum Pb content for the preparation of the (Bi,Pb)-2223 phase is between $x = 0.3$ and $x = 0.5$.

Zhu and Nicholson [34] found that besides the sintering temperature the heat treatment atmosphere is of importance for the formation of the (Bi,Pb)-2223 phase. They used nitrate powders with the initial cation ratio Bi : Pb : Sr : Ca : Cu = 1.84 : 0.34 : 1.91 : 2.03 : 3.06. The nitrate mixture was calcined at 800 °C for 48 h in air. The ground powder was pressed into pellets and heat-treated for 6 h in Ar/O_2 mixtures. A weight fraction of more than 50% of the 2223 phase was found for heat treatment at 830 °C and an oxygen partial pressure of 0.02 bar. An even larger weight fraction of 62% was achieved by heat treatment at 860 °C and an oxygen partial pressure of 0.13 bar.

The conditions reported for the synthesis of (Bi,Pb)-2223 powders are listed in Table 14.2 [35–52]. (Bi,Pb)-2223 can be prepared by solid-state reaction of Bi_2O_3, $SrCO_3$, $CaCO_3$, CuO, and PbO. The use of CaO instead of $CaCO_3$ allows the reduction of the carbon content of the initial powder. To get rid of the moisture typically present in CaO powders, they need to be carefully dried before weighing to get the correct calcium content in the initial powder. Pb_3O_4 can be used instead of PbO as the source of lead.

Instead of oxides and carbonates, metal nitrates can be used as starting materials for the preparation of (Bi,Pb)-2223 by solid-state reactions [48–52]. The chemical homogeneity can be enhanced by solution of the nitrate powders in nitric acid [49–52]. The cation ratios used for the preparation of (Bi,Pb)-2223 are almost the same for metal nitrate and oxide/carbonate precursors.

In general, it was found that the first step in the synthesis of the (Bi,Pb)-2223 phase is the formation of (Bi,Pb)-2212 as an intermediate phase. Thus, (Bi,Pb)-2223 can be prepared by a two-powder process starting from (Bi,Pb)-2212 and $(Sr,Ca)CuO_2$ powders [53–55]. The (Bi,Pb)-2212 is prepared by solid-state reaction of Bi_2O_3, PbO, $SrCO_3$, $CaCO_3$, and CuO. The mixed and milled precursor powders are calcined at 750 °C for 6 h in an oxygen atmosphere with a total pressure of 4 mbar. A second calcination for 24 h is performed at a temperature 840 °C and ambient pressure in CO_2-free air. The reaction products are $Bi_{1.8}Pb_{0.4}Sr_2CaCu_2O_8$, and Ca_2PbO_4.

$Sr_xCa_{1-x}CuO_2$ is prepared separately by the solid-state reaction of $SrCO_3$, $CaCO_3$, and CuO. The mixed and milled powders are calcined at 750 °C for 6 h in oxygen atmosphere with a total pressure of 4 mbar. A second heat treatment for 48 h is performed at a temperature of 900 °C and ambient pressure in CO_2-free air. For $x > 0.25$, the

Table 14.2 *Conditions for the synthesis of (Bi,Pb)-2223 powders (data taken from [1])*

Initial powders	Cation ratio Bi:Pb:Sr:Ca:Cu	Heat treatment conditions	Duration	Reference
Bi_2O_3, $CaCO_3$, $SrCO_3$, CuO, PbO	1.8 : 0.2 : 2 : 2 : 3	770 °C, air 860 °C, air	16 h 2 × 48 h	[35]
Bi_2O_3, $CaCO_3$, $SrCO_3$, CuO, PbO	1.8 : 0.2 : 2 : 2 : 3	770 °C, air 840 °C 830 °C pO_2 0.07 bar	80 h 400 h 128 h	[36]
Bi_2O_3, $SrCO_3$, CaO, CuO, PbO	1.8 : 0.33 : 1.93 : 1.93 : 3	800 °C 820 °C	3 h 16 h	[37]
Bi_2O_3, $CaCO_3$, $SrCO_3$, CuO, PbO	1.84 : 0.34 : 1.9 : 2 : 3.1	800 °C 835 °C	2 × 10 h 50 h	[38]
–	1.72 : 0.34 : 1.83 : 1.97 : 3.13	820 °C, air	3 × 24 h	[39]
Bi_2O_3, $SrCO_3$, CaO, CuO, PbO	1.8 : 0.4 : 2 : 2.2 : 3	800 °C 820 °C	3 h 16 h	[37]
Bi_2O_3, $SrCO_3$, CaO, CuO, PbO	1.8 : 0.4 : 2 : 2.5 : 3	840 °C	4 × 24 h	[40]
Bi_2O_3, $SrCO_3$, $CaCO_3$, CuO, PbO	1.7 : 0.4 : 2 : 2 : 3	810–820 °C, air	2 × 24 h	[41]
Bi_2O_3, Pb_3O_4, CuO $SrCO_3$, $CaCO_3$	1.8 : 0.43 : 1.71 : 2.14 : 3	760–860 °C flowing O_2 860 °C, air	15–48 h 90 h	[42]
Bi_2O_3, Pb_3O_4, CuO $SrCO_3$, $CaCO_3$	1.92 : 0.48 : 2 : 2 : 3.2	800 °C, air 850 °C, air	2 × 12 h 1–60 h	[43]
Bi_2O_3, Pb_3O_4, CuO $SrCO_3$, $CaCO_3$	1.4 : 0.6 : 2 : 2 : 3.6	800 °C, air 830–870 °C, air	17 h 40–100 h	[44]
Bi_2O_3, $CaCO_3$, $SrCO_3$, CuO, PbO	1.4 : 0.6 : 2 : 2 : 3.6	750 °C, air 800 °C, air 850 °C, air	6 h 17.5 h 12 h	[45]
Bi_2O_3, $SrCO_3$, CaO, CuO, PbO	1.4 : 0.6 : 2 : 2 : 3.6	800 °C 845 °C, air 700 °C	24 h 240 h 5 min	[46]
Bi_2O_3, $SrCO_3$, CaO, CuO, PbO	1.4 : 0.6 : 2 : 2 : 3.6	800 °C, air 845 °C, air 870 °C, air	11 h ≤ 700 h 12 h	[47]
Bi_2O_3, $Sr(NO_3)_2$, $Ca(NO_3)_2{\cdot}4H_2O$, $Cu(NO_3)_2{\cdot}2.5H_2O$	1.82 : 0.36 : 1.9 : 2.06 : 3.05	830–840 °C	3 × 60 h	[48]
Nitrates	1.85 : 0.35 : 1.9 : 2.03 : 3.05	820 °C 840 °C	10 h 20 h	[49]
Bi_2O_3, $SrCO_3$, $CaCO_3$, CuO, PbO dissolved in HNO_3	1.6 : 0.4 : 2 : 2 : 3	800 °C 850 °C, air 850 °C, air	10 h 120 h 48 h	[50]
Nitrates	1.85 : 0.35 : 1.9 : 2.05 : 3.05	830 °C, 845 °C	10 h, 20 h	[51]
Nitrates	1.8 : 0.4 : 2 : 2.2 : 3	800 °C 830–840 °C	2 h 3 × 15 h	[52]

reaction product is $(Sr_xCa_{1-x})CuO_2$. The (Bi,Pb)-2212 and the $(Sr,Ca)CuO_2$ powders are thoroughly mixed and milled before they are used for the manufacture of Ag/Bi-2223 tapes by the powder-in-tube method (see Chapter 15). The formation of phase-pure (Bi,Pb)-2223 powder by the two-powder process is quicker than the solid-state reaction starting from the carbonates and the binary oxides.

An alternative route for the rapid synthesis of single-phase (Bi,Pb)-2223, illustrated in Figure 14.4, is based on the use of freeze-dried nitrate solutions [56, 57]. The metal nitrates

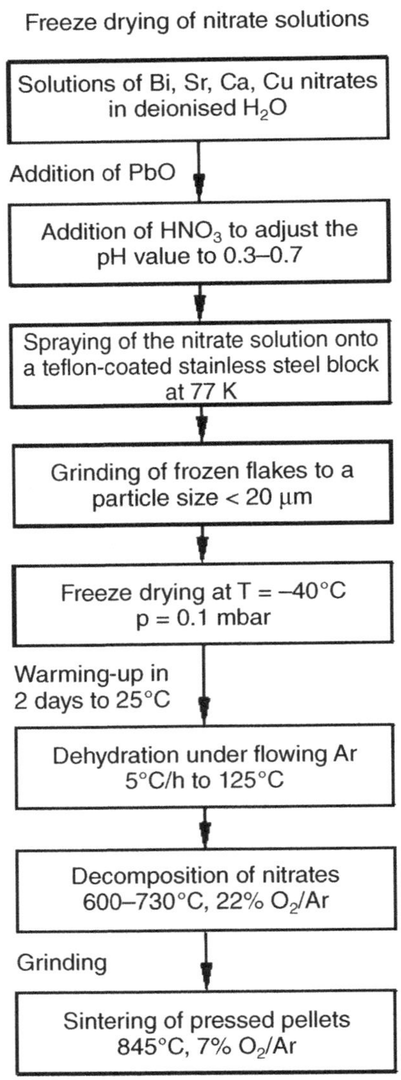

Figure 14.4 *Sketch of the synthesis of (Bi,Pb)-2223 by freeze-dried precursors as described by Krishnaraj et al. [56] (adapted from [1])*

with cation ratios of Bi : Pb : Sr : Ca : Cu = 1.8 : 0.4 : 2 : 2.2 : 3 or 1.8 : 0.4 : 2 : 2 : 3 are diluted in distilled and deionized water. The solution is diluted to a concentration 0.1 M in Bi (i.e. 0.1 g atom of bismuth per litre H_2O). By means of the addition of HNO_3 the pH of the solution is adjusted to values in the range of 0.3 to 0.7. Spraying the solution onto a Teflon-coated stainless steel cylinder provides flakes of frozen material, which are ground to a particle size of less than 20 μm. The nitrates, freeze-dried at −40 °C and 0.1 mbar, are dehydrated at a temperature of 125 °C under flowing argon. They are then decomposed to the oxides by heat treatment between 600 °C and 730 °C in a 22% O_2/78% Ar atmosphere. After grinding, the powder is pressed into pellets, which were sintered at 845 °C in a 7% O_2/93% Ar mixture.

Another synthesis route for formation of single-phase (Bi,Pb)-2223 is based on oxalate co-precipitation. Figure 14.5 shows a sketch of the oxalate precipitation route. The merits of this synthesis route are high reactivity, small particle size and mixing of the cations on an atomic scale [58–64]. In the first step, bismuth nitrate is dissolved in a 1 M aqueous solution of oxalic acid. After mixing and stirring, a solution of lead, strontium, calcium, and copper nitrates is added. Because of the dependence of the precipitation yield of Sr, Ca, and Cu on the pH value, its control is of crucial importance. The precipitation yield is the percentage ratio of the weights of the oxalates in the resulting filtrate to that in the starting solution. Typically, ammonium or sodium hydroxide is used to adjust the pH to the desired value. Copper oxalates are fully precipitated for pH values between 2 and 4 or above 10 [58], while bismuth and lead oxalates are almost completely precipitated for a pH above 1. The filtrate is washed, dried, and annealed at 600 °C to decompose the oxalates to the oxides. Pressed pellets are heat-treated for 12 h at 800 °C, reground, and again pressed into pellets, which are sintered for further 100 h at a temperature of 830 °C. The strong dependence of the precipitation yield on the pH-value is a disadvantage of oxalate co-precipitation starting from aqueous nitrate solutions. This problem can be avoided by the use of alcohol [59] or acetate solutions [60] in the oxalate co-precipitation process.

14.2.3 Thallium-based Superconductors

There exist two different families of thallium-based cuprate high-T_c superconductors. The Tl-22($n-1$)n family is characterized by high critical temperatures, reaching 128 K for $n = 3$. It is the thallium-based counterpart of the Bi-22($n-1$)n family. In the Tl-22($n-1$)n family, adjacent copper oxide blocks are separated by a TlO double layer. In contrast the copper oxide blocks of the Tl-12($n-1$)n cuprates are separated only by a single TlO layer. As a consequence of the smaller distance of adjacent copper oxide blocks, the anisotropy of the superconducting properties is less pronounced than in the corresponding Tl-22($n-1$)n compounds. As in the bismuth-based cuprate superconductors, synthesis leads to a mixture of several superconducting phases. Moreover, the preparation of thallium-based superconductors is difficult because of the toxicity and volatility of Tl_2O_3 and other thallium compounds. Single-phase material can be obtained only within small windows of initial composition and heat treatment conditions.

In general, long sintering at high temperatures leads to a loss of thallium. To avoid this, two-step synthesis routes have been developed. In the first step a precursor of the nonvolatile components Ba, Ca, and Cu is prepared. In the second step, Tl_2O_3 and the Ba-Ca-Cu-O precursor are mixed and pressed into pellets. To counteract the loss

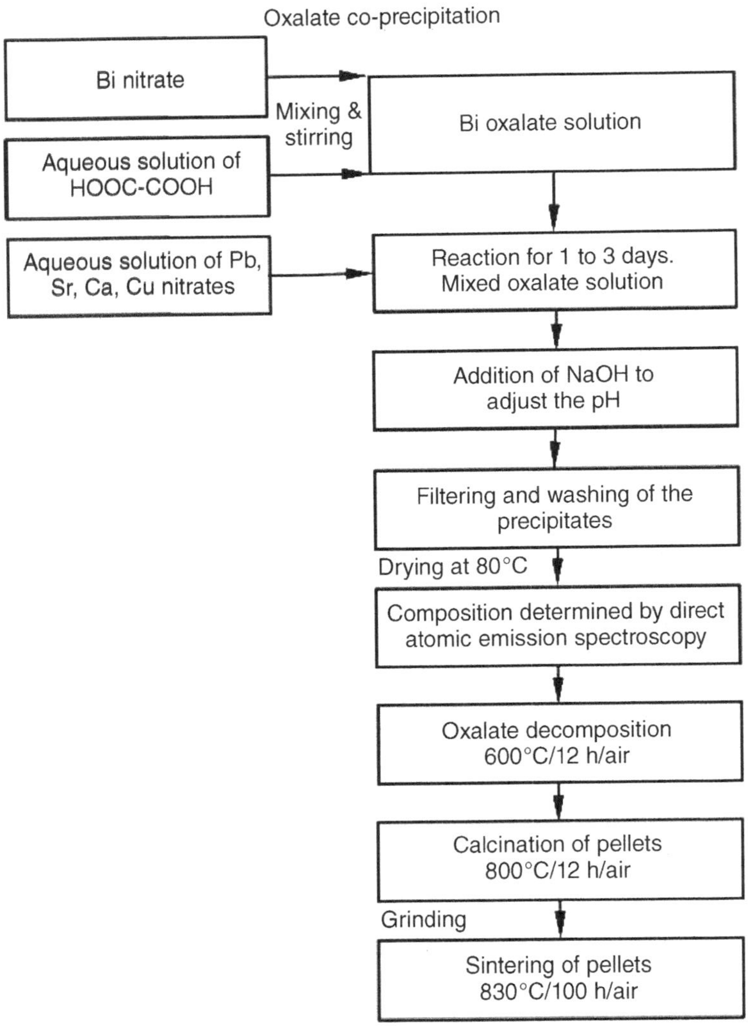

Figure 14.5 *Sketch of the synthesis of single phase (Bi,Pb)-2223 by oxalate co-precipitation as described by Hagberg et al. [58] (adapted from [1])*

of thallium, the pellets are wrapped in gold foils which are encapsulated in quartz tubes. Another way to avoid thallium loss is the preparation of thallium-based cuprate superconductors by means of high-pressure and high-temperature techniques. The partial substitution of Tl by Bi or Pb has been found to be beneficial for the formation of the Tl-1223 phase. We follow here the description of the preparation of Tl-based superconductors presented in [1].

The conditions reported for the preparation of $Tl_2Ba_2CuO_6$ (Tl-2201) are listed in Table 14.3. Some of the precursors (e.g., BaO) react strongly with CO_2 and moisture and must therefore be handled in a glove box filled with synthetic air or an inert gas.

Table 14.3 Conditions for the synthesis of Tl-2201 powders (data taken from [1])

Reactants	Comments Formed compounds	Heat treatment conditions	Duration	Reference
$BaCO_3$	Formation of BaO[a]	950 °C, vacuum	12 h	[65]
Tl_2O_3, BaO, CuO	Formation of $Tl_2Ba_2CuO_{6+\delta}$	820 °C ≈100 bar Ar	7 h	
Tl_2O_3, BaO_2, CuO Cu_2O or Cu	Reaction in silica ampoules	850 °C, 3 bar O_2 850 °C, 6 bar O_2		[66]
	Sintering	RT to 880 °C 880 °C	6 h 8 h	
	Optimized hole concentration	290 °C 10% H_2, 90% Ar	≤480 min	
$Tl_2Ba_2O_5$, CuO	Sintering	750 °C, 1 bar O_2	3 days	[67]
	Optimized hole concentration	850 °C 100 bar He	18 min	
$Ba(OH)_2$, $Cu(OH)_2$	Preparation of Ba_2CuO_x	400 °C, dynamic	10 h	[68]
	$Ba_2Cu_{1.25}O_x$, $Ba_2Cu_{1.5}O_x$	vacuum O_2 leak		
Tl_2O_3, Ba–Cu–O	$Tl_2Ba_2CuO_x$	550 °C, 80 bar Ar	10 h	
	$Tl_{1.75}Ba_2Cu_{1.25}O_x$	550 °C, 80 bar Ar	10 h	
	$Tl_{1.5}Ba_2Cu_{1.5}O_x$	550 °C, 80 bar Ar	10 h	

[a]Handling of powders in dry N_2

Synthesis procedures used to prepare Tl-2212 and Tl-2223 are presented in Tables 14.4 [68–71] and 14.5 [47, 69, 72–77]. In these compounds, the thallium site can be partly occupied by calcium ions. In the case of Tl-2223 this leads to the chemical formula $Tl_{2-x}Ba_2Ca_{2+x}Cu_3O_{10+\delta}$. As a consequence, precursor powders with reduced thallium and enhanced calcium content were found to be beneficial for the formation of the 2223 phase. Furthermore, the use of barium peroxide as the barium source was found to be advantageous for the formation of Tl-2223.

Nearly single-phase Tl-2234 was prepared by Presland *et al.* [78] by solid-state reaction starting from Tl_2O_3, BaO_2, CaO, and CuO. They used initial powders with a cation ratio Tl : Ba : Ca : Cu = (2 − x) : 2 : (4 + x) : 5, where x was 0, 0.25, and 0.5. Pressed pellets were placed into a gold tube and reacted for a few minutes at 904 °C under flowing oxygen until the formation of the Tl-2234 phase set in. To continue the growth of the 2234 phase, annealing at 850 °C for several hours was required. Ogborne and Weller [79] prepared Tl-2234 in a similar way, starting with the same reactants but a different cation ratio Tl : Ba : Ca : Cu = 1.5 : 2 : 3.5 : 4. The mixture was reacted in air for 20–30 h at 860 °C in a sealed gold tube. The critical temperature of the Tl-2234 was 114 K, and was not improved by additional annealing at 500 °C under flowing argon or in oxygen atmosphere.

The last aspect to be considered in this section is the synthesis of the single-layer thallium compounds Tl-1212 and Tl-1223. The extensive loss of thallium during long heat treatments at high temperatures can be avoided by two-step synthesis routes in which a precursor of the nonvolatile components (Sr, Ba, Ca, Cu) is first prepared by solid-state

Table 14.4 Conditions for the synthesis of Tl-2212 powders (data taken from [1])

Reactants	Comments Formed compounds	Heat treatment conditions	Duration	Reference
$Ba(OH)_2$, $Ca(OH)_2$ $Cu(OH)_2$	Preparation of $Ba_2CaCu_2O_x$	400 °C, dynamic vacuum O_2 leak	10 h	[68]
Tl_2O_3, $Ba_2CaCu_2O_x$	Tl : Ba : Ca : Cu = 2 : 2 : 1 : 2 Formation of 2201, 2212, minor phases	550 °C 100 bar Ar	10 h	
Tl_2O_3, $Ba_2CaCu_2O_x$	Tl : Ba : Ca : Cu = 1 : 2 : 1 : 2 Formation of 2212 and minor phases	550 °C 100 bar Ar	10 h	
$CaCO_3$, $BaCO_3$ CuO	Preparation of $CaBa_2CuO_4$	830–930 °C	60 h	[69]
Tl_2O_3, CuO $CaBa_2CuO_4$	Two-step synthesis of $Tl_2Ba_2CaCu_2O_x$	830 °C Pellets wrapped in Au foils	0.5 h, 14 h	
CaO, CuO	Preparation of $Ca_2Cu_3O_5$	850–880 °C	18–68 h	[70]
Tl_2O_3, BaO_2, $Ca_2Cu_3O_5$	Two-step synthesis Tl : Ba : Ca : Cu = 2 : 2 : 2 : 3 Formation of 2212 & minor phases	820 °C Pellets wrapped in Au foils	10 h	
Tl_2O_3, CaO, $BaCuO_2$	Tl : Ba : Ca : Cu = 2 : 2 : 1 : 2 Formation of 2212 $BaCuO_2$ as impurity	865 °C Pellets in sealed Au tubes	15 min	[70]
$Tl(CH_3OC_2H_4O)$ $Ba(CH_3OC_2H_4O)_2$ $Ca(CH_3OC_2H_4O)_2$ $Cu(C_2H_5O)_2$	Sol–gel technique Tl : Ba : Ca : Cu = 2 : 2 : 2 : 3 Formation of 2212 and minor phases	900 °C, O_2 Preheated furnace	30 min	[71]

reaction of carbonates and oxides. In a second step, Tl_2O_3 and the (Ba,Sr)–Ca–Cu–O precursor are pressed into pellets. For the synthesis of Tl-1212 or Tl-1223, these pellets are wrapped in gold foils and the reaction is performed in encapsulated quartz tubes or under enhanced pressure. Typical conditions for the synthesis of Tl-1212 and Tl-1223 are presented in Table 14.6 [80–83] and Table 14.7 [84–99], respectively.

In almost all of the examples presented in Table 14.7, thallium was partly substituted by lead and/or bismuth. In addition, barium was partly replaced by strontium. The preparation of Tl-1223 without bismuth, lead, and strontium substitution is possible if TlF is used as thallium source instead of Tl_2O_3 [84]. Moreover, because carbon impurities favor

Table 14.5 *Conditions for the synthesis of Tl-2223 powders (data taken from [1])*

Reactants	Comments Formed compounds	Heat treatment conditions	Duration	Reference
Tl_2O_3, CaO, BaO, CuO	Heat treatment in sealed quartz tubes Tl : Ba : Ca : Cu = 1 : 1 : 3 : 3	880 °C, 1 bar O_2 Pellets wrapped in Au foils	3 h	[72]
Tl_2O_3, CaO, BaO_2, CuO	Tl : Ba : Ca : Cu = 1.6 : 2 : 2.4 : 3	910 °C, O_2 Pellets wrapped in Au foils 5 °C/min to RT[a]	3 h	[73]
	Encapsulated in evacuated quartz tube	750 °C, 10^{-4} mbar	10 days	
	Rapidly quenched into liquid nitrogen Nearly single-phase 2223	600 °C 2% O_2 98% N_2	2 h	
Tl_2O_3, CaO, BaO_2, CuO	Tl : Ba : Ca : Cu = 1.7 : 2 : 2.3 : 3	890 °C, O_2 Wrapped in Au foils	5 h	[74]
	Encapsulated in evacuated quartz tube Nearly single-phase 2223	750 °C, 10^{-4} mbar	2–250 h	
Tl_2O_3, CaO, BaO_2, CuO	Tl : Ba : Ca : Cu = 1.8 : 2 : 2.2 : 3 Nearly single-phase 2223	910 °C, O_2 5 °C/min to RT Pellets in sealed Au foils	3 h	[75]
$BaCO_3$, $CaCO_3$, CuO	Preparation of $Ba_2Ca_2Cu_3O_x$	925 °C, air	2×	[76]
Tl_2O_3, $Ba_2Ca_2Cu_3O_x$	Tl : Ba : Ca : Cu = 2 : 2 : 2 : 3 Main phase Tl-2223	895 °C Pressed pellets Pt crucibles	5 min	
$BaCO_3$, CuO	Preparation of $Ba_2Cu_3O_x$	900 °C, air	8 h	[47]
Tl_2O_3, CaO, $Ba_2Cu_3O_x$	Tl : Ba : Ca : Cu = 2 : 2 : 2 : 3	910 °C 850 °C, O_2 Pressed bars	5 min 3 h	

(*continued overleaf*)

Table 14.5 *(continued)*

Reactants	Comments Formed compounds	Heat treatment conditions	Duration	Reference
$BaCO_3$, $CaCO_3$, CuO	Preparation of $CaBa_2CuO_4$	830–930 °C	60 h	[69]
Tl_2O_3, CaO, CuO, $CaBa_2CuO_4$	Tl : Ba : Ca : Cu = 2 : 2 : 2 : 3	830 °C, 870 °C Wrapped in Au foils	5 h, 2–10 h	
$BaCO_3$, $CaCO_3$, CuO	Preparation of $Ca_2Ba_2Cu_3O_x$	900 °C	3 × 12 h	[77]
Tl_2O_3, Ag_2O, $Ca_2Ba_2Cu_3O_x$	Tl:Ba:Ca:Cu:Ag = 2:2:2:3:1 Nearly single phase Tl-2223	900 °C, O_2 2 °C/min to RT Preheated furnace Sealed Au foil	6 min	

[a] RT: room temperature.

Table 14.6 *Conditions for the synthesis of Tl-1212 powders (data taken from [1])*

Reactants	Comments Formed compounds	Heat treatment conditions	Duration	Reference
Tl_2O_3, Bi_2O_3, SrO, CaO, CuO	Tl : Bi = 3 : 1 Formation of $(Tl,Bi)_{1.33}Sr_{1.33}Ca_{1.33}Cu_2O_{6.67+\delta}$	870 °C, O_2	12 h	[80]
Tl_2O_3, BaO_2, CaO, CuO	Tl : Ba : Ca : Cu = 2 : 2 : 1 : 2	885 °C, O_2 500 °C Quench in liquid N_2	1.5 h 0.5 h	[81]
$Ca(NO_3)_2$, $Sr(NO_3)_2$, Y_2O_3, La_2O_3, CuO, Tl_2O_3, PbO	Decomposition of nitrates $Tl_{0.5}Pb_{0.5}Sr_{2-x}La_xCaCu_2O_7$ $Tl_{0.5}Pb_{0.5}Sr_2Ca_{1-y}La_yCu_2O_7$ Sealed Au bags	700 °C 950 °C, O_2	1 h 3 h	[82]
	Optimization of T_c	700 °C, 1 bar O_2	1 h	
$SrCO_3$, BaO_2, CaO, Y_2O_3, CuO	Preparation of $Sr_{1.6}Ba_{0.4}Ca_{0.8}Y_{0.2}Cu_2O_5$ precursor	920 °C	4 × 10 h	[83]
Precursor, Tl_2O_3, PbO	Pellets between Au plates wrapped in Ag foils Formation of $Tl_{0.5}Pb_{0.5}(Sr,Ba)_2(Ca,Y)Cu_2O_y$	900–940 °C	3–10 h	

Table 14.7 *Conditions for the synthesis of Tl-1223 powders (data taken from [1])*

Reactants/ Formed compounds	Heat treatment conditions	Duration	Reference
TlF, $Ba_2Ca_2Cu_3O_x$ Formation of $TlBa_2Ca_2Cu_3O_z$	850 °C, flowing O_2	2 h	[84]
Tl_2O_3, PbO, SrO_2, CaO, CuO Formation of $(Tl_{0.5}Pb_{0.5})Sr_{1.6}Ca_{2.4}Cu_3O_9$	950 °C, flowing O_2 Pellets wrapped in Au foils 700–750 °C, 10^{-4} mbar Encapsulated quartz tubes	3 h 5–10 d	[85]
Tl_2O_3, PbO_2, CaO, SrO, CuO Formation of $Tl_{0.6}Pb_{0.5}Sr_2Ca_{2.4}Cu_{3.4}O_x$	850 °C, air Pellets crimped in Au tubes Quenched in liquid nitrogen	2 × 1 h	[86]
BaO, SrO, CaO, CuO Preparation of $Sr_{1.6}Ba_{0.4}Ca_2Cu_3O_x$ *Tl_2O_3, PbO, $(Sr,Ba)Ca_2Cu_3O_x$* Formation of $(Tl_{0.5}Pb_{0.5})(Sr,Ba)_2Ca_2Cu_3O_x$	870 °C 800–900 °C, pellets 950–1100 °C, partial melting 850–880 °C	20 h 10 h 1/6–2 h 10–50 h	[87, 88]
SrO, BaO_2, CaO, CuO $(Sr_{2-y}Ba_y)Ca_2Cu_3O_z$ (*y*: 0.5–1.5) *Tl_2O_3, Cr_2O_3, $(Sr,Ba)_2Ca_2Cu_3O_z$* $(Tl_{1-x}Cr_x)(Sr_{2-y}Ba_y)Ca_2Cu_3O_z$	840 °C, air 900–1000 °C, flowing O_2 Pellets in Al_2O_3 crucibles	3 × 24 h 5 h	[89]
Tl_2O_3, BaO_2, CaO, $SrCuO_2$ Formation of $Tl(Ba,Sr)_2Ca_2Cu_3O_x$	930 °C, slow cooling Evacuated quartz ampoules	6 h	[90]
$SrCO_3$, $BaCO_3$, $CaCO_3$, CuO Preparation of $Sr_{1.6}Ba_{0.4}Ca_2Cu_3O_7$ *Tl_2O_3, Bi_2O_3, $(Sr,Ba)_2Ca_2Cu_3O_7$* Formation of $(Tl_{0.9}Bi_{0.2})(Sr,Ba)_2Ca_2Cu_3O_x$	950 °C, air 920 °C, O_2 Pellets pressed in Au foils	60 h 6 h	[91]
$BaCO_3$, SrO, CaO, CuO Preparation of $Sr_{1.6}Ba_{0.4}Ca_2Cu_3O_7$ *Tl_2O_3, Bi_2O_3, $Sr_{1.6}Ba_{0.4}Ca_2Cu_3O_7$* Formation of $(Tl_{0.93}Bi_{0.22})(Sr,Ba)_2Ca_2Cu_3O_x$	905–920 °C 840 °C, Ag/Tl-1223 tapes	4 × 10 h	[92]

(continued overleaf)

Table 14.7 *(continued)*

Reactants/ Formed compounds	Heat treatment conditions	Duration	Reference
Tl_2O_3, PbO, or Bi_2O_3, $Sr_{1.6}Ba_{0.4}Ca_2Cu_3O_7$	880–920 °C, Tl vapor Tl source at 730–780 °C		[93]
Partially reacted Tl-1223			
Ag/(Tl,Pb)-1223 tapes	865–885 °C	10 h	
Ag/(Tl,Bi)-1223 tapes	865–885 °C	7 h	
$SrCO_3$, BaO_2, CaO, CuO	950 °C, precursor preparation	3 × 16 h	[94]
$Sr_{1.6}Ba_{0.4}Ca_2Cu_3O_7$ precursor			
Tl_2O_3, Bi_2O_3, $Sr_{1.6}Ba_{0.4}Ca_2Cu_3O_7$	940 °C, ≈ 95% Tl-1223	2 h	
$(Tl_{0.78}Bi_{0.22})(Sr,Ba)_2Ca_2Cu_3O_x$	In sealed Al_2O_3 crucibles		
$SrCO_3$, BaO_2, CaO, CuO	905–920 °C	4 × 10 h	[95]
$Sr_{1.6}Ba_{0.4}Ca_2Cu_3O_7$ precursor	Precursor preparation		
$(Tl_{0.93}Bi_{0.22})(Sr,Ba)_2Ca_2Cu_3O_x$			
Ag/Tl-1223 tapes	800–860 °C		
$SrCO_3$, $BaCO_3$, $CaCO_3$, CuO	920 °C, air	10 h	[96]
$Sr_{1.8}Ba_{0.2}Ca_2Cu_3O_7$ precursor	Precursor preparation		
Tl_2O_3, Bi_2O_3, PbO, $Sr_{1.8}Ba_{0.2}Ca_2Cu_3O_7$	920 °C, flowing O_2 Pellets wrapped in Au foils	3 h	
Formation of $(Tl_{0.6}Pb_{0.2}Bi_{0.2})(Sr,Ba)_2Ca_2Cu_3O_x$			
$SrCO_3$, BaO_2, CaO, CuO	905–920 °C	4 × 10 h	[97]
$Sr_{1.6}Ba_{0.4}Ca_2Cu_3O_7$ precursor	Precursor preparation		
Formation of $(Tl_{0.78}Bi_{0.22})(Sr,Ba)_2Ca_2Cu_3O_x$	840 °C, air		
Ag/Tl-1223 tapes			
$SrCO_3$, $BaCO_3$,$CaCO_3$, CuO	900 °C, flowing oxygen	24 h	[98]
$Sr_{1.8}Ba_{0.2}Ca_{1.9}Cu_3O_7$ precursor	980 °C, pellets	24 h	
Tl_2O_3, PbO, Bi_2O_3,	940 °C	3 h	
$Sr_{1.8}Ba_{0.2}Ca_{1.9}Cu_3O_7$	25 bar He, 0.5 bar O_2	5 h	
Formation of $(Tl_{0.7}Pb_{0.2}Bi_{0.2})(Sr,Ba)_2Ca_{1.9}Cu_3O_x$	870 °C, flowing O_2		
$SrCO_3$, $BaCO_3$,$CaCO_3$, CuO	900 °C, flowing O_2	24 h	[99]
$Sr_{1.8}Ba_{0.2}Ca_{1.9}Cu_3O_7$ precursor	980 °C, pellets	24 h	
Tl_2O_3, PbO, Bi_2O_3,	930–990 °C	3–6 h	
$Sr_{1.8}Ba_{0.2}Ca_{1.9}Cu_3O_7$	25 bar He, 0.5 bar O_2		
Formation of $(Tl_{0.7}Pb_{0.2}Bi_{0.2})(Sr,Ba)_2Ca_{1.9}Cu_3O_x$	890 °C, flowing O_2	3–5 h	

the formation of the Tl-2223 phase, low carbon content of the $Ba_2Ca_2Cu_3O_x$ precursor is necessary to make this alternative synthesis route effective.

14.2.4 Mercury-based Superconductors

To date the highest critical temperature of 135 K at ambient pressure has been achieved in the mercury-based cuprate $HgBa_2Ca_2Cu_3O_{8+\delta}$. This extraordinarily high value has attracted considerable attention. Disadvantages of mercury-based superconductors are the fact that mercury is as poisonous as, and even more volatile than, thallium.

Here, we follow the description of the preparation of mercury-based cuprate superconductors presented in [1]. The precursor powders need to be handled in a glove box filled with a dry inert gas to avoid contamination with H_2O and CO_2, which were found to be detrimental to the formation of Hg-12($n-1$)n cuprate superconductors. In order to counteract the loss of mercury at the relatively high temperatures required for the formation of Hg-12($n-1$)n superconductors, the precursor pellets are typically encapsulated in evacuated quartz tubes ($p < 10^{-4}$ mbar) during the heat treatments. The HgO used as a mercury source decomposes at relatively low temperature. There is a risk that the resulting mercury pressure is high enough to rupture the quartz tube. For safety reasons, the quartz tubes must be placed in stainless steel containers.

First, the synthesis of $HgBa_2CuO_{4+\delta}$ (Hg-1201) is briefly described [100–110]. Because the first member of the Hg-12($n-1$)n family does not contain calcium, the only superconducting phase that can be formed is Hg-1201. Typical synthesis procedures described in the literature are presented in Table 14.8. The 1201 phase can be prepared by solid-state reaction of the simple oxides

$$HgO + 2BaO + CuO \rightarrow HgBa_2CuO_4 \tag{14.7}$$

A final annealing is required for optimization of the oxygen content to obtain the maximum critical temperature of 96 K.

In another synthesis route a $Ba_2CuO_{3+\delta}$ precursor is prepared separately [100]. The metal nitrates $Ba(NO_3)_2$ and $Cu(NO_3)_2$ were decomposed in a two-step heat treatment at 600 °C and 800 °C in air. The final heat treatments were performed in flowing oxygen at 925 °C with two intermediate grindings, the second done in a dry box. In a second step, the precursor is thoroughly mixed with appropriate amounts of HgO in an agate mortar. The high-temperature synthesis (920 °C) was performed under a pressure of 18 kbar for times in the range of 0.6 to 2 h. The copper valence and the exact oxygen content in the synthesized $HgBa_2CuO_{4+\delta}$ can be controlled by their values in the $Ba_2CuO_{3+\delta}$ precursor. The oxygen content in the precursor can be adjusted by the heat treatment conditions [100].

The high vapor pressure of mercury originating from the decomposition of HgO causes the formation of $HgBaO_2$ as an impurity. Starting compositions deficient in mercury can suppress the formation of $HgBaO_2$.

Another synthesis route is based on the use of a Ba_2CuO_x precursor and unreacted Hg-1201 pellets as the mercury source [101–103]. An advantage of this preparation route is the reduced mercury pressure, which avoids the formation of $HgBaO_2$ impurity. Two Hg-1201 and one Ba_2CuO_x pellets are wrapped in gold foils, which are encapsulated in evacuated quartz tubes as illustrated in Figure 14.6.

Table 14.8 *Conditions for the synthesis of Hg-1201 powders (data taken from [1])*

Reactants	Heat treatments	Duration	Comments	Reference
HgO, BaO, CuO	Evacuated quartz tube 700–750 °C 300 °C, flowing Ar	1/4–1h 6 h	Mixing in dry box Pellet in Au tube Optimization of T_c	[104]
HgO, BaO, CuO	Evacuated quartz tube 750° 400 °C, flowing Ar	1 h 6 h	Mixing in dry box Bars Optimization of T_c	[105]
HgO, BaO, CuO	Sealed quartz tube 820 °C 820 °C → 200 °C 400 °C, flowing O_2	8 h 2 h 2–4 h	Ar filled dry box Preheated Furnace Optimization of T_c	[106]
HgO, BaO, CuO	Evacuated quartz tube 100 °C/h → 700–750 °C 700–750 °C 10 °C/h → 600 ° Furnace-cooling	 24 h	Ba_2CuO_3 + two Hg-1201 pellets wrapped in Au foil Ba_2CuO_3 pellet → No CuO, $HgBaO_2$, $BaCuO_2$ impurities	[101]
HgO, BaO, CuO	Evacuated quartz tube 750 °C, slow cooling	10 h	Ba_2CuO_3 + two Hg-1201 pellets wrapped in Au foil	[102]
BaO, CuO	860 °C, air	2 × 50 h	Preparation of Ba_2CuO_3 precursor	[107]
Ba_2CuO_3, HgO	Evacuated quartz tube RT → 800 °C 800 °C, 800 °C → RT	 5 h 5 h, 6 h	Powders handled in glove box 30% mercury loss	
BaO, CuO	930 °C, 20% O_2/80% Ar	24 h	Preparation of Ba_2CuO_3 precursor	[108]
Ba_2CuO_3, HgO	Evacuated quartz tube 845 °C 500 °C, flowing O_2	 8 h 24 h	 Preheated furnace Optimization of T_c	
BaO, CuO	925 °C, flowing O_2		Preparation of Ba_2CuO_x precursor	[109]
Ba_2CuO_x, HgO	Evacuated quartz tube 800 °C, high heating and cooling rates	 5–15 h	Pellets wrapped in Au foil	
BaO, CuO	930 °C, 20% O_2/80% Ar	24 h	Preparation of Ba_2CuO_x precursor Glove bag with Ar	[110]
Ba_2CuO_x, HgO	Evacuated quartz tube 800–830 °C, slow cooling	 8 h	Ba_2CuO_x pellet $Hg_yBa_2CuO_x$ pellets $y = 0.7$ → single-phase	
Ba_2CuO_x, HgO	Evacuated silica tube 790 °C	8 h	Ba_2CuO_x + two Hg-1201 pellets Ba_2CuO_x + HgO → Hg-1201	[103]

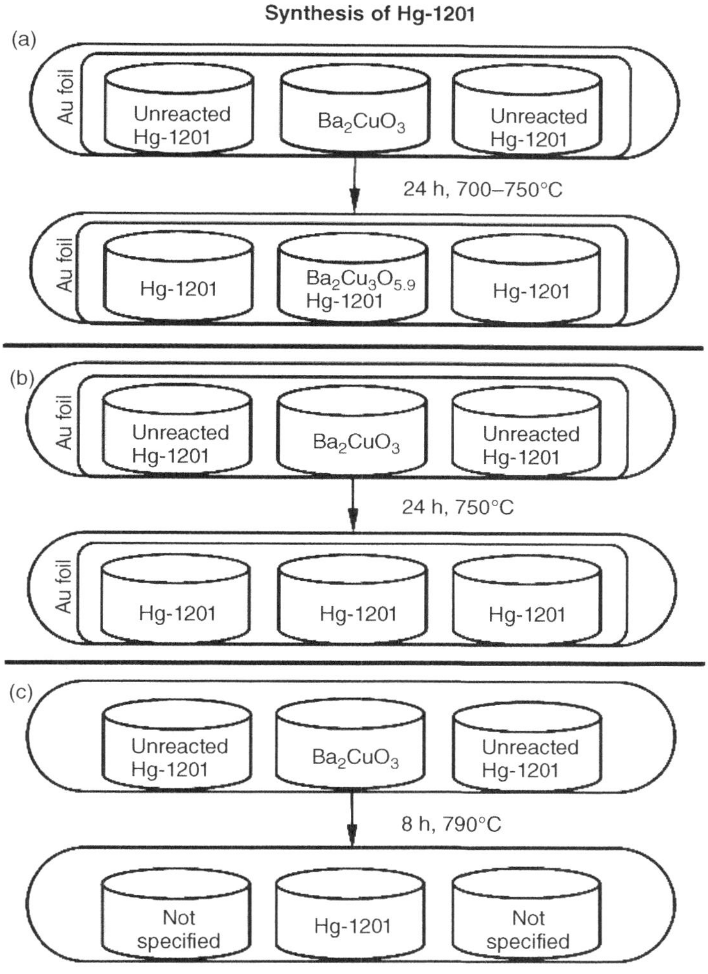

Figure 14.6 *Preparation of $HgBa_2CuO_{4+\delta}$ from Ba_2CuO_3 and unreacted $HgBa_2CuO_{4+\delta}$ pellets as mercury source (results from a) [101], b) [102], c) [103] (adapted from [1]))*

In all members of the Hg-12(n – 1)n family with $n \geq 2$, there are layers of Ca ions between the CuO_2 layers in the copper oxide blocks. In the Hg-1212 compound each copper oxide block consists of CuO_2–Ca–CuO_2 layers. The addition of a further Ca–CuO_2 unit in the copper oxide block leads to Hg-1223. As a consequence of their structural similarities, different members of the Hg-12(n – 1)n family are frequently formed simultaneously in a single sample. Careful control of the reaction conditions and the initial composition of the reactants are therefore required to obtain single-phase material. The conditions used for the preparation of Hg-1212 and Hg-1223 are listed in Table 14.9 [111–118] and Table 14.10 [119–135], respectively.

A description of the preparation of Hg-1234 can be found in [136–138]. There exist various cuprate high-T_c superconductors that can be prepared only under high pressure. A description of their synthesis can be found in [1] and the references therein.

Table 14.9 Conditions for the synthesis of Hg-1212 powders (data taken from [1])

Reactants	Heat treatments	Duration	Comments	Reference
HgO, BaO, CaO, CuO	Evacuated silica tube 680–750 °C	0.5–10 h	Mixing in glove box Pellet wrapped in Ag foil	[111]
	260 °C, O_2	6 h	Optimization of T_c	
HgO, BaO, CaO, CuO	Evacuated quartz tube 810 °C	8 h	Mixing in glove bag, pellets	[112]
HgO, BaO, CaO, CuO	Evacuated quartz tube 1–2 °C/min → 800 °C		Mixing in glove Bag	[113]
	800 °C	8 h		
$BaCuO_2$, CaO	900 °C, O_2	16 h	Preparation of $Ba_2CaCu_2O_5$	[114]
$Ba_2CaCu_2O_5$, HgO	Evacuated quartz tube 160 °C/h → 800 °C, 800 °C		Pellets	
	300 °C, flowing O_2	9 h	Optimization of T_c	
$Ba_2CaCu_2O_5$, HgO	Evacuated quartz tube 160 °C/h → 800 °C, 800 °C		Pellets	[115]
	Oxygenation at 300 °C	9 h	Optimization of T_c	
$BaCO_3$, CaO, CuO	900 °C, O_2	24–48 h	Preparation of $Ba_2CaCu_2O_x$	[116]
$Ba(NO_3)_2$, $Ca(NO_3)_2{\cdot}4H_2O$, $Cu(NO_3)_2{\cdot}4H_2O$	500–550 °C 600–620 °C 900 °C, flowing O_2	0.5 h 1 h 24–48 h	Nitrate Decomposition Preparation of $Ba_2CaCu_2O_x$	
$Ba_2CaCu_2O_x$, HgO	Evacuated quartz tubes Heating rate of 160 °C/h 750–820 °C	5 h	Pellets of $Ba_2CaCu_2O_x$ and Hg-1212	
	100–310 °C, flowing O_2	1–90 h	Optimization of T_c	
BaO_2, $CaCO_3$, CuO	800 °C, air	12 h	Preparation of $Ba_2CaCu_2O_x$	[117]
$Ba_2CaCu_2O_x$, HgO	800 °C, 60 bar	1 h	High-pressure synthesis	
$Ba(NO_3)_2$, $Ca(NO_3)_2{\cdot}4H_2O$, $Cu(NO_3)_2{\cdot}3H_2O$	Infrared evaporator 925 °C, flowing O_2	5 h 72 h	Preparation of $Ba_2CaCu_2O_x$	[118]
$Ba_2CaCu_2O_x$, HgO	880 °C, 18 kbar	1.5 h	Mixing in dry box High-pressure synthesis	
	300 °C, O_2	10 h	Optimization of T_c	

Table 14.10 Conditions for the synthesis of Hg-1223 powders (data taken from [1])

Reactants	Heat treatments	Duration	Comments	Reference
HgO, BaO, CaO, CuO	715–725 °C, nonclosed	4 × 3–4 min	Mixing in dry box	[119]
	Evacuated quartz tube		Bars wrapped in Au foil	
	750 °C	1–10 h		
	300 °C, flowing O_2	6 h	Optimization of T_c	
HgO, BaO, CaO, CuO	715–725 °C, nonclosed	4 × 3–4 min	Pellets wrapped in Au foil	[120, 121]
	Evacuated quartz tube			
	750 °C	10 h		
	300 °C, flowing O_2	8 h	Optimization of T_c	
HgO, BaO, CaO, CuO	Pellets in Al_2O_3 tubes		BaO, CaO	[122]
	Evacuated quartz tube		Prepared from carbonates	
	RT → 840–860 °C	8–10 h		
	840–860 °C	6 h	Mixing in dry box	
	840–860 °C → RT	8 h		
	300 °C, flowing O_2	6 h	Optimization of T_c	
HgO, BaO, CaO, CuO	Evacuated quartz tube		Mixing in dry box filled with Ar	[123]
	665 °C, quenched	70 h		
	300 °C, flowing O_2		Optimization of T_c	
HgO, BaO, CaO, CuO	Evacuated quartz tube		$BaCO_3 \rightarrow BaO + CO_2$	[124]
10% surplus of HgO	Preheated furnace		Mixing in dry box	
	660–670 °C	72 h	Pellets in Au foil	
	300 °C, flowing O_2	8 h	Optimization of T_c	
HgO, $Ba_2Ca_2Cu_3O_x$	Evacuated quartz tube		Pellets	[125]
	800 °C	5 h		
	300 °C, flowing O_2		Optimization of T_c	
HgO, $Ba_2Ca_2Cu_3O_x$	Sealed Au tube		Mixing in dry box	[126]
	850 °C, 25 kbar	1–3.5 h	Pellets	
$BaCuO_2$, CaO, CuO	900 °C, flowing O_2	16 h	Preparation of $Ba_2Ca_2Cu_3O_7$	[127]
HgO, $Ba_2Ca_2Cu_3O_7$	Evacuated quartz tube		Pellets	
	800–850 °C [a]	5 h		
	300 °C, flowing O_2		Optimization of T_c	

(continued overleaf)

Table 14.10 *(continued)*

Reactants	Heat treatments	Duration	Comments	Reference
$BaCO_3$, CaO,	900 °C	24 h	Preparation of	[128]
CuO	Evacuated quartz tube		$Ba_2Ca_2Cu_3O_7$	
HgO, $Ba_2Ca_2Cu_3O_7$	600–800 °C 300 °C, flowing O_2	8 h	 Optimization of T_c	
$Ba(NO_3)_2$, $Ca(NO_3)_2{\cdot}4H_2O$, $Cu(NO_3)_2{\cdot}3H_2O$ Dissolved in	Evaporation of H_2O RT → 530 °C 530 °C 530 °C → 620 °C, 620 °C	 4 h 2 h 2 h, 4 h	Green mixture Decomposition of nitrates	[129]
distilled H_2O	900 °C, in O_2	2 × 1–4 h	$Ba_2Ca_2Cu_3O_x$	
HgO, $Ba_2Ca_2Cu_3O_x$	Sealed quartz tube RT → 600 °C, 1 bar O_2 600 °C → 800 °C 800 °C 300 °C, O_2	 14 h 1.5 h 5 h	Pellets Furnace cooling Optimization of T_c	
$Ba(NO_3)_2$, $Ca(NO_3)_2{\cdot}4H_2O$, $Cu(NO_3)_2{\cdot}3H_2O$	600 °C, air 925 °C, O_2	12 h 72 h	Preparation of $Ba_2Ca_2Cu_3O_x$	[130]
HgO, $Ba_2Ca_2Cu_3O_x$	950 °C, 59 kbar	3 h	High-pressure synthesis	
$Ba(NO_3)_2$, $Ca(NO_3)_2{\cdot}4H_2O$, $Cu(NO_3)_2{\cdot}6H_2O$	60 °C/h → 600 °C, O_2 600 °C, O_2 60 °C/h → 900 °C 900 °C	 40 h 10 h	Preparation of $Ba_2Ca_2Cu_3O_x$	[131]
HgO, PbO, (4:1) $Ba_2Ca_2Cu_3O_x$	900 °C, 160 MPa	3 h	Hot isostatic pressing	
$Ba(NO_3)_2$, $Ca(NO_3)_2{\cdot}4H_2O$, $Cu(NO_3)_2{\cdot}3H_2O$ Ba-Ca-Cu-O,	200 °C 600 °C 900 °C, O_2 Pt or Au container	 72 h	Nitrate decomposition Ba–Ca–Cu-O Mixing in glove box	[132]
HgO	900 °C, 18 kbar	1–1.8 h		
BaO_2, $CaCO_3$	1000 °C dynamic vacuum		Preparation of BaO and CaO	[133]
HgO, BaO, CaO, CuO	Evacuated silica tube RT → 800 °C 800 °C 800 °C → RT 300 °C, flowing O_2	 5 h 20 h 16 h 5 h	Mixing in dry box Al_2O_3 crucibles $Ba_2Ca_2Cu_3O_7$+ 2Hg-1223 pellets Optimization of T_c	

Table 14.10 (continued)

Reactants	Heat treatments	Duration	Comments	Reference
$Ba_2Ca_2Cu_3O_7$, PbO or Bi_2O_3	$Pb_{0.2}Ba_2Ca_2Cu_3O_x$ or $Bi_{0.2}Ba_2Ca_2Cu_3O_x$		Precursors	[134, 135]
CaO, HgO	Evacuated quartz tube 775–800 °C	10 h	$CaHgO_2$ (Hg source)	
$CaHgO_2$,	Evacuated quartz tube		2 precursor+	
$Pb_{0.2}Ba_2Ca_2Cu_3O_x$	RT → 750–900 °C	2–3 h	Hg source pellets	
	750–900 °C	10–20 h	Hg,Pb-1223	
$Bi_{0.2}Ba_2Ca_2Cu_3O_x$	RT → 700–820 °C			
	700–820 °C	2–30 h	Hg,Bi-1223	

[a]Heating rate 160 °C/h

14.3 Bulk Cuprate High-T_c Superconductors

14.3.1 Introduction

In principle, bulk cuprate superconductors are interesting materials for various applications. Soon after the discovery of the weak-link problem in sintered cuprate superconductors, it was found that this problem can be strongly alleviated by melt processing. In bulk cuprate superconductors, moderately high critical current densities were achieved. Because of their extremely low thermal conductivity, bulk cuprate superconductors can be used in high-temperature superconducting current leads [139, 140]. Other possible applications are fault current limiters [141], bearings for magnetic levitation in flywheel energy storage [142], and magnet applications [143, 144]. The last are based on the fact that magnetic flux can be trapped in bulk high-temperature superconductors at temperatures well below T_c. The use of bulk high-T_c superconductors in motors and generators has been proposed [145].

In Section 14.3.2, the preparation and the properties of bismuth-based bulk superconductors are briefly described. Because of their relatively low irreversibility fields, Bi-2212 and (Bi,Pb)-2223 are not favored for magnet applications, which require a relatively large magnetic flux to be trapped at temperatures below T_c. In RE-123 (where RE is a rare earth element) bulk superconductors, characterized by much higher irreversibility fields than Bi-2212 and (Bi,Pb)-2223, the trapped flux can be as high as 17 T at 29 K [146]. The preparation and the properties of RE-123 high-T_c superconductors are described in Section 14.3.3.

14.3.2 Bi-2212 and (Bi,Pb)-2223 Bulk Superconductors

Bi-2212 bulk material can be manufactured by the melt cast process (MCP). The precursor oxides Bi_2O_3, CaO, SrO, and CuO are mixed in amounts leading to the cation ratio Bi : Sr : Ca : Cu = 2 : 2 : 1 : 2. The mixture of the oxides is melted for 15 min at a temperature of 1050 °C. The melt, which is cast in a copper mould, is slowly cooled to

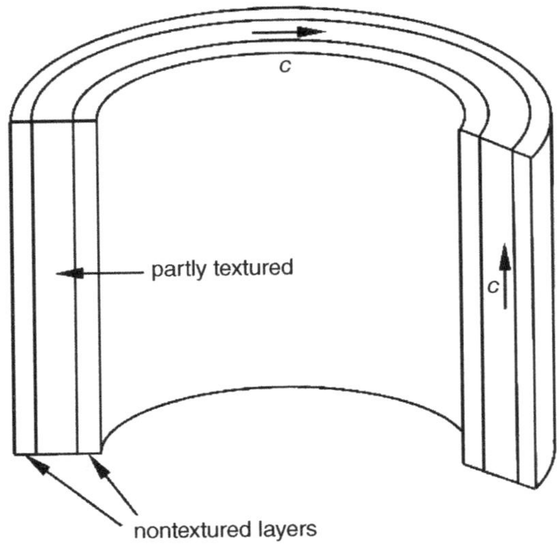

Figure 14.7 *Sketch of a Bi-2212 tube, which shows the nontextured surface layers and the partly textured central part of the wall. In the central part, the crystallographic c axes of the Bi-2212 grains are preferentially parallel to the tube axis or azimuthal direction (adapted from [148])*

room temperature. Nearly single-phase $Bi_2Sr_2CaCu_2O_8$ bulk material of high density is formed during annealing at 815 °C for 36 h [147]. In a slightly modified way, Bi-2212 rings and hollow cylinders can be manufactured. A homogenous melt of the metal oxides with the addition of 10% $SrSO_4$ is cast on the inner wall of a rotating steel mould. After solidification, the bulk material contains several different phases. Again an annealing at 750 °C to 850 °C promotes the formation of nearly single-phase Bi-2212 bulk material [148]. Typical dimensions of Bi-2212 rings manufactured by melt casting are 50 to 200 mm in length and 20 to 400 mm in diameter, with walls 2 to 8 mm thick [148]. A typical orientation of the Bi-2212 crystallites is illustrated in Figure 14.7. Thin layers at the inner and outer radii of the Bi-2212 bulk tubes are not textured, whereas in the central zone of the bulk material the crystallographic *c* axes of the Bi-2212 crystallites are preferentially parallel to the axial or the azimuthal directions. On the other hand, there exists no preferential orientation of the *c* axes parallel to the radial direction, which is the direction of the thermal gradient in the melt cast process [148].

As an alternative to Bi-2212, bulk (Bi,Pb)-2223 high-T_c superconductors were prepared by cold isostatic pressing (CIP) with repeated annealing steps. In sintered (Bi,Pb)-2223 rods of 10 mm diameter, critical current densities of 300 to 500 A/cm^2 were achieved at 77 K and zero applied magnetic field [140]. The initial powders used for the manufacture of (Bi,Pb)-2223 bulk material contained 90% of the 2223 phase. The best performance was found for bulk material prepared from partially reacted (Bi,Pb)-2223 powders [140]. The critical temperature of the (Bi,Pb)-2223 superconductor reaches a value of 105 K, as compared to only 90 K for Bi-2212. Moreover, the irreversibility field of (Bi,Pb)-2223 at 77 K

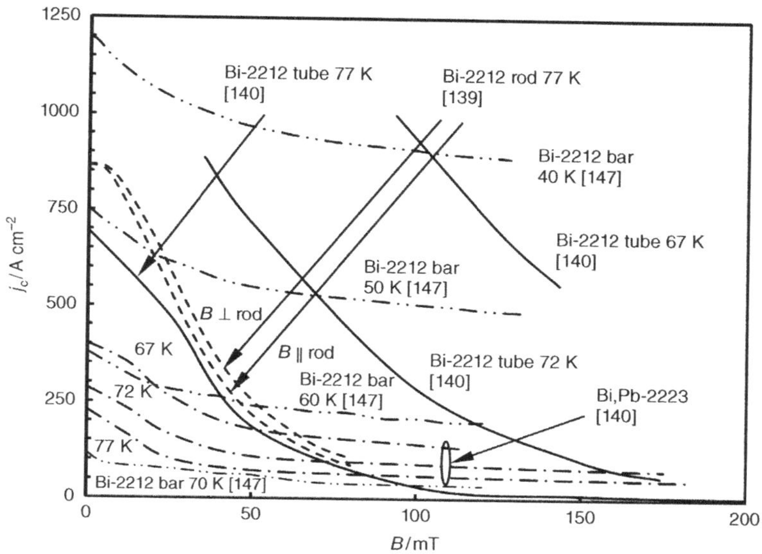

Figure 14.8 *Critical current densities versus field of various Bi-2212 and (Bi,Pb)-2223 bulk superconductors (results from [139, 140, 147])*

is significantly larger than that of Bi-2212. Unfortunately, in contrast to melt processing, the sintering process utilized for the manufacture of (Bi,Pb)-2223 bulk material provides only relatively low critical current densities.

Figure 14.8 shows the critical current densities of various bismuth-based high-T_c superconductors as a function of the applied magnetic field. The critical current of (Bi,Pb)-2223 bulk material reaches a value of $\approx$230 A/cm^2 at 77 K with no applied field. The relatively low critical current density is caused by the fact that in sintered (Bi,Pb)-2223 samples, the grain boundaries are strongly coupled in only a small fraction of the cross-section, while the majority of the grain boundaries act as weak links. A reduction of the temperature to 67 K increases the j_c in zero applied field to about 400 A/cm^2. In Bi-2212 rods, a critical current density of $\approx$870 A/cm^2 was achieved at 77 K without an applied field.

The j_c depends weakly on the direction of the applied magnetic field. The values measured in magnetic fields perpendicular to the axis are slightly higher than those for parallel fields. Bi-2212 tubes show a similar performance to the rods. The lowest critical current densities were reported for bar-shaped Bi-2212 samples. For all Bi-2212 samples, the dependence of j_c on the applied magnetic field is more pronounced than that found for the (Bi,Pb)-2223 bulk material. The reason for this difference is the lower irreversibility field of the Bi-2212.

Figure 14.9 shows the dependence of the critical current densities of (Bi,Pb)-2223 tubes [148], and Bi-2212 bars [147] and tubes [148] on temperature. The Bi-2212 bars and tubes show a similar dependence of j_c on temperature. However, the absolute values of j_c published for the tubes in 2010 [148] are much higher than those found for the

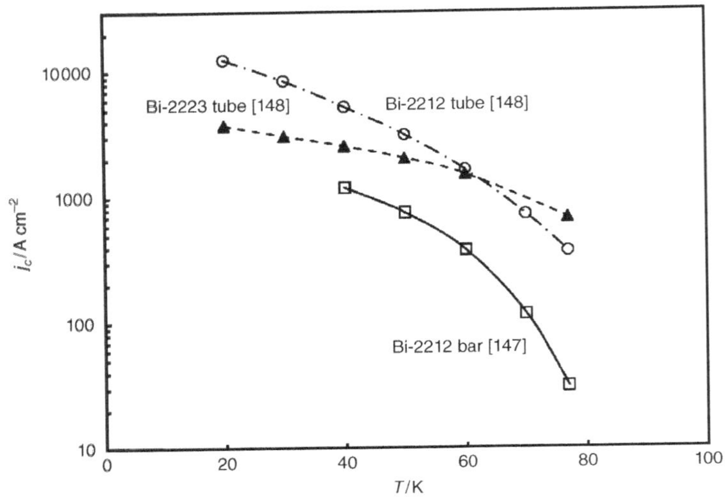

Figure 14.9 *Temperature dependence of the critical current density of Bi-2212 and (Bi,Pb)-2223 bulk superconductors (results from [147, 148])*

bars (cross-section of 1 mm × 1 mm) in 1989 [147]. The comparison indicates that the performance of Bi-2212 bulk material has been significantly improved since 1989. Due to the higher critical temperature of the (Bi,Pb)-2223 bulk material ($T_c^{2223} \approx 108$ K, $T_c^{2212} \approx 92$ K) the j_c of the 2223 tubes is higher than that of the 2212 tubes at temperatures above 60 K. On the other hand, the j_c of the 2212 tubes is significantly larger at lower temperatures. The better performance of the 2212 tubes originates from the fact that the connectivity of the grains in the melt cast 2212 is better than in the sintered 2223.

An interesting feature is the extremely low thermal conductivity of Bi-2212 and (Bi,Pb)-2223 bulk material. The thermal conductivity integrated over the temperature range of 4.2 to 77 K is as low as 80 W/m for Bi-2212 and 103 W/m for (Bi,Pb)-2223 [139]. In spite of relatively low critical current densities, the use of Bi-2212 or (Bi,Pb)-2223 for the low-temperature part of current leads would be very effective in reducing the heat load at the 4 K level due to heat conduction [139]. A drawback of the very low thermal conductivity is the difficulty of protecting the current lead against a burn-out in case of a quench (see, e.g., [149]).

14.3.3 RE-123 Bulk Superconductors

The most attractive bulk superconductors for applications requiring a high-trapped field are RE-123 compounds (where RE represents Y or another rare earth element). Soon after the discovery of superconductivity above 77 K in $YBa_2Cu_3O_{7-x}$ (Y-123), the grain boundaries in sintered material were found to act as weak links. Melt processing was found to strongly alleviate the weak link problem. All melt processing techniques are based on the peritectic reaction, in which the RE-123 phase melts incongruently over a wide compositional range to form RE_2BaCuO_5 (RE-211) and a Ba–Cu–O liquid phase. In the case of $YBa_2Cu_3O_{7-x}$ the peritectic temperature T_p is around 1015 °C. Slow cooling of this mixture below T_p

leads to the formation of the desired 123 phase according to the following chemical reaction [143, 150]:

$$Y_2BaCuO_5 + L(3BaCuO_2 + 2CuO) \rightarrow 2YBa_2Cu_3O_x \tag{14.8}$$

The strongly coupled grains formed during the melt processing lead to much higher transport critical current densities than in sintered material. Furthermore, the dependence of j_c on applied magnetic field is much less pronounced than in sintered Y-123.

The main steps of the melt texture growth (MTG) process of Y-123 [151, 152] are presented in Figure 14.10. The Y-123 powder is typically prepared by the solid-state

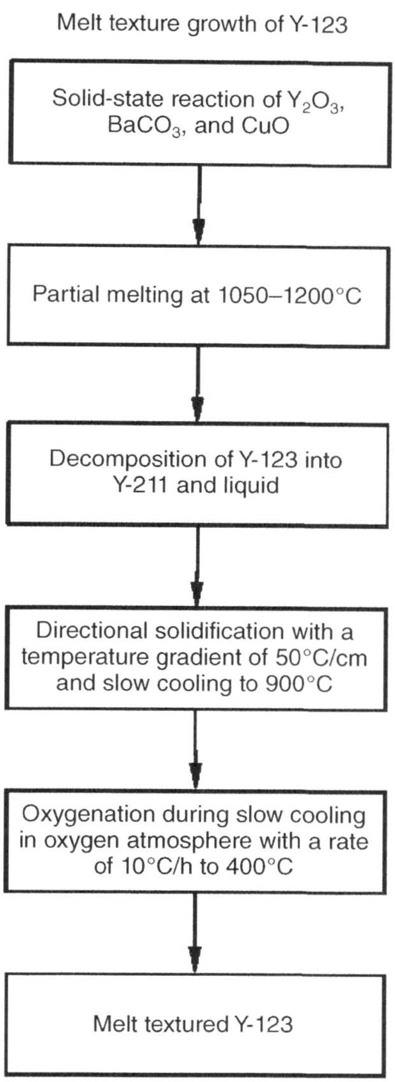

Figure 14.10 *Sketch of melt texture growth as described in [151, 152] (adapted from [1])*

reaction of Y_2O_3, $BaCO_3$, and CuO. Pressed pellets are heated above the peritectic temperature, leading to the decomposition of the Y-123 phase into solid Y-211 and an off-stoichiometric Ba–Cu–O liquid. During slow cooling through the peritectic temperature, the $YBa_2Cu_3O_{7-x}$ phase is re-formed according to reaction 14.8. The application of a thermal gradient of 50 °C/cm during the slow cooling to 900 °C leads to a textured microstructure. To reach the optimum critical temperature, the melt textured bulk material is cooled at a rate of 10 °C/h to 400 °C in oxygen.

A melt processing route without the application of a thermal gradient was developed by Salama *et al.* [153] (see Figure 14.11). Again the Y-123 powder is synthesized by the standard solid-state reaction. The powder is cold pressed into bar-shaped samples, which are inserted into a preheated furnace. The samples are rapidly cooled from 1100 °C to 1030 °C followed by slow cooling (1 °C/h) to 980 °C. After annealing for 8 h at 980 °C the material is furnace-cooled to 600 °C. The suppression of crack formation requires slow

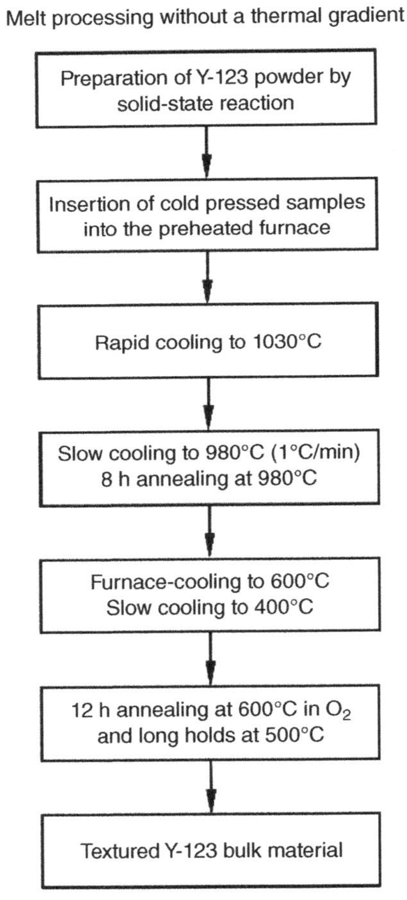

Figure 14.11 *Sketch of melt processing without the use of a thermal gradient as described in [153] (adapted from [1])*

cooling from 600 °C to 400 °C. The bulk material is oxygenated by annealing at 600 °C for 12 h in oxygen, followed by long holds at 500 °C and 400 °C.

Another way to prepare RE-123 bulk material is the melt powder melt growth (MPMG) process [143, 154]. We consider again the manufacture of Y-123 bulk material. A presintered Y–Ba–Cu–O sample is rapidly heated to a temperature above 1300 °C. In the peritectic reaction, the Y–Ba–Cu–O sample decomposes into a liquid and Y_2O_3. Using cold copper plates, the material is rapidly quenched to room temperature. In a second step, the material is heated to a temperature in the range of 1000–1200 °C, leading to re-melting of the sample. The Y_2O_3 reacts with the melt to form Y_2BaCuO_5, which reacts with the remaining liquid to form $YBa_2Cu_3O_{7-x}$. During these reactions, fine Y-211 particles of about 1 μm size are dispersed in the Y-123 matrix.

The formation of textured RE-123 bulk material can be facilitated by the use of seed crystals [143, 155–159]. We again consider $YBa_2Cu_3O_{7-x}$ as an example. A thin RE-123 single crystal, placed on top of a presintered Y-123 sample, acts as an effective nucleation site, which controls the nucleation of the Y-123 phase. The formation temperatures of RE-123 phases vary depending on the rare-earth element present in the compound. The RE-123 seed crystal must have a higher formation temperature than the presintered Y-123. The formation temperature of Y-123 is 1000 °C, while Sm-123, Eu-123, Gd-123, and Dy-123 nucleate at higher temperatures of 1060 °C, 1050 °C, 1030 °C, and 1010 °C, respectively [143]. Ho-123, Er-123, and Yb-123 are characterized by lower formation temperatures of 990 °C, 970 °C, and 900 °C, respectively [143]. As an alternative to thin RE-123 single crystals, novel thin films of Nd-123 grown on MgO crystals have been used to seed the formation of Nd-123, Sm-123, Eu-123, and Gd-123 [156]. The phenomenon of superheating has been observed for Nd-123 and other RE-123 thin films. This means that partial melting of the thin RE-123 films occurs at a temperature much higher than the peritectic temperature of the corresponding RE-123 bulk material. For the growth of large Y-123 bulk superconductors, the use of four seed crystals (multiseeding) has been successfully demonstrated by Wongsatanawarid *et al.* [157].

The critical current density of RE-123 bulk material depends on the microstructure of the material. RE-211 particles and the stresses around them, added nanoparticles, dislocations, stacking faults, and point defects can all act as pinning centres. On the other hand, cracks, and porosity obstruct the current flow. Diko *et al.* [155] studied crack formation in Y-123 bulk superconductors prepared by top-seeded melt growth. They found that post-growth treatments required to obtain the optimum oxygen content can introduce microcracks, which can significantly degrade the properties of the Y-123 bulk material. The formation of oxygenation cracks can be avoided by a high-pressure (160 bar) and high-temperature (750 °C) post-growth heat treatment.

The critical current densities achieved in melt-processed RE-123 bulk material are listed in Table 14.11. In the early days of high-temperature superconductivity, critical current densities of 17 000 A/cm^2 at 77 K and zero applied field were achieved in melt-processed Y-123, much higher than the values achieved in sintered material. In addition, the critical current density at a field of 1 T reached a value of 4000 A/cm^2 [151, 152]. The moderate decrease of j_c in the presence of an applied magnetic field is again in contrast to the very pronounced drop of j_c found in sintered Y-123. The use of seed crystals and improved pinning allowed the increase of j_c in present-day RE-123 bulk material to values of more than 300 000 A/cm^2 at 77 K and zero applied field [159].

Table 14.11 *Critical current densities achieved in melt processed bulk material*

Bulk material	T (K)	B (T)	j_c (A/cm^2)	Comments	Reference
$YBa_2Cu_3O_{7-x}$	77 77	0 1	17 000 ≈4 000	Melt textured, directional solidification	[151, 152]
$YBa_2Cu_3O_{7-x}$	77	0.6	37 000	Melt textured without thermal gradient	[153]
$YBa_2Cu_3O_{7-x}$	77 77	0 20[a]	71 000 11 000	Melt textured Y-123 doped with $BaSnO_3$	[160]
$YBa_2Cu_3O_7$, Y_2BaCuO_5	77	0	≈120000	Melt texture growth Annealing at 450 °C, $pO_2 = 100$ bar	[161]
(Nd,Eu,Gd)-123	77	0	≈72000	40 mol% (Nd,Eu,Gd)- 211	[158]
(Nd,Eu,Gd)- 123 + Gd-211	77 77	0 0	≈196000 ≈310000	0 mol% TiO_2 0.1 mol% TiO_2	[159]
$GdBa_2Cu_3O_{7-x}$	77	0	≈115000	Seeded infiltration and growth	[162]
Y-123	77	0	≈42000	$YBa_2(Cu_{1-x}Al_x)_3O_{7-\delta}$ $x = 0.0025$	[155]

[a] B parallel ab.

The field dependence of the critical current density in various RE-123 bulk superconductors is presented in Figure 14.12. For sample Y-123 A[1], the critical current density along the CuO_2 (ab) planes drops from ≈120000 A/cm^2 at 77 K and zero applied field to ≈20000 A/cm^2 in a field of 1 T parallel to the crystallographic c direction [161]. At a field of 2.8 T j_c is still around 10 000 A/cm^2. Because of the textured growth of the RE-123 bulk material, the critical current density depends on the direction of the applied magnetic field. This behavior is illustrated for sample Y-123 B (see note 1) [160]. At fields above 1.5 T the decrease of j_c with increasing field is much more pronounced for $B \,||\, c$ than for $B \,||\, ab$. An interesting feature of sample B, which has been heat-treated in 100 bar oxygen at 400 °C, is the improved pinning at fields of around 1 T parallel to the c direction. This has been attributed to stacking faults generated by the transformation of $YBa_2Cu_3O_7$ (Y-123) to $YBa_2Cu_4O_8$ (Y-124) [160]. In case of sample Y-123 C (see note 1) the effect of aluminum doping on flux pinning is illustrated. Up to a magnetic field of ≈2.7 T the j_c of the undoped sample is significantly lower than that measured in the sample $YBa_2(Cu_{1-x}, Al_x)_3O_{7-\delta}$ ($x = 0.0025$). The substitution of Al for Cu improves the pinning significantly for fields of 1 T parallel to the c direction. The achievable critical current density also depends on the rare earth elements present in the 123 compound in question. The results shown in Figure 14.12 suggest that the critical current densities in Gd-123 [162] and (Nd,Eu,Gd)-123 (NEG-123) [158, 159] are higher than those in Y-123. The critical current density of NEG-123 was found to depend on the preparation of the mixture of the NEG-123 and the NEG-211 powders. Ball-milling significantly increased the j_c of NEG-123 in fields of less than 5 T ($B||c$) [158]. The addition of 0.1 mol% TiO_2 to NEG-123/35 mol% Gd-211 was found to improve the performance of NEG-123 bulk material.

[1] Not the same sample as referred to previously.

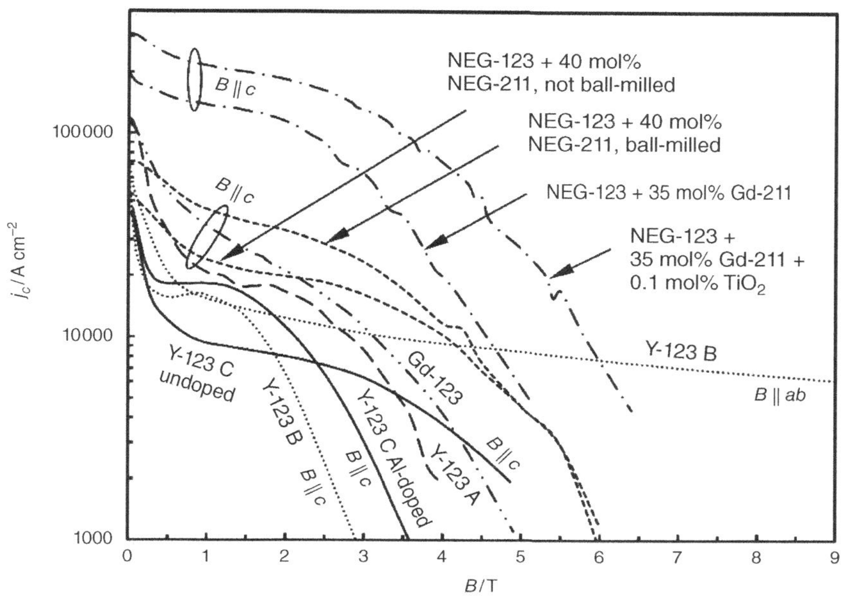

Figure 14.12 Critical current density versus field of various RE-123 bulk superconductors (results from [155, 158–162])

Finally, let us briefly consider the ability of RE-123 bulk material to trap magnetic flux. We follow here the discussion of the possible use of bulk cuprate superconductors for magnet applications by Campbell and Cardwell [143]. A bulk high-temperature superconductor can be magnetized by cooling from well above to below the transition temperature T_c in the presence of a magnetic field (field-cooling). Alternatively the bulk superconductor can be cooled to a temperature below T_c without an applied magnetic field. At low temperature a sufficiently large magnetic field is applied to magnetize the bulk superconductor (cold magnetization). The trapped magnetic field can be estimated using the Bean model, which assumes a constant j_c for the screening currents within the bulk superconductor. For simplicity, we consider a long cylinder where demagnetization effects can be omitted. At the penetration field ($B_p = \mu_0 j_c a$, where a is the radius of the cylinder) the applied magnetic field reaches the centre of the rod, i.e., the whole cylinder is filled with screening currents. To trap the maximum field it would be necessary to apply a magnetic field B_p in a field-cooling experiment. In case of cold magnetization a field twice as high as B_p needs to be applied. After removal of the external magnetic field, the superconducting cylinder traps a flux, which, for a cylinder fully saturated with current, reaches $B_p/2$ at the surface.

In Table 14.12, values of the trapped field reported in the literature are presented. The data indicate that the field trapped in the gap between two Y-123 disks of 24 mm diameter can reach around 1 T at 77 K [163]. At a reduced temperature of 51.5 K the trapped field can be enhanced to a value as high as 8.5 T. A record value of more than 17 T trapped field at 29 K has been reported for a Y-123 disk of 26.5 mm diameter. In general, the trapped field can be significantly increased at lower temperatures (see, e.g., the data for (Nd,Eu,Gd)-123 [158]). As expected the maximum achievable trapped field increases with

Table 14.12 *Trapped field in melt processed RE-123 bulk material*

Bulk material	*T* (K)	B_t (T)	Comments	Reference
$YBa_2Cu_3O_{7-x}$ (Y-123)	29	17.24	Diameter of 2.65 cm	[146]
$YBa_2Cu_3O_{7-x}$ (Y-123)	77	1[a]	Diameter of 2.4 cm	[163]
	51.5	8.5[a]		
$YBa_2Cu_3O_{7-x}$ (Y-123)	77	≈0.53	Diameter of 4.3 cm	[164]
$GdBa_2Cu_3O_{7-x}$ (Gd-123)	77	1.5	Ag addition, D = 3.2 cm	[165]
$GdBa_2Cu_3O_{7-x}$ (Gd-123)	77	0.32	Diameter of 2 cm	[162]
$GdBa_2Cu_3O_{7-x}$ (Gd-123)	77	≈0.9	Diameter of 2.4 cm	[156]
	77	1.35	Diameter of 4.5 cm	
$SmBa_2Cu_3O_{7-x}$ (Sm-123)	77	4.07	Diameter of 6 cm	[158]
(Nd,Eu,Gd)$Ba_2Cu_3O_{7-x}$	77	≈1.4	Diameter of 3 cm	[158]
((Nd,Eu,Gd)-123)	70	≈2.1		
	65	≈2.9		
	60	≈3.9		
	55	≈5.1		
	50	≈5.8		
	45	≈6.9		
	40	≈7.6		
	35	≈8.5		

[a]Trapped field in the gap between two samples.

the size of the bulk superconductor, as demonstrated for Gd-123 disks of 24 and 45 mm diameter [156]. In a Sm-123 disk of 6 cm diameter, a field of more than 4 T could be trapped even at a temperature of 77 K [158].

14.4 Summary

The synthesis of cuprate powders is an important step in the manufacturing of superconducting wires and tapes. RE-123, Bi-2212, and (Bi,Pb)-2223 powders can be prepared by solid-state reaction of binary oxides and carbonates. In order to ensure adequate homogeneity and low residual carbon content, several calcination steps with intermediate grindings are required. The detrimental effects of moisture and carbon dioxide can be minimized by handling the powders in a glove box filled with synthetic air or an inert gas. The main disadvantages of the superconductor powder preparation by solid-state reaction are long heat treatment times, time consuming intermediate grindings, and the unavoidable chemical inhomogeneity on the atomic scale. These disadvantages can be avoided by solution-based processing routes such as citrate pyrolysis and oxalate co-precipitation.

The synthesis of thallium- and mercury-based cuprate superconductors is more difficult because of their toxicity and volatility. To avoid substantial loss of thallium or mercury during long heat treatments at high temperatures, two-step reactions are used. In the first step, a precursor containing the nonvolatile constituents Ba, Ca, and Cu is prepared. In the second, the precursor reacts with Tl_2O_3 or HgO. For the synthesis of Hg-based high-temperature superconductors, nonreacted pellets of Hg–Ba–Ca–Cu–O are frequently used as a mercury source in place of HgO. By using unreacted Hg-12($n-1$)n pellets, mercury vapor pressure can be limited to acceptable values. Another way to avoid Hg or Tl loss is the synthesis of these cuprates under high pressure.

For some applications, superconducting bulk material is an interesting alternative to the use of superconducting wires and tapes. Bi-2212 bulk material can be easily manufactured by a melt cast process. Because of the low irreversibility field of Bi-2212 at temperatures close to 77 K, the material is suitable for low-field applications only. For magnetic levitation and magnet applications, RE-123 bulk material is more suitable because of the much higher irreversibility field at 77 K.

In applications of bulk superconductors, the possible occurrence of flux jumping has to be carefully investigated. For low-temperature superconductors, it is well known that the superconductor in a wire must be subdivided into thin filaments to avoid flux jump instability. Because of the much larger specific heat at elevated temperatures and the generally larger temperature margins provided by high-temperature superconductors, even bulk material may not show flux jump instabilities. A detailed discussion can be found in the literature [166–172].

References

1. R. Wesche, *High-Temperature Superconductors: Materials, Properties and Applications*, Kluwer Academic Publishers, Norwell, MA USA, 1998.
2. R.J. Cava, B. Battlog, R.B. van Dover, D.W. Murphy, S. Sunshine, T. Siegrist, J.P. Remeika, E.A. Rietman, S. Zahurak, and G.P. Espinosa, Bulk superconductivity at 91 K in single-phase oxygen-deficient perovskite $Ba_2YCu_3O_{9-\delta}$, *Phys. Rev. Lett.*, **58**, 1676–1679 (1987).
3. G.K. Bichile, D.G. Kuberkar, K.M. Jadhav, S.S. Shah, and R.G. Kulkarni, Influence of nickel substitution on flux pinning and critical currents in $YBa_2Cu_3O_{7-\delta}$, *Cryogenics*, **31**, 833–838 (1991).
4. V.G. Bessergenev, Yu.A. Kovalevskaya, V.N. Naumov, and G.I. Frolova, Electron and phonon characteristics of $YBa_2Cu_3O_{7-\delta}$, *Physica C*, **245**, 36–40 (1995).
5. M. Ishikawa, Y. Nakazawa, T. Takabatake, A. Kishi, R. Kato, and A. Maesono, Specific heat study on samples of $Ba_2YCu_3O_7$ with a double superconducting transition, *Solid State Commun.*, **66**, 201–204 (1988).
6. J. Heise, P. Gutsmiedl, K. Neumaier, Chr. Probst, H. Berndt, P. Müller, and K. Andres, On the influence of the heat treatment on the diamagnetic and thermodynamic properties of sintered $YBa_2Cu_3O_7$, *Physica C*, **153–155**, 1507–1508 (1988).
7. A. Manthiram and J.B. Goodenough, Factors influencing T_c in 123 copper-oxide superconductors, *Physica C*, **159**, 760–768 (1989).
8. H. Küpfer, I. Apfelstedt, W. Schauer, R. Flükiger, R. Meier-Hirmer, and H. Wühl, Critical current and upper critical field of sintered and powdered superconducting $YBa_2Cu_3O_7$, *Z. Phys. B*, **69**, 159–166 (1987).
9. A.K. Gangopadhyay and T.O. Mason, Solubility of Ag in $YBa_2Cu_3O_{6+y}$ and its effect on superconducting properties, *Physica C*, **178**, 64–70 (1991).
10. P. Müller, M. Schubert, Ch. Rodig, G. Fuchs, and K. Fischer, Heat treatment of wires on the basis of the high T_c superconductor $YBa_2Cu_3O_x$, *Appl. Phys. Lett.*, **55**, 917–918 (1989).

11. S.D. Peacor, C. Uher, A.H. Francis, and J. Shewchun, Physical and electromagnetic properties of Y-Ba-Cu-O superconductors synthesized with peroxides, *J. Appl. Phys.*, **67**, 7488–7492 (1990).
12. A. Junod, A. Bezinge, D. Cattani, M. Decroux, D. Eckert, M. François, A. Hewat, J. Muller, and K. Yvon, *Helvetica Physica Acta*, **61**, 460–469 (1988).
13. A. Junod, A. Bezinge, and J. Muller, Optimization of the specific heat jump at T_c and magnetic properties of the superconductor $YBa_2Cu_3O_7$, *Physica C*, **152**, 50--64 (1988).
14. S. Vieira, S. Bourgeal, R. Villar, A. Aguilo, M.A. Ramos, C. Moure, and P. Duran, *Solid St. Commun.*, **65**, 1555–1557 (1988).
15. Y. Idemoto, K. Shizuka, Y. Yasuda, and K. Fueki, Standard enthalpies of formation of member oxides in the Bi–Sr–Ca–Cu–O system, *Physica C*, **211**, 36–44 (1993).
16. J. Tenbrink, M. Wilhelm, K. Heine, and H. Krauth, Development of high-T_c superconductor wires for magnet applications, *IEEE Trans. Magn.*, **27**, 1239–1246 (1991).
17. W. Zhang and E.E. Hellstrom, The influence of carbon on melt processing Ag sheathed $Bi_2Sr_2CaCu_2O_8$ tape, *Physica C*, **234**, 137--145 (1994).
18. T. Hasegawa, T. Kitamura, H. Kobayashi, F. Takeshita, H. Kumakura, H. Kitaguchi, and K. Togano, Preparation of Bi-Sr-Ca-Cu-O superconducting tape by pyrolysis of organic acid salts, *Physica C*, **190**, 81–83 (1991).
19. P.E. Kazin, M. Jansen, A. Larrea, G.F. de la Funte, and Yu.D. Tretyakov, Flux pinning improvement in Bi-2212 silver sheathed tapes with submicron $SrZrO_3$ inclusions, *Physica C*, **253**, 391–400 (1995).
20. J. Kase, N. Irisawa, T. Morimoto, K. Togano, H. Kumakura, D.R. Dietderich, and H. Maeda, Improvement in critical current density of $Bi_2Sr_2Ca_1Cu_2O_x$ tapes synthesized by doctor-blade casting and melt growth, *Appl. Phys. Lett.*, **56**, 970–972 (1990).
21. J. Tenbrink, K. Heine, H. Krauth, A. Szulczyk, and M. Thöner, Entwicklung von Hoch-T_c-Supraleiterdrähten, *VDI Berichte*, **733**, 399–404 (1989).
22. J. Tenbrink, K. Heine, and H. Krauth, Critical currents and flux pinning in Ag stabilized high T_c superconductor wires, *Cryogenics*, **30**, 422–426 (1990).
23. W. Paul and Th. Baumann, Inductive measurements of intergranular critical currents in ceramic rings – A comparison between sintered and melt processed Y-, Bi- and Tl-based superconductors, *Physica C*, **175**, 102–110 (1991).
24. S. Nomura, H. Yoshino, and K. Ando, Cross-substitution of Bi and Sr in $Bi_2Sr_2CaCu_2O_{8+\delta}$, *Physica C*, **196**, 323–328 (1992).
25. K. Shibutani, T. Egi, S. Hayashi, R. Ogawa, and Y. Kawate, Investigation of heat-treating conditions for silver-sheathed Bi-2212 superconducting coils, *Japan. J. Appl. Phys.*, **30**, 3371–3376 (1991).
26. Th. Lang, D. Buhl, and L.J. Gauckler, Melting of Bi-2212 under controlled oxygen partial pressures with silver, *Physica C*, **275**, 284–292 (1997).
27. D. Buhl, T. Lang, M. Cantoni, T. Risold, B. Hallstedt, and L.J. Gauckler, Critical current densities in Bi-2212 thick films, *Physica C*, **257**, 151–159 (1996).
28. J.L. MacManus-Driscoll, J.C. Bravman, and R.B. Beyers, Pseudo-quarternary phase relations near $Bi_2Sr_2CaCu_2O_{8+x}$ in reduced oxygen pressures, *Physica C*, **251**, 71–88 (1995).

29. R. Wesche, A.M. Fuchs, B. Jakob, and G. Pasztor, Development and processing of Bi-2212/Ag superconducting wires and tapes, in *Processing of Long Lengths of Superconductors*, U. Balachandran, E.W. Collings, and A. Goyal (eds.), TMS, Warrendale, USA, 1994.
30. A.P. Baker and B.A. Glowacki, The dependence of $Bi_9Sr_{11}Ca_5$ oxide on the stoichiometry of Bi-2212 and its reversible phase transition in oxidising and non-oxidising annealing conditions, *Physica C*, **223**, 383–390 (1994).
31. P. Majewski, H.-L. Su, and F. Aldinger, The oxygen content of the high-temperature superconducting compound $Bi_{2+x}Sr_{3-y}Ca_yCu_2O_{8+d}$ as a function of the cation concentration, *Physica C*, **229**, 12–16 (1994).
32. J.F. Mitchell, J.A. Garcia, and D.G. Hinks, Rapid synthesis of 93 K Bi-2212 superconductor from hydroxide precursors, *Physica C*, **245**, 126–130 (1995).
33. L. Pierre, J. Schneck, D. Morin, J.C. Tolédano, J. Primot, C. Daguet, and H. Savary, Role of lead substitution in the production of 110-K superconducting single-phase Bi-Sr–Ca–Cu–O ceramics, *J. Appl. Phys.*, **68**, 2296–2303 (1990).
34. W. Zhu and P.S. Nicholson, Atmosphere–temperature–time relationships for the formation of 110 K phase in the Bi-(Pb)-Sr-Ca-Cu-O high T_c superconductor system, *Appl. Phys. Lett.*, **61**, 717–719 (1992).
35. Yu Mei, S.M. Green, C. Jiang, and H.L. Luo, Preparation of a single phase $(Bi_{0.9}Pb_{0.1})_2Sr_2Ca_2Cu_3O_y$ ceramic superconductor, *J. Appl. Phys.*, **66**, 1777–1781 (1989).
36. V. Plechâĉek, H. Hejdová, and Ž. Trejbalová, Pressing pressure dependence of critical current density in single-phase 2223 Bi, Pb-Sr-Ca-Cu-O polycrystalline superconductor, *Cryogenics*, **30**, 750–753 (1990).
37. M. Wilhelm, H.-W. Neumüller, and G. Ries, Fabrication and critical current densities of 2223-BiPbSrCaCuO silver sheathed tapes, *Physica C*, **185–189**, 2399–2400 (1991).
38. J.-H. Ahn, K.-H. Ha, S.-Y. Lee, J.-W. Ko, H.-D. Kim, and H. Chung, Alloying effect of Ag sheath on microstructure and superconducting properties of Bi-2223/Ag tape, *Physica C*, **235–240**, 3405–3406 (1994).
39. G. Grasso, A. Perin, B. Hensel, and R. Flükiger, Pressed and cold rolled Ag-sheathed Bi(2223) tapes: A comparison, *Physica C*, **217**, 335–341 (1993).
40. P. Haldar, J.G. Hoehn, Jr.,, J.A. Rice, and L.R. Motowidlo, Enhancement in critical current density of Bi-Pb-Sr-Ca-Cu-O tapes by thermomechanical processing: Cold rolling versus uniaxial pressure, *Appl. Phys. Lett.*, **60**, 495–497 (1992).
41. S.R. Shukla and R.G. Sharma, Transport critical current characteristics of Ag/B(P)SCCO tapes in magnetic field, *Physica C*, **247**, 83–90 (1995).
42. M.G. Smith, D.S. Phillips, D.E. Peterson, and J.O. Willis, *Physica C*, **224**, 168–174 (1994).
43. T. Hatano, K. Aota, H. Hattori, S. Ikeda, K. Nakamura, and K. Ogawa, Growth mechanism of high T_c phase in leaded Bi–Sr–Ca–Cu–O system, *Cryogenics*, **30**, 611–613 (1990).
44. M. Shiloh, I. Wood, M. Brown, F. Beech, and I.W. Boyd, Formation of $(Ca_{1-x}Sr_x)_2$ PbO_4 in Bi(Pb)-Sr-Ca-Cu-O system: Correlation with the formation of the 2223 high-T_c phase, *J. Appl. Phys.*, **68**, 2304–2307 (1990).

45. Y. Nishi, S. Moriya, and T. Manabe, Effect of annealing on T_c of liquid-quenched $Bi_{0.7}Pb_{0.3}SrCaCu_{1.8}O_x$, *J. Appl. Phys*, **65**, 2389–2391 (1989).
46. H.G. Lee, C.J. Kim, K.H. Lee, and D.Y. Won, Study on the stabilization of the high T_c phase in a BiSrCaCuO system, *Appl. Phys. Lett.*, **54**, 391–392 (1989).
47. K. Togano, H. Kumakura, and D.R. Dietderich, Critical currents and magnetic properties of Bi- and Tl-based new high T_c superconductors, *Cryogenics*, **29**, 286–290 (1989).
48. Q.Y. Hu, H.K. Liu, and S.X. Dou, Characterization of Ag-sheathed $(Bi,Pb)_2Sr_2Ca_2Cu_3O_{10-x}$ multifilamentary tapes, *Cryogenics*, **32**, 1038–1041 (1992).
49. H.K. Liu, Y.C. Guo, S.X. Dou, S.M. Cassidy, L.F. Cohen, G.K. Perkins, A.D. Caplin, and N. Savvides, Pinning mechanisms in Ag-sheathed Bi(Pb)SrCaCuO tapes, *Physica C*, **213**, 95–102 (1993).
50. C.J. Zhou, T.W. Li, and T.G. Chen, Out-diffusion of oxygen in Bi-system superconducting oxide, *Physica C*, **179**, 369–375 (1991).
51. S.X. Dou, Y.C. Guo, and H.K. Liu, Magnetic properties and microstructures in Ag-clad Bi-Pb-Sr-Ca-Cu-O wires, *Physica C*, **194**, 343–350 (1992).
52. Y.C. Guo, H.K. Liu, and S.X. Dou, Silver-doped $(Bi,Pb)_2Sr_2Ca_2Cu_3O_{10}$/Ag high-temperature superconducting composites, *Physica C*, **215**, 291–296 (1993).
53. R.H. Arend, M.F. Garbauskas, and P.J. Bednarczyk, An alternative preparation for $(Bi, Pb)_2Ca_2Sr_2Cu_3O_z$, *Physica C*, **176**, 126–130 (1991).
54. S.E. Dorris, B.C. Prorok, M.T. Lanagan, S. Sinha, and R.B. Poeppel, Synthesis of highly pure bismuth-2223 by a two-powder process, *Physica C*, **212**, 66–74 (1993).
55. M.G. Smith, J.O. Willis, D.E. Peterson, J.F. Bingert, D.S. Phillips, J.Y. Coulter, K.V. Salazar, and W.L. Hults, Synthesis of Ag-clad (Bi,Pb)-2223 tapes by an alternative precursor route, *Physica C*, **231**, 409–415 (1994).
56. P. Krishnaraj, M. Lelovic, N.G. Eror, and U. Balachandran, Synthesis and microstructure of $Bi_{1.8}Pb_{0.4}Sr_2Ca_{2.2}Cu_3O_x$ obtained from freeze-dried precursors, *Physica C*, **215**, 305–312 (1993).
57. P. Krishnaraj, M. Lelovic, T.A. Deis, N.G. Eror, and U. Balachandran, Synthesis and micro-structure of the 2223 Phase $(Bi_{2.0}Sr_{2.0}Ca_{2.0}Cu_{3.0}O_x)$ in the (Bi,Pb)-Sr-Ca-Cu-O system obtained from freeze-dried Precursors, in *Processing of Long Lengths of Superconductors*, U. Balachandran, E.W. Collings, and A. Goyal (eds.), TMS, Warrendale, USA, 1994.
58. J. Hagberg, A. Uusimäki, J. Levoska, and S. Leppävuori, Preparation of Bi-Pb-Sr-Ca-Cu-O high T_c superconducting material via oxalate route at various pH values, *Physica C*, **160**, 369–374 (1989).
59. Ch. Lang, S. Gauss, W. Becker, M. Rupich, and W. Carter, Precursor preparation by a modified oxalate coprecipitation and its suitability for manufacturing (Bi,Pb)SrCaCuO-conductors, in *Processing of Long Lengths of Superconductors*, U. Balachandran, E.W. Collings, and A. Goyal (eds.), TMS, Warrendale, USA, 1994.
60. G. Gritzner and K. Bernhard, $Pb_{0.3}Bi_{1.7}Ca_xSr_{4-x}Cu_3O_y$ and $Pb_{0.4}Bi_{1.6}Ca_xSr_{4-x}Cu_3O_y$ superconductors from coprecipitated oxalate precursors, *Physica C*, **181**, 201–205 (1991).
61. Cyril Chiang, C.Y. Shei, Y.T. Huang, W.H. Lee, and P.T. Wu, Preparation of high purity 110 K phase in the (Bi, Pb)-Sr-Ca-Cu-O superconductor system using a solution method, *Physica C*, **170**, 383–387 (1990).

62. H. Kumakura, K. Togano, H. Maeda, and M. Mimura, Bi(Pb)-Sr-Ca-Cu-O superconducting composite tapes prepared by the powder method using an Ag sheath, *J. Appl. Phys.*, **67**, 3443–3447 (1990).
63. Y. Idemoto, S. Kobayashi, and K. Fueki, Phase diagrams and ionic defects in the 2212 and 2201 phases of the bismuth system, *Physica C*, **229**, 47–58 (1994).
64. J.-C. Grivel and R. Flükiger, A study of the stability of the $(Bi,Pb)_2Sr_2Ca_2Cu_3O_x$ phase in Ag-sheathed tapes, *Physica C*, **235–240**, 505–506 (1994).
65. J.L. Wagner, O. Chmaissem, J.D. Jorgensen, D.G. Hinks, P.G. Radaelli, B.A. Hunter, and W.R. Jensen, Multiple defects in overdoped $Tl_2Ba_2CuO_{6+\delta}$: Effects on structure and superconductivity, *Physica C*, **277**, 170–182 (1997).
66. A. Maignan, C. Martin, M. Huve, J. Provost, M. Hervieu, C. Michel, and B. Raveau, The "2201" thallium cuprate: T_c's up to 92 K can be achieved by "hydrogen" annealing, *Physica C*, **170**, 350–360 (1990).
67. A. Mirmelstein, A. Junod, G. Triscone, K.-Q. Wang, and J. Muller, Specific heat of $Tl_2Ba_2CuO_6$ ("2201") 90 K superconducting ceramics in magnetic fields up to 14 T, *Physica C*, **248**, 335–342 (1995).
68. R. Mogilevsky, J.F. Mitchell, D.G. Hinks, D.N. Argyriou, and J.D. Jorgensen, Ternary and quaternary thallium-containing high-T_c superconductors synthesized at low temperature and elevated pressure, *Physica C*, **250**, 15–24 (1995).
69. N.-L. Wu, Y.D. Yao, S.-N. Lee, S.-Y. Wong, and E. Ruckenstein, Synthesis of $Tl_2Ca_nBa_2Cu_{n+1}O_{6+2n}$ from stoichiometric reactant mixtures, *Physica C*, **161**, 302–312 (1989).
70. K.D. Vernon-Parry, L.T. Romano, J.S. Lees, and C.R.M. Grovenor, The degradation process in $Tl_2Ba_2Ca_1Cu_2O_{10+x}$, *Physica C*, **170**, 388–394 (1990).
71. G. Kordas and M.R. Teepe, $Tl_2Ca_1Ba_2Cu_2O_{8+x}$ high T_c superconductors processed by the sol-gel technique, *Appl. Phys. Lett.*, **57**, 1461–1463 (1990).
72. S.S.P. Parkin, V.Y. Lee, E.M. Engler, A.I. Nazzal, T.C. Huang, G. Gorman, R. Savoy, and R. Beyers, Bulk superconductivity at 125 K in $Tl_2Ca_2Ba_2Cu_3O_x$, *Phys. Rev. Lett.*, **60**, 2539–2542 (1988).
73. R.S. Liu, J.L. Tallon, and P.P. Edwards, An efficient and reproducible approach for attaining superconductivity at 128 K in $Tl_2Ba_2Ca_2Cu_3O_{10-\delta}$, *Physica C*, **182**, 119–122 (1991).
74. T. Kaneko, H. Yamauchi, and S. Tanaka, Zero-resistance temperature of Tl-based "2223" superconductor increased to 127 K, *Physica C*, **178**, 377–382 (1991).
75. R.S. Liu and P.P. Edwards, Synthesis of monophasic $Tl_2Ba_2Ca_2Cu_3O_{10}$ ($T_c = 120$ K) superconductor, *Physica C*, **179**, 353–357 (1991).
76. C. Uher, S.D. Peacor, and J. Shewchun, Thermal conductivity of $Tl_2Ba_2Ca_2Cu_3O_{10}$ ceramics from 300 K down to 0.1 K, *Physica C*, **177**, 23–26 (1991).
77. H.S. Koo, W.-M. Hurng, W.H. Lee, T.Y. Tseng, M. Chen, and J.R. Lo, Effect of Ag_2O additive on microstructure and superconductivity of the $Tl_2Ba_2Ca_2Cu_3O_y$ compound, *Appl. Phys. Lett.*, **62**, 3354–3356 (1993).
78. M.R. Presland, J.L. Tallon, P.W. Gilberd, and R.S. Liu, Bulk single-superconducting-phase thallium "2234" superconductor – $Tl_{2-x}Ba_2Ca_{3+x}Cu_4O_{12-\delta}$, *Physica C*, **191**, 307–315 (1992).
79. D.M. Ogborne and M.T. Weller, Structure and oxygen stoichiometry in $Tl_2Ba_2Ca_3Cu_4O_{12-\delta}$ – A high-resolution powder neutron-diffraction study, *Physica C*, **223**, 283–290 (1994).

80. K. Chen, B. Maheswaran, P. Haldar, R.S. Markiewicz, and B.C. Giessen, *J. Appl. Phys.*, **65**, 3574–3578 (1989).
81. S. Nakajima, M. Kikuchi, Y. Syono, K. Nagase, T. Oku, N. Kobayashi, D. Shindo, and K. Hiraga, Improvement in superconductivity of $TlBa_2CaCu_2O_y$ system by introduction of oxygen loss, *Physica C*, **170**, 443–447 (1990).
82. M.R. Presland and J.L. Tallon, Superconductivity at 105 K in $Tl_{0.5}Pb_{0.5}CaSr_{2-x}La_x Cu_2O_7$, *Physica C*, **177**, 1–7 (1991).
83. Z.F. Ren and J.H. Wang, Superior flux pinning in *in situ* synthesized silver-sheathed superconducting tape of $Tl_{0.5}Pb_{0.5}Sr_{1.6}Ba_{0.4}Ca_{0.8}Y_{0.2}Cu_2O_y$, *Appl. Phys. Lett.*, **62**, 3025–3027 (1993).
84. Y.S. Sung, X.F. Zhang, P.J. Kostic, and D.J. Miller, The effects of carbon on $TlBa_2Ca_2Cu_3O_x$ phase formation in the TlF – $Ba_2Ca_2Cu_3O_z$ system, *Appl. Phys. Lett.*, **69**, 3420–3422 (1996).
85. R.S. Liu, S.F. Hu, D.A. Jefferson, and P.P. Edwards, Superconductivity at 124 K in (Tl, Pb)$Sr_2Ca_2Cu_3O_9$, *Physica C*, **198**, 318–322 (1992).
86. M. Paranthaman, M. Foldeaki, R. Tello, and A.M. Hermann, Synthesis and magnetic characterization of $(Tl_{0.5}Pb_{0.5})Sr_2Ca_2Cu_3O_9$ and $Tl_2Ba_2CaCu_2O_8$ bulk superconductors, *Physica C*, **219**, 413–419 (1994).
87. K. Aihara, T. Doi, A. Soeta, S. Takeuchi, T. Yuasa, M. Seido, T. Kamo, and S. Matsuda, Flux pinning in Tl-(1223) superconductor, *Cryogenics*, **32**, 936–939 (1992).
88. T. Doi, M. Okada, A. Soeta, T. Yuasa, K. Aihara, T. Kamo, and S.-P. Matsuda, Flux pinning in single Tl-layer 1223 superconductor, *Physica C*, **183**, 67–72 (1991).
89. Z.Y. Chen, Z.Z. Sheng, Y.Q. Tang, Y.F. Li, L.M. Wang, and D.O. Pederson, New high-T_c cuprate $(Tl_{1-x}Cr_x)(Sr_{2-y}Ba_y)Ca_2Cu_3O_9$, *Appl. Phys. Lett.*, **62**, 3034–3036 (1993).
90. C. Martin, M. Huvé, M. Hervieu, A. Maignan, C. Michel, and B. Raveau, A 116 K superconductor with "1223" structure: $TlBaSrCa_2Cu_3O_{9-\delta}$, *Physica C*, **201**, 362–368 (1992).
91. N.H. Hur, B.C. Chakoumakos, M. Paranthaman, J.R. Thompson, and D.K. Christen, Structure and superconducting properties of $(Tl_{0.8}Bi_{0.2})(Sr_{1.6}Ba_{0.4})Ca_2Cu_3O_{9-\delta}$, *Physica C*, **253**, 109–114 (1995).
92. D.Y. Jeong, M.H. Sohn, H.S. Kim, L.L. He, M. Cantoni, and S. Horiuchi, Microstructures in high-J_c Tl-1223/Ag tapes, *Physica C*, **269**, 279–285 (1996).
93. V. Selvamanickam, T. Finkle, K. Pfaffenbach, P. Haldar, E.J. Peterson, K.V. Salazaar, E.P. Roth, and J.E. Tkaczyk, *Physica C*, **260**, 313–320 (1996).
94. J.C. Moore, D.M.C. Hyland, C.J. Salter, C.J. Eastell, S. Fox, C.R.M. Grovenor, and M.J. Goringe, Microstructure and phase development in melt processed Tl-1223 tapes, *Physica C*, **276**, 202–217 (1997).
95. Z.F. Ren and J.H. Wang, Enhanced formation of 1223 phase by partial replacement of Bi for Tl in in-situ synthesized silver-sheathed superconducting tape of $Tl_{1-x}Bi_xSr_{2-y}Ba_yCa_2Cu_3O_{9-\delta}$, *Physica C*, **216**, 199–204 (1993).
96. R.S. Liu, S.F. Wu, D.S. Shy, S.F. Hu, and D.A. Jefferson, Improvement of phase purity and accelerated formation of the Tl-1223 phase from stoichiometric compositions $(Tl_{0.6}Pb_{0.2}Bi_{0.2})(Sr_{2-x}Ba_x)Ca_2Cu_3O_9$ ($x = 0.2$-0.3), *Physica C*, **222**, 278–282 (1994).

97. Z.F. Ren, C.A. Wang, J.H. Wang, D.J. Miller, and K.C. Goretta, Improvement of transport critical current density in in-situ synthesized silver-sheathed superconducting tape of $Tl_{0.78}Bi_{0.22}Sr_{1.6}Ba_{0.4}Ca_2Cu_3O_{9-\delta}$ by intermediate annealing and pressing, *Physica C*, **247**, 163–168 (1995).
98. G. Triscone, A. Junod, and R.E. Gladyshevskii, Magnetic and thermal properties of the 116 K superconductor Tl-1223, *Physica C*, **264**, 233–249 (1996).
99. R.E. Gladyshevskii, A. Perin, B. Hensel, R. Flükiger, R. Abraham, K. Lebbou, M.Th. Cohen-Adad, and J.-L. Jordan, Preparation by *in-situ* reaction and physical characterization of Ag(Au) and Ag(Pd) sheathed $(Tl,Pb,Bi)(Sr,Ba)_2Ca_2Cu_3O_{9-\delta}$ tapes, *Physica C*, **255**, 113-123 (1995).
100. S.M. Loureiro, ET. Alexandre, E.V. Antipov, J.J. Capponi, S. de Brion, B. Souletie, J.L. Tholence, M. Marezio, Q. Huang, and A. Santoro, Suppression of superconductivity and the overdoped region in $HgBa_2CuO_{4+\delta}$, *Physica C*, **243**, 1–9 (1995).
101. A. Bertinotti, V. Viallet, D. Colson, J.-F. Marucco, J. Hammann, G. LeBras, and A. Forget, Synthesis, crystal structure and magnetic properties of superconducting single crystals of $HgBa_2CuO_{4+\delta}$, *Physica C*, **268**, 257–265 (1996).
102. J.-F. Marucco, V. Viallet, A. Bertinotti, D. Colson, and A. Forget, Point defects, thermodynamic and superconducting properties of the non-stoichiometric $HgBa_2CuO_{4+\delta}$ phase, *Physica C*, **275**, 12–18 (1997).
103. P. Bordet, F. Duc, S. LeFloch, J.J. Capponi, E. Alexandre, M. Rosa-Nunes, S. Putilin, and E.V. Antipov, Single crystal X-ray diffraction study of the $HgBa_2CuO_{4+\delta}$ superconducting compound, *Physica C*, **271**, 189–196 (1996).
104. M. Itoh, A. Tokiwa-Yamamoto, S. Adachi, and H. Yamauchi, Increase of T_c in $HgBa_2CuO_y$ by reducing treatment, *Physica C*, **212**, 271–273 (1993).
105. R. Usami, A. Fukuoka, H. Kubota, and H. Yamauchi, The oxygen-content dependence of the magnetization anisotropy of $HgBa_2CuO_{4+\delta}$, *Physica C*, **243**, 19–23 (1995).
106. I.C. Chang, M.D. Lan, T.J. Goodwin, J.Z. Liu, P. Klavins, and R.N. Shelton, Magnetic and transport properties of $HgBa_2CuO_{4+\delta}$, *Physica C*, **223**, 207–212 (1994).
107. H.R. Khan, O. Loebich, P. Fabbricatore, A. Sciutti, and B. Zhang, Synthesis, X-ray diffraction and AC magnetic susceptibility investigations of $HgBa_2CuO_{4+\delta}$, *Physica C*, **229**, 165–168 (1994).
108. J.L. Wagner, P.G. Radaelli, D.G. Hinks, J.D. Jorgensen, J.F. Mitchell, B. Dabrowski, G.S. Knapp, and M.A. Beno, Structure and superconductivity of $HgBa_2CuO_{4+\delta}$, *Physica C*, **210**, 447–454 (1993).
109. A. Asab, A.R. Armstrong, I. Gameson, and P.P. Edwards, Single-step synthesis and crystal structure of $HgBa_2CuO_{4+\delta}$ with a T_c of 97 K, *Physica C*, **255**, 180–187 (1995).
110. Q. Xiong, F.T. Chan, Y.Y. Xue, and C.W. Chu, Synthesis and start cation-composition of $HgBa_2CuO_{4+\delta}$, *Physica C*, **253**, 329–333 (1995).
111. W.J. Zhu, Y.Z. Huang, L.Q. Chen, C. Dong, B. Yin, and Z.X. Zhao, Synthesis of the superconductors $HgBa_2CaCu_2O_{6+\delta}$ and $HgBa_2Ca_2Cu_3O_{8+\delta}$, *Physica C*, **218**, 5–7 (1993).
112. A.-K. Klehe, J.S. Schilling, J.L. Wagner, and D.G. Hinks, Hydrostatic pressure dependence of the superconducting transition temperature of $HgBa_2CaCu_2O_{6+\delta}$ and $HgBa_2Ca_2Cu_3O_{8+\delta}$, *Physica C*, **223**, 313–320 (1994).

113. P.G. Radaelli, J.L. Wagner, B.A. Hunter, M.A. Beno, G.S. Knapp, J.D. Jorgensen, and D.G. Hinks, Structure, doping and superconductivity in $HgBa_2CaCu_2O_{6+\delta}$ ($T_c \leq$ 128 K), *Physica C*, **216**, 29–35 (1993).
114. R.L. Meng, Y.Y. Sun, J. Kulik, Z.J. Huang, F. Chen, Y.Y. Xue, and C.W. Chu, Superconductivity at 112-117 K in $HgBa_2CaCu_2O_{6+\delta}$, *Physica C*, **214**, 307–312 (1993).
115. Z.J. Huang, Y.Y. Xue, R.L. Meng, and C.W. Chu, Irreversibility line of the $HgBa_2CaCu_2O_{6+\delta}$, *Phys. Rev. B*, **49**, 4218–4221 (1994).
116. R.L. Meng, L. Beauvais, X.N. Zhang, Z.J. Huang, Y.Y. Sun, Y.Y. Xue, and C.W. Chu, Synthesis of the high-temperature superconductors $HgBa_2CaCu_2O_{6+\delta}$ and $HgBa_2Ca_2Cu_3O_{8+\delta}$, *Physica C*, **216**, 21–28 (1993).
117. S.N. Putilin, E.V. Antipov, and M. Marezio, Superconductivity above 120 K in $HgBa_2CaCu_2O_{6+\delta}$, *Physica C*, **212**, 266–270 (1993).
118. E.V. Antipov, J.J. Capponi, C. Chaillout, O. Chmaissem, S.M. Loureiro, M. Marezio, S.N. Putilin, A. Santoro, and J.L. Tholence, Synthesis and neutron powder diffraction study of the superconductor $HgBa_2CaCu_2O_{6+\delta}$ before and after heat treatment, *Physica C*, **218**, 348–355 (1993).
119. H.M. Shao, L.J. Shen, J.C. Shen, X.Y. Hua, P.F. Yuan, and X.X. Yao, Synthesis of single-phase $HgBa_2Ca_2Cu_3O_{8+\delta}$ superconductor, *Physica C*, **232**, 5–9 (1994).
120. J.G. Wu, H.M. Shao, S.Y. Ding, X.X. Yao, J.C. Shen, and Z. Wu, Irreversibility and transport in the high T_c superconductor $HgBa_2Ca_2Cu_3O_{8+\delta}$, *Cryogenics*, **36**, 17–20 (1996).
121. C.C. Lam, J.C.L. Chow, P.C.W. Fung, and H.M. Shao, Observation of a screw dislocation in the Hg-1223 superconductor under a transmission electron microscope, *Physica C*, **258**, 253–260 (1996).
122. M. Paranthaman, Single-step synthesis of bulk $HgBa_2Ca_2Cu_3O_{8+\delta}$, *Physica C*, **222**, 7–12 (1994).
123. K. Isawa, A. Tokiwa-Yamamoto, M. Itoh, S. Adachi, and H. Yamauchi, Encapsulation method for the synthesis of nearly single-phase superconducting $HgBa_2Ca_2Cu_3O_{8+\delta}$ with $T_c \geq 135$ K, *Physica C*, **222**, 33–37 (1994).
124. Y.T. Wang and A.M. Hermann, Influence of fluorine in magnetic properties of Hg-1223 high-temperature superconductors, *Physica C*, **254**, 1–6 (1995).
125. A. Schilling, O. Jeandupeux, J.D. Guo, and H.R. Ott, Magnetization and resistivity study on the 130 K superconductor in the Hg–Ba–Ca–Cu–O system, *Physica C*, **216**, 6–11 (1993).
126. Y.S. Yao, Y.J. Su, W. Liu, Z.X. Liu, H. Gong, J.W. Li, G.H. Cao, F. Wu, and Z.X. Zhao, Superconductivity, microstructure and pressure effect on T_c of Hg-based compounds prepared under high pressure, *Physica C*, **224**, 91–98 (1994).
127. Z.J. Huang, R.L. Meng, X.D. Qiu, Y.Y. Sun, J. Kulik, Y.Y. Xue, and C.W. Chu, Superconductivity, structure and resistivity in $HgBa_2Ca_2Cu_3O_{8+\delta}$, *Physica C*, **217**, 1–5 (1993).
128. Y. Rui, H.L. Ji, H.M. Shao, X.N. Xu, M.J. Qin, J.C. Shen, X. Jin, X.X. Yao, X.S. Rong, B. Ying, G.C. Che, and Z.X. Zhao, Characterization of Hg-Ba-Ca-Cu-O 1223 superconductor, *Physica C*, **231**, 243–248 (1994).
129. A. Schilling, O. Jeandupeux, S. Büchi, H.R. Ott, and C. Rossel, Physical properties of $HgBa_2Ca_2Cu_3O_8$ with $T_c \approx 133$ K, *Physica C*, **235–240**, 229–232 (1994).

130. E.V. Antipov, S.M. Loureiro, C. Chaillout, J.J. Capponi, P. Bordet, J.L. Tholence, S.N. Putilin, and M. Marezio, The synthesis and characterisation of the $HgBa_2Ca_2Cu_3O_{8+\delta}$ and $HgBa_2Ca_3Cu_4O_{10+\delta}$, *Physica C*, **215**, 1–10 (1993).
131. W. Lechter, L. Toth, M. Osofsky, E. Skelton, R.J. Soulen Jr.,, S. Quadri, J. Schwartz, J. Kessler, and C. Wolters, One-step reaction and consolidation of Hg based high-temperature superconductors by hot isostatic pressing, *Physica C*, **249**, 213–219 (1995).
132. G. van Tendeloo, C. Chaillout, J.J. Capponi, M. Marezio, and E.V. Antipov, Atomic structure and defect structure of the superconducting $HgBa_2Ca_{n-1}Cu_nO_{2n+2+\delta}$ homologous series, *Physica C*, **223**, 219–226 (1994).
133. A. Maignan, S.N. Putilin, V. Hardy, Ch. Simon, and B. Raveau, Magnetic study of $HgBa_2Ca_2Cu_3O_{8+\delta}$ single crystals: Effect of doping on irreversibility and fishtail lines, *Physica C*, **266**, 173–177 (1996).
134. P.V.P.S.S. Sastry, K.M. Amm, D.C. Knoll, S.C. Peterson, and J. Schwartz, Synthesis and processing of $(Hg,Pb)_1Ba_2Ca_2Cu_3O_y$ superconductors, *Physica C*, **297**, 223–231 (1998).
135. P.V.P.S.S. Sastry, K.M. Amm, D.C. Knoll, S.C. Peterson, and J. Schwartz, Synthesis and processing of Bi-doped $Hg_1Ba_2Ca_2Cu_3O_y$ superconductors, *Physica C*, **300**, 125–140 (1998).
136. S.M. Loureiro, E.V. Antipov, E.M. Kopnin, M. Brunner, J.J. Capponi, and M. Marezio, Structure and superconductivity of the $HgBa_2Ca_3Cu_4O_{10+\delta}$ phase, *Physica C*, **257**, 117–124 (1996).
137. R. Usami, S. Adachi, M. Itoh, T. Tatsuki, A. Tokiwa-Yamamoto and K. Tanabe, Synthesis of $HgBa_2Ca_3Cu_4O_y$ under ambient pressure, *Physica C*, **262**, 21–26 (1996).
138. K.A. Lokshin, D.A. Pavlov, M.L. Kovba, E.V. Antipov, I.G. Kuzemskaya, L.F. Kulikova, V.V. Davydov, I.V. Morozov, and E.S. Itskevich, Synthesis and characterisation of overdoped Hg-1234 and Hg-1245 phases; the universal behavior of T_c variation in the $HgBa_2Ca_{n-1}Cu_nO_{2n+2+\delta}$ series, *Physica C*, **300**, 71–76 (1998).
139. P.F. Herrmann, C. Albrecht, J. Bock, C. Cottevieille, S. Elschner, W. Herkert, M.-O. Lafon, H. Lauvray, A. Leriche, W. Nick, E. Preisler, H. Salzburger, J.-M. Tourre, and T. Verhaege, European project for the development of high T_c current leads, *IEEE Trans. Appl. Supercond.*, **3**, 876–880 (1993).
140. C. Albrecht, J. Bock, P.F. Herrmann, and J.M. Tourre, Current leads – First applications of high T_c superconductors for power devices, *Physica C*, **235–240**, 205–208 (1994).
141. W. Paul, Th. Baumann, J. Rhyner, and F. Platter, Tests of 100 kW high-T_c superconducting fault current limiters, *IEEE Trans. Appl. Supercond.*, **5**, 1059–1062 (1995).
142. H.J. Bornemann and M. Sander, Conceptual system design of a 5 MWh/100 MW superconducting flywheel energy storage plant for power utility applications, *IEEE Trans. Appl. Supercond.*, **7**, 398–401 (1997).
143. A.M. Campbell and D.A. Cardwell, Bulk high temperature superconductors for magnet applications, *Cryogenics*, **37**, 567–575 (1997).
144. T. Oka, Processing and applications of bulk HTSC, *Physica C*, **463–465**, 7–13 (2007).

145. J.R. Hull and M. Strasik, Concepts for using trapped-flux bulk high-temperature superconductor in motors and generators, *Supercond. Sci. Technol.*, **23**, 124005 (2010).
146. M. Tomita and M. Murakami, High-temperature superconductor bulk magnets that can trap magnetic fields of over 17 tesla at 29 K, *Nature*, **421**, 517–520 (2003).
147. H.W. Neumüller and G. Ries, Critical currents and flux creep in melt processed $Bi_2Sr_2CaCu_2O_8$, *Physica C*, **162–164**, 363–364 (1989).
148. J.-F. Fagnard, S. Elschner, J. Bock, M. Dirickx, B. Vanderheyden, and P. Vanderbemden, Shielding efficiency and *E*(*J*) characteristics measured on large melt cast Bi-2212 hollow cylinders in axial magnetic fields, *Supercond. Sci. Technol.*, **23**, 095012 (2010).
149. R. Wesche and A.M. Fuchs, Design of superconducting current leads, *Cryogenics*, **34**, 145–154 (1994).
150. M. Murakami, K. Yamaguchi, H. Fujimoto, N. Nakamura, T. Taguchi, N. Koshizuka, and S. Tanaka, Flux pinning by non-superconducting inclusions in melt-processed YBaCuO superconductors, *Cryogenics*, **32**, 930–935 (1992).
151. S. Jin, T.H. Tiefel, R.C. Sherwood, M.E. Davis, R.B. van Dover, G.W. Kammlott, R.A. Fastnacht, and H.D. Keith, High critical currents in Y-Ba-Cu-O superconductors, *Appl. Phys. Lett.*, **52**, 2074–2076 (1988).
152. S. Jin, R.C. Sherwood, E.M. Gyorgy, T.H. Tiefel, R.B. van Dover, S. Nakahara, L.F. Schneemeyer, R.A. Fastnacht, and M.E. Davis, Large magnetic hysteresis in a melt-textured Y-Ba-Cu-O superconductor, *Appl. Phys. Lett.*, **54**, 584–586 (1989).
153. K. Salama, V. Selvamanickam, L. Gao, and K. Sun, High current density in bulk $YBa_2Cu_3O_x$ superconductor, *Appl. Phys. Lett.*, **54**, 2352–2354 (1989).
154. R. Ravinder Reddy, M. Murakami, S. Tanaka, and Venugopal Reddy , Elastic behavior of a Y-Ba-Cu-O sample prepared by MPMG method, *Physica C*, **257**, 137–142 (1996).
155. P. Diko, V. Antal, K. Zmorayová, M. Šefčiková, X. Chaud, J. Kováč, X. Yao, I. Chen, M. Eisterer, and H.W. Weber, The influence of post-growth thermal treatments on the critical current density of TSMG YBCO bulk superconductors, *Supercond. Sci. Technol.*, **23**, 124 002 (2010).
156. M. Muralidhar, K. Suzuki, A. Ishihara, M. Jirsa, Y. Fukumoto, and M. Tomita, Novel seeds applicable for mass processing of LRE-123 single-grain bulks, *Supercond. Sci. Technol.*, **23**, 124 003 (2010).
157. A. Wongsatanawarid, H. Seki, and M. Murakami, Growth of large bulk Y-Ba-Cu-O with multi-seeding, *Supercond. Sci. Technol.*, **23**, 045022 (2010).
158. H. Ikuta, T. Yamada, M. Yoshikawa, Y. Yanagi, Y. Itoh, B. Latha, and U. Mizutani, Melt processing of (rare earth)-Ba-Cu-O bulk superconductors, *Supercond. Sci. Technol.*, **18**, S119–S125 (2005).
159. M. Muralidhar, M. Jirsa, and M. Tomita, Flux pinning and superconducting properties of melt-textured NEG-123 superconductor with TiO_2 addition, *Physica C*, **470**, 592–597 (2010).
160. M. Lepropre, I. Monot, M.P. Delamare, M. Hervieu, Ch. Simon, J. Provost, G. Desgardin, B. Raveau, J.M. Barbut, D. Bourgalt, and D. Braithwaite, Critical currents up to 71 000 A cm^{-2} at 77 K in melt textured YBCO doped with $BaSnO_3$, *Cryogenics*, **34**, 63–67 (1994).

161. T. Puig, J. Plain, F. Sandiumenge, X. Obradors, J. Rabier, and J.A. Alonso, High oxygen pressure generation of flux-pinning centers in melt-textured $YBa_2Cu_3O_7$, *Appl. Phys. Lett.*, **75**, 1952–1954 (1999).
162. K. Iida, N.H. Babu, Y.H. Shi, D.A. Cardwell, and M. Murakami, Gd-Ba-Cu-O bulk superconductors fabricated by a seeded infiltration growth technique under reduced oxygen partial pressure, *Supercond. Sci. Technol.*, **19**, 641–647 (2006).
163. G. Fuchs, G. Krabbes, P. Schätzle, S. Gru*β*, P. Stoye, T. Staiger, K.-H. Müller, J. Fink, and L. Schultz, High trapped fields in bulk $YBa_2Cu_3O_{7-\delta}$ samples at temperatures around 50 K, *Appl. Phys. Lett.*, **70**, 117–119 (1997).
164. H.J. Bornemann, T. Burghardt, W. Hennig, and A. Kaiser, Processing technique for fabrication of advanced YBCO bulk materials for industrial applications, *IEEE Trans. Appl. Supercond.*, **7**, 1805–1808 (1997).
165. K. Iida, K. Nenkov, G. Fuchs, G. Krabbes, B. Holzapfel, B. Büchner, and L. Schultz, Effect of addition of planetary milled Gd-211 on the microstructure and superconducting properties of air-processed single grain Gd–Ba–Cu–O/Ag bulk superconductors, *Physica C*, **470**, 1153–1157 (2010).
166. T. Ogasawara, Conductor design issues for oxide superconductors. Part 1: Criteria of magnetic stability, *Cryogenics*, **29**, 3–5 (1989).
167. T. Ogasawara, Conductor design issues for oxide superconductors – Part 2: Exemplification of stable conductors, *Cryogenics*, **29**, 6–9 (1989).
168. E.W. Collings, Design considerations for high T_c ceramic superconductors, *Cryogenics*, **28**, 724–733 (1988).
169. S.L. Wipf, Review of stability in high temperature superconductors with emphasis on flux jumping, *Cryogenics*, **31**, 936–948 (1991).
170. K. Watanabe, S. Awaji, N. Kobayashi, S. Nimori, G. Kido, K. Kimura, and M. Hashimoto, Flux jumps in the magnetization of QMG processed $Y_1Ba_2Cu_3O_7$, *Cryogenics*, **32**, 959–963 (1992).
171. J.W. Burgoyne and J.H.P. Watson, Flux jump instabilities in a bulk $Bi_2Sr_2CaCu_2O_{8+x}$ flux tube, *Cryogenics*, **34**, 507–512 (1994).
172. A. Nabiałek and M. Niewczas, The critical state stability in textured $Bi_2Sr_2CaCu_2O_{8+\delta}$ superconductor, *Physica C*, **436**, 43–50 (2006).

15

First- and Second-Generation High-Temperature Superconductor Wires

15.1 Introduction

For power applications of superconductivity, long superconducting wires with sufficiently high critical current density are required. In addition, in magnet applications, wires need to be able to withstand the large Lorentz forces generated in the windings of a superconducting magnet. Another aspect of importance is the fact that the superconducting wires must be able to be wound into coils or high-current cables. Unfortunately, the cuprate superconductors are brittle ceramics. In order to achieve some flexibility for coil winding, superconducting wires are composites of the high-temperature superconductor (HTS) and a good normal conducting metal. The strain tolerance of the superconducting wire may be further improved by stainless steel laminations. A good normal conductor in parallel to the HTS is necessary not only for improved bending properties but also to ensure stability against disturbances, which would otherwise lead to a transition to the normal state. In the case of a quench, the superconducting magnet is discharged via an external dump resistor. During the discharge, the current decays exponentially with a time constant $\tau = L/R$, where L is the inductance of the magnet and R the resistance of the external dump resistor. The resistance of the dump resistor has to be selected in such a way that the discharge voltage, which may be as high as 10 kV, is well below the breakdown voltage of the electrical insulation. When a quench occurs, the current is transferred from the superconductor to the normally conducting stabilizer (e.g., copper or silver). The cross-section of the normal conducting stabilizer must be large enough to limit the maximum temperature in the winding (hot-spot temperature) to less than 200 K.

The merits of HTS are critical temperatures well above 77 K, and extremely high upper critical fields. For power applications (transformers, power transmission cables, fault current limiters, and generators), operation temperatures much higher than 4 K are required to

Physical Properties of High-Temperature Superconductors, First Edition. Rainer Wesche.

make them economically competitive with conventional technology. In principle, cooling by liquid nitrogen would be highly desirable for power applications of HTS. Another possible application is the generation of extremely high magnetic fields. The present-day work-horses in the field of superconducting magnets are NbTi ($T_c = 9.5$ K, $B_{c2}(T = 0) = 13$ T) [1] and Nb_3Sn ($T_c = 18$ K, $B_{c2}(T = 0) = 24$ T) [2]. Magnetic fields significantly higher than 20 T can be generated by the use of HTS. The most cost-effective way to generate magnetic fields well above 20 T is the use of a HTS insert coil, which increases the field generated by an outer Nb_3Sn main magnet by several tesla.

From a technical point of view, the engineering critical current density of a superconducting wire should reach 10^5 A/cm^2 at the envisaged operating field and temperature [3]. The engineering critical current density j_e is defined as the critical current divided by the total wire cross-section including the normal conducting matrix, whereas the critical current density j_c is based on the cross-section of the superconducting filaments. A high-engineering critical current density of more than 10^5 A/cm^2 is also desirable for economic reasons. In power applications, the use of HTS is in competition with conventional technology based on the use of copper. The widespread use of HTS in the grid would require the superconductor cost per unit length divided by the operating current density to be comparable to that of copper.

In this chapter, the development of first- (Ag/Bi-2212 and Ag/Bi-2223) and second-generation (RE-123-coated conductors, where RE represents Y or another rare earth element) HTS wires will be described. The factors limiting the critical current density will be discussed. Examples of the performance of industrially fabricated HTS wires will be presented.

15.2 First-Generation High-T_c Superconductor Wires and Tapes

15.2.1 Introduction

Soon after the discovery of high-temperature superconductivity above 77 K it was noticed that the transport critical current density in sintered $YBa_2Cu_3O_{7-x}$ is extremely low. In addition, the critical current density dropped by more than an order of magnitude in the presence of a magnetic field as small as 10 mT. Further investigations showed that the grain boundaries in HTS act as barriers for the transport current. Mannhart *et al.* [4] performed detailed studies of the transport critical current densities across individual grain boundaries. Their results suggest that the grain boundaries in cuprate superconductors behave as superconductor–normal–superconductor (S–N–S) type contacts. These weak links at the grain boundaries are responsible for extremely low transport critical current densities in cuprate superconductors, and also explain the high sensitivity to small applied magnetic fields. The fact that grain boundaries can act as weak links has been attributed to the extremely short coherence length in the cuprate superconductors especially along the crystallographic c direction (see Chapter 9). The effect of the misalignment angle between the crystallographic axes of two $YBa_2Cu_3O_{7-x}$ films, epitaxially grown on a $SrTiO_3$ bicrystal, on the grain boundary transport critical current density was studied by Dimos *et al.* [5, 6]. They found that the transport critical current density across the grain boundary declines exponentially with increasing misalignment angle. For misalignment angles of a few degrees, the grain boundaries do not block the supercurrents, which is in line with the high transport critical current densities found in epitaxial HTS films.

An important step in the development of the first generation of HTS wires and tapes was the discovery by Heine *et al.* in 1989 that strongly coupled grains and high critical current densities can be achieved in round Ag/Bi-2212 wires by a partial melting process [7]. A further breakthrough was the laboratory-scale manufacture of Ag/Bi-2223 tapes with a critical current density of 54 000 A/cm^2 at 77 K and zero applied field by Sato *et al.* [8–12] in the early 1990s. These tapes were manufactured by repeated pressing and sintering steps, which provided a *c*-axis texture of the Bi-2223 filaments. For the fabrication of long Ag/Bi-2223 tapes, the deformation needs to be performed by a rolling process. Due to the relatively high critical temperature of ≈110 K, relatively high critical current densities can be achieved in Ag/Bi-2223 tapes at 77 K and low magnetic fields. On the other hand, Ag/Bi-2212 wires offer high critical current densities at 4.2 K in fields well above 20 T. An important advantage of Bi-2212 is the fact that high critical current densities can be achieved not only in tapes but also in round wires.

15.2.2 Ag/Bi-2212 Wires and Tapes

The first aspect to be considered is the manufacture of Ag/Bi-2212 round wires by the powder-in-tube (PIT) method (see Figure 15.1). Silver or silver alloy tubes are filled with fully reacted Bi-2212 powder. For the manufacture of monocore wires, silver tubes of 6 mm inner and 8 mm outer diameter were used [7]. Round Ag/Bi-2212 monocore wires are obtained by drawing with a 10–15% reduction in the cross-section per pass. In the manufacture of multifilament wires, the initial silver tube filled with the Bi-2212 powder is drawn to hexagonally shaped single-core wires of 2–4 mm diameter. Typically, 19 or 37 of these monocore wires are stacked into a silver or silver alloy tube. This composite is deformed by drawing to a multifilament wire of 1 mm diameter. In monocore wires, the fraction of Bi-2212 in the total cross-section reaches 50%, while it is around 40% in multifilament wires [13].

In different parts of the heat treatment, oxygen is released or taken up by the Bi-2212. It is, therefore, essential that the silver matrix is permeable for oxygen. In addition, silver does not cause any chemical degradation of the Bi-2212 superconductor.

In order to reach high critical current densities, it is essential that the Bi-2212 is for a short time partially molten during the heat treatment. Figure 15.2 shows various heat treatment schedules used in the manufacture of Ag/Bi-2212 wires and tapes. A common feature of all of them is partial melting of the Bi-2212. The partial melting temperatures used in the different heat treatment schedules vary around 890 °C. The partial melting temperature was found to depend on the heat treatment atmosphere. It decreases with decreasing oxygen partial pressure. Holesinger *et al.* [14] determined the melting temperature of thick Bi-2212 films on silver substrates by means of differential thermal analysis (DTA). They found for the chemical composition $Bi_{2.09}Sr_{2.03}Ca_{0.78}Cu_{2.11}O_y$ melting temperatures of 856.5 °C for 1% O_2 and 99% Ar, 877.5 °C for 10% O_2 and 90% Ar and 889.5 °C in 100% O_2. The quoted values are based on the position of the endothermic peak in the DTA scans. The melting of the Bi-2212 phase sets in at a temperature which is 1.5–2.5 °C lower than the position of the peak in the DTA scan. Shibutani *et al.* [15] investigated the dependence of the melting temperature of Bi-2212 powder, estimated by DTA, on the oxygen partial pressure.

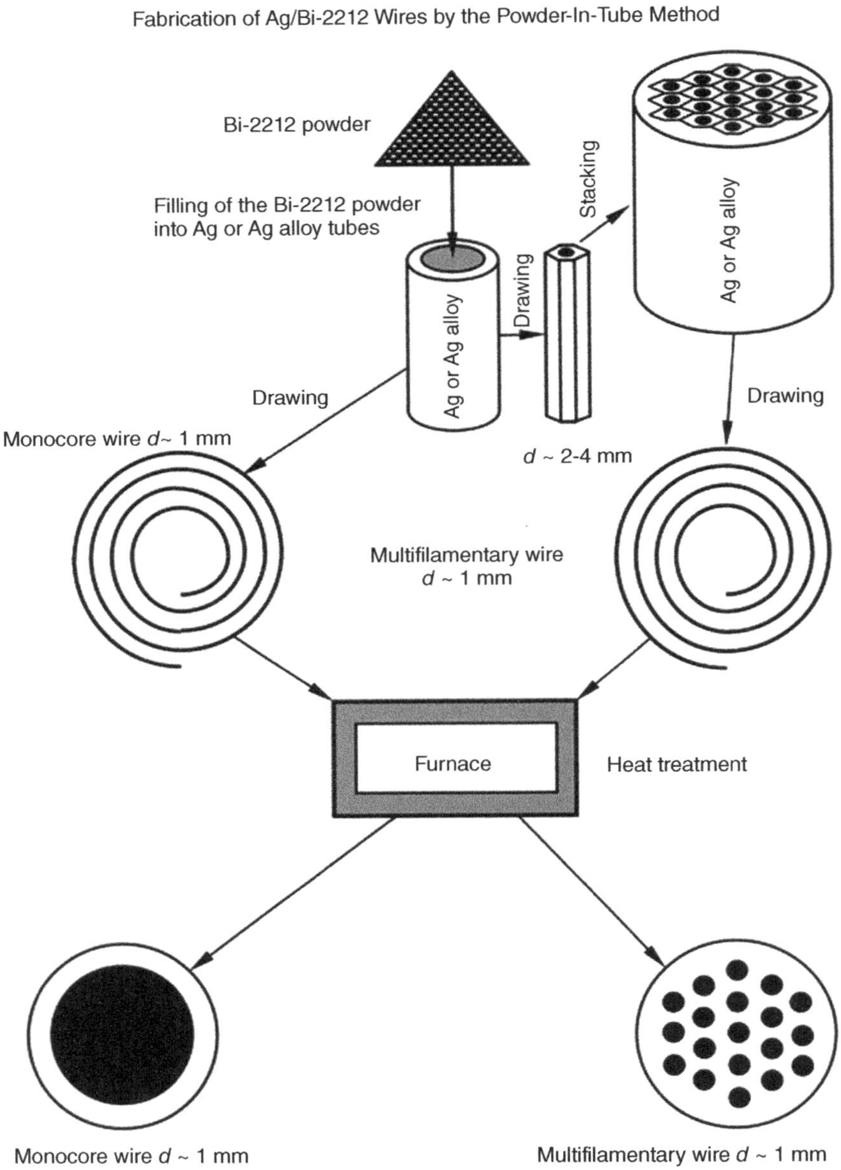

Figure 15.1 *Sketch of the manufacture of Ag/Bi-2212 round wires by the powder-in-tube (PIT) method*

For the chemical compound $Bi_{2.1}Sr_2CaCu_{1.9}Ag_{0.1}$, they found melting temperatures of 843 °C, 864 °C, 875 °C, and 881 °C at oxygen partial pressures of 0.01, 0.1, 0.2, and 0.5 bar, respectively. Furthermore, the melting point of Bi-2212 is reduced in the presence of silver. Hasegawa *et al.* [16] observed that the melting point of a Bi-2212 film on MgO is 883 °C, whereas melting of a Bi-2212 film on a silver substrate sets in at a temperature as low as 864 °C.

Figure 15.2a shows a heat treatment schedule developed for Ag/Bi-2212 wires of 1.5 mm diameter with seven filaments [17, 18]. The oxygen content of Bi-2212 depends on both temperature and oxygen partial pressure [19]. Below 800 °C, the oxygen content of Bi-2212 was found to be independent of oxygen partial pressure. In general, the oxygen content of Bi-2212 decreases with increasing temperature. At a temperature of 200 °C, the chemical composition of the 2212 phase is well described by the formula $Bi_2Sr_2CaCu_2O_{8.3}$, while at 700 °C the formula is $Bi_2Sr_2CaCu_2O_{8.15}$, reflecting the reduced oxygen content [19]. As a consequence, oxygen is released from the Bi-2212 filaments during heating to the partial melting temperature. For the heat treatment schedule presented in Figure 15.2a, all heat treatments were performed in air. The ramp rates for heating and the annealing at 880 °C are expected to lead to the optimum oxygen content before the Bi-2212 filaments are partially melted. It was found that the optimum annealing time, which is closely connected to oxygen diffusion through the silver matrix, depends on the wire and filament dimensions, and increases for thicker wires. The achievable critical current density is very sensitive to the maximum temperature during partial melting. For developmental Ag/Bi-2212 wires of 1.5 mm diameter, the highest short-sample critical current densities of 70 000 A/cm^2 (4.2 K, $B_a = 0$) were obtained for partial melting at 893 °C [17, 18]. Majewski [20] reported that, during the partial melting step, the Bi-2212 phase decomposes into Bi-2201, earth-alkali cuprates, and a liquid phase. After partial melting, the wires are furnace-cooled to room temperature. The final long-term annealing at 840 °C is necessary for renewed formation of the Bi-2212 phase. Again, heat treatment ends with furnace-cooling to room temperature.

Figure 15.2b shows a heat treatment schedule developed for Ag/Bi-2212 multifilament wires with filament diameters between 11 and 100 μm [21]. Independently of each other, Wesche *et al.* [22] (monocore wires) and Motowidlo *et al.* [23] (multifilament wires) discovered that the critical current density in Ag/Bi-2212 wires increases with decreasing filament diameter. Using the heat treatment schedule presented in Figure 15.2b, Motowidlo *et al.* [23] achieved a critical current density of 165 000 A/cm^2 at 4.2 K and $B_a = 0$ in Ag/Bi-2212 wires with 11 μm filaments. All heat treatments in the schedule presented in Figure 15.2b were performed in pure oxygen. A heating rate of 300 °C/h was used up to a temperature of 870 °C. Between 870 °C and the partial melting step at 885 °C, a slower heating rate was used. The renewed formation of the Bi-2212 phase occurred during slow cool-down to 870 °C and the following reduction in the temperature in 5 °C steps and hold times of 24 h for each temperature step. The slow cool-down between 885 °C and 870 °C is required to achieve better grain alignment, which is further improved during step solidification between 870 °C and 840 °C. The step solidification may also reduce the amount of minor phases in the superconducting filaments. The more rapid cool-down to room temperature starts at 840 °C. In general, it was found that heat treatment in pure oxygen instead of air is helpful in reaching high critical current densities, especially in thin filaments.

Figure 15.2c shows a heat treatment schedule for Ag/Bi-2212 tapes developed by Nomura *et al.* [24]. In air, the tapes were heated up at a rate of 5–60 °C/h to a temperature slightly below 885 °C. To prevent void swelling caused by diffusion of oxygen, the heating rate needs to be reduced to 10 °C/h just below the temperature of 885 °C at which the Bi-2212 is partially molten. During slow cooling at 5 °C/h to 825 °C, the Bi-2212 phase is again formed. Finally, the Ag/Bi-2212 tapes are furnace-cooled to room temperature. For tapes with a core thickness of 15 μm, a critical current density of 140 000 A/cm^2 at 4.2 K and 10 T was achieved.

Figure 15.2d shows a heat treatment schedule used by Endo and Nishikida for the preparation of Ag/Bi-2212 tapes of 1.5 cm width and a Bi-2212 core thickness of 40 μm [25]. All heat treatments were performed in pure oxygen. The tapes were heated at a rate of 10 °C/h to a temperature of 885 °C, at which the Bi-2212 core is partially molten. After 10 min at 885 °C, the tapes are cooled to a temperature of 820 °C at 5 °C/h. Slow cooling to 820 °C leads to the renewed formation of the 2212 phase. In addition, the low cooling rate improves the grain alignment in the Bi-2212 core. Finally, the tapes are cooled down to room temperature at 160 °C/h. Endo and Nishkida achieved a maximum critical current density of 160 000 A/cm^2 at 4.2 K and 1 T. They found that heat treatments performed in air with the same heat treatment schedule led to much lower critical current densities of only 10 000 A/cm^2 at 4.2 K and 1 T.

Sager *et al.* [26] varied the oxygen partial pressure for different steps of the heat treatment schedule (Figure 15.2e). They investigated the influence of the precursor calcination parameters on the properties of Bi-2212 ceramics. The heat treatment schedule presented in Figure 15.2e is applied to Bi-2212 pellets of 13 mm diameter and an initial thickness of 700 μm deposited on a silver substrate. After partial melting, the thickness of the Bi-2212 pellets is reduced to 400–500 μm. The sample is heated up to 895 °C at an initial rate of 180 °C/h, which is reduced to a value of 60 °C/h just below the partial melting temperature of 895 °C. After 2 h at 895 °C, the sample is cooled to 850 °C at a cooling rate of 45 °C/h. These heat treatments were performed at an oxygen partial pressure of 1 bar. After partial melting at 895 °C, the material recrystallizes during the cool-down to 850 °C. The following annealing steps at 850 °C for 20 h and 820 °C again for 20 h were performed at an oxygen partial pressure of 0.1 bar. These annealing steps reduce the amount of minor phases, and improve the connectivity of the Bi-2212 grains. Thereafter the samples were cooled to a temperature of 690 °C and the oxygen partial pressure was reduced to zero. The annealing for 2 h at 690 °C is needed to optimize the critical temperature of the Bi-2212 phase by adjusting the oxygen content, which determines the number of holes per CuO_2 unit. An important result of their studies is the fact that high critical current densities in the final material require a high degree of homogeneity in the initial Bi-2212 powder. Melting during the calcination of the initial powders was found to be detrimental to the homogeneity of the material, leading to reduced critical current densities.

Figure 15.2f shows a heat treatment schedule used by Nachtrab *et al.* [27] for the manufacture of Ag/Bi-2212 wires for magnet resonance imaging (MRI). The fraction of Bi-2212 (fill factor) in the total cross-section of Ag/Bi-2212 wires of 0.787 mm diameter reached ≈55% in a wire with 928 filaments. In the wire with the highest fill factor, an engineering critical current density of 71 600 A/cm^2 was achieved at 4.2 K without applied field. The critical current density reached a value of 131 300 A/cm^2 at 4.2 K and zero applied field. The heat treatments were carried out in pure oxygen (see Figure 15.2f). The partial melting was done at 890 °C. The wires were cooled down to 830 °C at the very low rate of 2 °C/h. The final annealing for 8 h at 830 °C was followed by furnace-cooling to room temperature.

Figure 15.2g shows a heat treatment schedule for melt processing of silver-sheathed tapes in pure oxygen atmosphere. Polak *et al.* [28] found that the highest critical current densities of 260 000 A/cm^2 at 4.2 K without an applied magnetic field result for melting temperatures between 894 °C and 896 °C. Temperatures during melting only 2 °C above or below the optimum reduce the critical current density by ≈25%. Polak *et al.* [28] observed that the highest density of the Bi-2212 core of 5.7 g/cm^3 and the highest Vickers microhardness (≈ 100 kg/mm^2) were obtained for the optimum melting temperature of ≈895°C. The X-ray density of Bi-2212 is ≈6.5 g/cm^3. As in many other heat treatment schedules, the Ag/Bi-2212 tapes are slowly cooled from the melting temperature of 895 °C to 840 °C at a rate of 10 °C/h. The temperature is held for 48 h at 840 °C and thereafter rapidly reduced (at 700 °C/h) to room temperature.

Because the density of the Bi-2212 core is lower than the X-ray density, there must be some porosity present. The formation of voids in the superconducting core of developmental Ag/Bi-2212 monocore wires was already observed in 1993 [29]. Typically the voids are non-uniformly distributed in the superconducting Bi-2212 filaments, and at some places large voids reduce the effective area of single Bi-2212 filaments to a small fraction of the nominal value. Large bubbles formed in the Bi-2212 core can even deform the silver sheath, especially in tape conductors. Possible sources for the bubble formation are the release of oxygen and carbon dioxide. In principle, the silver sheath is oxygen permeable, and hence bubble formation originating from oxygen loss should be controllable by an adequate heat treatment schedule.

In contrast to oxygen, CO_2, which is formed from residual carbon-containing materials, does not permeate through the silver sheath. It is necessary to use Bi-2212 powders with as little residual carbon as possible. Reeves *et al.* [30] studied the use of overpressure processing to reduce void formation in Ag/Bi-2212 tapes. Furthermore, they used initial Bi-2212 powders with a residual carbon content as low as 220 ppmwt (parts per million by weight). The residual carbon content of the Bi-2212 powder was lowered from 710 to 220 ppmwt by means of a heat treatment at 700 °C in partial vacuum with flowing oxygen for 1 day. Their heat treatment is presented in Figure 15.2h. The tapes were heated up to 835 °C and held at there for 1 h in pure oxygen ($p = 1$ bar). The following partial melting step at 895 °C was performed in a mixture of Ar and O_2 with a total pressure of 8.4 bar and an oxygen partial pressure of 1 bar. After annealing for 24 h at 840 °C, the tapes were cooled down to room temperature at 120 °C/h. Below 800 °C, the oxygen partial pressure was reduced to zero, and the tapes were cooled down to room temperature in an argon atmosphere of 8.4 bar. The overpressure processing was found to be effective in reducing the number of bubbles per tape length. The phase assembly found after processing at a few atmospheres overpressure is characterized by Bi-2212, small grains of $(Sr,Ca)_{14}Cu_{24}O_x$, very small grains of $Bi_2(Sr,Ca)_4O_x$, and some long thin grains of $Bi_2Sr_2CuO_x$ (Bi-2201). This is similar to that found after processing in pure oxygen ($p = 1$ bar). The main difference was a larger amount of Bi-2201 in these tapes [30].

The manufacture of magnets from Ag/Bi-2212 wires, in which the magnet is wound before heat treatment (wind and react), suffers from difficulties related to the partial melt process. Leakage of material from the oxide core through the silver alloy sheath has frequently been observed. Moreover, the critical current density of Ag/Bi-2212 wires is very sensitive to the maximum temperature during partial melting. The electrical insulation of wind and react Ag/Bi-2212 magnets must be compatible with the required oxygen and

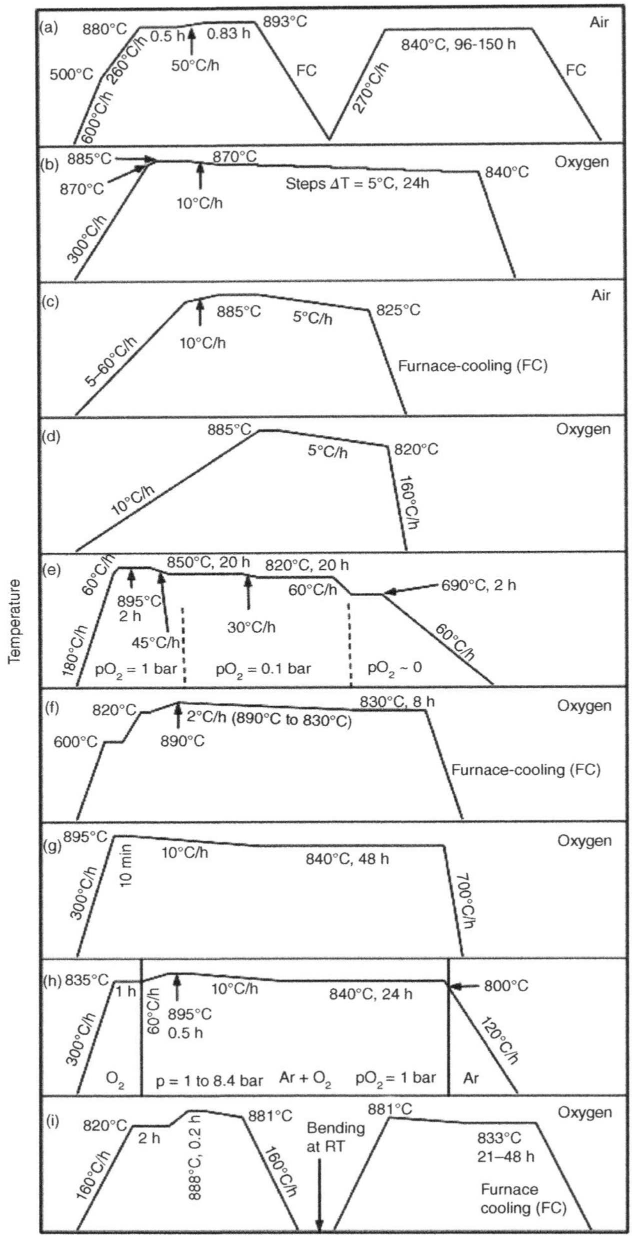

__Figure 15.2__ Various heat treatment schedules used in the manufacture of Ag/Bi-2212 wires and tapes. The details are described in the text (results from a [17, 18], b [21], c [24], d [25], e [26], f [27], g [28], h [30], and i [31, 32])

thermal diffusion. The necessity of heat treatment in pure oxygen at close to 900 °C also reduces the options for reinforcing materials [31]. On the other hand, there exist severe strain limitations for Ag/Bi-2212 magnets manufactured by the react and wind process, in which the magnet is wound using fully reacted Ag/Bi-2212 wire [31].

To avoid these problems, Liu *et al.* [31] and Shen *et al.* [32] have proposed the use of a react-wind-sinter process for the manufacture of Ag/Bi-2212 magnets. In this, the partial melt processing of the Ag/Bi-2212 wires is split into two parts, as illustrated in Figure 15.2i. The heat treatments were performed in pure oxygen. In the first heat treatment, the wires were heated up at a rate of 160 °C/h to 820 °C and held for 2 h at this temperature. The partial melting of the Bi-2212 filaments was performed at 888 °C, followed by slow cooling to 881 °C. Finally, the wires are cooled to room temperature at 160 °C/h, and wound into a coil. The winding may lead to the formation of cracks in the Bi-2212 filaments. The reaction continues during cooling from 881 °C, at which the heat treatment was interrupted.

Liu *et al.* found by means of DTA that the melting temperature of the Bi-2212 filaments is reduced after the reaction heat treatment [31]. Although the second heat treatment starts from only 881 °C, a small amount of liquid is formed, which is expected to be sufficient to heal cracks formed during coil winding. The wires are slowly cooled from 881 °C to 833 °C (at 1–2.5 °C/h), and held at this temperature for 21–48 h. The heat treatment is finished by furnace-cooling ($\approx$ 85 h) to room temperature. Even for bending diameters as small as 4 cm, the engineering critical current density of Ag/Bi-2212 wires of 0.78 mm diameter reached 75 000 A/cm^2 at 4.2 K without applied magnetic field. Even higher values of 145 000 A/cm^2 were achieved for wires of 0.4 mm diameter [31].

Other ways to manufacture Ag/Bi-2212 tapes are continuous dip-coating [33–37], tape-casting [33, 38–44], and electrophoretic deposition of Bi-2212 on silver [33]. In all these alternatives to the powder-in-tube method, Bi-2212 is deposited onto a metallic substrate in different ways. In contrast to the RE-123-coated conductors (second-generation HTS), the Bi-2212 grains of Ag/Bi-2212-coated conductors show only a c-axis texture, i.e., the crystallographic c-axis is oriented perpendicular to the broad face of the tapes, while there is no alignment of the grains in the ab planes.

Figure 15.3 shows a sketch of the continuous dip-coating technique. Yang *et al.* [34] mixed the $Bi_2Sr_2CaCu_2O_x$ powder with 10 wt% polyvinyl butyral binder and 90 wt% trichloroethylene solvent. In order to disperse the Bi-2212 powder, a small amount of phosphate ester was added. A silver or silver-alloy tape is passed through this slurry at a velocity of 20 cm/min. Yang *et al.* [34] used a silver substrate with 0.25 wt% Mg and 0.25 wt% Ni added. After heat treatment above 800 °C, fine Ni–Mg–O precipitates of 2 nm diameter are uniformly distributed throughout the silver alloy tape. The dispersed precipitates lead to a significant improvement in the mechanical properties as compared to pure silver. The uniformity of the thickness of the Bi-2212 coating can be controlled by the volatility of the solvent, the viscosity of the Bi-2212 slurry, and the coating speed [34]. The typical thickness of the Bi-2212 films on both sides of the silver substrate ranges from 10 to 50 μm after heat treatment. The dimensions of the silver substrate are several mm in width and 50 to 100 μm in thickness. The heat treatment will be discussed after the description of the tape-casting method. In short samples, critical current densities as high as 220 000 A/cm^2 ($B_a = 0$) and 70 000 A/cm^2 at 10 T were achieved at 4.2 K [35]. The j_c value of 70 000 A/cm^2 corresponds to an engineering critical current density $j_e = 29\,000$ A/cm^2 (4.2 K, 10 T).

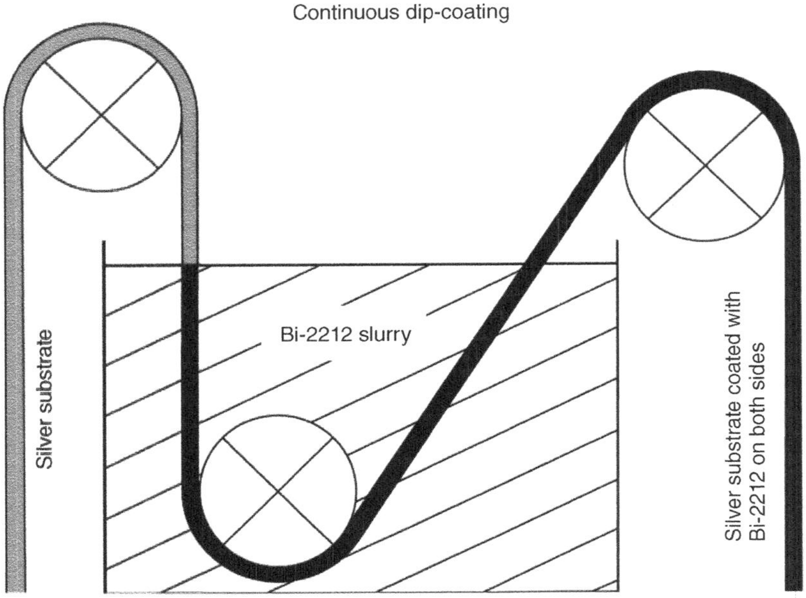

Figure 15.3 *Sketch of the continuous dip-coating technique as described in [34] (adapted from [33])*

Figure 15.4 shows a sketch of the tape-casting technique. Descriptions of this technique can be found in Refs. [33, 39, 40]. Heeb *et al.* [39] mixed Bi-2212 powder prepared by a solid-state reaction for 4 h with an organic solvent consisting of ethanol, triolein, polyethylene glycol, phtalic acid ester, and polyvinyl butyral. In this solution, triolein acts as a dispersant for the Bi-2212 powder, polyethylene glycol and phtalic acid ester were used as plasticizer, while polyvinyl butyral was used as the binder. The Bi-2212 slurry is tape-cast into tapes of 150 mm width and 50 μm thickness. Before partial melting, the tapes were put onto silver foils of 50 μm thickness.

Typical heat treatment schedules for dip-coating [37] and tape-casting [39, 44] are presented in Figure 15.5. In both processing routes, organic compounds and binders must be removed before the partial melting process is started. In the heat treatment schedule for dip-coating, the organic binder is burnt out by heat treatment at 400 °C. Partial melting is performed at 874–876 °C. This relatively low temperature seems to be required by a relatively large carbon content of the Bi-2212 slurry, originating from the polyvinyl butyral binder and the trichloroethylene solvent [37]. After the partial melting step, the tapes are slowly cooled to 840–845 °C and held at this temperature for several hours. This solidification step leads to the renewed formation and textured growth of the Bi-2212 phase.

In the case of tape-casting, the organic compounds are removed by heat treatment at 500 °C in air. The partial melting is performed in pure oxygen at 893 °C for thick films and 875 °C for thin films. The Bi-2212 phase is formed again and the grains are aligned during cooling to 820 °C at a rate of 5 °C/h. The tape-cast tapes are annealed for 1–50 h at 820 °C, again in pure oxygen. The partial melting-solidification step is finished by rapid

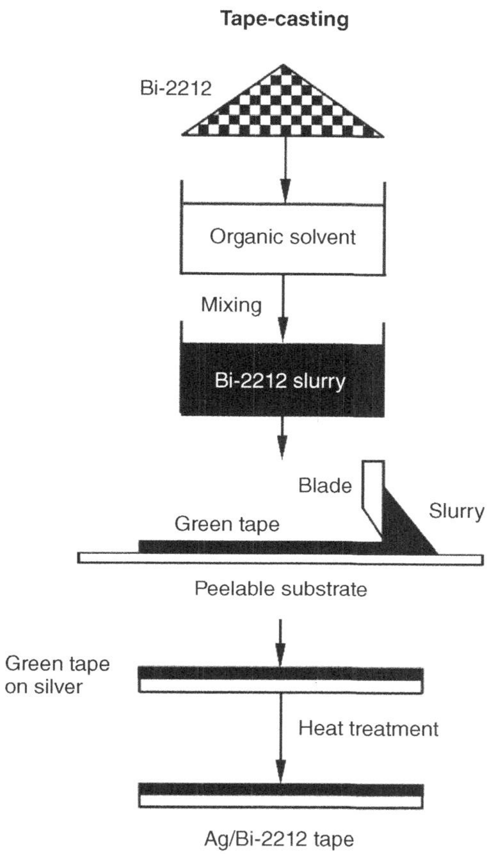

Figure 15.4 *Sketch of the tape-casting technique as described in [39, 40] (adapted from [33])*

cooling (500°C/h) to room temperature. Finally, the tapes are annealed at 500 °C for 10 h in nitrogen to reach the optimum critical temperature.

In 15 μm thick films deposited on silver substrates, the critical current density at 4.2 K exceeded 10^5 A/cm^2 for fields of 25 T applied parallel to the broad face of the tape [38]. Even for the unfavorable field direction perpendicular to the broad face of the tapes (i.e., $B||c$), the j_c is more than 10^5 A/cm^2 at 16 T. The quoted j_c values are based on an electric field criterion of 1 μV/cm.

The next aspect to be considered is the dependence of the critical current on applied magnetic field at different temperatures. Figure 15.6 shows the critical current densities normalized to the values at 4.2 K and zero applied field of two developmental monocore Ag/Bi-2212 wires [45, 46]. In both Ag/Bi-2212 wires, application of a magnetic field of 1 T reduced the j_c at 4.2 K to around 50% of the zero-field value. This reduction in j_c may be due to some weakly coupled grains in the Ag/Bi-2212 wires. For fields between 1 and 8 T, the decrease in j_c with increasing field is much less pronounced. As a consequence of the extremely high upper critical field of Bi-2212, the further reduction in j_c above 8 T is

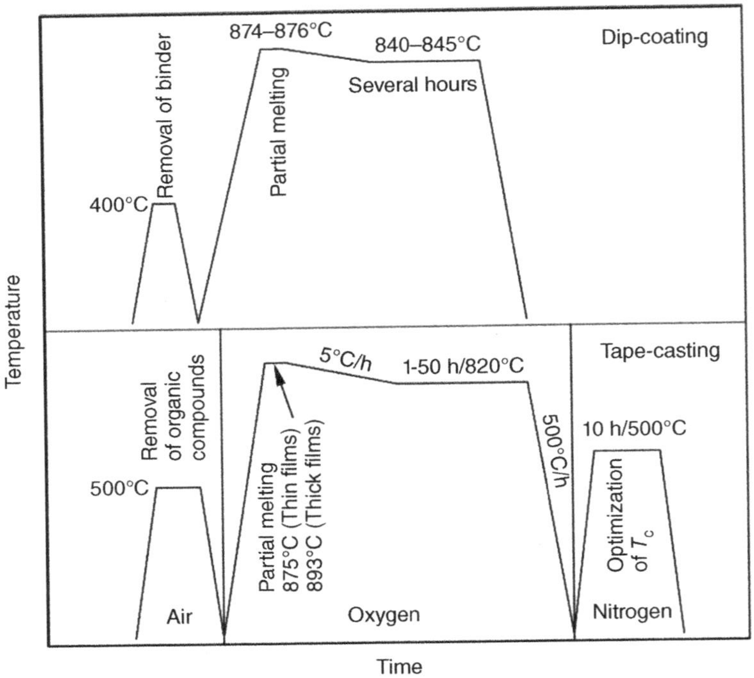

Figure 15.5 *Typical heat-treatment schedules for dip-coating (top) (after [37]) and tape-casting (bottom) (after [39, 44]). In both cases, the organic binders and compounds need to be removed before the partial melting step. For the dip-coated tape, the organic binder is removed by a heat treatment at 400 °C, while a temperature of 500 °C was necessary to remove the organic compounds from the tape manufactured by tape-casting. In order to reach high critical current densities, a partial melting process is required in both processing routes. The details are discussed in the text*

very small. Even at 4.2 K and 12 T, the j_c is still ≈30% of the zero-field value. At higher temperatures, the reduction in j_c with increasing applied magnetic field becomes more and more pronounced. The values of j_c achieved at 20 K and 12 T are between 6% and 10% of the j_c at 4.2 K with no field. At the highest temperature of 60 K, j_c drops to ≈1% of j_c (4.2 K, $B_a = 0$) at a field as low as 0.75 T. In general, the j_c declines exponentially with increasing fields for $B \geq 1$ T. This behavior can be attributed to thermally activated flux creep [33, 45, 46]. The performance of the Ag/Bi-2212 wires at temperatures above 60 K is very sensitive to the actual value of T_c. The critical temperature of the two developmental Ag/Bi-2212 wires is well above 80 K. The highest j_c values at 4.2 K in high magnetic fields are typically achieved in overdoped Ag/Bi-2212 wires with an excess of oxygen as compared to the content leading to the optimum critical temperature.

Critical current densities, which have been achieved in Ag/Bi-2212 conductors at 4.2 K, are listed in Table 15.1 [24, 26, 32, 47–57]. In wires as well as tapes, critical current densities of 4000 A/mm^2 have been reached at 4.2 K without applied field. In round wires, engineering critical current densities of 448 A/mm^2 have been obtained at 4.2 K even in a field of 25 T.

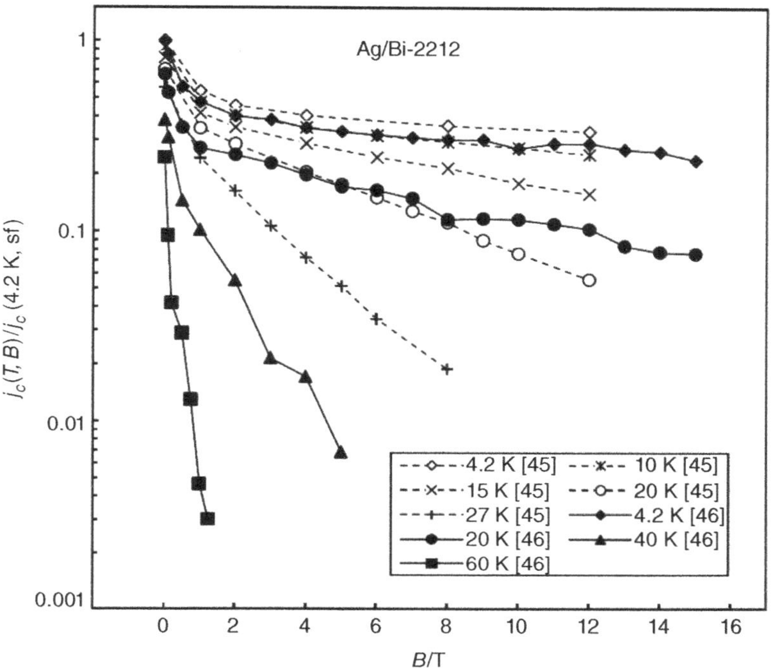

Figure 15.6 *Critical current densities of two developmental Ag/Bi-2212 monocore wires normalized to j_c (4.2 K, $B_a = 0$) versus field for temperatures in the range of 4.2–60 K (results from [45, 46])*

Figure 15.7 shows the critical current densities of various Ag/Bi-2212 conductors at 4.2 K as a function of applied magnetic field [24, 27, 32, 50, 52]. For textured Ag/Bi-2212 tapes, the j_c for fields parallel to the broad face of the tape ($B||ab$) is higher than for perpendicular fields ($B||c$). The results of Marken *et al.* [50] (see Figure 15.7) indicate that the ratio of j_c(4.2 K, 20 T $||ab$) to j_c(4.2 K, 20 T $||c$) is ≈1.65. This moderate anisotropy of j_c with respect to the field direction is not expected to be a problem for high-field insert coils.

The feasibility of manufacturing Rutherford cables with round Bi-2212 wires was successfully demonstrated by Hasegawa *et al.* [49]. A two-stage Rutherford cable was made of ten 1 × 6 cables. In the first-stage cable, a dispersion-strengthened silver-alloy core, used for reinforcement, is surrounded by six AgMgSb/Bi-2212 wires. A critical current of 9.5 kA was achieved in this Rutherford cable at 4.2 K without applied field. The improvement of the performance of Bi-2212 conductors is ongoing. Very recently, critical current densities of $\approx 10^5$ A/cm^2 at 4.2 K and 45 T have been reported [58].

15.2.3 Ag/Bi-2223 Tapes

The development of Ag/Bi-2223 tapes started with the pioneering work of Hikata *et al.* [8] in 1990. Bi-2223 tapes were the first HTS manufactured on an industrial scale in

Table 15.1 *Critical current densities of Ag/Bi-2212 conductors at 4.2 K*

Conductor type	B (T)	j_c (A/mm^2)	j_e (A/mm^2)	Comments	Reference
PIT wire	0	2490	858	$d=0.95$ mm 121 × 7 filaments	[47]
PIT wire	0	2750	1937	$d=0.74$ mm 55 × 7 filaments	[48]
PIT wire	0	5000	1042	$d=0.81$ mm 61 × 7 filaments Ag:Bi-2212 = 3.8:1	[49]
PIT wire	0	4000	1025	$d=0.81$ mm 61 × 7 filaments Ag:Bi-2212 = 2.9:1	[49]
PIT wire	18	1080	312	$d=0.8$ mm 91 × 7 filaments	[50]
PIT wire	0	1313	716	$d=0.787$ mm 928 filaments	[26]
PIT wire	25		380		[51]
PIT wire	0 19		692 215	$d=0.8$ mm 521 filaments	[32]
PIT wire	0 25		1330 448	$d=0.8$ mm 85 × 7 filaments	[52]
PIT tape	0	2150	650	6 mm × 0.2 mm	[53]
PAIR tape[a]	10 || *ab*	4000		≈4 mm width	[54]
PAIR tape[a]	10 || *ab*	3200		2.5–3.5 mm width	[55]
Tape[b]	0	2500		2 × 3.5 μm Bi-2212 on 50 μm thick Ag	[56]
PAIR tape	1 || *c*	2190		Bi-2212 core of 80 μm thickness	[57]
PIT tape	10 || *ab*		300	Double core tape 140 μm thickness	[24]
Tape	0	4107		Dip-coating	[52]

[a] PAIR process: pre-annealing and intermediate rolling process.
[b] MgO doped Bi-2212 electrolytic deposition on silver.

kilometer lengths (American Superconductor Corporation (AMSC), European Advanced Superconductors (EAS), and Sumitomo Electric Industries (SEI)). AMSC and EAS have stopped the production of Ag/Bi-2223 tapes and focused their research and development activities on RE-123 (RE: Y or another rare earth element) coated conductors (see Section 15.3.4), while SEI continued to fabricate Ag/Bi-2223 tapes with excellent properties.

An overview of the early activities to develop Ag/Bi-2223 tapes can be found in Ref. [33]. In this section, only some aspects of the early research and development activities will be presented. Ag/Bi-2223 tapes are typically manufactured by the powder-in-tube method. The manufacture of developmental Ag/Bi-2223 tapes described by Grasso *et al.* [59] is as follows. The $Bi_{1.72}Pb_{0.34}Sr_{1.83}Ca_{1.97}Cu_{3.13}O_x$ powder is filled into Ag tubes, and the composite is swaged and drawn to a wire with a final diameter between 1 and 2 mm. In contrast to the manufacture of Ag/Bi-2212 wires and tapes, the Bi-2223 powder is not fully reacted before filling. The powder with the above-mentioned overall composition is

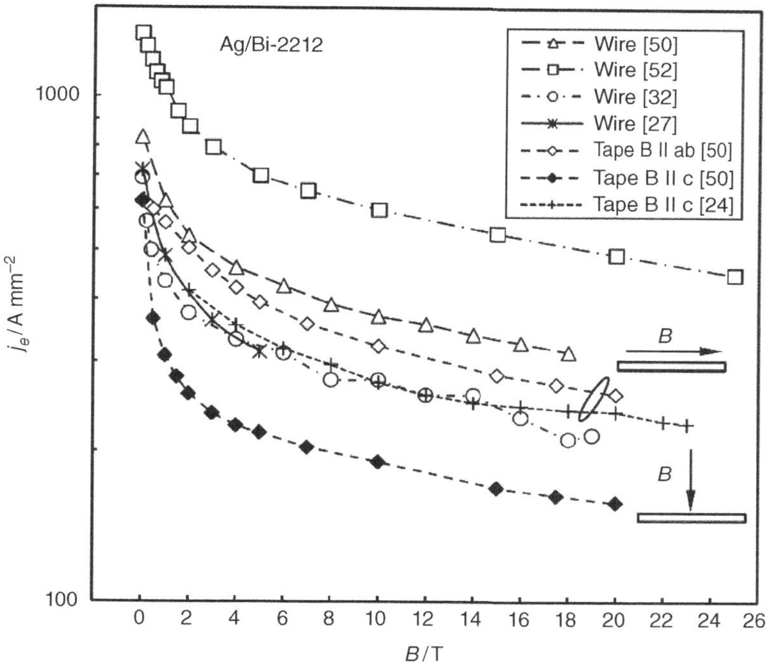

Figure 15.7 *Critical current density of various Ag/Bi-2212 conductors at 4.2 K versus magnetic field (results from [24, 27, 32, 50, 52])*

in fact a mixture of (Bi,Pb)-2212, Ca_2PbO_4, and CuO. The wire is rolled in steps with ≈10% thickness reduction per pass to a tape of ≈0.1 mm thickness. In an intermediate heat treatment at 837 °C in air (Bi,Pb)-2223 is formed. Due to the rolling process and the following heat treatment, the (Bi,Pb)-2223 crystallites grow in such a way that their crystallographic *c*-axis is almost perpendicular to the broad face of the tape. In a further rolling step, the tape thickness is reduced by around 15%, leading to a further enhanced core density. Some microcracks, formed by this final rolling step, need to be filled by a final heat treatment for 200 h at 837 °C. It is of crucial importance that, at the beginning of this final heat treatment, some unreacted (Bi,Pb)-2212, $CaPbO_4$, and CuO are present, which are able to form a liquid that fills the microcracks. The final heat treatment converts the (Bi,Pb)-2212 and the minor phases nearly completely to the desired (Bi,Pb)-2223 phase. The two-dimensional growth of the plate-like (Bi,Pb)-2223 grains, which is limited by the Bi-2223/Ag interface, further improves the *c*-axis texture.

In the case of multifilament tapes, the composites are swaged and drawn into hexagonally shaped wires, which are stacked in a further silver tube. This in turn is deformed by swaging, drawing, and rolling to a tape of 0.2–0.25 mm thickness.

Again an intermediate heat treatment is followed by a further rolling step and a final heat treatment. Grasso *et al.* [59] achieved critical current densities of 34 500 A/cm^2 (monocore tape) and 28 000 A/cm^2 (multifilament tape) at 77 K without applied magnetic field. The j_c is defined by a 1 μV/cm criterion.

The formation of the Bi-2223 phase is difficult without partial substitution of lead for bismuth. Lead substitution leads to an accelerated formation of the Bi-2223 phase. As a consequence, all initial powders used for the manufacture of Ag/Bi-2223 tapes contain 0.3–0.41 of lead [33]. Merchant *et al.* [60] studied the effect of lead substitution on the critical current density of Ag/$Bi_{1.8}Pb_xSr_{1.98}Ca_{1.97}Cu_{3.08}O_{10+\delta}$ monocore tapes. They found the highest j_c values for $x = 0.3$, close to the values quoted above. The bismuth content of the initial powders for tape manufacture is between 1.72 and 2 [33]. Thus the total content of Bi and Pb is in general slightly larger than 2. The selected values of the Sr content are between 1.78 and 2.01, while the Ca content is in the range of 1.9–2.2 [33]. In most of the initial powders used for tape manufacture, the Sr content is reduced, whereas the Ca content is increased as compared to the cation ratio of 2:2:2:3 in the stoichiometric composition. In general, the selected copper content is close to the stoichiometric value of 3. The reported values of the copper content are in the range of 3–3.13 [33].

Penny *et al.* [61] investigated the effect of the heat-treatment atmosphere on the critical current density. They heat-treated Ag/Bi-2223 tapes in atmospheres with oxygen partial pressures of 0.04, 0.075, and 0.21 bar. Their results indicate that the achievable j_c at 77 K without applied magnetic field is nearly independent of the oxygen partial pressure. However, the optimum total annealing time at an oxygen partial pressure of 0.04 bar is four times shorter than in air. The heat treatments of Ag/Bi-2223 tapes are typically performed in air or at a reduced oxygen partial pressure between 0.07 and 0.1 bar (7–10% O_2 balanced by Ar or N_2) [33]. The quoted values of the heat treatment times are between 100 and 240 h at temperatures in the range of 820–840 °C.

Next, some parameters limiting the critical current density of Ag/Bi-2223 tapes are briefly discussed. A first aspect of importance is the degree of the *c*-axis texture. In high-quality Ag/Bi-2223 tapes, the crystallographic *c*-axis is within ±10° of perpendicular to the broad face of the tapes. The degree of this *c*-axis texture can be determined from the full-width at half-maximum of X-ray rocking curves of major (0 0 l) reflections (e.g., (0 0 10), (0 0 12), (0 0 14) reflections). In Figure 15.8, the measurement of a rocking curve is illustrated. The Bragg condition for the (0 0 l) plane is fulfilled for the angle $2\theta_{00l}$ between the incident and the diffracted beams. For the measurement of the rocking curve, the angle between the incident and the diffracted beam is kept constant, while the angle ω between the incident beam and the sample surface is varied by rotating the sample under test. The intensity of the diffracted beam as a function of ω provides the rocking curve for the (0 0 l) reflection in question. For grains 1 and 2, the *c*-axis is exactly perpendicular to the surface of the Bi-2223 filament and the Bragg condition is fulfilled for $\omega = \theta_{00l}$, while for grain 3 the *c*-axis is slightly tilted to the surface normal, and hence the Bragg reflection can occur for $\omega \neq \theta_{00l}$ (see Figure 15.8). The full-width at half-maximum (FWHM) of the rocking curve, therefore, provides the typical misalignment angle.

Another way to determine the typical misalignment angle between the crystallographic *c*-axis of the grains and the surface normal was described by Hensel *et al.* [62]. At 77 K, the critical current density of textured Bi-2223 is highly anisotropic with respect to the direction of the applied magnetic field. For a magnetic field parallel to the *c*-axis ($\alpha = 0°$ see inset of Figure 15.9), j_c declines very rapidly, whereas for $B \perp c$ ($\alpha = 90°$) the decrease in j_c with increasing field is much less pronounced. It was found that the j_c of textured Bi-2223 films is limited by the perpendicular field component $B \cos\alpha$. The results of Hensel *et al.* [62], presented in Figure 15.9, show that the I_c data for different magnetic fields collapse

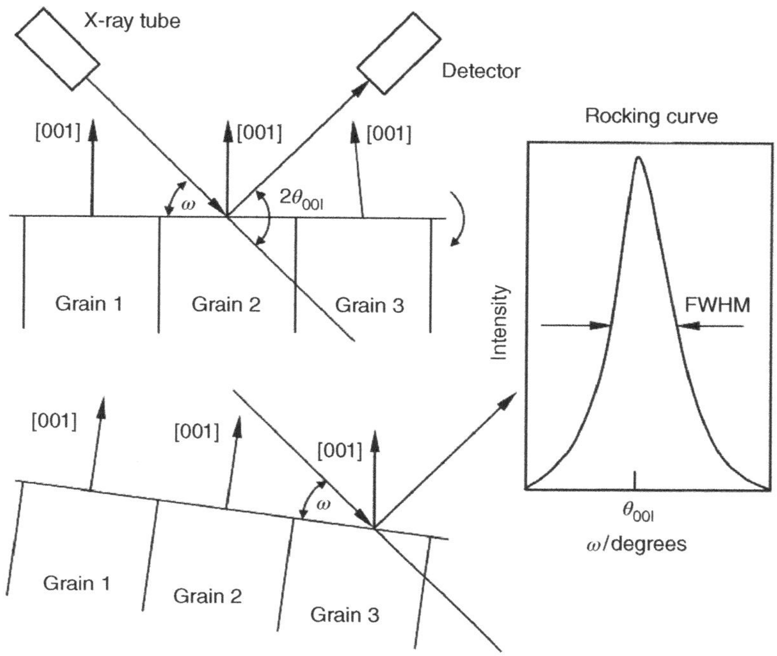

Figure 15.8 *Sketch of the measurement of a rocking curve (ω scan). The angle $2\theta_{00l}$ between the incident and the diffracted beam is kept constant, while the angle ω is varied by rotating the investigated sample. For grains 1 and 2, the c-axis is exactly perpendicular to the sample surface and hence Bragg reflection occurs for $\omega = \theta_{00l}$. In the case of grain 3, the c-axis is at a slight angle to the surface normal. Thus Bragg reflection is possible for $\omega \neq \theta_{00l}$ (adapted from [33])*

as expected onto a common I_c versus $B \cos\alpha$ curve. The deviations of the I_c data from this common curve at α values close to 90° originate from the misalignments of the *c*-axes of the individual grains. The intercepts of the plateau values with the common $I_c(B \cos\alpha)$ curve at $\alpha \approx 80°$ suggest that the typical misalignment angle is around 10°, which is in good agreement with the FWHM values of 11–14° of the X-ray rocking curves of the $00\underline{10}$, $00\underline{12}$, and $00\underline{14}$ reflections [62].

In textured Ag/Bi-2223 tapes, there exist colonies of plate-like Bi-2223 grains with 10–20 μm extension in the *ab* (CuO_2) plane and a thickness of only 1–2 μm along the crystallographic *c*-axis. In general, the *ab* planes of the grains are aligned parallel to the Ag/Bi-2223 interface. Colonies of very thin plate-like grains, in which the grains are rotated around a common *c*-axis, were found to be an important microstructural feature in the Bi-2223 filaments [63]. Li *et al.* [64] found that *c*-axis twist boundaries in Bi-2212 are strong links for the current flow irrespective of the twist angle, and allow the current to flow along the *c* direction within a single colony. On the other hand, amorphous phases present at the twist grain boundaries between adjacent colonies suggest that they are weakly coupled along the *c* direction. Small-angle grain boundaries between two grains of adjacent colonies in the *ab* plane facilitate current flow parallel to the CuO_2 planes [65]. As a consequence

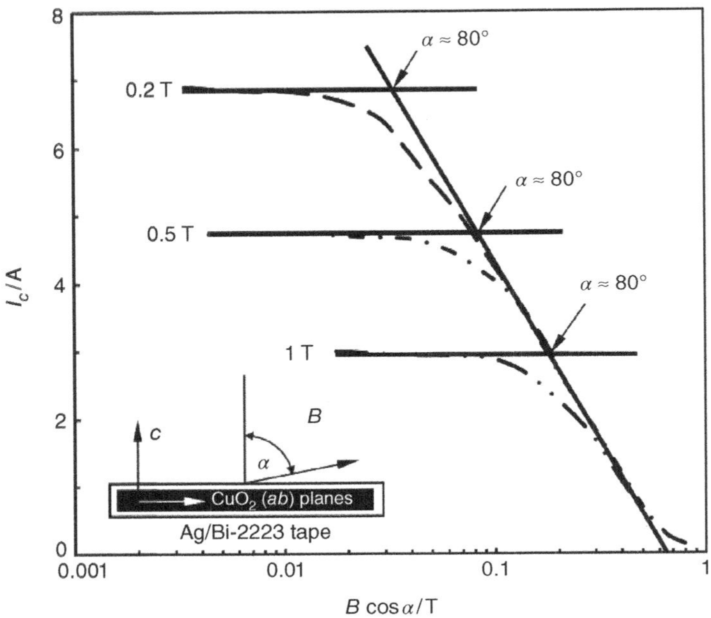

Figure 15.9 *Critical current versus the field component B cos α || c. The deviations from the common $I_c(B\cos\alpha)$ curve originate from the small misalignments of the individual grains (adapted from [62])*

of the combination of current transport across *c*-axis twist boundaries within single grain colonies, and current flow parallel to the CuO_2 planes via strongly coupled small angle grain boundaries between some grains of neighboring grain colonies, the *c*-axis texture is sufficient to reach high transport critical current densities in Ag/Bi-2223 tapes. This unique property of bismuth-based superconductors originates from the weak chemical bonding of the BiO double layers causing the observed colony structure.

The achievable critical current density depends also on the properties of the initial powders used for the manufacture of the Ag/Bi-2223 tapes. A requirement to reach high j_c values in Ag/Bi-2223 tapes is a residual carbon content of the initial powders as low as possible. For example, an increase in the residual carbon content from 0.034 to 0.057 wt% in the initial powder was found to reduce the j_c of Ag/Bi-2223 tapes from 16 000 to 7000 A/cm^2 at 77 K at zero field [66]. The CO_2 formed during the heat treatment may lead to the formation of voids in the ceramic core of Ag/Bi-2223 tapes. Furthermore, the residual carbon seems to be involved in the formation of secondary phases [66, 67]. Sources of the residual carbon content are the incomplete decomposition of $SrCO_3$ and $CaCO_3$, or extended milling of the powders in air leading to some uptake of CO_2. The carbon content of the initial powders can be reduced by heat treatment in oxygen or under vacuum. Moreover the powders may be handled in a glove box filled with an inert gas.

Another aspect of importance is the average particle size in the precursor powder, which affects their reactivity. Kim *et al.* [68] studied the effect of particle size in $Bi_{1.89}Pb_{0.41}Sr_{2.01}Ca_{2.23}Cu_{3.03}O_x$ on the critical current density of Ag/Bi-2223 tapes at

77 K and $B_a = 0$. The average particle size of 5.5 μm obtained after calcination was reduced to values as small as 1.5 μm by means of a planetary ball mill. Moisture and CO_2 taken up during extended milling was removed by an additional heat treatment at 738 °C for 1 h in 8 mbar oxygen. The j_c of Ag/Bi-2223 tapes prepared from precursor powders with average particle sizes in the range 1.5–5.5 μm showed a pronounced maximum ($\approx$ 19 000 A/cm^2) for a particle size of $\approx$2.8 μm. The use of powders with a larger particle size of 5.5 μm led to a j_c of only $\approx$9000 A/cm^2, while for the smallest particle size of 1.5 μm the achieved j_c was as small as 3000 A/cm^2. For the precursor powder with the optimum particle size, the reactivity is better than for a larger particle size. On the other hand, very small particles are amorphous, which seems to be unfavorable for the formation of the Bi-2223 phase.

The achievable critical current densities in Ag/Bi-2223 tapes depend strongly on the deformation process. Intermediate rolling steps enhance the density of the Bi-2223 filaments. On the other hand, they introduce cracks into the filaments. The main parameters for the rolling process are the roller diameter D_R and the reduction ratio defined as $R_r = (t_i - t_f)/t_i$, where t_i and t_f are the initial and final tape thicknesses, respectively. The deformation should be performed in such a way that the powder has enough time to flow along the tape in order to avoid cracking of the Bi-2223 filaments. Deformation studies by Zeimetz *et al.* [69] indicate that for $D_r = 234$ mm and a rolling speed of 4 cm/s, sausaging (i.e., a wavy Ag/Bi-2223 interface) can be avoided when R_r is smaller than 30%, 15%, and 10% for tape thicknesses above 0.8, above 0.4, and less than 0.4 mm, respectively.

The performance of Ag/Bi-2223 tapes can be significantly improved by means of high-pressure processing [70–79]. The research and development activities on high-pressure processing of Bi-2223 tapes performed until 2005 was reviewed in an article by Hellstrom *et al.* [70]. The random growth of plate-like (Bi,Pb)-2223 grains from a mixture of (Bi,Pb)-2212, $CaPbO_4$, and CuO leads to expansion of the tapes. In the filaments of Ag/Bi-2223 tapes, some pores and cracks are present after the first heat treatment at ambient pressure. To increase the density of the filaments, an intermediate rolling step is applied. However, this rolling step also introduces some additional microcracks. Even after the final heat treatment at ambient pressure, some porosity is still present and not all microcracks formed during intermediate rolling are completely healed. In principle, an enhanced filament density can be achieved by high-pressure processing performed at up to 300 bar in an Ar/O_2 mixture. The oxygen partial pressure needs to be kept in a range within which the Bi-2223 phase is stable. The results of Hellstrom *et al.* [70] indicate that in ambient pressure as well as in high-pressure synthesis, an intermediate rolling step is required to reach high critical current densities. The microcracks formed by the intermediate rolling seem to be beneficial for the conversion of residual Bi-2212 phase to the desired Bi-2223 phase [70]. They achieved a critical current density of 69 600 A/cm^2 at 77 K and zero applied field by high-pressure processing of already fully heat treated Ag/Bi-2223 tapes at 148 bar and an oxygen partial pressure in the range of 0.075–0.1 bar. The increased critical current density of high-pressure processed Ag/Bi-2223 tapes was attributed to higher filament density, a reduced number of cracks, better homogeneity, and a reduced content of the undesirable Bi-2212 phase [70]. Investigations of the local j_c by magneto-optical current reconstruction indicate that not only the average j_c is improved but also the highest values of the local j_c are increased from 200 000 A/cm^2 in ambient pressure processed tapes to 300 000 A/cm^2 after high-pressure heat treatment. The fact that the maximum local j_c is five times higher than the average value of 69 600 A/cm^2 suggests that there exists a large potential for further improvements [70].

For a few years, Sumitomo Electric Industries has been using a controlled high-pressure process in the industrial manufacturing of Ag/Bi-2223 tapes of kilometer length [73–78]. The tapes are sold under the brand name DI-BSCCO (DI, dynamically innovative). Figure 15.10 shows the critical current of two Ag/Bi-2223 tapes as a function of magnetic field parallel to the broad face of the tapes, for various temperatures. The tape with the higher I_c of 195 A at 77 K and zero field is 0.24 mm thick and 4.25 mm wide [76]. The second tape, with an I_c of only 76 A at 77 K and zero field, is reinforced by a stainless steel lamination [77]. The dimensions of the tape including the 0.02 mm thick stainless steel tapes are 2.7 mm in width and 0.25 mm in thickness. Thus the reinforced Ag/Bi-2223 tape is only 2.5 mm wide and 0.18 mm thick. The difference in the critical currents mainly reflects the different superconductor cross-sections. The relative reduction in the critical current with field and temperature is similar for both tapes. The critical current depends only weakly on the applied magnetic field at 4.2 K, and reaches a value of ≈640 A at 12 T for the unreinforced tape. The dependence of the critical current on magnetic fields

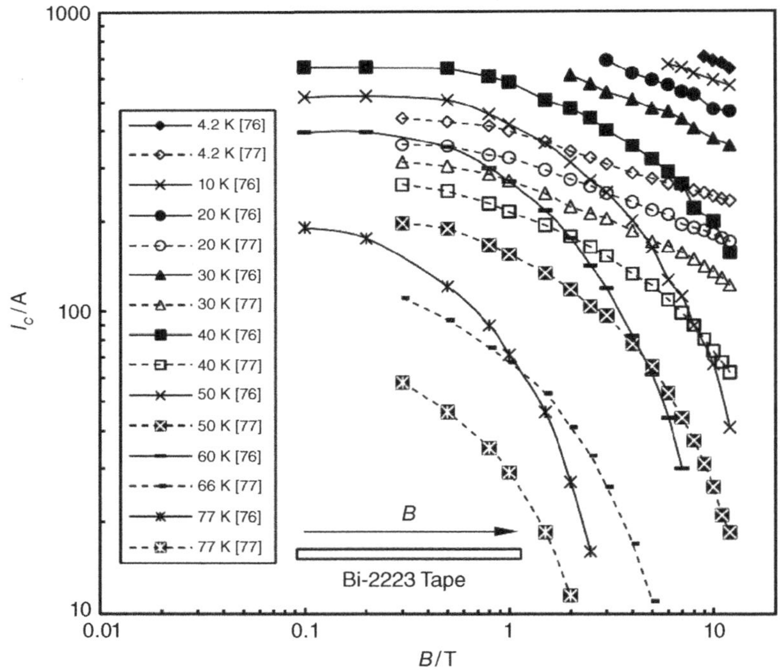

Figure 15.10 Critical current of two Ag/Bi-2223 tapes versus parallel field for various temperatures. The tape with the higher I_c of 195 A (77 K, $B_a = 0$) is 0.24 mm thick and 4.25 mm wide [76]. The second tape with an I_c of 76 A (77 K, $B_a = 0$) is stainless steel laminated [77]. The dimensions of the tape are 2.5 mm × 0.18 mm (without stainless steel lamination). The difference in the critical currents reflects mostly the different superconductor cross-sections (results from [76, 77])

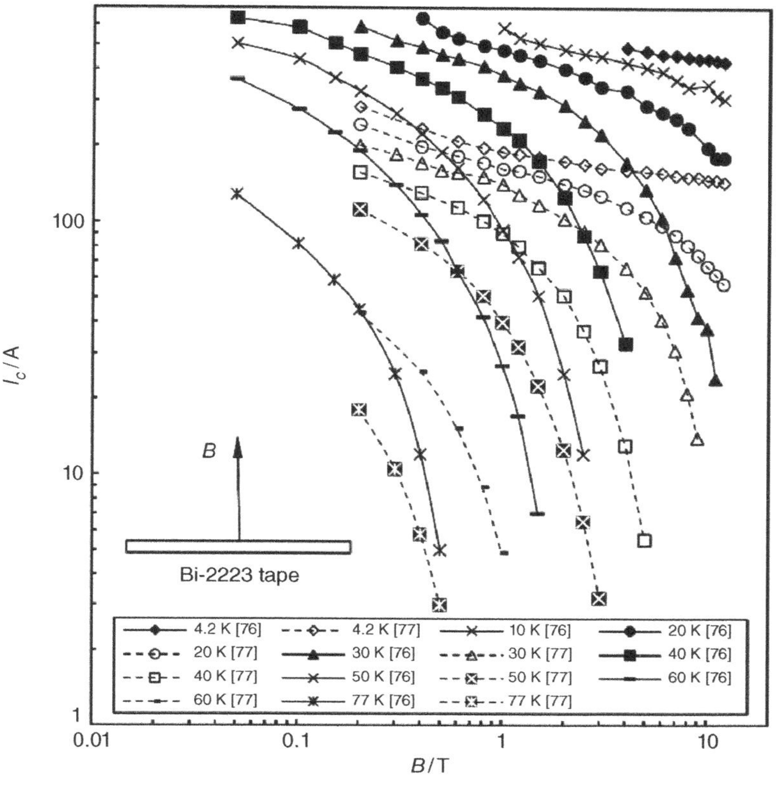

Figure 15.11 *Critical current of two Ag/Bi-2223 tapes versus perpendicular field for various temperatures. The tape with the higher I_c of 195 A (77 K, $B_a = 0$) is 0.24 mm thick and 4.25 mm wide [76]. The second tape with an I_c of 76 A (77 K, $B_a = 0$) is laminated with stainless steel [77]. The dimensions of the tape are 2.5 mm × 0.18 mm (without stainless steel lamination). The data for parallel fields have already been presented in Figure 15.10 (results from [76, 77])*

perpendicular to the broad face of these tapes is presented in Figure 15.11. A comparison with the data in Figure 15.10 indicates that the I_c declines more rapidly in perpendicular fields, especially at elevated temperatures. The normalized critical currents for parallel and perpendicular fields are compared in Figure 15.12. For clarity, the data for temperatures up to 30 K are presented in the top panel, and those for higher temperature in the bottom panel. The data indicate that the anisotropy of I_c with respect to the field direction increases with increasing temperature. In the design of superconducting magnets to be manufactured of Ag/Bi-2223 tapes, this anisotropy needs to be carefully considered.

The performance of industrially manufactured Ag/Bi-2223 tapes is still progressing. In October 2009, a critical current of 236 A (77 K, zero applied field, 1 μV/cm criterion) was achieved in a tape of 4.31 mm width and 0.257 mm thickness [73]. The corresponding critical current density is 59 800 A/cm^2. Recent studies of the current distribution in the tape cross-section by means of scanning Hall probe magnetic microscopy suggest that in tapes with an improved current uniformity, critical currents of 300 A would be feasible

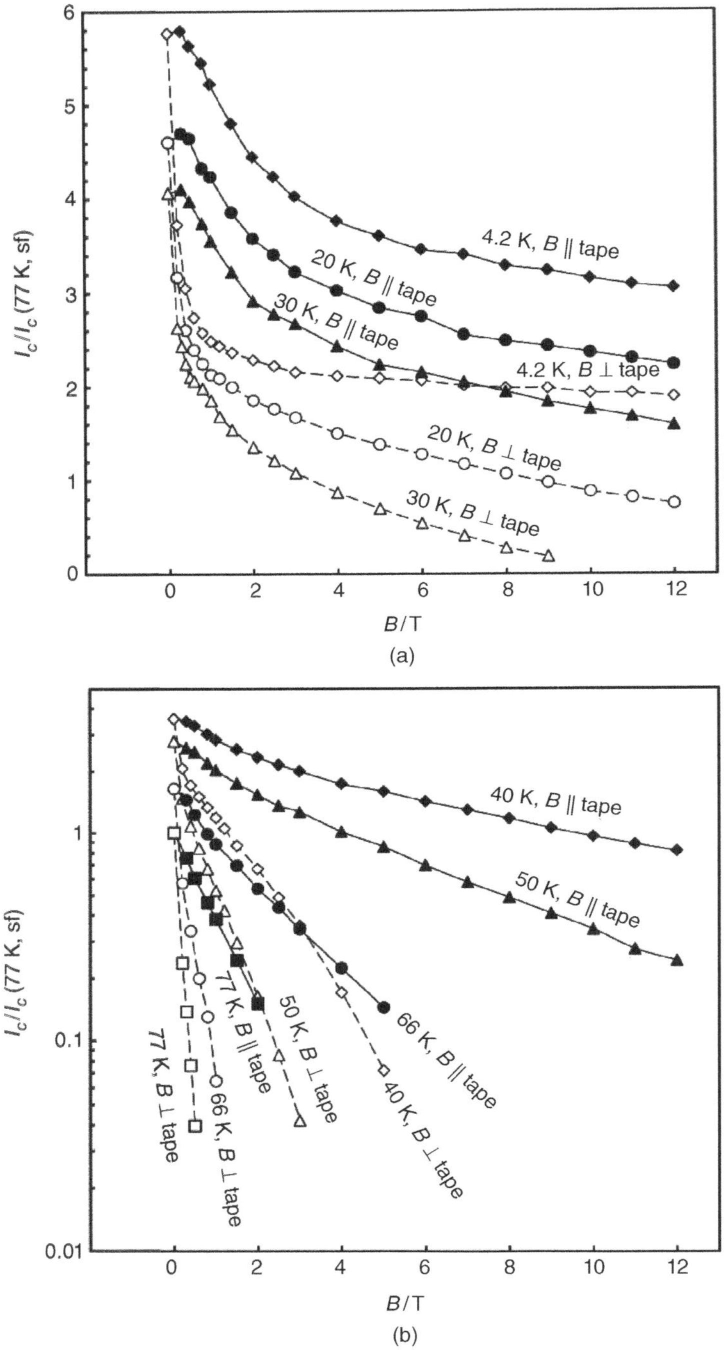

Figure 15.12 *Comparison of critical currents in the stainless steel laminated Ag/Bi-2223 tape for fields parallel (solid symbol) and perpendicular (open symbols) to the broad face of the tapes. The critical currents are normalized to the value of 76 A at 77 K and $B_a = 0$. The upper panel shows the data for temperatures between 4.2 and 30 K, while the lower panel provides the data for the range of 40–77 K (results from [77])*

[79]. In commercially available Ag/Bi-2223 tapes of up to 1500 m length, critical currents of 200 A (77 K, $B_a = 0$, 4.3 mm × 0.23 mm) have been achieved [80].

15.3 Second-Generation of High-T_c Superconductor Tapes

15.3.1 Introduction

In the Ag/Bi-2223 tapes of the first generation of HTS, a *c*-axis texture was found to be sufficient to achieve relatively high critical current densities. As described in the previous section, the required *c*-axis texture could be achieved in tapes manufactured by the powder-in-tube method. The rolling process included in their manufacture tends to align the Bi-2223 grains in such a way that the crystallographic *ab* planes are parallel to the Ag-Bi-2223 interface. For Y-123, a biaxial texture was found to be necessary to overcome the problem that high-angle grain boundaries act as barriers for the transport current (the weak-link problem). The difference between a *c*-axis texture and a biaxial alignment are schematically illustrated in Figure 15.13. In the case of the *c*-axis texture (Figure 15.13, left), the crystallographic *c*-axis is within a few degrees of perpendicular to the film surface, whereas the orientation of the crystallographic *a*-axis is random within the film plane. In a film with a biaxial alignment, the *c*-axes of all grains are again perpendicular to the film surface, while the crystallographic *a*-axis of the grains is also aligned within the film plane as illustrated in Figure 15.13 (right).

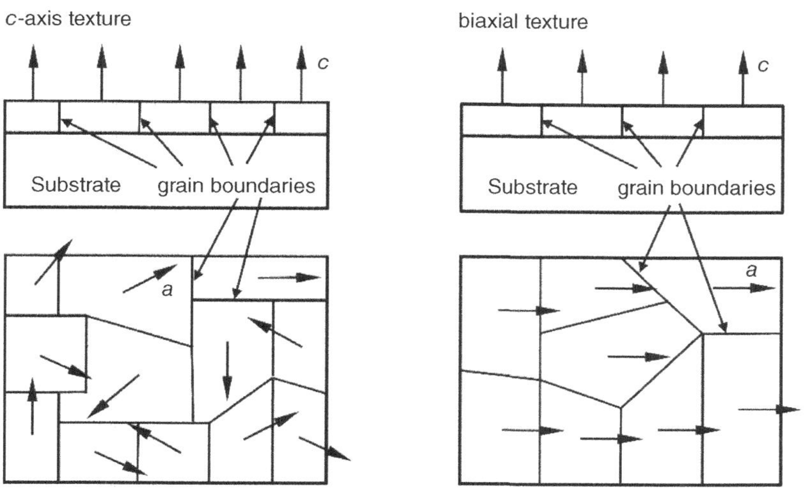

Figure 15.13 *Comparison of c-axis texture (left) and biaxial grain alignment (right). In the case of the c-axis texture, the crystallographic c-axis of all grains is almost perpendicular to the film surface as shown for the film cross-section (top left). The crystallographic a-axes of the grains are randomly oriented within the film surface (bottom left). For the biaxial texture, the c-axis of all grains is again almost perpendicular to the film surface (top right). In addition, the crystallographic a-axes of all grains are aligned (bottom right) (adapted from [33])*

15.3.2 Manufacturing Routes for Coated Conductors

The manufacture of RE-123-coated conductors (RE: Y or another rare earth element) requires biaxial alignment over lengths of several hundred meters. Three different procedures were developed to reach this target. The first of these procedures is ion-beam-assisted deposition (IBAD), which makes use of the discovery that yttria-stabilized zirconia (YSZ) can be grown with biaxial alignment on polycrystalline metallic substrates [81–88]. The growth of biaxially aligned YSZ was demonstrated for nickel [84], stainless steel [85], and nickel-based alloys (e.g., Hastelloy C-276) [81–83, 86–88]. A sketch of the IBAD process is presented in Figure 15.14. Argon ions of 1500 eV energy are used to sputter the YSZ film onto the metallic substrate at a temperature of 130 °C. The biaxial alignment originates from the use of an assisting Ar/O ion beam of 250 eV energy, at an angle of 35.3° to the substrate surface, and parallel to the [1 1 1] direction of the growing YSZ film. YSZ grains with other orientations are sputtered away by the assisting beam. As illustrated in Figure 15.14 (right), the [0 0 1] direction of the YSZ buffer layer is perpendicular to the film surface. A CeO_2 buffer layer, epitaxially grown onto the YSZ film, improves the matching of the lattice parameters of substrate and Y-123 [84]. The final Y-123 film can be deposited by pulsed laser deposition, electron beam evaporation, or chemical vapor deposition. Pulsed laser deposition is typically performed at an oxygen pressure of 400 mbar and a substrate temperature in the range of 700–800 °C. The degree of the in-plane alignment can be checked by X-ray ϕ-scans for selected Bragg peaks. For the YSZ (1 1 3) as well as the Y-123 (1 0 3) peaks, the in-plane alignment leads to four strong reflections separated by 90°. For a full-width at half-maximum of 7°of the Y-123 (1 0 3) peak, critical current densities of 10^6 A/cm^2 (77 K, $B_a = 0$) were achieved [85].

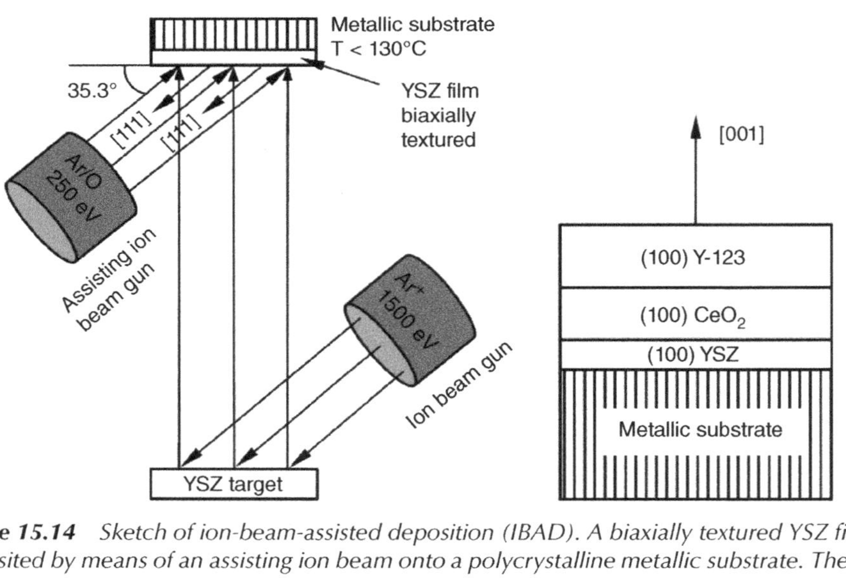

Figure 15.14 *Sketch of ion-beam-assisted deposition (IBAD). A biaxially textured YSZ film is deposited by means of an assisting ion beam onto a polycrystalline metallic substrate. The YSZ film grows in such a way that the [1 1 1] direction is parallel to the assisting ion beam (adapted from [33, 85, 88])*

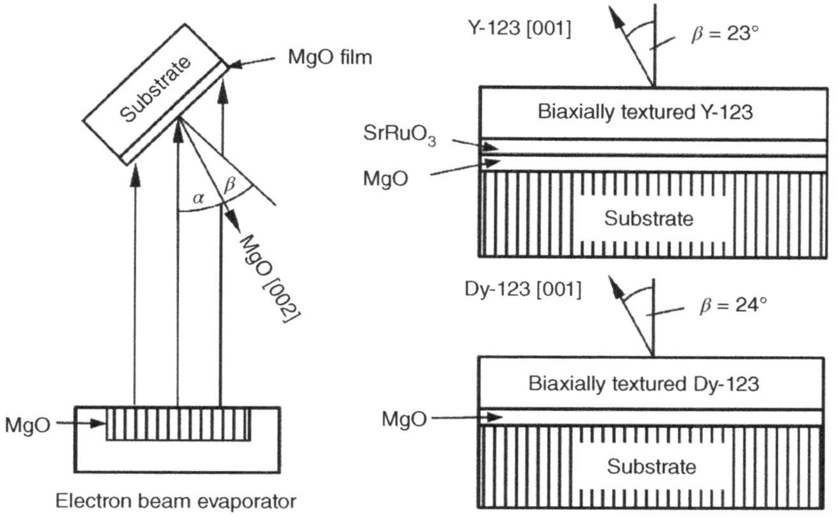

Figure 15.15 *Sketch of inclined substrate deposition (adapted from [89]). For inclination angles α of 35° and 55° a good biaxial texture of the MgO film was found. The crystallographic c direction of the MgO film is tilted to the surface normal by the angle β. Film sequences presented in [93] and [95] are shown on top right and bottom right, respectively*

A process closely related to IBAD is inclined substrate deposition (ISD) [89–95], shown in Figure 15.15. In contrast to the IBAD process, no assisting beam is required. The surface normal of the substrate is inclined at an angle α to the direction of evaporation. The biaxially textured MgO film is deposited by electron beam evaporation. The crystallographic c direction of the MgO film is tilted at an angle β to the normal to the film. Due to the inclination of the substrate, the MgO surface has a roof-tile structure. In order to reduce the surface roughness, a homoepitaxial MgO film is deposited on top of the ISD MgO film by pulsed layer deposition (PLD). Inclination angles α of 35° and 55° were found to provide the best in- and out-of-plane alignment of the MgO films [89]. The film sequences presented in Refs. [93, 95] are shown in Figure 15.15 (right).

The third way to prepare Y-123-coated conductors is the RABiTS (*R*olling *A*ssisted *Bi*axially *T*extured *S*ubstrate) process, which is based on the use of biaxially textured nickel substrates [88, 96–100]. The biaxial texture of the nickel substrates can be obtained by 90% reduction in the tape cross-section by means of cold rolling and subsequent annealing at temperatures of 400–1000 °C. The process leads to a texture of the nickel tape with the [1 0 0] direction parallel to the rolling direction and a (1 0 0) surface formed in the rolling plane [33, 96, 97]. In copper, such a cube texture can also be very easily obtained. However, the oxidation resistance of biaxially textured nickel or Ni–W alloys is much higher than that of copper, and hence the use of nickel or Ni–W alloys is preferred for the manufacturing of Y-123-coated conductors. A sketch of the preparation of cube-textured nickel, and a possible film sequence in a RABiTS Y-123-coated conductor are presented in Figure 15.16. Without the use of an assisting beam, a thin CeO_2 film can be epitaxially grown onto the metallic nickel substrate. In order to avoid oxidation of the nickel, the deposition of the

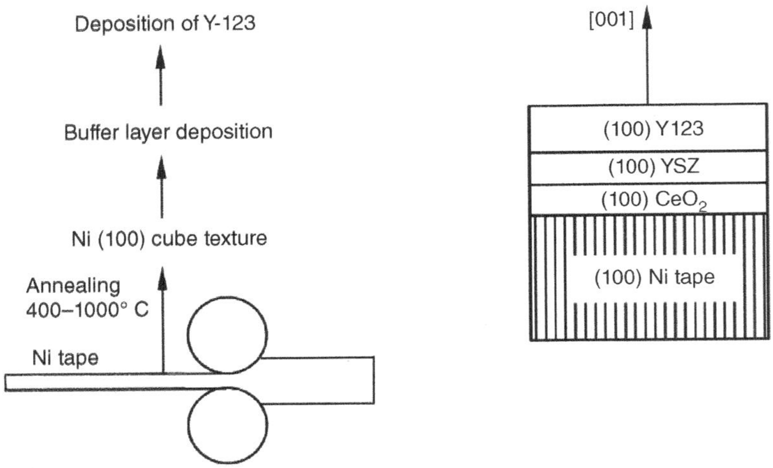

Figure 15.16 *Sketch of the RABiTS process. On cube-textured Ni (1 0 0) tapes, obtained by rolling and subsequent annealing, CeO_2 and YSZ buffer layers are grown epitaxially. The Y-123 film is deposited onto the YSZ buffer layer. For all deposited films and the nickel tape, the crystallographic c direction is perpendicular to the surface (adapted from [33, 88, 100])*

CeO_2 film is performed in a mixture of 4% hydrogen and 96% argon [98, 100]. An additional buffer layer of YSZ is deposited onto the CeO_2 buffer layer. These layers hinder the diffusion of nickel into the Y-123 film, which would degrade its superconductor properties. In addition, buffer layers can be used to reduce the mismatch of the lattice parameters of the biaxially textured substrate and the Y-123 film. The use of several buffer layers may reduce the surface roughness of the substrate, and contribute to an improved biaxial alignment of the epitaxially grown Y-123 film. Even in the early days of coated conductor development, critical current densities as high as 7×10^5 A/cm^2 (77 K, $B_a = 0$) were achieved with a Ni/CeO_2/YSZ architecture [100].

All the three described procedures aim to provide a biaxially textured buffer suitable as a substrate for the epitaxial growth of RE-123 (RE: Y or another rare earth element). Due to the absence of high-angle grain boundaries in the biaxially aligned Y-123 films, no weak links are present, and high transport critical current densities can be achieved in the superconducting layer. Various options for the type, sequence, and number of buffer layers described in the literature are presented in Table 15.2 [81, 83–87, 89, 91, 94, 97, 99–125]. For the ISD process, the nontextured metallic substrate is typically Hastelloy C 276. MgO can be grown with a biaxial texture onto the Hastelloy substrates for angles of 25–35° and 55° between the direction of evaporation and the normal to the substrate. The MgO film is deposited by electron beam evaporation at a substrate temperature between 20 °C and 50 °C [89]. In order to reduce the surface roughness, an additional MgO film may be deposited by homoepitaxy (direction of evaporation parallel to the substrate normal) on top of the biaxially textured ISD MgO film. Biaxially textured Y-123 or Dy-123 films can be grown directly on the MgO layer [89, 94]. The RE-123 film is typically deposited by PLD at substrate temperatures in the 700–800 °C range. Uprety *et al.* deposited an additional buffer layer of $SrRuO_3$ on top of the MgO layer by PLD [91]. The mismatch of the lattice

Table 15.2 *Buffer layer architectures used for the preparation of coated conductors (α, inclination of evaporation direction to surface normal; T_s, substrate temperature; p_O, oxygen partial pressure; ISD, inclined substrate deposition; IBAD, ion-beam-assisted deposition; RABiTS, rolling-assisted biaxially textured substrate; PLD, pulsed laser deposition, d_r, deposition rate, MOD, metal-organic deposition)[a]*

Sequence of layers	Process	Comments	Reference
Y-123/MgO/Hastelloy C	ISD $\alpha = 30°$ or $\alpha = 55°$	Mechanically polished Hastelloy e-beam evaporation of MgO $d_r = 2$–5 nm/s $T_s = 20$–50 °C PLD of Y-123 film at $T_s = 760$ °C $p_O \approx 130$–400 mbar	[89]
Y-123/$SrRuO_3$/MgO/ Hastelloy C276	ISD $\alpha = 35°$ or $\alpha = 55°$	e-beam evaporation of MgO PLD $SrRuO_3$, $T_s = 700$–800 °C PLD Y-123, $T_s = 700$–800 °C $p_O \approx 130$–400 mbar Annealing at 450 °C, $p_O \approx 1$ bar	[91]
Dy-123/MgO/Hastelloy C276	ISD $\alpha = 25$–$30°$	Electropolished and cleaned Hastelloy C276 e-beam evaporation of MgO $T_s = 400$ °C, $d_r > 4$ nm/s MgO thickness of 3 μm + 200 nm homo-epitaxial MgO cap layer e-beam evaporation of Dy-123	[94]
Y-123/CeO_2/Ag/Pd/Ni Y-123/YSZ/CeO_2/Pd/ Ni	RABiTS		[97]
Y-123/CeO_2/Pd/Ni	RABiTS	Ultrasonically cleaned Ni e-beam deposition of Pd $T_s = 100$–500 °C, $p \leq 10^{-6}$ mbar $d_r = 0.5$–1 nm/s	[99]
Y-123/YSZ/CeO_2/Ni	RABiTS	4% H_2/96% Ar $p \approx 1.3$ mbar Ni annealed at 700 °C for 1 h e-beam deposition of CeO_2 $T_s = 600$ °C, $p = 10^{-6}$ mbar $d_r = 0.1$ nm/s, <10 nm thick e-beam deposition of YSZ 4% H_2/96% Ar $p \approx 1.3$ mbar CeO_2/Ni annealed at 700 °C for 1 h $T_s = 650$–750 °C, $p = 2 \times 10^{-5}$ mbar $d_r = 0.1$ nm/s, 50–150 nm thick	[99]
Y-123/YSZ/CeO_2/Ni	RABiTS	Sputtering of CeO_2 in Ar/4% H_2 10–30 nm thick thereafter deposition in Ar/10% O_2 $T_s = 650$–750 °C YSZ deposition at 780 °C Ar/10% O_2 or Ar/4% H_2 PLD of Y-123	[100]

(continued overleaf)

Table 15.2 (continued)

Sequence of layers	Process	Comments	Reference
Y-123/CeO_2/YSZ/CeO_2/Ni	RABiTS	Reel-to-reel process	[104]
Y-123/$LaNiO_3$/Ni	RABiTS	Annealing of Ni tape at 900 °C for 1–4 h in 1 bar Ar/4% H_2 Deposition of $LaNiO_3$, $T_s = 400$ °C $d_r = 0.01$–0.2 nm/s, 400 nm thick Ar/4% H_2 sputtering gas PLD of Y-123, $T_s = 740$–780 °C $p_O \approx 250$ mbar	[102]
Y-123/YSZ/Ni	RABiTS	Cold rolled Ni tapes (50 μm thick) Annealing at 800–1000 °C, 1 h, 2.7×10^{-6} mbar in Ta envelope[b] YSZ deposition, 2.7×10^{-6} mbar $T_s = 800$ °C Y-123 deposition, $T_s = 750$ °C $p_O = 270$ mbar	[114]
Y-123/CeO_2/Ni-5% W[c]	RABiTS	PLD of CeO_2 and Y-123	[105]
Y-123/$LaMnO_3$/Ni-3% W[c]	RABiTS	Magnetron sputtering of $LaMnO_3$ $T_s = 575$–625 °C, 4 mbar Ar/4% $H_2 + 2.7$–6.7×10^{-5} mbar H_2O PLD Y-123, $T_s = 780$ °C, $p_O = 160$ mbar, 5 °C/min → 500 °C $T_s < 500$ °C, $p_O = 730$ mbar	[103]
Y-123/CeO_2/YSZ/Y_2O_3/Ni-5% W[c] 50 nm Y_2O_3 300 nm YSZ 30 nm CeO_2 1.2 μm Y-123	RABiTS	Ni-W tape 10 mm wide 75 μm thick Reel-to-reel deposition of epitaxial buffer layers e-beam evaporation of Y_2O_3 Sputtering deposition of YSZ CeO_2 MOD of Y-123[d] in H_2O/O_2 at $T_s < 400$ °C	[101]
Y-123/(La,Sr)TiO_3/TiN/Ni-3% W[c] 0.2 μm TiN 0.6–1.2 μm (La,Sr)TiO_3	RABiTS	PLD TiN layer, $T_s = 650$ °C in 1.3×10^{-4} mbar N_2 PLD (La,Sr)TiO_3 layer, $T_s = 750$ °C in vacuum PLD Y-123, $T_s = 780$ °C, $p_O = 0.13$ mbar	[111]
Y-123/CeO_2/Ni-5% W[c] 80 μm Ni-W 80–100 nm CeO_2 1 μm Y-123 Ag cap layer	RABiTS	Ni-W recrystallization at 700 °C in 10 mbar Ar/5% H_2[e] e-beam or thermal evaporation of CeO_2, $T_s \sim 690$ °C, $d_r = 0.25$ nm/s, 5×10^{-5} mbar Ar/5% H_2 Y-123, thermal co-evaporation, $T_s = 750$ °C, $d_r = 0.25$ nm/s, $p_O = 7\times 10^{-5}$ mbar 0.5 h annealing in $p_O = 500$ mbar	[106]

Table 15.2 *(continued)*

Sequence of layers	Process	Comments	Reference
Nd-123/CeO_2/YSZ/ Y_2O_3/Ni-3%W Ni-W 50 μm Y_2O_3 50 nm YSZ 200 nm CeO_2 30 nm	RABiTS	Deposition of Nd-123 by PLD $T_s = 760\,°C$, 1% O/Ar gas $p \approx 1060$ mbar, in situ annealing at 500 °C, $p_O = 665$ mbar Ex situ annealing at 450 °C for 1 h in flowing O_2	[107]
Ag/Y-123/CeO_2/YSZ/ Y_2O_3/Ni-W Ag 1 μm, Y-123 ~ 1 μm, CeO_2 75 nm, YSZ 75 nm, Y_2O_3 75 nm, Ni-W 50–75 μm	RABiTS	The three buffer layers were deposited by reactive sputtering MOD of the Y-123 film Cap layer of silver Y-123 tape industrially fabricated by AMSC[f]	[108]
Y-123/CeO_2/YSZ/Y_2O_3/ Ni-alloy	RABiTS	Typical architecture of RABiTS Y-123 tapes	[109]
Y-123/$Gd_2Zr_2O_7$/Ni-5% W	RABiTS	Deposition of $Gd_2Zr_2O_7$ by PLD $T_s = 750\,°C$, 5% H_2/Ar, $p = 3.6 \times 10^{-2}$ mbar Further deposition of $Gd_2Zr_2O_7$ $p_O = 3.3 \times 10^{-4}$ mbar Deposition of Y-123 by PLD $T_s = 810\,°C$, $p_O = 0.3$ mbar $p_O = 400$ mbar during cool-down	[110]
Y-123/YSZ/Hastelloy C276 YSZ 0.35–0.7 μm Y-123 1 μm	IBAD	IBAD YSZ, 300 eV assisting O + Ar beam 30–60° tilted to substrate normal $T_s = 100\,°C$, $d_r = 0.05$ nm/s PLD of Y-123, $T_s = 700\,°C$, $p_O = 266$ mbar	[81]
Y-123/Ag/YSZ/Hastelloy C276	IBAD	IBAD YSZ, CVD of Y-123	[83]
Y-123/CeO_2/YSZ/Ni YSZ 0.6–0.8 μm	IBAD	IBAD YSZ, 250 eV assisting beam PLD of CeO_2 and Y-123	[84]
Y-123/YSZ/stainless steel YSZ 0.6–3 μm Y-123 0.3 μm	IBAD	IBAD YSZ, 300 eV Ar + O beam 55° tilted to substrate normal $T_s \leq 130\,°C$, $d_r = 0.17$ nm/s Deposition of Y-123 at $T_s = 785\,°C$ Ar:O_2 = 1:1, $p = 0.2$ mbar	[85]
Y-123/CeO_2/YSZ/Ni alloy CeO_2 200 nm	IBAD	IBAD YSZ	[86]
Y-123/YSZ/Ni-alloy	IBAD	IBAD YSZ, Ar or O beam 30–60° tilted to substrate normal	[87]

(continued overleaf)

Table 15.2 (continued)

Sequence of layers	Process	Comments	Reference
Y-123/$SrTiO_3$/MgO/ Hastelloy C276 $SrTiO_3$ ≤360 nm Y-123 1.5 μm	IBAD	IBAD-MgO $SrTiO_3$ deposition, $T_s = 820\,°C$, $p_O = 400$ mbar, Y-123 deposition $T_s = 760\,°C$, $p_O = 270$ mbar	[115]
MgO/amorphous substrate	IBAD	IBAD-MgO, PLD of MgO Assisting Ar/O_2 beam (800 eV) 45° tilted to substrate normal	[116]
TiN/amorphous substrate	IBAD	IBAD TiN, PLD of TiN Assisting Ar/N_2 beam (800 eV) $d_r = 0.1$ nm/s, $p = 8 \times 10^{-4}$ mbar	[116]
Ag/Y-123/CeO_2/ $Gd_2Zr_2O_7$/Hastelloy C276 Ag cap layer 20 μm thick	IBAD	IBAD $Gd_2Zr_2O_7$, production speed 0.5–1 m/h, reel-to-reel process Assisting Ar^+ beam 200 eV 55° tilted to surface normal PLD of CeO_2 ($T_s = 400$–$600\,°C$) and Y-123 ($T_s = 900$–$1000\,°C$)	[112]
RE-123/CeO_2/$Gd_2Zr_2O_7$/ metal RE-123 films 0.25–2 μm	IBAD	IBAD $Gd_2Zr_2O_7$/PLD CeO_2 PLD of RE-123 films	[117]
Y-123/CeO_2/YSZ (1 0 0) single crystal	IBAD	Deposition of CeO_2 by the sol–gel method, MOD of Y-123	[118]
Y-123 or Gd-123/CeO_2/ $Gd_2Zr_2O_7$/Hastelloy	IBAD	IBAD $Gd_2Zr_2O_7$/PLD CeO_2 PLD of Y-123 or Gd-123	[119]
Y-123/CeO_2/YSZ/ stainless steel YSZ 4 μm thick CeO_2 10–30 nm thick	IBAD	IBAD YSZ MOD of CeO_2 MOD of Y-123 starting from trifluoroacetates	[120]
Gd-123/CeO_2/$Gd_2Zr_2O_7$/ Hastelloy $Gd_2Zr_2O_7$ 0.8 μm CeO_2 0.4 μm	IBAD	IBAD $Gd_2Zr_2O_7$/PLD CeO_2 PLD of Gd-123 at $T_s = 800$–$850\,°C$ $p_O = 800$ mbar	[121]
(Y,Sm)-123/$LaMnO_3$/ MgO/metal	IBAD	IBAD-MgO, deposition of homo-epitaxial MgO by magnetron sputtering, $LaMnO_3$ deposition by magnetron sputtering MOCVD[g] of (Y,Sm)-123	[122]
Sm-123/$LaMnO_3$/homo-epitaxial/MgO/ IBAD-MgO/Y_2O_3/ Al_2O_3/Hastelloy	IBAD	e-beam evaporation of Al_2O_3 and Y_2O_3 at ambient temperature IBAD of MgO < 10 nm, e-beam evaporation of 50 nm homo-epitaxial MgO at $T_s = 450\,°C$	[123]

Table 15.2 *(continued)*

Sequence of layers	Process	Comments	Reference
		PLD of $LaMnO_3$ at $T_s = 650–800\,°C$ $p_O = 13–800$ mbar, evaporation of Sm, Ba, Cu, Sm-123 formation at $T_s = 700–750\,°C$, $p_O = 6.7–26$ mbar	[123]
Y-123/$LaMnO_3$/ IBAD-MgO/metal Y-123 0.8 μm thick	IBAD	IBAD-MgO, MOCVD[g] of Y-123 Tetramethyl heptanedionate precursor solutions of Y, Ba, Cu Spraying of mixture + O_2 and Ar on substrate at $T_s = 920–960\,°C$	[124]
Gd-123/CeO_2/$LaMnO_3$/ IBAD-MgO/$Gd_2Zr_2O_7$/ Hastelloy	IBAD	$Gd_2Zr_2O_7$ deposition by ion beam sputtering, $T_s \sim 20\,°C$, IBAD-MgO $T_s \sim 20\,°C$, sputtering of $LaMnO_3$ onto IBAD-MgO, PLD of CeO_2 $T_s = 600–900\,°C$, PLD of Gd-123	[125]
Gd-123/CeO_2/IBAD-MgO/Y_2O_3/Al_2O_3/ Ni-alloy Al_2O_3 150 nm Y_2O_3 20 nm Gd-123 1 or 3.7 μm	IBAD	Ion beam sputtering of Al_2O_3 and Y_2O_3, IBAD-MgO at $T_s = 20\,°C$ assisting 800–1200 eV Ar^+ ion beam, 45° tilted to surface normal PLD of Gd-123	[113]
Y-123/$LaMnO_3$/IBAD-MgO/Nucleation layer/Ni alloy	IBAD	Typical architecture of IBAD Y-123 Tapes	[109]

[a]MOD: more precisely metal-organic decomposition.
[b]Tantalum envelope acts as oxygen getter;
[c]W content in atom %;
[d]trifluoroacetate-based precursor;
[e]0.5 h annealing for nickel oxide removal;
[f]American Superconductor Corporation;
[g]MOCVD, metal organic chemical vapor deposition.

parameters of $SrRuO_3$ (a = 0.393 nm (pseudocubic perovskite structure)) and Y-123 (a = 0.389) is smaller than that of MgO (a = 0.421 nm) and Y-123.

First attempts to manufacture Y-123-coated conductors by means of the RABiTS process were based on the use of cube-textured nickel tapes. Disadvantages of the use of pure nickel substrates are their magnetism and the relatively low tensile yield strength of 34 MPa at room temperature [101]. In Ni-5 atom% W, a cube texture can also be formed by a rolling and annealing process. The yield strength of Ni-5 atom% W of 145 MPa is around four times that of pure nickel. In addition, alloying with tungsten reduces the magnetism, which reduces alternating current losses [101]. Typically, the first buffer layer needs to be deposited in a reducing atmosphere (e.g., 4% H_2 + 96% Ar [99, 100, 102, 103]). Frequently CeO_2 [99, 100, 104–106] or Y_2O_3 films [101, 107–109] are grown as first buffer layers on top of the Ni or Ni alloy metallic substrate. The film sequences Y-123/YSZ/CeO_2/Ni [99, 100], Y-123/CeO_2/YSZ/CeO_2/Ni [104], RE-123/CeO_2/YSZ/Y_2O_3/Ni alloy [101, 107, 108], and Y-123/$Gd_2Zr_2O_7$/Ni alloy [110] have been successfully used in RABiTS tapes. The RE-123 films have been deposited by PLD [100, 102, 103, 107, 110, 111] or metal organic decomposition (MOD) [101, 108].

The use of an assisting ion beam allows the biaxially textured growth of YSZ, $Gd_2Zr_2O_7$, or MgO buffer layers on a polycrystalline metallic substrate (Table 15.2). YSZ and $Gd_2Zr_2O_7$ have the disadvantage that, for the evolution of the biaxial texture, film thicknesses of at least 1 μm are required. In contrast, a biaxial texture is already fully developed in an IBAD-MgO buffer layer of only 10 nm thickness. For cost reasons, it would be desirable to have a thin IBAD buffer layer. Examples of buffer layer sequences used in the IBAD process are Y-123/YSZ/Hastelloy C276 [81], Y-123/YSZ/stainless steel [85], Y-123/CeO_2/$Gd_2Zr_2O_7$/Hastelloy C276 [112], and Y-123/CeO_2/IBAD-MgO/Y_2O_3/Al_2O_3/Ni-alloy [113]. The RE-123 films were deposited by PLD, MOD, or metal organic chemical vapor deposition (MOCVD).

Good in- and out-of-plane alignment of the buffer layers is required for the architecture of coated conductors. Furthermore, the topmost buffer layer should allow the epitaxial growth of the RE-123 film. For industrial manufacture of long, coated conductors, cost reduction is crucial. Conductor architectures with as few buffer layers as possible are therefore preferred. Rapid development of the biaxial texture during IBAD is highly desirable. In the case of IBAD-MgO, a film thickness of 10 nm is sufficient to obtain a biaxial texture. In general, deposition of the RE-123 film by PLD leads to high performance, but relatively high production cost. Deposition by an all-chemical solution may significantly reduce the manufacturing cost but reduce the conductor performance. In Section 15.3.3, the critical current densities in coated conductors will be considered.

15.3.3 Critical Current Densities of Coated Conductors

Magnet and power applications require high engineering critical current densities of the order of 100 000 A/cm^2. In the case of the RE-123-coated conductors (RE: Y or another rare earth element), the superconducting layer of around 1 μm thickness is only a small fraction of the total conductor cross-section. The total thickness of a noninsulated coated conductor including a 40-μm thick copper stabilizer is around 100 μm. Thus the engineering critical current density $j_e = I_c/A_{\text{tot}}$ of a coated conductor (where I_c is the critical current, A_{tot} the total tape cross-section) is only around 1% of the critical current density $j_c = I_c/A_{\text{sc}}$ of a superconducting layer of cross-section A_{sc}. In a comparison of the performance of NbTi, Nb_3Sn, MgB_2, Ag/Bi-2212, and Ag/Bi-2223 with that of coated conductors, the engineering critical current densities need to be considered.

The critical current densities of various RE-123-coated conductors at 77 K are presented in Figure 15.17 [126–129]. The top panel shows the field dependence of the critical current density up to 8 T, while the bottom panel presents j_c data up to only 2 T. The critical current density at 77 K and zero applied field is in the range of 1.2–4 MA/cm^2. A total tape thickness of 100 μm (without electrical insulation), and a RE-123 layer thickness of 1 μm, a j_c of 4 MA/cm^2 would correspond to an engineering critical current density of 40 000 A/cm^2. The critical current densities of the three considered Y-123-coated conductors are considerably different. For example, at a field of 1 T ($B||c$), the lowest j_c is ≈0.16 MA/cm^2 (solid line), whereas the highest value is ≈0.35 MA/cm^2 (dash-dot-dot line). In principle, the critical current density in the RE-123 film is determined not only by the in- and out-of-plane alignment of the grains, but also by flux pinning. As long as the pinning force exceeds the Lorentz force acting on the flux lines, no flux motion, which would cause dissipation, will occur. The pinning force depends on the type and the number of defects present in the superconducting film. This aspect will be discussed later in more detail.

The data presented in Figure 15.17 suggest that the achievable j_c values depend also on the rare earth element present in the 123 compound. The j_c values achieved in Gd-123-coated conductors are higher than those measured in Y-123 tapes.

For one of the Gd-123 tapes [127], the critical current densities for magnetic fields applied parallel and perpendicular to the broad face of the tapes are compared in Figure 15.17. The critical current density at 2 T parallel to the CuO_2 (*ab*) planes reaches ≈0.66 MA/cm^2, while for $B||c$ j_c is only ≈0.25 MA/cm^2. This result reflects the well-known anisotropy of the current-carrying capacity of RE-123 films with respect to the field direction. The j_c $(B||c)$ of the Gd-123 film with ZrO_2 additions [126] is significantly higher than the values measured in Gd-123 films without ZrO_2 additions. Investigations of the microstructure by means of transmission electron microscopy (TEM) indicated that the addition of ZrO_2 led to the self-assembly of $BaZrO_3$ (BZO) nanorods perpendicular to the film surface. The BZO nanorods act as pinning centers for magnetic fields applied perpendicular to the film surface $(B||c)$.

Figure 15.18 shows the field dependence of j_c of Y-123-coated conductors with and without $BaZrO_3$ nanorods in magnetic fields up to 31 T [130]. For the Y-123 tape with BZO nanorods, j_c at 4.2 K and 30 T reaches values of ≈8.3 MA/cm^2 for $B||ab$ and ≈1.4 MA/cm^2 for $B||c$. These results in a factor of 5.9 for the anisotropy of j_c with respect to the direction of the applied field. For comparison, the anisotropy of j_c of Ag/Bi-2223 tapes at 4.2 K is less than a factor of 2 at fields above 10 T (Figures 15.10 and 15.11). The BZO nanorods improve the j_c at 4.2 K only at fields below 10 T parallel to the *c* direction. This result suggests that the flux-line pinning by the BZO nanorods is not efficient at 4.2 K and fields higher than 10 T. On the other hand, the j_c $(B||ab)$ in the film with BZO nanorods is lower than that achieved in the Y-123 film without added BZO.

For Gd-123 tapes with and without BZO nanorods [131], the field dependence $(B||c)$ of j_c is presented for temperatures of 20, 50, and 65 K in Figure 15.18. In the considered range of fields and temperatures, an improved critical current density was found for the tapes with added BZO. In general, the field dependence $(B||c)$ becomes more pronounced at higher temperatures.

In Figure 15.19, the field- and temperature-dependencies of the normalized critical current densities of Y-123- [108] and (Gd,Y)-123-coated conductors [132] are presented. Values have been normalized to j_c at 77 K and zero applied field for the material in question. The data for magnetic fields perpendicular to the broad face of the tapes are shown in the top panel, while that for parallel fields are indicated in the bottom panel. For simplicity, the ratio $j_c(T,B)/j_c(77\text{ K}, 0)$ will be called the "lift factor." For fields of more than 0.5 T perpendicular to the broad face of the tapes the lift factor found for (Gd,Y)-123 tapes is larger than that obtained for the Y-123 tapes. At the lowest temperatures of 20 and 30 K, and magnetic fields of less than 5.5 and 1.5 T $(B||ab)$, respectively, a higher lift factor was found for the Y-123 tape.

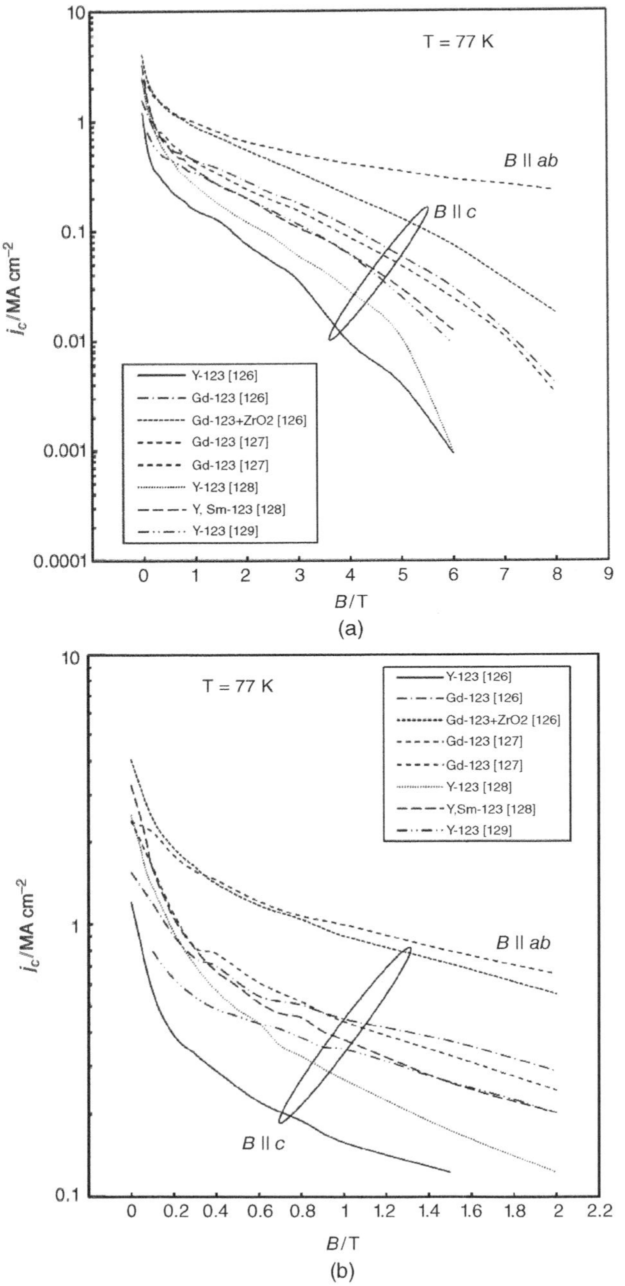

Figure 15.17 Critical current density at 77 K versus magnetic field of various RE-123 coated conductors. The top panel provides the j_c data up to 8 T, whereas in the bottom panel only the low field data up to 2 T are presented. The critical current densities at 77 K and zero applied magnetic field range from 1.2 to 4 MA/cm^2 (results from [126–129])

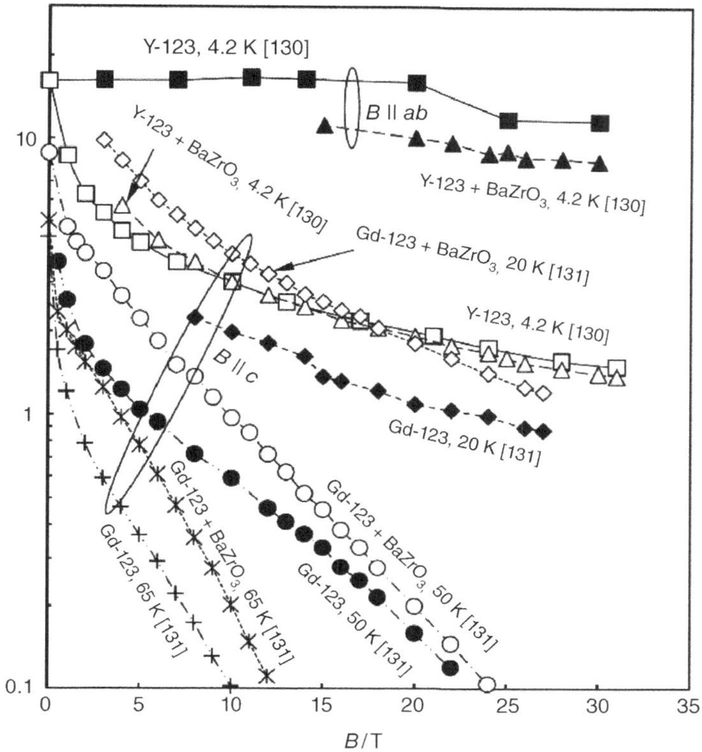

Figure 15.18 *Critical current density versus magnetic field of Y-123 and Gd-123 coated conductors, with and without added $BaZrO_3$. At 4.2 K, the critical current density of Y-123 exceeds 1 MA/cm^2 at magnetic fields as high as 30 T parallel to the crystallographic c direction. For fields parallel to the CuO_2 (ab) planes, the critical current density at 4.2 K and 30 T exceeds 8 MA/cm^2 for the Y-123 coated conductor with added $BaZrO_3$. The data for Gd-123 in the temperature range of 20–65 K indicate that the decrease in j_c with increasing magnetic field ($B \parallel c$) is much more pronounced at elevated temperatures (results from [130, 131])*

The next aspect to be considered is the dependence of the critical current density on the angle θ between the surface normal and the direction of the applied magnetic field. The angle dependence of j_c provides information on flux pinning in coated conductors. In addition, $j_c(\theta)$ is required for the design of superconducting magnets to be manufactured of coated conductors. Examples of the angle dependence of the critical current density of various RE-123-coated conductors are presented in Figure 15.20 [107, 130, 132–135]. In all measurements, a magnetic field of 1 T was applied at temperatures between 75 and 77 K. All considered conductors show a pronounced peak of the critical current density at an angle θ around 90°, i.e., B parallel to the CuO_2 planes. This originates from intrinsic pinning. The lowest j_c of $\approx$0.16 MA/cm^2 was found in a Y-123-coated conductor without artificial pinning centers (APC). Because of the absence of APC, no j_c maximum exists around $\theta = 0°$. The peak value of j_c reached 0.46 MA/cm^2 at $\theta = 89°$ [129]. A Y-123 sample with $BaZrO_3$ addition [130] showed a broad peak at θ values of around 10°. The lowest

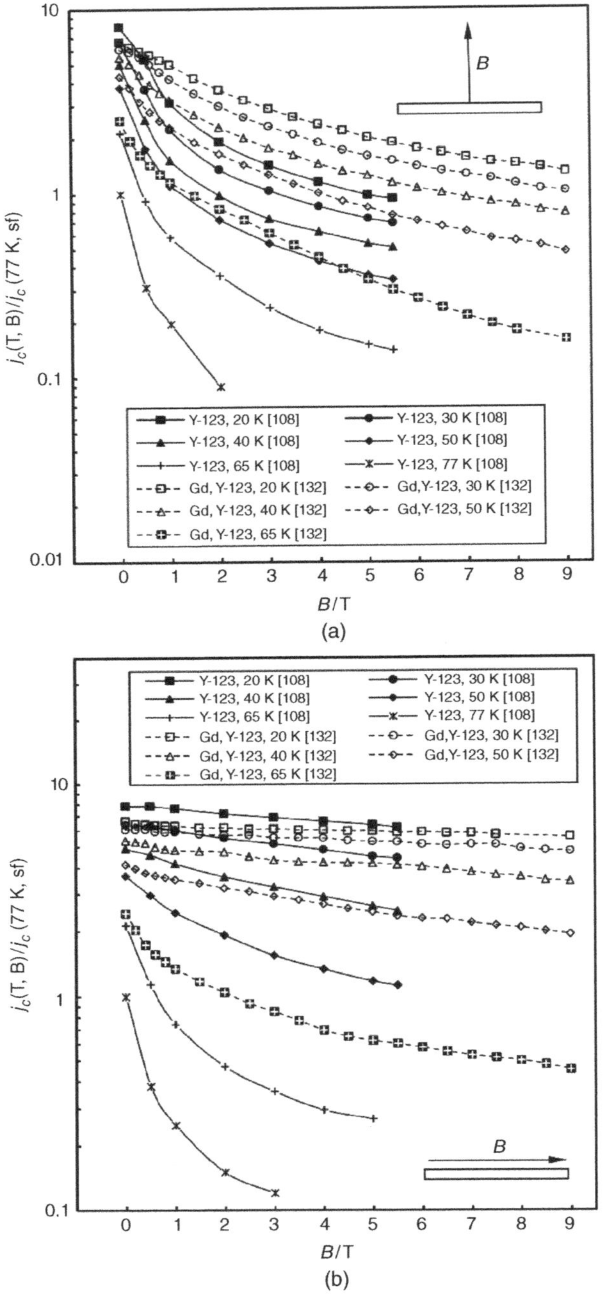

Figure 15.19 *Critical current densities of Y-123 and (Gd,Y)-123 coated conductors normalized to the j_c at 77 K and zero applied field. Especially for magnetic fields applied perpendicular to the broad face of the tapes (B || c), the normalized critical current density of the (Gd,Y)-123 coated conductor with artificial pinning centers is improved as compared to the Y-123 coated conductor (results from [108, 132])*

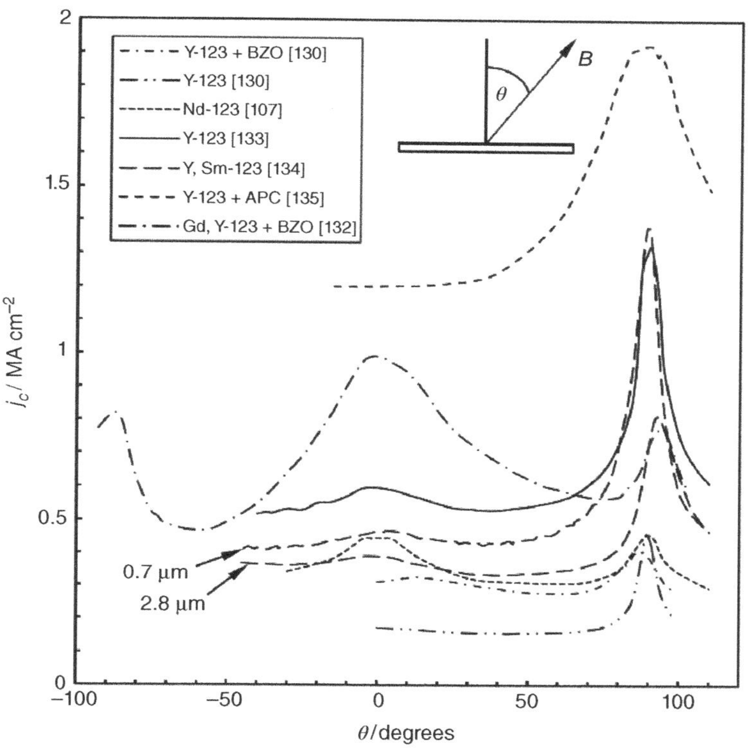

Figure 15.20 Angle dependence of the critical current density of various RE-123 coated conductors at temperatures close to 77 K and an applied magnetic field of 1 T. For $\theta = 0°$, the magnetic field is perpendicular to the broad face of the tape, i.e., B || c (results from [107, 130, 132–135])

j_c values of ≈0.28 MA/cm^2 occur at an angle of ≈60°. The maximum j_c of ≈0.41 MA/cm^2 at $\theta \approx$ 87° is slightly lower than the peak value found for Y-123 without added $BaZrO_3$ [130]. The anisotropy of j_c in the Y-123 film with added $BaZrO_3$ is clearly reduced as compared to the Y-123 without artificial pinning.

In Figure 15.20, the angle dependencies of j_c of two (Y,Sm)-123 layers of 0.7 and 2.8 μm thickness are compared. A broad maximum of j_c around $\theta = 0°$ exists in both conductors; however, the intrinsic pinning peak at $\theta \approx$ 90° is much higher for the thinner layer of only 0.7 μm thickness.

The most pronounced maximum of j_c around 0° was found for a (Gd,Y)-123-coated conductor with $BaZrO_3$ (BZO) APC [132]. The absolutely highest j_c values, presented in Figure 15.20, were measured in a Y-123 film with APC [135]. The j_c of this film varied between 1.2 ($\theta = 0°$) and 1.9 MA/cm^2 ($\theta = 90°$). In spite of high performance and low anisotropy, there exists no maximum of j_c around $\theta = 0°$. These results suggest that in this film, flux pinning is improved not only at $\theta = 0°$, but for a wide range of angles.

Figure 15.21 shows the angle dependence of j_c of a (Gd,Y)-123-coated conductor [132] at an applied magnetic field of 3 T and temperatures of 20, 30, 40, and 50 K. The comparison

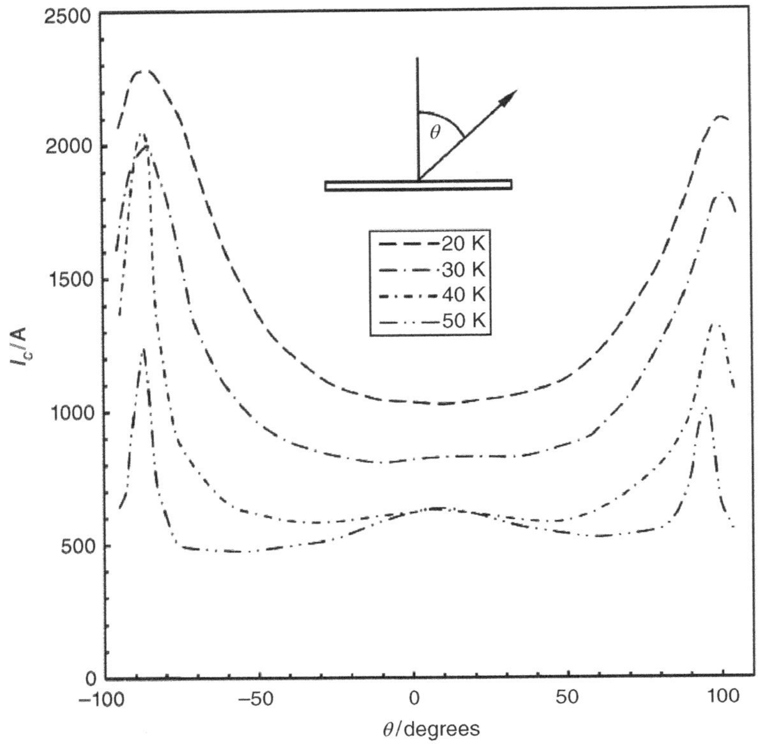

Figure 15.21 *Angle dependence of the critical current of (Gd,Y)-123 coated conductors of 12 mm width at a field of 3 T and temperatures of 20, 30, 40, and 50 K. The ratio of the largest to the lowest j_c value depends on temperature. The values of the ratio j_c^{max}/j_c^{min} are 2.22, 2.48, 3.51, and 2.61 for temperatures of 20, 30, 40, and 50 K, respectively (results from [132])*

of the curves indicates that the peak at $\theta = 0°$ is most pronounced at 50 K. At the lowest temperature of 20 K, this peak, due to pinning by c-axis-related defects, has completely vanished.

Next, the angle dependence of j_c at high magnetic fields is considered. In Figure 15.22, the angle dependence of j_c of a (Gd,Y)-123-coated conductor at 4.2 K and magnetic fields of 5, 10, 15, and 20 T is presented [136]. The film was doped with 7.5% $BaZrO_3$ in order to provide better pinning. The intrinsic pinning peak ($\theta = 90°$) was observed for all considered magnetic fields. The 4.2 K data do not show a significant peak at $\theta = 0°$ related to defects parallel to the crystallographic c-axis. The ratio of $j_c(90°)/j_c(0°)$ is ≈4, ≈5.7, ≈5.9, and ≈6.9 at fields of 5, 10, 15, and 20 T, respectively. The results suggest that the APCs are less efficient at higher fields.

Figure 15.22 also shows j_c versus angle of a $(Y_{0.77}Gd_{0.33})Ba_{1.5}Cu_3O_y$ {(Y,Gd)-123)} film measured at a temperature of 60 K and magnetic fields of 5, 9, 13, and 17 T [137]. The film contained $BaZrO_3$ nanoparticles acting as APCs. The broad peak of j_c around $\theta = 0°$ is most pronounced for the lowest considered magnetic field of 5 T. This peak becomes less and less pronounced with increasing magnetic field. For 5 and 9 T, the lowest j_c was found

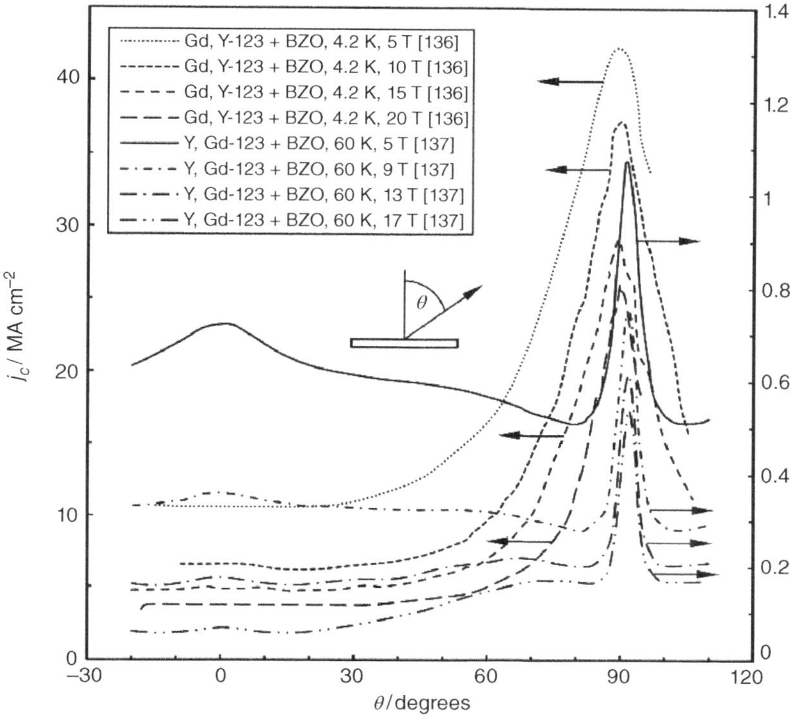

Figure 15.22 *Angular dependence of j_c at high magnetic fields and temperatures of 4.2 and 60 K. The (Y,Gd)-123 coated conductor shows a broad maximum of j_c for $\theta \approx 0°$ (B || c) at 60 K, indicating the presence of defects, parallel to the c direction, acting as pinning centers. A sharper and higher peak occurs for fields parallel to the ab planes (i.e., $\theta = 90°$). The j_c data of (Gd,Y)-123 at 4.2 K show only single peaks at $\theta = 90°$ (results from [136, 137])*

at an angle of ≈80°, whereas at the highest fields of 13 and 17 T an absolute minimum of j_c was observed near 15°.

Finally, the microstructure of RE-123 films with APCs will be considered. Figure 15.23 shows a transmission electron micrograph of the cross-section of a Y-123 film, which was deposited by PLD on a PLD CeO_2/IBAD-$Gd_2Zr_2O_7$/Hastelloy substrate [138]. The Y-123 target, used for the PLD of the Y-123 film, contained 2% by volume of YSZ ($Y_2O_3 + ZrO_2$). The transmission electron micrograph shows parallel $BaZrO_3$ (BZO) nanorods in the Y-123 film, which act as pinning centers for fields applied perpendicular to the broad face of the tape.

Figure 15.24 shows the transmission electron micrograph of the cross-section of a Y-123 film with BZO nanorods [126]. Because the appearance of the BZO nanorods was similar to that of bamboo, the term "bamboo structure" was coined by Japanese scientists.

Figure 15.25 shows TEM bright-field images of the cross-section of a Y-123 film with a high density of BZO nanorods, indicated by the tilted arrow (Figure 15.25a). In addition, small RE_2O_3 precipitates parallel to the *ab* planes are present (indicated by a horizontal arrow). In the plan-view bright-field image presented in Figure 15.25b, the ends of the BZO

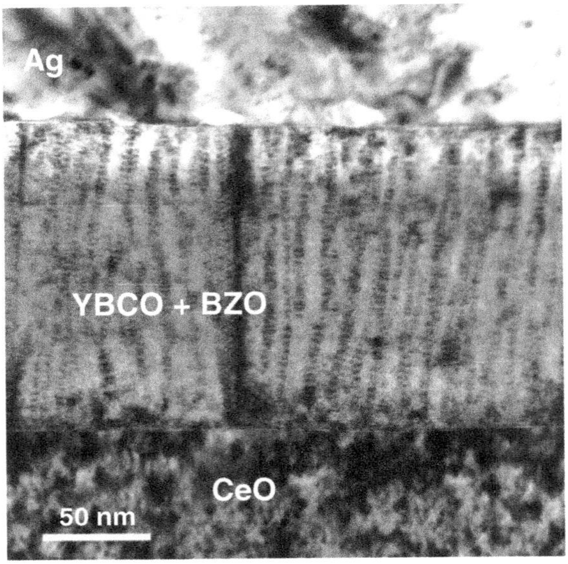

Figure 15.23 Transmission electron micrograph of the cross-section of a Y-123 film with a large number of $BaZrO_3$ nanorods. The Y-123 film was deposited by pulsed laser deposition on a PLD CeO_2/IBAD-$Gd_2Zr_2O_7$/Hastelloy substrate. The image shows the CeO_2 buffer layer at the bottom, the Y-123 film in the center, and the silver cap layer on top (Reprinted from [138] with kind permission from Elsevier B.V.)

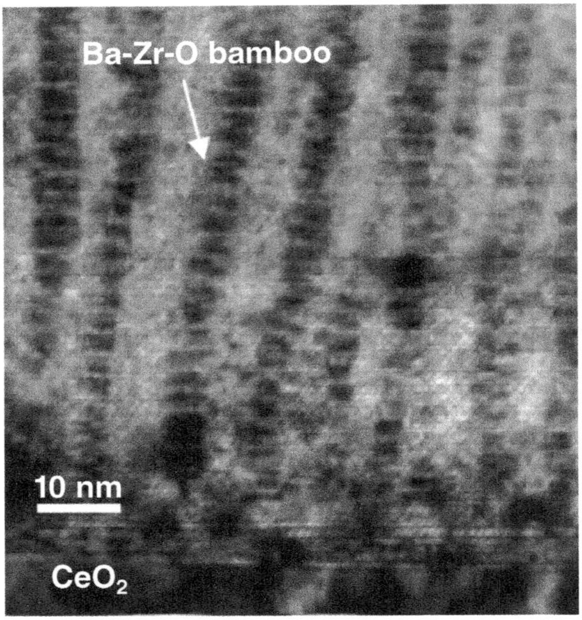

Figure 15.24 Transmission electron micrograph of the cross-section of a Y-123 film with BZO nanorods. Because the appearance of the BZO nanorods was similar to that of bamboo, the term "bamboo structure" was coined by Japanese scientists. (Reprinted from [126] with kind permission from Elsevier B.V.)

Figure 15.25 (a) TEM bright-field image of the cross-section of a Y-123 film with a high density of BZO nanorods (indicated by the tilted arrow). In addition, small RE_2O_3 precipitates parallel to the ab planes are present (indicated by a horizontal arrow). (b) Plan view bright-field image showing the ends of the BZO nanorods, which are parallel to the crystallographic c direction. (c) Magnified image of the BZO nanorods. (Reprinted from [136], doi: 10.1088/0953-2048/24/3/035 001 © IOP Publishing. Reproduced by permission of IOP Publishing. All rights reserved)

nanorods, which are parallel to the crystallographic *c* direction, are visible. Figure 15.25c shows a magnified image of the BZO nanorods [136].

15.3.4 Lengthy Coated Conductors

Power and magnet applications require superconducting wires with lengths of the order of 1000 m and a high, uniform critical current density over the whole wire. In the previous section, the current-carrying capability of coated conductors has been discussed. Nowadays, coated conductors several hundred meters long with critical currents of 250 A/cm-width at 77 K and zero applied field are commercially available. The length and performance of RE-123-coated conductors are still rapidly advancing [108, 112, 119, 121–123, 126–128, 139–142]. The critical current per unit width can be further enhanced by the deposition of thicker RE-123 layers. In the last few years, research and development has aimed at reaching critical current densities in thick films which are as high as those already obtained in 1 μm films. Feldmann *et al.* reported a critical current of 234 A/cm-width (75.6 K, 1 T) in a Y-123 film of 2 μm thickness [135]. The critical current density in the Y-123 layer reached a value as high as 5.2 MA/cm^2 at 75.6 K and zero applied field. Beyond that, they prepared a multilayer film of 1.8 μm total thickness consisting of three thin Y-123 layers separated by two Y_2O_3 layers. A critical current of 175 A/cm-width (75.6 K, 1 T) [135] has been achieved. Another field of intensive research and development is the introduction of APCs (e.g., $BaZrO_3$ nanorods) into the production tapes.

Table 15.3 *Milestones in the development of long coated conductors – the critical currents are typically based on a 1 μV/cm criterion (abbreviations: IBAD – ion-beam-assisted deposition, PLD – pulsed laser deposition, MOCVD – metal organic chemical vapor deposition, MOD – metal organic deposition)*

Composition	*L* (m)	*T* (K)	*B* (T)	I_c (A/cm-width)	Comments	Reference/Year
Y-123	105	77	0	126	IBAD $Gd_2Zr_2O_7$ PLD Y-123	[112]/2005
Y-123	212.6	77	0	245	IBAD $Gd_2Zr_2O_7$ PLD Y-123	[119]/2006
Gd-123	32	77	0	205	IBAD $Gd_2Zr_2O_7$	[121]/2006
	60.7	77	0	183	PLD Gd-123	
(Y,Sm)-123	203	77	0	191	IBAD-MgO MOCVD (Y,Sm)-123	[122]/2007
Y-123	56	77	0	250	IBAD $Gd_2Zr_2O_7$ MOD Y-123	[140]/2007
Sm-123	27	77	0	305	IBAD-MgO Cu, Sm, Ba metal sources	[123]/2008
Gd-123	368	77	0	304.8	IBAD $Gd_2Zr_2O_7$ PLD Gd-123	[126]/2008
Y-123	56	77	0	250	IBAD $Gd_2Zr_2O_7$ MOD Y-123	[128]/2008
Y-123	595	77	0	173	IBAD-MgO MOCVD Y-123	[141]/2008
Y-123	103	77	0	362	IBAD-MgO MOCVD Y-123	[141]/2008
Ho-123	200	77	0	205	RABiTS PLD Ho-123	[141]/2008
Y-123	94	77	0	350	RABiTS MOD Y-123	[141]/2008
Y-123	100	77	0	253	IBAD YSZ PLD Y-123	[141]/2008
Y-123	203	77	0	93	IBAD $Gd_2Zr_2O_7$ MOCVD Y-123	[141]/2008
Y-123	56	77	0	250	IBAD $Gd_2Zr_2O_7$ MOD Y-123	[141]/2008
Gd-123	201.5	77	0	318	IBAD $Gd_2Zr_2O_7$ PLD Gd-123	[127]/2008
Y-123	212.6	77	0	245	IBAD $Gd_2Zr_2O_7$ PLD Y-123	[142]/2008
Gd-123	215.6	77	0	220	PLD Gd123	
Y-123	200	77	0	250	RABiTS MOD Y-123	[108]/2009
Y-123	500	77	0	≥250	RABiTS MOD Y-123	[139]/2010

It appears worthwhile to describe briefly the progress made in the last few years in the performance of long, coated conductors. In their pioneering work, Iijima *et al.* [81] demonstrated as long ago as 1992 that high critical current densities can be achieved in Y-123 films deposited on polycrystalline nickel alloys with a biaxially textured layer of YSZ. The biaxial texture in the YSZ buffer layer was obtained by means of IBAD [81]. Further milestones in the development of long coated conductors are presented in Table 15.3. In 2005, Y-123-coated conductors over 100 m long with a critical current well above 100 A/cm-width at 77 K and zero applied field ($B_a = 0$) were demonstrated [112]. In 2008, a length of 368 m Gd-123-coated conductor with a critical current of 304.8 A (77 K, $B_a = 0$) was achieved [126]. By 2010, the piece length of industrially manufactured coated conductors reached 500 m; the critical current exceeded 250 A/cm-width (77 K, $B_a = 0$) [139].

Today, the performance of coated conductors is close to that required for magnet and power applications. For widespread use of these superconductors, the production capacity of the manufacturers would need to be scaled up significantly. The main hurdle for entering the big potential market in the power sector is economic. Thus a significant reduction in the superconductor cost would be highly desirable.

References

1. T.P. Orlando and K.A. Delin, *Foundations of Applied Superconductivity*, Addison-Wesley Publishing Company, Reading, MA, 1991.
2. J. Muller, A15-type superconductors, *Rep. Prog. Phys.*, **43**, 641–687 (1980).
3. D. Larbalestier, Critical currents and magnet applications of high-T_c superconductors, *Physics Today*, **44**, 74–82 (1991).
4. J. Mannhart, P. Chaudhari, D. Dimos, C.C. Tsuei, and T.R. McGuire, Critical currents in [001] grains and across their tilt boundaries in $YBa_2Cu_3O_7$ films, *Phys. Rev. Lett.*, **61**, 2476–2479 (1988).
5. D. Dimos, P. Chaudhari, J. Mannhart, and F.K. LeGoues, Orientation dependence of grain-boundary critical currents in $YBa_2Cu_3O_{7-\delta}$ bicrystals, *Phys. Rev. Lett.*, **61**, 219–222 (1988).
6. D. Dimos, P. Chaudhari, and J. Mannhart, Superconducting transport properties of grain boundaries in $YBa_2Cu_3O_7$ bicrystals, *Phys. Rev. B*, **41**, 4038–4049 (1990).
7. K. Heine, J. Tenbrink, and M. Thöner, High-field critical current densities in $Bi_2Sr_2Ca_1Cu_2O_{8+x}$/Ag wires, *Appl. Phys. Lett.*, **55**, 2441–2443 (1989).
8. T. Hikata, M. Ueyama, H. Mukai, and K. Sato, Magnetic field dependence of critical current density in Bi–Pb–Sr–Ca–Cu–O silver sheathed wire, *Cryogenics*, **30**, 924–929 (1990).
9. K. Sato, T. Hikata, H. Mukai, M. Ueyama, N. Shibuta, T. Kato, T. Masuda, M. Nagata, K. Iwata, and T. Mitsui, High-J_c silver-sheathed Bi-based superconducting wires, *IEEE Trans. Magn.*, **27**, 1231–1238 (1991).
10. K. Sato, T. Hikata, M. Ueyama, H. Mukai, N. Shibuta, T. Kato, and T. Masuda, Transport current properties of silver-sheathed BiPbSrCaCuO wire and coil, *Cryogenics*, **31**, 687–698 (1991).

11. K. Sato, N. Shibuta, H. Mukai, T. Hikata, M. Ueyama, and T. Kato, Silver-sheathed Bi(2223) wire and application, *Physica C*, **190**, 50–52 (1991).
12. K. Sato, N. Shibuta, H. Mukai, T. Hikata, M. Ueyama, T. Kato, and J. Fujikami, Bismuth superconducting wires and their applications, *Cryogenics*, **33**, 243–246 (1993).
13. J. Tenbrink and H. Krauth, Recent results on test windings and coils of Bi-2212/Ag HTSC wires, *Adv. Cryogen. Eng.*, **40**, 305–311 (1994).
14. T.G. Holesinger, D.S. Phillips, J.Y. Coulter, J.O. Willis, and D.E. Peterson, Isothermal melt processing of Bi-2212 thick film, *Physica C*, **243**, 93–102 (1995).
15. K. Shibutani, T. Hase, Y. Fukumoto, S. Hayashi, Y. Inoue, R. Ogawa, and Y. Kawate, Improvement of J_c in Bi-2212-Ag composite superconductor by controlling pO_2 on the partial melt process, *IEEE Trans. Appl. Supercond.*, **5**, 1849–1852 (1995).
16. T. Hasegawa, H. Kobayashi, H. Kumakura, H. Kitaguchi, and K. Togano, The effect of Ag on the formation of $Bi_2Sr_2Ca_1Cu_2O_y$ thick film, *Physica C*, **222**, 111–118 (1994).
17. R. Wesche, Development of Long Lengths Ag and AgNiMg/Bi-2212 Superconductors, *in High-Temperature Superconductors: Synthesis, Processing and Applications II*, U. Balachandran and P.J. McGinn (eds.), TMS, Warrendale, USA, 1997.
18. R. Wesche, A.M. Fuchs, and G. Pasztor, Fabrication and properties of superconducting AgNiMg and Ag/Bi-2212 wires, *in High-Temperature Superconductors: Synthesis, Processing, and Large Scale Applications*, U. Balachandran, P.J. McGinn, and J.S. Abell (eds.), TMS, Warrendale, USA, 1996.
19. P. Krishnaraj, M. Lelovic, N.G. Eror, and U. Balachandran, Oxygen stoichiometry, structure and superconductivity in $Bi_2Sr_2CaCu_2O_{8+x}$, *Physica C*, **246**, 271–276 (1995).
20. P. Majewski, The use of phase diagrams for the engineering of flux pinning centres in BSCCO ceramics, *in High-Temperature Superconductors: Synthesis, Processing, and Large Scale Applications*, U. Balachandran, P.J. McGinn, and J.S. Abell (eds.), TMS, Warrendale, USA, 1996.
21. L.R. Motowidlo, G. Galinski, G. Ozeryansky, W. Zhang, E.E. Hellstrom, M. Sumption, and T. Collings, The influence of filament size and atmosphere on the microstructure and J_c of round multifilament $Bi_2Sr_2Ca_1Cu_2O_x$ wires, *IEEE Trans. Appl. Supercond.*, **5**, 1162–1166 (1995).
22. R. Wesche, A.M. Fuchs, B. Jakob, and G. Pasztor, Diameter-dependent critical current densities for superconducting Bi-2212/Ag wires, *Cryogenics*, **34**, 805–811 (1994).
23. L.R. Motowidlo, G. Galinski, G. Ozeryanski, W. Zhang, and E.E. Hellstrom, Dependence of the critical current density on filament diameter in round multifilament Ag-sheathed $Bi_2Sr_2CaCu_2O_x$ wires processed in O_2, *Appl. Phys. Lett.*, **65**, 2731–2733 (1994).
24. K. Nomura, M. Seido, H. Kitaguchi, H. Kumakura, K. Togano, and H. Maeda, Fabrication conditions and superconducting properties of Ag-sheathed Bi-Sr-Ca-Cu-O tapes prepared by partial melting and slow cooling process, *Appl. Phys. Lett.*, **62**, 2131–2133 (1993).
25. A. Endo and S. Nishikida, Effects of heating temperature and atmosphere on critical current density of $Bi_2Sr_2Ca_1Cu_2Ag_{0.8}O_y$ Ag-sheathed tapes, *IEEE Trans. Appl. Supercond.*, **3**, 931–934 (1993).

26. D. Sager, L.P. Meier, L.J. Gauckler, and M. Chen, Influence of precursor calcination parameters on the critical current density of Bi-2212 superconductors, *Physica C*, **434**, 125–134 (2006).
27. W.T. Nachtrab, C.V. Renaud, Jr., T. Wong, X.T. Liu, T.M. Shen, U.P. Trociewitz, and J. Schwartz, Development of high superconductor fraction $Bi_2Sr_2CaCu_2O_x$/Ag wire for MRI, *IEEE Trans. Appl. Supercond.*, **18**, 1184–1187 (2008).
28. M. Polak, W. Zhang, A. Polyanskii, A. Pashitski, E.E. Hellstrom, and D.C. Larbalestier, The effect of the maximum processing temperature on the microstructure and electrical properties of melt processed Ag-sheathed $Bi_2Sr_2CaCu_2O_x$ tape, *IEEE Trans. Appl. Supercond.*, **7**, 1537–1540 (1997).
29. R. Wesche, B. Jakob, and G. Pasztor, Development and performance characteristics of Bi-2212/Ag superconducting wires, *IEEE Trans. Appl. Supercond.*, **3**, 927–930 (1993).
30. J.L. Reeves, M. Polak, W. Zhang, E.E. Hellstrom, S.E. Babcock, D.C. Larbalestier, N. Inoue, and M. Okada, Overpressure processing of Ag-sheathed Bi-2212 tapes, *IEEE Trans. Appl. Supercond.*, **7**, 1541–1543 (1997).
31. X.T. Liu, T.M. Shen, U.P. Trociewitz, and J. Schwartz, React-wind-sinter processing of high superconductor fraction $Bi_2Sr_2CaCu_2O_x$/AgMg round wire, *IEEE Trans. Appl. Supercond.*, **18**, 1179–1183 (2008).
32. T. Shen, X. Liu, U.P. Trociewitz, and J. Schwartz, Electromechanical behavior of $Bi_2Sr_2CaCu_2O_x$ conductor using a split melt process for react-wind-sinter magnet fabrication, *IEEE Trans. Appl. Supercond.*, **18**, 520–524 (2008).
33. R. Wesche, *High-Temperature Superconductors: Materials, Properties and Applications*, Kluwer Academic Publishers, Norwell, MA 1998.
34. M. Yang, R. Jenkins, H. Jones, M.J. Goringe, and C.R.M. Grovenor, Prototype BSCCO 2212 coils: Processing and properties, *Physica C*, **235–240**, 3435–3436 (1994).
35. W. Dai, K.R. Marken, Jr., S. Hong, L. Cowey, K. Timms, and I. McDougall, Fabrication of high T_c coils from BSCCO 2212 powder in tube and dip coated tape, *IEEE Trans. Appl. Supercond.*, **5**, 516–519 (1994).
36. M.S. Walker, C.M. Trautwein, L.R. Motowidlo, D.R. Dietderich, and F.A. List, Practical, coated BSCCO-2212 high T_c conductors, *IEEE Trans. Appl. Supercond.*, **5**, 1857–1859 (1995).
37. J.W. Burgoyne, C.G. Morgan, C.J. Eastell, R. Storey, D. Dew-Hughes, H. Jones, C.R.M. Grovenor, and M.J. Goringe, Continuous processing of BSCCO-2212/Ag dip-coated conductors, *in High-Temperature Superconductors: Synthesis, Processing, and Large Scale Applications*, U. Balachandran, P.J. McGinn, and J.S. Abell (eds.), TMS, Warrendale, USA, 1996.
38. H. Kumakura, K. Togano, H. Maeda, J. Case, and T. Morimoto, Anisotropy of critical current density in textured $Bi_2Sr_2Ca_1Cu_2O_x$ tapes, *Appl. Phys. Lett.*, **58**, 2830–2832 (1991).
39. B. Heeb, D. Buhl, Th. Lang, and L.J. Gauckler, Processing and properties of Bi-2212 thick films, in *Processing of Long Lengths of Superconductors*, U. Balachandran, E.W. Collings, and A. Goyal (eds.), TMS, Warrendale, USA, 1994.
40. J. Kase, T. Morimoto, K. Togano, H. Kumakura, D.R. Dietderich, and H. Maeda, Preparation of the textured Bi-based oxide tapes by partial melting process, *IEEE Trans. Magn.*, **27**, 1254–1257 (1991).

41. H. Kumakura, K. Togano, J. Kase, T. Morimoto, and H. Maeda, Superconducting properties of textured Bi–Sr–Ca–Cu–O tapes prepared by applying doctor blade casting, *Cryogenics*, **30**, 919–923 (1990).
42. H. Kumakura, K. Togano, D.R. Dietderich, H. Maeda, J. Kase, and T. Morimoto, J_c characteristics of textured Bi-based oxide tapes, *IEEE Trans. Magn.*, **27**, 1250–1253 (1991).
43. K. Nomura, J. Sato, S. Kuma, H. Kumakura, K. Togano, and N. Tomita, Characteristics of strengthened Ag substrates for $Bi_2Sr_2CaCu_2O_x$ doctor-bladed tapes, *Appl. Phys. Lett.*, **64**, 912–914 (1994).
44. Th. Lang, D. Buhl, and L.J. Gauckler, Influence of the maximum temperature during partial melt-processing of Bi-2212 thick films on microstructure and j_c, *Physica C*, **294**, 7–16 (1998).
45. R. Wesche, Temperature dependence of critical currents in superconducting Bi-2212/Ag wires, *Physica C*, **246**, 186–194 (1995).
46. C.M. Friend, J. Tenbrink, and D.P. Hampshire, Critical current density of $Bi_2Sr_2Ca_1Cu_2O_\delta$ monocore and multifilamentary wires from 4.2 K up to T_c in high magnetic fields, *Physica C*, **258**, 213–221 (1996).
47. S.-C. Kim, D.-W. Ha, S.-S. Oh, I.-Y. Han, J.-G. Oh, and H.-S. Sohn, Effect of filament configuration on critical current density of Bi2212/Ag wires with low Ag ratio, *IEEE Trans. Appl. Supercond.*, **18**, 1188–1191 (2008).
48. S.C. Kim, D.W. Ha, S.S. Oh, I.Y. Han, C.E. Bruzek, J.G. Oh, and H.S. Sohn, Study on thermo-mechanical treatment in fabrication of Bi2212/Ag HTS wire, *Cryogenics*, **47**, 437–441 (2007).
49. T. Hasegawa, N. Ohtani, T. Koizumi, Y. Aoki, S. Nagaya, N. Hirano, L. Motowidlo, R.S. Sokolowski, R.M. Scanlan, D.R. Dietderich, and S. Hanai, Improvement of superconducting properties of Bi-2212 round wire and primary test results of large capacity Rutherford cable, *IEEE Trans. Appl. Supercond.*, **11**, 3034–3037 (2001).
50. K.R. Marken, H. Miao, M. Meinesz, B. Czabaj, and S. Hong, BSCCO-2212 conductor development at Oxford Superconducting Technology, *IEEE Trans. Appl. Supercond.*, **13**, 3335–3338 (2003).
51. S. Hong, H. Miao, M. Meinesz, B. Czabaj, P. Noonan, A. Twin, R. Harrison, and K. Marken, Bi-2212 wire & high field insert development, presented at the Low Temperature Superconductor Workshop, Tallahassee, FL, USA, November 2006, unpublished.
52. H. Miao, K.R. Marken, M. Meinesz, B. Czabaj, and S. Hong, Development of Bi-2212 conductors for magnet applications, *Adv. Cryog. Eng.*, **50**, 603–611 (2004).
53. K.R. Marken, Jr., H. Miao, J.M. Sowa, J.A. Parrell, and S. Hong, Processing issues for high critical current density in BSCCO-2212 composites, *IEEE Trans. Appl. Supercond.*, **11**, 3252–3255 (2001).
54. H. Miao, H. Kitaguchi, H. Kumakura, K. Togano, T. Hasegawa, and T. Koizumi, Optimization of melt-processing temperature and period to improve critical current density of $Bi_2Sr_2CaCu_2O_x$/Ag multilayer tapes, *Physica C*, **320**, 77–86 (1999).
55. H. Kitaguchi, H. Miao, H. Kumakura, and K. Togano, Relationship between $Bi_2Sr_2CaCu_2O_x$ layer thickness and J_c enhancement by pre-annealing and intermediate rolling process, *Physica C*, **320**, 71–76 (1999).

56. L. Zani, M. Menget, P. Régnier, J. Le Bars, T. Dechambre, and P. Nicolas, Transport critical current at 4.2 K of pure and MgO-doped Bi-2212 tapes, *IEEE Trans. Appl. Supercond.*, **7**, 1683–1686 (1997).
57. W.P. Chen, H. Maeda, K. Watanabe, M. Motokawa, H. Kitaguchi, and H. Kumakura, Microstructures and properties of Bi2212/Ag tapes grown in high magnetic fields, *Physica C*, **324**, 172–176 (1999).
58. T. Shen, J. Jiang, F. Kametani, U.P. Trociewitz, D.C. Larbalestier, J. Schwartz, and E.E. Hellstrom, Filament to filament bridging and its influence on developing high critical current density in multifilamentary $Bi_2Sr_2CaCu_2O_x$ round wires, *Supercond. Sci. Technol.*, **23**, 025 009 (2010).
59. G. Grasso, F. Marti, Y. Huang, and R. Flükiger, Long lengths of Bi(2223) superconducting tapes, *Cryogenics*, **37**, 597–599 (1997).
60. N.N. Merchant, V.A. Maroni, A.K. Fischer, S.E. Dorris, W. Zhong, and N. Ashcom, Effect of lead content on phase evolution and microstructural development in Ag-clad Bi-2223 composite conductors, in *High-Temperature Superconductors: Synthesis, Processing, and Applications II*, U. Balachandran and P.J. McGinn (eds.), TMS, Warrendale, USA, 1997.
61. M. Penny, S. Feltham, C. Beduz, Y. Yang, R. Scurlock, and R. Wroe, Influence of partial oxygen pressure on the processing of Ag-sheathed BSCCO-2223 tapes, *Cryogenics*, **37**, 601–604 (1997).
62. B. Hensel, J.-C. Grivel, A. Jeremie, A. Perin, A. Pollini, and R. Flükiger, A model for the critical current in $(Bi,Pb)_2Sr_2Ca_2Cu_3O_x$ silver-sheathed tapes – The role of small-angle *c*-axis grain boundaries and of the texture, *Physica C*, **205**, 329–337 (1993).
63. D.P. Grindatto, B. Hensel, G. Grasso, H.-U. Nissen, and R. Flükiger, TEM study of twist boundaries and colony boundaries in $(Bi,Pb)_2Sr_2Ca_2Cu_3O_x$ silver-sheathed tapes, *Physica C*, **271**, 155–163 (1996).
64. Q. Li, Y.N. Tsay, Y. Zhu, M. Suenaga, G.D. Gu, and N. Koshizuka, Electromagnetic properties of high angle [0 0 1] twist grain boundaries in $Bi_2Sr_2CaCu_2O_{8+\delta}$ bicrystals, *IEEE Trans. Appl. Supercond.*, **7**, 1584–1587 (1997).
65. G.N. Riley, Jr., A.P. Malozemoff, Q. Li, S. Fleshler, and T.G. Holesinger, The freeway model: New concepts in understanding supercurrent transport in Bi-2223 tapes, *JOM*, **49**(10), 24–27, 60 (1997).
66. A. Jeremie, G. Grasso, and R. Flükiger, Effect of carbon impurities on Bi,Pb(2223) phase formation and critical current densities in silver-sheathed Bi,Pb(2223) tapes, *Physica C*, **255**, 53–60 (1995).
67. W. Zhang and E.E. Hellstrom, The influence of carbon on melt processing Ag sheathed $Bi_2Sr_2CaCu_2O_8$ tape, *Physica C*, **234**, 137–145 (1994).
68. W.-J. Kim, J.-K. You, S.-C. Kwon, H.J. Lee, H.-G. Lee, T.-H. Sung, J.-J. Kim, and G.-W. Hong, Effect of particle size reduction on the microstructure evolution and critical current density of Ag/Bi-2223 tapes, *in High-Temperature Superconductors: Synthesis, Processing, and Applications II*, U. Balachandran and P.J. McGinn (eds.), TMS, Warrendale, USA, 1997.
69. B. Zeimetz, A. Pan, and S.X. Dou, Effect of deformation parameters on interface morphology of silver-sheathed high-temperature superconductor tapes, *Physica C*, **250**, 170–174 (1995).

70. E.E. Hellstrom, Y. Yuan, J. Jiang, X.Y. Cai, D.C. Larbalestier, and Y. Huang, Review of overpressure processing Ag-sheathed $(Bi, Pb)_2Sr_2Ca_2Cu_3O_x$ wire, *Supercond. Sci. Technol.*, **18**, S325–S331 (2005).
71. T. Matsushita, Y. Himeda, M. Kiuchi, J. Fujikami, K. Hayashi, and K. Sato, Effects of controlled over-pressure process on the critical current properties in Bi-2223 tapes, *Supercond. Sci. Technol.*, **19**, 1110–1117 (2006).
72. M. Kiuchi, S. Takayama, E.S. Otabe, T. Matsushita, J. Fujikama, K. Hayashi, and K. Sato, Critical current properties in multifilamentary Bi-2223 tape produced by the over pressure processing, *Physica C*, **463–465**, 825–828 (2007).
73. N. Ayai, S. Kobayashi, M. Kikuchi, T. Ishida, J. Fujikami, K. Yamazaki, S. Yamade, K. Tatamidani, K. Hayashi, K. Sato, H. Kitaguchi, H. Kumakura, K. Osamura, J. Shimoyama, H. Kamijyo, and Y. Fukumoto, Progress in performance of DI-BSCCO family, *Physica C*, **468**, 1747–1752 (2008).
74. K. Takahashi, H. Kitaguchi, H. Kumakura, S. Kobayashi, T. Kato, and K. Sato, Homogeneity of superconducting properties in controlled overpressure processed Bi2223 tapes, *IEEE Trans. Appl. Supercond.*, **18**, 1167–1170 (2008).
75. S. Choi, T. Kiyoshi, S. Matsumoto, K. Itoh, T. Hase, K. Zaitsu, M. Hamada, and M. Sugano, The characteristics of Bi-2223/Ag conductor for high field application, *IEEE Trans. Appl. Supercond.*, **18**, 1159–1162 (2008).
76. S. Yamade, N. Ayai, J. Fujikami, S. Kobayashi, E. Ueno, K. Yamazaki, M. Kikuchi, T. Kato, K. Hayashi, K. Sato, H. Kitaguchi, and J. Shimoyama, Development of high performance DI-BSCCO wire with over 200 A critical current, *Physica C*, **463–465**, 821–824 (2007).
77. H. Kitaguchi, K. Takahashi, H. Kumakura, T. Hayashi, K. Fujino, N. Ayai, and K. Sato, Performance of an over-pressure processed Bi-2223/Ag/SUS tape and a small layer-winding coil, *Supercond. Sci. Technol.*, **22**, 045 005 (2009).
78. G. Osabe, N. Ayai, M. Kikuchi, K. Tatamidani, T. Nakashima, J. Fujikami, T. Kagiyama, K. Yamazaki, S. Yamade, E. Shizuya, S. Kobayashi, K. Hayashi, K. Sato, J. Shimoyama, H. Kitaguchi, and H. Kumakura, Recent progress of high performance Ag-sheathed Bi2223 wire, *Physica C*, **470**, 1365–1368 (2010).
79. Y. Honda, K. Higashikawa, M. Inoue, T. Kiss, N. Ayai, M. Kikuchi, K. Hayashi, and K. Sato, Study on critical current distribution in CT-OP Bi-2223 tape based on the scanning Hall probe magnetic microscopy, *Physica C*, **470**, 1377–1379 (2010).
80. Superconductivity Web Site, Sumitomo Electric Industries, http://global-sei.com/super/hts_e/type_h (2013).
81. Y. Iijima, N. Tanabe, O. Kohno, and Y. Ikeno, In-plane aligned $YBa_2Cu_3O_{7-x}$ thin films deposited on polycrystalline metallic substrates, *Appl. Phys. Lett.*, **60**, 769–771 (1992).
82. S. Aoki, T. Yamaguchi, Y. Iijima, A. Kagawa, O. Kohno, S. Nagaya, and T. Inoue, High J_c Y–Ba–Cu–O thin film on metal substrates prepared by chemical vapor deposition, *Jpn. J. Appl. Phys.*, **31**, L547–L549 (1992).
83. S. Aoki, T. Yamaguchi, Y. Iijima, O. Kohno, S. Nagaya, and T. Inoue, Characteristics of high J_c Y–Ba–Cu–O tape on metal substrate prepared by chemical vapor deposition, *IEEE Trans. Appl. Supercond.*, **3**, 1691–1694 (1993).
84. X.D. Wu, S.R. Foltyn, P. Arendt, J. Townsend, C. Adams, I.H. Campbell, P. Tiwari, Y. Coulter, and D.E. Peterson, High current $YBa_2Cu_3O_{7-\delta}$ thick films on flexible

nickel substrates with textured buffer layers, *Appl. Phys. Lett.*, **65**, 1961–1963 (1994).

85. A. Knierim, R. Auer, J. Geerk, G. Linker, O. Meyer, H. Reiner, and R. Schneider, High critical current densities of $YBa_2Cu_3O_{7-x}$ thin films on buffered technical substrates, *Appl. Phys. Lett.*, **70**, 661–663 (1997).
86. X.D. Wu, S.R. Foltyn, P.N. Arendt, W.R. Blumenthal, I.H. Campbell, J.D. Cotton, J.Y. Coulter, W.L. Hults, M.P. Maley, H.F. Safar, and J.L. Smith, Properties of $YBa_2Cu_3O_{7-\delta}$ thick films on flexible buffered metallic substrates, *Appl. Phys. Lett.*, **67**, 2397–2399 (1995).
87. R.P. Reade, P. Berdahl, R.E. Russo, and S.M. Garrison, Laser deposition of biaxially textured yttria-stabilised zirconia buffer layers on polycrystalline metallic alloys for high critical current Y–Ba–Cu–O thin films, *Appl. Phys. Lett.*, **61**, 2231–2233 (1992).
88. R. Hawsey and D. Peterson, Coated conductors: The next generation of high-T_c wires, *Supercond. Indus.*, **9**(3), 23–29 (1996).
89. B. Ma, M. Li, Y.A. Jee, R.E. Koritala, B.L. Fisher, and U. Balachandran, Inclined-substrate deposition of biaxially textured magnesium oxide thin films for YBCO coated conductors, *Physica C*, **366**, 270–276 (2002).
90. U. Balachandran, B. Ma, M. Li, B.L. Fisher, R.E. Koritala, and D.J. Miller, Development of coated conductors by inclined substrate deposition, *Adv. Cryogen. Eng. ICMC*, **50**, 637–644 (2004).
91. K.K. Uprety, B. Ma, R.E. Koritala, B.L. Fisher, S.E. Dorris, and U. Balachandran, Growth and properties of YBCO-coated conductors on biaxially textured MgO films prepared by inclined substrate deposition, *Supercond. Sci. Technol.*, **18**, 294–298 (2005).
92. B. Ma, K.K. Uprety, R.E. Koritala, B.L. Fisher, S.E. Dorris, D.J. Miller, V.A. Maroni, and U. Balachandran, Growth and properties of YBCO-coated conductors fabricated by inclined substrate deposition, *IEEE Trans. Appl. Supercond.*, **15**, 2970–2973 (2005).
93. B. Ma and U. Balachandran, Prospects for the fabrication of low aspect ratio coated conductors by inclined substrate deposition, *Supercond. Sci. Technol.*, **19**, 497–502 (2006).
94. W. Prusseit, C. Hoffmann, R. Nemetschek, G. Sigl, J. Handke, A. Lümkemann, and H. Kinder, Long length coated conductor fabrication by inclined substrate deposition and evaporation, *J. Phys.: Conf. Series*, **43**, 215–218 (2006).
95. M. Dürrschnabel, Z. Aabdin, M. Bauer, R. Semerad, W. Prusseit, and O. Eibl, $DyBa_2Cu_3O_{7-x}$ superconducting coated conductors with critical currents exceeding 1000 A cm^{-1}, *Supercond. Sci. Technol.*, **25**, 105 007 (2012).
96. V. Selvamanickam, A. Ivanova, D.B. Fenner, T. Thurston, M.S. Walker, A.E. Kaloyeros, and P. Haldar, Fabrication of biaxially-textured thick film Y–Ba–Cu–O superconductor by MOCVD on cube-textured metal substrates, in *High-Temperature Superconductors: Synthesis, Processing, and Applications II*, U. Balachandran and P.J. McGinn (eds.), TMS, Warrendale, USA 1997.
97. A. Goyal, D.P. Norton, J.D. Budai, M. Paranthaman, E.D. Specht, D.M. Kroeger, D.K. Christen, Q. He, B. Saffian, F.A. List, D.F. Lee, P.M. Martin, C.E. Klabunde, E. Hartfield, and V.K. Sikka, High critical current density superconducting tapes by

epitaxial deposition of $YBa_2Cu_3O_x$ thick films on biaxially textured metals, *Appl. Phys. Lett.*, **69**, 1795–1797 (1996).

98. D.P. Norton, A. Goyal, J.D. Budai, D.K. Christen, D.M. Kroeger, E.D. Specht, Q. He, B. Saffian, M. Paranthaman, C.E. Klabunde, D.F. Lee, B.C. Sales, and F.A. List, Epitaxial $YBa_2Cu_3O_7$ on biaxially textured nickel (0 0 1): An approach to superconducting tapes with high critical current density, *Science*, **274**, 755–757 (1996).
99. M. Paranthaman, A. Goyal, F.A. List, E.D. Specht, D.F. Lee, P.M. Martin, Q. He, D.K. Christen, D.P. Norton, J.D. Budai, and D.M. Kroeger, Growth of biaxially textured buffer layers on rolled-Ni substrates by electron beam evaporation, *Physica C*, **275**, 266–272 (1997).
100. Q. He, D.K. Christen, J.D. Budai, E.D. Specht, D.F. Lee, A. Goyal, D.P. Norton, M. Paranthaman, F.A. List, and D.M. Kroeger, Deposition of biaxially-oriented metal and oxide buffer-layer films on textured Ni tapes: New substrates for high-current, high-temperature superconductors, *Physica C*, **275**, 155–161 (1997).
101. D.T. Verebelyi, U. Schoop, C. Thieme, X. Li, W. Zhang, T. Kodenkandath, A.P. Malozemoff, N. Nguyen, E. Siegal, D. Buczek, J. Lynch, J. Scudiere, M. Rupich, A. Goyal, E.D. Specht, P. Martin, and M. Paranthaman, Uniform performance of continuously processed MOD-YBCO-coated conductors using a textured Ni-W tungsten substrate, *Supercon. Sci. Technol.*, **16**, L19–L22 (2003).
102. Q. He, D.K. Christen, R. Feenstra, D.P. Norton, M. Paranthaman, E.D. Specht, D.F. Lee, A. Goyal, and D.M. Kroeger, Growth of biaxially oriented conductive $LaNiO_3$ buffer layers on textured Ni tapes for high-T_c-coated conductors, *Physica C*, **314**, 105–111 (1999).
103. T. Aytug, M. Paranthaman, S. Kang, H.Y. Zhai, K.J. Leonard, C.E. Vallet, S. Sathyamurthy, H.M. Christen, A. Goyal, and D.K. Christen, $LaMnO_3$: A single oxide buffer layer for high-J_c $YBa_2Cu_3O_{7-\delta}$ coated conductors, *IEEE Trans. Appl. Supercond.*, **13**, 2661–2664 (2003).
104. X. Cui, F.A. List, D.M. Kroeger, A. Goyal, D.F. Lee, J.E. Matthis, E.D. Specht, P.M. Martin, R. Feenstra, D.T. Verebelyi, D.K. Christen, and M. Paranthaman, Continuous growth of epitaxial CeO_2 buffer layers on rolled Ni tapes by electron beam evaporation, *Physica C*, **316**, 27–33 (1999).
105. G. Celentano, E. Varesi, T. Petrisor, V. Boffa, L. Ciontea, V. Galluzzi, U. Gambardella, A. Mancini, A. Rufoloni, and A. Vannozzi, Influence of the substrate microstructure on the superconducting properties of YBCO coated conductors, *IEEE Trans. Appl. Supercond.*, **13**, 2591–2594 (2003).
106. L. Gianni, A. Baldini, M. Bindi, A. Gauzzi, S. Rampino, and S. Zannella, High J_c coated conductors with a simple buffer layer architecture, *Physica C*, **426–431**, 872–877 (2005).
107. S.H. Wee, A. Goyal, P.M. Martin, and L. Heatherly, High in-field critical current densities in epitaxial $NdBa_2Cu_3O_{7-\delta}$ films on RABiTS by pulsed laser deposition, *Supercond. Sci. Technol.*, **19**, 865–868 (2006).
108. S. Fleshler, D. Buczek, B. Carter, P. Cedrone, K. DeMoranville, J. Gannon, J. Inch, X. Li, J. Lynch, A. Otto, E. Podtburg, D. Roy, M. Rupich, S. Sathyamurthy, J. Schreiber, C. Thieme, E. Thompson, D. Tucker, K. Nagashima, and M. Ogata, Scale-up of 2G wire manufacturing at American Superconductor Corporation, *Physica C*, **469**, 1316–1321 (2009).

109. J.H. Durrell and N.A. Rutter, Importance of low-angle grain boundaries in $YBa_2Cu_3O_{7-\delta}$ coated conductors, *Supercond. Sci. Technol.*, **22**, 013 001 (2009).
110. L.L. Ying, F. Fan, B. Gao, Y.M. Lu, Z.Y. Liu, C.B. Cai, R. Hühne, and B. Holzapfel, Deposition of $Gd_2Zr_2O_7$ single buffer layers with different thickness for $YBa_2Cu_3O_{7-\delta}$ coated conductors on metallic substrates, *Physica C*, **470**, 543–546 (2010).
111. K. Kim, D.P. Norton, C. Cantoni, T. Aytug, A.A. Gapud, M.P. Paranthaman, A. Goyal, and D.K. Christen, (La, Sr)TiO_3 as a conductive buffer for high-temperature superconducting coated conductors, *IEEE Trans. Appl. Supercond.*, **15**, 2997–3000 (2005).
112. K. Kakimoto, Y. Sutoh, N. Kaneko, Y. Iijima, and T. Saitoh, Research of long IBAD-PLD coated conductors with high quality, *Physica C*, **426–431**, 858–865 (2005).
113. S. Hanyu, C. Tashita, T. Hayashida, Y. Hanada, K. Morita, Y. Sutoh, N. Nakamura, H. Kutami, M. Igarashi, K. Kakimoto, Y. Iijima, and T. Saitoh, Long-length IBAD-MgO buffer layers for high performance RE-123 coated conductors by a large ion beam source, *Physica C*, **470**, 1227–1229 (2010).
114. C. Park, D.P. Norton, D.T. Verebelyi, D.K. Christen, J.D. Budai, D.F. Lee, and A. Goyal, Nucleation of epitaxial yttria-stabilized zirconia on biaxially textured (0 0 1) Ni for deposited conductors, *Appl. Phys. Lett.*, **76**, 2427–2429 (2000).
115. H. Wang, S.R. Foltyn, P.N. Arendt, Q.X. Jia, Y. Li, and X. Zhang, Thickness effects of $SrTiO_3$ buffer layers on superconducting properties of $YBa_2Cu_3O_{7-\delta}$ coated conductors, *Physica C*, **433**, 43–49 (2005).
116. R. Hühne, S. Fähler, L. Schultz, and B. Holzapfel, Thin biaxially textured MgO and TiN films prepared by ion-beam assisted pulsed laser deposition for coated conductor applications, *Physica C*, **426–431**, 893–898 (2005).
117. K. Takahashi, M. Konishi, A. Ibi, H. Iwai, T. Muroga, S. Miyata, T. Watanabe, Y. Yamada, and Y. Shiohara, Magnetic field dependence of critical currents in PLD RE–Ba–Cu–O (RE = Er, Dy, Gd) film prepared on CeO_2 capped IBAD GZO layers, *Physica C*, **426–431**, 1001–1004 (2005).
118. S.S. Wang, L. Qiu, L. Wang, P. Du, S. Chen, and Z. Han, YBCO film growth on CeO_2-buffered YSZ substrate by the full-solution method, *Supercond. Sci. Technol.*, **18**, 1271–1274 (2005).
119. Y. Yamada, A. Ibi, H. Fukushima, R. Kuriki, S. Miyata, K. Takahashi, H. Kobayashi, S. Ishida, M. Konishi, T. Kato, T. Hirayama, and Y. Shiohara, Towards the practical PLD-IBAD coated conductor fabrication – Long wire, high production rate and J_c enhancement in a magnetic field, *Physica C*, **445–448**, 504–508 (2006).
120. A. Pomar, A. Cavallaro, M. Coll, J. Gàzquez, A. Palau, F. Sandiumenge, T. Puig, X. Obradors, and H.C. Freyhardt, All-chemical $YBa_2Cu_3O_7$ coated conductors on IBAD-YSZ stainless steel substrates, *Supercond. Sci. Technol.*, **19**, L1–L4 (2006).
121. A. Ibi, H. Fukushima, Y. Yamada, S. Miyata, R. Kuriki, K. Takahashi, and Y. Shiohara, Development of long GdBCO coated conductors using the IBAD/MPMT-PLD method, *Supercond. Sci. Technol.*, **19**, 1229–1232 (2006).
122. V. Selvamanickam, Y. Chen, X. Xiong, Y. Xie, X. Zhang, Y. Qiao, J. Reeves, A. Rar, R. Schmidt, and K. Lenseth, Progress in scale-up of second-generation HTS conductor, *Physica C*, **463–465**, 482–487 (2007).

123. S.S. Oh, H.S. Ha, H.S. Kim, R.K. Ko, K.J. Song, D.W. Ha, T.H. Kim, N.J. Lee, D. Youm, J.S. Yang, H.K. Kim, K.K. Yu, S.H. Moon, K.P. Ko, and S.I. Yoo, Development of long-length SmBCO coated conductors using a batch-type reactive co-evaporation method, *Supercond. Sci. Technol.*, **21**, 034 003 (2008).
124. T. Aytug, M. Paranthaman, L. Heatherly, Y. Zuev, Y. Zhang, K. Kim, A. Goyal, V.A. Maroni, Y. Chen, and V. Selvamanickam, Deposition studies and coordinated characterization of MOCVD YBCO films on IBAD-MgO templates, *Supercond. Sci. Technol.*, **22**, 015 008 (2009).
125. Y. Yamada, S. Miyata, M. Yoshizumi, H. Fukushima, A. Ibi, T. Izumi, Y. Shiohara, T. Kato, and T. Hirayama, Long IBAD-MgO and PLD coated conductor, *Physica C*, **469**, 1298–1302 (2009).
126. Y. Shiohara, M. Yoshizumi, T. Izumi, and Y. Yamada, Current status and future prospects of Japanese national project on coated conductor development and its applications, *Physica C*, **468**, 1498–1503 (2008).
127. H. Fuji, M. Igarashi, Y. Hanada, T. Miura, S. Hanyu, K. Kakimoto, Y. Iijima, and T. Saitoh, Long Gd-123 coated conductor by PLD method, *Physica C*, **468**, 1510–1513 (2008).
128. T. Izumi, M. Yoshizumi, M. Miura, Y. Sutoh, T. Nakanishi, A. Nakai, Y. Ichikawa, Y. Yamada, T. Goto, A. Yajima, Y. Aoki, T. Hasegawa, and Y. Shiohara, Research and development of reel-to-reel TFA-MOD process for coated conductors, *Physica C*, **468**, 1527–1530 (2008).
129. S. Kang, A. Goyal, J. Li, A.A. Gapud, P.M. Martin, L. Heatherly, J.R. Thompson, D.K. Christen, F.A. List, M. Paranthaman, and D.F. Lee, High-performance high-T_c superconducting wires, *Science*, **311**, 1911–1914 (2006).
130. A. Xu, J.J. Jaroszynski, F. Kametani, Z. Chen, D.C. Larbalestier, Y.L. Viouchkov, Y. Chen, Y. Xie, and V. Selvamanickam, Angular dependence of J_c for YBCO coated conductors at low temperature and very high magnetic fields, *Supercond. Sci. Technol.*, **23**, 014 003 (2010).
131. M. Inoue, R. Fuger, K. Higashikawa, T. Kiss, M. Namba, S. Awaji, K. Watanabe, A. Ibi, S. Miyata, Y. Yamada, and T. Izumi, Current transport property in GdBCO coated conductor with artificial pinning centers in a wide range of temperature, magnetic field up to 27 T, and field angle, *Physica C*, **470**, 1292–1294 (2010).
132. V. Selvamanickam, Y. Yao, Y. Chen, T. Shi, Y. Liu, N.D. Khatri, J. Liu, C. Lei, E. Galstyan, and G. Majkic, The low-temperature, high-magnetic-field critical current characteristics of Zr-added (Gd, Y)$Ba_2Cu_3O_x$ superconducting tapes, *Supercond. Sci. Technol.*, **25**, 125 013 (2012).
133. V. Matias, E.J. Rowley, Y. Coulter, B. Maiorov, T. Holesinger, C. Yung, V. Glyantsev, and B. Moeckley, YBCO films grown by reactive co-evaporation on simplified IBAD-MgO coated conductor templates, *Supercond. Sci. Technol.*, **23**, 014 018 (2010).
134. T.G. Holesinger, B. Maiorov, O. Ugurlu, L. Civale, Y. Chen, X. Xiong, Y. Xie, and V. Selvamanickam, Microstructural and superconducting properties of high current metal-organic chemical vapor deposition $YBa_2Cu_3O_{7-\delta}$ coated conductor wires, *Supercond. Sci. Technol.*, **22**, 045 025 (2009).

135. D.M. Feldmann, T.G. Holesinger, B. Maiorov, H. Zhou, S.R. Foltyn, J.Y. Coulter, and I. Apodoca, 1000 A cm^{-1} in a 2 μm thick $YBa_2Cu_3O_{7-x}$ film with $BaZrO_3$ and Y_2O_3 additions, *Supercond. Sci. Technol.*, **23**, 115 016 (2010).
136. V. Braccini, A. Xu, J. Jaroszynski, Y. Xin, D.C. Larbalestier, Y. Chen, G. Carota, J. Dackow, I. Kesgin, Y. Yao, A. Guevara, T. Shi, and V. Selvamanickam, Properties of recent IBAD-MOCVD coated conductors relevant to their high field, low temperature use, *Supercond. Sci. Technol.*, **24**, 035 001 (2011).
137. S. Awaji, M. Namba, K. Watanabe, M. Miura, M. Yoshizumi, T. Izumi, and Y. Shiohara, Flux pinning properties of TFA-MOD (Y, Gd)$Ba_2Cu_3O_x$ tapes with $BaZrO_3$ nanoparticles, *Supercond. Sci. Technol.*, **23**, 014 006 (2010).
138. H. Kobayashi, Y. Yamada, A. Ibi, S. Miyata, Y. Shiohara, T. Kato, and T. Hirayama, Investigation of in-field properties of YBCO multi-layer film on PLD/IBAD metal substrate, *Physica C*, **463–465**, 661–664 (2007).
139. M.W. Rupich, X. Li, C. Thieme, S. Sathyamurthy, S. Fleshler, D. Tucker, E. Thompson, J. Schreiber, J. Lynch, D. Buczek, K. DeMoranville, J. Inch, P. Cedrone, and J. Slack, Advances in second generation high temperature superconducting wire manufacturing and R&D at American Superconductor Corporation, *Supercond. Sci. Technol.*, **23**, 014 015 (2010).
140. T. Nakanishi, M. Yoshizumi, J. Matsuda, K. Nakaoka, Y. Kitoh, Y. Sutoh, T. Izumi, Y. Shiohara, Y. Yamada, T. Goto, and A. Yajima, Fabrication of long Y123 coated conductor tape by advanced TFA-MOD process, *Physica C*, **463–465**, 515–518 (2007).
141. Y. Shiohara, M. Yoshizumi, T. Izumi, and Y. Yamada, Present status and future prospect of coated conductor development and its application in Japan, *Supercond. Sci. Technol.*, **21**, 034 002 (2008).
142. A. Ibi, S. Miyata, R. Kuriki, H. Kobayashi, H. Fukushima, A. Kinoshita, T. Kato, T. Hirayama, Y. Yamada, and Y. Shiohara, Development of long REBCO coated conductors with artificial pinning centers by using MPMT-PLD method, *Physica C*, **468**, 1514–1517 (2008).

16

Cuprate Superconductor Films

16.1 Introduction

Applications of superconductivity in electronics and measurement techniques are frequently based on the use of superconducting films. In Section 15.3, the manufacture and the properties of second-generation high-temperature superconductor tapes have been presented. These coated conductors are based on RE-123 (RE: Y or another rare-earth element) films epitaxially grown on a suitable substrate. The thickness of these RE-123 films is typically in the range of 1–2 μm. The challenge in the manufacture of coated conductors is how to obtain a biaxially textured RE-123 film over a length of several hundred meters without defects that hinder the flow of the supercurrents.

The investigation of cuprate superconductor films was of importance to obtain a deeper insight into the factors limiting the transport critical current density. Soon after the discovery of superconductivity above 77 K in $YBa_2Cu_3O_{7-x}$, it was observed that the critical current density in sintered polycrystalline samples is disappointingly low. On the other hand, high critical current densities of the order of MA cm^{-2} were demonstrated in epitaxial films. The transport critical current densities achieved in cuprate superconductor films have already been presented in Section 9.5.4. Furthermore, the study of biaxially textured epitaxial films provided information on the anisotropy of the transport properties of cuprate high-T_c superconductors.

Many aspects of the manufacture of long coated conductors have been investigated using superconducting films of small area. Examples are investigations of the sequence and the conditions during deposition of various buffer layers on biaxially textured (RABiTS) or nontextured nickel alloy substrates (IBAD, ISD), in order to obtain substrates suitable for the epitaxial growth of RE-123 films. Another aspect addressed by studies of cuprate films is the introduction of artificial pinning centers into RE-123 films.

A comprehensive description of cuprate superconductor film deposition and the discussion of their properties are beyond the scope of this chapter. Besides an introduction to film deposition, this chapter will focus on the effects of the thickness of the superconducting

Physical Properties of High-Temperature Superconductors, First Edition. Rainer Wesche.

film in multilayer systems (Section 16.3) and the effect of epitaxial strain on the critical temperature (Section 16.4).

16.2 Film Deposition Techniques

16.2.1 Preparation of Bismuth-based Cuprate Superconductor Films

First, let us consider the preparation of bismuth-based high-T_c superconductor films. Conditions reported in the preparation of Bi-2212 [1–9] and Bi-2223 films [10–14] are listed in Table 16.1. Bi-2212 films were successfully deposited on (1 0 0) YSZ ($ZrO_2 + 9.5\%$ Y_2O_3) [1], (1 0 0) $SrTiO_3$ [4–6], and (1 0 0) MgO substrates [2, 3, 7–9]. The films were deposited by pulsed laser deposition (PLD) [3–5, 7, 9], electron beam evaporation [2], metal organic decomposition (MOD) [8], and sputtering [1]. Further details of the deposition conditions are presented in Table 16.1. In highly textured Bi-2212 films grown on (1 0 0) $SrTiO_3$ substrates, the critical current density exceeded 1 MA cm^{-2} at 4.2 K and zero applied field ($B_a = 0$) [4, 5]. The onset T_c of these Bi-2212 films reached 83 K.

Chen and Bhattacharya [10] reported the growth of biaxially textured Bi-2212 films by means of electrodeposition on (1 0 0) $LaAlO_3$ single-crystal substrates. The $LaAlO_3$ single crystals were coated by a 30-nm-thick silver film. The metal nitrates of the constituents were co-electrodeposited on the Ag-coated single-crystal substrates at room temperature. Typically, 2.0 g $Bi(NO_3)_3{\cdot}5H_2O$,1.0 g $Sr(NO_3)_2$, 0.6 g $Ca(NO_3)_2{\cdot}4H_2O$, and 0.9 g $Cu(NO_3)_2{\cdot}6H_2O$ were dissolved in 400 ml dimethyl sulfoxide. In a furnace preheated to 1000 °C, the Bi-2212 precursor films were melted and then rapidly quenched to room temperature. After melt quenching, Bi-2201 was the main phase in the precursor film. During annealing at 840–870 °C in air or oxygen, the Bi-2201 phase was converted to nearly single phase, biaxially textured Bi-2212 films. The critical current density in the Bi-2212 films reached 0.58 MA cm^{-2} at 4.2 K and the zero applied field. Magnetization measurements provided a critical temperature of 79 K [10].

(Bi,Pb)-2223 and Bi-2223 films were successfully grown on (1 0 0) MgO [11, 13, 14] and (1 0 0) $LaAlO_3$ [12]. (Bi,Pb)-2223 films were deposited by spray pyrolysis on (1 0 0) MgO [11]. Bi-2223 films without lead substitution for bismuth were deposited by metal-organic chemical vapor deposition (MOCVD) [12] and dc sputtering [13, 14]. The MOCVD Bi-2223 films grown on (1 0 0) $LaAlO_3$ substrates showed T_c(end) = 86–90 K and excellent critical current densities above 1 MA cm^{-2} at 4.2 K and $B_a = 0$ [12].

16.2.2 Preparation of Thallium-based Cuprate Superconductor Films

Conditions reported in the preparation of Tl-2212 [15–20], Tl-2223 [15, 16, 21–23], Tl-1212 [24], and Tl-1223 [25–30] are listed in Table 16.2. Tl-2212 films were successfully grown on MgO [15], $SrTiO_3$ [15], $LaAlO_3$ [16–20], CeO_2 buffered sapphire [15], and $LaNiO_3$ buffered $LaAlO_3$ [18]. The same substrates were used for the preparation of Tl-2223 films (MgO [15, 21], $SrTiO_3$ [15], CeO_2 buffered sapphire [15], and $LaAlO_3$ [16, 22, 23]).

A 5–10-µm-thick Tl-1212 film was prepared by MOCVD [24]. The Tl-1212 film showed a broad transition with an onset T_c of 110 K. Zero resistance was reached at 70 K. Tl-1223 films were deposited on yttria-stabilized zirconia (YSZ) [25–27], silver [28], $SrTiO_3$ [29], and $LaAlO_3$ [29, 30] substrates.

Table 16.1 Conditions used for the preparation of Bi-2212 and Bi-2223 high-T_c superconductor films (abbreviations: B_a – applied field, p – chamber pressure, p_O – oxygen partial pressure, MOD – metal organic decomposition, MOCVD – metal-organic chemical vapor deposition, T_s – substrate temperature, YSZ – yttria-stabilized zirconia)

Film/substrate	Deposition conditions/comments	Reference
Bi-2212/Al_2O_3 Bi-2212/$SrTiO_3$ Bi-2212/(1 0 0) YSZ YSZ: ZrO_2+ 9.5% Y_2O_3	Reactive magnetically enhanced triode sputtering $p = 4 \times 10^{-7}$ mbar, Sr, Bi–Cu and Ca sputter targets presputtered for 45 min at $p_{Ar} \approx 5$ mbar, combined deposition rate of all metals: 10 nm min^{-1}, $p_O = 2.7$ mbar 0.5-µm-thick film, annealed in tube furnace filled with O_2 28 °C min^{-1} → 890 °C, cool-down rate > 50 °C min^{-1} T_c (onset) = 85 K, T_c (end) = 74 K, linear dependence of resistance on temperature above 85 K	[1]
Bi-2212/(1 0 0) MgO	Sequential electron beam evaporation of Cu, SrF_2, CaF_2, and Bi, total of 16 layers 0.5-µm-thick Furnace annealing in O_2 and H_2O vapor (860 °C 5 min) 860 °C 10 min without H_2O vapor T_c (onset) = 88 K, T_c (end) = 68 K	[2]
Bi-2212/(1 0 0) MgO	Pulsed laser evaporation from a Bi-2212 ceramic disk $p = 2.7 \times 10^{-5}$ mbar, T_s close to room temperature ArF excimer laser (6.4 eV), 8 Hz repetition rate 10 J/cm^2 at target, deposition rate of 0.1 nm $pulse^{-1}$ 10 min annealing at 850 °C, 1–20% O_2 balanced by N_2 250-nm-thick films, T_c(end) ≥ 75 K	[3]
Bi-2212/(1 0 0) $SrTiO_3$	PLD from sintered Bi-2212 disks XeCl excimer laser (2 J pulse energy, 60 ns pulse duration, 5 Hz repetition rate), $p_O = 0.4$–0.8 mbar, T_s 5–10 °C below Bi-2212 decomposition temperature 50 min annealing at T_s, $p_O = 0.6$ mbar cool-down with 100 °C min^{-1} → highly textured Bi-2212 films, c⊥ surface T_c (onset) = 83 K, j_c (4.2 K, $B_a = 0$) > 1 MA cm^{-2}	[4, 5]
Bi-2212/(1 0 0) $SrTiO_3$	Flash evaporation of $Bi_2Sr_{2.5}Ca_2Cu_2O_y$ powders from 1900 °C hot Ta boat, $p = 10^{-5}$ mbar, $T_s = 150$ °C Annealing in tube furnace at 820–830 °C Single-phase Bi-2212 films, $T_c = 70$–81 K, c⊥ surface j_c (4.2 K, $B_a = 0$) ≈ 0.1 MA cm^{-2}	[6]

(continued overleaf)

Table 16.1 *(continued)*

Film/substrate	Deposition conditions/comments	Reference
Ag/Bi-2212/(1 0 0) MgO	PLD with a KrF excimer laser from a dense Bi-2212 ceramic target T_s = 500 °C, p = 18 mbar (flowing O_2) 400-nm-thick Bi-2212 films + 30 nm Ag Furnace annealing in Ar/7% O_2 at 872–876 °C *c*-axis-oriented Bi-2212 film with enlarged grain size due to the presence of the Ag cover layer	[7]
Bi-2212/(1 0 0) MgO 240-nm-thick film	MOD of metal carboxylates dissolved in xylene, the MOD solution directly coated on MgO substrates, furnace annealing at 500 °C 15 min^{-1}, 845 °C/2.5 h/O_2, j_c (4.2 K, B_a = 0) = 1.1 MA cm^{-2}	[8]
Bi-2212/(1 0 0) MgO	Infrared Nd:YAG laser with 100–226 mJ pulse energy Sintered Bi-2212 target, 10^3 mbar $\leq p \leq$ 10^1 mbar and nonheated substrate, no O_2 Heat treatments outside the vacuum chamber in air at 840–850 °C for several times, some samples partially melted at 870–880 °C and annealing at 850 °C for 10 h *c*-axis-oriented Bi-2212 films with T_c (onset) = 58 K Stoichiometry of the target is transferred to the film	[9]
Bi-2212/(1 0 0) $LaAlO_3$ 30-nm-thick Ag film on $LaAlO_3$	Co-electrodeposition of metal nitrates (electrolyte bath composition: 2.0 g $Bi(NO_3)_3 \cdot 5H_2O$, 1.0 g $Sr(NO_3)_2$, 0.6 g $Ca(NO_3)_2 \cdot 4H_2O$, and 0.9 g $Cu(NO_3)_2 \cdot 6H_2O$ dissolved in 400 ml dimethyl sulfoxide), melting of Bi-2212 precursor films in a furnace preheated to 1000 °C, quenching to RT leads to Bi-2201 as a main phase, annealing at 840 to 870 °C in air or O_2 converts Bi-2201 to Bi-2212 j_c (4.2 K, B_a = 0) = 0.58 MA cm^{-2}, T_c = 79 K	[10]
Bi,Pb-2223/(1 0 0) MgO	Spray pyrolysis of carboxylates, Bi, Sr, Ca, Pb, and Cu oxides dissolved in 1.5 molar solution of $C_3H_6O_3$ in H_2O Solution was sprayed on freshly cleaved (1 0 0) MgO at T_s = 80–400 °C, furnace annealing at 860 °C, 12 h in air Optimal metal composition $Bi_{1.7}Pb_{0.4}Sr_{1.7}Ca_{2.5}Cu_{3.5}$ T_c (end) of the films varied between 84 and nearly 100 K	[11]

Table 16.1 (continued)

Film/substrate	Deposition conditions/comments	Reference
Bi-2223/(1 0 0) $LaAlO_3$	MOCVD of $Bi(C_6H_5)_3$, $Sr(DPM)_2$, $Ca(DPM)_2$, and $Cu(DPM)_2$ (DPM: dipivaloyl methane) 60–80-nm-thick c-axis-oriented single-phase films T_c (end) = 86–90 K, j_c (4.2 K, B_a = 0) > 1 MA cm^{-2}	[12]
Bi-2223/(1 0 0) MgO	dc sputtering from a single cylindrical target of Bi-2234 composition, mixture of O_2/Ar = 0.8–1 as sputtering gas p = 0.9–1.6 mbar, T_s = 750–760 °C, 80–110-nm-thick films Cool-down to 600° in 3 min, 10 min annealing at 600 °C in 600 mbar O_2	[13], [14]

The preparation of Tl-based high-T_c superconductor films is typically done in two steps. In the first, a thallium-free precursor film is deposited using one of the several deposition techniques (sputtering, MOCVD, PLD, or spray pyrolysis of metal nitrates). In the second step, the precursor film is thallinated, for example, in a quartz tube or a small alumina vessel. Sources of thallium are Tl_2O_3 or Tl–Ba–Ca–Cu–O pellets. The two-step synthesis avoids the poisoning of vacuum chambers, pumps, and other equipment with thallium.

16.2.3 Preparation of Mercury-based Cuprate Superconductor Films

Next, the preparation of mercury-based cuprate high-T_c superconductor films is briefly considered. Various procedures used to prepare Hg-based cuprate superconductor films are listed in Table 16.3 [31–46]. Hg-1201 [31], Hg-1212 [32, 33, 35, 40], and Hg-1223 films [41–46] were successfully grown on $SrTiO_3$ single-crystal substrates. $LaAlO_3$ [34, 37, 39] and MgO [36] were also used as substrates for the growth of Hg-1212 films. Hg-1223 films were also prepared on CeO_2-buffered sapphire substrates. Hg-based high-T_c superconductor films are very sensitive to air exposure. The chemical stability of the films can be improved by partial substitution of Re or Pb for Hg.

Frequently, the preparation of Hg-1212 and Hg-1223 starts from a mercury-free precursor film, which may be deposited by RF magnetron sputtering [32, 42, 46], PLD [35, 39, 40], evaporation [36], or spray pyrolysis [44, 45]. The reaction to the desired Hg-1212 or Hg-1223 phase is performed by heat treatment in the presence of mercury vapor. Unreacted pellets of Hg–Ba–Ca–Cu–O and $CaHgO_2$ [39] have been used as mercury sources. The composition of the unreacted Hg–Ba–Ca–Cu–O pellets depends on the phase desired.

In other procedures, layers of HgO and Ba–Ca–Cu–O are sequentially deposited [33, 34]. The multilayer film is protected by a HgO cap layer. Finally, the precursor films are made to react to the desired phase in an enclosed quartz tube in the presence of mercury vapor. Hg-1212 pellets were used as mercury sources [34].

Table 16.2 *Conditions used in the preparation of Tl-2212, Tl-2223, Tl-1212, and Tl-1223 high-T_c superconductor films (abbreviations: B_a – applied field, d_r – deposition rate, p – chamber pressure, p_O – oxygen partial pressure, MOCVD – metal-organic chemical vapor deposition, T_a – annealing temperature, T_s – substrate temperature, YSZ – yttria-stabilized zirconia)*

Film/substrate	Deposition conditions/comments	Reference
Tl-2212 films on (1 0 0) MgO, $SrTiO_3$, CeO_2/sapphire $T_c = 100\,K$	$Ba_2Ca_1Cu_2$ precursor deposited by reactive high-rate sputtering from a metal alloy target, $d_r = 50\,nm\,min^{-1}$, annealing together with a Tl–Ba–Ca–Cu–O ceramic at $T_a = 860–880\,°C$, rapid heating to T_a, cool-down with 1 K/min in O_2, thallination in a small container avoiding contamination of the vacuum chamber with Tl	[15]
Tl-2223 films on (1 0 0) MgO, $SrTiO_3$, CeO_2/sapphire $T_c = 115\,K$	$Ba_2Ca_2Cu_3$ precursor deposited by reactive high-rate sputtering from a metal alloy target, $d_r = 50\,nm\,min^{-1}$, annealing together with a Tl–Ba–Ca–Cu–O ceramic at $T_a = 860–880\,°C$, rapid heating to T_a, cool-down with 1 K/min in O_2, thallination in a small container avoiding contamination of the vacuum chamber with Tl	[15]
Tl-2212/$LaAlO_3$ $j_c >$ 1 MA cm^{-2} at 20 K, $B_a = 0$	Films grown by sputtering and post annealing Amorphous Tl–Ba–Ca–Cu–O film, low temperature to convert the precursor film to Tl-2212	[16]
Tl-2223/$LaAlO_3$ $j_c >$ 1 MA cm^{-2} at 40 K, $B_a = 0$	Films grown by sputtering and post annealing Post annealing in Tl atmosphere	[16]
Tl-2212/(0 0 1) $LaAlO_3$ T_c(end) = 105 K $j_c = 0.12\,MA\,cm^{-2}$ 77 K, $B_a = 0$	MOCVD of Ba–Ca–Cu–O(F) thin films Post annealing in the presence of Tl_2O_3 vapor Precursor film together with Tl-2212 pellet sealed in Au foil, annealing at 867 °C for 1 h in pure O_2	[17]
Tl-2212/$LaNiO_3$/$LaAlO_3$ $j_c \approx$ 1 MA cm^{-2} at 5 K, $B_a = 0$	PLD of $LaNiO_3$ on $LaAlO_3$ single crystal in ≈ 320–480 mbar O_2, Tl-free precursor film deposited by sputtering at ambient temperature Furnace thallination of the precursor films	[18]
Tl-2212/$LaAlO_3$ $j_c \approx$ 10 MA cm^{-2} 5 K, $B_a = 0$	Prepared together with the Tl-2212/$LaNiO_3$/$LaAlO_3$ film	[18]
Tl-2212/(0 0 1) $LaAlO_3$ 100-nm-thick film $T_c = 105\,K$ $j_c = 2.33\,MA\,cm^{-2}$	Deposition of amorphous Tl–Ba–Ca–Cu–O precursor film on $LaAlO_3$ substrate dc magnetron sputtering in 4:1 Ar/O_2 mixture at $p = 1.5 \times 10^{-2}$ hPa Precursor films annealed at 730–760 °C for 1–4 h	[19]

Table 16.2 *(continued)*

Film/substrate	Deposition conditions/comments	Reference
Tl-2212/$LaAlO_3$ 300-nm-thick film	Deposition of an amorphous Tl–Ba–Ca–Cu–O film by dc sputtering (Tl:Ba:Ca:Cu = 2.1:2:1:2), annealing at 820 °C for 1 h in flowing O_2 in a closed crucible	[20]
Ag doped Tl-2223 on MgO substrates $j_c = 23000\,A\,cm^{-2}$ at 77 K, $B_a = 0$ $T_c \approx 120\,K$	Spray pyrolysis of $Ba(NO_3)_2{\cdot}4H_2O$, $Ca(NO_3)_2$, $Ag(NO_3)_2$, $Cu(NO_3)_2{\cdot}3H_2O$ on MgO ($T_s = 300\,°C$), Ba:Ca:Cu:Ag = 2 : 2 : 3 : 0.3, 20–30 μm thick precursor films, $T_a = 885\,°C$ 3 min in an Au foil $Tl_2Ba_2Ca_3Cu_3O_y$ bulks as a Tl source	[21]
Tl-2223/$LaAlO_3$ 1–1.9-μm-thick film $T_c = 109–112\,K$	Sputter deposition of $Ba_2Ca_2Cu_3O_x$ precursor films from a metallic $Ba_2Ca_2Cu_3$ alloy target, $T_a = 1161–1168\,°C$	[22]
Tl-2223/(1 0 0) $LaAlO_3$ $T_c = 118.5–121\,K$ (mid-point)	Spray pyrolysis of $Ba_2Ca_2Cu_3O_y$ precursor films starting from Ba, Ca, Cu nitrates dissolved in H_2O (0.3 mol l^{-1}) Solution sprayed on $LaAlO_3$ at $T_s = 850\,°C$ within 15 min Tl source 0.5 g $Tl_{1.8}Ba_2Ca_2Cu_3O_z$ in contact with precursor film sealed in a quartz tube, $p_O = 0.5\,bar$ $T_a = 900\,°C$ 30 min, mainly biaxially textured Tl-2223	[23]
Tl-1212/(1 0 0) $LaAlO_3$ 5–10-μm-thick film T_c(onset) = 110 K T_c(end) = 70 K	MOCVD of Tl-1212 films in the presence of Tl_2O vapor in a small alumina vessel MOCVD at 400 °C/1 h/≈7 hPa Annealing at 820 °C for 1 h, $p_O = 400\,mbar$	[24]
Tl-1223/YSZ 7–8-μm-thick film $T_c = 105–111\,K$ $j_c \approx 60\,kA\,cm^{-2}$ 77 K, $B_a = 0$	Spray pyrolysis of metal nitrates (5 mol% Ag in precursor films), decomposition in air → oxides Thallination at 853–868 °C/1 h in a two-zone furnace under flowing O_2, Tl_2O_3 source at 730 °C	[25]
Tl-1223/YSZ	Aqueous solution of Ca, Ba, Cu, Ag nitrates sprayed on a heated polycrystalline YSZ substrate $Ba_2Ca_2Cu_3Ag_{0.37}O_x$ film, $T_a = 650\,°C$ 5 min cool-down $T_a = 845\,°C$ in O_2 (nitrates → oxides) thallination in a two-zone furnace, film at 860 °C 20 min, Tl source at 730 °C	[26]

(continued overleaf)

Table 16.2 *(continued)*

Film/substrate	Deposition conditions/comments	Reference
Tl-1223/YSZ $j_c = 60\,\text{kA cm}^{-2}$ $77\,\text{K}$, $B_a = 0$ 2.5–5-μm-thick films	Spray deposition of metal nitrates followed by decomposition at 650 °C in air, 845 °C in O_2 → $Ba_2Ca_2Cu_3Ag_{0.05}O_7$ precursor film of 2.5 μm thickness Thallination in a two-zone furnace at 858 °C (Tl_2O_3 source at 730 °C)	[27]
Tl-1223/Ag $j_c(77\,\text{K}, B_a = 0) =$ $25\,\text{kA cm}^{-2}$ 10–30-μm-thick films	Spin coating of $Ba_2Ca_2Cu_3Ag_{0.37}O_7$ precursor films on as-rolled Ag substrates, 600 °C for 6 h in air (burn out of organic binder), thallination in a two-zone furnace (precursor film at 860 °C, Tl_2O_3 source at 730 °C, 1–4 h) Annealing at 600 °C in flowing O_2 for 10 h	[28]
Tl-1223/(0 0 1) $SrTiO_3$ Tl-1223/(1 0 0) $LaAlO_3$	Off-axis laser ablation from a stoichiometric target Ba:Ca:Cu = 2:2:3 or Bi:Ba:Ca:Cu = 0.25:2:2:3 $p_O \leq 20\,\text{mbar}$, $T_s = 650$–720 °C, Tl_2O_3 as source of Tl_2O Tl_2O partial pressure of 0.5–1 mbar, $T_{Tl2O3} = 800$–850 °C Directly after deposition, p_O is increased to 600 mbar Bi substitution improved the film microstructure	[29]
Tl-1223/(0 0 1) $LaAlO_3$ 100–125-nm-thick film $T_c(\text{onset}) = 103\,\text{K}$	On-axis radio-frequency sputtering of $Ba_2Ca_2Cu_3O_x$ $p = 5 \times 10^{-7}$ mbar in a deposition chamber, 10^{-2} mbar Ar during sputtering, thallination in a sealed quartz tube $Tl_{0.8}Ba_2Ca_2Cu_3O_x$ as a Tl source, $T_a = 860$ °C, 3 h, $p_O = 500$ mbar	[30]

Stelzner and Schneidewind [37] used Tl–Hg cation exchange to prepare Hg-1212 films. In this special method, Tl-2212 films were first grown on $LaAlO_3$ or CeO_2-buffered sapphire substrates. The conversion of the Tl-2212 to Hg-1212 films was done in quartz ampoules at a temperature of 790–810 °C for 30–45 min in the presence of mercury vapor and oxygen. Ceramic pellets of composition $Tl_{0.002}Hg_{1.15}Ba_2Ca_2Cu_3O_x$ were used as the mercury source. The Hg-1212 films obtained by cation exchange exhibited transition temperatures between 121 and 124 K. The critical current achieved in these films is as high as $7\,\text{MA cm}^{-2}$ at 77 K and the zero applied field.

Valeriánová *et al.* [46] deposited mercury-based high-T_c superconductor films on CeO_2-buffered sapphire. Their studies focused on the influence of the CeO_2 buffer layer thickness on the formation of Hg-12($n-1$)n films ($n = 2$ and 3). For epitaxial growth,

Table 16.3 *Conditions used for the preparation of Hg-1201, Hg-1212, and Hg-1223 high-T_c superconductor films (abbreviations: B_a – applied field, d_r – deposition rate, p – chamber pressure, PLD – pulsed laser deposition, p_O – oxygen partial pressure, RF – radio frequency, RT – room temperature, T_a – annealing temperature, T_s – substrate temperature*

Film/substrate	Deposition conditions/comments	Reference
Hg-1201/(1 0 0) $SrTiO_3$ $T_c \approx$ 85 K	RF planar magnetron sputtering of a mixture of Ba_2CuO_3, CuO, HgO (Hg:Ba:Ca:Cu = 2.5:2:2.5:4) Sputtering in 0.5×10^{-2} mbar Ar input RF power of 50 W 1-μm-thick amorphous film after 2 h Annealing at 650–700 °C in flowing N_2/1% O_2 for 0.5–1 h $Hg(Ba,Ca)_2CuO_4$ film (*c*-axis perpendicular to the substrate) Remaining Ca–Cu–O as impurity phases	[31]
Hg-1212/$SrTiO_3$ T_c(mid) ~ 120 K j_c ~ 10 MA cm^{-2} (10 K)	RF-sputtering of $Ba_2CaCu_2O_x$ precursor film, target of BaO_2, CaO, CuO, Ba : Ca : Cu = 2 : 1.2 : 2.1 p ~ 10^{-6} mbar, 4 : 1 Ar/O_2 mixture ~ 110 mbar, reaction of the precursor film with Hg vapor in a sealed quartz tube (Hg source: pellet of Hg-1212), 160°C/h → T_a, T_a = 800 °C/5 h	[32]
Hg-1212/(1 0 0) $SrTiO_3$ $T_c \approx$ 120 K $j_c \geq$ 10 MA cm^{-2} 5 K, small fields $B \| c$	Sequential layers of HgO and $Ba_2CaCu_2O_x$ deposited by PLD from two separate targets at RT in vacuum 0.25-μm-thick film covered by a 40-nm-thick HgO layer 800 °C/1 h in an encapsulated quartz tube (film, Hg–Cu–O, precursor pellets) → *c*-axis-oriented high quality epitaxial Hg-1212 film	[33]
(Hg,Re)-1212/(1 0 0) $LaAlO_3$ $T_c \approx$ 120 K *c*-axis-oriented epitaxial film	Sequential deposition of HgO (5 nm thick) and $Re_{0.1}Ba_2CaCu_2O_y$ (25 nm) repeated several times 180-nm-thick HgO cap layer, all depositions at RT in vacuum ($\approx 10^{-7}$ mbar), total film thickness of 0.5–1 μm Annealing in a sealed quartz tube together with 0.4 g $Ba_2CaCu_2O_z$ and 1 g Hg-1212 pellets, T_a = 845 °C, 0.75 h, finally quenched in air	[34]
Hg-1212/(1 0 0) $SrTiO_3$ T_c(end) = 121 K	$Ba_2CaCu_2O_x$ film deposited by PLD from a $Ba_2CaCu_2O_x$ target prepared from $Ba(NO_3)_2$, $Ca(NO_3)_2 \cdot 4H_2O$, and CuO T_s = 673 K, p_O = 0.06 mbar, 0.4 nm s^{-1} deposition rate Hg-free precursor film and 2 Hg-1212 pellets placed into a sealed and evacuated quartz tube Annealing at 1043 K for 5 h and cooling to RT	[35]

(continued overleaf)

Table 16.3 *(continued)*

Film/substrate	Deposition conditions/comments	Reference
(Hg,Pb)-1212/(1 0 0) MgO Hg-1212/(1 0 0) MgO Both film types T_c(end) 105–117 K	Sequential evaporation of BaF_2, Cu, CaF_2 with a cation ratio of Ba : Ca : Cu = 2 : 2 : 3 for Hg-1212 Sequential evaporation of BaF_2, Cu, Pb, CaF_2 with cation ratio of Pb : Ba : Ca : Cu = 0.5 : 2 : 2 : 3 for (Hg,Pb)-1212 $T_s = 700\,°C$, *ex situ* annealing in 1.3×10^{-6} mbar dry O_2 Followed by increase in p_O to ambient pressure at $T_s = 700\,°C$ to remove fluorine Compacted pellets of unreacted $Hg_{1.2}Pb_{0.3}Ba_2Ca_2Cu_3O_x$ as a Hg source, 3 K min^{-1} to 800 °C for 1 h, 3 K min^{-1} to RT	[36]
Hg-1212/$LaAlO_3$ $T_c = 121$–124 K $j_c \leq 7$ MA cm^{-2} 77 K, $B_a = 0$ ≈ 300-nm-thick films	Ba–Ca–Cu–O precursor film sputtered from a metallic target, annealing together with Tl–Ba–Ca–Cu–O pellet in the Au/Pt container at 870 °C under flowing O_2 → Tl-2212 precursor film, conversion to Hg-1212 at 790–810 °C for 0.5 to 0.75 h in Hg/O_2 vapor in quartz ampoules Post annealing at 350 °C for 3 h in flowing O_2.	[37]
(Hg,Re)-1212/LSAT ≈ 300-nm-thick films $T_c = 117$–120 K $j_c = 4$–6 MA cm^{-2} 77 K, $B_a = 0$	(1 0 0) LSAT: $(LaAlO_3)_{0.3}$-$(SrAl_{0.5}Ta_{0.5}O_3)_{0.7}$ Preparation of a 80-nm-thick seed layer PLD of $Re_{0.1}Ba_2CaCu_2O_z$ in vacuum at T_s = RT Deposition of a 25-nm-thick HgO cap layer Precursor film and nonreacted pellets of $HgBa_2CaCu_2O_x$ and $Ba_2CaCu_2O_y$ encapsulated in a sealed quartz tube Heat treatment at 725 °C for 5 h PLD of a 220-nm-thick (Hg,Re)-1212 film, heat treatment together with pellets of $Hg_{0.8}Ba_2CaCu_2O_x$ and $Ba_2CaCu_2O_y$ at 725 °C for 100 h	[38]
(Hg,Re)-1212/$LaAlO_3$	PLD of precursor films from a $Re_{0.2}Ba_2Ca_2Cu_3O_x$ target 133 mbar O_2, $T_s = 250\,°C$, laser energy of 200 mJ Reaction in vacuum-sealed quartz tubes at 780 °C for 10 h using $CaHgO_2$ as a Hg source	[39]
(Hg,Re)-1212/(1 0 0) $SrTiO_3$ $T_c = 124$ K	$(Hg_{0.9}Re_{0.1})Ba_2CaCu_2O_{6+\delta}$ target of $Re_{0.1}Ba_2CaCu_2O_y$ (Mixture of ReO_2 and $Ba_2CaCu_2O_x$), $p = 1.33 \times 10^{-5}$ mbar PLD, 650 mJ/pulse, 8 Hz, 1 h, RT in 0.3 mbar O_2	[40]

Table 16.3 *(continued)*

Film/substrate	Deposition conditions/comments	Reference
	Precursor films and pellets of (Hg,Re)-1212 (film side) and 212 (substrate side) sealed into quartz tubes evacuated to 1 mbar, 1 h → 750 °C, 0.5 h → 850 °C, 850 °C/1 h, cool-down to RT in 6 h	
Hg-1223/(1 0 0) $SrTiO_3$ T_c(onset) ≈ 131 K j_c = 85000 A cm^{-2} 77 K, $B_a = 0$	RF sputtering of $Ba_2Ca_2Cu_3O_x$ T_s = 25–600 °C, 67–93 mbar Ar, *in situ* deposition of the HgO cap layer, 100–500-nm-thick precursor film Thickness ratio HgO : $Ba_2Ca_2Cu_3O_x$ = 0.2 : 1 T_a = 750–850 °C/ ≤ 5 h in an evacuated quartz tube	[41]
Hg-1223/(1 0 0) $SrTiO_3$ T_c(end) ≈ 130 K j_c = 23 MA cm^{-2} 5 K, $B_a = 0$ j_c = 2 MA cm^{-2} 77 K, $B_a = 0$	$Ba_2Ca_2Cu_3O_x$ radio-frequency sputtered from $Ba_2Ca_2Cu_3O_x$ target at RT in 0.07 mbar Ar Target synthesized from $Ba(NO_3)_2$, CaO, CuO at 900 °C for 17 h in O_2 atmosphere, $p = 1.3 \times 10^{-5}$ mbar, d_r = 150 nm h^{-1}, Hg-1223 pellets sealed in the evacuated quartz tube, annealing at 850 °C for 0.5 h	[42]
Hg-1223/(1 0 0) $Y_{0.15}Zr_{0.85}O_{1.93}$ T_c ≈ 135 K j_c = 10 MA cm^{-2} 10 K, $B_a = 0$	Sol–gel process, solution of $Cu(NO_3)_2{\cdot}3H_2O$, $Ca(NO_3)_2{\cdot}4H_2O$, and $Ba(NO_3)_2$ in a mixture of H_2O and glycerin Calcination at 900 °C → $Ba_2Ca_2Cu_3$ precursor film Annealing at 870 °C/1.5 h in Hg vapor in quartz tubes	[43]
Hg-1223/(1 0 0) $SrTiO_3$ T_c(end) = 130 °C j_c ~ 0.44 MA cm^{-2} 77 K, $B_a = 0$	Spray pyrolysis of $Ba(NO_3)_2$, $Ca(NO_3)_2{\cdot}4H_2O$, $Cu(NO_3)_2{\cdot}4H_2O$ dissolved in distilled H_2O, T_s = 100–120 °C, T_a = 650 °C for 6 h in flowing O_2 Precursor film and unreacted Hg–Ba–Ca–Cu–O pellet in an evacuated quartz tube, annealing at 775 °C/24 h Post annealing at 320 °C for 50 h in flowing O_2	[44]
Hg-1223/(1 0 0) $SrTiO_3$ 1.5–10 μm thick T_c(end) = 130 K j_c = 0.1 MA cm^{-2} 100 K, $B_a = 0$	Spray pyrolysis of a solution of Ba, Ca, Cu nitrates in distilled H_2O (Ba : Ca : Cu = 2 : 2 : 3), T_s = 100–120 °C Precursor film annealed at 650 °C/6 h in O_2 Film and pellet of unreacted $HgBa_2Ca_3Cu_4O_x$ heat treated at 700–775 °C for 24 h	[45]
Hg-1223/sapphire 10-nm-thick CeO_2 buffer layer 300-nm-thick precursor film	Preparation of a 10-nm-thick CeO_2 buffer layer and Re:Ba:Ca:Cu = 0.15:2:2:3 precursor films (10 Pa Ar, T_s = RT) by radio-frequency magnetron sputtering *Ex situ* annealing of film and unreacted pellet of (Hg,Re)-1223 wrapped in an Au foil in a closed quartz tube Reaction at 800 °C for 3 h	[46]
Thicker CeO_2 buffer	Favors the formation of Hg-1212 films	

the mismatch of the lattice parameters of the substrate and the film should be less than 10%. The lattice mismatch between CeO_2 and Hg-12($n-$ 1)n is only 0.3%, while it can reach up to 20% for sapphire and the mercury-based high-T_c films. A CeO_2 buffer layer only 10 nm thick on sapphire is heavily strained, and hence the lattice parameter is different from that in CeO_2 bulk material or thick films. A c lattice parameter of 0.541 nm was found for the 10-nm-thick film by X-ray diffraction. In thicker films, the strain caused by different lattice parameters is relaxed. For a CeO_2 buffer layer of 120 nm thickness, the c lattice parameter is 0.539 nm. The Re–Ba–Ca–Cu–O precursor films, deposited on the CeO_2 buffered sapphire, were heat treated at 800 °C in Hg vapor. In this way, superconducting films of 300 nm thickness were obtained. The films were mixtures of c-axis-oriented Hg-1212 and Hg-1223. The c lattice parameters were found to be independent of the buffer layer thickness, whereas the ratio of Hg-1223 to Hg-1212 decreased with increasing buffer layer thickness. The results suggest that the thickness of the CeO_2 buffer layer influences the growth of the Hg-12($n-$ 1)n film. The formation of Hg-1223 starts with the initial formation of Hg-1212, which is converted to Hg-1223 by the intercalation of Ca–Cu–O [46]. For films on thin-strained CeO_2 buffer layers, more dislocations are present in the initial Hg-1212 film. This promotes diffusion of Ca, Cu, and oxygen into the layer, leading to the accelerated formation of the Hg-1223 phase. Beyond that, the epitaxial growth of Hg-12($n-$ 1)n films is influenced by the strain present in the buffer layer [46].

16.2.4 Preparation of RE-123 Superconductor Films

In Chapter 15, the preparation of RE-123-coated conductors has already been described. Coated-conductor fabrication is based on the epitaxial growth of RE-123 films on flexible metallic substrates, which have been coated with suitable buffer layers in a reel-to-reel process. In this section, a few examples of the deposition of RE-123 films on single-crystal substrates are presented [47–66]. Published preparation conditions are listed in Table 16.4.

RE-123 films were deposited on $SrTiO_3$ [47–49, 55, 56, 61, 63, 64], YSZ [48], MgO [50, 53, 62], silicon wafers [51], and $LaAlO_3$ [52, 54, 58, 59, 65, 66]. In some cases, a buffer layer, deposited on the single crystal, was used as a substrate for the growth of the RE-123 films. CeO_2 buffer layers were deposited on $SrTiO_3$ [57], $LaAlO_3$ [58], and sapphire [60]. $LaAlO_3$ single crystals buffered with MgO [54] were also used as a substrate for the growth of RE-123 films. In a Y-123 film eight inches in diameter, deposited on a silicon wafer, the critical current density reached 20 MA cm^{-2} at 4.2 K and $B_a = 0$ [51]. The critical current density of RE-123 films was typically in the range of 1.6–2.7 MA cm^{-2} at 77 K and $B_a = 0$ [63–66]. The RE-123 films were deposited by various methods, including PLD [49, 52, 54, 56, 57, 63], various sputtering techniques [47, 51, 55, 61], and MOD [58–60, 65, 66]. Critical current densities of the order of 2 MA cm^{-2} (77 K, $B_a = 0$) were reported for RE-123 films prepared by MOD [60, 65, 66].

MOD is an interesting alternative to vacuum deposition techniques. Frequently, MOD starts from a solution of the Y, Ba, and Cu TFA, which is (for example) spin coated on the single-crystal substrate [65]. The substrate with the spin-coated solution is slowly heated to 400 °C in the presence of humid oxygen to decompose the TFA. To obtain films of 1–2 μm thickness, the process needs to be repeated several times. Finally, the precursor film is fired at 800 °C in low-pressure oxygen diluted by argon in the presence of water

Table 16.4 *Conditions used for the preparation of RE-123 high-T_c superconductor films on single-crystal substrates (abbreviations: B_a – applied field, d_r – deposition rate, MOD – metal organic decomposition, p – chamber base pressure, PLD – pulsed laser deposition, p_O – oxygen partial pressure, RF – radio frequency, RE Y – or another rare earth element, RT – room temperature, T_a – annealing temperature, T_s – substrate temperature, YSZ – yttria-stabilized zirconia*

Film/substrate	Deposition conditions/comments	Reference
Y-123/(1 0 0) $SrTiO_3$ T_c(onset) = 90 K j_c = 0.1 MA cm^{-2} 78 K, $B_a = 0$ j_c = 2 MA cm^{-2} 4.2 K, $B_a = 0$	Reactive magnetron cosputtering from Y, Ba, Cu metal targets in Ar/O_2 atmosphere $p = 1.33 \times 10^{-6}$ mbar, $p_{Ar} = 2.67 \times 10^{-3}$ mbar at Ba and Y sources, O_2 injection at the substrate (p_O = 1.33×10^{-4} mbar in the chamber) T_s = 400 °C, d_r = 0.4–0.5 nm s^{-1}, furnace annealing in flowing O_2, 650 °C/6 h, 750 °C/1 h, 850 °C/1 h	[47]
Y-123/$SrTiO_3$ or Y-123/YSZ T_c = 77–79 K (YSZ) T_c = 81 K ($SrTiO_3$) j_c = 1 MA cm^{-2} 4.2 K, $B_a = 0$	Y and Ba evaporated by two electron beam evaporators Cu deposited from a thermal source, d_r = 0.45 nm s^{-1}, p_O = 0.87–1.33 mbar, T_s = 700 °C, 700 to 200 °C in 20 min, 1.33 mbar oxygen during cool-down	[48]
Y-123/(1 0 0) $SrTiO_3$ j_c = 8–14 MA cm^{-2} 77 K, $B_a = 0$	Laser ablation from Y-123 + 5 wt% Ag pellets 100–200-nm-thick films	[49]
(0 0 1) Y-123/(0 0 1) MgO	Pulsed organometallic beam epitaxy	[50]
Y-123/Si wafer T_c = 87.5 K j_c = 20 MA cm^{-2} 4.2 K, $B_a = 0$	Deposition of 8 inch diameter Y-123 films by a 90° off-axis sputtering technique, 267 mbar 80% Ar + 20% O_2 T_s = RT	[51]
Y-123/(1 0 0) $LaAlO_3$	Pulsed laser ablation from Y-123 + 5 wt% Ag pellets p_O = 267 mbar, T_s = 700 °C, 100–400-nm-thick films	[52]
Y-123/MgO T_c = 84 K j_c = 1 MA cm^{-2}	Thermal coevaporation of Y, Ba, Cu from resistively heated metal boats, d_r = 0.25 nm s^{-1}, T_s = 650 °C, cool-down in 100 mbar O_2, 200-nm-thick Y-123 film	[53]
Y-123/MgO/$LaAlO_3$ T_c = 87 K	PLD of a 100-nm-thick MgO buffer layer Laser ablation of Y-123 at T_s = 750 °C, p_O = 250 mbar Y-123 films are c-axis oriented and epitaxial	[54]
Nd-123/(1 0 0) $SrTiO_3$ T_c(end) = 90 K	Off-axis RF sputtering method, 740 °C ≤ T_s ≤ 780 °C, p_{Ar}/p_O = 4–2 (0.12–0.18 mbar), 200-nm-thick Y-123 film	[55]

(continued overleaf)

Table 16.4 *(continued)*

Film/substrate	Deposition conditions/comments	Reference
(Y,Ca)-123/(0 0 1) $SrTiO_3$	PLD from a $Y_{1-x}Ca_xBa_2Cu_3O_{7-\delta}$ target, $T_s = 850\,°C$, $p_O = 400\,mbar$, at the end of deposition slow cool-down to 450 °C, annealing at 450 °C in 30 mbar O_2 for 90 h	[56]
Y-123/CeO_2/$SrTiO_3$	PLD of CeO_2 on $SrTiO_3$, 248 nm laser wave length, 108 mJ laser energy, 1 Hz repetition rate, $T_s = 650\,°C$, $p_O = 0.2\,mbar$, MOD of trifluoro-acetates (TFA), spin coating at RT, furnace annealing to 400 °C in 1l/min humid O_2 flow, $T_a = 840\,°C/70\,min$ in 2.1% humid Ar/O_2 (200–400 ppm O_2)	[57]
Y-123/$LaAlO_3$ Y-123/CeO_2/$LaAlO_3$	Solution of the TFA of Y and Ba and Cu naphtenate is spin coated on $LaAlO_3$ or CeO_2 buffered $LaAlO_3$, calcination of the precursor film at 400 °C in humid O_2, coating and calcination repeated for several times, calcination of the multilayer film at 500 °C, crystallization at 760 °C in humid Ar-0.1% O_2	[58]
Y-123/(1 0 0) $LaAlO_3$ $I_c = 105\,A\,cm^{-1}$-width 1.17-μm-thick film $j_c = 0.76$–$0.9\,MA\,cm$–^{2} (77 K, $B_a = 0$)	TFA MOD, Y_2BaCuO_x and $Ba_3Cu_5O_8$ dissolved in a ratio of 1:2, heat treatment at 80 °C/6 h, obtained blue residue is dissolved in methanol and dip-coated on $LaAlO_3$ single crystal, drying at 150 °C for 2 h and $p = 1.33 \times 10^{-3}\,mbar$, film calcined at 460 °C in humid O_2, 775 °C/4 h in 20% H_2O Ar-0.1% O_2 Annealing 450 °C/10 h in dry O_2	[59]
Y-123/CeO_2/sapphire $j_c = 3\,MA\,cm^{-2}$	Y, Ba, and Cu acetyl-acetonates dissolved in pyridine and propionic acid, evaporation of the solvents, dissolution of residual metal complexes in methanol spin coating on CeO_2 buffered sapphire, fired at 500 °C, 765 °C/1.5 h in flowing nitrogen and air (0.1 % O_2), total pressure of 100 mbar in the furnace, $p_O = 0.1\,mbar$, below 550 °C 1 bar pure O_2	[60]
Nd-123/(1 0 0) $SrTiO_3$	Deposition by high oxygen pressure sputtering	[61]
Nd-123/MgO $T_c = 90.4$–$91.6\,K$	Molecular-beam epitaxy (MBE), $T_s = 665\,°C$, 250-nm-thick films	[62]
Y-123/Nd-123 multilayers on $SrTiO_3$ ~ 2.7 MA cm^{-2} 77 K, $B_a = 0$	PLD of Y-123/Nd-123 multilayers on $SrTiO_3$ Y-123 layers of 300 nm thickness Nd-123 layers of 50 nm thickness Total thickness of ~ 1000 nm	[63]

Table 16.4 (continued)

Film/substrate	Deposition conditions/comments	Reference
Sm-123/$SrTiO_3$ I_c = 83.3 A cm^{-1}-width 77 K, $B_a = 0$, 1 μV cm^{-1} j_c = 2.1 MA cm^{-2}	Coevaporation of Sm, Ba, and Cu with deposition rates of 0.61 nm s^{-1} (Sm), 1.46 nm s^{-1} (Ba), and 0.45 nm s^{-1} (Cu), T_s = 700 °C, in an evaporation chamber $p < 4 \times 10^{-5}$ mbar, in a reaction chamber $p_O = 6.7 \times 10^{-3}$ mbar, after deposition p_O = 1 bar, cool-down to RT	[64]
Y-123 multilayer films on $LaAlO_3$ 5-layer film of 1 μm thickness $j_c \geq$ 1.6 MA cm^{-2} I_c = 160 A cm^{-1}-width 77 K, $B_a = 0$	Ultrasonic cleaning of the $LaAlO_3$ single-crystal substrate in an ultrasonic bath of (1) acetone and of (2) methanol, annealing at 800 °C in pure O_2 Y, Ba, Cu dissolved in trifluoroacetic acid and H_2O Several cycles of drying and dissolving in methanol Spin coating of the solution on the substrate (0.2-μm-thick layer), slow heating to 400 °C in humid O_2 to decompose the TFA Process is repeated until the multilayer film is 1–2-μm-thick, firing at 800 °C in humid 0.1% O_2/Ar mixture	[65]
Gd-123/(1 0 0) $LaAlO_3$ $j_c \approx$ 2.1 MA cm^{-2} 77 K, $B_a = 0$	MOD of Ba–TFA and fluorine-free Gd and Cu sources Deposition by dip coating, pyrolysis at 400 °C/3 h in humid O_2, firing at 775–800 °C in humid Ar/O_2 atmosphere	[66]

vapor. The rate of the reaction leading to the desired Y-123 film is controlled by the oxygen and the water vapor pressure [65].

16.3 Multilayers of Ultrathin Films

In this section, the properties of multilayers of ultrathin high-T_c superconductor films are considered. In the layered crystal structures of the cuprates, the superconducting CuO_2 planes are separated by insulating layers, which act as charge carrier reservoirs. Triscone *et al.* [67] used multilayers of ultrathin $YBa_2Cu_3O_7$/$PrBa_2Cu_3O_7$ films to study the effect of the thickness of the insulating and nonsuperconducting $PrBa_2Cu_3O_7$ layers on the superconducting coupling between the CuO_2 planes of adjacent layers of $YBa_2Cu_3O_7$. They prepared $YBa_2Cu_3O_7$/$PrBa_2Cu_3O_7$ superlattices by dc magnetron sputtering using stoichiometric targets of Y-123 and Pr-123. Multilayer films with a total thickness of 150 nm were deposited on heated (1 0 0) MgO substrates. A sketch of three multilayer films with alternating layers of $YBa_2Cu_3O_7$ and $PrBa_2Cu_3O_7$ is shown in Figure 16.1. In all multilayer films, each of the Y-123 layers is 1.2 nm thick, while the thickness of the Pr-123 layers differs in the different superlattices. Triscone *et al.* [67] studied multilayers with

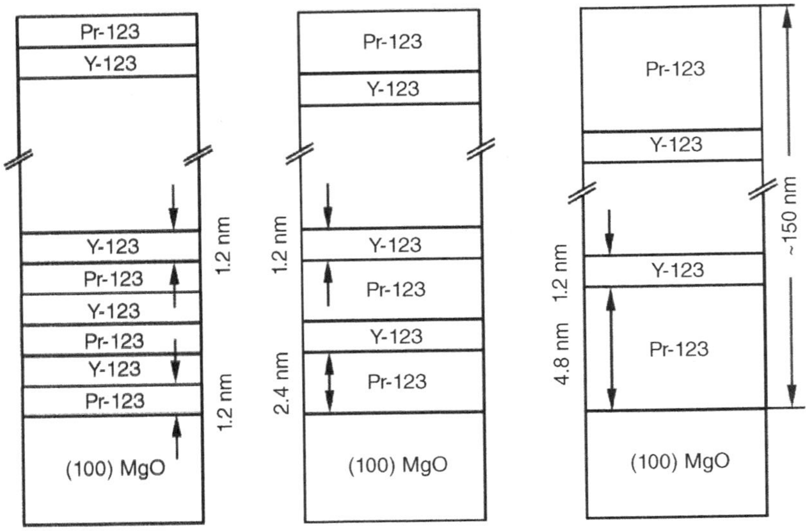

Figure 16.1 *Sketch of Pr-123/Y-123 multilayer films with different thicknesses (1.2, 2.4, and 4.8 nm) of the nonsuperconducting Pr-123 layers. The total thickness of each multilayer is approximately 150 nm. To study the effect of the Josephson coupling, the thickness of the Pr-123 layers, which separate 1.2-nm-thick Y-123 films, was varied. A thickness of 1.2 nm corresponds to the height of one unit cell of Y-123 or Pr-123 along the crystallographic c-direction (adapted from [67])*

Pr-123 thicknesses in the range of 1.2–7.2 nm. The height of a single unit cell of Y-123 or Pr-123 is approximately 1.2 nm along the crystallographic *c*-direction. In the Y-123 compound, two CuO_2 planes, separated by a layer of yttrium ions, form the conductive copper oxide blocks. Adjacent copper oxide blocks are separated by charge carrier reservoirs with the layer sequence BaO–CuO–BaO. For comparison, Triscone *et al.* [67] prepared single Y-123 and $(Y_{0.5}Pr_{0.5})Ba_2Cu_3O_7$ films of 150 nm thickness. In the $(Y_{0.5}Pr_{0.5})Ba_2Cu_3O_7$ film, they measured a critical temperature of $\approx$20 K, which is more than a factor of 2 less than the value of $\approx$50 K found for the multilayer film with alternating layers of Pr-123 and Y-123, each 1.2 nm thick. The quoted T_c values are based on the end of the resistive transition. For multilayer films with one unit cell thick Y-123 layers embedded between Pr-123 layers of 1.2–7.2 nm thickness, the critical temperature was found to decrease from $\approx$50 K (1.2-nm-thick Pr-123) to $\approx$ 13 K (7.2-nm-thick Pr-123).

Terashima *et al.* [68] studied the effect of the thickness of ultrathin Y-123 films embedded between two Pr-123 films of six unit cells thickness. As a substrate for the trilayer film system, they used (1 0 0) $SrTiO_3$ single-crystal substrates. The films were grown by reactive coevaporation of the metals in oxygen. The layer growth was checked by the observation of strong intensity oscillations of reflection high-energy electron diffraction, which correspond to the height of a single unit cell. The first Pr-123 film is used as a buffer layer for the growth of the epitaxial Y-123 film. The lattice mismatch between Pr-123 and Y-123 is as small as 1.5%. It was found that without a cap layer of Pr-123, the ultrathin Y-123 films are non-superconducting. The Pr-123 cap layer was found to transfer a number of holes

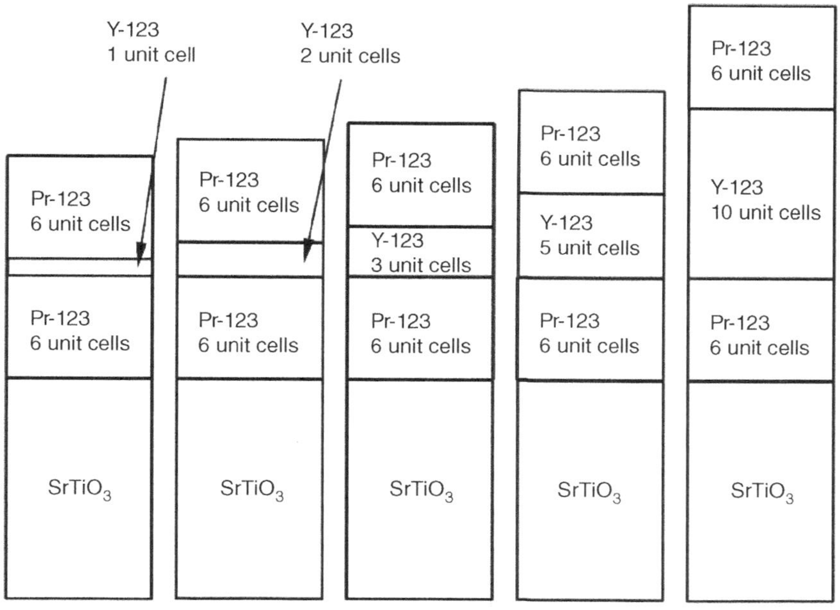

Figure 16.2 *Examples of Pr-123/Y-123/Pr-123 trilayers grown on* $SrTiO_3$*. From left to right, the thickness of the Y-123 layer increases from 1 to 10 unit cells (adapted from [68])*

to the Y-123 film, which is sufficient to establish superconductivity [68]. A sketch of the studied trilayer film systems is presented in Figure 16.2.

The variation of the critical temperature with the thickness of the Y-123 layer in Pr-123 (six unit cells)/Y-123 (n unit cells)/Pr-123 (six unit cells) trilayers is presented in Figure 16.3 (results from [68]). For comparison, the dependence of T_c on the thickness of the Pr-123 layer in Pr-123/Y-123 superlattices is also shown in Figure 16.3. All T_c values are based on 10% of the normal-state resistance. The results of Terashima *et al.* [68] indicate that a single unit cell of Y-123 embedded between Pr-123 layers (each 6 unit cells thick) is superconducting. The measured T_c of ~ 41 K is much lower than the critical temperature of slightly more than 90 K achieved in a 100-nm-thick Y-123 film. The critical temperature was found to increase with increasing thickness of the Y-123 layer. The data for the Pr-123/Y-123 superlattices indicate that the critical temperature of one unit cell thick Y-123 layers decreases nearly linearly with the thickness of the Pr-123 layers. The extrapolation to vanishing thickness of the Pr-123 layer leads to a critical temperature below 60 K, which is less than the T_c of ~ 84 K of a Y-123 film prepared under identical conditions (see Figure 16.3). This result suggests that T_c is determined by two factors, namely the superconducting coupling of adjacent Y-123 layers and a reduction of T_c in a Y-123 film of only one unit cell thickness. The onset T_c (beginning of the resistive transition) of ultrathin Y-123 layers in Pr-123/Y-123/Pr-123 trilayers is much less affected than the offset T_c. For the Y-123 layers of two, three, and four unit cells thickness, the onset T_c of around 91 K is only slightly lower than that of a 100-nm-thick film, while for a Y-123 layer of only one unit cell thickness, the onset T_c is reduced

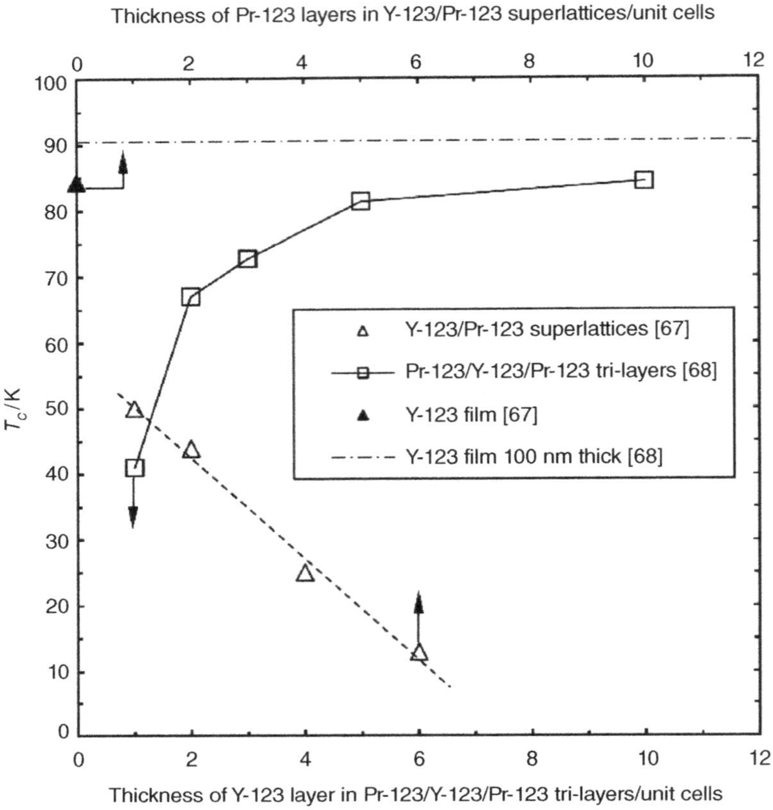

Figure 16.3 *Critical temperature of Pr-123/Y-123/Pr-123 trilayers vs. the thickness of the Y-123 layer (open squares). T_c increases with increasing thickness of the Y-123 layer from 41 K (one unit cell thick Y-123) to 84 K (10 unit cells thick Y-123). The critical temperature of a 100-nm-thick Y-123 film reaches ~90.5 K. The results indicate that a Y-123 layer one unit cell thick is superconducting. Critical temperature of Pr-123/Y-123 superlattices vs. the thickness of the Pr-123 layer separating adjacent Y-123 layers each 1.2 nm thick (open triangles). The T_c of these superlattices decreases with increasing thickness of the Pr-123 layers. In a 150-nm-thick film, a critical temperature of around 84 K was achieved (solid triangle) (results from [67, 68])*

to ≈85 K. Measurements of the Hall voltage indicated that the charge carrier density in the thinnest Y-123 layer is smaller than that in a thick film. For an increased number of holes in this layer, the onset T_c may reach 90 K [68].

Logvenov *et al.* [69] investigated superconductivity in a single CuO_2 layer. They prepared $La_{1.65}Sr_{0.45}CuO_4/La_2CuO_4$ bilayers by means of molecular-beam epitaxy (MBE). The deposition process was monitored by atomic absorption spectroscopy. The real-time observation of the atomic fluxes is used to control pneumatic shutters, enabling atomic layer-by-layer film deposition. Furthermore, the MBE system allows 3% Zn doping of a single predetermined layer of La_2CuO_4 or $(La,Sr)_2CuO_4$. A sketch of the film system under study is presented in Figure 16.4. The bilayer consisted of six layers of metallic

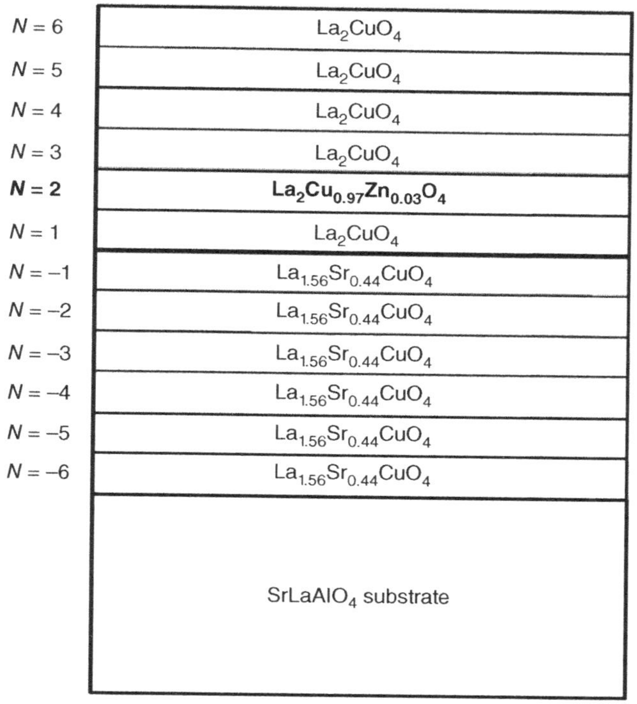

Figure 16.4 *Six unit cells thick metallic, non-superconducting $La_{1.56}Sr_{0.44}CuO_4$ film, grown on the $SrLaAlO_4$ substrate, covered with a six unit cells thick insulating $LaCuO_4$ film. In one of the layers, 3% of the copper atoms are replaced by Zn (in this example layer 2). The critical temperature was measured for different positions of the Zn-doped layer (see Table 16.5) (adapted from [69])*

$La_{1.56}Sr_{0.44}CuO_4$ and six layers of insulating La_2CuO_4. Outside such bilayers, none of the material superconducts. Several studies of superlattices [70] or bilayers of insulating and overdoped La–Sr–Cu–O [71, 72] revealed that these systems do superconduct. The results suggest that, at the interface of the overdoped metal and the insulator, the concentration of mobile holes is in a range that allows superconductivity to occur. In the experiments of Logvenov *et al.* [69], a single Zn-doped layer was introduced at different positions in the bilayer (i.e., $N = -6$–6). The measured T_c values are presented in Table 16.5.

Zn doping of copper in the CuO_2 layers is known to reduce T_c, an effect attributed to pair breaking by in-plane scattering [69]. For $La_{1.65}Sr_{0.45}CuO_4/La_2CuO_4$ bilayers without Zn doping in any layer, the zero-resistance temperature is typically 34.4 K. As can be seen in Table 16.5, Zn doping in the $N = 2$ layer reduces T_c as compared with doping in any other layer. In particular, the values for Zn doping in the neighboring layers $N = 1$ and 3 are also in the range of 32 ± 4 K, taking into account the sample-to-sample spread of T_c. This spread is due to small variations in the surface roughness of the substrate and the deposited film, interdiffusion of cations, and local changes in the charge carrier density. In contrast, T_c is significantly reduced to $16.7 - 23.1$ K, when the second, but only the second,

Table 16.5 *Measured zero-resistance critical temperature for different positions N of the Zn-doped layer [69]*

Position N	Measured T_c (K)
−6	35.1
−5	33.8
−4	34.2
−3	35.3
−2	29.1–33.9
−1	28.3–35.9
1	29–35.6
2	16.7–23.1
3	29.1–30
4	31.7
5	33.6
6	33.2

CuO_2 layer ($N = 2$) is Zn-doped. This result suggests that the second layer is responsible for the superconductivity of the bilayer. The T_c of all other layers is less than 18 K if they are superconducting at all.

16.4 Strain Effects

The critical temperature of cuprate high-T_c superconductors was found to depend on hydrostatic pressure, as discussed in Section 7.4. In this section, the effect of epitaxial strain on the critical current of $(La,Sr)_2CuO_4$ and $(La,Ce)_2CuO_4$ thin films is briefly discussed. A characteristic feature of epitaxial growth is the transfer of the crystallography of the substrate to the film. In the case of sufficiently thin films, the lattice parameters of the film are equal to that of the substrate. In the case of a mismatch of the lattice constants of the substrate and the bulk high-T_c superconductor, an epitaxial strain can be induced in a very thin superconductor film. Locquet *et al.* [73] deposited $La_{1.9}Sr_{0.1}CuO_4$ on both $SrLaAlO_4$ (lattice parameter $a = 0.3754$ nm) and $SrTiO_3$ (lattice parameter $a = 0.3905$) substrates in order to study the epitaxial strain effect. They deposited the $(La,Sr)_2CuO_4$ films of 10–15 nm thickness at a substrate temperature of 750 °C by means of block-by-block MBE. The selected thickness is large enough to avoid a reduction in T_c characteristic of ultrathin films. On the other hand, the film is so thin that the nucleation of misfit dislocations is minimized, which in thick films relieves epitaxial strain. Differential thermal expansion also contributes to epitaxial strain at low temperatures. The coefficients of thermal expansion of $(La,Sr)_2CuO_4$, $SrTiO_3$, and $SrLaAlO_4$ are 8.5×10^{-6} K^{-1}, 9×10^{-6} K^{-1}, and 10.5×10^{-6} K^{-1}, respectively. For $(La,Sr)_2CuO_4$ epitaxial films on $SrLaAlO_4$, the coefficient of thermal expansion of the film is smaller than that of the substrate, which enhances the compressive stress in the CuO_2 planes. Under epitaxial growth conditions, a compressive strain in the CuO_2 planes of $(La,Sr)_2CuO_4$ is typically accompanied by tensile strain in the c direction. Both in-plane compressive and out-of-plane tensile strains enhance the T_c of $(La,Sr)_2CuO_4$.

The expected change in the critical temperature of the films is given by [73]

$$T_c = T_c(0) + 2\frac{\delta T_c}{\delta \varepsilon_{ab}}\varepsilon_{ab} + \frac{\delta T_c}{\delta \varepsilon_c}\varepsilon_c \tag{16.1}$$

where $\varepsilon = (d_{\mathrm{bulk}} - d_{\mathrm{strained}})/d_{\mathrm{bulk}}$, with d being a lattice parameter. Locquet *et al.* [73] measured the actual strain values by X-ray diffraction. The strain values (positive sign for tensile strain and negative sign for compressive strain), found for $(La,Sr)_2CuO_4$ on $SrTiO_3$, are $\varepsilon_{ab} = 0.63\%$ and $\varepsilon_c = -0.76\%$. The corresponding strain values for the $(La,Sr)_2CuO_4$ film on $SrLaAlO_4$ are $\varepsilon_{ab} = -0.54\%$ and $\varepsilon_c = 0.35\%$. Sekitani *et al.* [74] performed similar experiments with strained $La_{1.85}Sr_{0.15}CuO_4$ films deposited on $SrTiO_3$ and $SrLaAlO_4$. The resistance versus temperature curves of Loquet *et al.* [73] and Sekitani *et al.* [74] are presented in Figure 16.5. As expected, the critical temperature of $(La,Sr)_2CuO_4$ films on $SrLaAlO_4$ is higher than the T_c of films deposited on $SrTiO_3$. The critical temperature of $La_{1.9}Sr_{0.1}CuO_4$ bulk material is 25 K, which has to be compared to a T_c as high as 49 K for the $La_{1.9}Sr_{0.1}CuO_4$ film on $SrLaAlO_4$ with the CuO_2 planes under compressive strain [73]. On the other hand, the CuO_2 planes of the $La_{1.9}Sr_{0.1}CuO_4$ film on $SrTiO_3$ experience a tensile strain leading to a T_c as low as 10 K [73]. Furthermore, the

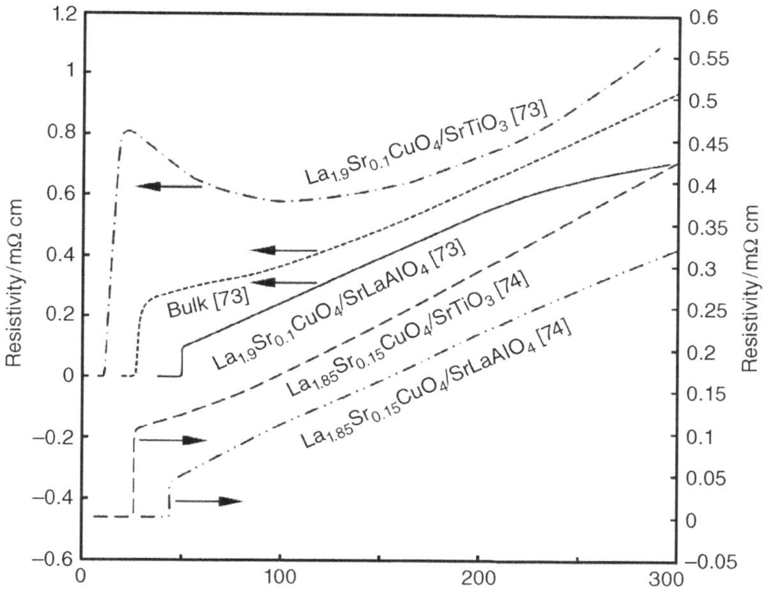

Figure 16.5 *Resistivity vs. temperature of $La_{1.9}Sr_{0.1}CuO_4$ films on $SrTiO_3$ and $SrLaAlO_4$. The differences of the T_c values of the films from the bulk value are caused by different epitaxial strains. For the $SrTiO_3$ substrate, the in-plane strain is tensile, whereas it is compressive for $La_{1.9}Sr_{0.1}CuO_4$ films on $SrLaAlO_4$. For comparison, the data of $La_{1.85}Sr_{0.15}CuO_4$ on the same substrates are also presented (results from [73, 74])*

Table 16.6 *Lattice constants of the $SrLaAlO_4$ (SLAO) buffer layer and the critical temperature measured in 80-nm-thick $La_{1.85}Sr_{0.15}CuO_4$ (LSCO) films on the buffered $SrTiO_3$ substrate [75]*

Thickness of SLAO (nm)	Lattice constant a (nm)	Lattice constant c (nm)	T_c of 80-nm-thick LSCO film (K)
10	3.8077	1.255	–
15	3.7828	1.267	–
20	3.7728	1.270	–
40	3.7682	1.272	35.3
60	3.7718	1.272	37.4
80	3.7611	1.276	39.1
100	3.7577	1.277	39.5

dependence of the normal-state resistivities on temperature is remarkably different. Similar to underdoped high-T_c superconductors, the film under tensile in-plane strain shows an upturn of the normal-state resistance at temperatures below 100 K. The critical temperature of $La_{1.85}Sr_{0.15}CuO_4$ bulk material (lattice constants $a = 0.3777$ nm, $c = 1.323$ nm [74]) is 36.7 K. As expected, the critical temperature of the $La_{1.85}Sr_{0.15}CuO_4$ film on $SrTiO_3$ (lattice constants $a = 0.3837$ nm, $c = 1.318$ nm [74]) of 26 K is lower, while the T_c of 44 K of the film on $SrLaAlO_4$ (lattice constants $a = 0.3762$ nm, $c = 1.329$ nm [74]) is higher than that of the bulk material.

Si *et al.* [75] varied the epitaxial strain by deposition of $SrLaAlO_4$ buffer layers of different thickness on $SrTiO_3$ substrates. The lattice constant c increases with increasing thickness of the buffer layer, whereas the lattice constant a decreases. In other words, the thicker the buffer layer, the higher is the in-plane compressive strain. Their results are presented in Table 16.6.

Yuan *et al.* [76] studied the effect of epitaxial strain on the critical temperature of the electron-doped superconductor $La_{1.89}Ce_{0.11}CuO_4$. They deposited $La_{1.89}Ce_{0.11}CuO_4$ films of identical thickness on $LaAlO_3$ ($a = 0.3821$ nm [76]), $SrTiO_3$ ($a = 0.3900$ nm [76]), and MgO ($a = 0.4216$ nm [76]) substrates. For some films, $BaTiO_3$ buffer layers were deposited on the substrates. A compressive in-plane strain lowers the T_c of electron-doped cuprate superconductors, whereas it enhances the critical temperature in hole-doped high-T_c superconductors. The resistivity versus temperature curves of Yuan *et al.* [76] are presented in Figure 16.6. The lowest zero-resistance critical temperature of 7.5 K was measured for a $La_{1.89}Ce_{0.11}CuO_4$ film directly deposited on $LaAlO_3$. As expected, this is lower than the T_c of 18 K found for $La_{1.89}Ce_{0.11}CuO_4$ on $SrTiO_3$. A 20-nm-thick $BaTiO_3$ buffer layer deposited on the $LaAlO_3$ substrate increased the zero-resistance T_c of the $La_{1.89}Ce_{0.11}CuO_4$ film to 15 K. The critical temperature of $La_{1.89}Ce_{0.11}CuO_4$ films on MgO substrates buffered with 20- and 40-nm-thick $BaTiO_3$ layers is 18.5 and 25 K, respectively. As expected, the T_c values are higher than that found for the $BaTiO_3$ buffered $LaAlO_3$. However, the T_c of only 15 K of a $La_{1.89}Ce_{0.11}CuO_4$ film directly deposited on MgO is lower than that found for $La_{1.89}Ce_{0.11}CuO_4$ on $SrTiO_3$, which is in contrast to expectations. The results are not due to different film thicknesses, which are the same in all considered $La_{1.89}Ce_{0.11}CuO_4$ films. The relatively low T_c of the nonbuffered

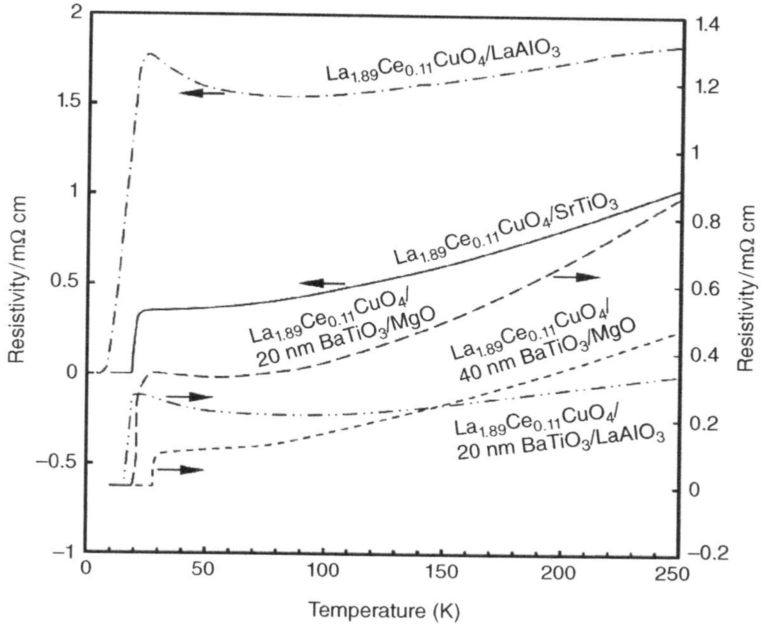

Figure 16.6 *Resistivity vs. temperature of $La_{1.89}Ce_{0.11}CuO_4$ electron-doped cuprate superconductor films on $LaAlO_3$, $SrTiO_3$, and $BaTiO_3$ buffered MgO. For electron-doped cuprate superconductors, an in-plane tensile strain is expected to increase T_c. The critical temperature of the films is determined not only by epitaxial strain but also by the perfection of the film growth. As expected, the T_c of 7.5 K of $La_{1.89}Ce_{0.11}CuO_4$ on $LaAlO_3$ (a = 0.3821 nm [76]) is lower than the T_c of 18 K found for $La_{1.89}Ce_{0.11}CuO_4$ on $SrTiO_3$ (a = 0.3900 nm [76]). A 20-nm-thick $BaTiO_3$ buffer layer on the $LaAlO_3$ substrate reduces the lattice mismatch and, together with improved film quality, leads to a T_c of 15 K. The highest T_c of all considered $La_{1.89}Ce_{0.11}CuO_4$ films of 25 K was obtained on MgO (a = 0.4216 nm [76]) with a 40-nm-thick $BaTiO_3$ buffer layer. For smaller buffer layer thickness, the lattice mismatch seems to be too large to allow high-quality films to be grown, as reflected by reduced T_c values (results from [76])*

MgO substrate suggests that the lattice mismatch of 5.16% is too large for the growth of high-quality $La_{1.89}Ce_{0.11}CuO_4$ films, leading to a lower T_c [76]. The lattice mismatch is reduced with increasing thickness of the $BaTiO_3$ buffer layer. For the 40-nm-thick $BaTiO_3$ buffer layer, the lattice mismatch is small enough to allow high-quality $La_{1.89}Ce_{0.11}CuO_4$. The improved film quality and increased in-plane tensile strain, as compared to the $La_{1.89}Ce_{0.11}CuO_4/SrTiO_3$ system, lead to a T_c of 25 K, significantly larger than the 18 K found for the film on $SrTiO_3$.

Hühne *et al.* [77] used piezoelectric (0 0 1) $Pb(Mg_{0.333}Nb_{0.667})_{0.72}Ti_{0.28}O_3$ (PMN-PT) substrates for the deposition of $YBa_2Cu_3O_{7-x}$ (Y-123) films. The epitaxial growth of the Y-123 films was monitored by X-ray diffraction. Pronounced cube-on-cube epitaxial growth of the Y-123 film ($a = 0.3823$ nm, $b = 0.388$ nm) on the PMN-PT substrate ($a = b = 0.4022$ nm) was observed. To allow the epitaxial strain to be controlled, a

NiCr/Au electrode on the back of the PMN-PT substrate and an Au cap layer on top of the Y-123 film are used to apply an electric voltage up to 500 V, leading to an electric field of less than 12.5 kV cm^{-1} across the piezocrystal of 0.4 mm thickness. Applied voltages of ±200 V were found to change the transition temperature reversibly by 0.1 K. A voltage of 400 V changed the in-plane strain by 0.05% [77]. The results indicate that piezoelectric single-crystal substrates can be used to study the influence of strain on the superconducting transition of Y-123 films.

16.5 Summary

In this chapter, the growth conditions and the properties of cuprate high-T_c superconductor films, except the special case of RE-123 coated conductors, have been described. Films can be used to study the current-carrying capacity of weak-link-free cuprate superconductors. In addition, the effect of defects of nanometer size on flux pinning can be investigated. Cuprate high-T_c superconductor films were deposited by various methods, including PLD, electron beam evaporation, MBE, magnetron sputtering, MOCVD, and MOD. In the case of thallium- or mercury-based cuprate superconductors, reactions with poisonous vapor can be performed in a separate furnace, for example, in an evacuated quartz ampoule, to avoid the need for decontamination of vacuum chambers and other equipment.

Various single-crystal substrates were used for the preparation of biaxially textured high-T_c superconductor films. Frequently used single-crystal substrates are $SrTiO_3$, YSZ, MgO, $LaAlO_3$, and sapphire. For high-quality epitaxial Y-123 films, the critical current density without applied field exceeds 10 MA cm^{-2} at 4.2 K and 1 MA cm^{-2} at 77 K. The critical temperature, achieved in thick epitaxial films, is comparable to the T_c value for the bulk superconductor in question.

In Pr-123/Y-123 multilayers, the critical temperature of extremely thin Y-123 layers was found to decrease with increasing thickness of the non-superconducting Pr-123 layers, indicating that Josephson coupling of adjacent Y-123 layers increases T_c. Furthermore, the results suggest that a Y-123 film, only a single unit cell in thickness, is superconducting. The critical temperature of extremely thin high-T_c superconductor films was found to be reduced as compared to the values achieved in thick films or bulk material.

Interesting effects were observed at the interface of insulating La_2CuO_4 and overdoped metallic $(La,Sr)_2CuO_4$ films. In spite of the fact that both are non-superconducting as individual films, high-T_c superconductivity was observed at their interface. The results suggest that holes are transferred from the overdoped metal to the insulator providing at their interface a concentration of mobile holes appropriate for the occurrence of high-T_c superconductivity.

The mismatch of the lattice parameters of the substrate and the high-temperature superconductor in question causes strains in thin epitaxial films. The transition temperature was found to be changed by epitaxial strain. A similar effect is the dependence of T_c on the applied pressure. Using piezocrystal substrates, the strain in thin Y-123 films can be controlled by the application of a voltage to the piezocrystal. Changes in the voltage were found to be accompanied by reversible changes of the critical temperature.

References

1. D.W. Face, M.J. Neal, M.M. Matthiesen, J.T. Kucera, J. Crain, J.M. Graybeal, T.P. Orlando, and D.A. Rudman, Preparation of superconducting thin films of bismuth strontium calcium copper oxides by reactive sputtering, *Appl. Phys. Lett.*, **53**, 246–248 (1988).
2. A. Mogro-Campero, L.G. Turner, M.F. Garbauskas, and R.W. Green, Superconducting thin films of Bi–Ca–Sr–Cu–O by sequential evaporation, *Appl. Phys. Lett.*, **53**, 327–328 (1988).
3. C. Richard Guarnieri, R.A. Roy, K.L. Saenger, S.A. Shivashankar, D.S. Yee, and J.J. Cuomo, Thin-film Bi–Sr–Ca–Cu–O high-temperature superconductors using pulsed laser evaporation from sintered disks, *Appl. Phys. Lett.*, **53**, 532–533 (1988).
4. P. Schmitt, L. Schultz, and G. Saemann-Ischenko, Electrical properties of $Bi_2Sr_2CaCu_2O_x$ thin films prepared in situ by pulsed laser deposition, *Physica C*, **168**, 475–478 (1990).
5. L. Schultz, B. Roas, P. Schmitt, P. Kumeth, and G. Saemann-Ischenko, Comparison of the critical current anisotropy in YBaCuO and BiSrCaCuO films, *IEEE Trans. Magn.*, **27**, 990–996 (1991).
6. C. Stölzel, M. Huth, and A. Adrian, *c*-axis oriented thin $Bi_2Sr_2CaCu_2O_{8+\delta}$-films prepared by flash evaporation, *Physica C*, **204**, 15–20 (1992).
7. M. Haruta and S. Sakai, Large grain growth by annealing of Ag-covered $Bi_2Sr_2CaCu_2O_{8+\delta}$ thin films and its application in the fabrication of intrinsic Josephson junctions, *Supercond. Sci. Technol.*, **23**, 115006 (2010).
8. K. Hamanaka, T. Tachiki, and T. Uchida, Investigation of characteristics for different film thickness of Bi-2212/MgO fabricated by metal-organic decomposition, *Physica C*, **470**, 1457–1460 (2010).
9. J.C. De Vero, J.L.F. Gabayno, W.O. Garcia, and R.V. Sarmago, Growth evolution of $Bi_2Sr_2CaCu_2O_{8+\delta}$ thin films deposited by infrared (1064 nm) pulsed laser deposition, *Physica C*, **470**, 149–154 (2010).
10. J. Chen and R.N. Bhattacharya, Growth of 1-2 μm thick biaxially textured Bi-2212 films on (100) $LaAlO_3$ single crystal substrates by electrodeposition, *Physica C*, **399**, 171–177 (2003).
11. M. Schieber, T. Tsach, M. Maharizi, M. Levinsky, B.L. Zhou, M. Golosovsky, and D. Davidov, Films of BSCCO superconductors prepared by spray pyrolysis of carboxylates, *Cryogenics*, **30**, 451–454 (1990).
12. H. Yamasaki, K. Endo, S. Kosaka, M. Umeda, S. Misawa, S. Yoshida, and K. Kajimura, Magnetic-field angle dependence of the critical current density in high-quality $Bi_2Sr_2Ca_2Cu_3O_x$ thin films, *IEEE Trans. Appl. Supercond.*, **3**, 1536–1539 (1993).
13. M. Ohkubo, J. Geerk, G. Linker, and O. Meyer, Phase intergrowth in $Bi_2Sr_2Ca_2Cu_3O_x$ superconducting thin films prepared by single cylindrical-sputtering gun, *Appl. Phys. Lett.*, **67**, 2403–2405 (1995).
14. M. Ohkubo, E. Brecht, G. Linker, J. Geerk, and O. Meyer, Different epitaxial growth modes of $Bi_2Sr_2Ca_2Cu_3O_x$ on MgO, *Appl. Phys. Lett.*, **69**, 574–576 (1996).

15. E. Steinbeiß, M. Manzel, M. Veith, H. Bruchlos, T. Eick, S. Huber, K. Steenbeck, W. Brodkorb, W. Morgenroth, G. Bruchlos, H.G. Schmidt, S. Bornmann, T. Köhler, L. Redlich, H.J. Fuchs, K. Schlenga, G. Hechtfischer, and P. Müller., Epitaxial Tl–Ba–Ca–Cu–O films – preparation, properties and applications, *Vacuum*, **47**, 1117–1122 (1996).
16. G. S. Hosseinali, W. Straif, B. Starchl, K. Kundzins, H.W. Weber, S.L. Yan, M. Manzel, E. Stangl, S. Proyer, D. Bäuerle, and E. Mezzetti., Critical currents in Tl-2212 and Tl-2223 thin films, *Physica C*, **268**, 307–316 (1996).
17. X.F. Zhang, Y.S. Sung, D.J. Miller, B.J. Hinds, R.J. McNeely, D.L. Studebaker, and T.J. Marks, New structural aspects of $Tl_2Ba_2CaCu_2O_y$ epitaxial thin films grown by MOCVD on $LaAlO_3$, *Physica C*, **275**, 146–154 (1997).
18. C.M. Carlson, P.A. Parilla, M.P. Siegal, D.S. Ginley, Y.-T. Wang, R.D. Blaugher, J.C. Price, D.L. Overmyer, and E.L. Venturini, $LaNiO_3$ buffer layers for high critical current density $YBa_2Cu_3O_{7-\delta}$ and $Tl_2Ba_2CaCu_2O_{8-\delta}$ films, *Appl. Phys. Lett.*, **75**, 2479–2481 (1999).
19. R.T. Lu, S.L. Yan, L. Fang, M. He, X.J. Zhao, Y.C. Ma, and W. Li, Fabrication and superconductivity of very thin $Tl_2Ba_2CaCu_2O_x$ films, *Physica C*, **399**, 43–47 (2003).
20. H. Zhao and J.Z. Wu, Converting Hg-1212 to Tl-2212 via Tl-Hg cation exchange in combination with Tl cation intercalation, *Supercond. Sci. Technol.*, **20**, 327–330 (2007).
21. H.S. Koo, H.K. Ku, M. Chen, and T. Kawai, Preparation and physical properties of superconducting films in the Tl–Ba–Ca–Cu–Ag_x–O system prepared by spray pyrolysis and Tl diffusion, *Physica C*, **455**, 71–75 (2007).
22. S. Huber, M. Manzel, H. Bruchlos, S. Hensen, and G. Müller, Thallium-based high-T_c films with very low surface impedance, *Physica C*, **244**, 337–340 (1995).
23. Z. Matkovičová, H. Nguyen Xuan, Ph. Galez, S. Beauquis, V. Štrbík, and G. Plesch, The influence of the reaction conditions on the formation of Tl-2223 superconducting films deposited by spray pyrolysis, *J. Phys. Chem. Solids* **68**, 1230–1233 (2007).
24. G. Malandrino, L.M.S. Perdicaro, I.L. Fragala, A. Cassinese, and A. Prigiobbo, Novel MOCVD approach to the low pressure in situ growth of $TlBa_2CaCu_2O_7$ films, *Physica C*, **408–410**, 894–895 (2004).
25. A. Mogro-Campero, P.J. Bednarcyk, Y. Gao, R.B. Bolon, J.E. Tkaczyk, and J.A. DeLuca, Critical current density and microstructure of 7–8 μm thick spray pyrolyzed films of $TlBa_2Ca_2Cu_3O_y$, *Physica C*, **269**, 325–329 (1996).
26. A. Goyal, E.D. Specht, D.M. Kroeger, J.E. Tkacyzk, C.L. Briant, and J.A. DeLuca, Crystallization of colonies of locally aligned grains during thallination of spray-pyrolized $TlBa_2Ca_2Cu_3O_x$ thick films, *Appl. Phys. Lett.*, **67**, 2563–2565 (1995).
27. A. Mogro-Campero, P.J. Bednarczyk, J.E. Tkaczyk, and J.A. DeLuca, Critical current as a function of film thickness for spray pyrolyzed films of $TlBa_2Ca_2Cu_3O_y$, *Physica C*, **247**, 239–242 (1995).
28. Q. He, D.K. Christen, C.E. Klabunde, J.E. Tkaczyk, K.W. Lay, M. Paranthaman, J.R. Thompson, A. Goyal, A.J. Pedraza, and D.M. Kroeger, High-current superconducting $Tl_1Ba_2Ca_2Cu_3O_y$ thick films on polycrystalline Ag by spin coating, *Appl. Phys. Lett.*, **67**, 294–296 (1995).
29. J.D. O'Connor, D. Dew-Hughes, N. Reschauer, W. Brozio, H.H. Wagner, K.F. Renk, M.J. Goringe, C.R.M. Grovenor, and T. Kaiser, The transmission electron microscopy

investigation and electrical properties of epitaxial Tl-1223 and Bi-substituted Tl-1223 thin films grown by in-situ laser ablation, *Physica C*, **302**, 277–289 (1998).
30. T. Guerfi, S. Beauquis, P. Galez, and M.F. Mosbah, Investigations on $TlBa_2Ca_2Cu_3O_x$ superconducting thin films formation on $LaAlO_3$ substrate, *Thin Solid Films*, **518**, 4986–4988 (2010).
31. H. Adachi, T. Satoh, and K. Setsune, Highly oriented Hg–Ba–Ca–Cu–O superconducting thin films, *Appl. Phys. Lett.*, **63**, 3628–3629 (1993).
32. Y.Q. Wang, R.L. Meng, Y.Y. Sun, K. Ross, Z.J. Huang, and C.W. Chu, Synthesis of preferred-oriented $HgBa_2CaCu_2O_{6+\delta}$ thin films, *Appl. Phys. Lett.*, **63**, 3084–3086 (1993).
33. L. Krusin-Elbaum, C.C. Tsuei, and A. Gupta, High current densities above 100 K in the high-temperature superconductor $HgBa_2CaCu_2O_{6+\delta}$, *Nature*, **373**, 679–681 (1995).
34. C. Gasser, Y. Moriwaki, T. Sugano, K. Nakashini, X.-J. Wu, S. Adachi, and K. Tanabe, Orientation control of *ex-situ* $(Hg_{1-x}Re_x)Ba_2CaCu_2O_y$ ($x \approx 0.1$) thin films on $LaAlO_3$, *Appl. Phys. Lett.*, **72**, 972–974 (1998).
35. J.D. Guo, G.C. Xiong, D.P. Yu, Q.R. Feng, X.L. Xu, G.J. Lian, and Z.H. Hu, Preparation of superconducting $HgBa_2CaCu_2O_x$ films with a zero-resistance transition temperature of 121 K, *Physica C*, **276**, 277–280 (1997).
36. G. Plesch, Š. Chromik, V. Štrbík, M. Mair, G. Gritzner, Š. Beňačka, I. Sargánková, and A. Bučkuliaková, Thin $(Hg,Pb)Ba_2CaCu_2O_y$ films prepared from thermally evaporated precursors by post annealing in Hg-atmosphere, *Physica C*, **307**, 74–78 (1998).
37. Th. Stelzner and H. Schneidewind, Double sided high critical current density $HgBa_2CaCu_2O_{6+\delta}$ films on lanthanum aluminate and sapphire substrates, *Physica C*, **398**, 37–40 (2003).
38. A. Ogawa, N. Inoue, T. Sugano, S. Adachi, and K. Tanabe, Fabrication of $(Hg,Re)Ba_2CaCu_2O_y$ thin films on LSAT substrates by long heat treatment, *Physica C*, **385**, 443–448 (2003).
39. J.H. Su, P.V.P.S.S. Sastry, and J. Schwartz, Fabrication and morphology of (Hg,Re)-1212 thin films on $LaAlO_3$, *Physica C*, **386**, 309–313 (2003).
40. A. Salem, G. Jakob, and H. Adrian, Preparation, scaling behavior of activation energy, Hall effect, and flux-flow anisotropy of $(Hg_{0.9}Re_{0.1})Ba_2CaCu_2O_{6+\delta}$ HTS thin films, *Physica C*, **204**, 354–364 (2004).
41. S.H. Yun, J.Z. Wu, B.W. Kang, A.N. Ray, A. Gapud, Y. Yang, R. Farr, G.F. Sun, S.H. Yoo, Y. Xin, and W.S. He, Fabrication of *c*-oriented $HgBa_2Ca_2Cu_3O_{8+\delta}$ superconducting thin films, *Appl. Phys. Lett.*, **67**, 2866–2868 (1995).
42. S.H. Yun and J.Z. Wu, Superconductivity above 130 K in high-quality mercury-based cuprate thin films, *Appl. Phys. Lett.*, **68**, 862–864 (1996).
43. Y. Tsabba and S. Reich, Giant mass anisotropy and high critical current in Hg-1223 superconducting films, *Physica C*, **269**, 1–4 (1996).
44. Y. Moriwaki, T. Sugano, C. Gasser, A. Fukuoka, K. Nakanishi, S. Adachi, and K. Tanabe, Epitaxial $HgBa_2Ca_2Cu_3O_y$ films on $SrTiO_3$ substrates prepared by spray pyrolysis technique, *Appl. Phys. Lett.*, **69**, 3423–3425 (1996).
45. Y. Moriwaki, T. Sugano, C. Gasser, K. Nakanishi, S. Adachi, and K. Tanabe, Highly *c*-axis oriented $HgBa_2Ca_2Cu_3O_y$ thick and thin films with $T_c > 130\,K$, *Physica C*, **282–287**, 643–644 (1997).

46. M. Valeriánová, P. Odier, Š. Chromik, V. Štrbík, M. Polák, and I. Kostič, Influence of the buffer layer on the growth of superconducting films based on mercury, *Supercond. Sci. Technol.*, **20**, 900–903 (2007).
47. K. Char, A.D. Kent, A. Kapitulnik, M.R. Beasley, and T.H. Geballe, Reactive magnetron sputtering of thin-film superconductor $YBa_2Cu_3O_{7-x}$, *Appl. Phys. Lett.*, **51**, 1370–1372 (1987).
48. D.K. Lathrop, S.E. Russek, and R.A. Buhrman, Production of $YBa_2Cu_3O_{7-y}$ superconducting thin films *in situ* by high-pressure reactive evaporation and rapid thermal annealing, *Appl. Phys. Lett.*, **51**, 1554–1556 (1987).
49. D. Kumar, M. Sharon, R. Pinto, P.R. Apte, S.P. Pai, S.C. Purandare, L.C. Gupta, and R. Vijayaraghavan, Large critical currents and improved epitaxy of laser ablated Ag-doped $YBa_2Cu_3O_{7-\delta}$ thin films, *Appl. Phys. Lett.*, **62**, 3522–3524 (1993).
50. D.B. Buchholz, J.S. Lei, S. Mahajan, P.R. Markworth, R.P.H Chang, B. Hinds, T.J. Marks, J.L. Schindler, C.R. Kannewurf, Y. Huang, and K.L. Merkle, In-plane orientation control of (0 0 1) $YBa_2Cu_3O_{7-\delta}$ grown on (001) MgO by pulsed organometallic beam epitaxy, *Appl. Phys. Lett.*, **68**, 3037–3039 (1996).
51. R.A. Rao, Q. Gan, C.B. Eom, Y. Suzuki, A.A. McDaniel, and J.W.P. Hsu, Uniform deposition of $YBa_2Cu_3O_7$ thin films over an 8 inch diameter area by a 90° off-axis sputtering technique, *Appl. Phys. Lett.*, **69**, 3911–3913 (1996).
52. P. Selvam, E.W. Seibt, D. Kumar, R. Pinto, and P.R. Apte, Enhanced J_c and improved grain-boundary properties in Ag-doped $YBa_2Cu_3O_{7-\delta}$ fims, *Appl. Phys. Lett.*, **71**, 137–139 (1997).
53. B. Utz, S. Rieder-Zecha, and H. Kinder, Continuous $YBa_2Cu_3O_{7-\delta}$ film deposition by optically controlled reactive thermal co-evaporation, *IEEE Trans. Appl. Supercond.*, **7**, 1181–1184 (1997).
54. R.J. Kennedy, W.P. Tucker, and P.A. Stampe, Growth of $YBa_2Cu_3O_7$ on silicon and $LaAlO_3$ with MgO buffer layers, *Physica C*, **314**, 69–72 (1999).
55. W.H. Tang and J. Gao, The relation between *c*-axis lattice parameter and superconducting transition temperature of $NdBa_2Cu_3O_y$ thin films, *Physica C*, **313**, 115–120 (1999).
56. C. Cancellieri, A. Augieri, V. Boffa, G. Celentano, L. Ciontea, F. Fabbri, V. Galluzzi, U. Gambardella, G. Grassano, T. Petrisor, and R. Tebano., Deposition and characterization of $Y_{1-x}Ca_xBa_2Cu_3O_{7-\delta}$ epitaxial films, *IEEE Trans. Appl. Supercond.*, **15**, 3038–3041 (2005).
57. X.M. Cui, G.Q. Liu, J. Wang, Z.C. Huang, Y.T. Zhao, B.W. Tao, and Y.R. Li, Enhancement of critical current density of $YBa_2Cu_3O_{7-\delta}$ thin films by nanoscale CeO_2 pretreatment of substrate surfaces, *Physica C*, **466**, 1–4 (2007).
58. T. Tanaka, K. Tada, N. Mori, K. Yamada, R. Teranishi, M. Mukaida, Y. Shiohara, T. Izumi, and J. Matsuda, Deposition process, microstructure and properties of YBCO film fabricated by advanced TFA-MOD method, *Physica C*, **463–465**, 527–531 (2007).
59. J.H. Lim, S.H. Jang, K.T. Kim, S.M. Hwang, J. Joo, H.-J. Lee, H.-G. Lee, and G.-W. Hong, The effect of film thickness on critical properties of YBCO film fabricated by TFA-MOD using 211-process, *Physica C*, **463–465**, 532–535 (2007).
60. I. Yamaguchi, M. Sohma, K. Tsukada, W. Kondo, K. Kamiya, T. Kumagai, and T. Manabe, Preparation of large-size Y-123 films on CeO_2-buffered sapphire substrates by MOD using a low-cost vacuum technique, *Physica C*, **463–465**, 549–553 (2007).

61. M. Saluzzo, G.M. de Luca, D. Marrè, M. Putti, M. Tropeano, U. Scotti di Uccio, and R. Vaglio, Thickness effect on the structure and superconductivity of $Nd_{1.2}Ba_{1.8}Cu_3O_{7+x}$ epitaxial films, *Physica C*, **460–462**, 724–725 (2007).
62. S. Karimoto, H. Sato, H. Shibata, and T. Makimoto, Deposition of $NdBa_2Cu_3O_{7-\delta}$ thin films on large area of over 5 in. diameter, *Physica C*, **463–465**, 927–929 (2007).
63. A.V. Pan, S.V. Pysarenko, D. Wexler, S. Rubanov, M. Ionescu, and S.X. Dou, Multi-layered deposition and its role in the enhancement of $YBa_2Cu_3O_7$ film performance, *Physica C*, **460–462**, 1379–1380 (2007).
64. H.S. Kim, H.S. Ha, T.H. Kim, J.S. Yang, R.K. Ko, K.J. Song, D.W. Ha, N.J. Lee, S.S. Oh, J. Youm, and Chan Park , The deposition of $Sm_1Ba_2Cu_3O_{7-\delta}$ on $SrTiO_3$ using co-evaporation method, *Physica C*, **460–462**, 1361–1362 (2007).
65. S. Ghalsasi, Y.X. Zhou, J. Chen, B. Lv, and K. Salama, MOD multi-layer YBCO films on single-crystal substrate, *Supercond. Sci. Technol.*, **21**, 045015 (2008).
66. J.W. Lee, G.M. Shin, S.H. Moon, and S.I. Yoo, High-J_c $GdBa_2Cu_3O_{7-\delta}$ films fabricated by the metal-organic deposition process, *Physica C*, **470**, 1253–1256 (2010).
67. J.-M. Triscone, Ø. Fischer, O. Brunner, L. Antognazza, A.D. Kent, and M.G. Karkut, $YBa_2Cu_3O_7/PrBa_2Cu_3O_7$ superlattices: Properties of ultra-thin superconducting layers separated by insulating layers, *Phys. Rev. Lett.*, **64**, 804–807 (1990).
68. T. Terashima, K. Shimura, Y. Bando, Y. Matsuda, A. Fujiyama, and S. Komiyama, Superconductivity of one-unit-cell thick $YBa_2Cu_3O_7$ thin film, *Phys. Rev. Lett.*, **67**, 1362–1365 (1991).
69. G. Logvenov, A. Gozar, and I. Bozovic, High-temperature superconductivity in a single copper-oxygen plane, *Science*, **326**, 699–702 (2009).
70. S. Smadici, J.C.T. Lee, S. Wang, P. Abbamonte, G. Logvenov, A. Gozar, C. Deville Cavellin, and I. Bozovic, Superconducting transition at 38 K in insulating-overdoped La_2CuO_4-$La_{1.64}Sr_{0.36}CuO_4$ superlattices: Evidence for interface electronic redistribution from resonant soft X-ray scattering, *Phys. Rev. Lett.*, **102**, 107004 (2009).
71. A. Gozar, G. Logvenov, L. Fitting Kourkoutis, A.T. Bollinger, L.A. Giannuzzi, D.A. Muller, and I. Bozovic, High-temperature interface superconductivity between metallic and insulating copper oxides, *Nature*, **455**, 782–785 (2008).
72. O. Yuli, I. Asulin, O. Millo, D. Orgad, L. Iomin, and G. Koren, Enhancement of the superconducting transition temperature of $La_{2-x}Sr_xCuO_4$ bilayers: Role of pairing and phase stiffness, *Phys. Rev. Lett.*, **101**, 057005 (2008).
73. J.-P. Locquet, J. Perret, J. Fompeyrine, E. Mächler, J.W. Seo, and G. Van Tendeloo, Doubling the critical temperature of $La_{1.9}Sr_{0.1}CuO_4$ using epitaxial strain, *Nature*, **394**, 453–456 (1998).
74. T. Sekitani, H. Sato, M. Naito, and N. Miura, High-field magnetotransport in strained $(La,Sr)_2CuO_4$ films, *Physica C*, **378–381**, 195–198 (2002).
75. W. Si, H.-C. Li, and X.X. Xi, Strain and oxygenation effects on superconductivity of $La_{1.85}Sr_{0.15}CuO_4$ thin films, *Appl. Phys. Lett.*, **74**, 2839–2841 (1999).
76. J. Yuan, S.J. Zhou, K. Jin, H. Wu, B. Xu, X.G. Qiu, and B.R. Zhao, In-plane stress effect on the electron-doped superconductor $La_{1.89}Ce_{0.11}CuO_4$ thin films, *Physica C*, **468**, 1876–1878 (2008).
77. R. Hühne, D. Okai, K. Dörr, S. Trommler, A. Herklotz, B. Holzapfel, and L. Schultz, Dynamic investigations on the influence of epitaxial strain on the superconducting transition in $YBa_2Cu_3O_{7-x}$, *Supercond. Sci. Technol.*, **21**, 075020 (2008).

17

MgB_2 – An Intermediate-Temperature Superconductor

17.1 Introduction

The discovery by Akimitsu *et al.* [1] in 2001 that MgB_2 is superconducting at a remarkably high temperature of 39 K was a big surprise. Before the discovery of MgB_2, the critical temperature of 23.2 K, found for Nb_3Ge in 1974 [2–4], was the highest known T_c-value in a binary intermetallic compound. The discovery of the superconductivity of this compound was unexpected because it had been known since the 1950s. In an article in "Science" soon after the discovery, Campbell [5] used the title "How could we miss it?", reflecting the surprise of the scientific community.

Immediately after the discovery of superconductivity in MgB_2, the question of whether its grain boundaries are weak links, as in the cuprate high-T_c superconductors, was addressed by Larbalestier *et al.* [6]. They found that the current-carrying capacity of polycrystalline MgB_2 samples is dominated by flux pinning, with $j_c^{0.5}H^{0.25}$ being linear in H over a wide range of temperature (j_c critical current density, H magnetic field), as in Nb_3Sn. Together with the absence of grain-boundary weak links, the results of Larbalestier *et al.* [6] suggest that MgB_2 is more similar to low-T_c metallic superconductors than to high-T_c cuprate superconductors.

The low material cost and the absence of grain-boundary weak links make MgB_2 an interesting material for applications, in spite of the only moderately high critical temperature of 39 K. As a consequence of the absence of grain-boundary weak links, no texture is required to reach high transport critical current densities; hence, the manufacture of MgB_2 conductors is much easier than that of coated conductors with biaxial alignment of the grains in the superconducting RE-123 layer. If manufacturing costs are sufficiently low, MgB_2 may be competitive with NbTi at operating temperatures close to 4 K. In addition, MgB_2 may be interesting at relatively low fields (~2–4 T) and operating temperatures of 20 K.

Physical Properties of High-Temperature Superconductors, First Edition. Rainer Wesche.

17.2 Physical Properties of MgB_2

The first aspect to be considered is the crystal structure of MgB_2. A sketch of the hexagonal crystal structure (space group *P6/mmm*) with the lattice parameters $a = 0.3086$ nm and $c = 0.3524$ nm is presented in Figure 17.1 [1]. The space group symbol indicates a primitive unit cell, with sixfold rotational symmetry and three mirror planes. In the crystal structure of MgB_2, layers of magnesium and boron alternate along the crystallographic c direction. In the boron layer, the atoms are arranged in the same way as the carbon atoms in graphite, i.e., each boron atom is surrounded by three other boron atoms at the same distance. In the magnesium layer, each magnesium atom is in the center of a regular hexagon formed by six other Mg atoms. The layered crystal structure causes anisotropic material properties.

A sketch of the typical X-ray powder diffraction pattern is shown in Figure 17.2. The intensities and line positions are based on the results of measurements using Cu $K\alpha$ radiation published by Akimitsu *et al.* [1].

In conventional metallic superconductors, the pairing of two single electrons to a Cooper pair originates from the electron–phonon interaction (see Chapter 3). According to the BCS theory, the critical temperature T_c is proportional to the Debye frequency ω_D (see Equation 3.8, Chapter 3), the cut-off frequency of the phonon spectrum in a solid. Considering the atoms in a solid as harmonic oscillators, the phonon frequency is expected to be proportional to $M^{-1/2}$ for isotopes of the same element. In fact, this isotope effect on T_c is found in many simple metallic superconductors (see Chapter 3). For this reason, the isotope effect on T_c was studied in isotopically pure $Mg^{10}B_2$ and $Mg^{11}B_2$ by Bud'ko *et al.* [7]. They measured transition temperatures (2% onset criterion) of 39.2 K and 40.2 K for $Mg^{11}B_2$ and

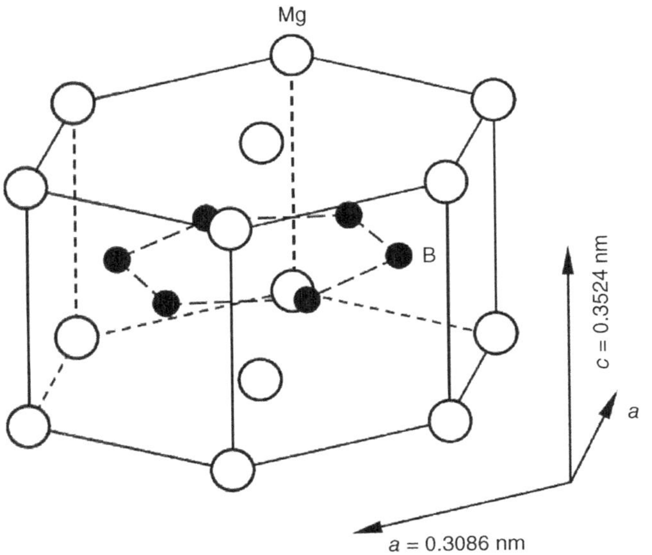

Figure 17.1 Sketch of the hexagonal crystal structure of MgB_2 (space group P6/mmm) with alternating layers of boron and magnesium. The lattice parameters of MgB_2 are $a = 0.3086$ nm and $c = 0.3524$ nm (adapted from [1])

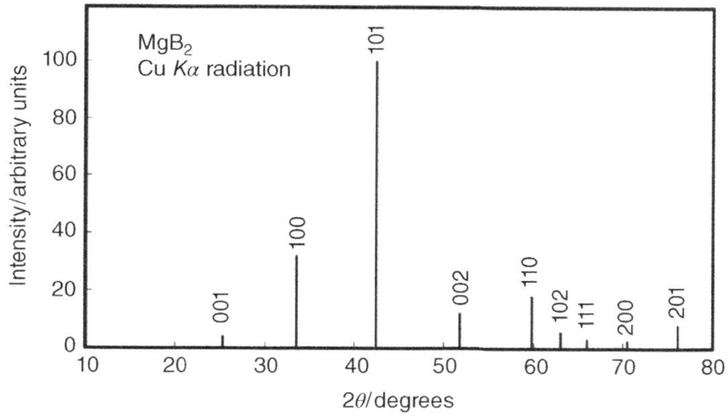

Figure 17.2 *Sketch of the Cu Kα diffraction pattern of MgB_2 (results from [1])*

$Mg^{10}B_2$, respectively. As expected, the critical temperature of the compound containing the lower mass isotope is enhanced. Next, the partial boron isotope-effect exponent, defined as $\beta_B = \Delta \ln T_c / \Delta \ln M_B$ (where M_B is the atomic mass of boron), is considered. From the experimental data, a value of $\beta_B = 0.26 \pm 0.03$ was found [7]. The exponent β_B indicates that boron phonons contribute to the superconductivity of MgB_2. Hinks *et al.* [8] studied the magnesium isotope effect using isotopically pure $^{26}MgB_2$ and $^{24}MgB_2$ samples. The measured difference of the critical temperatures is only 0.1 K, leading to a partial magnesium isotope effect $\beta_{Mg} = 0.02$. This very small effect suggests that the pairing of the electrons is dominated by boron phonons. Thus, MgB_2 seems to be a conventional metallic superconductor. The unusually high critical temperature of MgB_2 seems to be the consequence of the occurrence of high phonon frequencies because of the small atomic mass of boron.

Although the pairing of the electrons is mediated by the electron–phonon interaction, MgB_2 shows more complex physical properties than simple metallic superconductors because of the existence of two superconducting energy gaps. The electronic structure of MgB_2 is similar to that of graphite. MgB_2 may be considered as metallic layers of boron with intercalated layers of Mg^{2+} ions (see Figure 17.1) [9]. The highest occupied electronic states at the Fermi level are mainly σ-(sp_xp_y) or π-bonding boron orbitals (p_z) [9–11]. A unique feature of MgB_2 is that the two σ-bands are incompletely filled. The electrons of the σ-bands are confined in the boron planes. The electrons of the boron p_z orbitals fill two three-dimensional metallic π-bands. As compared to graphite, the energy of the nonbonding boron $\pi(p_z)$ bands is lowered relative to the bonding boron $\sigma(sp_xp_y)$ bands. As a consequence, charge is transferred from the σ- to π-bands, forming hole-doped σ-bands. The results of band calculations indicate that the superconductivity of MgB_2 is dominated by the σ-band holes [9]. The σ-band states and the in-plane vibrations of the boron atoms are strongly coupled, leading to strong pairing with an energy gap of around 7 meV [10]. This strong pairing, confined to the boron planes, occupies only parts of the Fermi surface. In addition, much weaker electron pairs are formed by the π-band electronic states on the remaining parts of the Fermi surface. The energy gap related to the π-band is around ~2 meV [10]. Calculations of Choi *et al.* [12] provide partial isotope-effect exponents $\beta_B = 0.32$ and $\beta_{Mg} = 0.03$, which are in good agreement with the measured values of $\beta_B = 0.26 \pm 0.03$ [7] and $\beta_{Mg} = 0.02$ [8]. The reduction of β_B as compared to the expected value of 0.5 is caused by inharmonic effects among the phonons involved [12]. Since the discovery of superconductivity at 39 K in MgB_2, a large number of calculations have been performed, providing evidence for two superconducting gaps in MgB_2 [10–16]. Experimental evidence for a two-gap scenario was found by angle-resolved photo electron spectroscopy [17, 18], scanning tunneling spectroscopy [19], point-contact Andreev-reflection spectroscopy (PCAR) [20–25], the specific heat jump at the normal-superconductor transition [26–28], and the behavior of B_{c2} versus T close to T_c [29–31].

Figure 17.3 shows the half-widths Δ of the two gaps in MgB_2, as measured by tunneling spectroscopy [19] and PCAR [22–25]. A description of PCAR can be found in a review article by Daghero and Gonnelli [21]. The solid symbols show the values of the half-width Δ of the energy gap related to the σ-band. The measured data indicate that $\Delta_\sigma(T = 0)$ is between 6.5 and 8 meV, which leads to $2\Delta_\sigma(T = 0)/k_BT_c \approx 4.2$ for $\Delta \approx 7$ meV and $T_c =$ 39 K. This value is slightly higher than the value of $2\Delta(0)/k_BT_c = 3.5$ predicted by the BCS theory and frequently observed in elemental metallic superconductors (see Figure 3.4 in Chapter 3). The crosses and the open symbols represent the measured values of the half-width Δ of the energy gap related to the π-band. From the data, a value between 1.9 and 2.8 meV can be estimated for $\Delta_\pi(0)$. For the weakly bonded pairs originating from

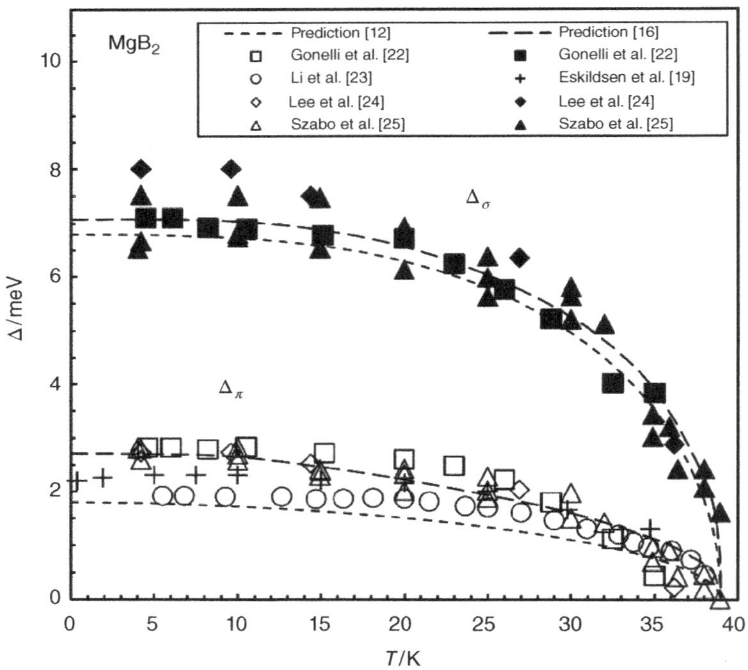

Figure 17.3 Comparison of the predicted temperature dependences [12, 16] of the half-widths of the σ- and π-band gaps $\Delta_\sigma(T)$ and $\Delta_\pi(T)$ with the measured data [19, 22–25]

the π-bands, the ratio of $2\Delta_\pi(0)/k_BT_c$ is between 1.1 and 1.7. For comparison, theoretical predictions of the temperature dependence of the half-width of the energy gap Δ published by Choi *et al.* [12] and by Brinkman *et al.* [16] are also shown in Figure 17.3. The measured data agree more or less with the theoretical predictions. The values of the half-width of the gap calculated by Choi *et al.* [12] can be approximated by the expression

$$\Delta(T) = \Delta(0)\left(1 - \left(\frac{T}{T_c}\right)^p\right)^{1/2} \tag{17.1}$$

The fit parameters for the σ gap are $p = 2.9$ and $\Delta(0) = 6.8\,\text{meV}$, and $p = 1.8$ and $\Delta(0) = 1.8\,\text{meV}$ for the π gap. An interesting aspect is the fact that despite the significantly different energy scales for the two gaps, they close at the same temperature. The reason for this behavior is the coupling between the σ- and π-bands, which leads to a single T_c, and to values of the ratio $2\Delta(0)/k_BT_c$ which are less than and greater than the BCS value of 3.5, for the π- and σ-bands, respectively [21].

The existence of two gaps in MgB_2 causes a non-BCS shape of the electronic specific heat. In Figure 17.4, calculations of the electronic specific heat C_{se} of MgB_2 in the superconducting state, for single- and two-gap scenarios (results from [10]), are compared. The data are normalized to the electronic specific heat $C_{ne} = \gamma T$ in the normal state with $\gamma = 2.6\,\text{mJ mole}^{-1}\,\text{K}^{-2}$ [10]. The temperature is normalized to the critical temperature of 39.4 K used in the calculations. Due to the existence of the smaller π-band gap, the specific heat below 0.5 T_c is considerably enhanced in the two-gap scenario. On the other hand, the jump in the electronic specific heat at T_c is smaller than that expected for a single-gap BCS

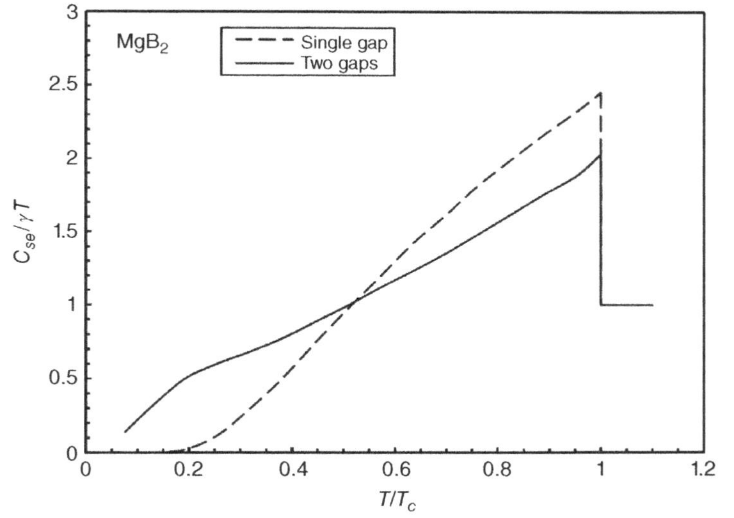

Figure 17.4 *Comparison of the calculated electronic specific heats C_{se} of MgB_2 normalized to the normal state electronic specific heat $C_{ne} = \gamma T$ ($\gamma = 2.62\,mJ\,mole^{-1}\,K^{-2}$) for single- and two-gap scenarios. The existence of the second smaller π gap leads to a considerably enhanced specific heat below 0.5 T_c, whereas the jump in the specific heat at T_c is reduced as compared to a single-gap scenario (results from [10])*

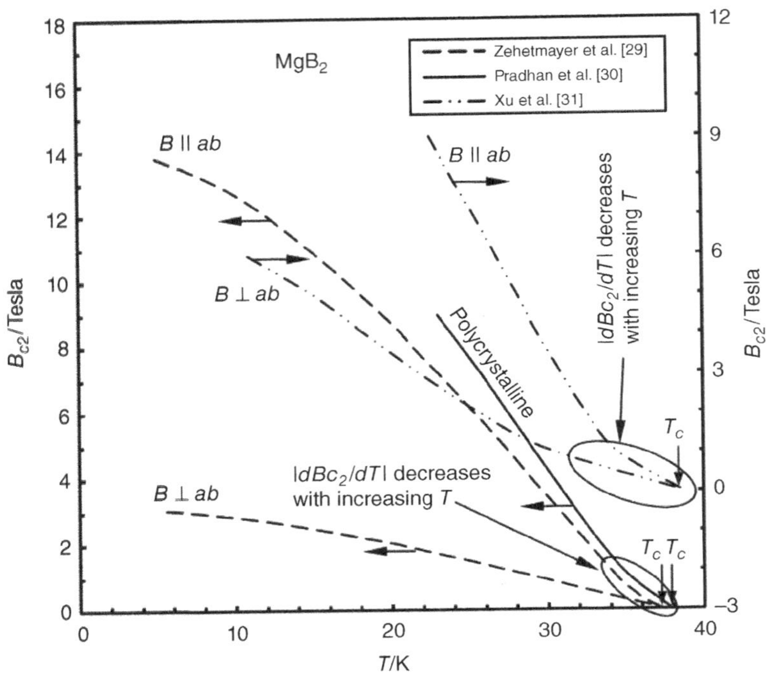

Figure 17.5 *Upper critical field of MgB_2 vs. temperature for dense polycrystalline bulk material and two single crystals. The upper critical field at $T = 0$ is for $B \parallel ab$ a factor 2.6 [31] to 4.5 [29] higher than that for $B \perp ab$. An interesting feature is the decrease of the absolute value of dB_{c2}/dT with increasing temperature close to T_c, which has been attributed to the presence of two gaps in MgB_2 (results from [29–31])*

superconductor. Measured specific heat data show these two features, which are characteristic of a two-gap superconductor.

Figure 17.5 shows the upper critical field of dense polycrystalline bulk material and two MgB_2 single crystals as a function of temperature [29–31]. As expected, the upper critical field depends on the direction of the applied magnetic field with respect to the boron planes (*ab* planes). Zehetmayer *et al.* [29] estimated that the upper critical field of MgB_2 at $T = 0$ is 14.5 T for $B||ab$, but only 3.2 T for $B||c$. The resulting ratio of $B_{c2,ab}$ to $B_{c2,c}$ is ~4.5.

For a single-gap superconductor, the temperature dependence of the thermodynamic critical field B_c is typically well described by the expression $B_c(T) = B_c(0)(1 - (T/T_c)^2)$ (Equation 2.25 Chapter 2), which leads to $dB_c(T)/dT = -2T(B_c(0)/T_c^{\,2})$. The absolute value of the temperature derivative of the critical field increases with temperature, and reaches its highest value at T_c. The upper critical field and the thermodynamic critical field are related by the expression $B_{c2} = \kappa B_c \sqrt{2}$ (Equation 4.20 Chapter 4), where κ is the Ginzburg–Landau parameter. In contrast to expectation, the absolute value of the temperature derivative of B_{c2} of MgB_2 decreases close to T_c as illustrated in Figure 17.5. This special feature of the B_{c2} versus T curves is a further evidence for the existence of two gaps in MgB_2.

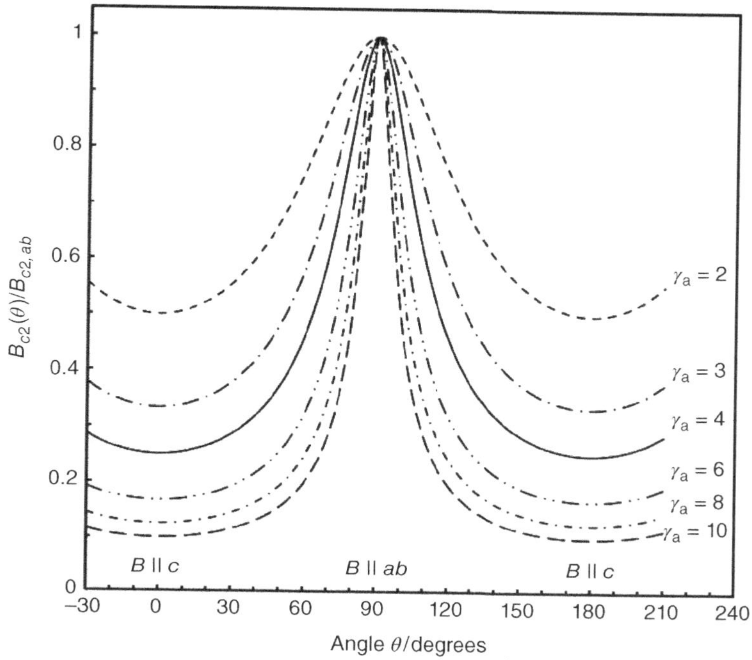

Figure 17.6 Dependence of the upper critical field on the angle between the direction of the applied magnetic field and the crystallographic c-axis calculated with Equation 17.2 for γ_a values between 2 and 10

Next, the angular dependence of the upper critical field of MgB_2 will be considered. Using the ratio $\gamma_a = B_{c2,ab}/B_{c2,c}$, the angle dependence of B_{c2} can be well described by the formula [32]

$$B_{c2}(\theta) = \frac{B_{c2,ab}}{\sqrt{\gamma_a{}^2\cos^2(\theta) + \sin^2(\theta)}} \tag{17.2}$$

Here, θ is the angle between the direction of the applied magnetic field and the crystallographic c-axis. Figure 17.6 shows the angular dependence of B_{c2} calculated with Equation 17.2 for values of γ_a in the range of 2–10. The calculated values of B_{c2} are normalized to $B_{c2,ab}$. Values of $B_{c2,ab}$, $B_{c2,c}$, and the anisotropy factor γ_a reported in the literature are listed in Table 17.1. The anisotropy factor decreases with increasing temperature, and hence is lowest close to the critical temperature. At temperatures below 15 K, the values of the anisotropy factor are in the range of 1.4–6. Values of 2 or less were reported for MgB_2 films. In pure MgB_2, the upper critical field is between 15 and 20 T. Much higher values of B_{c2} were reported for carbon-doped material and MgB_2 films (see Figure 17.7).

As in the case of the cuprate high-T_c superconductors, the critical current density of MgB_2 drops to zero at the irreversibility field B_{irr}, well below the upper critical field B_{c2}.

Table 17.1 Measured upper critical fields $B_{c2,ab}$, $B_{c2,c}$, and anisotropy factor γ_a of MgB_2

$B_{c2,ab}$ (T)	$B_{c2,c}$ (T)	$\gamma_a = B_{c2,ab}/B_{c2,c}$	Comments	Reference
14.5	3.18	4.6	Estimation for $T = 0$	[29]
~15	~4.8	~3.2	Estimation for $T = 0$	[30]
19.8	7.7	2.6	Estimation for $T = 0$	[31]
~17	~3.5	5 (low T), 2 close to T_c	Estimated B_{c2} values at $T = 0$	[33]
~23	–	6 at 15 K, 2.8 at 35 K	Estimated B_{c2} value at $T = 0$	[34]
25.5	9.2	2.8	Linear extrapolation to $T = 0$	[35]
26.4	14.6	1.8	MgB_2 film estimation for $T = 0$	[36]
19.6	13.8	1.4	MgB_2 film estimation for $T = 0$	[36]

At the irreversibility field the flux line pinning vanishes, and hence the critical current density drops to zero. Figure 17.7 shows the irreversibility fields and the upper critical fields of various MgB_2 conductors. The measured irreversibility field depends on the criterion used to define a vanishing critical current density, i.e., a j_c small enough to be considered as zero. For magnetic fields parallel to the boron (*ab*) planes, the irreversibility fields and the upper critical fields of MgB_2 single crystals [30, 37, 38] are presented in Figure 17.7 (a). The irreversibility field reaches ~70–85% of the upper critical field at 25 K. The irreversibility fields and the upper critical fields for magnetic fields perpendicular to the boron planes are presented in Figure 17.7 (b). The ratio B_{irr}/B_{c2} ranges from 55% to 80% at 15 K [30, 37, 38]. Figure 17.7 (c) shows the irreversibility and upper critical field data of MgB_2 tapes [39, 40], wires [41], and bulk material [42]. A comparison of pure and 5% carbon-doped MgB_2 tapes indicates that the carbon doping enhances both the upper critical and the irreversibility field, whereas the critical temperature is reduced. Additions of C, SiC, and multiwalled carbon nanotubes are found to increase the upper critical field as well as the irreversibility field (see Figure 17.7).

Impurities decrease the mean free path L_{el} of the electrons, which causes an enhanced normal-state resistivity. The relations between the upper critical field, the normal-state resistivity, and the mean free path of the electrons in MgB_2 have been discussed in a review article by Eisterer [32]. The upper critical field and the mean free path L_{el} of the electrons are connected by Equation 17.3 [32]

$$B_{c2} = B_{c2}{}^{clean}\left(1 + \frac{\xi_0}{L_{el}}\right) \tag{17.3}$$

where $B_{c2}{}^{clean}$ is the upper critical field in the clean limit and ξ_0 is the BCS value of the coherence length:

$$\xi_0 = \frac{\hbar v_F}{\pi \Delta} \tag{17.4}$$

Here, v_F is the Fermi velocity of the electrons and Δ is the superconducting energy gap. Because the superconductivity in the π-band is strongly suppressed at high magnetic fields, the upper critical field of MgB_2 is mainly determined by the properties of the σ-band [32]. Equation 17.3 indicates that the shortening of the mean free path caused by impurities enhances the upper critical field. For a quantitative description of the upper critical field,

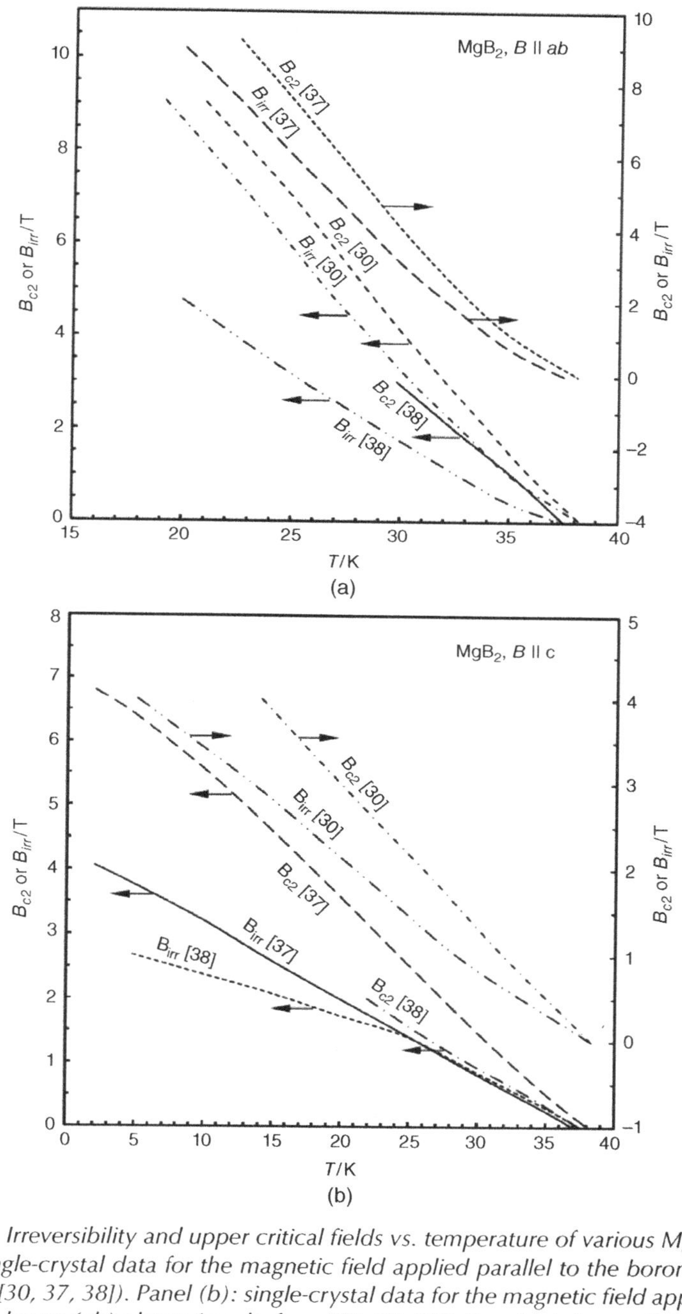

Figure 17.7 *Irreversibility and upper critical fields vs. temperature of various MgB_2 samples. Panel (a): single-crystal data for the magnetic field applied parallel to the boron (ab) planes (results from [30, 37, 38]). Panel (b): single-crystal data for the magnetic field applied perpendicular to the boron (ab) planes (results from [30, 37, 38]). Panel (c): irreversibility and upper critical fields of MgB_2 wires, tapes, and bulk material (results from [39–42])*

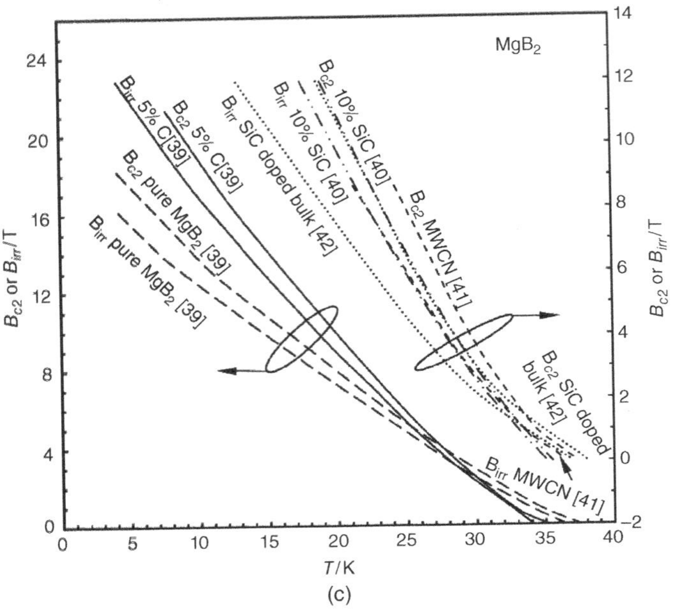

Figure 17.7 *(continued)*

the decrease of the critical temperature of MgB_2 with increasing disorder needs to be taken into account, leading to

$$B_{c2} = B_{c2}^{\text{clean}} \left[\left(\frac{T_c}{T_c^{\text{clean}}} \right)^2 + \frac{\xi_0}{L_{\text{el}}} \frac{T_c}{T_c^{\text{clean}}} \right] \tag{17.5}$$

with $T_c^{\text{clean}} = 39.43\,\text{K}$ and $B_{c2,ab} = 13.8\,\text{T}$ in the clean limit [32]. The B_{c2} value of 14.5 T, measured by Zehetmayer *et al.* [29], is close to the estimated value for the clean limit.

The main physical properties of MgB_2 are listed in Table 17.2. MgB_2 is a Type II superconductor with a Ginzburg–Landau parameter of the order of 30. The coherence length is of the order of ~2.5 nm along the *c*-direction and around 7 nm in the *ab* planes. The penetration depth λ_c (screening current along the *c*-direction) reaches a value of 370 nm, whereas λ_{ab} (screening current flow within the *ab* planes) is only around 80 nm. In polycrystalline samples, penetration depths between 110 and 180 nm have been reported. The upper critical field of "dirty" MgB_2 films can reach values as high as 60 T. In general, the upper critical fields in MgB_2 wires, tapes, and bulk material are much lower than in "dirty" MgB_2 films. In any case, the practical use of MgB_2 is limited by the irreversibility field B_{irr}, which can exceed 20 T at 4.2 K in MgB_2 tapes (see Table 17.2), but is still not high enough to allow Nb_3Sn high-field superconductors to be replaced. On the other hand, the extremely high upper critical fields of "dirty" films suggest that there is still some potential for further improvements of MgB_2 wires and tapes.

Table 17.2 Physical properties of MgB_2

Parameter	Value	Comments	Reference
Critical temperature			
T_c (K)	39	MgB_2, natural mixture of ^{10}B and ^{11}B isotopes	[1]
T_c (K)	40.2	$Mg^{10}B_2$, isotope pure sample	[7]
T_c (K)	39.2	$Mg^{11}B_2$, isotope pure sample	[7]
Crystal structure and lattice parameters			
a (nm)	0.3086	Hexagonal crystal structure	[1]
c (nm)	0.3524	Space group P6/mmm	[1]
Lower and upper critical fields, irreversibility field			
$B_{c1,ab}$(mT)	22	Single crystal $T = 0$, $B\|\|ab$	[29]
$B_{c1,c}$(mT)	63	Single crystal $T = 0$, $B\|\|c$	[29]
B_{c1}(mT)	18	Polycrystalline $T = 0$	[43]
B_{c1}(mT)	24	Polycrystalline $T = 0$	[38]
$B_{c2,ab}$(T)	14.5	Single crystal $T = 0$, $B\|\|ab$	[29]
$B_{c2,c}$(T)	3.18	Single crystal $T = 0$, $B\|\|c$	[29]
B_{c2}(T)	14	Polycrystalline $T = 0$	[43]
B_{c2}(T)	16.4	MgB_2 wires, $T = 0$	[44]
B_{c2}(T)	43	Magnesium-rich MgB_2 thin film	[45]
B_{c2}(T)	~60	MgB_2 thin film, C-doped, $T = 0$	[46]
$B_{c2,ab}$(T)	38.6	MgB_2 thin film, $T = 0$, $T_c = 30.4$ K	[47]
$B_{c2,c}$(T)	19.5	MgB_2 thin film, $T = 0$, $T_c = 30.4$ K	[47]
B_{irr}(T)	22.9	Fe-sheathed MgB_2 tapes, C-doped, $T = 4.2$ K	[39]
B_{irr}(T)	14.5	MgB_2 tapes with CuNi/Nb sheath, $T = 4.2$ K	[48]
B_{irr}(T)	25.5	MgB_2/Fe tape, SiC-doped, $T = 4.2$ K	[49]
Coherence length and penetration depth			
ξ_{ab}(nm)	10.2	Single crystal, $T = 0$	[29]
ξ_c(nm)	2.3	Single crystal, $T = 0$	[29]
ξ_{ab}(nm)	8.2	Single crystal, $T = 0$	[30]
ξ_c(nm)	2.7	Single crystal, $T = 0$	[30]
ξ_{ab}(nm)	6.5	Single crystal, $T = 0$	[31]
ξ_c(nm)	2.5	Single crystal, $T = 0$	[31]
ξ_{ab}(nm)	6.8	Single crystal, $T = 0$	[37]
ξ_c(nm)	2.3	Single crystal, $T = 0$	[37]
ξ(nm)	4.9	Polycrystalline, $T = 0$	[43]
ξ_{ab}(nm)	7.0	Aligned MgB_2 crystals, $T = 0$	[50]
ξ_c(nm)	4.0	Aligned MgB_2 crystals, $T = 0$	[50]
ξ(nm)	5.38–6.06	MgB_2 bulk samples, $T = 20$ K	[51]
λ_{ab}(nm)	82	Single crystal, $T = 0$	[29]
λ_c(nm)	370	Single crystal, $T = 0$	[29]
λ(nm)	180	Polycrystalline, $T = 0$	[43]
λ(nm)	110–140	$T = 0$	[52]
Ginzburg–Landau parameter $\kappa = \lambda/\xi, \kappa_{ab} = (\lambda_{ab}\lambda_c/\xi_{ab}\xi_c)^{0.5}$, $\kappa_c = \lambda_{ab}/\xi_{ab}$			
κ_{ab}	37.1	$T = 0$	[29]
κ_c	8.1	$T = 0$	[29]
κ	38	$T = 0$	[43]
κ	~25	$T = 0$	[53]

The resistivity versus temperature curves of various types of MgB_2 conductors are presented in Figure 17.8 [31, 54–62]. The normal-state resistivity of MgB_2 is typically characterized by the residual resistivity ρ_0 at a temperature of 42 K just above T_c, and the residual resistivity ratio (RRR) is defined as $\rho(300\,K)/\rho_0$. As in other metals, the resistivity increases with increasing amount of impurities and defects present in the material. The top panel of Figure 17.8 shows the resistivity data of a MgB_2 single crystal [31], two polycrystalline samples [55], and two MgB_2 films [54, 56]. The residual resistivity of the MgB_2 single crystal is ρ_0~2.5 μΩcm, while the resistivity at 270 K increases to around 16 μΩcm. Extrapolation of the data to 300 K leads to a value of $\rho(300\,K)$ ~18 μΩcm leading to a residual resistivity ratio of ~7.2. The resistivity of the two polycrystalline samples [55] is to a first approximation a factor of 2 different in the considered temperature range of 40–300 K. The difference of the resistivity of the two samples originates from handling the MgB_2 powders under different atmospheric conditions. Milling of the MgB_2 powder in dry nitrogen reduces the resistivity as compared to the sample made of powders milled in moist air. The sensitivity of MgB_2 powders to handling in different atmospheres can explain different results for polycrystalline samples, which are otherwise identical. The values of the resistivity of the two MgB_2 films presented in the top panel of Figure 17.8 are relatively high with ρ_0 values of 16 μΩcm [56] and 30 μΩcm [54], and residual resistivity ratios of ~1.8 and ~1.5, respectively. Both the high ρ_0 values and the small residual resistivity ratios suggest that many defects and perhaps some oxygen impurities are present in these films. The data [57–62], presented in the bottom panel of Figure 17.8, indicate that the residual resistivity of undoped MgB_2 films can reach values as low as 0.29 μΩcm [57]. In general, the residual resistivity ratio was found to be higher for MgB_2 films with a lower residual resistivity. For example, the RRR reached a value of ~15 for a 150-nm-thick MgB_2 film with ρ_0 ~0.58 μΩcm [60], while the RRR value is as small as 2.3 for a 400-nm-thick MgB_2 film on Al_2O_3, characterized by ρ_0 ~6.2 μΩcm [62].

In order to increase the upper critical field, MgB_2 is frequently doped with carbon. Zhuang *et al.* [60] investigated the effect of carbon doping on the resistivity of MgB_2 films. The residual resistivity ρ_0 of the MgB_2 films at 42 K increases from ~0.58 μΩcm for an undoped film to more than 10 000 μΩcm for a carbon content of 37 at%. The enhancement of the residual resistivity is accompanied by a reduction of the critical temperature. Carbon doping of $\geq$ 39 at% leads to nonsuperconducting MgB_2 films. The RRR decreases from a value around 15 for the undoped film to values close to 1 for carbon contents above 15 at%.

Specific heat data for the zero applied field and temperatures between 10 and 300 K are presented in Figure 17.9. The specific heat reaches a value of around 48 J mole^{-1} K^{-1} at a temperature of 300 K [43, 63], which is only 2/3 of the Dulong–Petit equipartition value of $3 \times 3R = 9R$ (74.83 J mole^{-1} K^{-1}). The value of 48 J mole^{-1} K^{-1} at 300 K suggests that the effective Debye temperature reaches an extremely high value of 920 K [43]. The factor of 3 takes into account that one mole of MgB_2 is equivalent to three gram atoms. Collings *et al.* [63] reported Debye temperatures between 600 and 745 K for doped and undoped MgB_2 samples. For the purpose of better readability, the low temperature specific heat data of Rao *et al.* [64] are presented in Figure 17.9 using secondary temperature and specific heat axes (see for a comparison the specific heat data of cuprate high-T_c superconductors in Chapter 13).

The thermal conductivity data of a MgB_2 single crystal [65], two bulk samples [66], and one polycrystalline sample [67] are compared in Figure 17.10. The thermal conductivity

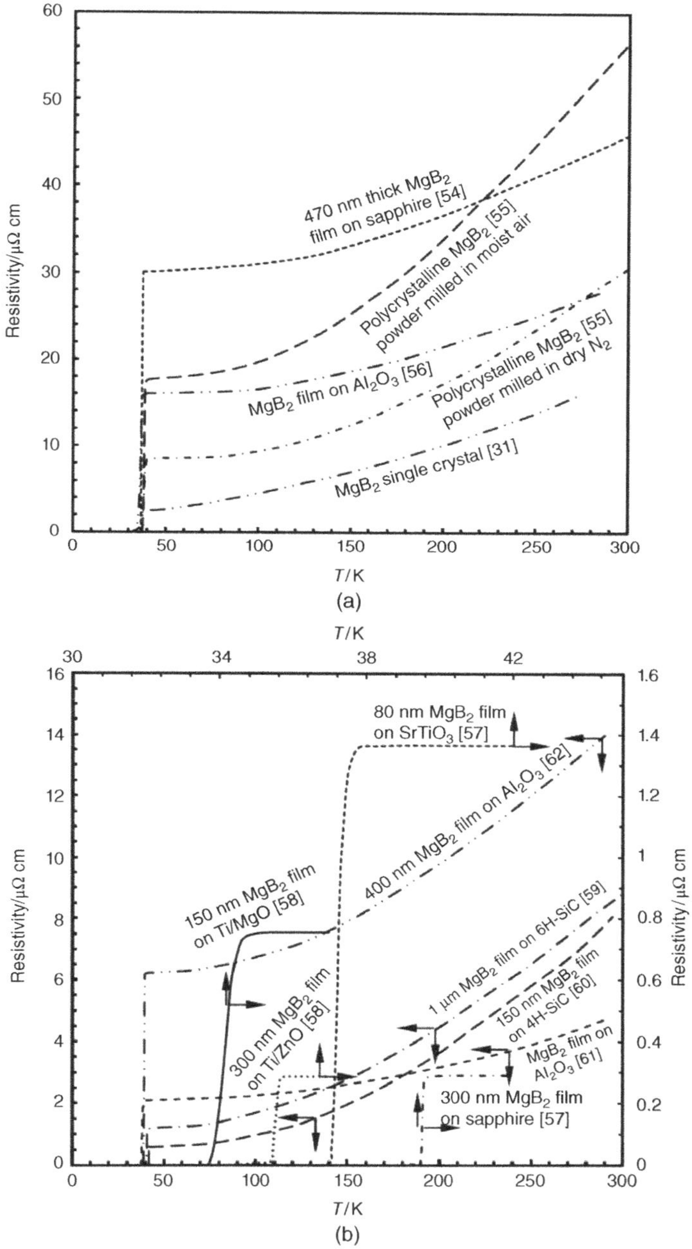

Figure 17.8 *Resistivity vs. temperature of various MgB_2 conductors. Top panel: MgB_2 single crystal [31], two polycrystalline samples [55], and two MgB_2 films with relatively large residual resistivity [54, 56]. Bottom panel: resistivity of various MgB_2 films. The details are described in the text. The arrows indicate which axes are to be used (results from [57–62])*

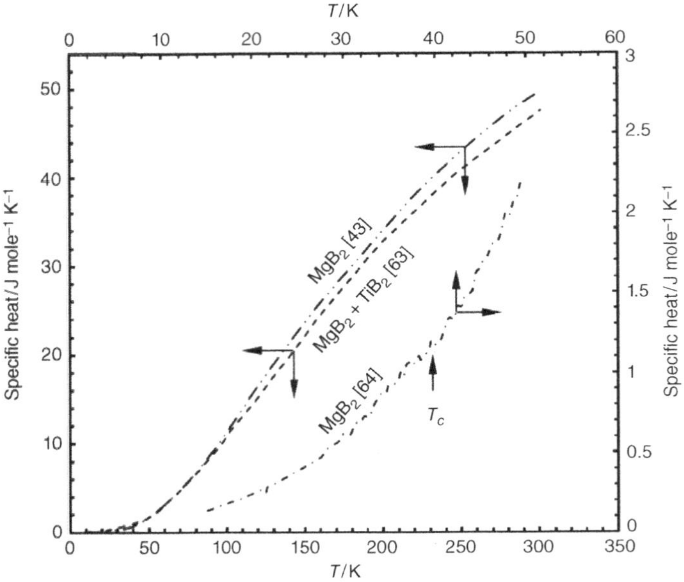

Figure 17.9 Specific heat of MgB_2 for temperatures between 10 and 300 K (results from [43, 63, 64])

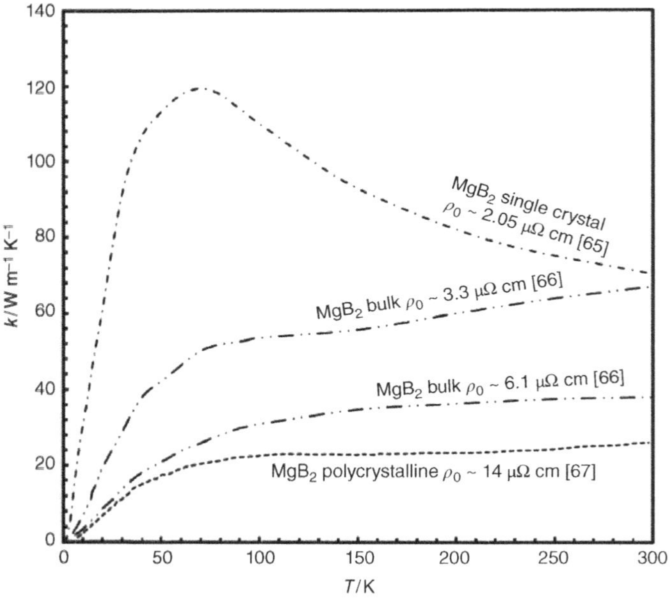

Figure 17.10 Thermal conductivity of MgB_2 as a function of temperature. The thermal conductivity is closely related to the electrical resistivity of the sample in question. Samples with lower electrical resistivity show higher thermal conductivity (results from [65–67])

increases with decreasing residual resistivity. For the single crystal with $\rho_0 = 2.05\,\mu\Omega\,cm$, the thermal conductivity reaches a maximum value of $120\,W\,m^{-1}\,K^{-1}$ at a temperature of 70 K [65]. On the other hand, the thermal conductivity of the polycrystalline sample with $\rho_0 = 14\,\mu\Omega\,cm$ shows no pronounced maximum at low temperatures, and is less than $26\,W\,m^{-1}\,K^{-1}$ over the whole temperature range considered. In general, the thermal conductivity data show no pronounced anomaly at the critical temperature. The electronic and the phonon contributions to the thermal conductivity are comparable in the normal conducting state, and below $0.4T_c$ the lattice contribution dominates [67].

17.3 MgB_2 Wires and Tapes

MgB_2 wires and tapes are typically manufactured by the powder-in-tube (PIT) method. In the *ex situ* route, reacted MgB_2 powder is used for wire or tape manufacture, whereas the *in situ* route starts from a mixture of unreacted magnesium and boron powders. Figure 17.11 (a) shows the manufacture of mono-core and multifilament MgB_2 wires by the *in situ* PIT process as described by Jiang *et al.* [68] and Huŝek and Kováĉ [69]. In the manufacture of the mono-core wire [68], a stoichiometric mixture of MgH_2 and amorphous boron powders was filled into pure iron tubes of 6.2 mm outer and 3.35 mm inner diameter. To improve the pinning properties of the MgB_2 wires, 5 or 10 mol% nano-SiC powder with a particle size of around 30 nm was added to the powder mixture. The composite was groove-rolled into a rectangular wire of 2 mm × 2 mm cross section. Next, the MgB_2/Fe composites were drawn to a round wire of 1.45 mm diameter, which was preannealed at a temperature of 600 °C for 15 min under flowing argon. The preannealed MgB_2/Fe wire was drawn to a final diameter of 1.1 mm and heat treated at 600 °C for 1 h under flowing argon. The heat treatment is required to form the desired MgB_2 compound from the initial powder mixture. The formation of MgB_2 is accompanied by shrinkage of the wire core originating from the fact that the X-ray density of MgB_2 of $2.62\,g/cm^3$ is larger than the densities of MgH_2, Mg, and B of 1.45, 2.34, and $1.74\,g/cm^3$, respectively. Huŝek and Kováĉ [69] started the manufacture of MgB_2/Ti/Cu/stainless steel multifilament wires from a powder mixture of Mg, B, and 5 wt% SiC. In the first step, a Ti/Cu composite tube was filled with the powder mixture. The inner Ti tube acts as a diffusion barrier to avoid a reaction of MgB_2 and copper. The copper is required for electrical stabilization of the MgB_2 multifilament superconductor. The composite was drawn to form mono-core wires of 1.15 mm diameter, 19 of which were inserted into a stainless steel tube of 6 mm inner and 8 mm outer diameter, which was cold drawn to a final diameter of 0.86 mm. The desired MgB_2 compound was formed during heat treatment at 800 °C for 0.5 h in pure argon atmosphere.

Figure 17.11 (b) shows the manufacture of MgB_2 single- and multifilament wires by the *ex situ* route as described by Suo *et al.* [70] and Malagoli *et al.* [71]. In contrast to the *in situ* route, the *ex situ* route starts with reacted MgB_2 powders. Suo *et al.* [70] used iron tubes of 5 mm inner and 8 mm outer diameter for the manufacture of MgB_2 mono-core wires. The MgB_2 powder was filled into the iron tubes and the composite was drawn to a wire of 2 mm diameter, which was deformed to a tape of 0.375 mm thickness by cold rolling. This MgB_2/Fe tape was heat treated at 950 °C for half an hour in argon. The heat treatment improves the connectivity of the MgB_2 grains in the tape core necessary to achieve high critical current densities. Malagoli *et al.* [71] filled the reacted MgB_2 powder into a niobium tube of 13 mm diameter and a wall thickness of 1 mm, which was inserted into a cupronickel (Cu70Ni30) alloy tube of 13 mm inner and 16 mm outer diameter. The niobium barrier avoids reactions of the MgB_2 powder with the cupronickel alloy tube. The composite was drawn to a wire of 3.5 mm diameter. Twelve of these wires were stacked, together with a high-conductivity copper core, into a Monel (Cu30Ni70 alloy) tube. To avoid poisoning of the pure copper by nickel, the copper core is surrounded by a thin niobium barrier. The Monel is harder than cupronickel, allowing a more uniform deformation of the inner filaments [71]. The composite was groove-rolled and drawn to a round wire

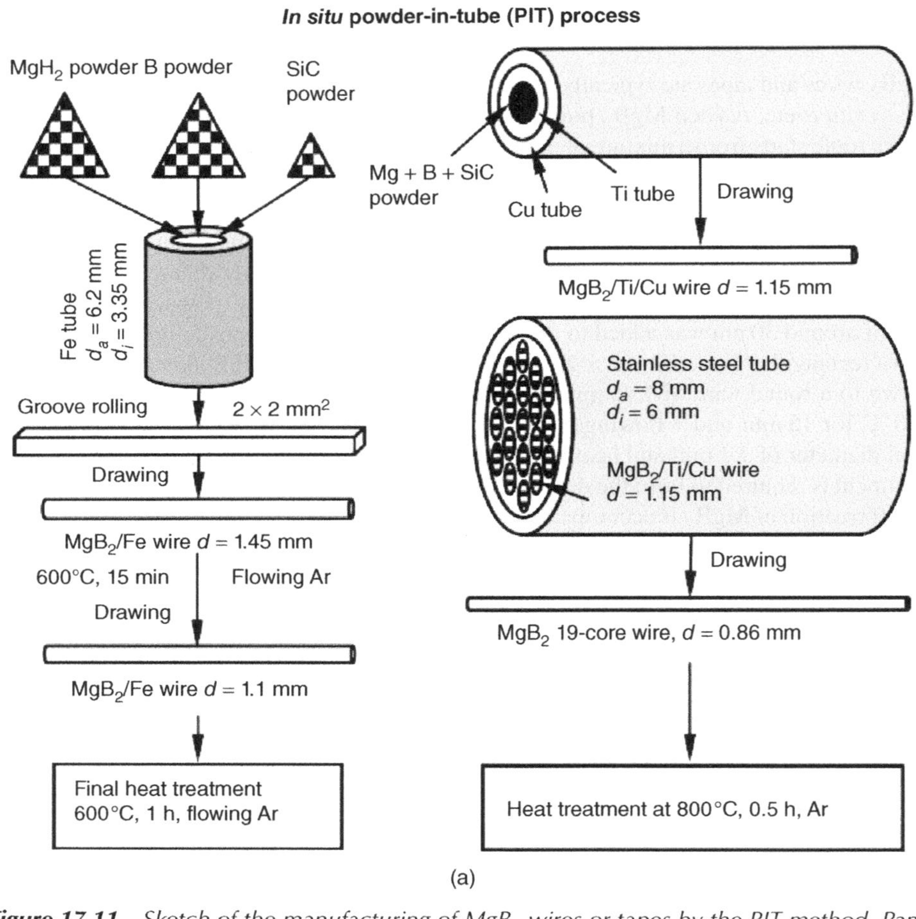

Figure 17.11 *Sketch of the manufacturing of MgB_2 wires or tapes by the PIT method. Panel (a) shows the manufacture of mono-core and multifilament MgB_2 conductors by the in situ route starting from magnesium and boron powders. For carbon doping, SiC powder may be added [68, 69]. Panel (b) shows the manufacture of mono-core and multifilament MgB_2 conductors by the ex situ route starting from already reacted MgB_2 powder [70, 71]. The details are described in the text*

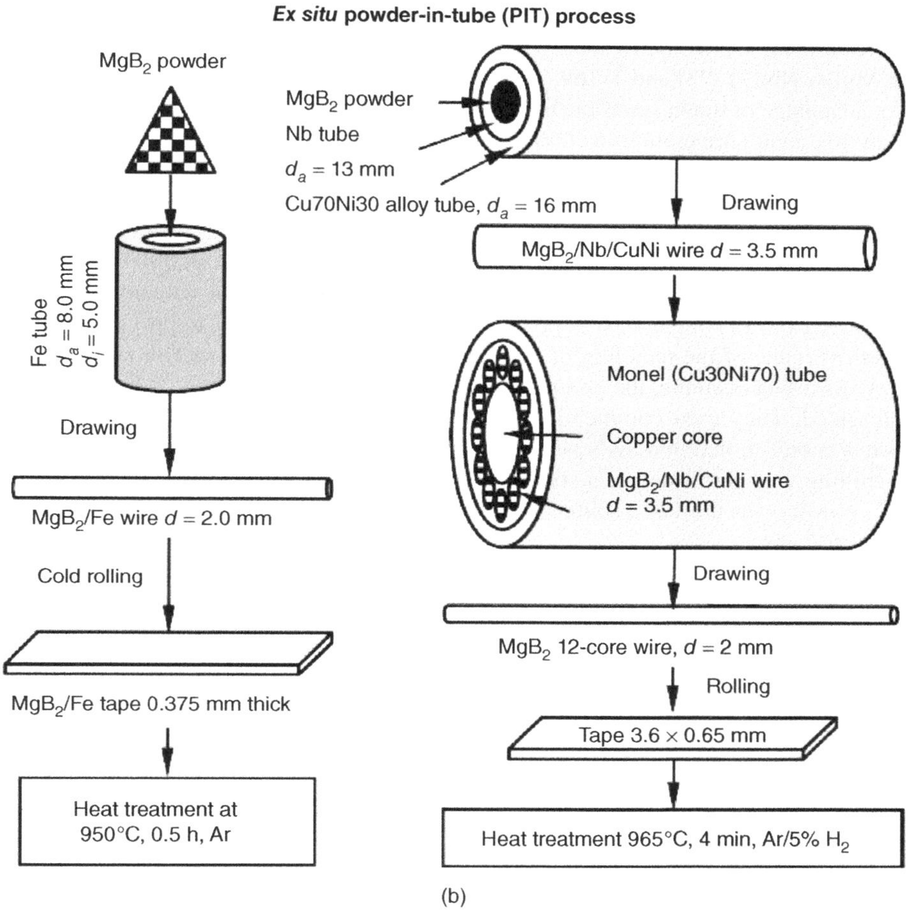

Figure 17.11 (continued)

of 2 mm diameter. The multifilament wire was flat-rolled to a tape of 3.6 mm width and 0.65 mm thickness. The tapes were heat treated at 965 °C for 4 min in a flow of a mixture of Ar and 5% H_2.

An important aspect of the manufacture of MgB_2 wires by the PIT method is the selection of a sheath material that does not react with magnesium at the reaction and sintering temperatures. In the *ex situ* route, annealing at more than 900 °C is necessary for the formation of well-connected MgB_2 grains. Magnesium is volatile and forms solid solutions or intermetallic compounds, with lowered melting points, with many metals [72]. Metals that do not react with magnesium are Fe and the refractory metals Mo, Nb, V, Ta, Hf, and W [72]. The manufacture of thin wires by the PIT method requires a ductile sheath material. Iron is a favorite sheath or barrier material because it does not react with magnesium and is much more ductile than the refractory metals. Superconducting wires for magnet applications need to be electrically stabilized by a highly conductive metal in parallel to the superconducting filaments. To avoid poisoning of the pure copper stabilizer due to reactions with MgB_2, Nb [71, 73] and Ti [69, 74] barriers have been successfully used.

An advantage of the *in situ* route is the fact that good grain connectivity can be achieved at heat treatment temperatures as low as 650 °C. As a consequence of these lower temperatures, the reaction layers formed at the interface with less inert sheath materials may thin enough to allow sufficient current to be carried. Fe, Nb, Ta, Ti, Ni, Cu, Ag, stainless steel, cupronickel, Monel, or combinations of different materials have been successfully used for the manufacture of MgB_2 wires and tapes by the PIT method (see, e.g., [70–75]).

Another important aspect is the handling of the powders used for wire and tape manufacturing by the PIT method. It is beyond the scope of this chapter to provide more than a brief overview of the handling of the precursor powders for the *ex situ* and the *in situ* routes. As a first example, the *ex situ* route powder handling described by Fujii *et al.* [76] is presented. They used commercially available MgB_2 powder supplied by Alfa Aesar, which was ball-milled in a Si_3N_4 jar with planetary balls made of the same material. The ball-milling was performed at a speed of 300 rpm for 5–100 h. Mineral oil added to the MgB_2 powder was used as a source of carbon doping. The carbon doping can be summarized as follows:

$$MgB_2 + xC \rightarrow MgB_{2-x}C_x + xB \tag{17.6}$$

The formula indicates that boron is partly substituted by carbon, leading to a significant enhancement of the upper critical field as well as the irreversibility field. The excess boron released from the MgB_2 powder can act as a barrier for the supercurrent flow [76]. In order to counteract this effect, magnesium powder was added in the ratio MgB_2:Mg of 100:5 or 100:10. The added mineral oil prevents the oxidation of the MgB_2 powder during ball milling in air. Before being filled into iron tubes, the powder was dried at 200 °C for 2 h in vacuum. The use of silicon oil was found to be a less efficient carbon source [77]. Very recently chitosan, a linear polysaccharide, has been used to prepare C-doped boron powder [78]. The amorphous C-doped boron powder was used as starting material for the synthesis of $MgB_{2-x}C_x$ powder for the *ex situ* manufacture of MgB_2 wires or tapes.

For the preparation of the starting powders used for the *in situ* manufacture of MgB_2 wires or tapes, the methods of Yamada *et al.* [79], Lee *et al.* [80], and Kario *et al.* [81] are presented here. As starting materials, Yamada *et al.* [79] used commercial MgH_2 and amorphous boron powders. A partial substitution of boron by carbon improves the upper critical field and the critical current density of MgB_2 wires and tapes in high fields. They, therefore, added 10 mol% ethyltoluene and/or 10 mol% SiC to the powder mixture as carbon sources. The powders were mixed for 1 h by ball milling with tungsten carbide in high-purity argon. Lee *et al.* [80] used lauric acid as a carbon source. The boron powder was dissolved in methyl alcohol and mixed with up to 5 wt% lauric acid. The slurry was dried at 100 °C in flowing argon. Before filling into the iron tubes, the boron and magnesium powders were mixed together. Kario *et al.* [81] used a combination of *in situ* and the *ex situ* processes for the manufacture of Glidcop® (dispersion-strengthened copper) sheathed MgB_2 wires. Magnesium and boron powders were mechanically alloyed in a planetary ball-mill for 5 h in argon. The *in situ* precursor powder was pressed into a cylinder of 6 mm diameter by cold isostatic pressing at 0.4 GPa. This cylinder was placed into the center of a Glidcop®

tube (10 mm inner, 13.8 mm outer diameter). The gap of 2 mm width between the Glidcop® tube and the central *in situ* MgB_2 cylinder was filled with commercially available MgB_2 powder.

Next, we consider the critical current densities achieved in MgB_2 wires and tapes. Figure 17.12 shows the critical current densities at 4.2 K of various MgB_2 wires and tapes manufactured by the *ex situ* PIT route [71, 73, 75, 77, 82–84]. The performance of the MgB_2 wires depends strongly on the details of the wire manufacture. As illustrated in Figure 17.12, MgB_2 wires manufactured by the *ex situ* route are superconductive even without heat treatment [75]. However, the connectivity of the MgB_2 grains and, therefore, the critical current density is significantly improved after heat treatment with a maximum temperature slightly above 900 °C in a mixture of 95% Ar and 5% H_2. Frequently, the performance of MgB_2 wires or tapes is characterized by the magnetic field at which the current density reaches 10 000 A cm^{-2}. The j_c value of the *ex situ* wire without heat treatment exceeds 10 000 A cm^{-2} below 0.25 T, while after heat treatment the same j_c level is reached at a field of ~4.8 T. The j_c also depends on the sheath material. In a commercial tape with a nickel sheath and an iron barrier around the copper stabilizer, j_c reaches 10 000 A cm^{-2} at a field of ~3.8 T [71], whereas in a CuNi/Nb/MgB_2 tape the same j_c level is achieved at a field of ~5.3 T [71]. In both cases, the magnetic field is applied perpendicular to the broad face of the multifilament MgB_2 tapes. The

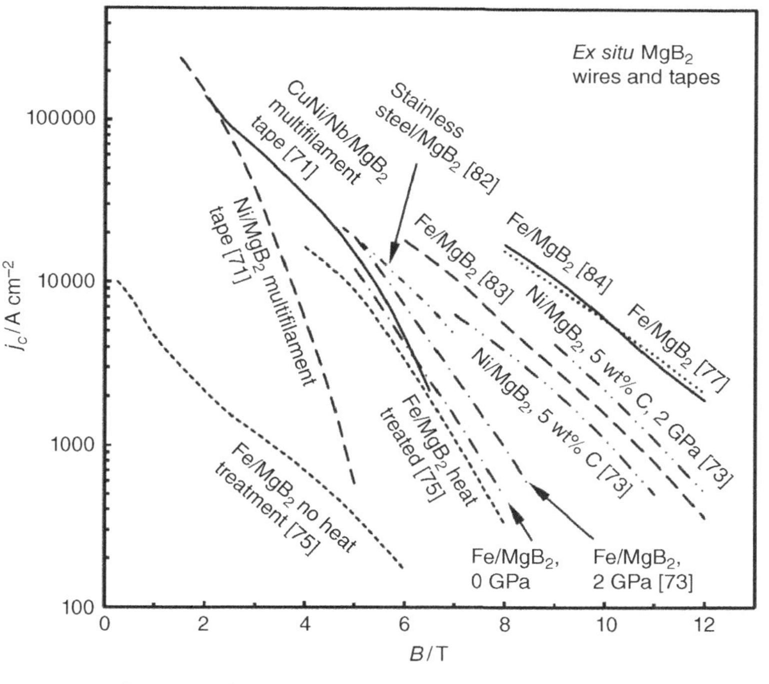

Figure 17.12 *Critical current density at 4.2 K vs. applied magnetic field of various MgB_2 conductors manufactured by the ex situ PIT process (results from [71, 73, 75, 77, 82–84])*

connectivity of the grains can be improved by the densification of the MgB_2 filaments, as demonstrated by cold high-pressure deformation [73]. A j_c of 10 000 A cm^{-2} was achieved at ~5.2 T without cold high-pressure deformation, while after the application of 2 GPa, the same j_c was reached at a field of ~5.7 T. Furthermore, Kulich *et al.* [73] demonstrated that the performance of a MgB_2 wire doped with 5 wt% C can be improved by the application of 2 GPa. As expected, the wires with carbon doping show an improved high-field performance. The carbon doping increases both the upper critical field and the irreversibility field, and, hence, also the j_c at high fields. Good performance was reported for Fe/MgB_2 tapes with carbon-doped MgB_2 powders [77, 84]. The MgB_2 powders were ball-milled with additions of mineral oil [77] or benzene and benzoic acid [84] to provide the desired carbon doping. The j_c level of 10 000 A cm^{-2} was reached at fields of 8.9 T [77] and 9.1 T [84].

Figure 17.13 shows the critical current densities at 4.2 K of various MgB_2 wires and tapes manufactured by the *in situ* PIT route [69, 75, 85–92]. The comparison of the performances of the mono-core and multifilament MgB_2 wires indicates that a higher critical current density was achieved in the multifilament wire. In both wires, 5 wt% SiC was added to the Mg and B powders. The better performance of the stainless steel/Cu/Ti/MgB_2 multifilament wire seems to originate from the stainless steel outer sheath, which is not present

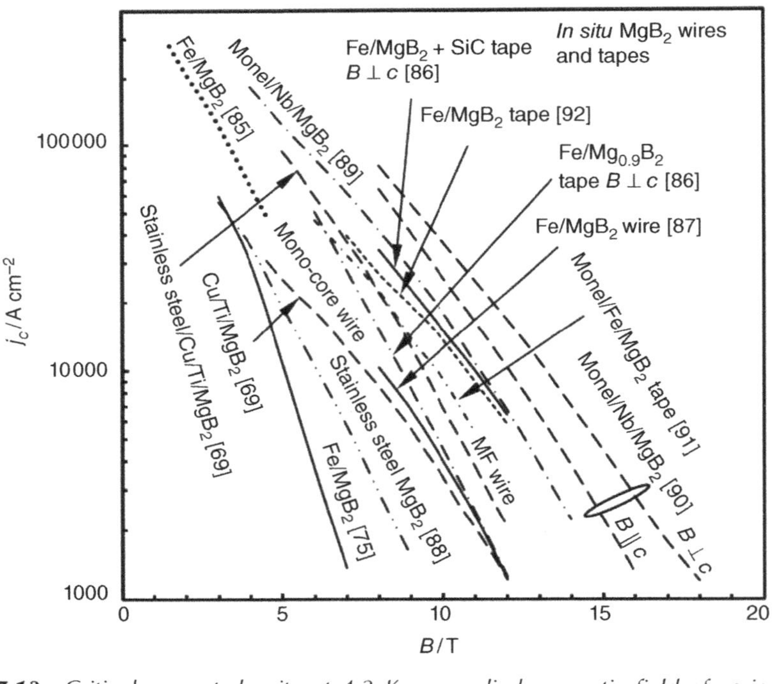

Figure 17.13 Critical current density at 4.2 K vs. applied magnetic field of various MgB_2 conductors manufactured by the in situ PIT process (results from [69, 75, 85–92])

in the mono-core wires, and which allows better densification of the MgB_2 filament. A j_c of 10000 A cm^{-2} was achieved at ~9.5 T and 4.2 K.

For a Monel/Nb/MgB_2 tape [90], the anisotropy with respect to the field direction is presented. As a carbon source, malic acid ($C_4H_6O_5$) was used. In addition, the density of the MgB_2 filaments was increased by the application of 1.48 GPa (cold high-pressure densification). A critical current density of 10 000 A cm^{-2} was achieved at a field of 13.2 T applied perpendicular to the crystallographic *c* direction and 12.2 T for fields parallel to the *c* axis [90]. As compared to cuprate high-T_c superconductors, the anisotropy of j_c with respect to the field direction is not very pronounced. An advantage of the *in situ* route is the easier doping of MgB_2 by carbon, which is reflected in improved performance at high magnetic fields.

Figure 17.14 shows the field dependence of various MgB_2 conductors at 20 K [93–97]. Depending on the manufacturing process, a critical current density of 10000 A cm^{-2} was achieved for magnetic fields between 4 T [95] and 7 T [93]. Figure 17.15 shows the dependence of j_c on temperature for fields of 5 and 7 T [93]. The data indicate that the operation of MgB_2 wires at 20 K could be an interesting option. For the developmental conductor considered, the j_c at 20 K and 5 T is of the order of 50 000 A cm^{-2}. The critical current density exceeds 10000 A cm^{-2} even at 7 T and 20 K.

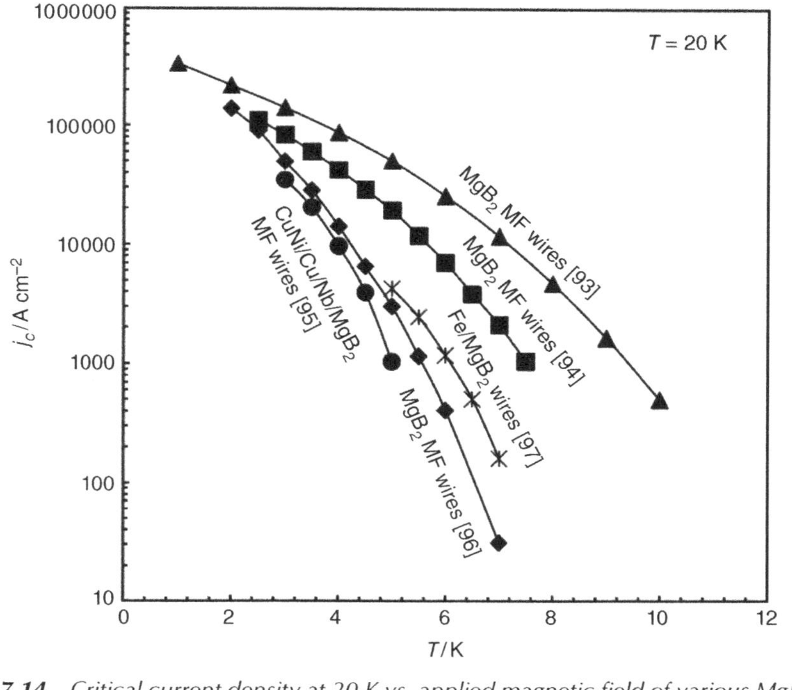

Figure 17.14 *Critical current density at 20 K vs. applied magnetic field of various MgB_2 conductors (results from [93–97])*

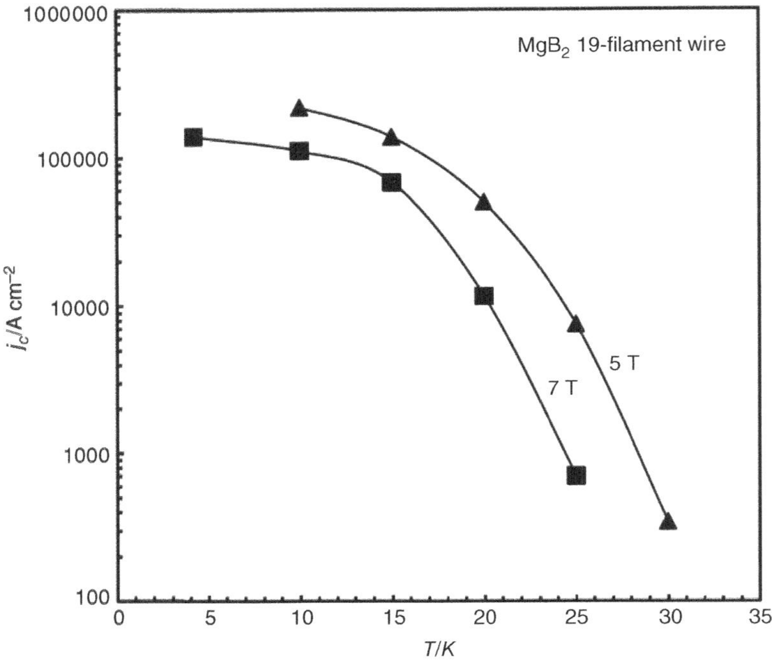

Figure 17.15 *Critical current density of MgB_2 vs. temperature at applied magnetic fields of 5 and 7 T (results from [93])*

17.4 MgB_2 Bulk Material

Bulk samples have frequently been used to study the physical properties of MgB_2. Zhu *et al.* [98] prepared bulk material starting from the commercial MgB_2 powder. The powder was pressed into MgB_2 pellets of 6 mm diameter, which were sealed together with pure magnesium in Al_2O_3 tubes. The pellets were sintered at 1050 °C for 24 h and a magnesium vapor pressure of 400 hPa. The samples were furnace-cooled to room temperature. Zhou *et al.* [99] prepared Si- and C-doped MgB_2 pellets by solid-state reactions. Mg and B powders were mixed with Si or C particles, forming $MgB_2Si_{0.1}$ or $MgB_2C_{0.1}$ samples. The powders, hand grounded for 0.5 h in a mortar, were compacted under high pressure to pellets of 10 mm diameter. These pellets were enclosed in iron tubes closed at both ends, sintered at 750 °C for 0.5 h in a pure argon atmosphere, and furnace-cooled to room temperature. Xu *et al.* [100] synthesized Mg_xB_2 samples ($0.5 \leq x \leq 1.3$) by a solid-state reaction, starting with well-ground mixtures of magnesium and amorphous boron powder. Pressed pellets were wrapped in tantalum foils heated at a rate of 2 °C min^{-1} from room temperature to 800 °C. The heat treatment was performed for 1 h in a tube furnace under argon, after which the pellets were furnace-cooled to room temperature. Shimoyama *et al.* [101] studied the effect of added silver on the formation of MgB_2 bulk material starting from magnesium and boron powders. They found that ~3% Ag reduced the reaction temperature from well above 600 °C to 550 °C, with no degradation of the critical temperature. Senkowicz *et al.*

[102] used mechanically alloyed carbon-doped MgB_2 powder in the preparation of pellets, which were sealed into evacuated stainless steel tubes. After heat treatment for 200 min at 1000 °C and a pressure of ~2070 bar in a hot isostatic press (HIP), the density of the samples reached 90% of the X-ray density.

An interesting route for the manufacture of MgB_2 bulk material as well as MgB_2 wires is reactive liquid magnesium infiltration, developed by Giunchi *et al.* [103–107], which may be considered as a special variant of the *in situ* route. The reaction of the liquid Mg and B powder is performed in a stainless steel container. To manufacture a MgB_2 disk, a solid block of Mg with a central cylindrical hole is inserted into a stainless steel container. The hole is filled with the B powder with 50% of the bulk density. At the top and the bottom, the boron powder is in contact with the stainless steel container. The container is sealed with a welded steel lid so that no empty space remains in it.

Heat treatment at temperatures between 750 °C and 950 °C leads to infiltration of the boron powder by liquid magnesium and the formation of MgB_2. In an intermediate step, the boron-rich phase Mg_2B_{25} is formed [106, 107]. Finally, the container is machined away, leaving MgB_2 bulk material and some residual Mg. Giunchi *et al.* [106] have demonstrated the manufacture of MgB_2 rods and cylinders in this way.

In the application of low-temperature superconductors, it is well known that the superconductor must be in the form of thin filaments to avoid flux jumping. The conditions for the occurrence of flux jumps in high-T_c superconductors were discussed by Wipf [108]. The lowest field B_{fj} at which flux jumping can occur in a superconducting slab is

$$B_{fj} \geq \sqrt{3\mu_0 C(T_c - T_{op})} \tag{17.7}$$

Here μ_0 is the permeability of free space, C is the specific heat per unit volume, T_c is the critical temperature, and T_{op} is the operating temperature. In the derivation of Equation 17.7, it was assumed that j_c depends linearly on temperature. Equation 17.7 indicates that the flux jump field increases with increasing specific heat and temperature margin. At low temperatures, the specific heat of solids is proportional to T^3. Thus, the specific heat at 20 K is by more than a factor of 100 larger than at 4.2 K. As a consequence, operation of high-T_c bulk superconductors is possible without flux jumping at elevated temperatures. At temperatures of 5 and 20 K, flux jumps have been observed in MgB_2 bulk material [109]. On the other hand, MgB_2 cylinders have been successfully used for magnetic shielding even at a temperature of 4.2 K with no observed flux jumps [110].

Figure 17.16 shows the critical current densities of various MgB_2 bulk superconductors at 20 K [100, 101, 110–112]. The data indicate that a minimum critical current density of 10 kA cm^{-2} was achieved in the different MgB_2 bulk samples at fields between 2 and 4 T. Thus, MgB_2 bulk material is a potential superconductor for operation at 20 K in magnetic fields up to 4 T. High critical current densities were achieved in slightly Mg-deficient bulk material [100]. For the composition $Mg_{0.8}B_2$, the highest irreversibility field of 5.2 T was achieved at 20 K. In the stoichiometric compound MgB_2, the irreversibility field at 20 K reached a value of only 4.5 T. The improved irreversibility field and the higher critical current densities at higher fields were attributed to the formation of MgB_4 nanoparticles found by transmission electron microscopy (TEM) [100]. An even better high-field performance was achieved in MgB_2 bulk material doped with 1.5 wt% of nanodiamond [112].

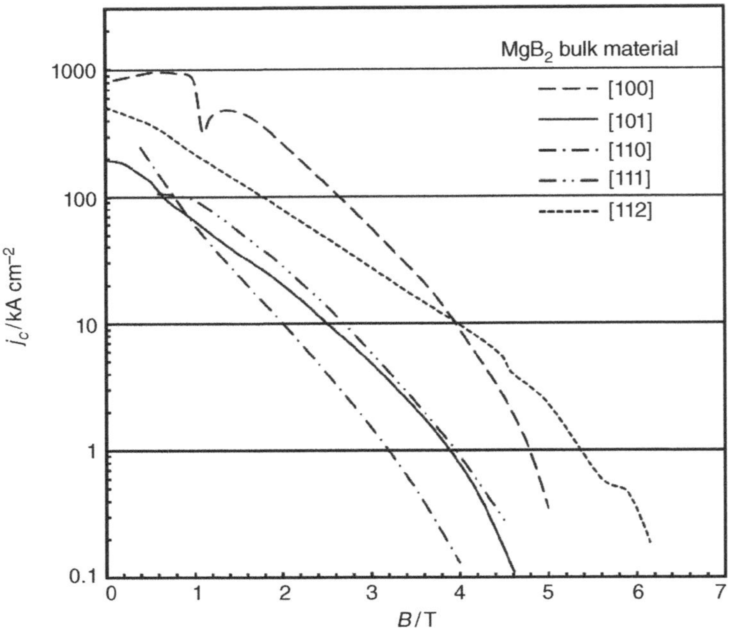

Figure 17.16 *Critical current density vs. field of various MgB_2 bulk superconductors measured at 20 K (results from [100, 101, 110–112])*

17.5 MgB_2 Films

For completeness, the preparation and the superconducting properties of MgB_2 films are briefly described. It is beyond the scope of the present section to provide a comprehensive discussion of MgB_2 film properties. Conditions used for the preparation of MgB_2 films are presented in Table 17.3 [113–124]. Different cuts of sapphire (α-Al_2O_3) single crystals were frequently used as substrates for the growth of thin MgB_2 films [114, 115, 117, 120, 122–124]. Other substrates used for the growth of MgB_2 films are (1 1 1) $SrTiO_3$ [113], (1 0 0) MgO [116], Ti-buffered (1 0 0) ZnO [116], Al_2O_3-buffered (1 0 0) Si [121] and 6H-SiC [118, 119]. For MgB_2 film deposition, pulsed laser deposition [113, 120], electron beam evaporation [114, 115], molecular beam epitaxy (MBE) [116, 117], sputtering [122, 123], thermal evaporation of Mg [123], and hybrid physical–chemical vapor deposition (HPCVD) [119, 121, 124] have all been successfully used.

As an example, a brief description of HPCVD is presented (see, e.g., [119, 121, 124]). The deposition is based on the reaction of magnesium vapor with diborane (B_2H_6). The reaction can be performed in an evacuated quartz tube [121]. The substrate temperature during film deposition is typically kept between 590 °C and 620 °C. The substrate is surrounded by Mg ingots during film deposition [119], and a mixture of 95% H_2 and 5% B_2H_6 is used as boron source. The diborane is thermally decomposed in the vicinity of the substrate surface and reacts with the magnesium vapor to form MgB_2. The thickness of the MgB_2 film can be controlled by varying the deposition time.

Table 17.3 Preparation of MgB_2 films. Abbreviations: PLD – pulsed laser deposition, p_b – base pressure, T_s – substrate temperature, T_a – annealing temperature, B_a – applied magnetic field, RT – room temperature

Substrate	T_c (K)	j_c ($A\,cm^{-2}$)	Deposition conditions	Reference
(1 1 1) $SrTiO_3$ single			PLD, $p_b = 3\times10^{-6}$ Pa, $p_{Ar} = 0.3$ Pa	[113]
crystals			during deposition, T_s = RT	
Each film around			Annealing together with Mg pellet	
500 nm thick	36		Film 1: evacuated Nb tube,	
			$T_a = 850°C$, 15 min	
	31	$>10^6$	Film 2: evacuated quartz tube,	
		1 T, 4.2 K	$T_a = 750°C$, 30 min	
	35		Film 3: Ta envelope in evacuated	
			Nb tube $T_a = 750°C$, 30 min	
Sapphire C(0 0 0 1)	33.12	4.7 to	Electron beam evaporation,	[114]
single crystal	To	7.1×10^{-6}	$p_b < 10^{-6}$ Pa, $T_s = 543°C$, growth	
Films of 300 and	33.9	4.2 K,	rate of MgB_2 layer of 1 nm s	
280 nm thickness		$B_a = 0$		
(1 0 0) α-Al_2O_3			Electron beam evaporation,	[115]
			$p_b < 10^{-6}$ Pa, $T_s = 543°C$, growth	
			rate of MgB_2 layer of 1 nm s^{-1}	
r-cut sapphire			Electron beam evaporation of	
			B/Mg multilayers	
			$p_b = 10^{-6}-10^{-7}$ mbar, T_s = RT	
			$T_a = 700°C$, 30 min in Mg vapor	
(1 0 0) MgO	33.6		Molecular beam epitaxy,	[116]
Ti/(1 0 0) ZnO	34.9	5.8×10^5	$p_b < 5.3\times10^{-9}$ hPa, $T_s = 200°C$,	
20-nm-thick Ti		10 K, 5 T	0.03 nm s^{-1} B, 0.24 nm s^{-1} Mg	
buffer layer				
300-nm-thick *c*-axis				
oriented thin films				
(0 0 0 1) sapphire			Molecular beam epitaxy,	[117]
150–200 nm thick			$p_b < 6.7\times10^{-10}$ hPa, $T_s = 300°C$,	
(0 0 0 1) 6H-SiC			Hybrid physical–chemical vapor	[118]
7.5-nm-thick film	32.8		deposition, Mg source heated to	
10-nm-thick film	35.5	10^6	650–700°C, B supplied from the	
		20 K	thermal decomposition of B_2H_6 gas	
			H_2 pressure of 5 kPa	
c-cut 6H-SiC	40.8	7.9×10^6	Hybrid physical–chemical vapor	[119]
100–150-nm-thick		5 K	deposition, $T_s = 620°C$, substrate	
Films		$B_a = 0$	surrounded by 2–4 g Mg ingots	
5% Si doping		2.7×10^5	5% B_2H_6/95% H_2 as B source	
		5 K, 3 T		

(continued overleaf)

Table 17.3 (continued)

Substrate	T_c (K)	j_c ($A cm^{-2}$)	Deposition conditions	Reference
Sapphire Al_2O_3-c(0 0 0 1)	34.2		Off-axis PLD, $p_b = 10.7 \times 10^{-7}$ hPa Deposition $p_{Ar} = 0.16$ hPa, $T_s = 250°C$, Mg capping layer, $T_a = 650°C$, 9 min	[120]
Al_2O_3/(1 0 0) Si 80-nm- or 300-nm-thick buffer layer	39		Atomic layer deposition of Al_2O_3 buffer layer, hybrid physical vapor deposition of MgB_2, Mg + B_2H_6 vapor $T_s = 500–600°C$, 33–67 hPa	[121]
c-plane sapphire	22		DC magnetron sputtering of Mg Radio frequency sputtering of B $p_b = 2.7 \times 10^{-7}$ hPa, $p_{Ar} = 0.2$ Pa, T_s = RT, Mg + B capping layer $T_a = 620°C$, 5–60 min, $p_{Ar} \sim 1$ Pa	[122]
r-plane sapphire	28		Radio frequency sputtering of B Thermal evaporation of Mg $p_b = 6.7 \times 10^{-7}$ hPa, film growth under 3 Pa Ar, $T_s = 800°C$, 30 min	[123]
(0 0 0 1) Al_2O_3 440-nm-thick film		2.5×10^6 20 K 0.2 T	Hybrid physical–chemical vapor deposition, MgB_2 films reacted from Mg metal vapor and B_2H_6 gas $T_s = 590°C$	[124]

In MgB_2 films, the critical temperature reached values up to 40.8 K [119], which is comparable to the maximum values achieved in wires, tapes and bulk material. Figure 17.17 shows the critical current densities of various MgB_2 films [114–116, 125, 126]. The critical current densities were measured at temperatures between 4.2 and 10 K. The critical current of the MgB_2 films reached values up to $\sim$10 MA cm^{-2} in the absence of an applied magnetic field. For magnetic fields between 4 and 7 T, the critical current densities range from 0.1 to 1.8 MA cm^{-2}. In some cases, the critical current was measured for fields parallel and perpendicular to the film surface. The results of Kitaguchi and Doi [114] indicate that below 7 T, the j_c for perpendicular fields is higher than that for parallel fields, whereas at higher fields, the j_c is higher in the parallel field. A similar behavior was reported by Sosiati *et al.* [115] and Haruta *et al.* [116]. The high j_c values found for B perpendicular to the film surface have been attributed to the columnar structure of the films. The grain boundaries perpendicular to the film surface seem to act as efficient pinning centers up to fields of 7–8 T at 4.2 K [114].

The last aspect to be considered is the dependence of the critical temperature on the thickness of the MgB_2 layer. Zhuang *et al.* [127] prepared ultrathin MgB_2 films on (0 0 0 1) 6H-SiC single-crystal substrates. The film thicknesses ranged from 10 to 200 nm. The T_c of the 100- and 200-nm-thick films is 40.8 K, dropping to 37 K for the thinnest film of 10 nm thickness (see Figure 17.18). Doi *et al.* [128] prepared multilayers with alternating

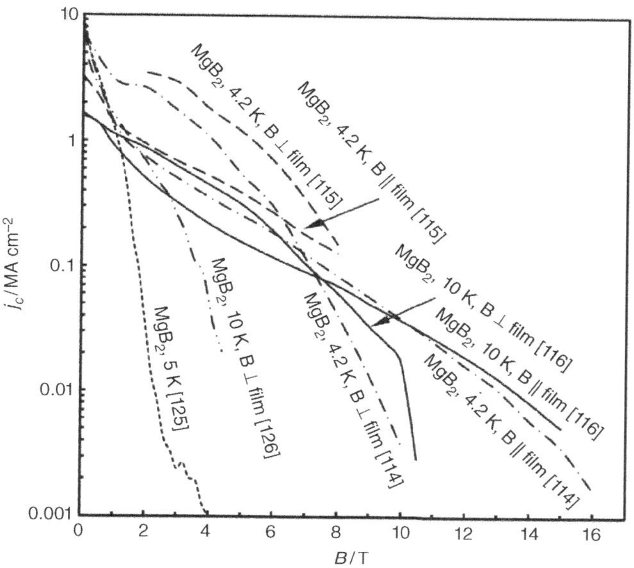

Figure 17.17 Critical current density vs. field of various MgB_2 films measured at temperatures between 4.2 and 10 K. In the absence of an applied magnetic field, the critical current density reaches values close to $10\,MA\,cm^{-2}$. Further details are discussed in the text (results from [114–116, 125, 126])

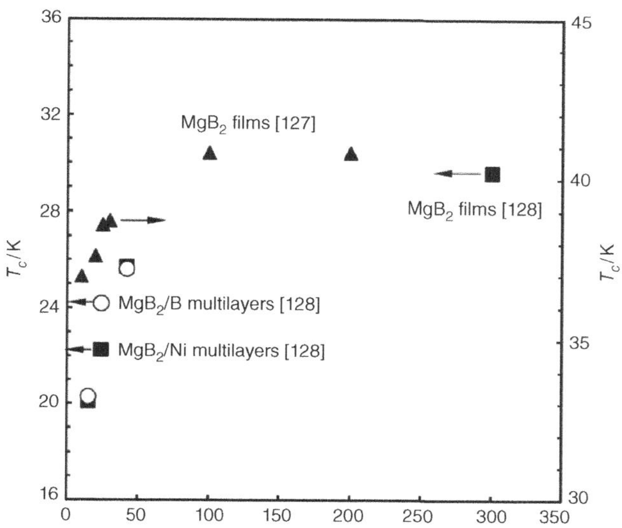

Figure 17.18 Dependence of the critical temperature of MgB_2 films or MgB_2/M multilayers (M represents Ni or B) on the thickness of the MgB_2 layer. For the films as well as the multilayers, the critical temperature decreases with decreasing thickness of the MgB_2 layer (results from [127, 128])

MgB_2 and Ni or MgB_2 and B layers. In the MgB_2/Ni multilayers, adjacent MgB_2 layers are separated by 0.2-nm-thick Ni layers, while in the MgB_2/B multilayers, the boron layers are 5 nm thick. The drop in the critical temperature with decreasing MgB_2 layer thickness was found to be independent of the interlayer material (see Figure 17.18).

17.6 Summary

MgB_2 is an intermediate-temperature superconductor with a critical temperature of 39 K. The pairing of the electrons to form Cooper pairs is mediated by the electron–phonon interaction. The high critical temperature seems to be caused by high-frequency boron phonons. This picture is supported by the observed boron isotope effect. A special feature of MgB_2 is the presence of two superconducting energy gaps. The smaller energy gap ($\Delta(T = 0) \cong$ 2 meV) originates from the π electrons, while the larger gap ($\Delta(T = 0) \cong 7$ meV) is related to the σ electrons. Despite its critical temperature of 39 K, much lower than that of cuprate superconductors, MgB_2 seems to be an interesting superconductor for applications, especially at 20 K and moderate fields up to 4 T. In contrast to cuprate superconductors, the grain boundaries of MgB_2 do not act as weak links for the supercurrents. Nevertheless, it is of importance to ensure that the MgB_2 grains are well connected, and that the current flow is not interrupted by the presence of impurity phases at the grain boundaries. The performance of MgB_2 wires at 4.2 K is still not competitive with Nb_3Sn because the irreversibility field is too low. Thin-film data indicate that the j_c of MgB_2 can reach 10 MA cm^{-2} at 4.2 K with no magnetic field. Furthermore, high upper critical fields were found in dirty MgB_2 films, suggesting that the high-field performance of MgB_2 wires and tapes can be further improved.

References

1. J. Nagamatsu, N. Nakagawa, T. Muranaka, Y. Zenitani, and J. Akimitsu, Superconductivity at 39 K in magnesium diboride, *Nature*, **410**, 63–64 (2001).
2. J.R. Gavaler, Superconductivity in Nb–Ge films above 22 K, *Appl. Phys. Lett.*, **23**, 480–482 (1973).
3. L.R. Testardi, J.H. Wernick, and W.A. Royer, Superconductivity with onset above 23 K in Nb–Ge sputtered films, *Solid State Commun.*, **15**, 1–4 (1974).
4. J.R. Gavaler, M.A. Janocko, and C.K. Jones, Preparation and properties of high-T_c Nb-Ge films, *J. Appl. Phys.*, **45**, 3009–3013 (1974).
5. A.M. Campbell, How could we miss it? *Science*, **292**, 65–66 (2001).
6. D.C. Larbalestier, L.D. Cooley, M.O. Rikel, A.A. Polyanskii, J. Jiang, S. Patnaik, X.Y. Cai, D.M. Feldmann, A. Gurevich, A.A. Squitieri, M.T. Naus, C.B. Eom, E.E. Hellstrom, R.J. Cava, K.A. Regan, N. Rogado, M.A. Hayward, T. He, J.S. Slusky, P. Khalifah, K. Inumaru, and M. Haas, Strongly linked current flow in polycrystalline forms of the superconductor MgB_2, *Nature*, **410**, 186–189 (2001).
7. S.L. Bud'ko, G. Lapertot, C. Petrovic, C.E. Cunningham, N. Anderson, and P.C. Canfield, Boron isotope effect in superconducting MgB_2, *Phys. Rev. Lett.*, **86**, 1877–1880 (2001).

8. D.G. Hinks, H. Claus, and J.D. Jorgenson, The complex nature of superconductivity in MgB_2 as revealed by the reduced total isotope effect, *Nature*, **411**, 457–460 (2001).
9. J.M. An and W.E. Pickett, Superconductivity of MgB_2: Covalent bonds driven metallic, *Phys. Rev. Lett.*, **86**, 4366–4369 (2001).
10. H.J. Choi, D. Roundy, H. Sun, M.L. Cohen, and S.G. Louie, The origin of the anomalous superconducting properties of MgB_2, *Nature*, **418**, 758–760 (2002).
11. I.I. Mazin and V.P. Antropov, Electronic structure, electron–phonon coupling, and multiband effects in MgB_2, *Physica C*, **385**, 49–65 (2003).
12. H.J. Choi, M.L. Cohen, and S.G. Louie, Anisotropic Eliashberg theory of MgB_2: T_c, isotope effects, superconducting energy gaps, quasiparticles, and specific heat, *Physica C*, **385**, 66–74 (2003).
13. O.C. Abah, G.C. Asomba, and C.M.I. Okoye, The effect of interband interactions of phonon and charge fluctuation on the superconducting parameters of MgB_2, *Supercond. Sci. Technol.*, **23**, 045031 (2010).
14. A.Y. Liu, I.I. Mazin, and J. Kortus, Beyond Eliashberg superconductivity in MgB_2: Anharmonicity, two-phonon scattering, and multiple gaps, *Phys. Rev. Lett.*, **87**, 087005 (2001).
15. J. Kortus, Current progress in the theoretical understanding of MgB_2, *Physica C*, **456**, 54–62 (2007).
16. A. Brinkman, A.A. Golubov, H. Rogalla, O.V. Dolgov, J. Kortus, Y. Kong, O. Jepsen, and O.K. Andersen, Multiband model for tunneling in MgB_2 junctions, *Phys. Rev. B*, **65**, 180517(R) (2002).
17. S. Souma, Y. Machida, T. Sato, T. Takahashi, H. Matsui, S.-C. Wang, H. Ding, A. Kaminski, J.C. Campuzano, S. Sasaki, and H. Kadowaki., The origin of multiple superconducting gaps in MgB_2, *Nature*, **423**, 65–67 (2003).
18. S. Souma, Y. Machida, T. Sato, T. Takahashi, H. Matsui, S.-C. Wang, H. Ding, A. Kaminski, J.C. Campuzano, S. Sasaki, and H. Kadowaki, Direct observation of superconducting gaps in MgB_2 by angle-resolved photoemission spectroscopy, *Physica C*, **408–410**, 102–103 (2004).
19. M.R. Eskildsen, M. Kugler, G. Levy, S. Tanaka, J. Jun, S.M. Kazakov, J. Karpinski, and Ø. Fischer, Scanning tunneling spectroscopy on single crystal MgB_2, *Physica C*, **385**, 169–176 (2003).
20. D. Daghero, G.A. Ummarino, M. Tortello, D. Delaude, R.S. Gonnelli, V.A. Stepanov, M. Monni, and A. Palenzona, Effect of Li–Al co-doping on the energy gaps of MgB_2, *Supercond. Sci. Technol.*, **22**, 025012 (2009).
21. D. Daghero and R.S. Gonnelli, Probing multiband superconductivity by point-contact spectroscopy, *Supercond. Sci. Technol.*, **23**, 043001, (2010).
22. R.S. Gonnelli, D. Daghero, G.A. Ummarino, V.A. Stepanov, J. Jun, S.M. Kazakov, and J. Karpinski, Direct evidence for two-band superconductivity in MgB_2 single crystals from directional point-contact spectroscopy in magnetic fields, *Phys. Rev. Lett.*, **89**, 247004 (2002).
23. Z.-Z. Li, H.-J. Tao, Y. Xuan, Z.-A. Ren, G.-C. Che, and B.-R. Zhao, Andreev reflection spectroscopy evidence for multiple gaps in MgB_2, *Phys. Rev. B*, **66**, 064513 (2002).
24. S. Lee, Z.G. Khim, Y. Chong, S.H. Moon, H.N. Lee, H.G. Kim, B. Oh, and E.J. Choi, Measurement of the superconducting gap of MgB_2 by point contact spectroscopy, *Physica C*, **377**, 202–207 (2002).

25. P. Szabó, P. Samuely, J. Kačmarčik, T. Klein, J. Marcus, D. Fruchart, S. Miraglia, C. Marcenat, and A.G.M. Jansen, Evidence for two superconducting gaps in MgB_2 by point-contact spectroscopy, *Phys. Rev. Lett.*, **87**, 137005 (2001).
26. F. Bouquet, R.A. Fisher, N.E. Phillips, D.G. Hinks, and J.D. Jorgensen, Specific heat of $Mg^{11}B_2$: Evidence for a second energy gap, *Phys. Rev. Lett.*, **87**, 047001 (2001).
27. A.A. Golubov, J. Kortus, O.V. Dolgov, O. Jepsen, Y. Kong, O.K. Andersen, B.J. Gibson, K. Ahn, and R.K. Kremer, Specific heat of MgB_2 in a one- and a two-band model from first-principles calculations, *J. Phys.: Condens. Matter*, **14**, 1353–1360 (2002).
28. F. Bouquet, Y. Wang, I. Sheikin, P. Toulemonde, M. Eisterer, H.W. Weber, S. Lee, S. Tajima, and A. Junod, Unusual effects of anisotropy on the specific heat of ceramic and single crystal MgB_2, *Physica C*, **385**, 192–204 (2003).
29. M. Zehetmayer, M. Eisterer, J. Jun, S.M. Kazakov, J. Karpinski, A Wisniewski, and H.W. Weber, Mixed-state properties of superconducting MgB_2 single crystals, *Phys. Rev. B*, **66**, 052505 (2002).
30. A.K. Pradhan, Z.X. Shi, M. Tokunaga, T. Tamegai, Y. Takano, K. Togano, H. Kito, and H. Ihara, Electrical transport and anisotropic superconducting properties in single crystalline and dense polycrystalline MgB_2, *Phys. Rev. B*, **64**, 212509 (2001).
31. M. Xu, H. Kitazawa, Y. Takano, J. Ye, K. Nishida, H. Abe, A. Matsushita, N. Tsujii, and G. Kido, Anisotropy of superconductivity from MgB_2 single crystals, *Appl. Phys. Lett.*, **79**, 2779–2781 (2001).
32. M. Eisterer, Topical review – Magnetic properties and critical currents of MgB_2, *Supercond. Sci. Technol.*, **20**, R47–R73 (2007).
33. L. Lyard, P. Samuely, P. Szabo, T. Klein, C. Marcenat, L. Paulius, K.H.P. Kim, C.U. Jung, H.-S. Lee, B. Kang, S. Choi, S.-I. Lee, J. Marcus, S. Blanchard, A.G.M. Jansen, U. Welp, G. Karapetrov, and W.K. Kwok, Anisotropy of the upper critical field and critical current in single crystal MgB_2, *Phys. Rev. B*, **66**, 180502(R) (2002).
34. M. Angst, R. Puzniak, A. Wisniewski, J. Jun, S.M. Kazakov, J. Karpinski, J. Roos, and H. Keller, Temperature and field dependence of the anisotropy of MgB_2, *Phys. Rev. Lett.*, **88**, 167004 (2002).
35. M. Xu, H. Kitazawa, Y. Takano, J. Ye, K. Nishida, H. Abe, A. Matsushita, N. Tsujii, and G. Kido, Single crystalline MgB_2 superconductor, *J. Phys. Soc. Japan*, **71**, 320–322 (2002).
36. C. Ferdeghini, V. Ferrando, G. Grassano, W. Ramadan, V. Braccini, M. Putti, P. Manfrinetti, and A. Palenzona, Transport properties of *c*-oriented MgB_2 thin films grown by pulsed laser deposition, *Physica C*, **372–376**, 1270–1273 (2002).
37. Yu. Eltsev, S. Lee, K. Nakao, N. Chikumoto, S. Tajima, N. Koshizuka, and M. Murakami, Anisotropic superconducting properties of MgB_2 single crystals, *Physica C*, **378–381**, 61–64 (2002).
38. Z.X. Shi, A.K. Pradhan, M. Tokunaga, K. Yamazaki, T. Tamegai, Y. Takano, K. Togano, H. Kito, and H. Ihara, Magnetic properties and flux pinning in single crystalline and dense polycrystalline MgB_2, *Physica C*, **378–381**, 550–553 (2002).
39. Y. Ma, X. Zhang, S. Awaji, D. Wang, Z. Gao, G. Nishijima, and K. Watanabe, Large irreversibility field in nanoscale C-doped MgB_2/Fe tape conductors, *Supercond. Sci. Technol.*, **20**, L5–L8 (2007).
40. H. Kumakura, H. Kitaguchi, A. Matsumoto, and H. Yamada, Upper critical field, irreversibility field, and critical current density of powder-in-tube-processed MgB_2/Fe tapes, *Supercond. Sci. Technol.*, **18**, 1042–1046 (2005).

41. J.H. Kim, W.K. Yeoh, X. Xu, S.X. Dou, P. Munroe, M. Rindfleisch, and M. Tomsic, Superconductivity of MgB_2 with embedded multiwall carbon nanotube, *Physica C*, **449**, 133–138 (2006).
42. S.C. Yan, G. Yan, Y.F. Lu, and L. Zhou, The upper critical field in micro-SiC doped MgB_2 fabricated by a two-step reaction, *Supercond. Sci. Technol.*, **20**, 549–553 (2007).
43. Y. Wang, T. Plackowski, and A. Junod, Specific heat in the superconducting and the normal state (2–300 K, 0–16 T), and magnetic susceptibility of the 38 K superconductor MgB_2: Evidence for a multicomponent gap, *Physica C*, **355**, 179–193 (2001).
44. S.L. Budko, C. Petrovic, G. Lapertot, C.E. Cunningham, P.C. Canfield, M.-H. Jung, and A.H. Lacerda, Magnetoresistivity and $H_{c2}(T)$ in MgB_2, *Phys. Rev. B*, **63**, 220503 (2001).
45. Y. Shen, R. Gandikota, R.K. Singh, F.L. Hunte, J. Jaroszynski, D.C. Larbalestier, J.M. Rowell, and N. Newman, A novel technique for synthesizing MgB_2 thin films with high upper critical fields, *Supercond. Sci. Technol.*, **21**, 085009 (2008).
46. Y. Zhu, A.V. Pogrebnyakov, R.H. Wilke, K. Chen, X.X. Xi, J.M. Redwing, C.G. Zhuang, Q.R. Feng, Z.Z. Gan, R.K. Singh, Y. Shen, N. Newman, J.M. Rowell, F. Hunte, J. Jaroszynski, D.C. Larbalestier, S.A. Baily, F.F. Balakirev, and P.M. Voyles, Nanoscale disorder in pure and doped MgB_2 thin films, *Supercond. Sci. Technol.*, **23**, 095008 (2010).
47. S. Noguchi, A. Kuribayashi, T. Oba, H. Iriuda, Y. Harada, M. Yoshizawa, S. Miki, H. Shimakage, Z. Wang, K. Satoh, T. Yotsuya, and T. Ishida, Systematic characterization of upper critical fields for MgB_2 thin films by means of the two-band superconducting theory, *Supercond. Sci. Technol.*, **22**, 055004 (2009).
48. W. Häßler, M. Herrmann, C. Rodig, M. Schubert, K. Nenkov, and B. Holzapfel, Further increase of the critical current density of MgB_2 tapes with nanocarbon-doped mechanically alloyed precursor, *Supercond. Sci. Technol.*, **21**, 062001 (2008).
49. A. Matsumoto, H. Kitaguchi, and H. Kumakura, Superconducting properties of MgB_2/Fe tape heavily doped with nanosized SiC, *Supercond. Sci. Technol.*, **21**, 065007 (2008).
50. O.F. de Lima, R.A. Ribeiro, M.A. Avila, C.A. Cardoso, and A.A. Coelho, Anisotropic superconducting properties of aligned MgB_2 crystallites, *Phys. Rev. Lett.*, **86**, 5974–5977 (2001).
51. T. Matsushita, M. Kiuchi, A. Yamamoto, J. Shimoyama, and K. Kishio, Essential factors for the critical current density in superconducting MgB_2: Connectivity and flux pinning by grain boundaries, *Supercond. Sci. Technol.*, **21**, 015008 (2008).
52. F. Simon, A. Jánossy, T. Fehér, F. Murányi, S. Garaj, L. Forró, C. Petrovic, S.L. Bud'ko, G. Lapertot, V.G. Kogan, and P.C. Canfield, Anisotropy of superconducting MgB_2 as seen in electron spin resonance and magnetization data, *Phys. Rev. Lett.*, **87**, 047002 (2001).
53. P.C. Canfield, S.L. Bud'ko, and D.K. Finnemore, An overview of the basic physical properties of MgB_2, *Physica C*, **385**, 1–7 (2003).
54. H. Shimakage, M. Tatsumi, and Z. Wang, Ultrathin MgB_2 films fabricated by the co-evaporation method at high Mg evaporation rates, *Supercond. Sci. Technol.*, **21**, 095009 (2008).

55. B.J. Senkowicz, R. Pérez Moyet, R.J. Mungall, J. Hedstrom, O.N.C. Uwakweh, E.E. Hellstrom, and D.C. Larbalestier, Atmospheric conditions and their effect on ball-milled magnesium diboride, *Supercond. Sci. Technol.*, **19**, 1173–1177 (2006).
56. L.P. Chen, C. Zhang, Y.B. Wang, Y. Wang, Q.R. Feng, Z.Z. Gan, J.Z. Yang, and X.G. Li, A sol–gel method for growing superconducting MgB_2 films, *Supercond. Sci. Technol.*, **24**, 015002 (2011).
57. C. Zhuang, T. Tan, Y. Wang, S. Bai, X. Ma, H. Yang, G. Zhang, Y. He, H. Wen, X.X. Xi, Q. Feng, and Z. Gan, Clean MgB_2 thin films on different types of single-crystal substrate fabricated by hybrid physical–chemical vapor deposition, *Supercond. Sci. Technol.*, **22**, 025002 (2009).
58. K. Yonekura, A. Kugo, T. Fujiyoshi, T. Sueyoshi, Y. Harada, M. Yoshizawa, T. Ikeda, S. Awaji, and K. Watanabe, Flux pinning properties of MgB_2 thin films on Ti buffered substrate prepared by molecular beam epitaxy, *Physica C*, **470**, 1461–1464 (2010).
59. M. Hanna, S. Wang, J.M. Redwing, X.X. Xi, and K. Salama, Thickness dependence of critical current density in MgB_2 films fabricated by *ex situ* annealing of CVD-grown B films in Mg vapor, *Supercond. Sci. Technol.*, **22**, 015024 (2009).
60. C.G. Zhuang, S. Meng, H. Yang, Y. Jia, H.H. Wen, X.X. Xi, Q.R. Feng, and Z.Z. Gan, Significant improvements of the high-field properties of carbon-doped MgB_2 films by hot-filament-assisted hybrid physical–chemical vapor deposition using methane as the doping source, *Supercond. Sci. Technol.*, **21**, 082002 (2008).
61. W.N. Kang, H.-J. Kim, E.-M. Choi, C.U. Jung, and S.-I. Lee, MgB_2 superconducting thin films with a transition temperature of 39 Kelvin, *Science*, **292**, 1521–1523 (2001).
62. H.-J. Kim, W.N. Kang, E.-M. Choi, M.-S. Kim, K.H.P. Kim, and S.-I. Lee, High current-carrying capability in *c*-axis-oriented superconducting MgB_2 thin films, *Phys. Rev. Lett.*, **87**, 087002 (2001).
63. E.W. Collings, M.D. Sumption, M. Bhatia, M.A. Susner, and S.D. Bohnenstiehl, Prospects for improving the intrinsic and extrinsic properties of magnesium diboride superconducting strands, *Supercond. Sci. Technol.*, **21**, 103001 (2008).
64. A. Rao, B. Gahtori, S.K. Agarwal, T. Chakraborty, C.K. Sarkar and A. Das, Low temperature specific heat (zero field and with field) of Fe and Mn-doped MgB_2, *Physica C*, **469**, 64–69 (2009).
65. A.V. Sologubenko, J. Jun, S.M. Kazakov, J. Karpinski, and H.R. Ott, Thermal conductivity of single crystalline MgB_2, *arXiv: cond-mat*[0201517v3] (2002).
66. T. Cavallin, E.A. Young, C. Beduz, Y. Yang, and G. Giunchi, Thermal conductivity of bulk MgB_2 produced by infiltration of different boron powders, *IEEE Trans. Appl. Supercond.*, **17**, 2770–2773 (2007).
67. E. Bauer, Ch. Paul, St. Berger, S. Majumdar, H. Michor, M. Giovannini, A. Saccone, and A Bianconi, Thermal conductivity of superconducting MgB_2, *arXiv: cond-mat*[0104203v2] (2001).
68. C.H. Jiang, T. Nakane, H. Hatakeyama, and H. Kumakura, Enhanced J_c property in nano-SiC doped thin MgB_2/Fe wires by a modified in situ PIT process, *Physica C*, **422**, 127–131 (2005).
69. I. Huŝek and P. Kováĉ, Mechanical properties, interface reactions and transport current densities of multi-core MgB_2/Ti/Cu/SS wire, *Supercond. Sci. Technol.*, **23**, 075012 (2010).

70. HongLi Suo, C. Beneduce, M. Dhallé, N. Musolino, J.-Y. Genoud, and R. Flükiger, Large transport critical currents in dense Fe- and Ni-clad MgB_2 superconducting tapes, *Appl. Phys. Lett.*, **79**, 3116–3118 (2001).
71. A. Malagoli, C. Bernini, V. Braccini, C. Fanciulli, G. Romano, and M. Vignolo, Fabrication and superconducting properties of multifilamentary MgB_2 conductors for AC purposes: Twisted tapes and wires with very thin filaments, *Supercond. Sci. Technol.*, **22**, 105017 (2009).
72. S. Jin, H. Mavoori, C. Bower, and R.B. van Dover, High critical currents in iron-clad superconducting MgB_2 wires, *Nature*, **411**, 563–565 (2001).
73. M. Kulich, R.L. Flükiger, C. Senatore, M. Tropeano, and R. Piccardo, Effect of cold high pressure deformation on the properties of *ex situ* MgB_2 wires, *Supercond. Sci. Technol.*, **26**, 105019 (2013).
74. L. Kopera, P. Kováĉ, I. Huŝek, and T. Meliŝek, Rutherford cable made of single-core MgB_2 wires, *Supercond. Sci. Technol.*, **26**, 125007 (2013).
75. S.I. Schlachter, W. Goldacker, J. Reiner, S. Zimmer, B. Liu, and B. Obst, Influence of the preparation process on microstructure, critical current density and T_c of MgB_2 powder-in-tube wires, *IEEE Trans. Appl. Supercond.*, **13**, 3203–3206 (2003).
76. H. Fujii, K. Ozawa, and H. Kitaguchi, Improved critical current density in *ex situ* processed carbon-substituted MgB_2 tapes by Mg addition, *Supercond. Sci. Technol.*, **25**, 065008 (2012).
77. H. Fujii, K. Ozawa, and H. Kitaguchi, Effect of the carbon addition to filling powder ball-milled with oils on the critical current density in *ex situ* processed MgB_2 tapes, *Supercond. Sci. Technol.*, **25**, 105008 (2012).
78. G. Bovone, M. Vignolo, C. Bernini, S. Kawale, and A.S. Siri, An innovative technique to synthesize C-doped MgB_2 by using chitosan as the carbon source, *Supercond. Sci. Technol.*, **27**, 022001 (2014).
79. H. Yamada, N. Uchiyama, A. Matsumoto, H. Kitaguchi, and H. Kumakura, The excellent superconducting properties of *in situ* powder-in-tube processed MgB_2 tapes with both ethyltoluene and SiC powder added, *Supercond. Sci. Technol.*, **20**, L30–L33 (2007).
80. C.M. Lee, S.M. Lee, G.C. Park, J. Joo, J.H. Lim, W.N. Kang, J.H. Yi, B.-H. Jun, and C.J. Kim, Investigation of lauric acid dopant as a novel carbon source in MgB_2 wire, *Physica C*, **470**, 1438–1441 (2010).
81. A. Kario, A. Morawski, W. Häßler, K. Nenkov, M. Schubert, M. Herrmann, B. Ringsdorf, S.I. Schlachter, W. Goldacker, B. Holzapfel, and L. Schultz, *Ex situ* MgB_2 barrier behavior of monofilament *in situ* MgB_2 wires with Glidcop® sheath material, *Supercond. Sci. Technol.*, **23**, 115007 (2010).
82. L. Civale, A. Serquis, D.L. Hammon, X.Z. Liao, J.Y. Coulter, Y.T. Zhu, T. Holesinger, D.E. Peterson, and F.M. Mueller, High critical currents in powder in tube MgB_2 wires: Influence of microstructure and heat treatments, *IEEE Trans. Appl. Supercond.*, **13**, 3347–3350 (2003).
83. H. Fujii and K. Ozawa, Effect of temperature and concentration of solution in chemical treatment for MgB_2 powder on the J_c-B property of *ex situ* processed MgB_2 superconducting tapes, *Physica C*, **470**, 326–330 (2010).
84. H. Fujii, K. Ozawa, and H. Kitaguchi, The effect of refinement of carbon-substituted MgB_2 filling powder using various starting powders on the microstructure and critical

current density of *ex situ* processed superconducting tapes, *Supercond. Sci. Technol.*, **27**, 035002 (2014).
85. C.H. Jiang, H. Hatakeyama, and H. Kumakura, Effect of nanometer MgO addition on the in situ PIT processed MgB_2/Fe tapes, *Physica C*, **423**, 45–50 (2005).
86. C.H. Jiang and H. Kumakura, Stoichiometry dependence of the critical current density in pure and nano-SiC doped MgB_2/Fe tapes, *Physica C*, **451**, 71–76 (2007).
87. A. Xu, Y. Ma, D. Wang, Z. Gao, X. Zhang, and K. Watanabe, Improved flux pinning and high critical current density in Fe-sheathed MgB_2 wires and tapes by carbon nanotube doping, *Physica C*, **466**, 190–195 (2007).
88. A. Jung, S.I. Schlachter, B. Runtsch, B. Ringsdorf, H. Fillinger, H. Orschulko, A. Drechsler, and W. Goldacker, Influence of Ni and Cu contamination on the superconducting properties of MgB_2 filaments, *Supercond. Sci. Technol.*, **23**, 095006 (2010).
89. M.A. Susner, T.W. Daniels, M.D. Sumption, M.A. Rindfleisch, C.J. Thong, and E.W. Collings, Drawing induced texture and the evolution of superconductive properties with heat treatment time in powder-in-tube *in situ* processed MgB_2 strands, *Supercond. Sci. Technol.*, **25**, 065002 (2012).
90. M.S.A. Hossain, C. Senatore, R. Flükiger, M.A. Rindfleisch, M.J. Tomsic, J.H. Kim, and S.X. Dou, The enhanced J_c and B_{irr} of *in situ* MgB_2 wires and tapes alloyed with $C_4H_6O_5$ (malic acid) after cold high pressure densification, *Supercond. Sci. Technol.*, **22**, 095004 (2009).
91. C. Wang, D. Wang, X. Zhang, C. Yao, C. Wang, Y. Ma, H. Oguro, S. Awaji, and K. Watanabe, Improved J_c-B properties of MgB_2 multifilamentary wires and tapes, *Supercond. Sci. Technol.*, **25**, 125001 (2012).
92. M. Takahashi and H. Kumakura, The enhancement of critical current density of powder-in-tube processed MgB_2 tapes by pre-heating of Mg and B mixed powder, *Supercond. Sci. Technol.*, **26**, 075007 (2013).
93. K. Togano, J. Hur, A. Matsumoto, and H. Kumakura, Microstructures and critical currents of single- and multi-filamentary MgB_2 superconducting wires fabricated by an internal Mg diffusion process, *Supercond. Sci. Technol.*, **23**, 085002 (2010).
94. K. Togano, J.M. Hur, A. Matsumoto, and H. Kumakura, Fabrication of seven-core multi-filamentary MgB_2 wires with high critical current density by an internal Mg diffusion process, *Supercond. Sci. Technol.*, **22**, 015003 (2009).
95. W. Häßler, H. Hermann, M. Herrmann, C. Rodig, A. Aubele, L. Schmolinga, B. Sailer and B. Holzapfel, Influence of milling energy transferred to the precursor powder on the microstructure and the superconducting properties of MgB_2 wires, *Supercond. Sci. Technol.*, **26**, 025005 (2013).
96. S. Oh, J.H. Kim, C. Lee, H. Choi, C.-J. Kim, S.X. Dou, M. Rindfleisch, and M. Tomsic, Field, temperature and strain dependence of the critical current for multi-filamentary MgB_2 wire, *Physica C*, **468**, 1821–1824 (2008).
97. R. Flükiger, C. Senatore, M. Cesaretti, F. Buta, D. Uglietti, and B. Seeber, Optimization of Nb_3Sn and MgB_2 wires, *Supercond. Sci. Technol.*, **21**, 054015 (2008).
98. Y. Zhu, L. Wu, V. Volkov, Q. Li, G. Gu, A.R. Moodenbaugh, M. Malac, M. Suenaga, and J. Tranquada, Microstructure and structural defects in MgB_2 superconductor, *Physica C*, **356**, 239–253 (2001).
99. S.H. Zhou, A.V. Pan, M.J. Qin, X.L. Wang, H.K. Liu, and S.X. Dou, Effects of Si and C doping on the superconducting properties of MgB_2, *Adv. Cryogen. Eng. – ICMC*, **50**, 554–560 (2004).

100. G.J. Xu, R. Pinholt, J. Bilde-Sørensen, J.-C. Grivel, A.B. Abrahamsen, and N.H. Andersen, Effect of starting composition and annealing temperature on irreversibility field and critical current density in Mg_xB_2, *Physica C*, **434**, 67–70 (2006).
101. J. Shimoyama, K. Hanafusa, A. Yamamoto, Y. Katsura, S. Horii, K. Kishio, and H. Kumakura, Catalytic effect of silver addition on the low temperature phase formation of MgB_2, *Supercond. Sci. Technol.*, **20**, 307–311 (2007).
102. B.J. Senkowicz, R.J. Mungall, Y. Zhu, J. Jiang, P.M. Voyles, E.E. Hellstrom, and D.C. Larbalestier, Nanoscale grains, high irreversibility field and large critical current density as a function of high-energy ball milling time in C-doped magnesium diboride, *Supercond. Sci. Technol.*, **21**, 035009 (2008).
103. G. Giunchi, S. Ginocchio, S. Raineri, D. Botta, R. Gerbaldo, B. Minetti, R. Quarantiello, and A. Matrone, High density MgB_2 bulk materials of different grains size: Supercurrents instability and losses in variable magnetic fields, *IEEE Trans. Appl. Supercond.*, **15**, 3230–3233 (2005).
104. G. Giunchi, S. Ceresara, G. Ripamonti, A. DiZenobio, S. Rossi, S. Chiarelli, M. Spadoni, R. Wesche, and P.L. Bruzzone, High performance new MgB_2 superconducting hollow wires, *Supercond. Sci. Technol.*, **16**, 285–291 (2003).
105. G. Giunchi, S. Raineri, R. Wesche, P.L. Bruzzone, The voltage–current relations for MgB_2 obtained by reactive liquid infiltration, *Physica C*, **401**, 310–315 (2004).
106. G. Giunchi, G. Ripamonti, T. Cavallin, and E. Bassani, The reactive liquid Mg infiltration process to produce large superconducting bulk MgB_2 manufacts, *Cryogenics*, **46**, 237–242 (2006).
107. G. Giunchi, C. Orecchia, L. Malpezzi, and N. Masciocchi, Analysis of the minority crystalline phases in bulk superconducting MgB_2 obtained by reactive Mg infiltration, *Physica C*, **433**, 182–188 (2006).
108. S.L. Wipf, Review of stability in high temperature superconductors with emphasis on flux jumping, *Cryogenics*, **31**, 936–948 (1991).
109. Y. Kimishima, S. Takami, T. Okuda, M. Uehara, T. Kuramoto, and Y. Sugiyama, Complete flux jump in bulk MgB_2 sintered under high pressure, *Physica C*, **463–465**, 281–285 (2007).
110. J.J. Rabbers, M.P. Oomen, E. Bassani, G. Ripamonti, and G. Giunchi, Magnetic shielding capability of MgB_2 cylinders, *Supercond. Sci. Technol.*, **23**, 125003 (2010).
111. Y. Yang, C.H. Cheng, L. Wang, H.H. Sun, and Y. Zhao, Effect of sorbic acid doping on flux pinning in bulk MgB_2 with the percolation model, *Physica C*, **470**, 1100–1102 (2010).
112. Y. Zhao, Y. Yang, C.H. Cheng, and Y. Zhang, Doping and effect of nano-diamond and carbon-nanotubes on flux pinning properties of MgB_2, *Physica C*, **463–465**, 165–169 (2007).
113. C.B. Eom, M.K. Lee, J.H. Choi, L.J. Belenky, X. Song, L.D. Cooley, M.T. Naus, S. Patnaik, J. Jiang, M. Rikel, A. Polyanskii, A. Gurevich, X.Y. Cai, S.D. Bu, S.E. Babcock, E.E. Hellstrom, D.C. Larbalestier, N. Rogado, K.A. Regan, M.A. Hayward, T. He, J.S. Slusky, K. Inumaru, M.K. Haas, and R.J. Cava, High critical current density and enhanced irreversibility field in superconducting MgB_2 thin films, *Nature*, **411**, 558–560 (2001).
114. H. Kitaguchi and T. Doi, High-temperature and high-field performance of MgB_2 films with J_c of 10^6 A cm^{-2} (4.2 K, 4 T), *Supercond. Sci. Technol.*, **18**, 489–493 (2005).

115. H. Sosiati, S. Hata, N. Kuwano, Y. Tomokiyo, H. Kitaguchi, T. Doi, H. Yamamoto, A. Matsumoto, K. Saitoh, and H. Kumakura, Relationship between microstructure and J_c property in MgB_2/α-Al_2O_3 film fabricated by *in situ* electron beam evaporation, *Supercond. Sci. Technol.*, **18**, 1275–1279 (2005).
116. M. Haruta, T. Fujiyoshi, S. Kihara, T. Sueyoshi, K. Miyahara, Y. Harada, M. Yoshizawa, T. Takahashi, H. Iriuda, T. Oba, S. Awaji, K. Watanabe, and R. Miyagawa, High critical current density under magnetic fields in as-grown MgB_2 thin films deposited by molecular-beam epitaxy, *Supercond. Sci. Technol.*, **20**, L1–L4 (2007).
117. R.K. Singh, Y. Shen, R. Gandikota, J.M. Rowell, and N. Newman, Effect of stoichiometry on oxygen incorporation in MgB_2 thin films, *Supercond. Sci. Technol.*, **21**, 015018 (2008).
118. Y. Wang, C. Zhuang, X. Sun, X. Huang, Q. Fu, Z. Liao, D. Yu, and Q. Feng, Ultrathin epitaxial MgB_2 superconducting films with high critical current density and T_c above 33 K, *Supercond. Sci. Technol.*, **22**, 125015 (2009).
119. C.Y. Zhang, Y.B. Wang, W.W. Hu, and Q.R. Feng, The effect of Si addition in MgB_2 thin films by hybrid physical–chemical vapor deposition using silane as the doping source, *Supercond. Sci. Technol.*, **23**, 065017 (2010).
120. Y.S. Wu, Y. Zhao, D. Wexler, J.H. Kim, and S.X. Dou, Optimization of *in situ* annealing conditions for off-axis PLD MgB_2 films, *Physica C*, **468**, 218–222 (2008).
121. T.G. Lee, S.W. Park, W.K. Seong, J.Y. Huh, S.-G. Jung, B.K. Lee, K.-S. An, and W.N. Kang, Growth of MgB_2 thin films on alumina-buffered Si substrates by using hybrid physical–chemical vapor deposition method, *Physica C*, **468**, 1888–1891 (2008).
122. A. O'Brian, B. Villegas, and J.Y. Gu, Sputtered magnesium diboride thin films: Growth conditions and surface morphology, *Physica C*, **469**, 39–43 (2009).
123. S.G. Kang, S.C. Park, J.-K. Chung, D.G. Jeong, C.-J. Kim, and C.J. Kim, Microstructure of MgB_2 thin film deposited on *r*-plane sapphire substrate by co-evaporation method, *Physica C*, **469**, 1571–1573 (2009).
124. D.H. Kim, J.-K. Chung, T.J. Hwang, W.N. Kang, and K.C. Chung, A possible role of amorphous regions at grain-boundary vertex points as linear defects on flux pinning in MgB_2 films with columnar grain structure, *Physica C*, **469**, 1950–1953 (2009).
125. A.T. Matveev, J. Albrecht, M. Konuma, G. Christiani, Y. Krockenberger, U. Starke, G. Schütz, and H.U. Habermeier, Synthesis of MgB_2 films in Mg vapour flow and their characterization, *Supercond. Sci. Technol.*, **19**, 299–305 (2006).
126. Z.X. Shi, Y.X. Zhang, H. Lv, M. Xu, E.-M. Choi, S.-I. Lee, Flux pinning in *c*-axis-oriented MgB_2 thin film, *Physica C*, **467**, 101–105 (2007).
127. C. Zhuang, K. Chen, J.M. Redwing, Q. Li, and X.X. Xi, Surface morphology and thickness dependence of the properties of MgB_2 thin films by hybrid physical–chemical vapor deposition, *Supercond. Sci. Technol.*, **23**, 055004 (2010).
128. T. Doi, H. Kitaguchi, S. Hata, K. Fukuyama, K. Masuda, K. Takahashi, T. Yoshidome, Y. Hakuraku, and N. Kuwano, Monotonic decrease of T_cs with thinning of the superconducting MgB_2 layer in MgB_2/Ni and MgB_2/B alternately-layered thin films, *Supercond. Sci. Technol.*, **20**, 1223–1227 (2007).

18

Iron-Based Superconductors – A New Class of High-Temperature Superconductors

18.1 Introduction

Before the discovery of superconductivity in the iron-pnictides [1], the cuprates were the only known class of materials showing high-temperature superconductivity. Very recently a record T_c value of 58.1 K has been reported for the iron-pnictide superconductor $Sm(O_{1-x}F_x)FeAs$ (Sm-1111) [2]. The iron-based superconductors are characterized by layered crystal structures similar to the cuprate high-T_c materials. There exist four main families of iron-based superconductors, which are distinguished by their layering schemes and their chemical compositions. The layering schemes of the (1 1 1 1), (1 1 1), (1 2 2), and (1 1) families are presented in Figure 18.1, together with the c lattice parameters of La-1111 [1], Li-111 [3], Ba-122 [4], and FeTe [5]. In the parent compounds of the (1 1 1 1) family (LnOFeAs (Ln = lanthanide)), layers of LnO^{1+} and $FeAs^{1-}$ alternate in the crystal structure along the crystallographic c direction. The parent compounds of the (1 1 1) family are described by the chemical formula AFeAs, where A is Li or Na [6]. In the (1 1 1) compounds, the $FeAs^{1-}$ layer is embedded between layers of A^{1+} ions, with the c-axis perpendicular to the layers. In the (1 2 2) family ($AEFe_2As_2$ (AE = alkaline earth)), layers of $Fe_2As_2^{2-}$ alternate with layers of AE^{2+}($c \perp$ layers) [6]. The parent compounds are non-superconducting metals showing structural and magnetic instability at low temperatures. In the cuprate high-T_c superconductors, the parent compounds show antiferromagnetic ordering. For example, at a temperature of ~160 K LaOFeAs (La-1111) shows a transition from a tetragonal to an orthorhombic [7] or monoclinic crystal structure [8], and <140 K an antiferromagnetic ordering of the iron moments. Partial replacement of oxygen by fluorine (doping) suppresses both the transition and the ordering, and superconductivity occurs [7]. In the nondoped $BaFe_2As_2$ (Ba-122) compound the crystal structure changes

Physical Properties of High-Temperature Superconductors, First Edition. Rainer Wesche.

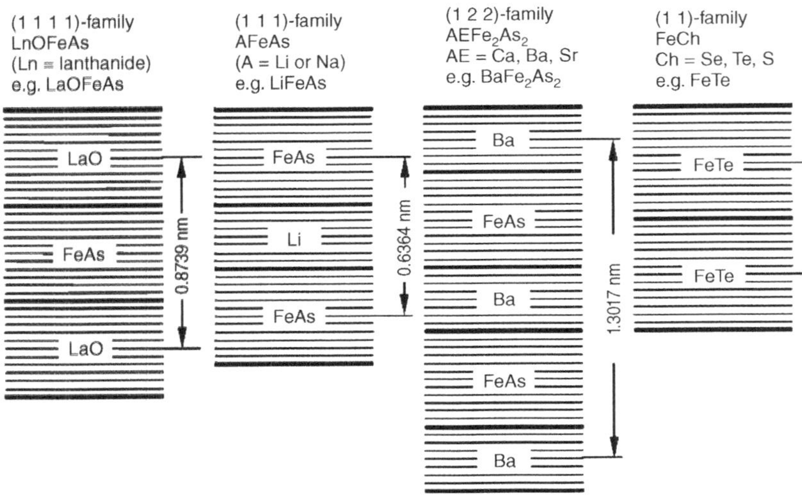

Figure 18.1 *Sequence of layers along the crystallographic c direction in the four main families of iron-based superconductors. In the (1 1 1 1) family (left), LnO (Ln = lanthanide) layers alternate with FeAs layers. The c lattice parameter of the LaOFeAs shown is 0.8739 nm at room temperature [1]. In the (1 1 1) family (second from left), the* FeAs *layers are embedded between layers of alkali ions. The c lattice parameter of the considered LiFeAs compound is 0.6364 nm [3]. The chemical formula for the (1 2 2) family (second from right) is* $AEFe_2As_2$ *(AE = alkaline earth atom). In the (1 2 2) crystal structure, AE and FeAs layers alternate along the c direction. The c lattice parameter of* $BaFe_2As_2$ *is 1.3017 nm [4]. In contrast to the iron-pnictides, there exist no charge carrier reservoir layers between the iron chalcogenide layers of the (1 1) family (FeCh (Ch = Se, Te, S). The c lattice parameter of tetragonal FeTe is 0.6252 nm at 80 K [5]*

from tetragonal to orthorhombic at ~140 K [4]. The crystallographic phase transition is accompanied by antiferromagnetic ordering of the iron moments [4]. Again, structural and magnetic instability can be suppressed by doping, e.g., partial replacement of Ba by K. In the (1 1 1 1), (1 1 1), and (1 2 2) families shown in Figure 18.1, superconductivity is enabled by the FeAs layer in a way similar to the CuO_2 planes in the cuprate high-T_c superconductors. The layers separating the FeAs resemble the charge carrier reservoirs in the cuprates (see Chapter 5, Figures 5.1–5.3). In the (1 1) family of iron-based superconductors, the FeAs layers are replaced by iron-chalcogenide layers, and there exist no blocking layers between adjacent FeCh layers (Ch = Se, Te, or S) [6]. An important difference between iron-based and cuprate high-T_c superconductors is the fact that in the former the parent compounds are metals and not Mott insulators. As in cuprate high-T_c superconductors, the upper critical fields of the iron-based superconductors are extremely high, making them candidates for the generation of very high magnetic fields.

Like the cuprates, the iron-pnictides show a large number of interesting physical phenomena. This new class of high-temperature superconductors therefore attracted great attention, as reflected in the several hundred articles published since their discovery in 2008. This chapter will provide an overview of their properties. A comprehensive discussion of their physical properties and the investigations of the pairing mechanism in these complex materials is beyond the scope of this introduction. In the next section, the critical temperatures of the iron-based superconductors will be considered.

18.2 Critical Temperatures of Iron-based Superconductors

The critical temperatures of various iron-based superconductors are listed in Table 18.1. As early as 2006, superconductivity at ~4 K was discovered in LaOFeP [9]. A much higher critical temperature of 26 K was found in the As analog $La(O_{1-x}F_x)FeAs$ ($x = 0.05$–0.1) [1]. The highest known critical temperature of 58.1 K was achieved in Sm-1111 [2]. For the (1 2 2) family, the highest critical temperature of 38 K has been found in $Ba_{0.6}K_{0.4}Fe_2As_2$ [6]. The critical temperatures reported for the (1 1 1) and (1 1) families are typically <30 K.

In the nonsuperconducting parent compound LaOFeAs, the charges per formula unit are +1 in the LaO layer and −1 in the FeAs layer. The charges of La, O, and F are +3, −2, and −1, respectively. Replacement of 10% of the oxygen atoms by fluorine (resulting in $LaO_{0.9}F_{0.1}FeAs$) leads to a charge of +1.1 per formula unit in the LaOF layer. For charge neutrality, an increased negative charge of −1.1 must be present in the FeAs layer, i.e., the FeAs layer is doped with electrons.

Further the charges of barium and potassium are +2 and +1, respectively. In $BaFe_2As_2$, the charges per formula unit are +2 in the Ba layer and −2 in the Fe_2As_2 layer. Replacement of 40% of the barium atoms by potassium leads to a charge of +1.6 per formula unit in the BaK layer, and hence the charge per Fe_2As_2 needs to be −1.6. The negative charge of the Fe_2As_2 layer is reduced as compared to the parent compound, i.e., holes have been doped into the FeAs layer.

Panel (a) of Figure 18.2 shows the dependence of the critical temperature of fluorine-doped La-1111 [1, 6], Ce-1111 [5], and Sm-1111 [2] on the doping level, while panel (b) shows the effect of potassium [4] or phosphorus doping [30] on the critical temperature of Ba-122. The fluorine doping adds electrons to the FeAs layers of the (1 1 1 1) compounds. In La-1111, the maximum critical temperature of ~26 K has been achieved when 8–11% of the oxygen atoms are replaced by fluorine atoms. For Ce-1111, the highest T_c of ~41 K has been reached at 16% fluorine doping. In the case of Sm-1111, the replacement of 16 to 18% of the oxygen atoms leads to a maximum T_c of ~54 K. The quoted values are based on the end-point of the resistive transition (i.e., $T_c(R = 0)$), which are a few kelvin lower than the onset T_c. The data indicate that the maximum T_c has been achieved at slightly different fluorine doping levels in the three (1 1 1 1) compounds considered. A maximum T_c of ~39 K has been achieved in $Ba_{1-x}K_xFe_2As_2$ at $x = 0.37$. Substitution of potassium for barium leads to hole doping into the FeAs layer. Interestingly, partial replacement of arsenic by phosphorous in $BaFe_2(As_{1-x}P_x)_2$ leads to a maximum T_c of ~30 K. Because As and P have the same valence, the partial replacement of As by P does not introduce additional charge carriers. This means that the effect of doping on superconductivity cannot be simply explained by the addition of charge carriers. Superconductivity in the iron-based superconductors seems to be very sensitive to changes in the crystal structure caused by doping.

Table 18.1 *Critical temperatures of various iron-based superconductors*

Compound	Symbol	T_c(K)	Comments	Reference
LaOFeP	–	≈4	P analog to La-1111	[9]
$LaO_{1-x}F_xFeAs$	La-1111	26	$x = 0.05–0.1$	[1]
$La_{0.87}Sr_{0.13}OFeAs$	La-1111	25	La partly replaced by Sr	[10]
$LaO_{0.8}F_{0.2}FeAs_{1-x}Sb_x$	La-1111	30.1	$x = 0.05$	[11]
$LaO_{1-x}F_xFeAs$	La-1111	43	Under a pressure of 4 GPa	[12]
$SmO_{0.85}F_{0.15}FeAs$	Sm-1111	43	O partly replaced by F	[13]
$SmO_{0.8}F_{0.2}FeAs$	Sm-1111	56.1	O partly replaced by F	[14]
$SmO_{0.74}F_{0.26}FeAs$	Sm-1111	58.1	O partly replaced by F	[2]
$SmO(Fe_{0.9}Co_{0.1})As$	Sm-1111	15.2	Fe partly replaced by Co	[15]
$CeO_{0.84}F_{0.16}FeAs$	Ce-1111	41	O partly replaced by F	[16]
$Ce_{0.6}Gd_{0.4}O_{0.84}F_{0.16}FeAs$	Ce-1111	47.5	Ce partly replaced by Gd	[17]
$PrO_{0.89}F_{0.11}FeAs$	Pr-1111	52	O partly replaced by F	[18]
$PrO_{1-\delta}FeAs$	Pr-1111	44	Oxygen deficiency	[19]
$NdO_{0.89}F_{0.11}FeAs$	Nd-1111	52	O partly replaced by F	[18, 20]
$NdO_{0.75}F_{0.25}FeAs$	Nd-1111	55	O partly replaced by F	[21]
$NdO_{0.85}FeAs$	Nd-1111	46	Oxygen deficiency	[22]
$GdO_{0.83}F_{0.17}FeAs$	Gd-1111	36	O partly replaced by F	[18]
$GdO_{1-\delta}FeAs$, $\delta = 0.85$	Gd-1111	53.5	Oxygen deficiency	[23]
$Gd_{0.8}Th_{0.2}OFeAs$	Gd-1111	56	Gd partly replaced by Th	[24]
$TbO_{0.9}F_{0.1}FeAs$	Tb-1111	46	O partly replaced by F	[16]
$DyO_{0.9}F_{0.1}FeAs$	Dy-1111	45	O partly replaced by F	[18]
$CaF(Fe_{0.9}Co_{0.1})As$	Ca-1111	22	Fe partly replaced by Co	[25]
$Sr_{0.5}Sm_{0.5}FFeAs$	Sr,Sm-1111	56	50% of Sr replaced by Sm	[26]
$Sr_{0.8}La_{0.2}FFeAs$	Sr-1111	36	Sr partly replaced by La	[26]
LiFeAs	Li-111	~18	Both superconducting	[3]
NaFeAs	Na-111	12–25	without doping	[3]
$Ca_{0.5}Na_{0.5}Fe_2As_2$	Ca-122	~20		[27]
$BaFe_2As_2$	Ba-122	29	Pressure of 40 kbar	[28]
$SrFe_2As_2$	Sr-122	27	Pressure of 30 kbar	[28]
$Ba_{0.6}K_{0.4}Fe_2As_2$	Ba-122	38	Ba partly replaced by K	[6]
$Sr_{0.6}K_{0.4}Fe_2As_2$	Sr-122	38	Sr partly replaced by K	[6]
$Ba_{0.84}Rb_{0.1}Sn_{0.09}Fe_2As_{0.96}$	Ba-122	22–24		[29]
$Ba(Fe_{0.8}Co_{0.2})_2As_2$	Ba-122	22	Co doping in FeAs layer	[6]
$BaFe_2(As_{1-x}P_x)_2$	Ba-122	30	As partly replaced by P	[30]
α-FeSe	Fe,Se-11	8–27		[6]
$Fe(Se_{0.5}Te_{0.5})$	Fe,Se-11	~15.5	Se partly replaced by Te	[6]
$Fe(Se_{0.8}S_{0.2})$	Fe,Se-11	~15.5	Se partly replaced by S	[3]
$Fe(Te_{0.8}S_{0.2})$	Fe,Te-11	10	Te partly replaced by S	[3]

Zhao *et al.* [31] studied the relations between structure and high-temperature superconductivity in the (1 1 1 1) and (1 2 2) families. They found an interesting relation between the Fe-As(P)-Fe bond angle (see the inset of Figure 18.3) and the maximum achieved T_c. Figure 18.3 shows the maximum T_c versus the Fe-As(P)-Fe bond angle. The highest critical temperatures were found in (1 1 1 1) compounds with a bond angle close to 109.47°, characteristic for the ideal Fe-As-Fe tetrahedron [31].

Mizuguchi *et al.* [32] found a relation between critical temperature and the anion height in the iron-pnictide or chalcogenide layer. The maximum T_c of ~55 K in the iron-based superconductors occurs for an anion height of 0.138 nm. For anion heights smaller or larger than this value, the T_c drops in a nearly symmetric way. In fluorine-doped LaOFeP with a T_c of 9 K, the anion height is 0.114 nm, while in $FeTe_{0.53}Se_{0.47}$ with a T_c of 13.9 K the anion height is 0.162 nm [32]. The observed relation seems to be valid for (1 1), (1 1 1), (1 2 2), and (1 1 1 1) compounds, both at ambient pressure and with applied pressure. The relation is not valid for compounds with doping in the FeAs layer [32].

Figure 18.4 illustrates the effect of chemical pressure on the critical temperature of $Ce_{1-x}Gd_xO_{0.84}F_{0.16}FeAs$. Yang *et al.* [17] studied the effect of the substitution of Ce by Gd on the critical temperature. The ionic radius of Gd is smaller than that of Ce, and hence the *a* and *c* lattice parameters of $Ce_{1-x}Gd_xO_{0.84}F_{0.16}FeAs$ decrease linearly with increasing

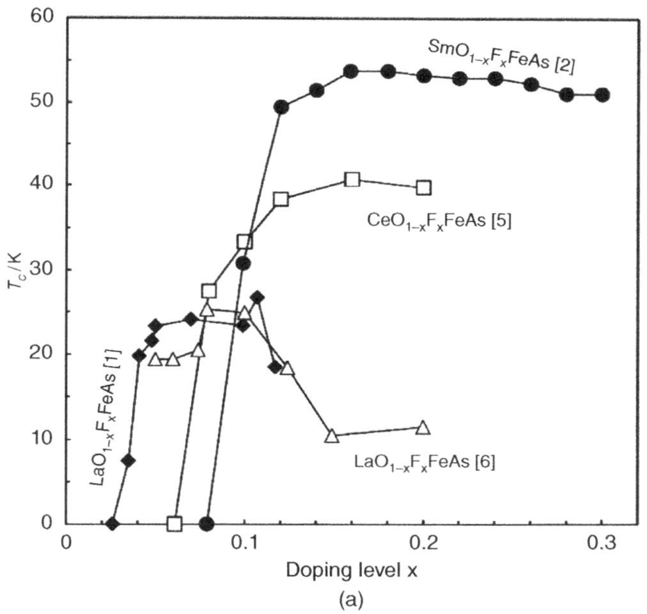

Figure 18.2 *Panel (a): critical temperature of La-1111, Ce-1111, and Sm-1111 versus the level x of fluorine doping. In the (1 1 1 1) compounds $LaO_{1-x}F_xFeAs$, $CeO_{1-x}F_xFeAs$, and $SmO_{1-x}F_xFeAs$, superconductivity sets in for x values of 0.026, 0.061, and 0.079, respectively. In La-1111, maximum critical temperatures of ~26 K have been found for x between 0.08 and 0.11. Higher fluorine content reduces the critical temperature (e.g., $T_c = 10.5$ K at $x = 0.15$). For Ce-1111, a maximum T_c of 40.7 K has been found at $x = 0.16$. A critical temperature as high as 54 K has been achieved in Sm-1111 for x between 0.16 and 0.18 (results from [1, 2, 5, 6]). Panel (b): T_c of $Ba_{1-x}K_xFe_2As_2$ and $BaFe_2(As_{1-x}P_x)_2$ versus doping level x. Superconductivity occurs in the potassium-doped Ba-122 for $x \geq 0.11$, and T_c reaches a maximum of 39 K at $x = 0.37$. In contrast to potassium doping, the substitution of arsenic by phosphorous does not introduce additional charge carriers. Nevertheless, P doping leads to a superconducting dome with a maximum T_c of ~30 K at $x = 0.32$, suggesting that the occurrence of superconductivity is closely related to changes in the crystal structure (results from [4, 30])*

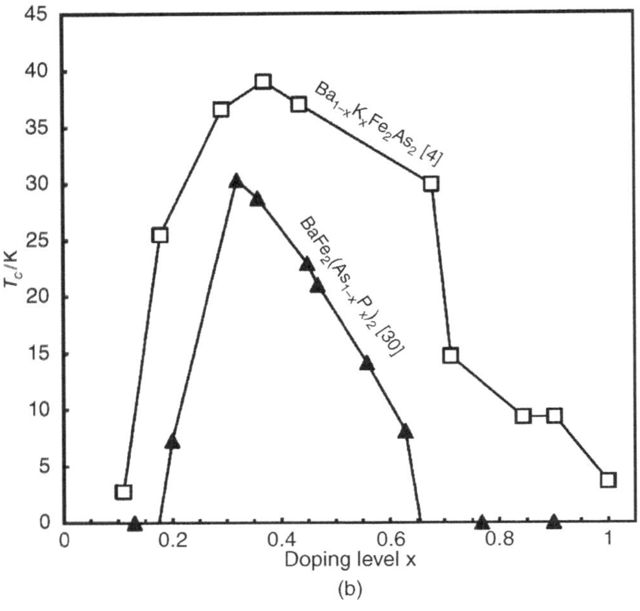

Figure 18.2 *(continued)*

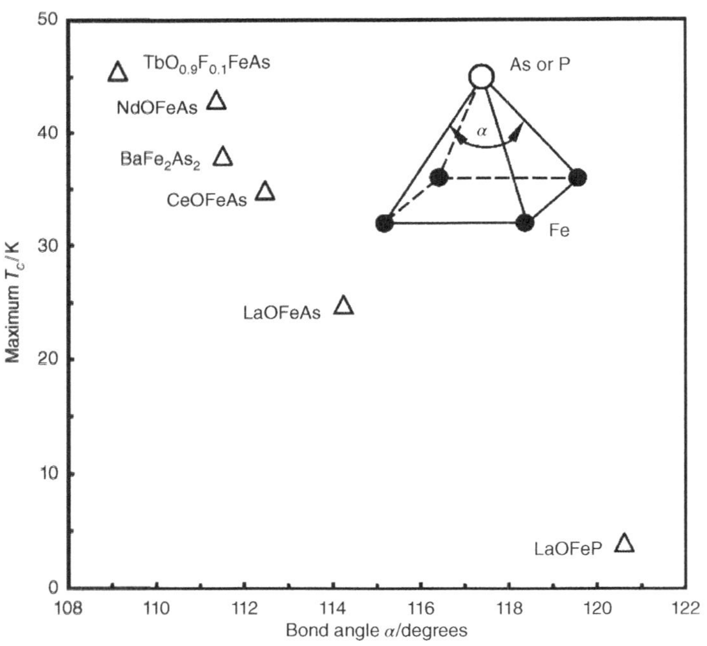

Figure 18.3 *Relation between the critical temperature of iron-pnictide superconductors and the Fe-As(P)-Fe bonding angle. An ideal Fe–As tetrahedron with a bonding angle of 109.47° leads to the highest critical temperatures (adapted from [31])*

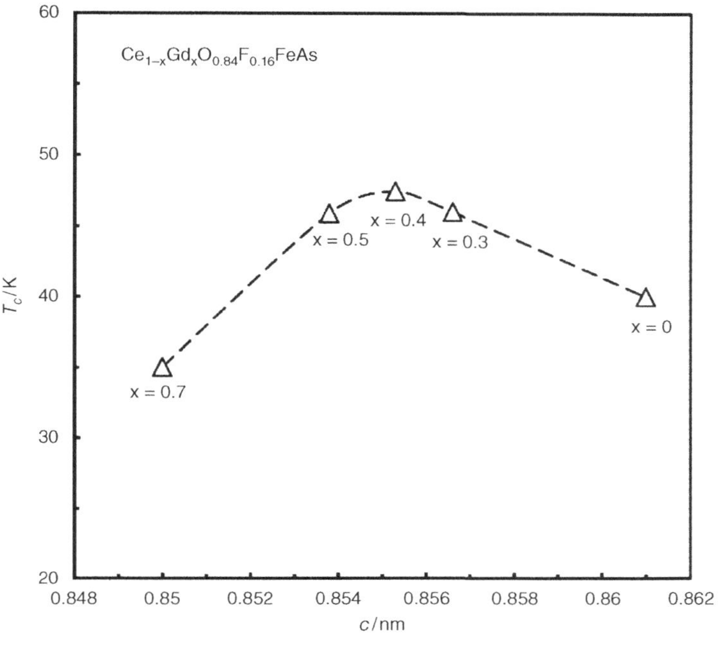

Figure 18.4 *Dependence of the critical temperature of $Ce_{1-x}Gd_xO_{0.84}F_{0.16}FeAs$ on the c lattice parameter. The decrease of the lattice parameters a and c with increasing gadolinium-doping is a consequence of the fact that the ionic radius of Gd is smaller than that of Ce (results from [17])*

Gd content x (chemical pressure). Figure 18.4 indicates that the maximum T_c is reached for $x = 0.4$, leading to lattice parameters $c = 0.8553$ nm and $a = 0.3964$ nm.

Further evidence for the correlation of the critical temperature of iron-based superconductors with structural features is the observed dependence of T_c on pressure. For example, $CaFe_2As_2$ shows superconductivity at ~12 K without doping under a hydrostatic pressure of 5.5 kbar [33]. Nondoped FeSe is not magnetically ordered and is a superconductor with a T_c of ~9 K [34, 35]. The critical temperature of FeSe reaches a maximum value of ~37 K under an applied pressure of 8.9 GPa [35]. Hardy *et al.* [36] found a large dependence of the critical temperature of $Ba(Fe_{0.92}Co_{0.08})_2As_2$ single crystals on uniaxial pressure [36]. They obtained values of $dT_c/dp_a = 3.1 \pm 0.1$ K/GPa and $dT_c/dp_c = -7.0 \pm 0.2$ K/GPa. The pressure derivatives of the critical temperature were deduced from measurements of the thermal expansion and the specific heat. Application of hydrostatic pressure leads to $dT_c/dp = 2dT_c/dp_a + dT_c/dp_c = -0.9$ GPa/K [36], i.e., the uniaxial pressure dependences nearly cancel each other.

Next, let us consider the phase diagram of various compounds in the (1 1 1 1) family. Figure 18.5 shows the boundaries between the room temperature tetragonal phase and the low temperature orthorhombic phase (T_s), between the magnetically nonordered phase and the antiferromagnetic phase (T_N), and the superconducting region (T_c). For the nondoped La-1111 parent compound (solid lines), the crystal structure changes from tetragonal to

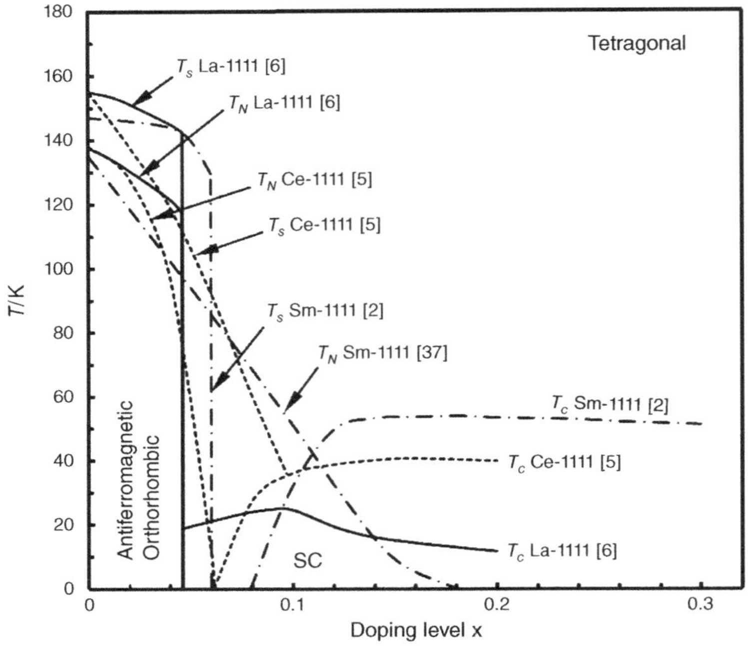

Figure 18.5 *Phase diagrams of La-1111 (solid lines), Ce-1111 (dashed lines), and Sm-1111 (dash-dot lines). The T_s lines indicate the temperature of the structural transition from a tetragonal to an orthorhombic structure. Below the T_N lines, antiferromagnetic ordering sets in. The superconducting state occurs below the T_c line in question. In contrast to La-1111 and Ce-1111, superconductivity and antiferromagnetic ordering can coexist in Sm-1111 for doping levels x between 0.079 and 0.18. All lines have been smoothed (results from [2, 5, 6, 37])*

orthorhombic at ~155 K [6]. Below ~138 K, the spin density wave (SDW) magnetic order has been observed. The SDW transition is accompanied by a steep decrease of the resistivity below ~150 K [6] (see also Figure 18.17). The parent compound is an antiferromagnetic, nonsuperconducting metal. Above ~4.6% doping of fluorine for oxygen, the structural phase transition and the spin density wave magnetic ordering are suppressed, and $LaO_{1-x}F_xFeAs$ (La-1111) shows superconductivity with a maximum $T_c \sim 26$ K at a doping level of $x \sim 0.1$. The dashed lines indicate the phase boundaries for Ce-1111. In the nondoped $CeO_{1-x}F_xFeAs$ parent compound, the crystal structure changes from tetragonal to orthorhombic below ~155 K. Antiferromagnetic ordering occurs < 138 K. Fluorine doping at 6% suppresses antiferromagnetic ordering, and the compound becomes superconducting. In contrast to La-1111, superconductivity can develop not only in the tetragonal but also in the orthorhombic phase. The dash-dotted lines show the phase boundaries for $SmO_{1-x}F_xFeAs$ (Sm-1111). Again, fluorine doping suppresses the transition to an orthorhombic crystal structure and the onset of antiferromagnetic ordering. Using muon-spin spectroscopy, Amato *et al.* [37] observed that spin-density wave ordering and superconductivity can coexist for x between ~0.08 and ~0.15 in Sm-1111, but not in La-1111 or Ce-1111.

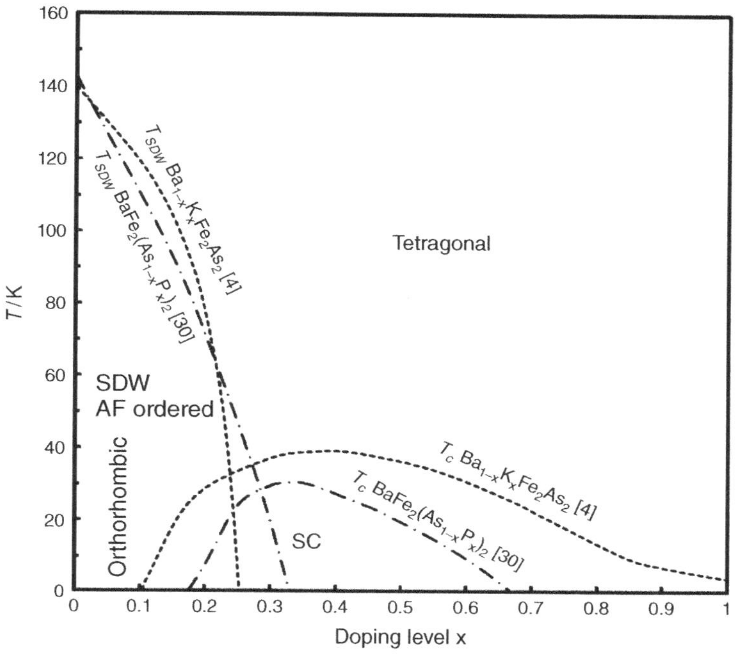

Figure 18.6 *Phase diagrams of potassium- (dotted lines) and phosphorous-doped (dash-dot lines) Ba-122. At the temperature T_{SDW}, antiferromagnetic spin density wave ordering sets in. At the same temperature, the crystal structure changes from tetragonal to orthorhombic. Superconductivity occurs below the T_c line in question. In Ba-122 compounds, superconductivity can coexist with spin density wave ordering and can be present in the orthorhombic crystal structure (results from [4, 30])*

Figure 18.6 shows the phase diagrams of $Ba_{1-x}K_xFe_2As_2$ [4] and $BaFe_2(As_{1-x}P_x)_2$ [30]. In the nonsuperconducting parent compound ($x = 0$), simultaneous structural and magnetic transitions have been observed at ~140 K. The crystal structure changes from tetragonal to orthorhombic. In the orthorhombic phase an SDW antiferromagnetic ordering has been found. Potassium doping introduces additional charge carriers into the FeAs layers and suppresses the structural transition and the SDW ordering. Although phosphorous doping for arsenic does not introduce additional charge carriers, it does suppress the structural and magnetic phase transitions at low temperatures. In both cases, doping leads to superconductivity. An interesting feature is the coexistence of superconductivity and SDW order in the orthorhombic structure at lower doping levels.

18.3 Crystal Structures of Iron-based Superconductors

First, the X-ray powder diffraction patterns of various iron-based superconductors are presented. In the synthesis, X-ray powder diffraction patterns are used to identify the impurity phases present in addition to the desired superconducting compound, and hence may be used to optimize the synthesis conditions. Moreover, X-ray powder diffraction provides the lattice parameters and further information on the crystal structure.

Sketches of the Cu Kα X-ray diffraction patterns of $LaO_{0.8}F_{0.2}FeAs$ (La-1111), $SmO_{0.85}F_{0.15}FeAs$ (Sm-1111), and $CeO_{0.84}F_{0.16}FeAs$ (Ce-1111) are presented in Figures 18.7, 18.8, and 18.9, respectively. The sketches are based on the results of Shekhar *et al.* (La-1111) [38], Chen *et al.* (Sm-1111) [13], and Chen *et al.* (Ce-1111) [16]. The compounds of the (1 1 1 1) family crystallize in a tetragonal structure with the space group (*P4/nmm*) [5]. In the space group symbol, *P* stands for a primitive unit cell, 4 for a fourfold rotational symmetry, *n* for a diagonal glide plane, and *m* for a mirror plane (for more details, see [39]). The lattice parameters deduced from X-ray powder diffraction are collected in Table 18.2. For the three (1 1 1 1) compounds, both the *a* and the *c* lattice

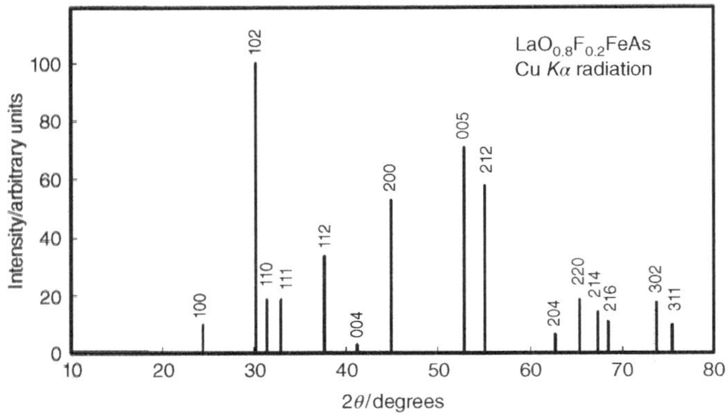

Figure 18.7 Sketch of the Cu Kα X-ray diffraction pattern of $LaO_{0.8}F_{0.2}FeAs$ based on the results of Shekhar et al. [38]. The reflexes are indexed based on a tetragonal unit cell (space group P4/nmm) with the lattice parameters a = 0.4030 nm and c = 0.8716 nm [38]

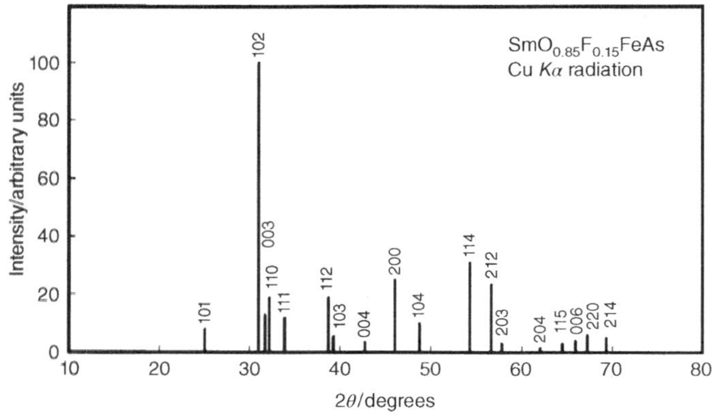

Figure 18.8 Sketch of the Cu Kα X-ray diffraction pattern of $SmO_{0.85}F_{0.15}FeAs$ based on the results of Chen et al. [13]. The reflexes are indexed based on a tetragonal unit cell (space group P4/nmm) with the lattice parameters a = 0.3932 nm and c = 0.8490 nm [13]

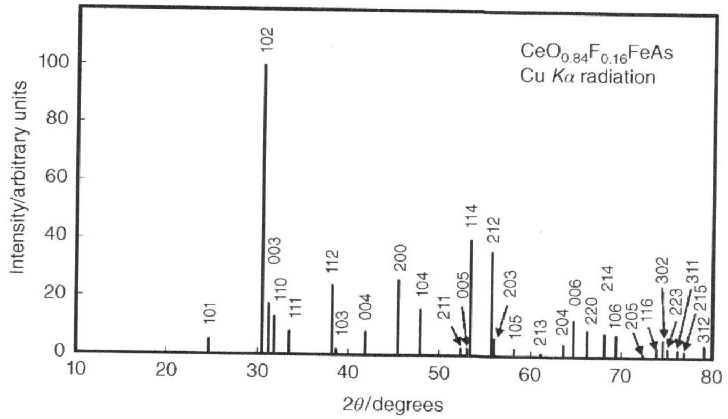

Figure 18.9 *Sketch of the Cu Kα X-ray diffraction pattern of $CeO_{0.84}F_{0.16}FeAs$ based on the results of Chen et al. [16]. The reflexes are indexed based on a tetragonal unit cell (space group P4/nmm) with the lattice parameters a = 0.3989 nm and c = 0.8631 nm [16]*

Table 18.2 *Lattice parameters of selected members of the (1 1 1 1) family*

Compound	a (nm)	c (nm)	Reference
LaOFeAs	0.4039	0.8742	[38]
$LaO_{0.8}F_{0.2}FeAs$	0.4030	0.8716	[38]
SmOFeAs	0.3940	0.8501	[13]
$SmO_{0.85}F_{0.15}FeAs$	0.3932	0.8490	[13]
CeOFeAs	0.3996	0.8648	[16]
$CeO_{0.84}F_{0.16}FeAs$	0.3989	0.8631	[16]

parameters decrease slightly with fluorine doping. The *a* lattice parameter of the (1 1 1 1) compounds is close to 0.4 nm, while the *c* lattice parameter varies between 0.849 and 0.8742 nm.

Sketches of the Cu Kα X-ray powder diffraction patterns of $BaFe_{1.87}Co_{0.13}As_2$ [40] and $Ba_{0.6}K_{0.4}Fe_2As_2$ [41] are presented in Figures 18.10 and 18.11, respectively. The (1 2 2) compounds crystallize in a tetragonal structure of the space group *I4/mmm* [40]. The *a* lattice parameters of $BaFe_2As_2$ ($a = 03964$ nm) and $BaFe_{1.87}Co_{0.13}As_2$ ($a = 0.3963$ nm) are nearly identical, whereas the *c* lattice parameter of the Co-doped compound ($c = 1.2991$ nm) is slightly shorter than that of nondoped Ba-122 ($c = 1.3022$ nm) [40].

Figure 18.12 shows the Cu Kα X-ray powder diffraction pattern of LiFeAs (Li-111), which is based on the results of Tapp *et al.* [42]. The reflexes are indexed based on a tetragonal unit cell (space group *P4/nmm*) with the lattice parameters $a = 0.3791$ nm and $c = 0.6364$ nm [42].

Cu Kα X-ray powder diffraction patterns of the iron chalcogenides FeTe and FeSe are presented in Figures 18.13 and 18.14, respectively. The sketches are based on the results of Yeh *et al.* [43]. The Miller indices are based on a tetragonal unit cell. The cell dimensions of FeSe are $a = 0.3775$ nm and $c = 0.5512$ nm. Due to the larger ionic radius of tellurium,

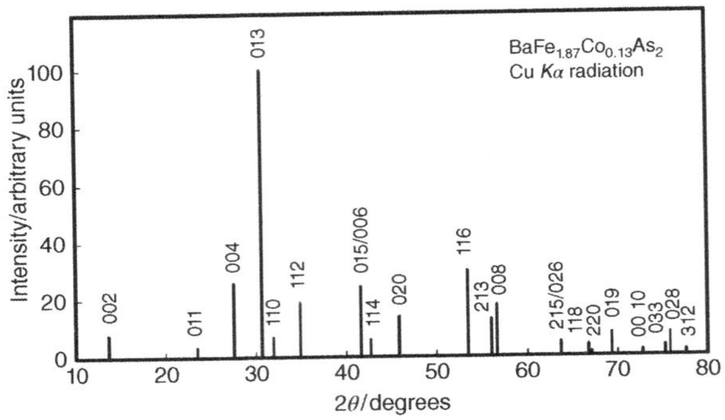

Figure 18.10 *Sketch of the Cu Kα X-ray diffraction pattern of $BaFe_{1.87}Co_{0.13}As_2$ based on the results of Sun et al. [40]. The reflexes are indexed based on a tetragonal unit cell (space group I4/mmm) with the lattice parameters a = 0.3963 nm and c = 1.2991 nm [40]*

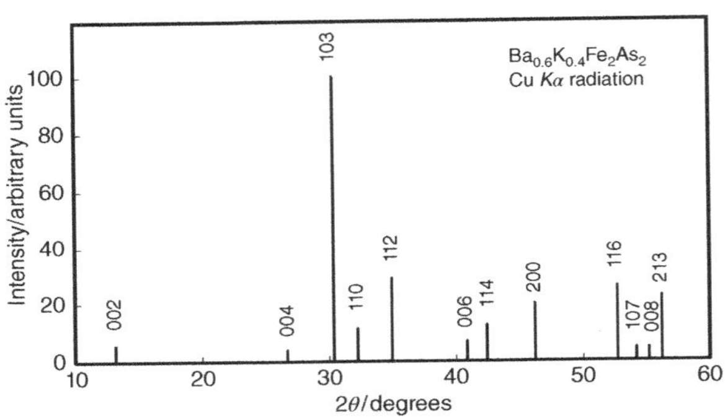

Figure 18.11 *Sketch of the Cu Kα X-ray diffraction pattern of $Ba_{0.6}K_{0.4}Fe_2As_2$ based on the results of Togano et al. [41]*

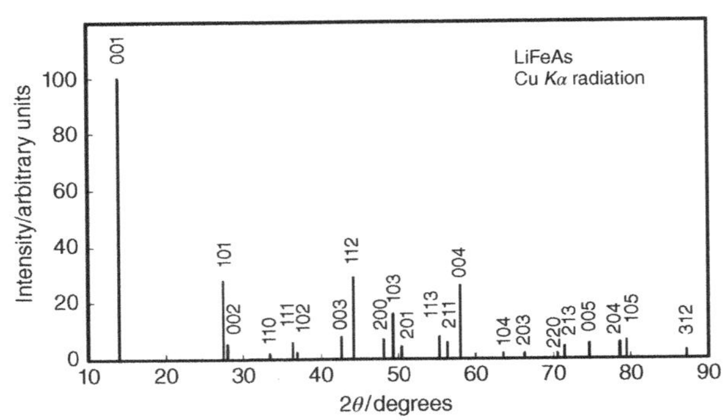

Figure 18.12 *Sketch of the Cu Kα X-ray diffraction pattern of LiFeAs based on the results of Tapp et al. [42]. The reflexes are indexed based on a tetragonal unit cell (space group P4/nmm) with the lattice parameters a = 0.3791 nm and c = 0.6364 nm [42]*

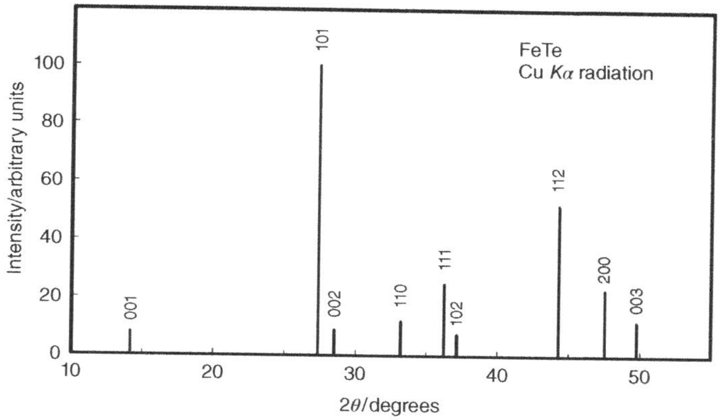

Figure 18.13 Sketch of the Cu Kα X-ray diffraction pattern of FeTe based on the results of Yeh et al. [43]. The Miller indices are based on a PbO-type tetragonal structure [43]

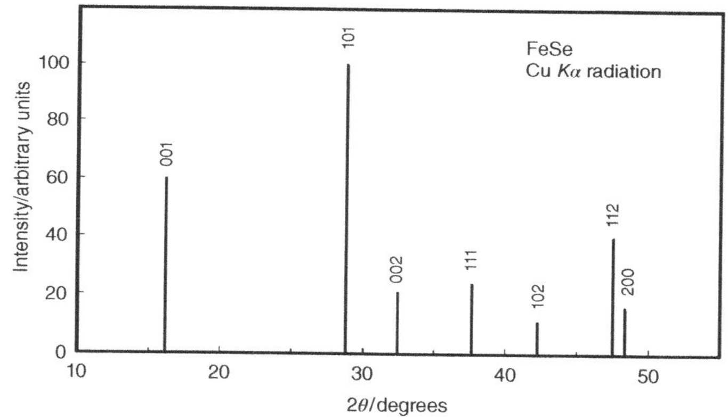

Figure 18.14 Sketch of the Cu Kα X-ray diffraction pattern of FeSe based on the results of Yeh et al. [43]. The Miller indices are based on a PbO-type tetragonal structure [43]

the lattice parameters of FeTe are larger than those of FeSe. The increase of the lattice parameter a is only 1.4%, whereas c is 13.6% longer than in FeSe. The increased lattice parameters are reflected in the smaller diffraction angles from the FeTe.

Figure 18.15 shows the tetragonal crystal structure of LaOFeAs (La-1111) (space group *P4/nmm*). The number of atoms in the unit cell corresponds to two formula units. As can be seen for the Fe atom in the center of the unit cell, each iron atom is associated with four arsenic atoms.

In Figure 18.16, the crystal structure of $CaFe_2As_2$ (Ca-122) is presented [27, 44]. The (1 2 2) compounds crystallize in a tetragonal structure of the space group *I4/mmm*.

18.4 Physical Properties of Iron-based Superconductors

In the present section, some of the physical properties of iron-based superconductors will be summarized. The first aspect to be considered is the superconducting energy gap. A

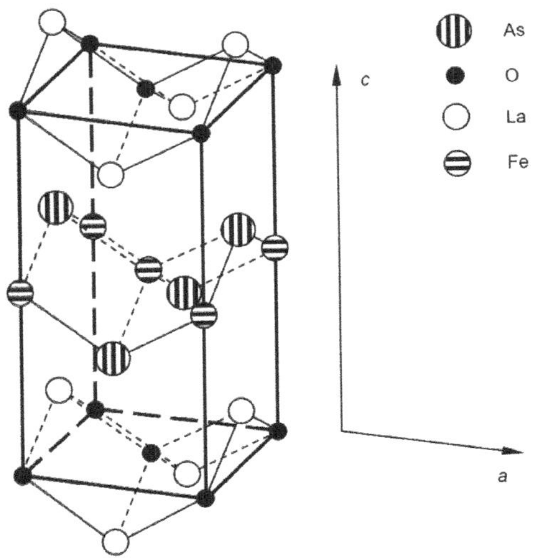

Figure 18.15 *Sketch of the tetragonal crystal structure of LaOFeAs (La-1111) (space group P4/nmm). The lattice parameters of the undoped compound are* $a = 0.4033$ *nm and* $c = 0.8741$ *nm at 300 K [7]. The z positions of the atoms are* $z = 0$ *and* $z = c$ *for oxygen,* $z = 0.141\,c$ *and* $z = 0.859\,c$ *for La,* $z = 0.5\,c$ *for Fe, and* $z = 0.348\,c$ *and* $z = 0.652\,c$ *for As [7] (adapted from [24, 44])*

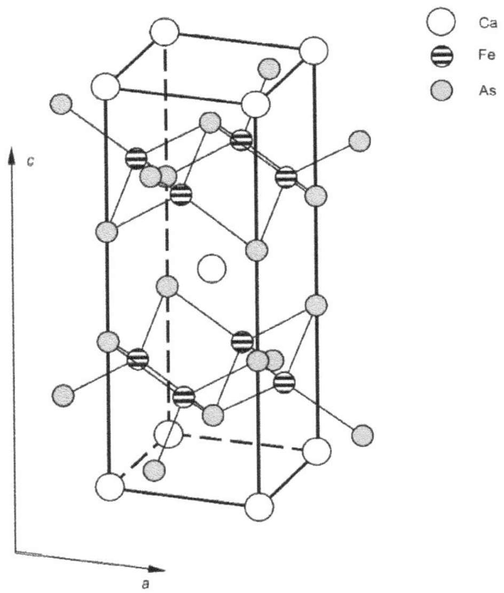

Figure 18.16 *Sketch of the tetragonal crystal structure of* $CaFe_2As_2$ *(Ca-122) (space group I4/mmm). The lattice parameters of the undoped compound are* $a = 0.3872$ *nm and* $c = 1.1730$ *nm at 297 K [27]. The z positions of the atoms are* $z = 0$, $z = 0.5\,c$ *and* $z = c$ *for Ca,* $z = 0.25\,c$ *and* $z = 0.75\,c$ *for Fe, and* $z = 0.133\,c$, $z = 0.367\,c$, $z = 0.633\,c$ *and* $z = 0.867\,c$ *for As [27] (adapted from [27, 44])*

Table 18.3 *Superconducting energy gaps of iron-based superconductors*

Compound	T_c(K)	Δ(meV)	$2\Delta/k_BT_c$	Comments	Reference
$LaO_{0.9}F_{0.1-\delta}FeAs$	27	3.9 ± 0.7	3.35 ± 0.6	Point contact spectroscopy	[45]
$SmO_{0.8}F_{0.2}FeAs$	51.2	6.45 ± 0.25	2.95 ± 0.15	PCAR[a]	[29]
		16.6 ± 1.6	7.65 ± 0.85	2 gaps	
$SmO_{0.85}F_{0.15}FeAs$	42	6.67	3.68	PCAR[a]	[46]
$NdO_{0.85}FeAs$	45.5	7	3.57	PCAR[a]	[47]
$BaFe_{1.8}Co_{0.2}As_2$	23.6	1.64 ± 0.2	1.6 ± 0.2	2 gaps	[48]
		6.2 ± 0.2	6.1 ± 0.2		
$Ba_{0.6}K_{0.4}Fe_2As_2$	37	6	3.76	ARPES[b]	[49]
		12	7.53	2 gaps	
LiFeAs	17	3 ± 0.2	4.1 ± 0.3	Single gap ARPES [b]	[50]
$FeSe_{0.5}Te_{0.5}$	19.1	4.1 ± 0.4	4.9 ± 0.5		[51]
$Fe_{1.03}Te_{0.7}Se_{0.3}$	13	~4	~7	Single crystal ARPES [b]	[52]

[a]PCAR – point-contact Andreev-reflection spectroscopy.
[b]ARPES – angle-resolved photoemission spectroscopy.

discussion of the pairing symmetry is beyond the scope of the present introduction. The measured energy gaps of various iron-based superconductors are presented in Table 18.3.

For Sm-1111 and Ba-122, the existence of two energy gaps has been reported [29, 48, 49], while only a single gap has been found in LiFeAs [50]. The values of $2\Delta/k_BT_c$ for Sm-1111 and Ba-122 are between 6.1 and 7.65 for the larger gap and in the range of 1.6–3.76 for the smaller gap. The $2\Delta/k_BT_c$ values for the larger gap are significantly larger than the single gap value of 3.5 provided by the BCS theory.

Next the lower and upper critical fields of iron-based superconductors are considered. In general, the large values of the derivative dB_{c2}/dT measured at the critical temperature indicate that the upper critical fields of iron-based superconductors can reach values of the order of 100 T at zero temperature. Frequently the values of B_{c2} at zero temperature have been estimated using the Werthamer–Helfand–Hohenberg (WHH) formula (see footnote to Table 18.4). It should be noted that these extrapolated values have to be considered as only a rough estimate of B_{c2}. A more detailed discussion of the limitations of the upper critical field values can be found in a review article by Fuchs *et al.* [53]. The lower and upper critical fields of various iron-based superconductors are listed in Table 18.4.

Due to the layered crystal structures of the iron-based superconductors, their upper critical fields are anisotropic. For Sm-1111 [29], a moderate anisotropy factor $B_{c2}{}^{ab}/B_{c2}{}^{c}$ of ~7.3 has been found at temperatures close to T_c. An even smaller anisotropy factor of 2.9 has been reported for the upper critical field of (Ba,K)-122 close to T_c [29]. The upper critical fields of Sm-1111, Nd-1111, Ba-122, Ca-122, Sr-122, and $K_{0.58}Fe_{1.56}Se_2$ reach or exceed 100 T at zero temperature.

The characteristic length scales of various iron-based superconductors are listed in Table 18.5. The penetration depth $\lambda_{ab}(T = 0)$ for screening currents flowing in the planes of the FeAs layers are in the range of 115–320 nm, while $\lambda_c(0)$ lies between 304 and 870 nm. The coherence length in the planes of the FeAs layers is of the order of a few

Table 18.4 *Lower and upper critical fields of iron-based superconductors*

Compound	T_c(K)	B_{c2}(T)	B_{c1}(mT)	Comments	Reference
La-1111		36		$B_{c2}(T=0)$ estimated with WHH formula[a]	[6]
La-1111		63–65	5		[54]
$LaO_{0.8}F_{0.2}FeAs_{1-x}Sb_{0.05}$	30.1	~73		Polycrystalline	[11]
$LaO_{0.9}F_{0.1-\delta}FeAs$	28.9	56		$B_{c2}(T=0)$	[55]
Sm-1111		150		$B_{c2}(T=0)$ estimated with WHH formula[a]	[6]
$SmO_{0.65}F_{0.35}FeAs$	52	~120		$B_{c2}(T=0)$ estimated with WHH formula[a]	[56]
$SmO_{0.7}F_{0.25}FeAs$	49.7	13.5 13.5		$B \parallel c$, ~40 K $B \parallel ab$, ~46.9 K $B_{c2}{}^{ab}/B_{c2}{}^{c} \sim 7.3$, $T \approx T_c$	[29]
$SmO_{0.9}F_{0.1}FeAs$	52		9.33	$B_{c1}(0)$	[57]
Nd-1111		204		$B_{c2}(T=0)$ estimated with WHH formula[a]	[6]
$NdO_{0.82}F_{0.18}FeAs$	50.5	250		$B_{irr}(0) \approx 150$ T	[58]
Pr-1111		72		$B_{c2}(T=0)$ estimated with WHH formula[a]	[6]
Ce-1111		107		$B_{c2}(T=0)$ estimated with WHH formula[a]	[6]
$CeO_{0.9}F_{0.1}FeAs$	38.4	94		$B_{c2}(T=0)$ estimated with WHH formula[a]	[59]
$CeO_{0.9}F_{0.1}FeAs$	38.6	185		$B_{c2}(T=0)$ $\xi(0) = 3.34$ nm	[60]
$(Ba,Rb)Fe_2As_2$	23.35	13.5 13.5		$B \parallel c$, ~19.3K $B \parallel ab$, ~21.2K $B_{c2}{}^{ab}/B_{c2}{}^{c} \sim 2.9$, $T \approx T_c$	[29]
$Ba_{0.6}K_{0.4}Fe_2As_2$	36.2		59	$B_{c1}(0)$	[57]
$Ba_{0.6}K_{0.4}Fe_2As_2$	38	75		$B_{c2}{}^{ab}/B_{c2}{}^{c} \sim 1$–3	[4]
$(Ba,K)Fe_2As_2$		100		$B_{c2}(T=0)$ estimated with WHH formula[a]	[6]
$(Ba,K)Fe_2As_2$	28	70		$B_{c2}{}^{ab}/B_{c2}{}^{c} \sim 1.4$ Measurements of resistivity up to 60 T	[61]
$Ca_{0.77}La_{0.23}Fe_2As_2$ Single crystals	~36	102.7 16.8		$B_{c2}{}^{ab}$ $B_{c2}{}^{c}$ $B_{c2}(T=0)$ estimated with WHH formula[a]	[62]
$Sr_{0.6}K_{0.4}Fe_2As_2$	34	170		$B_{c2}(T=0)$ estimated with WHH formula[a]	[63]
$FeSe_{0.89}$	9.91	33.4	0.95	$B_{c2}(T=0)$ estimated with WHH formula[a]	[64]
Fe(Te,Se)		87		$B_{c2}(T=0)$ estimated with WHH formula[a]	[6]
$K_{0.58}Fe_{1.56}Se_2$ Single crystal	31.9	276 84		$B_{c2}{}^{ab}$ $B_{c2}{}^{c}$ $B_{c2}(T=0)$ estimated with WHH formula[a]	[65]

[a]Werthamer–Helfand–Hohenberg (WHH) formula $B_{c2}(0) = -0.693T_c[dB_{c2}/dT]$ at $T = T_c$.

Table 18.5 *Penetration depth and coherence length of iron-based superconductors*

Compound	T_c(K)	λ(nm)	ξ(nm)	Comments	Reference
$LaO_{0.89}F_{0.11}FeAs$			5	$\xi_{ab}(0)$	[54]
			1.2	$\xi_c(0)$	
$LaO_{0.9}F_{0.1}FeAs$	26	254		$\lambda_{ab}(0)$	[66]
$LaO_{0.8}F_{0.2}FeAs_{1-x}Sb_{0.05}$	30.1		2.2	Polycrystalline	[11]
$CeO_{0.8}F_{0.2}FeAs$	42.5		2.7		[59]
$NdO_{0.82}F_{0.18}FeAs$	47		3.1	$\xi_{ab}(0)$	[67]
			0.78	$\xi_c(0)$	
Nd-1111	47.4		1.8	$\xi_{ab}(0)$	[68]
			0.45	$\xi_c(0)$	
$SmO_{0.9}F_{0.1}FeAs$	52	205		$\lambda_{ab}(0)$	[57]
$Ca_{0.77}La_{0.23}Fe_2As_2$	~36		4.43	$\xi_{ab}(0)$	[62]
			0.72	$\xi_c(0)$	
$Ba_{0.6}K_{0.4}Fe_2As_2$	34.6	125	1.25	$\lambda_{ab}(0)$, $\xi_{ab}(0)$	[67]
		304	0.45	$\lambda_c(0)$, $\xi_c(0)$	
$Ba_{0.6}K_{0.4}Fe_2As_2$	36.2	115		$\lambda_{ab}(0)$	[57]
(Ba,K)-122	32	320		$\lambda_{ab}(0)$, μSR [a]	
		270		$\lambda_{ab}(0)$, ARPES [b]	
Ba-122	22		2.4	$\xi_{ab}(0)$	[68]
			1.2	$\xi_c(0)$	
$BaFe_{1.8}Co_{0.2}As_2$		169		$\lambda_{ab}(0)$	[48]
$SrFe_{1.75}Co_{0.25}As_2$	13.3	316		$\lambda_{ab}(0)$	[69]
		870		$\lambda_c(0)$	
LiFeAs	17	210	2–4	$\lambda_{ab}(0)$, $\xi_{ab}(0)$	[50]
$FeSe_{0.89}$	9.91	1008	3.14	$\lambda(0)$, $\xi(0)$	[64]
Fe(Se,Te)	14.5		1.2	$\xi_{ab}(0)$	[68]
			0.35	$\xi_c(0)$	

[a] μSR – muon spin rotation.
[b] ARPES – angle-resolved photoemission spectroscopy.

nanometers, whereas ξ_c is as small as 1.2 nm or even less. In general, the coherence length is much smaller than the penetration depth, indicating that the iron-based superconductors are extreme Type II superconductors.

Finally, the resistivity of iron-based superconductors and their parent compounds is considered. The top panel of Figure 18.17 shows the resistivity of various members of the (1 1 1 1) family. The resistivity of LaOFeAs (La-1111) reaches ~4.5 mΩ cm at 300 K. The resistivity anomaly at a temperature of ~140 K is caused by the occurrence of a spin density wave and antiferromagnetic ordering. The nondoped La-1111 is not superconducting, while the fluorine-doped compound $LaO_{0.8}F_{0.2}FeAs$ is superconducting. In addition, fluorine doping suppresses magnetic ordering, and hence the resistivity anomaly at ~140 K has vanished. The room temperature resistivity of ~1.5 mΩ cm is around three times lower than that of the undoped La-1111. Resistivity anomalies are also visible in the $\rho(T)$ curves of the nonsuperconducting compounds SrFeAsF and $SmO_{0.98}F_{0.02}FeAs$, whereas the superconducting compounds $Sr_{0.5}Sm_{0.5}FeAsF$ and $SmO_{0.74}F_{0.26}FeAs$ show no anomaly, and resistivity drops nearly linearly till the transition to superconductivity at the T_c in question. The bottom panel of Figure 18.17 shows the resistivity versus temperature curves of

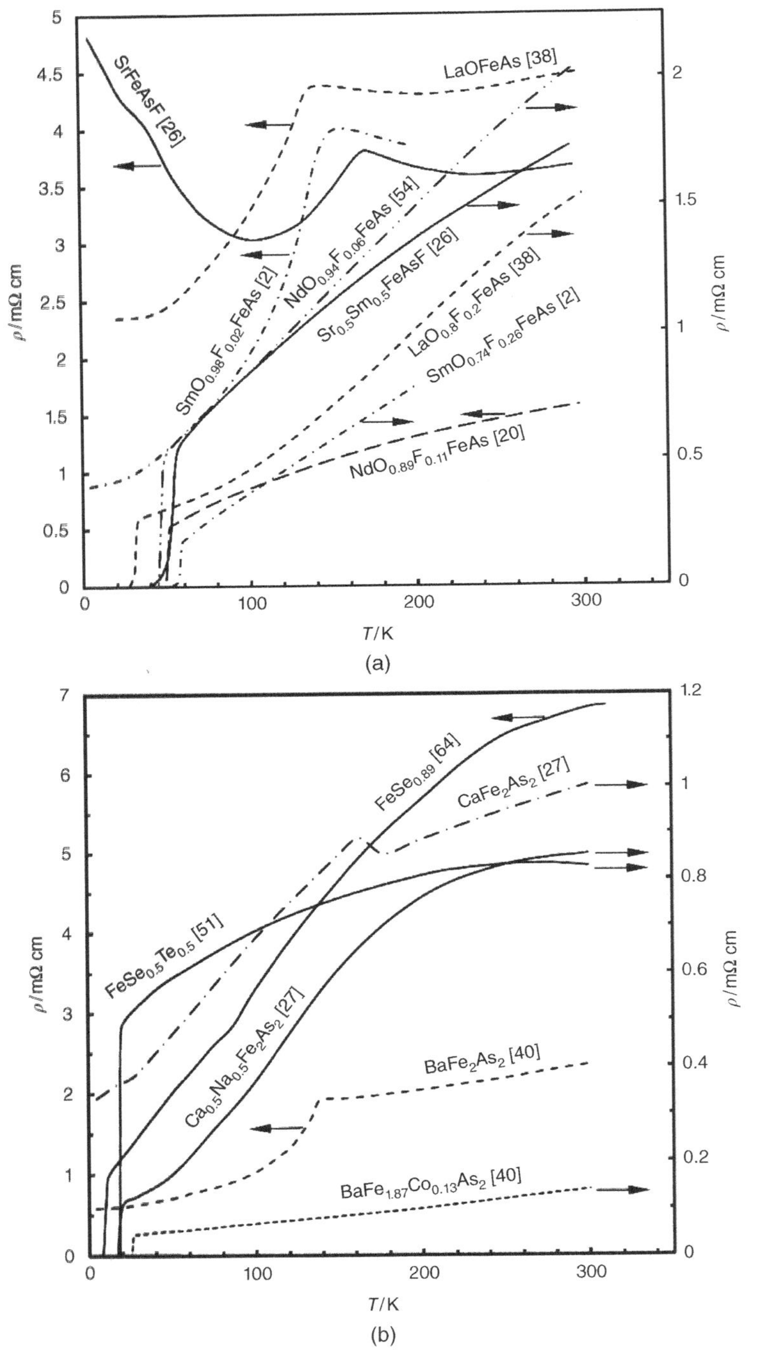

Figure 18.17 *Resistivity versus temperature of various iron-based superconductors. Top panel: resistivity of (1 1 1 1) superconductors and their undoped parent compounds (Results from [2, 20, 26, 38, 54]). Bottom panel: resistivity of (1 1) and (1 2 2) superconductors. (Results from [27, 40, 51, 64])*

$FeSe_{0.89}$, $FeSe_{0.5}Te_{0.5}$, $CaFe_2As_2$, $Ca_{0.5}Na_{0.5}Fe_2As_2$, $BaFe_2As_2$, and $BaFe_{1.87}Co_{0.13}As_2$. Again, the $\rho(T)$ curves of the nonsuperconducting parent compounds $CaFe_2As_2$ (Ca-122) and $BaFe_2As_2$ (Ba-122) show a resistivity anomaly related to spin density waves and magnetic ordering. The room temperature resistivity of these iron-based superconductors varies between 0.14 and 6.8 mΩcm. For comparison, the room temperature resistivity of copper is only 1.67 μΩ cm.

18.5 Synthesis of Iron-based Superconductors

Examples of the synthesis conditions used for the preparation of various iron-based superconductors are gathered in Table 18.6. In the preparation of iron-pnictide superconductors, the toxicity and volatility of arsenic need attention. Some dopants (e.g., fluorine, alkali metals) are also very volatile. Because of sensitivity to oxygen and moisture, the starting materials must be handled in a glove box filled with highly pure argon or nitrogen. The main techniques for the preparation of iron-pnictide superconductors are solid-state reaction, high pressure synthesis, and the flux method for the growth of single crystals (see, e.g., [6]). In the solid-state reaction technique, arsenic is provided by arsenic compounds (e.g., LaAs, SmAs, BaAs, KAs, FeAs, and Fe_2As) because the vapor pressure of elemental arsenic is so high. The initial materials need to be carefully mixed and pressed into pellets. The reaction is typically performed in evacuated quartz tubes. Heat treatment conditions for the synthesis of various iron-pnictides by the solid-state reaction are shown in Table 18.6. The advantage of the high-pressure method is the suppression of arsenic and fluorine loss from compounds such as $LaO_{1-x}F_xFeAs$ at high reaction temperatures. A publication by Karpinski *et al.* [29] provides a short description of the growth of $LnO_{1-x}F_xFeAs$ (Ln represents La, Pr, Nd, Sm, or Gd) and $(Ba,Rb)Fe_2As_2$ single crystals (see also Table 18.6).

18.6 Critical Current Densities in Iron-based Superconductors

The weak-link behavior of large-angle grain boundaries is a severe disadvantage of cuprate high-T_c superconductors. To overcome the blocking of the supercurrents by large-angle grain boundaries, the superconducting RE-123 (RE = rare earth) film of coated conductors needs to be biaxially textured. Due to the biaxial texture, most of the individual grains are connected by low-angle grain boundaries, which do not block the supercurrents. The coherence length in iron-based superconductors is nearly as short as in cuprate superconductors, which may cause weak-link behavior of clean grain boundaries as in the cuprates.

Several research groups [54, 70–74] have studied the intergrain and intragrain critical current densities in iron-based superconductors. In polycrystalline Sm-1111 and Nd-1111, Yamamoto *et al.* [54] found intragrain j_c values of $\approx 5 \times 10^6$ A/cm^2 at 5 K, whereas the intergrain j_c values reached only 1000 to 10000 A/cm^2. However, the large difference of intergrain and intragrain critical current densities is partly caused by wetting of the grain boundaries by FeAs. Very recently, Hong *et al.* [74] have reported weak coupling at junctions across natural grain boundaries in Ba-122.

Figure 18.18 shows the critical current densities achieved in a Ba-122-coated conductor [75] and an epitaxial $FeTe_{0.5}\ Se_{0.5}$ film [76]. The critical current density in the

Table 18.6 *Conditions used for the synthesis of iron-based superconductors (symbols: RT room temperature)*

Compound	Starting materials	Short description	Reference
$LaO_{1-x}F_xFeAs$ $0.05 \leq x \leq 0.12$	LaAs, FeAs, La_2O_3 (dehydrated)	Reaction in an Ar-filled silica tube at 1250 °C for 40 h	[1]
	Addition of LaF_3 and La with a 1:1 ratio	Fluorine doping	
$LaO_{0.9}F_{0.1-\delta}FeAs$	Fe powder, As grains mixed and ground in 1:1 ratio	Two-step solid-state reaction Pressed pellets heat treated at 700 °C for 10 h in evacuated quartz tube	[55]
	Addition of LaF_3, La_2O_3, La grains	Pressed pellets heat treated in evacuated quartz tube 940 °C/2 h and 1150 °C/48 h Slow cooling to RT	
$LaO_{1-x}F_xFeAs$ $0 \leq x \leq 0.4$	La, Fe, As mixed in the ratio 1:3:3 → LaAs, FeAs, Fe_2As	Two-step solid-state reaction Pressed pellets heat treated in evacuated quartz tube at 900 °C for 12 h	[38]
	Addition of La_2O_3, La, and LaF_3	Ground powder is pressed into pellets and heat treated at 1150 °C for 60 h in evacuated quartz tube	
$LaO_{0.8}F_{0.2}FeAs$	Fe chips and As powder → FeAs	950 °C for 24 h	[11]
	La_2O_3, LaF_3, La, FeAs, $FeSb_2$, Fe	Reaction in evacuated silica ampoules 950 °C/48 h Pellets wrapped in Ta foil 1150 °C/48 h in evacuated silica ampoules	
$SmO_{1-x}F_xFeAs$ $x = 0.15$	SmAs, SmF_3, Fe, Fe_2O_3	Sm + As → SmAs 600 °C/3 h, 900 °C/5 h Pressed pellets wrapped in Ta foils, 1160 or 1200 °C for 40 h in evacuated quartz tube	[13]
$SmO_{0.85}FeAs$	SmAs, Fe, Fe_2O_3	Mixed powders pressed into pellets, 1250 °C/2 h under 6 GPa	[54]
$SmO_{1-x}F_xFeAs$ $0 \leq x \leq 0.3$	Sm, Fe, As → $SmFe_3As_3$ (1 3 3), $Sm_2Fe_3As_3$ (2 3 3)	Precursor preparation at 850 °C for 10 h	[2]
	Sm_2O_3, SmF_3, 133, 233	Mixed powders pressed into Pellets 980 °C/40 h/5 °C/h to 600 °C in evacuated quartz tube	
$NdO_{1-x}F_xFeAs$ $x = 0.11$	Nd, As, Fe, Fe_2O_3, FeF_3	Pressed pellets sealed in BN crucibles. Synthesis at 1300 °C/2 h/6 GPa	[20]

Table 18.6 *(continued)*

Compound	Starting materials	Short description	Reference
$NdO_{0.94}F_{0.06}FeAs$	NdAs, Fe, Fe_2O_3, FeF_2	Mixed powders pressed into pellets, 1250 °C/2 h/6 GPa	[54]
$CeO_{1-x}F_xFeAs$ x = 0.1, 0.2	Ce, CeO_2, CeF_3, FeAs	Fe + As → FeAs 900 °C/24 h Mixed powders sealed in evacuated silica ampoules sintered at 1000 °C/48 h Pellets wrapped in Ta foil in evacuated silica ampoules 1180 °C/48 h	[59, 60]
$Sr_{0.5}Sm_{0.5}FeAsF$	SrF_2, SrAs, SmAs, Fe_2As	Sr + As → SrAs (873 K[a]) Sm + As → SmAs (1073 K[a]) 2Fe + As → Fe_2As (1073 K[a]) Ground and mixed powders. Pressed pellets wrapped in Ta foil, 1173 K, 40 h[b] After grinding second heat treatment at 1273 K, 20 h[b]	[26]
$Ba_{1-x}K_xFe_2As_2$ $0 \le x \le 1$	Ba, Fe, As Ba purified by distillation As purified by sublimation K doping	$BaFe_2As_2$ parent compound 650–900 °C in alumina crucibles in pure argon Repeated mixing and annealing required to obtain phase pure samples Prereaction of K with As at low temperatures	[4]
$BaFe_2(As_{1-x}P_x)_2$ $0 \le x \le 1$	Ba, Fe, As, P	Mixing and pressing of pellets in glove box filled with Ar. All heat treatments performed in evacuated quartz tubes 600 K/12 h and 1173 K/40 h Regrinding and pelletizing 1273 K for 24 h	[30]
$Ba_{0.6}K_{0.4}Fe_2As_2$	BaAs, KAs, Fe_2As Composition $Ba_{0.6}K_{0.42}Fe_2As_{2.02}$	Alumina crucible sealed in a stainless steel tube. N_2 atmosphere, 800 °C/20 h	[41]
LiFeAs	Li, Fe, As	Starting materials sealed in Nb tubes welded in Ar Reaction in evacuated quartz containers at 740 °C for 24 h	[42]
LiFeAs	Li, FeAs Handling in glove box filled with purified Ar.	Reaction in evacuated quartz containers at 750 °C, 4 days, heating rate 1 °C/min, cooling to 150 °C with 0.05 °C/min	[3]

(continued overleaf)

Table 18.6 (continued)

Compound	Starting materials	Short description	Reference
NaFeAs	Na, FeAs	Analogously to Li-111 5% excess of Na	[3]
$LnO_{1-x}F_xFeAs$ Ln: La, Pr, Nd, Sm, Gd Single crystals	LnAs, FeAs, Fe, Fe_2O_3, LnF_3 Handling in glove box. NaCl/KCl flux dissolved in H_2O.	NaCl/KCl flux, precursor:flux ratio of 1:1 and 1:3 Ground and mixed precursor powders and flux placed in BN crucible 3 GPa, RT → T_a in 1 h T_a = 1350–1450 °C for 4–85 h, T_a → RT in 1–24 h followed by pressure release	[29]
$Ba_{1-x}Rb_xFe_2As_2$ Single crystals	Ba, Rb, Fe_2As, As, Sn	Sn flux (Fe : Sn = 1 : 24) Placed in alumina crucibles and sealed in silica tubes 1/3 bar Ar, 850 °C/3 h/50 h → 500 °C Dissolution of Sn in liquid Hg at RT and removal of Hg at 190 °C in vacuum	[29]

[a]For 10 h in evacuated quartz tubes.
[b]In evacuated quartz tubes.

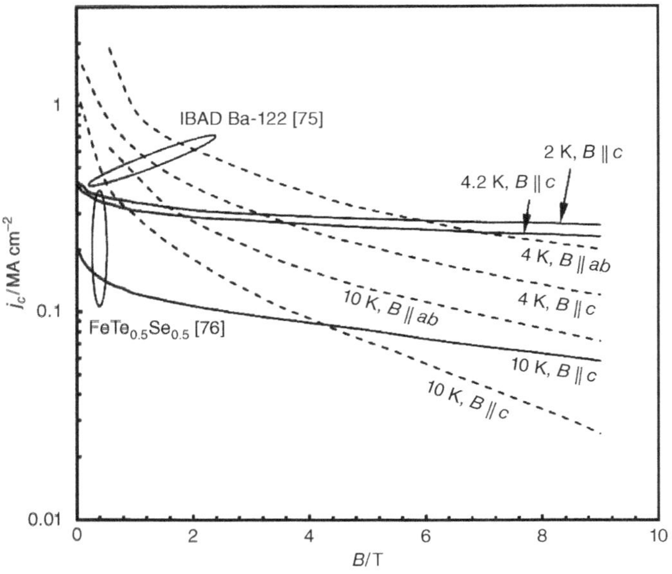

Figure 18.18 *Field dependences of critical current densities (1 μV/cm criterion) in an epitaxial $FeTe_{0.5}Se_{0.5}$ film deposited on a (1 0 0) CaF_2 single crystal and in a Ba-122-coated conductor at temperatures between 2 and 10 K. The sequence of the materials in the Ba-122-coated conductor is C276 Hastelloy™, Y_2O_3, IBAD (ion-beam-assisted deposition, see Chapter 15) MgO, Fe buffer layer, and $BaFe_{1.8}Co_{0.2}As_2$ (Results from [75, 76])*

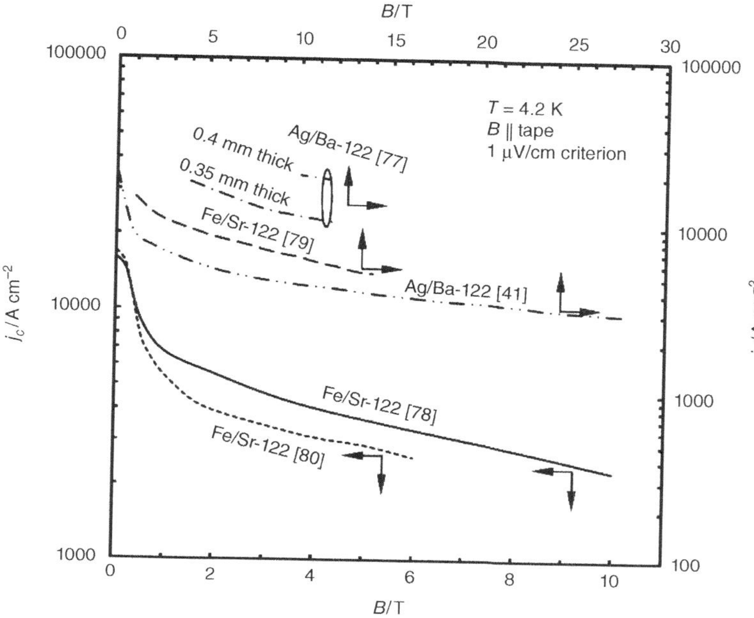

Figure 18.19 *Critical current densities at 4.2 K versus field of various Ba-122 and Sr-122 tapes prepared by the ex situ powder-in-tube (PIT) technique (Results from [41, 77–80])*

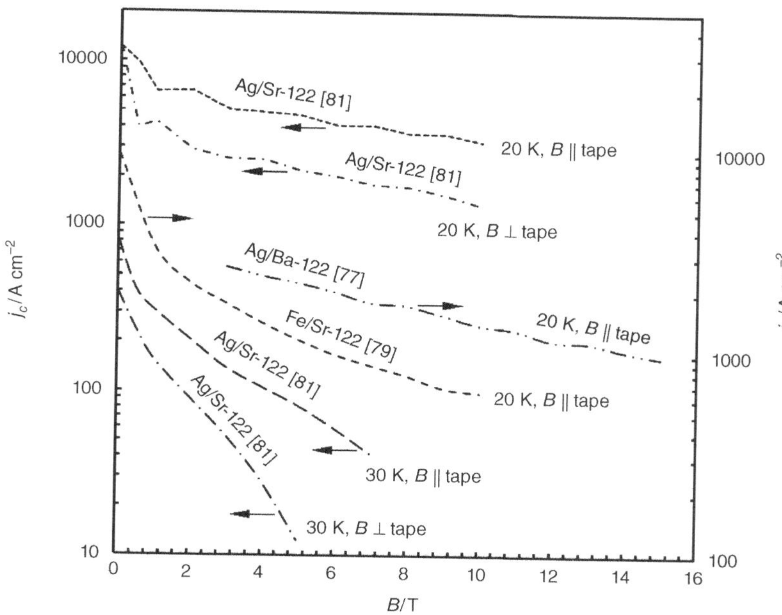

Figure 18.20 *Critical current densities of various (1 2 2) tapes versus field for temperatures in the range of 20–30 K (Results from [77, 79, 81])*

Ba-122-coated conductor exceeds 1 MA/cm^2 at a zero applied field and a temperature of 10 K. The anisotropy of j_c with respect to the applied magnetic field is smaller than that in RE-123-coated conductors. The ratio of the critical current densities for fields parallel ($B \,||\, ab$) and perpendicular to the FeAs layers ($B \,||\, c$) is less than a factor of 2 at 4 K and 9 T. The j_c in the epitaxial $FeTe_{0.5}Se_{0.5}$ film reaches 0.41 MA/cm^2 without applied field at 4.2 K, and 0.23 MA/cm^2 at 9 T and 4.2 K.

Figure 18.19 shows the critical current densities of Ba-122 [41, 77] and Sr-122 tapes [78–80] prepared by the *ex situ* powder-in-tube (PIT) technique. In this technique, fully reacted (1 2 2) powder is filled into a metal tube (e.g., Ag and Fe) and drawn and rolled into a tape. In uniaxially pressed Ag/Ba-122 tapes 0.4 mm thick, the j_c exceeded 20000 A/cm^2 at 4.2 K and 10 T parallel to the broad face of the tapes. For an Ag/Ba-122 tape, j_c measurements, at up to 28 T, indicate that the iron-based superconductors are potentially high-field superconductors.

Figure 18.20 shows the field dependence of j_c of Ag/Ba-122 [77], Fe/Sr-122 [79], and Ag/Sr-122 [81] tapes at elevated temperatures. For Ag/Sr-122 [81], the anisotropy with respect to the direction of the applied magnetic field increases with increasing temperature.

18.7 Summary

The iron-based superconductors with critical temperatures up to 58 K in Sm-1111 [2] form a second class of high-temperature superconductors. In the iron-pnictides, the supercurrents flow in the FeAs layers, which are separated by layers of (for example) LnO (Ln = lanthanide) in the (1 1 1 1) compounds or layers of Ca, Ba or Sr ions in the (1 2 2) compounds. The nonsuperconducting parent compounds show antiferromagnetic ordering. In contrast to the cuprates, these parent compounds are metals and not Mott insulators. Doping suppresses the spin density wave antiferromagnetic ordering, and superconductivity occurs. In the iron-pnictides, superconductivity can also be induced by doping into the FeAs layer, whereas doping into the CuO_2 planes of cuprates destroys superconductivity. As in the cuprates, the upper critical fields are extremely high, reaching values of the order of 100 T at zero temperature. Advantages of the (1 2 2) family are the high critical current density, and the very small anisotropy of the upper critical field with respect to the direction of the applied magnetic field. The coherence length perpendicular to the FeAs layers is extremely short (typically < 1 nm). The iron-based superconductors are extreme Type II superconductors with penetration depths much larger than the coherence length.

References

1. Y. Kamihara, T. Watanabe, M. Hirano, and H. Hosono, Iron-based layered superconductor $La[O_{1-x}F_x]FeAs$ ($x = 0.05$-0.12) with $T_c = 26$ K, *J. Am. Chem. Soc.*, **130**, 3296–3297 (2008).
2. M. Fujioka, S.J. Denholme, T. Ozaki, H. Okazaki, K. Deguchi, S. Demura, H. Hara, T. Watanabe, H. Takeya, T. Yamaguchi, H. Kumakura, and Y. Takano, Phase diagram and superconductivity at 58.1 K in α-FeAs-free $SmFeAsO_{1-x}F_x$, *Supercond. Sci. Technol.*, **26**, 085023 (2013).

3. C.W. Chu, F. Chen, M. Gooch, A.M. Guloy, B. Lorenz, B. Lv, K. Sasmal, Z.J. Tang, J.H. Tapp, and Y.Y. Zue, The synthesis and characterization of LiFeAs and NaFeAs, *Physica C*, **469**, 326–331 (2009).
4. D. Johrendt and R. Pöttgen, Superconductivity, magnetism and crystal chemistry of $Ba_{1-x}K_xFe_2As_2$, *Physica C*, **469**, 332–339 (2009).
5. J.W. Lynn and P. Dai, Neutron studies of the iron-based family of high T_c magnetic superconductors, *Physica C*, **469**, 469–476 (2009).
6. P.M. Aswathy, J.B. Anooja, P.M. Sarun, and U. Syamaprasad, An overview on iron based superconductors, *Supercond. Sci. Technol.*, **23**, 073001 (2010).
7. T. Nomura, S.W. Kim, Y. Kamihara, M. Hirano, P.V. Sushko, K. Kato, M. Takata, A.L. Shluger, and H. Hosono, Crystallographic phase transition and high-T_c superconductivity in LaFeAsO:F, *Supercond. Sci. Technol.*, **21**, 125028 (2008).
8. C. de la Cruz, Q. Huang, J.W. Lynn, J. Li, W. Ratcliff II,, J.L. Zarestky, H.A. Mook, G.F. Chen, J.L. Luo, N.L. Wang, and P. Dai, Magnetic order close to superconductivity in the iron-based layered $LaO_{1-x}F_xFeAs$ system, *Nature*, **453**, 899–902 (2008).
9. Y. Kamihara, H. Hiramatsu, M. Hirano, R. Kawamura, H. Yanagi, T. Kamiya, and H. Hosono, Iron-based layered superconductor: LaOFeP, *J. Am. Chem. Soc.*, **128**, 10012–10013 (2006).
10. H.-H. Wen, G. Mu, L. Fang, H. Yang, and X. Zhu, Superconductivity at 25 K in hole-doped $(La_{1-x}Sr_x)OFeAs$, *Europhys. Lett.*, **82**, 17009 (2008).
11. S.J. Singh, J. Prakash, S. Patnaik, and A.K. Ganguli, Enhancement of the superconducting transition temperature and upper critical field of $LaO_{0.8}F_{0.2}FeAs$ with antimony doping, *Supercond. Sci. Technol.*, **22**, 045017 (2009).
12. H. Takahashi, K. Igawa, K. Arii, Y. Kamihara, M. Hirano, and H. Hosono, Superconductivity at 43 K in an iron-based layered compound $LaO_{1-x}F_xFeAs$, *Nature*, **453**, 376–378 (2008).
13. X.H. Chen, T. Wu, G. Wu, R.H. Liu, H. Chen, and D.F. Fang, Superconductivity at 43 K in $SmFeAsO_{1-x}F_x$, *Nature*, **453**, 761–762, (2008).
14. C. Wang, Z. Gao, L. Wang, Y. Qi, D. Wang, C. Yao, Z. Zhang, and Y. Ma, Low-temperature synthesis of $SmO_{0.8}F_{0.2}FeAs$ superconductor with $T_c = 56.1$ K, *Supercond. Sci. Technol.*, **23**, 055002 (2010).
15. Y. Qi, Z. Gao, L. Wang, D. Wang, X. Zhang, and Y. Ma, Superconductivity in Co-doped SmFeAsO, *Supercond. Sci. Technol.*, **21**, 115016 (2008).
16. G.F. Chen, Z. Li, D. Wu, G. Li, W.Z. Hu, J. Dong, P. Zheng, J.L. Luo, and N.L. Wang, Superconductivity at 41 K and its competition with spin-density-wave instability in layered $CeO_{1-x}F_xFeAs$, *Phys. Rev. Lett.*, **100**, 247002 (2008).
17. Y. Wang, W.J. Ren, D. Li, W.J. Hu, B. Li, and Z.D. Zhang, Superconductivity modulated by internal pressure in $Ce_{1-x}Gd_xFeAsO_{0.84}F_{0.16}$ compounds, *Supercond. Sci. Technol.*, **23**, 025003 (2010).
18. M.A. McGuire, R.P. Hermann, A.S. Sefat, B.C. Sales, R. Jin, D. Mandrus, F. Grandjean, and G.J. Long, Influence of the rare-earth element on the effects of the structural and magnetic phase transitions in CeFeAsO, PrFeAsO and NdFeAsO, *New J. Phys.*, **11**, 025011 (2009).
19. M. Ishikado, S. Shamoto, H. Kito, A. Iyo, H. Eisaki, T. Ito, and Y. Tomioka, Growth of single crystal $PrFeAsO_{1-y}$, *Physica C*, **469**, 901–904 (2009).

20. Z.-A. Ren, J. Yang, W. Lu, W. Yi, X.-L. Shen, Z.-C. Li, G.-C. Che, X.-L. Dong, L.-L. Sun, F. Zhou, and Z.X. Zhao, Superconductivity in the iron-based F-doped layered quaternary compound [$NdO_{1-x}F_x$]FeAs, *Europhys. Lett.*, **82**, 57002 (2008).
21. A. Kursumovic, J.H. Durrell, S.K. Chen, and J.L. MacManus-Driscoll, Ambient/low pressure synthesis and fast densification to achieve 55 K T_c superconductivity in $NdFeAsO_{0.75}F_{0.25}$, *Supercond. Sci. Technol.*, **23**, 025022 (2010).
22. J.D. Moore, L.F. Cohen, Y. Yeshurun, A.D. Caplin, K. Morrison, K.A. Yates, C.M. McGilvery, J.M. Perkins, D.W. McComb, C. Trautmann, Z.A. Ren, J. Yang, W. Lu, X.L. Dong, and Z.X. Zhao, The effect of columnar defects on the pinning properties of $NdFeAsO_{0.85}$ conglomerate particles, *Supercond. Sci. Technol.*, **22**, 125023 (2009).
23. J. Yang, Z.-C. Li, W. Lu, W. Yi, X.-L. Shen, Z.-A. Ren, G.-C. Che, X.-L. Dong, L.-L. Sun, F. Zhou, and Z.-X. Zhao, Superconductivity at 53.5 K in $GdFeAsO_{1-\delta}$, *Supercond. Sci. Technol.*, **21**, 082001 (2008).
24. C. Wang, L. Li, S. Chi, Z. Zhu, Z. Ren, Y. Li, Y. Wang, X. Lin, Y. Luo, S. Jiang, X. Xu, G. Cao, and Z. Xu, Thorium-doping-induced superconductivity up to 56 K in $Gd_{1-x}Th_xFeAsO$, *Europhys. Lett.*, **83**, 67006 (2008).
25. S. Matsuishi, Y. Inoue, T. Nomura, Y. Kamihara, M. Hirano, and H. Hosono, Effect of 3d transition metal doping on the superconductivity in quaternary fluoroarsenide CaFeAsF, *New J. Phys.*, **11**, 025012 (2009).
26. G. Wu, Y.L. Xie, H. Chen, M. Zhong, R.H. Liu, B.C. Shi, Q.J. Li, X.F. Wang, T. Wu, Y.J. Yan, J.J. Ying, and X.H. Chen, Superconductivity at 56 K in samarium-doped SrFeAsF, *J. Phys.: Condens. Matter*, **21**, 142203 (2009).
27. G. Wu, H. Chen, T. Wu, Y.L. Xie, Y.J. Yan, R.H. Liu, X.F. Wang, J.J. Ying, and X.H. Chen, Different resistivity response to spin-density wave and superconductivity at 20 K in $Ca_{1-x}Na_xFe_2As_2$, *J. Phys.: Condens. Matter*, **20**, 422201 (2008).
28. P.L. Alireza, Y.T. Chris Ko, J. Gillett, C.M. Petrone, J.M. Cole, G.G. Lonzarich, and S.E. Sebastian, Superconductivity up to 29 K in $SrFe_2As_2$ and $BaFe_2As_2$ at high pressures, *J. Phys.: Condens. Matter*, **21**, 012208 (2009).
29. J. Karpinski, N.D. Zhigadlo, S. Katrych, Z. Bukowski, P. Moll, S. Weyeneth, H. Keller, R. Puzniak, M. Tortello, D. Daghero, R. Gonnelli, I. Maggio-Aprile, Y. Fasano, Ø. Fischer, K. Rogacki, and B. Batlogg, Single crystals of $LnFeAsO_{1-x}F_x$ (Ln = La, Pr, Nd, Sm, Gd) and $Ba_{1-x}Rb_xFe_2As_2$: Growth, structure and superconducting properties, *Physica C*, **469**, 370–380 (2009).
30. S. Jiang, H. Xing, G. Xuan, C. Wang, Z. Ren, C. Feng, J. Dai, Z. Xu, and G. Cao, Superconductivity up to 30 K in the vicinity of the quantum critical point in $BaFe_2(As_{1-x}P_x)_2$, *J. Phys.: Condens. Matter*, **21**, 382203 (2009).
31. J. Zhao, Q. Huang, C. de la Cruz, S. Li, J.W. Lynn, Y. Chen, M.A. Green, G.F. Chen, G. Li, Z. Li, J.L. Luo, N.L. Wang, and P. Dai, Structural and magnetic phase diagram of $CeFeAsO_{1-x}F_x$ and its relation to high-temperature superconductivity, *Nat. Mater.*, **7**, 953–959 (2008).
32. Y. Mizuguchi, Y. Hara, K. Deguchi, S. Tsuda, T. Yamaguchi, K. Takeda, H. Kotegawa, H. Tou, and Y. Takano, Anion height dependence of T_c for the Fe-based superconductor, *Supercond. Sci. Technol.*, **23**, 054013 (2010).
33. M.S. Torikachvili, S.L. Bud'ko, N. Ni, and P.C. Canfield, Pressure induced superconductivity in $CaFe_2As_2$, *Phys. Rev. Lett.*, **101**, 057006 (2008).

34. T. Imai, K. Ahilan, F.L. Ning, T.M. McQueen, and R.J. Cava, Why does undoped FeSe become a high-T_c superconductor under pressure?, *Phys. Rev. Lett.*, **102**, 177005 (2009).
35. S. Medvedev, T.M. McQueen, I. Trojan, T. Palasyuk, M.I. Eremets, R.J. Cava, S. Naghavi, F. Casper, V. Ksenofontov, G. Wortmann, and C. Felser, Superconductivity at 36 K in β-$Fe_{1.01}Se$ with the compression of the interlayer separation under pressure, arXiv:0903.2143 cond-mat.supra-con (2009).
36. F. Hardy, P. Adelmann, T. Wolf, H. V. Löhneysen, and C. Meingast, Large anisotropic uniaxial pressure dependencies of T_c in single crystalline $Ba(Fe_{0.92}Co_{0.08})_2As_2$, *Phys. Rev. Lett.*, **102**, 187004 (2009).
37. A. Amato, R. Khasanov, H. Luetkens, and H.-H. Klauss, Probing the ground state properties of iron-based superconducting pnictides and related systems by muon-spin spectroscopy, *Physica C*, **469**, 609–613 (2009).
38. C. Shekhar, S. Singh, P.K. Siwach, H.K. Singh, and O.N. Srivastava, Synthesis and microstructural studies of iron oxypnictide $LaO_{1-x}F_xFeAs$ superconductors, *Supercond. Sci. Technol.*, **22**, 015005 (2009).
39. T. Hahn (ed.), *International Tables for Crystallography, Brief Teaching Edition of Volume A – Space-Group Symmetry*, D. Reidel Publishing Company, Dordrecht, The Netherlands, 1988.
40. D.L. Sun, Y. Liu, J.T. Park, and C.T. Lin, Growth of large single crystals of $BaFe_{1.87}Co_{0.13}As_2$ using a nucleation pole, *Supercond. Sci. Technol.*, **22**, 105006 (2009).
41. K. Togano, Z. Gao, H. Taira, S. Ishida, K. Kihou, A. Iyo, H. Eisaki, A. Matsumoto, and H. Kumakura, Enhanced high-field transport critical current densities observed for *ex situ* PIT processed Ag/(Ba, K)Fe_2As_2 thin tapes, *Supercond. Sci. Technol.*, **26**, 065003 (2013).
42. J.H. Tapp, Z. Tang, B. Lv, K. Sasmal, B. Lorenz, P.C.W. Chu, and A.M. Guloy, LiFeAs: An intrinsic FeAs-based superconductor with $T_c = 18$ K, *Phys. Rev. B*, **78**, 060505(R) (2008).
43. K.-W. Yeh, T.-W. Huang, Y.-L. Huang, T.-K. Chen, F.-C. Hsu, P.M. Wu, Y.-C. Lee, Y.-Y. Chu, C.-L. Chen, J.-Y. Luo, D.-C. Yan, and M.-K. Wu, Tellurium substitution effect on superconductivity of the α-phase iron selenide, *Europhys. Lett.*, **84**, 37002 (2008).
44. T. Yildirim, Frustrated magnetic interactions, giant magneto-elastic coupling, and magnetic phonons in iron-pnictides, *Physica C*, **469**, 425–441 (2009).
45. L. Shan, Y. Wang, X. Zhu, G. Mu, L. Fang, C. Ren, and H.-H. Wen, Point-contact spectroscopy of iron-based layered superconductor $LaO_{0.9}F_{0.1-\delta}FeAs$, *Europhys. Lett.*, **83**, 57004 (2008).
46. T.Y. Chen, Z. Tesanovic, R.H. Liu, X.H. Chen, and C.L. Chien, A BCS-like gap in the superconductor $SmFeAsO_{0.85}F_{0.15}$, *Nature*, **453**, 1224–1227 (2008).
47. K.A. Yates, L.F. Cohen, Z.-A. Ren, J. Yang, W. Lu, X.-L. Dong, and Z.-X. Zhao, Point contact Andreev reflection spectroscopy of $NdFeAsO_{0.85}$, *Supercond. Sci. Technol.*, **21**, 092003 (2008).
48. K.-Y. Choi, S.H. Kim, C. Choi, M.-H. Jung, X.F. Wang, X.H. Chen, J.D. Noh, and S.I. Lee, Two S-wave gap symmetry for single crystals of the superconductor $BaFe_{1.8}Co_{0.2}As_2$, *Physica C*, **470**, S506–S507 (2010).

49. H. Ding, P. Richard, K. Nakayama, K. Sugawara, T. Arakane, Y. Sekiba, A. Takayama, S. Souma, T. Sato, T. Takahashi, Z. Wang, X. Dai, Z. Fang, G.F. Chen, J.L. Luo, and N.L. Wang, Observation of Fermi-surface-dependent nodeless superconducting gaps in $Ba_{0.6}K_{0.4}Fe_2As_2$, *Europhys. Lett.*, **83**, 47001 (2008).
50. D.S. Inosov, J.S. White, D.V. Evtushinsky, I.V. Morozov, A. Cameron, U. Stockert, V.B. Zabolotnyy, T.K. Kim, A.A. Kordyuk, S.V. Borisenko, E.M. Forgan, R. Klingeler, J.T. Park, S. Wurmehl, A.N. Vasiliev, G. Behr, C.D. Dewhurst, and V. Hinkov, Weak superconducting pairing and a single isotropic energy gap in stoichiometric LiFeAs, *Phys. Rev. Lett.*, **104**, 187001 (2010).
51. C. Bonavolontà, L. Parlato, G.P. Pepe, C. de Lisio, M. Valentino, E. Bellingeri, I. Pallecchi, M. Putti, and C. Ferdeghini, Ultrafast quasiparticle relaxation dynamics in high quality epitaxial $FeSe_{0.5}Te_{0.5}$ thin films, *Supercond. Sci. Technol.*, **26**, 075018 (2013).
52. K. Nakayama, T. Sato, P. Richard, T. Kawahara, Y. Sekiba, T. Qian, G.F. Chen, J.L. Luo, N.L. Wang, H. Ding, and T. Takahashi, Angle-resolved photoemission spectroscopy of the iron-chalcogenide superconductor $Fe_{1.03}Te_{0.7}Se_{0.3}$: Strong coupling behavior and the universality of interband scattering, *Phys. Rev. Lett.*, **105**, 197001 (2010).
53. G. Fuchs, S.-L. Drechsler, N. Kozlova, M. Bartkowiak, J.E. Hamann-Borrero, G. Behr, K. Nenkov, H.-H. Klauss, H. Maeter, A. Amato, H. Luetkens, A. Kwadrin, R. Khasanov, J. Freudenberger, A. Köhler, M. Knupfer, E. Arushanov, H. Rosner, B. Büchner, and L. Schultz, Orbital and spin effects for the upper critical field in As-deficient disordered Fe pnictide superconductors, *New J. Phys.*, **11**, 075007, (2009).
54. A. Yamamoto, A.A. Polyanskii, J. Jiang, F. Kametani, C. Tarantini, F. Hunte, J. Jaroszynski, E.E. Hellstrom, P.J. Lee, A. Gurevich, D.C. Larbalestier, Z.A. Ren, J. Yang, X.L. Dong, W. Lu, and Z.X. Zhao, Evidence for two distinct scales of current flow in polycrystalline Sm and Nd iron oxypnictides, *Supercond. Sci. Technol.*, **21**, 095008 (2008).
55. X. Zhu, H. Yang, L. Fang, G. Mu, and H.-H. Wen, Upper critical field, Hall effect and magnetoresistance in the iron-based layered superconductor $LaFeAsO_{0.9}F_{0.1-\delta}$, *Supercond. Sci. Technol.*, **21**, 105001 (2008).
56. Z. Gao, L. Wang, Y. Qi, D. Wang, X. Zhang, Y. Ma, H. Yang, and H. Wen, Superconducting properties of granular $SmFeAsO_{1-x}F_x$ wires with $T_c = 52$ K prepared by the powder-in-tube method, *Supercond. Sci. Technol.*, **21**, 112001 (2008).
57. C. Ren, Z.S. Wang, H.Q. Luo, H. Yang, L. Shan, and H.H. Wen, Temperature dependence of the lower critical field H_{c1} in $SmFeAsO_{0.9}F_{0.1}$ and $Ba_{0.6}K_{0.4}Fe_2As_2$ iron-arsenide superconductors, *Physica C*, **469**, 599–605 (2009).
58. M. Pissas, D. Stamopoulos, Z.-A. Ren, X.-L. Shen, J. Yang, and Z.X. Zhao, Magnetic measurements of superconducting $NdFeAsO_{0.82}F_{0.18}$ oxypnictide, *Supercond. Sci. Technol.*, **22**, 055008 (2009).
59. J. Prakash, S.J. Singh, S. Patnaik, and A.K. Ganguli, Superconductivity in $CeO_{1-x}F_xFeAs$ with upper critical field of 94 T, *Physica C*, **469**, 82–85 (2009).
60. M. Shahbazi, X.L. Wang, C. Shekhar, O.N. Srivastava, and S.X. Dou, Upper critical field, critical current density and thermally activated flux flow in fluorine doped CeFeAsO superconductors, *Supercond. Sci. Technol.*, **23**, 105008 (2010).

61. H.Q. Yuan, J. Singleton, F.F. Balakirev, S.A. Baily, G.F. Chen, J.L. Luo, and N.L. Wang, Nearly isotropic superconductivity in (Ba,K)Fe_2As_2, *Nature*, **457**, 565–568 (2009).
62. W. Zhou, F.F. Yuan, J.C. Zhuang, Y. Sun, Y. Ding, L.J. Cui, J. Bai, and Z.X. Shi, Upper critical fields and anisotropy of $Ca_{1-x}La_xFe_2As_2$ single crystals, *Supercond. Sci. Technol.*, **26**, 095003 (2013).
63. Z. Zhang, Y. Qi, L. Wang, Z. Gao, D. Wang, X. Zhang, and Y. Ma, Effects of heating conditions on the microstructure and superconducting properties of $Sr_{0.6}K_{0.4}Fe_2As_2$, *Supercond. Sci. Technol.*, **23**, 065009 (2010).
64. S.B. Zhang, X.D. Zhu, H.C. Lei, G. Li, B.S. Wang, L.J. Li, X.B. Zhu, Z.R. Yang, W.H. Song, J.M. Dai, and Y.P. Sun, Superconductivity of $FeSe_{0.89}$ crystal with hexagonal and tetragonal structures, *Supercond. Sci. Technol.*, **22**, 075016 (2009).
65. Z. Gao, Y. Qi, L. Wang, C. Yao, D. Wang, X. Zhang, and Y. Ma, Upper fields and critical current density of $K_{0.58}Fe_{1.56}Se_2$ single crystals grown by one step technique, *Physica C*, **492**, 18–20 (2013).
66. H. Luetkens, H.-H. Klauss, R. Khasanov, A. Amato, R. Klingeler, I. Hellmann, N. Leps, A. Kondrat, C. Hess, A. Köhler, G. Behr, J. Werner, and B. Büchner, Field and temperature dependence of the superfluid density in $LaFeAsO_{1-x}F_x$ superconductors: A muon spin relaxation study, *Phys. Rev. Lett.*, **101**, 097009 (2008).
67. U. Welp, G. Mu, R. Xie, A.E. Koshelev, W.K. Kwok, H.Q. Luo, Z.S. Wang, P. Cheng, L. Fang, C. Ren, and H.-H. Wen, Specific heat and phase diagrams of single crystal iron pnictide superconductors, *Physica C*, **469**, 575–581 (2009).
68. M. Putti, I. Pallecchi, E. Bellingeri, M.R. Cimberle, M. Tropeano, C. Ferdeghini, A. Palenzona, C. Tarantini, A. Yamamoto, J. Jiang, J. Jaroszynski, F. Kametani, D. Abraimov, A. Polyanskii, J.D. Weiss, E.E. Hellstrom, A. Gurevich, D.C. Larbalestier, R. Jin, B.C. Sales, A.S. Sefat, M.A. McGuire, D.C. Larbalestier, R. Jin, B.C. Sales, A.S. Sefat, M.A. McGuire, D. Mandrus, P. Cheng, Y. Jia, H.H. Wen, S. Lee, and C.B. Eom, New Fe-based superconductors: Properties relevant for applications, *Supercond. Sci. Technol.*, **23**, 034003 (2010).
69. R. Khasanov, A. Maisuradze, H. Maeter, A. Kwadrin, H. Luetkens, A. Amato, W. Schnelle, H. Rosner, A. Leithe-Jasper, and H.-H. Klauss, Superconductivity and field-induced magnetism in $SrFe_{1.75}Co_{0.25}As_2$, *Phys. Rev. Lett.*, **103**, 067010 (2009).
70. E.S. Otabe, M. Kiuchi, S. Kawai, Y. Morita, J. Ge, B. Ni, Z. Gao, L. Wang, Y. Qi, X. Zhang, and Y. Ma, Global and local critical current density in superconducting $SmFeAsO_{1-x}F_x$ measured by two methods, *Physica C*, **469**, 1940–1944 (2009).
71. F. Kametani, A.A. Polyanskii, A. Yamamoto, J. Jiang, E.E. Hellstrom, A. Gurevich, D.C. Larbalestier, Z.A. Ren, J. Yang, X.L. Dong, W. Lu, and Z.X. Zhao, Combined microstructural and magneto-optical study of current flow in polycrystalline forms of Nd and Sm Fe-oxypnictides, *Supercond. Sci. Technol.*, **22**, 015010 (2009).
72. M. Eisterer, M. Zehetmayer, H.W. Weber, J. Jiang, J.D. Weiss, A. Yamamoto, E.E. Hellstrom, D.C. Larbalestier, N.D. Zhigadlo, and J. Karpinski, Disorder effects and current percolation in FeAs-based superconductors, *Supercond. Sci. Technol.*, **23**, 054006 (2010).
73. S. Haindl, M. Kidszun, A. Kauffmann, K. Nenkov, N. Kozlova, J. Freudenberger, T. Thersleff, J. Hänisch, J. Werner, E. Reich, L. Schultz, and B. Holzapfel, High upper

critical fields and evidence of weak-link behavior in superconducting $LaFeAsO_{1-x}F_x$ thin films, *Phys. Rev. Lett.*, **104**, 077001 (2010).

74. S.-H. Hong, N.H. Lee, W.N. Kang, and S.-G. Lee, Observation of weak coupling effects in $Ba_{0.6}K_{0.4}Fe_2As_2$ junctions patterned across a naturally formed grain boundary, *Supercond. Sci. Technol.*, **27**, 055007 (2014).
75. S. Trommler, J. Hänisch, V. Matias, R. Hühne, E. Reich, K. Iida, S. Haindl, L. Schultz, and B. Holzapfel, Architecture, microstructure and J_c anisotropy of highly oriented biaxially textured Co-doped $BaFe_2As_2$ on Fe/IBAD-MgO-buffered metal tapes, *Supercond. Sci. Technol.*, **25**, 084019 (2012).
76. P. Mele, K. Matsumoto, K. Fujita, Y. Yoshida, T. Kiss, A. Ichinose, and M. Mukaida, Fe–Te–Se epitaxial thin films with enhanced superconducting properties, *Supercond. Sci. Technol.*, **25**, 084021 (2012).
77. K. Togano, Z. Gao, A. Matsumoto, and H. Kumakura, Enhancement in transport critical current density of *ex situ* PIT Ag/(Ba, K)Fe_2As_2 tapes achieved by applying a combined process of flat rolling and uniaxial pressing, *Supercond. Sci. Technol.*, **26**, 115007 (2013).
78. H. Lin, C. Yao, X. Zhang, H. Zhang, Q. Zhang, D. Wang, and Y. Ma, Enhanced transport critical current density in textured $Sr_{0.6}K_{0.4}Fe_2As_2$ + Ag tapes by an intermediate sintering process of precursors, *Physica C*, **490**, 37–42 (2013).
79. Y. Ma, C. Yao, X. Zhang, H. Lin, D. Wang, A. Matsumoto, H. Kumakura, Y. Tsuchiya, Y. Sun, and T. Tamegai, Large transport critical currents and magneto-optical imaging of textured $Sr_{1-x}K_xFe_2As_2$ superconducting tapes, *Supercond. Sci. Technol.*, **26**, 035011 (2013).
80. C. Yao, H. Lin, X. Zhang, D. Wang, Q. Zhang, Y. Ma, S. Awaji, and K. Watanabe, Microstructure and transport critical current in $Sr_{0.6}K_{0.4}Fe_2As_2$ superconducting tapes prepared by cold pressing, *Supercond. Sci. Technol.*, **26**, 075003 (2013).
81. A. Matsumoto, Z. Gao, K. Togano, and H. Kumakura, Critical current density enhancement of uniaxially pressed Ag-sheathed (Sr, K) Fe_2As_2 superconducting wires, *Supercond. Sci. Technol.*, **27**, 025011 (2014).

19

Outlook

19.1 Introduction

In this book, the physical properties of high-temperature superconductors have been described. A comparison with simple metallic low-temperature superconductors shows some special feature characteristics of the cuprate high-T_c superconductors. Their layered crystal structures cause highly anisotropic physical properties in both the normal and the superconducting states. New phenomena are the dependence of the critical temperature on the number of mobile charge carriers per CuO_2 unit and the existence of an irreversibility line well below the upper critical field. The irreversibility line separates a superconducting state without flux pinning and with vanishing critical current density from a region with pinned vortices and current transport without resistance. Due to an extremely short coherence length, the current transport across high-angle grain boundaries is strongly suppressed; today this is known as the weak-link problem. Another interesting aspect is the evolution of high-temperature superconductivity from an antiferromagnetic Mott insulator by means of doping (see Chapter 8). In the last few years, it has been established that the superconducting order parameter of cuprate high-T_c superconductors has d-wave symmetry (see Chapter 8). In contrast to simple metallic superconductors, in which the superconducting energy gap is independent of the wave vector, the width of the energy gap of high-T_c superconductors depends on the direction of the wave vector and vanishes at the positions where the sign of the superconducting order parameter changes.

Especially in underdoped cuprate superconductors close to the transition to the insulating state, a pronounced pseudogap has been observed. Recent results seem to show that the pseudogap state is in competition with superconductivity and is not a precursor of the superconducting state. For the time being, no generally accepted theory of the pairing mechanism in cuprate high-T_c superconductors has been developed.

Since the discovery of high-temperature superconductivity by J.G. Bednorz and K.A. Müller, the intermediate-temperature superconductor MgB_2 (see Chapter 17) and the iron-based superconductors (see Chapter 18) have been discovered. It has become clear that the pairing of the electrons in MgB_2 is mediated by the electron–phonon interaction, as in simple metallic superconductors. Due to the presence of two superconducting gaps, MgB_2 shows a more complex behavior than simple metallic superconductors.

In the following sections, future investigations of the physical properties of novel superconducting materials, directions in the development of superconducting wires and tapes, and possible future applications will be briefly considered.

19.2 The Investigation of Physical Properties

Future investigations of the physical properties of cuprate and iron-based high-T_c superconductors will be focused on the clarification of the pairing mechanism and on materials-science aspects of importance in applications. In order to identify the pairing mechanism in cuprate high-T_c superconductors, detailed investigations of the superconducting energy gap and the pseudogap are still a topic of research. Using scanning tunneling spectroscopy, a spatial variation of the size of the superconducting energy gap has been observed in Bi-2212 [1–4]. The observed inhomogeneity of high-T_c superconductivity on the nanometer scale may be of importance for the identification of the pairing mechanism. The reasons for the existence of the pseudogap have not yet been fully clarified (see Chapter 8).

The investigation of the symmetry of the superconducting gap (superconducting order parameter) of iron-based superconductors is at a much earlier stage than that of the cuprate superconductors. For the iron-based superconductors, various theoretical studies [5–10] suggest an extended s-wave gap ($s^{\pm}$ symmetry) without nodes. The absence of nodes means that there exists no direction in the momentum space at which the superconducting gap vanishes. A special feature is a sign reversal of the superconducting order parameter between different Fermi surface sheets (hole and electron pockets [5, 11]). Comparisons of angle-resolved photoemission spectroscopy (ARPES) data with results of full-potential linearized plane wave calculations show general similarities in the case of NdOFeAs (Nd-1111) and $Ba_{1-x}K_xFe_2As_2$ (Ba-122) [12]. The observed nodeless gap of Nd-1111 points toward s-wave or extended s-wave symmetry of the order parameter. For Ba-122, the ARPES results indicating a nodeless gap are in contradiction to measurements of the penetration depth indicating the existence of nodes. Nagi *et al.* [13] investigated nuclear magnetic relaxation and the superfluid density (number density of Cooper pairs) in iron-pnictide superconductors. They show that the experimental data are consistent with a fully gapped anisotropic $s^{\pm}$-wave energy gap. ARPES measurements of the energy gap of $Ba_{0.6}K_{0.4}Fe_2As_2$ are consistent with extended s-wave symmetry [14]. On the other hand, Reid *et al.* [15] reported that the symmetry of the order parameter is d-wave in KFe_2As_2, whereas it is s-wave in $Ba_{0.6}K_{0.4}Fe_2As_2$. For Pr-1111 [16], Sm-1111 [17], Nd-1111 [18], $FeSe_{1-x}$ [19], Ba-122 [20], and Sr-122 [20], experimental evidence for a two- or a multigap scenario has been reported.

Measurements of the phonon density of states (PDOS) of $LaO_{1-x}F_xFeAs$ [21] by means of inelastic neutron scattering show that the PDOS of the nonsuperconducting parent compound ($x = 0$) and the optimally fluorine-doped sample ($x = 0.1$) are very similar. These results reported by Christianson *et al.* [21] are not consistent with electron–phonon mediated superconductivity. Calculations of the electron–phonon coupling in $LaO_{1-x}F_xFeAs$, performed by Boeri *et al.* [22] indicate that the calculated coupling constant is 5–6 times too small to explain the observed critical temperature of 26 K. Theoretical considerations as well as experimental results favor an unconventional pairing mechanism.

For applications of cuprate high-T_c superconductors, iron-based superconductors, and MgB_2, investigations of the relations between electronic anisotropy, efficiency of flux pinning, and the position of the irreversibility line with respect to the upper critical field are all of importance. A further cardinal aspect is the presence or absence of weak links. The investigation of artificial grain boundaries in cuprate high-T_c superconductors revealed that high-angle grain boundaries can act as weak links for the supercurrents. On the other hand, it was found that clean grain boundaries of MgB_2 do not act as barriers for the supercurrents [23]. This means that an intrinsic weak-link problem is absent in MgB_2. Nevertheless, porosity or impurity phases may reduce the connectivity of adjacent grains. Investigations of iron-based superconductors suggest weak coupling of adjacent grains. However, for the time being, it is not clear to what extent weak coupling is an intrinsic problem, or if the observed difference of intergranular and intragranular critical current densities is caused by impurity phases at the grain boundaries.

Investigations of cuprate high-T_c superconductors indicated that in RE-123 with a moderate electronic anisotropy, the irreversibility line is closer to the upper critical field than in highly anisotropic Bi-22($n-1$)n or Tl-22($n-1$)n superconductors. Tl-12($n-1$)n and Hg-12($n-1$)n superconductors show a less pronounced anisotropy than the 22($n-1$)n compounds, and hence their irreversibility lines are typically closer to the upper critical field (see Chapter 10).

Artificial pinning centers are very efficient in the moderately anisotropic RE-123 superconductors but not in the highly anisotropic Bi-2212. The reason for this difference is the subdivision of the flux lines into pancake vortices. In (1 2 2) iron-based superconductors, and perhaps also in MgB_2, the electronic anisotropy is smaller than in RE-123; in the (1 1 1 1) family of iron-based superconductors, it is comparable. The typical upper critical fields of MgB_2 wires and tapes are smaller than that of Nb_3Sn and are a limiting factor for their use in magnet applications. On the other hand, iron-based (1 2 2) superconductors provide a relatively small electronic anisotropy along with moderately high critical temperatures of up to 38 K.

19.3 Conductor Development

Even for the first generation Ag/Bi-2212 round multifilament wires, there is still progress in the conductor critical current density. A few years ago, the current carrying capacity of Ag/Bi-2223 tapes could be significantly improved by heat treatment under high pressure. Very recently, the critical current density of Bi-2212 round multifilament wires has also been considerably improved by the use of an over-pressure partial melting process [24–26]. The partial melting process performed under an over-pressure of $\approx$ 100 bar leads to densification of the Bi-2212 filaments, avoiding the formation of large bubbles during the partial melting process. The absence of large bubbles seems to be the main reason for the improved transport critical current density.

In the case of RE-123-coated conductors, conductor development is focused on improved flux pinning, increased conductor unit length, and the subdivision of the RE-123 film into thinner filaments (see Chapter 15). Improved flux pinning, especially for fields parallel to the crystallographic *c*-axis, reduces the undesirable anisotropy of the critical current density with respect to the direction of the magnetic field. In addition, improved pinning, thinner substrates, and thicker RE-123 films all provide higher overall tape critical currents.

An important issue for applications is the production of long RE-123 tapes. All chemical solution deposition techniques may open the way to reduced conductor cost. To reduce AC losses, the superconductor must be subdivided into thin filaments. Developmental RE-123 conductors with striations provide quasitwisting of the filaments.

In the development of MgB_2 conductors (see Chapter 17), improvement of the flux pinning, well-connected grains (absence of voids and secondary phases at the grain boundaries), and an increase of the upper critical field in wires and tapes would all be beneficial. The development of iron-based conductors is still at an early stage. First results for (1 2 2) wires prepared by the powder-in-tube method seem to be promising (see Chapter 18). As in other superconductors, well-connected grains and good flux pinning are required.

19.4 Magnet and Power Applications

Research in the field of superconductivity is stimulated by applications in magnet technology and the power sector. Advantages provided by the use of superconductivity are zero resistance and high operation current densities, allowing the reduction of volume and weight. The most important applications of low-temperature superconductors are for accelerators, nuclear fusion, magnetic resonance imaging (MRI), and nuclear magnetic resonance (NMR). Magnet applications of superconductivity are economic even at operating temperatures of 4.2 K. In the power sector, the use of superconducting cables, transformers, and generators is only economically competitive with conventional technology at elevated operating temperatures (30–77 K).

The optimum operation temperature of superconducting devices depends strongly on the required magnetic field. Low-field applications ($B \approx 0.1$ T) are superconducting current leads [27–32] and power transmission cables [33, 34]. The envisaged operating temperature of superconducting power cables is around 77 K. Transformers and MRI require fields of 1 to 3 T, while for generators and superconducting magnetic energy storage, the typical fields are in the range of 3–10 T. For these intermediate field applications, the optimum operation temperatures are expected to be 30–50 K. Even higher magnetic fields of 12 to 16 T are envisaged for the magnets of future fusion reactors. High-T_c superconductors operated at 4.5 K might be used only for the highest field section of the fusion magnet, while the lower field sections might be made of low-temperature superconductors (Nb_3Sn, NbTi). If wire or tape costs could be significantly reduced, the whole fusion magnet could be fabricated out of high-T_c superconductors, allowing operating temperatures above 20 K. At present, cable concepts for high current conductors made of coated conductor tapes are under development [35–39]. Magnetic fields above 25 T would significantly improve the resolution of NMR spectrometers [40]. At fields well above 20 T, the high-T_c superconductor will most likely be operated at 4.2 K, perhaps enabling the lower field sections to be made of low-temperature superconductors (Nb_3Sn, NbTi). Laboratory magnets generating fields well above 20 T by means of high-T_c superconductor insert coils are under development [41–44].

A further interesting application is the superconducting fault-current limiter [34], which uses the strongly nonlinear current–voltage characteristic of the superconductor at the transition to the normal state to limit fault currents in the power grid.

The main hurdle for a widespread use of superconductivity is economic. To make superconducting cables, transformers, motors, and generators economically competitive with the conventional technology, the price of high-T_c superconductors (RE-123-coated conductors, Ag/Bi-2223, Ag/Bi-2212) needs to be reduced significantly. Mass production could be expected to strongly reduce coated-conductor cost. The large investments necessary for this will be made only when there is an adequate demand for large amounts of these materials.

References

1. K.M. Lang, V. Madhavan, J.E. Hoffman, E.W. Hudson, H. Eisaki, S. Uchida, and J.C. Davis, Imaging the granular structure of high-T_c superconductivity in underdoped $Bi_2Sr_2CaCu_2O_{8+\delta}$, *Nature*, **415**, 412–416 (2002).
2. K.K. Gomes, A.N. Pasupathy, A. Pushp, S. Ono, Y. Ando, and A. Yazdani, Visualizing pair formation on the atomic scale in the high-T_c superconductor $Bi_2Sr_2CaCu_2O_{8+\delta}$, *Nature*, **447**, 569–572 (2007).
3. A.N. Pasupathy, A. Pushp, K.K. Gomes, C.V. Parker, J. Wen, Z. Xu, G. Gu, S. Ono, Y. Ando, and A. Yazdani, Electronic origin of the inhomogeneous pairing in high-T_c superconductor $Bi_2Sr_2CaCu_2O_{8+\delta}$, *Science*, **320**, 196–201 (2008).
4. T.S. Nunner, B.M. Andersen, A. Melikyan, and P.J. Hirschfeld, Dopant-modulated pair interaction in cuprate superconductors, *Phys. Rev. Lett.*, **95**, 177003 (2005).
5. I.I. Mazin, D.J. Singh, M.D. Johannes, and M.H. Du, Unconventional superconductivity with a sign reversal in the order parameter of $LaFeAsO_{1-x}F_x$, *Phys. Rev. Lett.*, **101**, 057003 (2008).
6. D. Zhang, Nonmagnetic impurity resonances as a signature of sign-reversal pairing in FeAs-based superconductors, *Phys. Rev. Lett.*, **103**, 186402 (2009).
7. P. Ghaemi, F. Wang, and A. Vishwanath, Andreev bound states as a phase-sensitive probe of the pairing symmetry of the iron pnictide superconductors, *Phys. Rev. Lett.*, **102**, 157002 (2009).
8. K. Kuroki and H. Aoki, Unconventional pairing originating from disconnected Fermi surfaces in the iron-based superconductor, *Physica C*, **469**, 635–639 (2009).
9. Z.-J. Yao, J.-X. Li, and Z.D. Wang, Spin fluctuations, interband coupling and unconventional pairing in iron-based superconductors, *New J. Phys.*, **11**, 025009 (2009).
10. M.A.N. Araújo, M. Cardoso, and P.D. Sacramento, Single vortex structure in two models of iron pnictide $s^{\pm}$ superconductivity, *New J. Phys.*, **11**, 113008 (2009).
11. T. Hanaguri, S. Niitaka, K. Kuroki, and H. Takagi, Unconventional s-wave superconductivity in Fe(Se,Te), *Science*, **328**, 474–476 (2010).
12. C. Liu, T. Kondo, A.D. Palczewski, G.D. Samolyuk, Y. Lee, M.E. Tillman, N. Ni, E.D. Mun, R. Gordon, A.F. Santander-Syro, S.L. Bud'ko, J.L. McChesney, E. Rotenberg, A.V. Fedorov, T. Valla, O. Copie, M.A. Tanatar, C. Martin, B.N. Harmon, P.C. Canfield, R. Prozorov, J. Schmalian, and A. Kaminski, Electronic properties of iron arsenic high temperature superconductors revealed by angle resolved photoemission spectroscopy (ARPES), *Physica C*, **469**, 491–497 (2009).

13. Y. Nagai, N. Hayashi, N. Nakai, H. Nakamura, M. Okumura, and M. Machida, Nuclear magnetic relaxation and superfluid density in Fe-pnictide superconductors: An anisotropic ±s-wave scenario, *New J. Phys.*, **10**, 103026 (2008).
14. K. Nakayama, T. Sato, P. Richard, Y.-M. Xu, Y. Sekiba, S. Souma, G.F. Chen, J.L. Luo, N.L. Wang, H. Ding, and T. Takahashi, Superconducting gap symmetry of $Ba_{0.6}K_{0.4}Fe_2As_2$ studied by angle-resolved photoemission spectroscopy, *Europhys. Lett.*, **85**, 67002 (2009).
15. J.-Ph. Reid, A. Juneau-Fecteau, R.T. Gordon, S. René de Cotret, N. Doiron-Leyraud, X.G. Luo, H. Shakeripour, J. Chang, M.A. Tanatar, H. Kim, R. Prozorov, T. Saito, H. Fukazawa, Y. Kohori, K. Kihou, C.H. Lee, A. Iyo, H. Eisaki, B. Shen, H.-H. Wen, and L. Taillefer, From d-wave to s-wave pairing in the iron-pnictide superconductor (Ba, K)Fe_2As_2, *Supercond. Sci. Technol.*, **25**, 084013 (2012).
16. K. Matano, Z.A. Ren, X.L. Dong, L.L. Sun, Z.X. Zhao, and Guo-quing Zheng, Spin-singlet superconductivity with multiple gaps in $PrFeAsO_{0.89}F_{0.11}$, *Europhys. Lett.*, **83**, 57001 (2008).
17. Y.-L. Wang, L. Shan, L. Fang, P. Cheng, C. Ren, and H.-H. Wen, Multiple gaps in $SmFeAsO_{0.9}F_{0.1}$ revealed by point-contact spectroscopy, *Supercond. Sci. Technol.*, **22**, 015018 (2009).
18. P. Samuely, P. Szabó, Z. Pribulová, M.A. Tillman, S.L. Bud'ko, and P.C. Canfield, Possible two-gap superconductivity in $NdFeAsO_{0.9}F_{0.1}$ probed by point-contact Andreev-reflection spectroscopy, *Supercond. Sci. Technol.*, **22**, 014003 (2009).
19. R. Khasanov, M. Bendele, A. Amato, K. Conder, H. Keller, H.-H. Klauss, H. Luetkins, and E. Pomjakushina, Evolution of two-gap behavior of the superconductor $FeSe_{1-x}$, *Phys. Rev. Lett.*, **104**, 087004 (2010).
20. Z.G. Chen, T. Dong, R.H. Ruan, B.F. Hu, B. Cheng, W.Z. Hu, P. Zheng, Z. Fang, X. Dai, and N.L. Wang, Measurement of the c-axis optical reflectance of AFe_2As_2 (A = Ba, Sr) single crystals: Evidence of different mechanisms for the formation of two energy gaps, *Phys. Rev. Lett.*, **105**, 097003 (2010).
21. A.D. Christianson, M.D. Lumsden, O. Delaire, M.B. Stone, D.L. Abernathy, M.A. McGuire, A.S. Sefat, R. Jin, B.C. Sales, D. Mandrus, E.D. Mun, P.C. Canfield, J.Y.Y. Lin, M. Lucas, M. Kresch, J.B. Keith, B. Fultz, E.A. Goremychkin, and R.J. McQueeny, Phonon density of states of $LaFeAsO_{1-x}F_x$, *Phys. Rev. Lett.*, **101**, 157004 (2008).
22. L. Boeri, O.V. Dolgov, and A.A. Golubov, Is $LaFeAsO_{1-x}F_x$ an electron–phonon superconductor? *Phys. Rev. Lett.*, **101**, 026403 (2008).
23. D.C. Larbalestier, L.D. Cooley, M.O. Rikel, A.A. Polyanskii, J. Jiang, S. Patnaik, X.Y. Cai, D.M. Feldmann, A. Gurevich, A.A. Squitieri, M.T. Naus, C.B. Eom, E.E. Hellstrom, R.J. Cava, K.A. Regan, N. Rogado, M.A. Hayward, T. He, J.S. Slusky, P. Khalifah, K. Inumaru, and M. Haas, Strongly linked current flow in polycrystalline forms of the superconductor MgB_2, *Nature*, **410**, 186–189 (2001).
24. D.C. Larbalestier, J. Jiang, U.P. Trociewitz, F. Kametani, C. Scheuerlein, M. Dalban-Canassy, M. Matras, P. Chen, N.C. Craig, P.J. Lee, and E.E. Hellstrom, Isotropic round-wire multifilament cuprate superconductor for generation of magnetic fields above 30 T, *Nat. Mater.*, **13**, 375–381 (2014).

25. T. Shen, A. Ghosh, L. Cooley, and J. Jiang, Role of internal gases and creep of Ag in controlling the critical current density of Ag-sheathed $Bi_2Sr_2CaCu_2O_x$ wires, *J. Appl. Phys.*, **113**, 213901 (2013).
26. T. Kanai and N. Inoue, Development of gas pressure melting (GPM) method for Ag-sheathed Bi-2212 wires, *J. Mater. Sci.*, **30**, 3200–3206 (1995).
27. A. Ballarino, HTS current leads for the LHC magnet powering system, *Physica C*, **372–376**, 1413–1418 (2003).
28. R. Wesche, P. Bruzzone, C. Fiamozzi Zignani, L. Affinito, S. Chiarelli, R. Freda, A. Formichetti, M Marchetti, H Ehmler, J. Heinrich, and P. Smeibidl, Results of the test of a pair of 20 kA HTS currents leads, *J. Phys.: Conf. Series*, **507**, 032056 (2014).
29. R. Wesche, M. Bagnasco, P. Bruzzone, R. Felder, M. Guetg, M. Holenstein, M. Jenni, S. March, F. Roth, and M. Vogel, Test results of the 18 kA EDIPO HTS current leads, *Fus. Eng. Des.*, **86**, 1414–1417 (2011).
30. R. Heller, W.H. Fietz, A. Kienzler, and R. Lietzow, High temperature superconductor current leads for fusion machines, *Fus. Eng. Des.*, **86**, 1422–1426 (2011).
31. R. Heller, S.M. Darweschsad, G. Dittrich, W.H. Fietz, S. Fink, W. Herz, F. Hurd, A. Kienzler, A. Lingor, I. Meyer, G. Nother, M. Susser, V.L. Tanna, A. Vostner, R. Wesche, F. Wuchner, and G. Zahn, Experimental results of a 70 kA high temperature superconductor current lead demonstrator for the ITER magnet system, *IEEE Trans. Appl. Supercond.*, **15**, 1496–1499 (2005).
32. R. Wesche, P. Bruzzone, S. March, C. Müller, M. Vogel, H. Quack, M. Börsch, E. Iten, N. Magginetti, D. Oertig, and F. Holdener, Results of the test of industrially manufactured HTS current leads with novel design features, *IEEE Trans. Appl. Supercond.*, **24**, 4800205 (2014).
33. F. Schmidt and A. Allais, Superconducting cables for power transmission cables – A review, http://cds.cern.ch/record/962751/files/p232.pdf.
34. AmpaCity – technologische Weltpremiere in Essen, http://www.rwe.com/web/cms/de/1301026/rwe-deutschland-ag/energiewende/intelligente-netze/ampacity/
35. M. Takayasu, L. Chiesa, L. Bromberg, and J.V. Minervini, HTS twisted stacked-tape cable conductor, *Supercond. Sci. Technol.*, **25**, 014011 (2012).
36. S.I. Schlachter, W. Goldacker, F. Grilli, R. Heller, and A. Kudymow, Coated conductor Rutherford cables (CCRC) for high-current applications: Concepts and properties, *IEEE Trans. Appl. Supercond.*, **21**, 3021–3024 (2011).
37. D.C. van der Laan, X.F. Lu, and L. Goodrich, Compact $GdBa_2Cu_3O_{7-\delta}$ coated conductor cables for electric power transmission and magnet applications, *Supercond. Sci. Technol.*, **24**, 042001 (2011).
38. D. Uglietti, N. Bykovsky, R. Wesche, and P. Bruzzone, Fabrication of a prototype HTS conductor for fusion magnets, Paper 2LOr2B-01 presented at the ASC 2014, Charlotte, USA, August 10 –15, 2014.
39. N. Bykovsky, D. Uglietti, R. Wesche, and P. Bruzzone, Strain management in HTS high current cables, Paper 3LPo1C-04 presented at the ASC 2014, Charlotte, USA, August 10–15, 2014.
40. J. Bascuñán, S. Hahn, D.K. Park, Y. Kim, and Y. Iwasa, On the 600 MHz HTS insert for a 1.3 GHz NMR magnet, *IEEE Trans. Appl. Supercond.*, **22**, 4302104 (2012).

41. H.W. Weijers, U.P. Trociewitz, K. Marken, M. Meinesz, H. Miao, and J. Schwartz, The generation of 25.05 T using a 5.11 T $Bi_2Sr_2CaCu_2O_x$ superconducting insert magnet, *Supercond. Sci. Technol.*, **17**, 636–644 (2004).
42. K. Ohkura, T. Okazaki, and K. Sato, Large HTS magnet made by improved DI-BSCCO tapes, *IEEE Trans. Appl. Supercond.*, **18**, 556–559 (2008).
43. W.D. Markiewicz, D.C. Larbalestier, H.W. Weijers, A.V. Voran, K.W. Pickard, W.R. Sheppard, J. Jaroszynski, A. Xu, R.P. Walsh, J. Lu, A.V. Gavrilin, and P.D. Noyes, Design of a superconducting 32 T magnet with REBCO high field coils, *IEEE Trans. Appl. Supercond.*, **22**, 4300704 (2012).
44. D. Uglietti, R. Wesche, and P. Bruzzone, Construction and test of a non-insulated insert coil using coated conductor tape, *J. Phys.: Conf. Series*, **507**, 032052 (2014).

Author Index

Subject Index

Printed and bound by CPI Group (UK) Ltd, Croydon, CR0 4YY
17/09/2026
14964270-0006